Lecture Notes in Electrical Engineering

Volume 476

** Indexing: The books of this series are submitted to ISI Proceedings, EI-Compendex, SCOPUS, MetaPress, Springerlink **

Lecture Notes in Electrical Engineering (LNEE) is a book series which reports the latest research and developments in Electrical Engineering, namely:

- Communication, Networks, and Information Theory
- Computer Engineering
- Signal, Image, Speech and Information Processing
- Circuits and Systems
- Bioengineering
- Engineering

The audience for the books in LNEE consists of advanced level students, researchers, and industry professionals working at the forefront of their fields. Much like Springer's other Lecture Notes series, LNEE will be distributed through Springer's print and electronic publishing channels.

For general information about this series, comments or suggestions, please use the contact address under "service for this series".

To submit a proposal or request further information, please contact the appropriate Springer Publishing Editors:

Asia:

China, *Jessie Guo, Assistant Editor* (jessie.guo@springer.com) (Engineering)

India, *Swati Meherishi, Senior Editor* (swati.meherishi@springer.com) (Engineering)

Japan, *Takeyuki Yonezawa, Editorial Director* (takeyuki.yonezawa@springer.com) (Physical Sciences & Engineering)

South Korea, *Smith (Ahram) Chae, Associate Editor* (smith.chae@springer.com) (Physical Sciences & Engineering)

Southeast Asia, *Ramesh Premnath, Editor* (ramesh.premnath@springer.com) (Electrical Engineering)

South Asia, *Aninda Bose, Editor* (aninda.bose@springer.com) (Electrical Engineering)

Europe:

Leontina Di Cecco, Editor (Leontina.dicecco@springer.com) (Applied Sciences and Engineering; Bio-Inspired Robotics, Medical Robotics, Bioengineering; Computational Methods & Models in Science, Medicine and Technology; Soft Computing; Philosophy of Modern Science and Technologies; Mechanical Engineering; Ocean and Naval Engineering; Water Management & Technology)

(christoph.baumann@springer.com)

(Heat and Mass Transfer, Signal Processing and Telecommunications, and Solid and Fluid Mechanics, and Engineering Materials)

North America:

Michael Luby, Editor (michael.luby@springer.com) (Mechanics; Materials)

More information about this series at http://www.springer.com/series/7818

Vijay Nath · Jyotsna Kumar Mandal
Editors

Proceeding of the Second International Conference on Microelectronics, Computing & Communication Systems (MCCS 2017)

 Springer

Editors
Vijay Nath
Department of Electronics and
  Communication Engineering
Birla Institute of Technology, Mesra
Ranchi, Jharkhand
India

Jyotsna Kumar Mandal
Department of Computer
  Science and Engineering
University of Kalyani
Kalyani, West Bengal
India

ISSN 1876-1100   ISSN 1876-1119 (electronic)
Lecture Notes in Electrical Engineering
ISBN 978-981-10-8233-7   ISBN 978-981-10-8234-4 (eBook)
https://doi.org/10.1007/978-981-10-8234-4

Library of Congress Control Number: 2018935883

This Springer imprint is published by the registered company Springer Nature Singapore Pte Ltd.
The registered company address is: 152 Beach Road, #21-01/04 Gateway East, Singapore 189721, Singapore

Preface

Modern complex ICs' design is complicated. As on average up to 2015 in desktop, more than 1 billion of transistors existed. For complete design having more than 500 rules and it's in statistical nature. As per the demand of the market for sophisticated systems, designers/engineers are playing with automated EDA software tools for efficient and bulk design. For these purposes, different automated EDA software have been launched such as cadence and mentor graphics. From these tools, designers are preparing full-custom and semicustom ICs. These tools consider the length of the transistor in manometer range. These range-designed ICs required good support of nanomaterials and its related chemicals. Testing of ICs is also a major issue in sophisticated systems. Without software support, testing is difficult. Nanotechnology allowed the integration of purely electronic devices, chips, circuits and systems. The nanoscale dimensions of nanoelectronic components for systems for giga-scale complexity are measured on a chip or in a package. Nanotechnology improves the capabilities of electronic components by reducing the size of transistors in ICs, increases the density of memory chips, reduces the power consumption and improves the display screen, thickness, etc. In global automation, control and functional environment IoT is playing a major role today. Now, you can imagine a single example—cashless world. The computing systems which are playing behind the cashless world is more reliable, robust, correct and highly accurate. Now, this system is going to adopt smart education system for making a global education hub. This setup already adopted in global factory operation, control and monitoring for perfect and optimized products. This conference provides an excellent forum for young researchers, engineers and professors to work together and share their knowledge. It also gives the ideas on how to work in electronic media safely and securely. Manufacturing companies/industries and education universities have well contribution in the development of any countries. However, they face several challenges such as rapid product development, flexibility, low-to-medium volume, transportation and low cost. Many advanced/unconventional technologies/tools/software are being developed worldwide to face these challenges. Among recent technologies IC design and manufacturing, Internet of Things has become more popular due to the ability of precise work. For the research, development, sharing knowledge and exchange of ideas in the current trends, the Second International Conference on Microelectronics, Computing & Communication Systems (MCCS-2017) has been organized by Indian Society for VLSI Education (ISVE) Ranchi at Advanced Regional Telecom Training Centre (ARTTC) near Jumar River Hazaribagh Road Ranchi from 13 to 14 May 2017 and 2nd IC-NCCS-2016 organized from 11 to 12 December 2016 at the same venue. These conferences cover advancement in MEMS and nanoelectronics, wireless communications, optical communication, instrumentation, signal processing, image processing, bioengineering, green energy, hybrid vehicles, environmental science, weather forecasting, cloud computing, renewable energy, RFID, CMOS sensors, actuators, transducers, telemetry systems, embedded systems, sensor network applications in mines, etc. Population of the countries are increasing day by day, then the duty of the society to provide all types of facility to the new users for serving their life efficiently. It is possible with the help of research and development only.

In this conference 350 papers and 50 chapters in pedagogy (Washington Accord) was received, in which 67 blindly peer-reviewed, registered and presented papers were accepted for publication in conference proceeding of Springer Scopus book series Lecture Notes in Electrical Engineering (LNEE) and 10 outstanding papers were selected for SCI Journals like *Microsystem Technologies* and *IETE Technical Review & Technical Research*, etc. In parallel conference, excellent reviewer's team guided authors for the extended version of the research articles, i.e. 70–80% improvement/changes with new content and title forward for the publication in listed SCI and Scopus journals. All articles are blindly peer-reviewed by at least three reviewers and details comments have been passed to concern authors with decisions. In presentation session, six expert reviewers' team evaluated their work. They also guided the authors for IPR/patents and new innovative project for funding in their area of research. These series of conferences (NCCS & MCCS) organized by ISVE & IETE Ranchi in support of ARTTC BSNL Ranchi gave a unique platform to young researchers, scientists, engineers and professors for the presentation of their work worldwide. This platform provides outcome-based learning and research strategy. Pedagogy method gives new ideas to the learner to enhance their knowledge in scientific research. This technique is highly beneficial to students, researchers, professors and industrial people to recognize or evaluate the value of their work.

In the conference on dais, Chief Guest Dr. R. K. Pandey, VC, Ranchi University, Ranchi; Guest of Honour Dr. K. K. Thakur, CGMT, BSNL Ranchi; Guest of Honour Dr. A. A. Khan, Former VC, Ranchi University, Ranchi; Dr. R. K. Singh, Former Chairman, IETE Ranchi; Guest of Honour Sh. Sanjay Kumar Jha, Past Chairman, IETE, Ranchi, and Chief Executive Engineer, Government of Jharkhand; Guest of Honour Dr. P. R. Thakura, Professor, BIT Mesra, and Executive Member, ISVE & IETE, Ranchi; Sh. Ajay Kumar, Chairman, IETE, Ranchi & AGM(Admin), ARTTC BSNL, Ranchi; Dr. Anand Kumar Thakur, Faculty, SSMC RU & Organizing Secretary, MCCS-2017; Dr. Raj Kumar Singh, Faculty, RLSYC RU & Convenor MCCS-2017, Ranchi; Dr. Vijay Nath, Faculty, BIT Mesra & General Chair, MCCS-2017; Keynote Speaker Dr. A. Srinivasulu,

Professor, Vignan University, Guntur (AP) were present. The conference began with welcome address by Dr. Vijay Nath, General Chair of the conference, and keynote address by Prof. A. Srinivasulu, Vignan University Guntur, AP, on the topic "VLSI Design and Challenges". All other dignitaries gave their views for the growth of society, quality of publications, research innovations and challenges.

In this conference, the authors are invited for the original papers submission and the quality of papers has been selected for presentation. The authors described their articles in above-mentioned domain very well. Mohan Kashyap et al. explain their research article "Optimal Placement of Distributed Generation Using Genetic Algorithm Approach". Piyush Kumar Ojha et al. demonstrate their work on "Analysis of Voltage Source Boost Inverter". Rahul Sharma et al. explain their research article "Reduction of Redundant Frames in Active Wireless Capsule Endoscopy". Pritha Roy et al. explain their research work on "A Study on Filter Design Aspects of Single Phase Inverter with Various Modulation Schemes". Maninder Kaur et al. demonstrate their work on "Design, Analysis and Testing of Low Voltage CMOS OTA". Sumit Srivastava et al. demonstrate their research article "Speech Based Access to Price of Different Agricultural Commodities Using MFCC, GMM and Naïve Bayes Classifier". Shahid Aziz et al. demonstrate their work as per the current demand of market "Dual Axis Solar Tracker for Solar Panel with Wireless". Naghma Khatoon et al. demonstrate their work on "A Node Stability Based Multi-metric Weighted Clustering Algorithm for Mobile Ad-Hoc Networks". Kumari Mamta et al. explain their work on "Design and Development of Microstrip Patch Antenna for Millimeter-Wave Application". Chitrita Saha et al. introduce their work on "Review on Fault Tolerant Control (FTC) and Fault Detection & Isolation (FDI) Schemes of Wind Turbine". Ankita et al. explain their work on "A Comprehensive survey on Computational Grid Resource Management". Kanchan Bala et al. introduce their research work on "Thunderstorm Prediction Using Soft Computing and Wavelet". Mohammad Javeed et al. define their work on "Design of a Clock Distribution Network Using Low Power Prescaler and Fused P & S Counters". K. Rama Devi et al. demonstrate their research article "Design of RFID Tag Antenna with Impedance Matching Techniques at UHF Band". Chandrashekhar Azad et al. introduce in their research article "Decision Tree and Genetic Algorithm Based Intrusion Detection System". B. Jayalakshmi et al. explain "A Novel Method of Designing Using Non-logical Algorithm". Annu Priya et al. explain in their research article "A Novel Multimedia Encryption and Decryption Technique Using Binary Tree Traversal". Malladi Lakshmi Lavanya et al. demonstrate their research work on "ZC-CDTA Based Integrator Circuit Using Single Passive Component". Saka Harshavardhan et al. explain "Road and Traffic Sign Detection Using Colour Segmentation". Pawan Kumar et al. explain "Switch in Based Trimmed Median Filter for Noise Removal from Medical Image". Surya Gupta et al. introduce their research article "Characterization of Interfering Signal in S Band of IRNSS". Sandip Kumar Singh Modak et al. present their article "Enhancing Multibiometric System Security Using ECC Based on Score Level Fusion". Santhoshi Rupa Gayatri Neralla et al. present their article "Generation of Photographic Mosaic Using Apache Spark and Scalding for Image Processing".

Nibha Rani et al. introduce "Mitigation of Congestion in Transmission Line Using Series Smart Wire". Naghma Khatoon et al. explain their research article "Mobility Aware Distributed Clustering and Routing Algorithm Based on A* Search for Mobile Ad-Hoc Networks". Meena Singh presents her research article "Floating Admittance Matrix Model Development of Active Devices and Circuits". Ravi Devesh et al. demonstrate "An Efficient Approach for Monuments Image Retrieval Using Multi-visual Descriptors". Ruchira et al. present "Comparison of ANN based MPPT Controller and Incremental Conductance for Photovoltaic System". Sneha Mangalwedhe et al. introduce "Low Power Implementation of 32-Bit RISC Processor with Pipelining". Bollampally Joy Persis et al. present their research work on "A Novel SINR Based Cooperative Radio Resource Allocation Mechanism (SBC-RRAM) for LTE/Wi-Fi Radio Access System in Smart Home Environment". Pallavi Choudekar et al. present their work on "Optimal Location of Thyristor Controlled Series Compensator to Assuage Congestion in Transmission Network". Priyanka Gupta et al. explain "An Efficient Brightness Preserving Contrast Enhancement Technique Using Discrete Wavelet Transform and Singular Value Decomposition". Ruhi Dubey et al. introduce their research article "Computer Assisted Valuation of Descriptive Answers Using WEKA with Random Forest Classification". Shaik Qadeer et al. present a chapter on "VLSI Signal Processing". Debaprasad De et al. present a chapter on "Ramanujan Sums and Signal Processing: An Overview". Akshaya R. et al. introduce "Analysis and Design of Bandgap Reference (BGR)". Avireni Srinivasulu et al. introduce the chapter "FinFET Based Negative Rectifiers for Low-Power Analog Applications". Sonali B. Wankhede et al. introduce "Denial of Service Attacks". Trilok Kumar Parashar et al. introduce "Analytical Modelling of Room Temperature GaAs/InAs$_{0.3}$Sb$_{0.7}$ Detector for H$_2$S Gas Detection". Om Prakash et al. describe the "Fuzzy Prediction Model for Water Temperature in Scheffler Solar Reflector". Anil Pinapati et al. introduce "A Reversible Data Hiding Using Difference-Histogram Modification on Multi-directional in Two-Dimensional Histogram". Chandan Kumar et al. describe "A Novel Single-phase Multilevel Inverter Topology with Reduced Component Count". K. Srilakshmi et al. demonstrate "Energy Efficient 64-Bit Asynchrobatic Adder". Gourab Das et al. demonstrate "PV Array's Resistance & Temperature Sensitivity Analysis with Shading Effects". Sukesha et al. explain the "Effect of Location of Piezoelectric Sensor Over a Smart Structure". Dipti Kumari et al. introduce "A Systematic Approach Towards Development of Universal Software Fault Prediction Model Using Object-Oriented Design Measurement". Varun Bohra et al. introduce "Design and Implementation of a Reaction Timer Using CMOS Logic". Shubham Goswami et al. describe "A Novel Hybrid Resource Allocation Scheme for Maximum Fairness Among Multiple Services". Aditya S. Sengupta describes the "Modelling Equivalent Circuit for Supercapacitor Module Voltage Decay". Rajinder Tiwari et al. demonstrate "An Innovative Design Approach of SoC Based Smart CMOS Sensor for Mixed Signal Processing Based Applications". Priyanka Parihar et al. describe the "Investigation of MTCMOS 6T SRAM Cell for Ultra Low Power Application". Neha Gupta et al. describe "Novel Approach for Sleep Transistor Sizing to Suppress Power and Ground Bounding Noise in

MTCMOS Clustering Technique". Mukesh Patidar et al. describe "Efficient Design and Simulation of Novel Exclusive-OR Gate Based on Nano-electronics Using Quantum Dot Cellular Automata". Anup Tiwari et al. demonstrate the "Design Strategy for Smart Toll-Billing Systems". Sanket P. Singhania et al. describe the "Novel Cell Search Method in Long Term Evolution System". Sukesha Sharma et al. describe the "Various Feature Extraction and Classification Techniques". Abhinav Kumar et al. explain the "Path Tracking Method of ALV Based on ADRC Strategy and Differential Flatness Theory". Ravneet Kaur et al. describe the "Performance Analysis of Conventional SRAM with Higher Order SRAM Topologies". Shashank Shekhar et al. demonstrate the "Performance Analysis of Time-Reversal Division Multiple Access Under Multi-path Rician Fading Channels". Amninder Singh et al. introduce the "Analysis of Software Development Life Cycle Models". Sumit Singh et al. explain the "Design of Narrowband 2.69 GHz CMOS Low Noise Amplifier for WiMAX Application". Sarita Kumari et al. describe the "Sensitivity Analysis of Various Magneto-Optic Materials Based on Faraday Rotation Principle". Anil Kumar et al. describe "A Secure Three—Way Handshake Authentication Process in IEEE 802.11i". Gireesh Joshi et al. explain the "BrowserGuard2: A Solution for Drive-By-Download Attacks". Nishat Aafreen et al. demonstrate the "GA Based Energy Optimization in Traffic Grooming WDM Optical Mesh Network". Oshin Garg et al. define the "Piezoelectric Energy Harvesting: A Developing Scope for Low Power Applications". Deepak Prasad et al. explain "An Overview of CMOS Temperature Sensors". Abhishek Pandey et al. demonstrate "A 3.65 mW, Op-Amp Based Up-Conversion Mixer for Zigbee Front-End Transmitter". Vijay Kumar Karan et al. explain "Adaptive Compensation Algorithm for Flux Estimation of PM BLDC Motor Drives". A. Srinivasulu et al. demonstrate the "Performance Analysis of Inter-Satellite Optical Wireless Communication Using 12 and 24 Transponders". These articles are helpful for the product development of smart ICs.

The authors and editors have taken utmost care in presenting the information and acknowledging the original sources whenever necessary. The editors express their gratitude towards the authors, organizers of IC-MCCS and staff of Springer, India, for making the publication of this research book/proceeding possible. Readers are requested to provide their valuable feedback on the quality of presentation and inadvertent error or omission of information if any. We expect that the book will be welcomed by students as well as practicing engineers/researchers/professors.

Ranchi, India   Vijay Nath
Kolkata, India   Jyotsna Kumar Mandal

Organizing Committee

Patron

- Dr. A. K. S. Chandele, President, IETE, New Delhi

Chairman Committee

- Sh. V. B. Pandey, DOT BSNL Term Cell1 & Chairman, ISVE, Ranchi
- Sh. Ajay Kumar, AGM(HR) BSNL & Chairman, IETE, Ranchi
- Prof. P. S. Neelakanta, C. Engg., Fellow IEE, Florida Atlantic University (FAU), USA.
- Prof. J. K. Mondal, Professor, Kalyani University, WB
- Prof. C. K. Sarkar, Chairman, IEEE Kolkata Section, WB
- Prof. Vinay Gupta, Professor, Delhi University, Delhi
- Sh. Sanjay Kumar Jha, Past Chairman, IETE, Ranchi
- Prof. Subir Sarkar, Professor, Jadavpur University, WB

Co-chairman Committee

- Dr. Umesh Yadav, DDU GU
- Sh. Rajan Kumar Ram, ARTTC BSNL, Ranchi
- Dr. R. K. Singh Founder Chairman, IETE, Ranchi
- Dr. Anshuman Sarkar, Kalyani Government Engineering College, Kalyani, WB

Organizing Secretary

- Dr. Anand Kr. Thakur, SSMC RU & Treasurer, IETE, Ranchi

General Chair

- Dr. Vijay Nath, BIT Mesra & Secretary, IETE, Ranchi

Convener

- Dr. Raj Kumar Singh, RLYC RU & Executive Mem IETE Ranchi

International Advisory Committee

- Prof. Bernd Michel, Micro Materials Centre (MMC) Berlin, Germany
- Prof. Bharath Bhushan, Ohio Eminent Scholar & The Howard D. Winbigler Professor, Director NBLL, The Ohio State University, Columbus, Ohio, USA
- Smt. Smriti Dagur, Former President, IETE, New Delhi
- Dr. A. A. Khan, Former VC, Ranchi University, Ranchi
- Dr. M. K. Mishra, VC, BIT Mesra, Ranchi
- Dr. K. K. Thakur, CGM, BSNL, Ranchi
- Dr. Ramgopal Rao, Professor, Director, IIT Delhi
- Dr. P. K. Barhai, Former VC, BIT Mesra, Ranchi
- Dr. S. Pal, Professor & SD, Satellite Navigation, ISRO, Bangalore
- Dr. M. S. Kori, Chairman IETE, TPC, New Delhi
- Sh. R. K. Gupta, Former President, IETE, New Delhi
- Dr. Rajendra Prasad, Professor, IIT Roorkee
- Dr. Labh Singh, Former CGM, ARTTC, BSNL, Ranchi
- Sh. R. Mishra, Former CMD, HEC, Ranchi
- Dr. Yogesh Singh Chauhan, Associate Professor, Nano Lab, Department of EE, IIT Kanpur
- Dr. S. N. Verma, Former CMD, EDC, Ltd. Jharkhand
- Sh. S. C. Thakur, Chief Engineer, Rural Electrification Energy Distribution Corporation Limited Jh.
- Sh. Ravindra Kr. Rakesh, Editor, Dainik Bhaskar, Jharkhand
- Sh. Gopal Jha, Journalist, New Delhi
- Dr. A. N. Mishra, VC, Central University, Jharkhand
- Dr. R. Pandey, VC, RU, Ranchi
- Dr. A. Chakrabarty, Professor, IIT Kharagpur
- Dr. S. Banerjee, Professor, IIT Kharagpur
- Dr. Nandita Das Gupta, Professor, IIT Chennai
- Dr. L. K. Singh, Former Professor, Dr. RML AU, Faizabad
- Dr. B. S. Rai, Professor, MMMUT, Gorakhpur
- Dr. D. Samathanam, Former Adviser & Head TDT, DST New Delhi
- Dr. P. Chakrabarty, Professor, IIT BHU
- Dr. G. A. Murthy, Scientist-G, DRDO, Hyderabad
- Dr. M. Srinivasa, Scientist-G, DRDO, Hyderabad
- Dr. S. C. Bose, Scientist-G, CEERI, Pilani
- Dr. Jamir Akhatar, Sr. Scientist, CEERI, Pilani
- Dr. Arokiaswami ALPHONES, Vice-Chairman, IEEE Singapore Section & Professor, NTU, Singapore
- Dr. K. Rajasekhar, Dy. Director General, NIC, DEIT, Mo CIT, Government of India, Hyderabad

- Dr. N. V. Kalyankar, Principal, Yeshwant Mahavidyalaya, Nanded
- Dr. R. P. Panda, Professor, VSSUT, Burla, Odisha
- Dr. Allen Klinger, Professor, University of California
- Dr. Hisao Ishibuchi, Professor, Osaka Prefecture University, Japan
- Dr. T. K. Bhattacharya, Professor, IIT Kharagpur
- Dr. N. Gupta, Professor, BIT Mesra, Ranchi & Fellow Member, IETE, New Delhi
- Dr. V. R. Gupta, Professor, BIT Mesra, Ranchi & Fellow Member, IETE, New Delhi
- Dr. M. Chakrabarty, Professor, IIT Kharagpur
- Dr. A. S. Dhar, Professor, IIT Kharagpur
- Dr. D. K. Sharma, Professor, IIT Bombay
- Dr. Nandita Das Gupta, Professor, IIT Chennai
- Dr. B. Mishra, Professor, BIT Mesra, Ranchi
- Dr. Swaroop Gosh, Assistant Professor, University of South Florida
- Dr. S. P. Maity, Professor, IIEST, Shibpur
- Dr. S. K. Ghorai, Professor, BIT Mesra, Ranchi & Executive Member, IETE, Ranchi Centre
- Dr. M. Bhuyan, Professor, Tezpur University, Assam
- Dr. S. Hosimin Thilangar, Professor, Anna University, Chennai
- Dr. V. N. Mani, Senior Scientist, CMET, Hyderabad
- Dr. V. Kumar, Professor, ITT-ISM, Dhanbad
- Dr. S. K. Paul, Professor, ISM, Dhanbad
- Dr. J. P. Gupta, Former Pro-VC, DDU Gorakhpur University
- Dr. H. C. Prasad, Former Professor, DDU Gorakhpur University
- Dr. S. Bhaumik, Associate Professor, NIT, Tripura
- Dr. P. D. Kashyap, Professor, NIT, Arunachal Pradesh
- Dr. J. Akhatar, Senior Scientist, CEERI, Pilani
- Dr. S. Ahmad, Former Director, CEERI, Pilani
- Dr. P. Kapoor, Former Director, CSIO, Chandigarh
- Dr. Uma Maheshwari, Professor, Anna University
- Dr. Sandip Rakshit, Professor, Kaziranga University, Assam
- Dr. Abhijit Biswas, Professor, Institute of Radio Physics & Electronics, Calcutta University
- Dr. Vikash Patel, SAC ISRO, Ahmedabad
- Dr. Parul Patel, SAC ISRO, Ahmedabad

National Advisory Committee

- Dr. Gaurav Trivedi, Assistant Professor, IIT Guwahati
- Dr. B. K. Kaushik, Associate Professor, IIT Roorkee
- Dr. K. K. Khatua, Associate Professor, NIT Rourkela
- Dr. M. Bhaskar, Associate Professor, NIT Trichy
- Dr. P. Kumar, Associate Professor, IIT Patna
- Dr. Soumya Pandit, Assistant Professor, Kolkata University

- Dr. Soma Berman, Assistant Professor, University of Calcutta
- Dr. K. B. Raja, Professor, Bangalore College of Engineering, Bangalore
- Dr. R. P. Panda, Professor, VSSUT, Burla, Odisha
- Dr. P. R. Thakua, Associate Professor, BIT Mesra, Ranchi
- Dr. S. S. Solanki, Associate Professor, BIT Mesra, Ranchi
- Dr. Mahesh Chandra, Associate Professor, BIT Mesra, Ranchi
- Dr. D. K. Malik, Associate Professor, BIT Mesra, Ranchi
- Dr. Nutan Lata, Associate Professor, BIT Mesra, Ranchi
- Dr. K. K. Senapati, Assistant Professor, BIT Mesra, Ranchi
- Dr. K. K. Patnaik, Associate Professor, IIITM Gwalior
- Dr. M. Goswami, Associate Professor, IIIT Allahabad
- Dr. Sukalayam Chakraborty, Assistant Professor, BIT Mesra, Ranchi
- Dr. Lallan Yadav, Associate Professor, DDU University, Gorakhpur
- Dr. S. Chakrabarty, Associate Professor, BIT Mesra, Ranchi
- Dr. D. Devaraj, Professor, Kalasalingam University, Tamil Nadu
- Dr. J. S. Roy, Professor, KIIT, Bhubaneswar
- Dr. N. K. Kamila, Professor, CVRCE, Bhubaneswar, Odisha
- Dr. B. K. Ratha, Associate Professor, Utkal University, Odisha
- Dr. A. Srinivasulu, Professor, Vignan University, Andhra Pradesh
- Dr. Manish Prateek, Professor, Petroleum University, Dehradun
- Dr. Vijay Laxmi, Associate Professor, BIT Mesra, Ranchi
- Dr. V. K. Jha, Associate Professor, BIT Mesra, Ranchi
- Dr. R. K. Lal, Associate Professor, BIT Mesra, Ranchi
- Dr. L. B. Singh, Professor, RPSIT, Patna
- Sh. H. S. Gupta, Senior Scientist ISRO, Bangalore
- Dr. N. S. Rao, Associate Professor, MECS, Hyderabad
- Dr. Usha Mehta, Professor, Nirma Institute of Technology, Ahmedabad

Technical Programme Committee

- Dr. Kota Solomon Raju, Scientist-F, CEERI, Pilani
- Dr. Amalin Prince, Associate Professor, BITS, Pilani, Goa Campus
- Dr. M. Mishra, Assistant Professor, DDU University, Gorakhpur
- Dr. J. B. Sharma, Associate Professor, Rajasthan Technical University, Kota
- Dr. Prabir Saha, Assistant Professor, NIT Meghalaya
- Dr. S. P. Tiwari, Assistant Professor, IIT Jodhpur
- Dr. Santosh Vishvakerma, Associate Professor, IIT Indore
- Dr. S. N. Shukla, Professor, Dr. RML Avadh University, Faizabad
- Dr. B. N. Sinha, Associate Professor, SSMC, Ranchi
- Dr. V. S. Rathore, Assistant Professor, BIT Mesra, Ranchi
- Dr. Manish Kumar, Assistant Professor, NERIST
- Dr. A. N. Jadhav, Professor, Y.M. R.T. Marathwada University, Nanded

Joint Secretary

- Prof. D. Acharya, PIET, Rourkela
- Prof. Rajeev Ranjan, ISM, Dhanbad
- Prof. Amar Prakash Sinha, BIT, Sindri
- Prof. Jayant Pal, NIT, Agarpara, Kolkata
- Prof. Adesh Kumar, Energy & Petroleum University, Dehradun
- Prof. J. Dinesh Reddy, BMS College of Engineering, Bangalore
- Prof. N. Srinivasa Rao, BMS College of Engineering, Bangalore
- Prof. P. Kumar, CIT, Ranchi
- Sh. Ramkrishna Kundu, IBM, Bangalore
- Sh. Dipayan Gosh, GM Aircel, Kolkata
- Sh. S. Chakrabarty, IBM, Bangalore
- Prof. Jyoti Kumari, RBS, Bangalore
- Sh. Rahul Kumar Singh, ST Microelectronics, Noida
- Sh. Suraj Kumar, NIT Agartala
- Prof. Anand Kr. Signh, GNIT, Ghaziabad

Treasurer

- Dr. Anand Kr. Thakur, IETE, Ranchi
- Sh. S. K. Saw, MCCS-2017
- Smt. Saroj, ISVE, Ranchi

Editorial Acknowledgements

We extend our thanks to all the authors for contributing to this book/proceeding by sharing their valuable research findings. We specially thank a number of reviewers for promptly reviewing the papers submitted to the conference. We are grateful to the volunteers, invited speakers, session chairs, sponsors, subcommittee members, members of the international advisory committee, members of the national advisory committee, members of the technical program committee, members of joint secretary and members of the scientific advisory committee for the successful conduct of the conference. The editors express their heartfelt gratitude towards Dr. A. K. S. Chandelle, President, IETE, New Delhi; Smt. Srimati Dagur, Past President, IETE, New Delhi; Sh. Sanjay Kumar Jha, Past Chairman, IETE, Ranchi, and Executive Engineer, Government of Jharkhand; Sh. Prasad Vijay Bhushan Pandey, DTO Term Cell1 BSNL, Ranchi, and Chairman, ISVE Ranchi; Prof. A. A. Khan, Former VC, Ranchi University; Prof. M. K. Mishra, VC, BIT Mesra; Dr. K. K. Thakur, CGMT, BSNL Ranchi; Prof. R. K. Pandey, VC, Ranchi University; Prof. P. K. Barhai, Former VC, BIT Mesra; Sh. R. Mishra, Former CMD, HEC Ranchi; Dr. M. Chakraborty, Professor, IIT Kharagpur; Dr. Ramgopal Rao, Professor, IIT Bombay, and Director IIT Delhi; Dr. P. Chakraborty, Professor, IIT BHU; Dr. Abhijit Biswas, Professor, Kolkata University; Dr. Subir Kumar Sarkar, Professor, Jadavpur University; Dr. Gaurav Trivedi, Associate Professor, IIT Guwahati; Dr. Y. S. Chauhan, Associate Professor, IIT Kanpur; Dr. B. K. Kaushik, Professor, IIT Roorkee; Dr. Shree Prakash Tiwari, Faculty, IIT Jodhpur; Dr. P. Kumar, Associate Professor, IIT Patna; Dr. M. Bhaskar, Professor, NIT Trichy; Dr. Adesh Kumar, Faculty, UPES University, Dehradun; Dr. Manish Kumar, Associate Professor, MMMUT Gorakhpur; Dr. Manish Mishra, Associate Professor, DDU University, Gorakhpur; Dr. Umesh Yadav, Professor, DDU University, Gorakhpur; Dr. J. K. Mandal, Professor, Kalyani University; Prof. D. Acharjee, President, ISTM Kolkata; Dr. N. Gupta, Professor, BIT Mesra Ranchi; Dr. Vibha Rani Gupta, Professor, BIT Mesra; Dr. B. K. Mishra, Professor, BIT Mesra; Dr. V. K. Jha, BIT Mesra; Sh. Ajay Kumar, AGM (admin), ARTTC BSNL Ranchi, and Chairman, IETE Ranchi;

Dr. P. R. Thakura, Executive Member, IETE and ISVE Ranchi, and Professor, BIT Mesra Ranchi; Dr. M. Chandra, Executive Member IETE Ranchi, and Professor, BIT Mesra Ranchi; Dr. S. K. Ghorai, Executive Member, IETE Ranchi, and Professor, BIT Mesra Ranchi; Dr. B. Chakraborty, Executive Member, IETE Ranchi, and Executive Engineer, Mecon, Ranchi; Dr. S. Chakraborty, Executive Member, IETE Ranchi, and Professor, BIT Mesra Ranchi; Dr. S. S. Solanki, Professor, BIT Mesra Ranchi; Dr. S. Pal, Professor, BIT Mesra Ranchi; Dr. S. Kumar, Executive Member, IETE Ranchi, and Associate Professor, BIT Mesra Ranchi; Dr. B. K. Bhattacharya, Professor, NIT Agartala; Dr. Anand Kumar Thakur, Treasurer, IETE Ranchi; Dr. Raj Kumar Singh, Executive Member, IETE Ranchi, and Faculty, RLSYC Ranchi University; Dr. R. K. Lal, Associate Professor, BIT Mesra Ranchi; Smt. Saroj, Treasurer, ISVE Ranchi; Prof. Jyoti Singh, Joint Secretary, ISVE Ranchi; Prof. A. K. Pandey, Secretary, ISVE Ranchi; Sh. Suraj Kumar Saw; Sh. Subro Chakraborty; Sh. Dipayan Ghosh; Sh. Ramkrishna Kundu, Executive Member, ISVE Ranchi; Sh. Deepak Prasad; Sh. Sumit Singh; Sh. H. Kar; Sh. Rajanish Yadav; Sh. Anup Tirkey for their endless support, encouragement, motivation to organize such prestigious event that paved the way for this book on Microelectronics, Computing & Communication Systems (MCCS). At last, we express our sincere gratitude towards the staff members of Springer, India, who helped in publishing this book.

Contents

About the Editors

Dr. Vijay Nath was born in Gorakhpur (UP), India, in 1976. He received his bachelor's degree in Physics and master's degree in Electronics from DDU Gorakhpur University, India, in 1998 and 2001, respectively. He also received PGDCN (GM) from MMMUT, Gorakhpur, in 1999. He received his Ph.D. degree in VLSI Design & Technology from Dr. RML Avadh University, Faizabad, in association with CEERI, Pilani, in 2008. From 2000 to 2001, he was Project Trainee in IC Design Group, CEERI, Pilani, under the guidance of Dr. K. S. Yadav (senior scientist). From 2002 to 2006, he severed as Faculty in the Department of Electronics, DDU University, Gorakhpur. In 2006, he joined as a Faculty in the Department of Electronics and Communication Engineering, Birla Institute of Technology Mesra Ranchi (JH), India. He is Professor-in-charge of VLSI Design Lab, BIT Mesra, Ranchi. He is Faculty Advisor of NAPS: News & Publication Society, Advisor IEI Student Chapter (ECE) BIT Mesra and Honorary Secretary of IETE Ranchi Centre. He is a recipient of Vivekananda Techno Fiesta Award 2002, Young Scientist Award 2004, Cadence Design Contest 2013, 2014, CCSN Best Paper Award 2013, IEI Technical Contest Best Paper Award 2013. His research interests include analog, digital, mixed CMOS VLSI circuits, low-power VLSI circuits, ADC, DAC, PTAT, CMOS bandgap voltage reference, ASICs, embedded systems designs, smart cardiac pacemaker, smart grids, Internet of Things and early-stage cancer detection. He has to his credit around 110 publications in reputed Scopus and SCI journals and conferences. He has successfully completed three R&D projects funded by DST New Delhi, DRDL Hyderabad and MHRD New Delhi, and one project is in ongoing stage funded by RESPOND ISRO Ahmedabad. He developed e-learning course of VLSI Design in Pedagogy pattern funded by MHRD New Delhi. Now, the complete course is available in IIT Kharagpur official website and open for all IITs, NITs, BITs and technical universities. He is Editor of Proceeding of International Conference on Nanoelectronics, Circuits & Communication Systems (NCCS 2015) of Lecture Notes of Electrical Engineering, Springer. He is a member of several professional societies and academic bodies including IETE, ISTE, ISVE and IEEE.

Prof. Jyotsna Kumar Mandal received his M.Sc. in Physics from the Jadavpur University in 1986 and M. Tech. in Computer Science from the University of Calcutta. He was awarded Ph.D. in Computer Science and Engineering by the Jadavpur University in 2000. Presently, he is Professor of Computer Science and Engineering and former Dean, Faculty of Engineering, Technology and Management, Kalyani University, Kalyani, Nadia, West Bengal, for two consecutive terms. He started his career as Lecturer at NERIST, Arunachal Pradesh, in September 1988. He has teaching and research experience of 28 years. His research interests include coding theory, data and network security, remote sensing and GIS-based applications, data compression, error correction, visual cryptography, steganography, security in MANET, wireless networks and unify computing. He has guided 15 Ph.D. students, 2 submitted (2015–16) and 8 ongoing. He has supervised 3 M.Phil. and more than 50 M.Tech. dissertations. He has life member of the Computer Society of India since 1992, CRSI since 2009, ACM since 2012, IEEE since 2013 and Fellow member of IETE since 2012, Honorary Chairman of CSI Kolkata Chapter. He has chaired more than 30 sessions in various international conferences and delivered more than 50 expert/invited lectures during the last 5 years. He has acted as program chair of many international conferences and edited more than 15 volumes of proceedings from Springer Series, ScienceDirect, etc. He is reviewer of various international journals and conferences. He has over 360 articles and 6 books published to his credit. He is one of the editors for Springer AISC Series, FICTA 2014, CSI 2013, IC3T 2015, INDIA 2015 and INDIA 2016. He is also the corresponding editor of CIMTA 2013 (Procedia Technology, Elsevier), INDIA 2015 (AISC Springer) and ICIC2 2016 (AISC Springer).

Reduction of Redundant Frames in Active Wireless Capsule Endoscopy

Rahul Sharma, Rampal Bhadu, Surender Kumar Soni and Nithin Varma

1 Introduction

Wireless capsule endoscopy (WCE) is a technique that enables a clinician to examine the interior of the whole gastrointestinal (GI) tract of a human body. The capsule endoscope is a disposable plastic capsule which weighs 3.7 g and measures 11 mm in diameter and 26 mm in length. At one end of the capsule, there is a miniaturized color camera and an optical dome with four white light-emitting diodes (LEDs). A color camera that captures two images (256×256 pixels) a second and these images are compressed using JPEG [1, 2]. It contains a battery of lifetime 8 h and a wireless transmitter that relayed the images to a data recorder worn by the patient on a belt using a radio frequency signal. After the 8 h procedure, about 50,000 images are obtained and uploaded to a workstation where the doctor can view and examine endoscopic images.

In WCE, a miniaturized pill size wireless camera is swallowed by a patient and motion of the capsule is mainly driven by the gastrointestinal peristalsis that leads to nonuniform motion of the camera [3] as shown in Fig. 1. In active WCE, locomotion strategy is used for controlling the motion of capsule. It uses a permanent magnet inside the capsule and an external rotating magnetic field to rotate

R. Sharma (✉) · R. Bhadu · S. K. Soni · N. Varma
Electronics & Communication Engineering Department,
National Institute of Technology, Hamirpur, Hamirpur, India
e-mail: sharma03071992@gmail.com

R. Bhadu
e-mail: errampal.bhadu@gmail.com

S. K. Soni
e-mail: surender.soni@gmail.com

N. Varma
e-mail: nithinvarma.a3@gmail.com

© Springer Nature Singapore Pte Ltd. 2019   1
V. Nath and J. K. Mandal (eds.), *Proceeding of the Second International Conference on Microelectronics, Computing & Communication Systems (MCCS 2017)*, Lecture Notes in Electrical Engineering 476, https://doi.org/10.1007/978-981-10-8234-4_1

Fig. 1 Motion of WCE in
gastrointestinal tract [3]

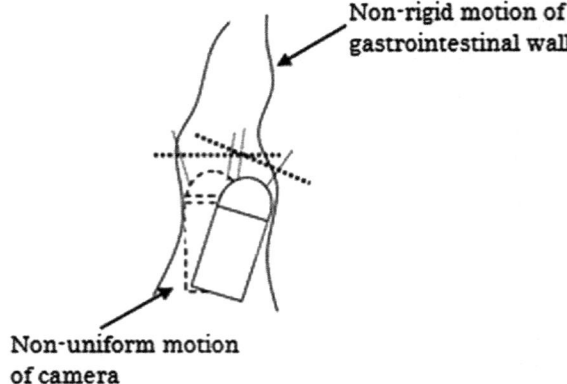

**Non-rigid motion of
gastrointestinal wall**

**Non-uniform motion
of camera**

the capsule [4]. Although natural peristalsis is present in gastrointestinal tract, endoscopic capsule slows down and even can stop. Therefore, several frames are redundant in active WCE video.

This paper presents a reduction scheme that includes two stages. The first stage applied skip-prediction [5] in which intensity based sum of absolute difference (SAD) value is used for detection of similar frames. In second stage, Harris algorithm is used for corner detection and after that RANSAC is applied to find out the redundant frames.

2 Redundant Frames Detection

2.1 Similar Frames Detection Using Skip-Prediction

Skip-prediction is a technique that identifies the neighboring macroblocks in the same image, likely to be skipped without consequential data loss. We skipped the neighboring frames by comparing original frame. Consider a sequence of image frames of a WCE video as follows:

$$F = [f_1, f_2 \cdots f_n] \tag{1}$$

where n is number of frames. From f_1 to f_n, measure the similarity of consecutive frames through SAD value.

Commercially, WCE images are received at receiver in YUV color space. Since Y is luminance, most of the information is contained in Y-component [6]. Therefore, we applied skip-prediction only on Y-component to save cost and time.

SAD is used as an algorithm that can measure the similarity between two images. It takes the absolute difference of two intensity values at corresponding pixels of two images for comparison and summed up the differences at all the pixels to create SAD value for frame similarity [7].

$$\text{SAD} = \sum_i \sum_j |I_k(i,j) - I_{k+1}(i,j)| \qquad (2)$$

where I_k and I_{k+1} are the value of intensity level of kth and $(k + 1)$th frame at the (i, j)th pixel. If SAD is smaller than the threshold, the two frames considered are similar. After that, similarity between $(k + 1)$th and $(k + 2)$th frames is detected.

There is no significant change in the visualization effect if the value of intensity level at all pixels of image is decremented or incremented by two [8]. Therefore, we compare two images of resolution of $M \times N$ using SAD with threshold smaller than $M \times N \times 2$.

Algorithm 1: Skip-Prediction

Step 1 Input images are subsequent images I_k and I_{k+1}.
Step 2 Find differences in the intensity values at each pixel of I_{k+1} and corresponding pixels of I_k.
Step 3 Compute the sum of absolute value of the differences (SAD).
Step 4 Compare the SAD value with the threshold.
Step 5 Find the frames below the threshold as redundant frames.

Since skip-prediction is not invariant to illumination, brightness, rotation, and noise, the SAD value can be greater than the threshold for some similar frames due to illumination variation and reflection of light from the gastrointestinal wall. We applied Harris algorithm for corner detection and RANSAC for detecting these types of the similar frames in the next stage of this paper.

2.2 Similar Frames Detection Using RANSAC Combined with Harris Algorithm for Corner Detection

There are three types of features that an image can contain: Edges, corners, and flat regions. A feature proved good if it can be located in different views of a scene without any ambiguity. Edge and flat region extraction lead to aperture problem [9]. Therefore, in this study, we find the corner points for similarity matching of subsequent WCE frames.

Corners are regions in an image with large variation in intensity in all direction. It represents the high variation of gradient in the image in several directions. Initially, Moravec [10] proposed a corner detector that recognizes corner points by shifting a window locally in any image in various direction which leads to large changes in intensity. Harris and Stephens [9] used the directional derivatives in place of shifted windows, which proved to be the more robust corner detector. Harris algorithm is invariant to illumination variation, rotation, and image noise [11].

Harris corner detector basically finds the difference in intensity $E(u, v)$ for (u, v) displacement in all directions.

$$E(u,v) = \sum_{x,y} W(x,y)[I(x+u, y+v) - I(x,y)]^2 \tag{3}$$

where $I(x, y)$ represents intensity at (x, y) and $I(x + u, y + v)$ is intensity at moved window. $W(x, y)$ is window function that applies weight to each pixel. It can be uniform weight or Gaussian weight. For detection of corners, we maximize above Eq. (3) that is approximated and simplified in [12] and can be written in matrix form as follows:

$$E(u,v) \approx [u,v]M \begin{bmatrix} u \\ v \end{bmatrix} \tag{4}$$

M is a 2×2 matrix and it is computed as given in Eq. (5) using image derivatives,

$$M = \sum_{x,y} W(x,y) \begin{bmatrix} I_x^2 & I_x I_y \\ I_x I_y & I_y^2 \end{bmatrix} \tag{5}$$

and I_x, I_y are the first derivatives of the image in x and y directions respectively. The decision of whether a pixel location is corner or not depends upon eigenvalues λ_1 and λ_2 of the matrix M through corner response R.

$$R = (\lambda_1 . \lambda_2) - k(\lambda_1 + \lambda_2)^2 \tag{6}$$

The value of R is large for a corner, negative with large magnitude for an edge and small for a flat region. Therefore, the pixel for which R value is large with empirical constant, k is considered as corner point.

Algorithm 2: Harris Corner Detector

Step 1   Compute derivatives of the image in horizontal and vertical directions: I_x and I_y.

Step 2   Compute three images corresponding to three terms in matrix M.

Step 3   Convolve the three images with a larger Gaussian window.

Step 4   Compute the scalar corner response, R.

Step 5   Find local maxima above some threshold as detected corner point.

After detection of corner points in two subsequent WCE images, we applied normalized cross-correlation [13] for matching the appropriate corner point in second image corresponding to a corner point in first image and labeled by same number in both the images. Normalized cross-correlation is basically a comparison of a little picture to all possible local regions in another image. For normalization, the pixels in a window are subtracted by mean of the patch intensities and divided by the standard deviation as shown in Eqs. (7) and (8).

$$\hat{f} = \frac{f - \bar{f}}{\sqrt{\sum (f - \bar{f})^2}} \tag{7}$$

$$\hat{g} = \frac{g - \bar{g}}{\sqrt{\sum (g - \bar{g})^2}} \tag{8}$$

$$C_{fg}\left(\hat{f}, \hat{g}\right) = \sum_{i,j} \hat{f}(i,j)\hat{g}(i,j) \tag{9}$$

where $C_{fg}\left(\hat{f}, \hat{g}\right)$ is normalized cross-correlation (NCC) between corner window f and corner window g. The value of NCC varies from -1 to 1 where 1 represents perfecting matching and -1 shows completely anticorrelated.

After matching of corresponding corner points in subsequent images, we apply random sample consensus (RANSAC). Fischer and Bolles [14] introduced random sample consensus (RANSAC) in 1981 for solving the problem of location determination using automated image analysis. RANSAC is a method of iteration that contains mainly two steps: In the first step, it randomly selects a sample subset of minimal data elements from input dataset and estimates a model and computes the parameters of the model using the elements of the subset. The second step checks the all the elements from input dataset that are fitted with estimated model within tolerable error in model parameters. The elements that are fitted with estimated model form the consensus set and all elements of consensus set are called inliers. The elements that are not fitted with model parameters considered as outliers. RANSAC algorithm reestimates the model until the consensus set has enough inliers.

Fundamental matrix F is used as mathematical model and estimated by using Eq. (10).

$$P_1^{\mathrm{T}} F P_2 = 0 \tag{10}$$

where P_1, P_2 are the points of randomly selected 8-point subset from image 1 and 2. After calculating the set of corner points of image 1 in the form of matrix as $P1$ and set of corner points of image 2 in the form of matrix as $P2$ applied in equation 10 while F is the fundamental matrix. The points for which residual value is smaller than threshold are considered as inliers otherwise outliers.

The number of iterations, N required [14] for maximum number of inliers depends upon probability, p that at least one of the sets of random samples includes all inliers, v that represents the probability that any selected data point is an outlier and number of elements, m in randomly selected sample subset.

$$N = \frac{\log(1 - p)}{\log(1 - (1 - v)^m)} \tag{11}$$

Table 1 Reduction of redundant frames in the active WCE videos

Videos	Total number of frames	Number of redundant frames	Frame reduction (%)
Video 1	27,000	8640	32
Video 2	32,000	12,480	39
Video 3	25,000	8250	33
Video 4	29,000	12,180	42
Video 5	23,000	9430	41

Table 2 Table for comparison of results with the previous work done

	Frame reduction % (slow motion)	Frame reduction % (backward + forward motion)	Total frame reduction %
Lee et al. [15]	26.5	25.8	52.3
Chen et al. [16]	37	17.7	54.7
Proposed	38.2	–	38.2

On the basis of fraction of inliers, we can detect the frames that are similar to the previous subsequent frame if the fraction of inliers is larger than the threshold.

$$\text{Fraction of inliers} = \frac{\text{Number of inliers}}{\text{Total number of data points}} \qquad (12)$$

3 Experiments and Results

We applied our reduction scheme to several active WCE videos from different patients provided by Gastrolab.com after fragmenting the videos into frames. We divided our reduction scheme into two stages. From first stage, we took the threshold for SAD value smaller than $M \times N \times 2$. If the SAD value is between $M \times N \times 2$ and $M \times N \times 4$ then take the threshold for the fraction of inliers greater than 0.4 from second stage for detecting similar frames. In this paper, we are using MATLAB R2016a for simulation. We found that every active WCE video contains several redundant frames that can be reduced as shown in Table 1 and without any loss of diagnostic information. We compare our results with previous work done in Table 2.

4 Conclusion

In this study, we present two different techniques for reduction of redundant frames in active WCE videos. The reduction scheme of active WCE videos is first time explored in this study, while for the passive WCE has been studied before.

The redundant frames in the active WCE are created mainly due to natural peristalsis and we use Harris algorithm for corner detection and RANSAC algorithm that is robust to illumination, brightness, scale, and noise. Based on our experimental results, we can reduce the diagnosis time from 40 to 50% with no loss of diagnostic information.

References

1. M. Q.-H. Meng, T. M. T. Mei, J. P. J. Pu, C. H. C. Hu, X. W. X. Wang, and Y. C. Y. Chan, "Wireless robotic capsule endoscopy: state-of-the-art and challenges," *Fifth World Congr. Intell. Control Autom. (IEEE Cat. No.04EX788)*, vol. 6, pp. 1–5, 2004.
2. B. Li and M. Q. H. Meng, "Computer-based detection of bleeding and ulcer in wireless capsule endoscopy images by chromaticity moments," *Comput. Biol. Med.*, vol. 39, no. 2, pp. 141–147, 2009.
3. H. Liu, N. Pan, H. Lu, E. Song, Q. Wang, and C.-C. Hung, "Wireless Capsule Endoscopy Video Reduction Based on Camera Motion Estimation," *J. Digit. Imaging*, vol. 26, no. 2, pp. 287–301, 2013.
4. G. Ciuti, A. Menciassi, and P. Dario, "Capsule Endoscopy : From Current Achievements to Open Challenges," *IEEE Trans. Biomed. Eng.*, vol. 4, pp. 59–72, 2011.
5. C. Sampath Kannangara *et al.*, "Low-complexity skip prediction for H.264 through lagrangian cost estimation," *IEEE Trans. Circuits Syst. Video Technol.*, vol. 16, no. 2, pp. 202–207, 2006.
6. R. C. Gonzalez and R. E. Woods, *Digital Image Processing (3rd Edition)*. Upper Saddle River, NJ, USA: Prentice-Hall, Inc., 2006.
7. M. Bhat, P. Kapoor, and B. L. Raina, "Application of Sad Algorithm in Image Processing for Motion Detection and Simulink Blocksets for Object Tracking," no. 3, pp. 731–736, 2012.
8. X. Xiang, G.-L. Li, and Z. Wang, "Low-complexity and high-efficiency image compression algorithm for wireless endoscopy system," *J. Electron. Imaging*, vol. 15, no. 2, pp. 23015–23017, 2006.
9. C. Harris and M. Stephens, "A Combined Corner and Edge Detector," *Proceedings Alvey Vis. Conf. 1988*, pp. 147–151, 1988.
10. H. P. Moravec, "Obstacle avoidance and navigation in the real world by a seeing robot rover.," DTIC Document, 1980.
11. C. Schmid, R. Mohr, and C. Bauckhage, "Evaluation of Interest Point Detectors," *Int. J. Comput. Vis.*, vol. 37, no. 2, pp. 151–172, 2000.
12. N. D. B. Bruce and P. Kornprobst, "Harris corners in the real world: A principled selection criterion for interest points based on ecological statistics," *2009 IEEE Comput. Soc. Conf. Comput. Vis. Pattern Recognit. Work. CVPR Work. 2009*, pp. 2160–2167, 2009.
13. F. Zhao, Q. Huang, and W. Gao, "Graduate School of the Chinese Academy of Sciences, Beijing, China," pp. 729–732, 2006.
14. M. a. Fischler and R. C. Bolles, "Random sample consensus: a paradigm for model fitting with applications to image analysis and automated cartography," *Commun. ACM*, vol. 24, no. 6, pp. 381–395, 1981.
15. H. G. Lee, M. K. Choi, B. S. Shin, and S. C. Lee, "Reducing redundancy in wireless capsule endoscopy videos," *Comput. Biol. Med.*, vol. 43, no. 6, pp. 670–682, 2013.
16. C. Yi, L. Yihua, and R. Haozheng, "Trimming the wireless capsule endoscopic video by removing redundant frames," *2012 Int. Conf. Wirel. Commun. Netw. Mob. Comput. WiCOM 2012*, 2012.

Analysis of Voltage Source Boost Inverter

Piyush Kumar Ojha and P. R. Thakura

1 Introduction

The great concern of today's society is the depletion of fossil fuels and to protect the environment from pollution caused by conventional energy sources. Therefore, there is demand for application of nonconventional energy sources systems. The power generated from nonconventional energy sources such as solar, wind, etc., are not adequate enough to deliver the high power load demand. Thus, one has to use the power conditioning units in between the nonconventional energy sources and load in the application area. The voltage source inverter used in many applications has the ability to generate ac power output from the dc supply input but this VSI cannot be used in such applications where the required ac output power is greater from DC input power. For such applications, one may use boost inverter topology circuit which will generate greater ac output power from dc power input [1, 2]. The major advantages of using single stage dc-ac boost inverter as compared to two-stage inverter topology configuration are reduced volume, weight, and cost of the system and improved efficiency [1–4]. In this paper topology of boost inverter, rating estimation and experimental simulation analysis is presented.

P. K. Ojha (✉) · P. R. Thakura
Department of Electrical & Electronic Engineering, Birla Institute
of Technology, Mesra, Ranchi, India
e-mail: piyushkojha@gmail.com

P. R. Thakura
e-mail: prthakura@bitmesra.ac.in

© Springer Nature Singapore Pte Ltd. 2019
V. Nath and J. K. Mandal (eds.), *Proceeding of the Second International Conference on Microelectronics, Computing & Communication Systems (MCCS 2017)*, Lecture Notes in Electrical Engineering 476, https://doi.org/10.1007/978-981-10-8234-4_2

2 Topology of Proposed Boost Inverter

The dc to ac boost converter which is proposed here obtains dc-ac conversion as shown in Fig. 1. The blocks 1 and 2 represent dc to ac converters. The above converters will generate output which is sinusoidal in nature and also DC biased. Each converter modulation is being 180° out of phase with respect to each other, which creates maximum voltage excursion across the load [1]. The connection of load across the converter output is differential in nature. The boost converter is the basic configuration of boost inverter that is proposed here. In this configuration of boost inverter, two inductors are used that are energy storing elements, two capacitors, and four power semiconductor switches are used. The resistive load is considered here and that is connected across load terminals differentially. Here, MOSFET has been used as a power semiconductor switch in the circuit of boost inverter [5].

The operation of proposed dc-ac boost inverter can be explained in an easy manner by considering the circuit operation of the current bidirectional boost dc-dc converter [2]. The circuit operation of the current bidirectional boost dc-dc converter as shown in Fig. 2 may be given in two modes. In Mode I when lower switch SW1 conducts the path of current conduction is from input dc source, inductor and through lower switch SW1, in this way the inductor L stores the energy itself, at the same time upper switch SW2 is switched off and capacitor supplies energy to the resistive load. In Mode II when lower switch SW1 is off and upper switch SW2 is on, the current will flow from the source through inductor L, upper switch SW2 and capacitance and resistance in parallel. In Mode II, the load resistance and capacitance connected across load terminals are being supplied with energy from the source and the energy stored in inductor also both. Thus, it can be seen that in this way, the output terminal has the more voltage level as compared with input terminal. The two current bidirectional boost dc-dc converter may be connected in such a manner that is shown in Fig. 2 to produce output voltage as an ac voltage and having greater magnitude also as compared to input voltage.

Fig. 1 Voltage source boost inverter

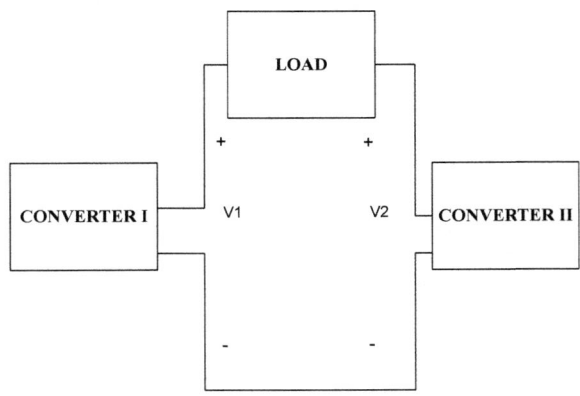

Fig. 2 The current
bidirectional boost converter

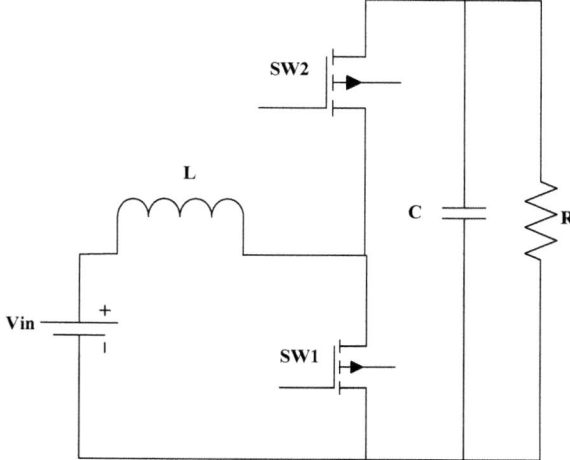

The proposed dc-ac boost voltage source inverter is shown in Fig. 3. The circuit operation may be explained in two modes in the following manner:

Mode 1: In Mode 1, when switch SW1 and Switch SW3 will be in turn on condition and switch SW2 and switch SW4 will be in turn off condition. Inductance $L2$ stores energy in it through switch SW3 from input supply. Supply and

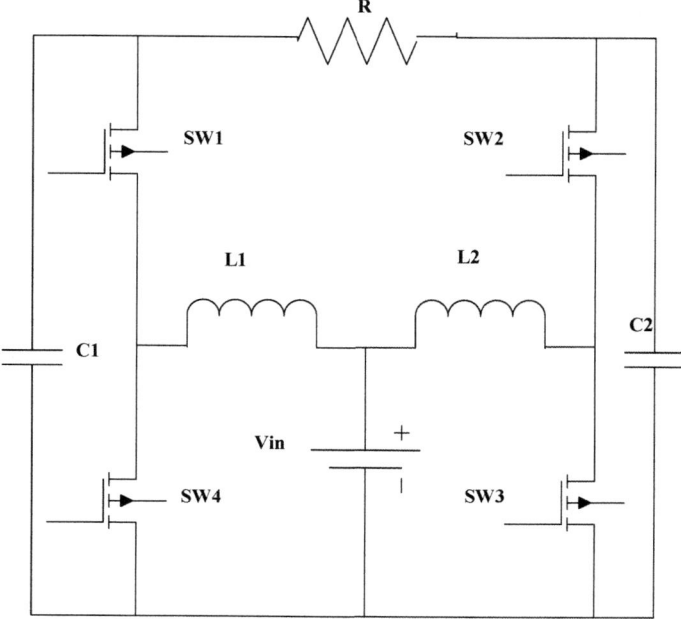

Fig. 3 The proposed dc-ac boost inverter

inductance $L1$ deliver energy to the capacitor $C1$ for the purpose to charge the capacitor $C1$ and at the same time, capacitor $C2$ discharges via the load resistance R and the capacitor $C1$.

Mode 2: In Mode 2, when switch SW2 and Switch SW4 will be in turn on condition and switch SW1 and switch SW3 will be in turn off condition. Inductance $L1$ stores energy in it through switch SW4 from input supply. Supply and inductance $L2$ delivers energy to the capacitor $C2$ for the purpose to charge the capacitor $C2$ and at the same time, capacitor $C1$ discharges via the load resistance R and the capacitor $C2$.

One can derive the voltage equation for boost converter as given below for the continuous conduction mode by using the averaging concept:

$$\frac{V_1}{V_{in}} = \frac{1}{1 - \Delta} \tag{1}$$

where Δ is the duty ratio/duty cycle.

$$V_o = V_1 - V_2 = \frac{V_{in}}{1 - \Delta} - \frac{V_{in}}{\Delta} \tag{2}$$

$$\frac{V_o}{V_{in}} = \frac{2\Delta - 1}{\Delta(1 - \Delta)} \tag{3}$$

3   Simulation Results and Discussion

Circuit parameters specifications are as follows:

V_{in} = 100 V (input voltage);
V_o = 300 V (output voltage);
f_s = 50 Hz (switching frequency);
f_o = 50 Hz (output voltage frequency);
R = 1000 Ω;
$L1 = L2$ = 10 mH;
$C1 = C2$ = 600 μF;

Figure 4 shows the output voltage waveform across the load resistance, its peak value is more than 300 V and ac voltage is obtained with switching frequency of 50 Hz by and applying 100 V dc source input and the output frequency of 50 Hz is also obtained.

Figure 5 shows the output current waveform across the load, since load is resistive, the phase of load current is the same as load voltage.

Figure 6 shows the voltage across the capacitor, it is unidirectional and its peak value is more than 300 V. Figure 7 shows the voltage across inductor. Waveforms

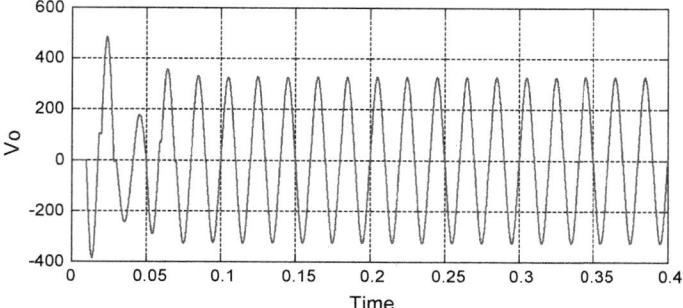

Fig. 4 Output voltage of boost inverter

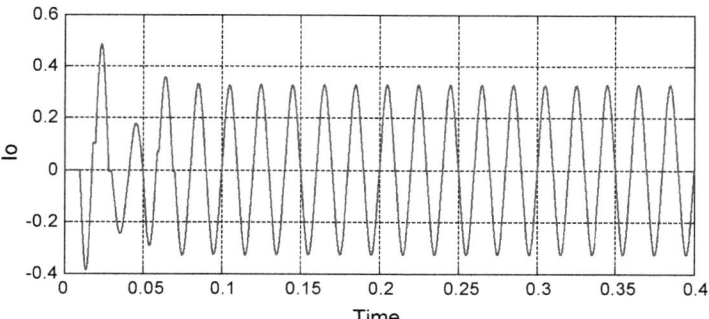

Fig. 5 Output current of boost inverter

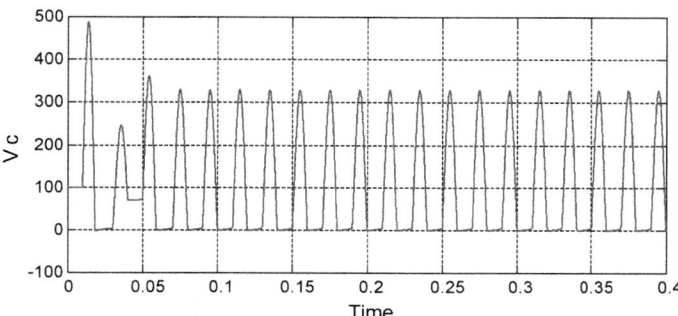

Fig. 6 Capacitor voltage

of different circuit parameters such as inductor voltage, input current of boost inverter circuit, and inductor current are also obtained in MATLAB and also shown in figures. Figure 10 shows the waveform and their FFT analysis of output current of boost inverter. The THD of the output current of boost inverter is 3.89% and which is desirable (Figs. 8 and 9).

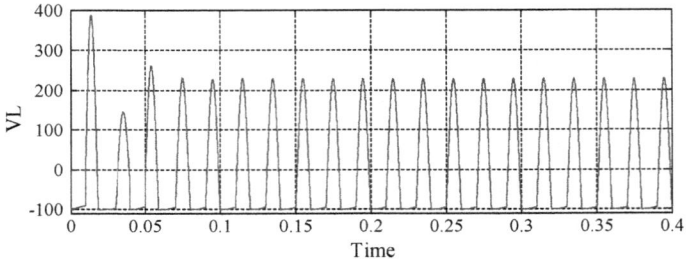

Fig. 7 Inductor voltage

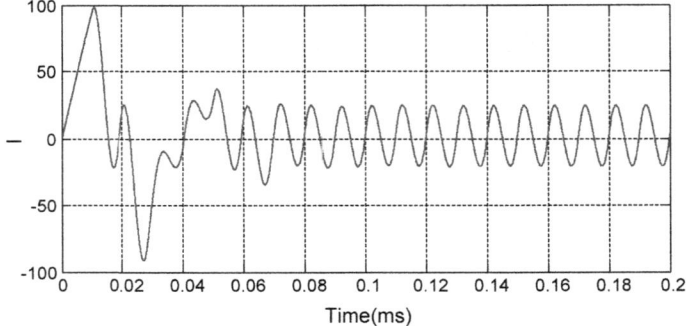

Fig. 8 Input current of boost inverter

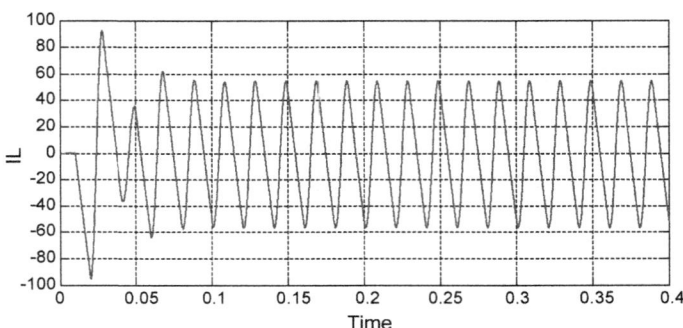

Fig. 9 Inductor current

Signal   Selected signal: 20 cycles. FFT window (in red): 1 cycles

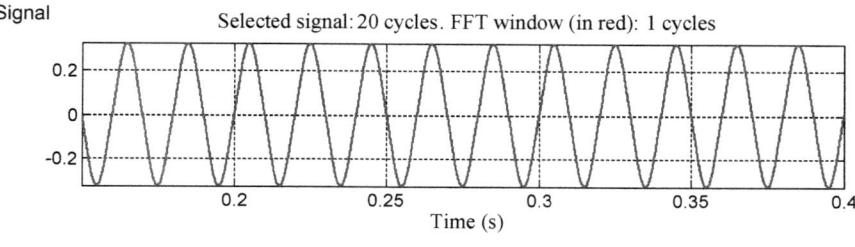

FFT analysis

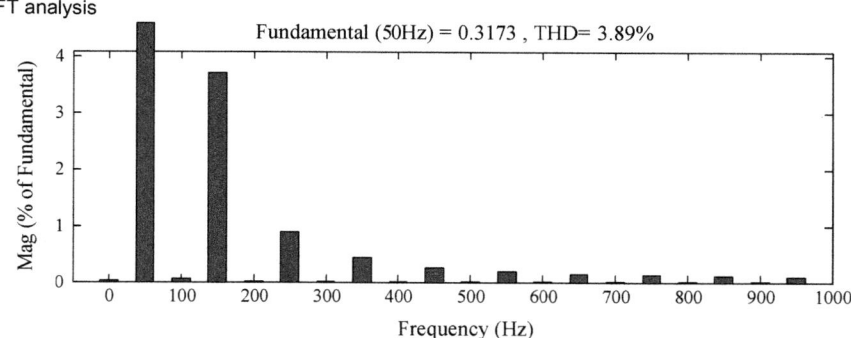

Fig. 10 FFT analysis of output current of boost inverter

4 Comparative Study Table

S. No.	Quantity	Value in Ref. [1]	Value in proposed work
1	Input voltage	96 V	100 V
2	Output voltage across load	200 V	300 V
3	Switching frequency	20 kHz	50 Hz
4	Output frequency	60 Hz	50 Hz

The abovementioned table reflects the comparison between proposed work on voltage source boost inverter versus the boost inverter output result data depicted in Ref. [1].

From the above study table, one may have the idea that the ratio of output voltage to input voltage (boost ratio = 3 times) is more in proposed work as compared to the output result data given in Ref. [1] where the ratio of output voltage to input voltage (boost ratio) is 2.083.

Another import aspect of the work carried out in this paper is the value of switching frequency that is 50 Hz, or one may say that it is the lower value of switching frequency as compared to used in Ref. [1]. Therefore, the lower value of switching losses will occur.

5   Conclusion

In this paper, a different topology of inverter is presented. The boost dc-ac inverter produces ac output voltage whose average value is greater than the input dc voltage. This configuration of dc-ac boost inverter may be applied as power converter in the field of electric vehicles and hybrid electric vehicles to drive the power train of vehicles. The single phase dc-ac boost inverter is able to interface with induction motor and other ac motor drives also.

The tools which are used for simulations has been associated with SimPowerSystem section of Simulink Library of MATLAB.

Acknowledgements The authors acknowledge UGC for providing major research project to the Department of Electrical and Electronic Eengineering BIT Mesra, Ranchi on boost inverter.

References

1. R. Caceres and I. Barbi, "A boost dc-ac converter: Operation, analysis, control and experimentation," in Proc. Int. Conf. Industrial Electronics, Control and instrumentation (IECON'95), Nov. 1995, pp. 546–551.
2. R.O. Caceres, I. Barbi, "A Boost DC-AC Converter: Analysis design, and Experimentation." IEEE Transactions on Power Electronics, pp. 134–140, 1999.
3. R. C´aceres, "DC–AC converters family, derived from the basic dc–dc converters," Ph.D. dissertation, Federal Univ. Santa Catarina, Brazil, 1997 (in Portuguese).
4. V. VorpCrian, "Simplified Analysis of PWM Converters Using the Model of the PWM Switch **Part** I: Continuous Conduction", Proceeding of the VPEC seminar, Blacksburg, **VA,** pp 1–9, 1989.
5. M.H. Rashid. Power electronics: Circuits, Devices and Applications. 3rd Edition, Printice Hall India.

A Study on Filter Design Aspects of Single-Phase Inverter with Various Modulation Schemes

Pritha Roy, J. N. Bera, G. Sarkar and S. Chowdhuri

1 Introduction

Inverter is one of the most important objects for micro-grid model. We can use inverters for both in stand-alone mode and grid-connected mode [1–3]. The main advantage of inverter is the elimination of the low-frequency harmonics. The different techniques like PWM, SPWM, SVM, etc., are helpful for the elimination of low-frequency harmonics during inverter switching. But the inverter output contains the high-frequency harmonics [4]. Filters are necessary to eliminate the high-frequency harmonics. Different authors have considered different types of filters. The output LC filter mainly performs two duties. One is attenuation of ripple in output voltage and the second one is the high-frequency ripple current minimization of the inverter switches. But the selection of LC filter is not very straightforward. The switching frequency voltage attenuation at output node is highly depended on the filter cutoff frequency. Other than the cutoff frequency, the other important matter is the determination of the values of inductor as well as capacitor inside the filter. The proper measurement of L and C inside an LC filter are dependent on the output impedance of the inverter, transient response, efficiency and cost of the inverter [5, 6].

P. Roy (✉) · J. N. Bera · G. Sarkar · S. Chowdhuri
Department of Applied Physics, Calcutta University, Kolkata, India
e-mail: roypritha09@gmail.comd

J. N. Bera
e-mail: jitendrabera@rediffmail.com

G. Sarkar
e-mail: gautamgs2010@yahoo.in

S. Chowdhuri
e-mail: sumana_cu05@rediffmail.com

© Springer Nature Singapore Pte Ltd. 2019   17
V. Nath and J. K. Mandal (eds.), *Proceeding of the Second International Conference on Microelectronics, Computing & Communication Systems (MCCS 2017)*, Lecture Notes in Electrical Engineering 476, https://doi.org/10.1007/978-981-10-8234-4_3

2   Materials and Methods

2.1   *Adopted Modulation Schemes of the Inverter for Comparison*

The sinusoidal pulse width modulation (SPWM) technique is adopted here. This SPWM technique is capable of minimizing the level of THD inside the output waveform. This technique involves a comparison in between a triangular carrier wave with desired ac waveform which is called the modulation waveform. Frequency and amplitude of the carrier wave are fixed. The amplitude of modulation waveform is generally a sinusoid. For example, if A_m is the amplitude of modulating sinusoidal and A_c is the amplitude of the triangular carrier, then the modulation index (MI) will be the ratio of A_m/A_c. MI varies in between 0 to 1 ($0 \leq MI \leq 1$). The magnitude of the output voltage can be adjusted by controlling this MI [7–9].

The bipolar modulation and unipolar modulation are the two basic schemes of SPWM. The main difference between them, with respect to their output voltage is the presence of harmonic content near the carrier frequency as shown in Fig. 1. The high-frequency harmonics of Fig. 1 can be eliminated by selecting the values of inductor and capacitor of LC filter at the output of SPWM inverter properly.

2.2   *Filter Design Procedure*

For comparison between the bipolar and unipolar SPWM schemes with respect to their output LC filter design, the main characteristics of an inverter are mentioned in Table 1.

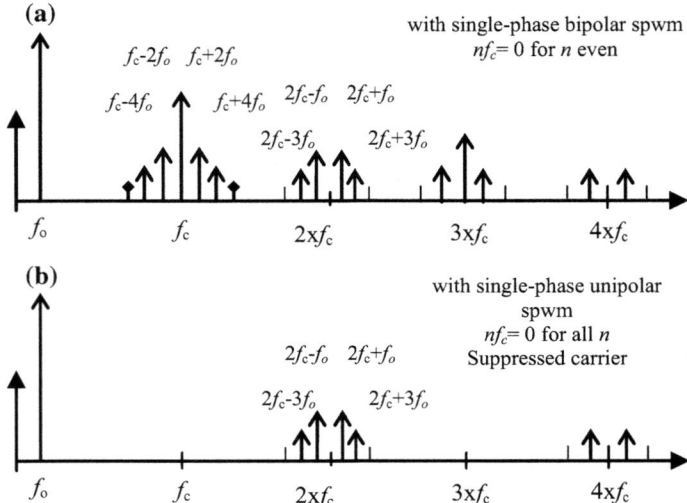

Fig. 1 Harmonic content for **a** bipolar; **b** unipolar SPWM

Table 1 Specification of the inverter

Parameter	Value
V_{DC}	450 V
V_O	240 V_{RMS}
S_{OUT}	2 KVA
f_S	10 kHz
f_1	50 Hz

Filter at the output of the inverter should be properly designed to attenuate ripple at the output voltage and to minimize the ripple current of the inverter switches which is mainly due to high-frequency switching. Here, cutoff frequency of the LC filter is selected depending on the Fourier series expansion of output voltage of the inverter [10, 11]. Evaluation of the parameter of the LC filter will be erroneous by considering only the total harmonics of the capacitor voltage. The other criterion is also used to specify these parameters which are based on the least amount of reactive power consumption of LC filter. The attenuation which is done by the filter depends on the value of load cutoff frequency and also the values of the inductor or capacitor. The proper measurement of inductor and capacitor inside the LC filter depend upon the output impedance of the inverter, transient response, efficiency and cost of the inverter. We can also able to select the values of L and C depending on the reactive power minimization technique. But this technique will lead a large inductor value. So this approach increases the size as well as the cost of the filter. The inductor mainly controls the ripple current of the inverter. The maximum allowable ripple current and the inverter switching frequency are also responsible to find out the minimum acceptable value of the inductor [12]. The proposed system is designed and analyzed using the environment of MATLAB®/SIMULINK.

3 Description of Proposed System

Figure 2 shows the schematic diagram of the proposed system. The inverter is designed on SPWM technique with an attached DC source. An LC filter is connected in between the inverter and the load for rejection of the high-frequency disturbance. Equations (1) and (2), as given below, are used to design the filter, where I_L is the current flowing through the inductor and V_L is the voltage drop across the inductor.

$$\Delta I_L = \frac{V_L}{L} \Delta t \tag{1}$$

Fig. 2 Schematic diagram of the proposed system

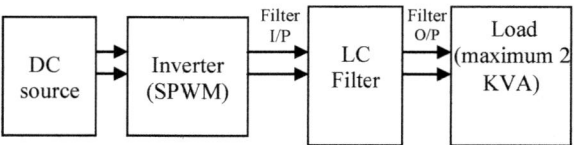

$$V_L = V_{DC} - V_{OM} \sin(\omega_1 t) \tag{2}$$

In Eq. (2), V_{DC} is the modulated output of the inverter when the switches are on and $V_{OM} \sin(\omega_1 t)$ is the peak value of the instantaneous voltage across the load. Δt is the duration of the switches when they are on and f_s is the switching frequency

$$\Delta t = \frac{m \times \sin(\omega_1 t)}{f_s} \tag{3}$$

Combining (1), (2), (3) and dividing by the inductor fundamental current:

$$\frac{\Delta I_L}{I_L} = \frac{m\omega_1 \times (V_{DC} - V_{OM} \sin(\omega_1 t)) \times \sin(\omega_1 t)}{f_s I_L L \omega_1} \tag{4}$$

$$\frac{\Delta I_L}{I_L} = \frac{m\omega_1 \times (V_{DC} - V_{OM} \sin(\omega_1 t)) \times \sin(\omega_1 t)}{f_s |V_L(\omega_1)|} \tag{5}$$

The percentage of ripple current can be verified using Eq. (4) or (5). The inductor voltage is considered as $V_L(\omega_1)$ at fundamental frequency. $V_L(\omega_1)$ is taken as the fraction of the output voltage:

$$|V_L(\omega_1)| = \alpha \times |V_O(\omega_1)| \tag{6}$$

Here α is the fraction of the output voltage ($V_O(\omega_1)$) at fundamental frequency. It depends on the switching frequency at the same time on the maximum allowable ripple current [12].

Now the inductor and capacitor values are

$$L = \frac{R_{Lm}}{\omega_1} \sqrt{\left(\alpha^2 - \frac{\omega_1^4}{\omega_r^4} \right)} \tag{7}$$

$$C = \frac{1}{R_{Lm}} \sqrt{\left(\frac{\omega_1^2}{\alpha^2 \omega_r^4 - \omega_1^4} \right)} \tag{8}$$

R_{Lm} is the maximum load of the inverter. ω_r is the cutoff frequency of the LC filter. Equation (9) will be satisfied at the time of cutoff frequency calculation of the LC filter.

$$\omega_r > \frac{\omega_1}{\sqrt{\alpha}} \tag{9}$$

Table 2 is showing that for the same values of α, inductor values are remaining the same though the cutoff frequency is changing.

Plotting in Fig. 3 is done using the values in Table 3. This is showing linear characteristics between α (fraction of output voltage) and L (value of inductor). This

Table 2 Observation of the values of Inductor and capacitor for designing the filter

α (fraction of output voltage)	ω_r in rad	L in Henry	C in Farad
0.03	12,560	0.0027510	0.00000230426
0.03	18,840	0.0027510	0.00000102390
0.03	25,120	0.0027510	0.00000057595
0.03	25,182	0.0027510	0.00000057308
0.03	25,246	0.0027510	0.00000057023
0.03	25,434	0.0027510	0.00000056181
0.03	31,400	0.0027510	0.00000036860
0.03	37,680	0.0027510	0.00000025597
0.04	25,120	0.0036688	0.00000043196
0.04	25,182	0.0036688	0.00000042981
0.04	25,246	0.0036688	0.00000042767
0.04	25,308	0.0036688	0.00000042555

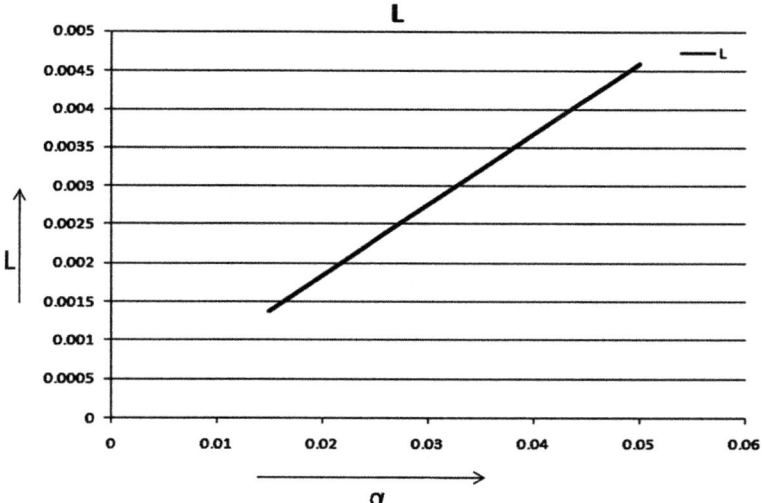

Fig. 3 Plotting of α versus inductor value

Table 3 Observation for filter designing	α (fraction of output voltage)	L in Henry
	0.015	0.0013757 H
	0.030	0.0027510 H
	0.040	0.0036688 H
	0.042	0.0038522 H
	0.045	0.0042274 H
	0.050	0.0045860 H

is depicted in Fig. 3. By selecting a very low value of α, an inductor with a very small value can be chosen.

4   Case Studies with Simulated Diagram and Results

Figure 4 is the experimental model of the single-phase inverter in MATLAB®/ SIMULINK. The values inductor and capacitor and cutoff frequency of this model are shown in Table 4.

Figure 5 is showing the bode characteristics of the filter, the cutoff frequency is verified using Bode plot which is exactly same as the calculated value.

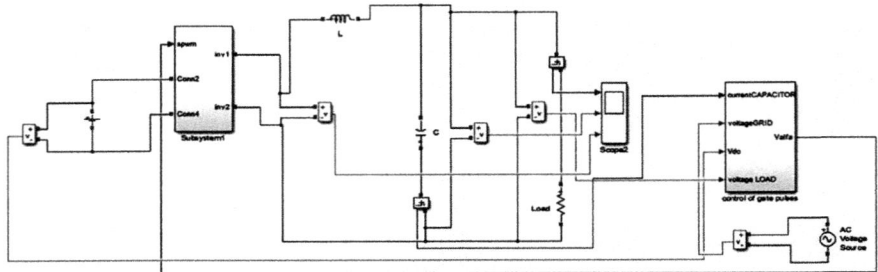

Fig. 4 MATLAB model of the proposed system in closed loop

Table 4 Filter characteristic

Cutoff frequency (f_r)	L in Henry	C in Farad
4020 Hz	0.003688 H	0.00000042767 F

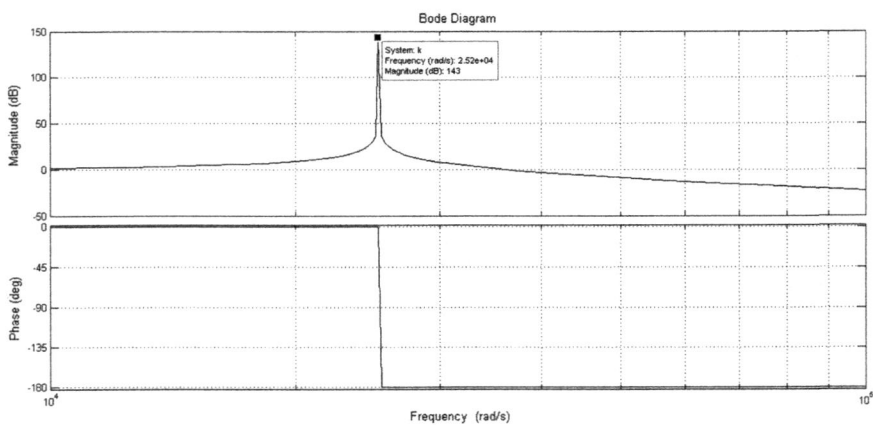

Fig. 5 Bode plot of the designed filter

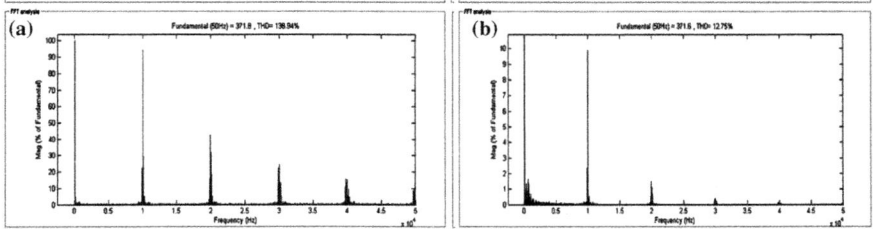

Fig. 6 FFt analysis of the bipolar SPWM **a** before the filter (THD (138.94%)) **b** after the filter (THD (12.75%))

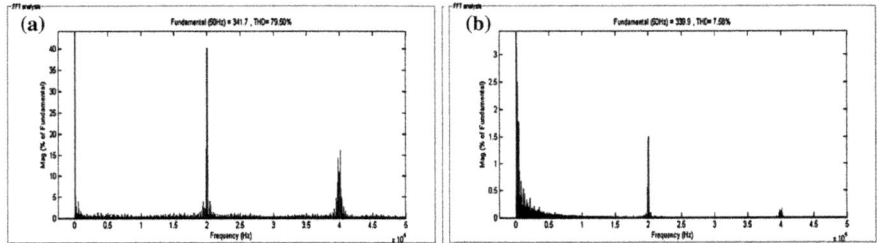

Fig. 7 FFt analysis of the unipolar SPWM **a** before the filter (THD (79.50%)) **b** after the filter (THD (7.58%))

The output waveforms for bipolar and unipolar modulation scheme are captured and the harmonic components are verified. Figures 6 and 7 are representing the harmonic components inside the voltage waveform before and after using the filter for bipolar and unipolar SPWM respectively.

5 Conclusion

A study of the design aspects of filter design for the inverter is made. To eliminate the error between the reference voltage and instantaneous output voltage, PR controller is used and to eliminate the error of current feedback loop, PI controller is used.

As mentioned in the IEEE standard 1547, the total harmonics distortion should be within 5% and the corresponding harmonic distortion is also mentioned in the standard, as given in Table 5.

As seen from Figs. 6 and 7, using the bipolar SPWM the total harmonic distortion is 12.75% but for unipolar modulation technique, it is Σ 7.59%. It is seen that the unipolar modulation creates less harmonic distortion than bipolar modulation and it is very much nearer to the IEEE standard.

Table 5 Maximum harmonic voltage distortion as per the IEEE Standard 1547

Individual harmonic order	H < 11	$11 \leq H < 17$	$17 \leq H < 23$	$23 \leq H < 35$	$35 \leq H$	Total harmonic distortion
Percentage (%)	4.0	2.0	1.5	0.6	0.3	5.0

The difference between the values of LC for bipolar and unipolar modulation is noted and this justifies the theoretical expression for both the modulation schemes.

Acknowledgements The authors acknowledge the University Grant Commission (UGC), India, for providing necessary support and infrastructural facility under the auspices of UGC SAP-DRS-II project.

References

1. Mohammad Monfared, Saeed Golestan and Josep M. Guerrero "Analysis, Design, and Experimental Verification of a Synchronous Reference Frame Voltage Control for Single-Phase Inverters" IEEE TRANSACTIONS ON INDUSTRIAL ELECTRONICS, VOL. 61, NO. 1, JANUARY 2014.
2. J. J. Pollack, "Advanced pulse width modulated inverter techniques." IEEE Trans. Ind. Appl.. vol. IAS-8. pp. 145–154, Mar./Apr. 1972.
3. Nikos Hatziargyriou, Hiroshi Asano, Reza Iravani, and Chris Marnay "Microgrid - An Overview of Ongoing Research, Development, and Demonstration Projects" in IEEE power & energy magazine, July-August 2007.
4. Pritha Roy, J.N. Bera, S. Chowdhuri and G. Sarkar "Synchronization Aspects of Single Phase SPWM Inverters for Microgrid Mode of Operation" 2016 2nd International Conference on Control, Instrumentation, Energy & Communication (CIEC), 28–30 Jan. 2016, IEEE.
5. Hyosung Kim, S.K.S., Analysis on Oupput LC filter for PWM inverter" IPEMC2009, 2009: p. 6.
6. J. Kim, S.M., IEEE, J. Choi, Member, IEEE, H. Hong, Student Member, IEEE, Output LC Filter Design of Voltage Source Inverter Considering the Performance of Controller. 2000: p. 6.
7. Pankaj H Zope, Pravin G. Bhangale, Prashant Sonare, S. R. Suralkar "Design and Implementation of carrier based Sinusoidal PWM Inverter" International Journal of Advanced Research in Electrical, Electronics and Instrumentation Engineering Vol. 1, Issue 4, October 2012, ISSN: 2278 – 8875.
8. Anuja Namboodiri, Harshal S. Wani "Unipolar and Bipolar PWM Inverter" IJIRST – International Journal for Innovative Research in Science & Technology| Volume 1 | Issue 7 | December 2014 ISSN (online): 2349–6010.
9. J. Holtz, "Pulse width modulation – A survey", *IEEETrans. Ind. Electr.*, v. 39, pp. 410–419, 1992.
10. S. B. Dewan, P.D.Z., *Optimum Filter Design for a Single Phase Solid State UPS System.* IEEE Transaction on Industrial Application, 1975. **IA-21**(3): p. 6.
11. H. van der Broeck, Untersuchung des Oberschwingungsverhaltens eines hochtaktenden Vierquadmntstellers, ET2 Archiv., Vol. 8, No.6, June 1986, pp. 195–199.
12. Ahmed Ale Ahmed, Abid Abrishamifar and Mohammad Farzi, "A New Design Procedure for Output LC Filter of Single Phase Inverter" 2010 3rd international Conference on Power Electronics and Intelligent Transportation System, IEEE.

Design, Analysis, and Testing of Low-Voltage CMOS OTA

Maninder Kaur and Jasdeep Kaur

1 Introduction

Very large scale integrated circuit design (mixed and analog VLSI) and testing with high performance undergoes many difficulties due to power consumption, cost, area overhead, and power supply. Needs for low-voltage low-power analog integrated circuit with little or no reduction in performance exists and are widely used in many applications such as telecommunication, remote computing, medical, etc. [1, 2]. Also as the level of integration increases, testing of the low-voltage VLSI circuits has become the critical portion of the circuit designing and implementation. It also increases the overall cost of the circuits. Integrated circuit has been tested for critical parameters such as slew rate, gain, gain bandwidth, signal to noise ratio, linearity, etc., which result in high cost, large testing time, and poor fault coverage [3, 4].

In this paper, a well-defined design and test procedure based on main parameters such as slew rate, load capacitance, gain bandwidth, output swing, input common mode range, etc., is presented for two-stage low-voltage OTA. The OTA is an essential building block in mixed and analog circuit design with linear input–output characteristics. It is used for implementing digital to analog converters (DAC), analog to digital converters (ADC), filters, voltage controlled oscillators, buffers, etc. [5].

M. Kaur (✉)
Guru Tegh Bahadur Institute of Technology (Affiliated to Guru Gobind Singh Indraprastha University), New Delhi, India
e-mail: mindersaini82@gmail.com

J. Kaur
Indira Gandhi Delhi Technical University for Women, Kashmere Gate, New Delhi, India
e-mail: jasdeepkaur@yahoo.com

© Springer Nature Singapore Pte Ltd. 2019
V. Nath and J. K. Mandal (eds.), *Proceeding of the Second International Conference on Microelectronics, Computing & Communication Systems (MCCS 2017),* Lecture Notes in Electrical Engineering 476, https://doi.org/10.1007/978-981-10-8234-4_4

Two-stage OTA is very robust and simple topology that provides reliable performance in terms of its electrical parameters like dc gain, linearity, common mode rejection ratio (CMRR), slew rate, output swing, etc. [6].

I_{DDQ} testing is also presented in this paper to detect both open and short defects introduced in OTA circuit using fault injection transistors. It is a current-based test method based on calculating the current on power supply and compared it with threshold limit. The fault is said to be detected if measured quiescent current exceeds threshold limit. CUT draws a very low quiescent current but due to the presence of faults, this current may rise to large value [7, 8].

The format of this paper is as follows; two-stage operational transconductance amplifiers are discussed in Sect. 2. Section 2 introduces I_{DDQ} testing. Simulation results and discussion are given in Sects. IV and V contain conclusions.

2 Two-Stage Operational Transconductance Amplifier Circuit Description

OTA is a voltage-controlled current source because it converts input voltage difference into current. The OTA is an operational amplifier with output buffer. It replaces operational amplifier as of its high voltage swing, high signal to noise ratio, bandwidth, high input impedance, and low-power dissipation even at low power and low voltage. The amplifier's transconductance is controlled by additional input current which is amplifier bias current [9, 10]. The OTA has all terminals at low impedance except the input and output terminal. It is used to drive capacitive loads. The output current is a linear function of the differential input voltage as shown in Eq. 1.

$$I_{out} = g_m(Vin^+ - Vin^-) \qquad (1)$$

where I_{out}, Vin^+, Vin^-, and g_m are output current, non-inverting, and inverting input voltages and transconductance, respectively. The transconductance (g_m) is proportionality constant between the output current and input differential voltage. g_m also depends on temperature. It provides more reliable performance at higher frequencies due to its current mode operation.

The schematic diagram of two-stage CMOS OTA is shown in Fig. 1. The first stage comprises of differential gain stage which converts differential input into single output with high gain. The driver stage of the circuit is formed by the transistors M1 and M2 due to improved slew rate and reduced power supply rejection ratio. Transistors M3 and M4 is current mirror which act as active load to provide high output resistance and reduces power supply rejection ratio. The second gain stage consists of transistor M6 and M7 to provide additional gain in the circuit. It is basically a current sink load inverter to provide high voltage swing and high gain. The input resistance of the first stage as well the second stage of the OTA

Fig. 1 Schematic diagram of two-stage CMOS OTA (CUT)

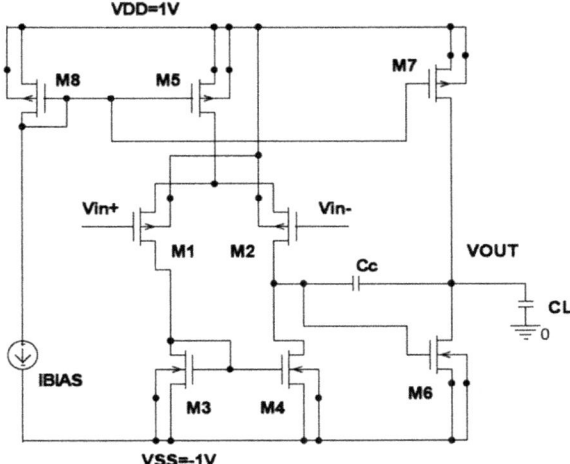

circuit is infinite [11]. The output resistance (R_{out}) of the second stage of the circuit is represented by Eq. 2:

$$R_{out} = R_{06}R_{07}/R_{06} + R_{07} \qquad (2)$$

where R_{06} and R_{07} are the internal resistance of M6 and M7, respectively [12].

The high gain of the second stage and an overall gain for the complete circuit depend upon output resistance. The bias current (IBIAS) of the circuit flows through current mirror formed by transistors M5, M7, and M8 which are P-type MOSFETs [13]. The aspect ratio of transistors is designed to produce bias current of 113 μA to generate sufficient gain, gain bandwidth, output voltage swing, and slew rate. The value of C_c and C_L is 1 and 3 pf, respectively.

Table 1 tabulated the values of channel Width (W) in μm for all transistors when channel length is 0.18 μm in Figs. 1 and 4.

Table 1 Values of channel width (W) for all transistors ($L = 0.18$ μm)

Parameters	Values
Channel length (μm)	0.18
$W_{1,2}$ (μm)	25
$W_{3,4}$ (μm)	10
W_5 (μm)	20
W_6 (μm)	11
W_7 (μm)	40
W_8 (μm)	20

2.1 Fundamental Specifications

Specifications considered for designing of the two-stage low-voltage OTA are supply voltage (VDD), load capacitance (C_L), gain margin, phase margin, gain bandwidth, common mode rejection ration, slew rate, output voltage swing, power dissipation, and offset. The performance of an OTA is classified based on time domain and frequency domain parameters [14, 15]. The frequency domain parameters are bandwidth, quality factor, gain, and phase. Slew rate is an important parameter that influences the circuit design in its time and frequency domain.

(1) *Slew Rate*: It is the rate of change of output voltage, whose rate of change is limited due to the electronic circuitry inside the OTA. It supplies small current to charge and discharge the capacitor. It can be obtained as shown in Eq. 3:

$$\frac{dv_0}{dt} = \frac{i}{c} = \frac{g_m V_{in}}{c} \tag{3}$$

where v_0, V_{in}, g_m, and c are output voltage, input voltage, transconductance and capacitance, respectively.

(2) *Gain*: The ratio of output voltage to the input voltage of the circuit is known as gain of the circuit. It is a measure of the ability of the circuit to increase the amplitude of a signal from the input to the output. It is measured in dB. The open loop gain of an amplifier determines the precision of the feedback system. A high open loop gain is necessary to suppress nonlinearity. The open loop gain is given by Eq. 4:

$$A_v dB = 20 \log_{10} \left(\frac{V_{out}}{V^+ - V^-} \right) \tag{4}$$

(3) *Power dissipation*: Power dissipation is easily calculated from the supply voltage and current when the output is open circuited. When current flows into a load, it is easy to calculate the total dissipation and then subtract the load dissipation to obtain the device dissipation. Unity gain frequency and slew rate of the OTA circuit can be reduced by increasing the load capacitance. Settling behaviour of the OTA can be maintained constant by increasing load capacitance as a result which increases the power consumption.

2.2 Design Steps Followed for Two-Stage CMOS OTA

See Fig. 2.

Fig. 2 Design steps of Two-Stage CMOS Operational Transconductance Amplifier

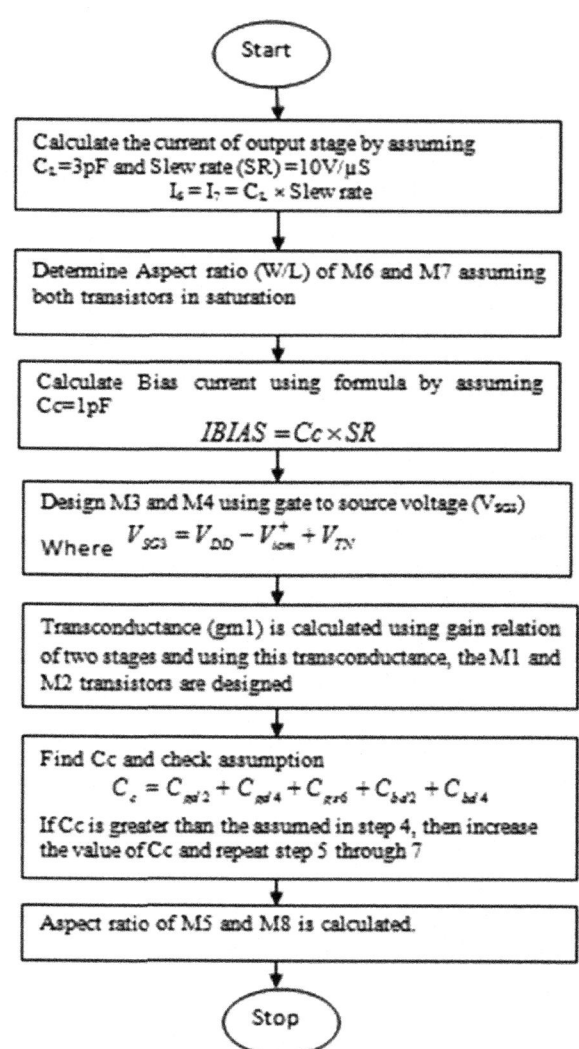

Start

Calculate the current of output stage by assuming
C_L=3pF and Slew rate (SR)=10V/μS
$I_6 = I_7 = C_L \times$ Slew rate

Determine Aspect ratio (W/L) of M6 and M7 assuming both transistors in saturation

Calculate Bias current using formula by assuming Cc=1pF
$$IBIAS = Cc \times SR$$

Design M3 and M4 using gate to source voltage (V_{SG3})
Where $V_{SG3} = V_{DD} - V_{icm}^{+} + V_{TN}$

Transconductance (gm1) is calculated using gain relation of two stages and using this transconductance, the M1 and M2 transistors are designed

Find Cc and check assumption
$$C_c = C_{gd2} + C_{gd4} + C_{gr6} + C_{bd2} + C_{bd4}$$
If Cc is greater than the assumed in step 4, then increase the value of Cc and repeat step 5 through 7

Aspect ratio of M5 and M8 is calculated.

Stop

3 I_{DDQ} Testing

I_{DDQ} testing is the testing technique based upon determination of quiescent current for testing both digital as well as analog VLSI circuit. In the presence of fault, the values of quiescent current may be decreased or increased. Thus, the quiescent current fluctuations are monitored by built-in current sensor (BICS) to accomplished fault detection. The presence of physical defects in the circuit would be indicated by the current which is above quiescent current [16, 17]. The block diagram of I_{DDQ} testing in Fig. 3 indicates that BICS is connected in series with ground lines of the CUT.

During quiescent state condition, the current of the power supply (VDD) or ground (GND) terminals is monitored after the inputs have changed and before next input change and is compared with threshold value. FIT which is n-MOS transistor is used to induce the fault in the circuit. The circuit operates in normal mode, when FIT is inactive while circuit work in test mode by applying voltage across the FIT. FIT prevents performance degradation and permanent damage to the CUT [18]. In the present work, total 14 faults (seven short and seven open faults) have been introduced in CUT with BICS as shown in Fig. 4.

Seven short faults are shown in the Fig. 4 whereas open faults are introduced in CUT by connecting gate of transistor to the VSS. Both short and open faults on CUT are detected using simple BICS without performance degradation of circuit. BICS uses less area as there is no requirement of external voltage or current source. The CUT operates in normal as well as in test mode. During normal mode, the BICS is totally disconnected from the circuit while in the test mode BICS is connected to the circuit. The proposed BICS design transforms the faulty current (I_{DD}) and the reference current (I_{REF}) difference to a voltage which depicts faulty circuit and fault-free circuit. The existence of fault in the circuit is indicated when I_{DD} is greater than I_{REF} that set the output signal (PASS/FAIL) of BICS to 1 otherwise output signal (PASS/FAIL) of BICS is 0 [19–21].

Fig. 3 Block diagram of I_{DDQ} testing

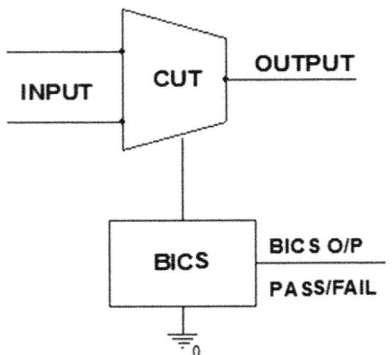

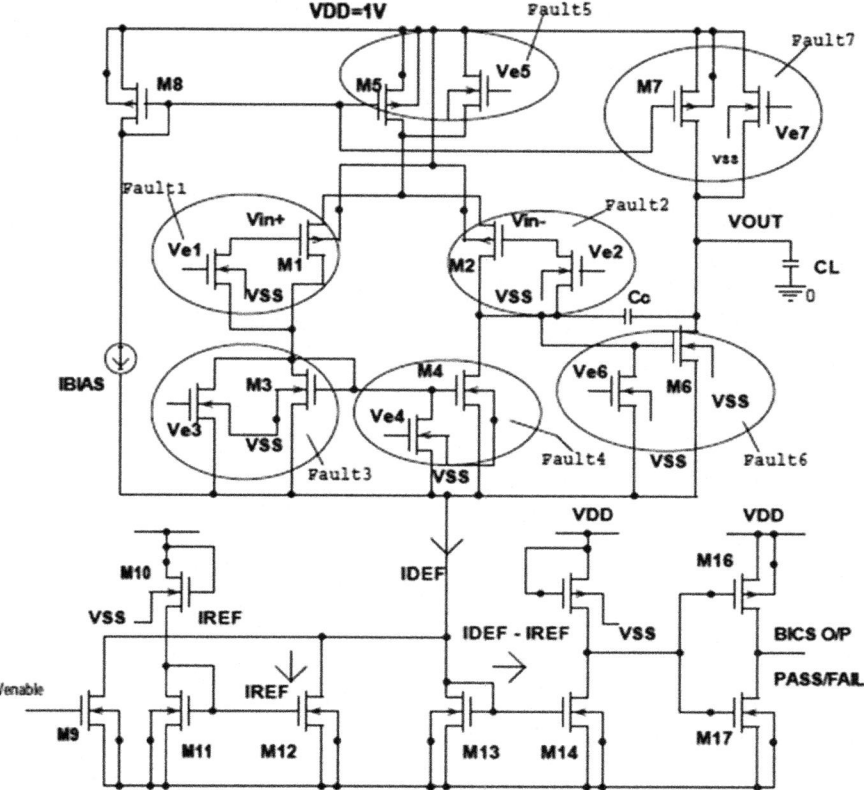

Fig. 4 Schematic diagram of CUT with built in current sensor (BICS) and seven short faults

4 Simulation Results and Discussions

The CUT is simulated using CMOS 0.18 μm technology.

4.1 Transient Analysis

In this type of analysis, a sinusoid signal of 5 mV is applied to the positive input terminal of OTA and output voltage swing of 1.3–1.4 V is obtained as shown in Fig. 5.

4.2 DC Analysis

Using DC analysis, the transfer characteristics is achieved as shown in Fig. 6. From the results, the calculated offset voltage at the input is 0.5 mV.

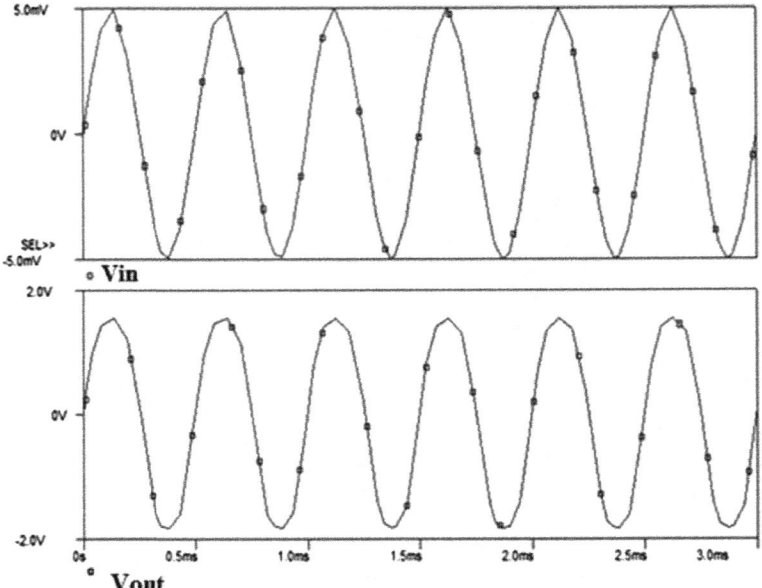

Fig. 5 Transient response of CMOS OTA

4.3 AC Analysis

AC analysis performed to determine the value of gain, gain bandwidth, and phase margin which is shown in Fig. 7 for sinusoidal input with start frequency = 1 kHz and stop frequency = 1 GHz. The gain margin and DC gain are obtained as 30 and 62 dB, respectively, from the simulated results.

Figure 8 illustrates simulated output frequency response which shows 3 dB Gain of 59 dB with a large GBW of 29.5 and 0.5 MHz of 3 dB Bandwidth.

A step response of CMOS OTA is depicted in Fig. 9 which gives the slew rate as 11 V/μS. The accurate agreement between expected and simulated slew rate is apparent.

Table 2 compiles the performance comparison of the proposed OTA with the previous work. Due to different design technologies of OTAs, explicit comparison cannot be obtained but few important results can be concluded from Table 2. The proposed design consumes less voltage and power than previous published design. It is also better in terms of gain, gain margin and slew rate. Therefore, it is discernible that proposed design is better than previously designed OTA's.

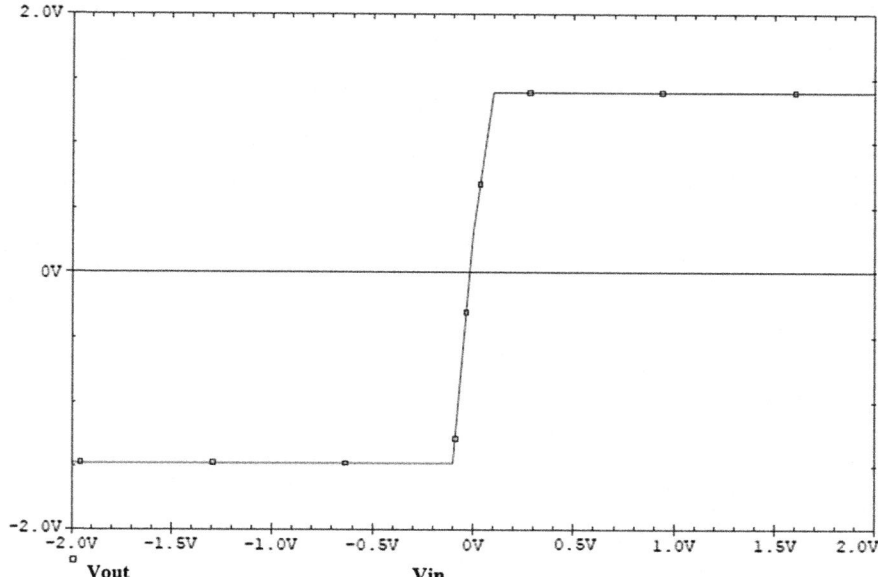

Fig. 6 DC sweep of CMOS OTA

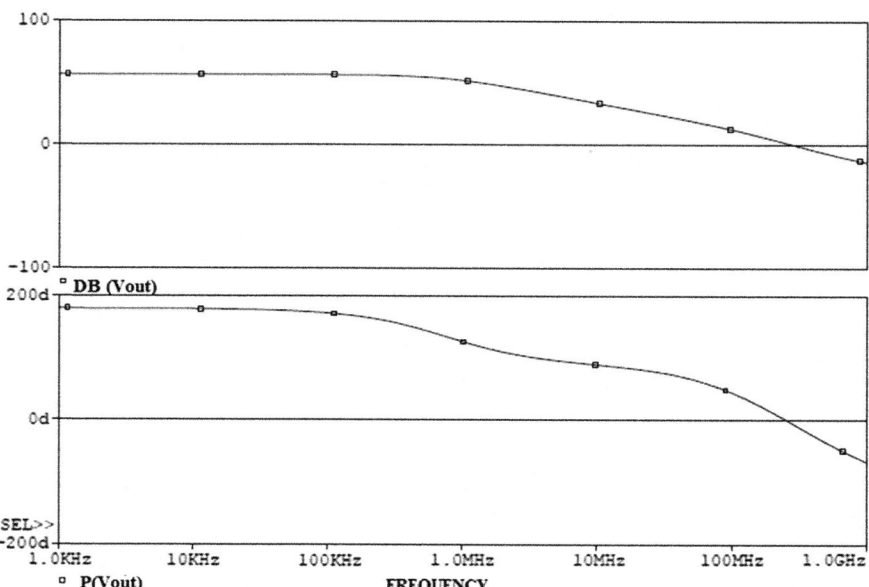

Fig. 7 AC response of CMOS OTA (amplitude and phase)

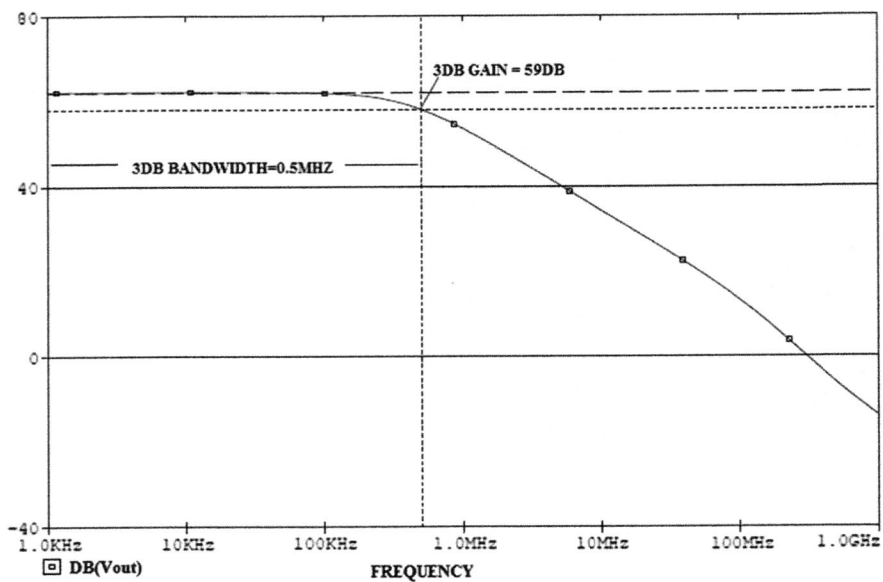

Fig. 8 Simulated frequency response characteristic of OTA

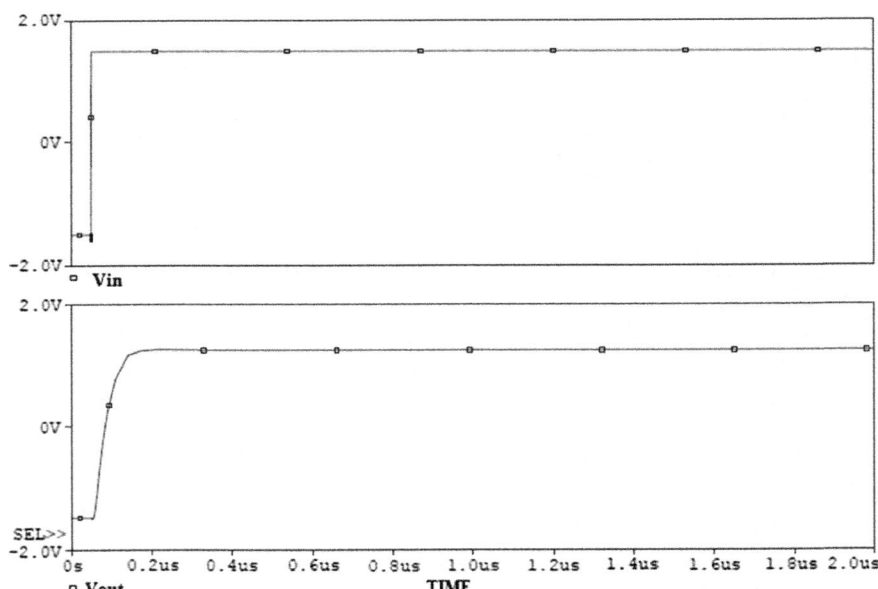

Fig. 9 Step response of CMOS OTA

Table 2 Comparison of proposed OTA with previous work

Technology used (μm)	Power supply (V)	Power dissipation (mW)	Slew rate (V/μS)	Gain (DB)	Gain margin (DB)
1.2 [6]	2	–	11.5	67	–
0.35 [13]	3.3	3.4	26	48	50
0.35 [12]	3.3	3.2	31.67	46.75	16.94
0.18 [12]	1.8	1.3	37.58	47.8	15.4
This work (0.18)	1	1.1	11	59	30

4.4 Simulated Results Using I$_{DDQ}$ Testing

I$_{DDQ}$ test approach based on the simulated results obtained from PSPICE simulations using n-well 0.18 μm technology. Figure 5 exhibits the simulated output response of CUT without BICS whereas Fig. 10 shows the output of CUT with BISC when no fault is introduced in the circuit. Since the output of CUT is 1.2 V

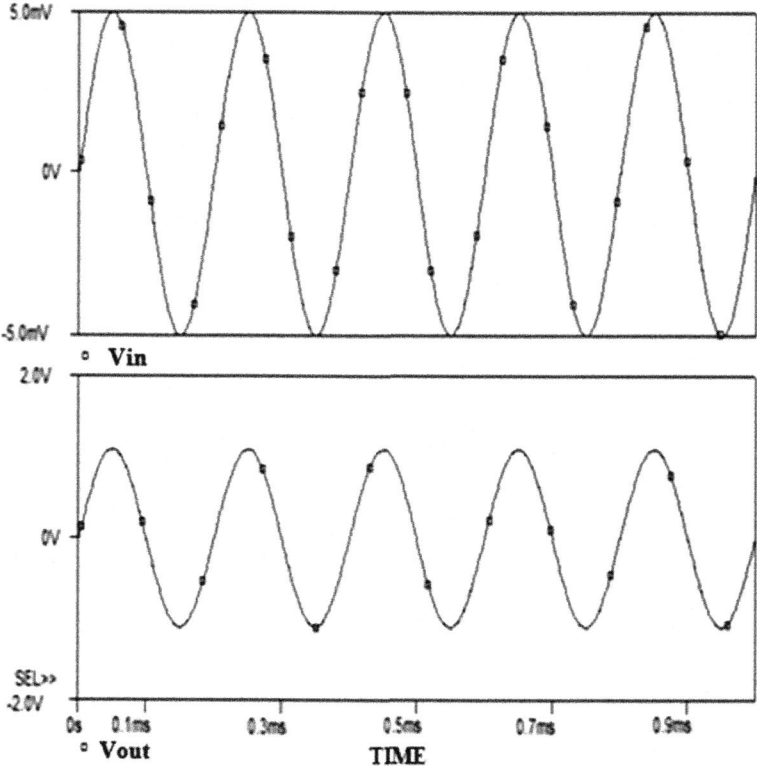

Fig. 10 Simulated input and output response of CUT (without fault)

Table 3 Simulated I_{DDQ} ($I_{REF} = 400$ μA)

Component	Fault type	Simulated I_{DDQ} (μA)	Fault type	Simulated I_{DDQ} (μA)
M1	GDS	500	OPEN	990
M2	GDS	1000	OPEN	650
M3	DSS	870	OPEN	1100
M4	GDS	990	OPEN	800
M5	DSS	910	OPEN	150
M6	GDS	600	OPEN	1800
M7	DSS	550	OPEN	200

GDS gate–drain short, *DSS* drain–source short

Table 4 Fault analysis by I_{DDQ} testing

Type of fault	Number of faults	Number of fault detected	Fault coverage (%)
Short	7	7	100
Open	7	5	71.42
Total	14	12	85.71

peak to peak when BICS is present, we can see that there is little performance degradation of the CUT with BICS.

Total 14 faults are injected in the CUT. Simulated quiescent current obtained is shown in Table 3. The results obtained show that seven short faults and five open faults out of fourteen faults are detected by this methodology.

The simulated fault coverage by I_{DDQ} test methodology is shown in Table 4. It is concluded from the simulated results that total 12 faults out of 14 faults are detected which leads to the fault coverage of 85.71%.

5 Conclusion

This paper provides novel design procedure and I_{DDQ} testing method for two-stage CMOS OTA using CMOS 0.18 μm technology. The proposed method regnant in comparison with previous work in terms of slew rate, gain (dB), GBW, output voltage swing, and power consumption. It has low-power consumption, better output voltage swing, improved gain and GBW. It is also suitable for high-frequency operation with good performance at low-voltage and low-power dissipation. Also, simulation test results show that I_{DDQ} testing technique is a valuable tool to achieve high fault coverage and low cost. Out of 14 faults which include 7 bridging and 7 open faults, 12 faults have been detected by this test methodology. Thus, I_{DDQ} testing methodology provides 85.71% fault coverage. It is exhibited that BICS design with the use of a FIT has notably improved the testing of mixed signal VLSI circuits. It is concluded that proposed design and test method are a valid aid for analog and mixed signal integrates circuit designers.

References

1. P.R Gray and R.G Mayer, "Analysis and Design of Analog Integrated Circuits", New York: Wiley, 2001.
2. Sayan Bandyopadhyay, Deep Mukherjee and Rajdeep Chatterjee, "Design of Two Stage CMOS Op-Amp in 180 nm Technology with low power & high CMRR," Int. J. of Recent Trends in Engineering & Technology, ACEEE, vol. 11, pp. 239-247, June 2011.
3. M. L. Bushnell and V. D. Aggarwal, "Essentials of Electronic Testing for Digital, Memory and Mixed -Signal VLSI circuits", Kluwer Academic Publication.
4. Stanley L. Hurst, VLSI Testing: Digital and Mixed Analogue/Digital Techniques, Institution of Engineering and Technology, 1998.
5. Randall L Geigir and Edgar Sanchez Sineccio, "Active Filter design using Operational Transconductance Amplifier: A tutorial," IEEE Circuit and Devices Magazines, Vol. 1, Issue 2, pp 20–32, 1985.
6. G. Palmisano, G. Palumba and S. Pennisi, "Design Procedure for Two-stage CMOS Transconducatnce Operational Amplifier: A Tutorial", article in Analog Integrated Circuits and Signal Processing, 27, 179–189, 2001.
7. Linda S. Milor, Member, IEEE, "A Tutorial Introduction to Research on Analog and Mixed-Signal Circuit Testing," IEEE Transactions on circuits and system-II: Analog and digital Signal processing, vol 45, no. 10 october 1998.
8. R. K. Gulati and C. F Hawkins, "IDDQ testing of VLSI circuits", Kluwer Academic, The Netherland, 1993.
9. O. M. Saravanakumar, N. Kaleeswari and K. Rajendran, "Design and Analysis of Two-Stage Operational Amplifier (OTA) using Candence Tool", International Journal of Emerging Technology and Advanced Engineering, Volume 4, Issue 4, April 2014.
10. Soolmaz Abbasalizadeh, Samad Sheikhaei and Behjat Forouzandeh, " A 0.9 V Supply OTA in 0.18 μm CMOS Technology and Its Application in Realizing a Tunable Low-Pass Gm-C Filter for Wireless Sensor Networks", proceeding of Circuits and Systems, 4, 34–43, 2013.
11. Dhaval Modi and Jayesh Patel, "Design and simulation for CMOS OTA with 1.0 V, 55db Gain & 5PF Load," International Journal of Managing Public Sector Information and Communication Technologies (IJMPICT), Vol. 5, No. 2, June 2014.
12. Hitesh Modi and Nilesh D. Patel, "Design and Simulation of Two Stage OTA Using 0.18 μm and 0.35 μm Technology", International Journal of Engineering and Advanced Technology (IJEAT), Volume-2, Issue-3, February 2013.
13. H. Bh. Soni and R. N. Dhavse, "Design of OTA using 0.35 μm Technology', International Journal of Wisdom Based Computing, vol. 1,2011.
14. P.E. Allen and D.R. Holberg, "CMOS Analog Circuit Design" Oxford University Press, 2nd edition.
15. B. Razavi, "Design of Analog CMOS Integrated Circuits," Tata McGraw Hill, 2002.
16. A. L. Crouch, et al., "Design-for-Test for Digital IC's and Embedded Core Systems", Prentice Hall PTR, New Jersey, 1999 (Chapter 2).
17. R. Rajsuman et al., 'IDDQ testing for CMOS VLSI' Proceeding of IEEE, vol. 88, Issue 4, pp 544–566, 2000.
18. S. Matakias, Yiorgos, A. Arapoyanni and T. Haniotakis, "A current monitoring technique for IDDQ testing in Digital integrated circuits", Integration, The VLSI journal 50, pp 48–60, 2015.
19. P. Engelke, I. Polian, M. Renovell and B. Becker, "Simulating resistive bridging and stuck-at faults", International Test conference, pp 1051–1059, 2003.
20. Jeong Beom kim, Sung Je Hong and Jong Kim, "Design of a Built in current sensor for Iddq testing", IEEE Journal of Solid State Circuits, vol. 33, pp. 1266–1272, 1998.
21. M.L Ali and N. H. Khamis, "Design of a current sensor for IDDQ testing of CMOS IC", American Journal of applied Sciences 2, pp. 682–687, 2005.

Speech-Based Access to Price of Different Agricultural Commodities Using MFCC, GMM, and Naïve Bayes Classifier

Sumit Srivastava, Arvind Kumar, Mahesh Chandra and G. Sahoo

1 Introduction

Automatic speech recognition is not a new field and many researchers have devoted their time in improving the recognition accuracy over the period of time. The main challenges are background noises, variation in user's vocal tract, echo and channel noise [1–3]. The application of ASR is unlimited in day-to-day life and for a creative and curious mind, the sky is only the limit. These days, it is widely used in automation, robotics, medical science, military application, and many more fields [4–7]. It is the most convenient way of taking input from a user as one does not need any expertise for using it. A system built on ASR can effectively interact with an illiterate or a handicapped via their speech. This paper proposes a system which aims at improving the lives of Indian farmers by empowering them with knowledge of the current market price of their products. Once familiar with the device, it will help them to understand the market rate and prevent them from getting manipulated by middlemen who takes their product in less value and earns the major chunk of the profit. For ASR, different features and classifier have been proposed by different researcher over the period of time. Out of these, MFCC feature is used along with

S. Srivastava · G. Sahoo
Department of Computer Science & Engineering, Birla Institute of Technology,
Mesra, Ranchi, India
e-mail: sumit.srivs88@gmail.com

G. Sahoo
e-mail: gsahoo@bitmesra.ac.in

A. Kumar (✉) · M. Chandra
Department of Electronics and Communication, Birla Institute of Technology,
Mesra, Ranchi, India
e-mail: arvind9835@gmail.com

M. Chandra
e-mail: shrotiya@bitmesra.ac

© Springer Nature Singapore Pte Ltd. 2019
V. Nath and J. K. Mandal (eds.), *Proceeding of the Second International Conference on Microelectronics, Computing & Communication Systems (MCCS 2017)*, Lecture Notes in Electrical Engineering 476, https://doi.org/10.1007/978-981-10-8234-4_5

GMM and Naïve Bayes Classifier and their comparative study is shown for Hindi word detection for five different commodities. The main advantage of going for MFCC is that it nullifies the effect of variation in time delay introduced by the user while speaking as it is a cepstral feature [5]. GMM and Bayes classifier work accurately with cepstral feature unlike ANN which performs better for statistical data [8, 9].

2 MFCC and GMM

2.1 MFCC

Cepstral feature is commonly used feature in speech processing. Short-term power spectrum of a sound is illustrated in MFCC. It is based on a discrete cosine transform of a log power spectrum mapped on a nonlinear Mel scale of frequency. MFCC coefficients try to replicate the properties of the human auditory system.

Block diagram of MFCC calculation is shown in Fig. 1. Preprocessed framed and windowed signal is taken to frequency domain by evaluating FFT. The power of the spectrum obtained is mapped onto the Mel scale by using triangular overlapping windows. Logs of the powers at each of the Mel frequencies are found. Discrete cosine transform (DCT) of the list of Mel log powers is then taken. MFCCs are the amplitude of the resulting spectrum. Out of the many values, we obtain only the first 13 coefficients that are saved as it contains most of the information. Linear frequency scale is converted to the Mel scale frequency m_f, using the following equation:

$$m_f = 2595\log_{10}\left(1 + \frac{f}{700}\right) \tag{1}$$

where f is frequency (hertz) in linear scale.

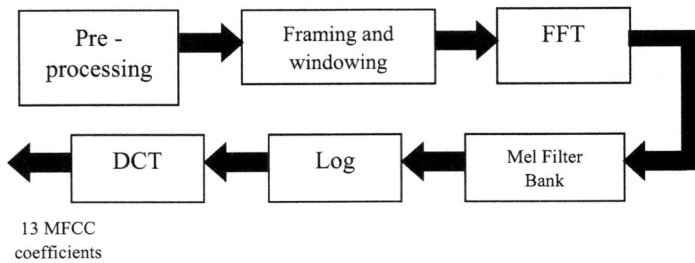

Fig. 1 MFCC feature calculation

2.2 Gaussian Mixture Model

GMM is a probabilistic and parametric method used widely for the application for speech recognition. Feature vectors are clustered and plotted in a d-dimensional feature space, a pattern similar to Gaussian distribution is seen where each corresponding cluster is seen as a Gaussian probability distribution and features associated with the clusters can be symbolized by their probability values (Fig. 2). Figure 2 graphically represents the feature space and corresponding Gaussian model for a set of data.

A Gaussian mixture density is the weighted sum of M component densities and given by Eq. 2.

$$p\left(\vec{x}|\lambda\right) = \sum_{i=1}^{M} p_i b_i(\vec{x}) \tag{2}$$

where x denotes the set of input feature vectors, $b_i(x)$ is the probability distribution of the ith element and p_i represents mixture weight of ith element.

2.3 Naïve Bayes Classifier

Naïve Bayes classifier is based on the application of Bayes Theorem. These are simple probabilistic classifiers based on assumptions that there is strong independence between features. Using Bayes rule

$$P(Y/X_1, \ldots, X_n) = \frac{P\left(\frac{X_1, \ldots, X_n}{Y}\right) P(Y)}{P(X_1, \ldots, X_n)}$$

where

$P\left(\frac{X_1, \ldots, X_n}{Y}\right)$ Likelihood Probability

$P(Y/X_1, \ldots, X_n)$ Posterior Probability

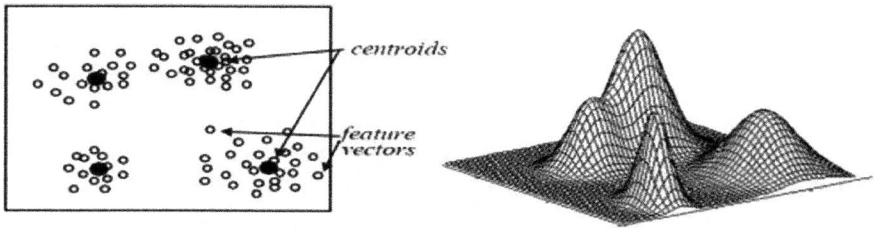

Fig. 2 Feature space and corresponding Gaussian model

$P(Y)$ Prior Probability

$X_1, \ldots, X_n$ Set of feature vectors

Assuming that all the features are independent,

$$P(X_1, \ldots, X_n | Y) = \prod_{i=1}^{n} P(X_i | Y) \tag{3}$$

This reduces computation complexity. Using these equations, a Naïve Bayes probability model is generated which is combined with a decision rule. One such rule is to select the most probable outcome which is also known as maximum a posteriori rule.

3 Building GUI

MATLAB GUIDE is used to build the graphical user interface for the application [10]. It primarily has two main steps which are discussed below.

3.1 Designing the Front End Layout

Initially, an interface is designed for the front end layout for the user to interact with the application. This is done by keying "guide" command on MATLAB command line. This pops up a new window which has tools for efficient designing of the layout. It includes various objects to be dragged and placed on the GUI. Various alignment tools are also present to modify the position of the placed objects. The property window, used for modifying various properties of an object, can be accessed by double clicking the objects. In this GUI, 4 push buttons, 2 axes, 2 panels, and 6 static texts are used. For better appearance, images can also be added via the axis panel. Figure 3 displays the implemented GUI with all the required components.

3.2 Designing the Backend

Smooth execution of the application depends on the backend operations which handles various procedures on receiving commands from the front end interface. Pressing a button on the GUI will redirect the program to the callback routine of the object where various commands are written as per desired results. All the supporting and necessary files should be included in the current directory to avoid any error. The GUI in total has four push buttons which are described below:

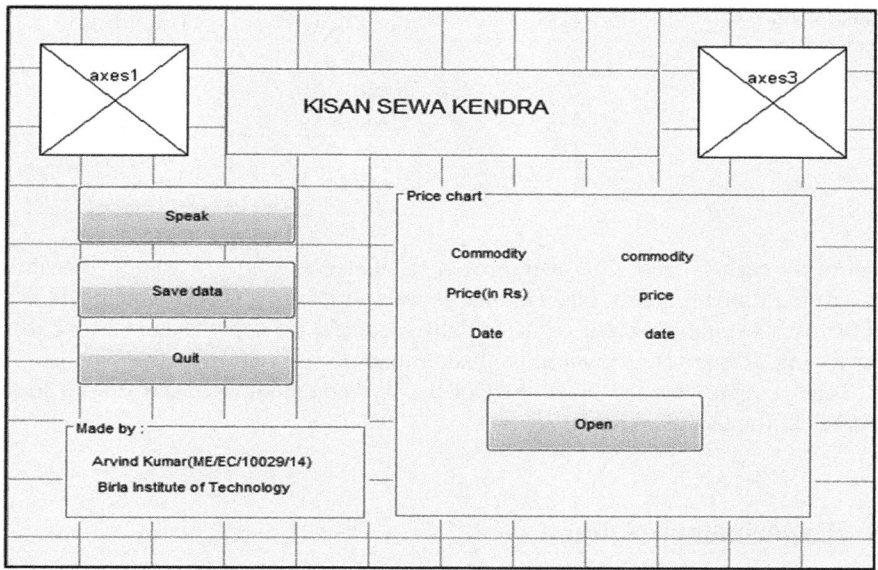

Fig. 3 GUI implementation

(1) *Speak button*

User needs to feed his voice input into the system to start the application. Speak button prompts the user to utter the name of a commodity into the microphone. A time slot of 2 s is given to a user to speak the same. This step is followed by feature extraction and classification in the background.

(2) *Save Data and Open button*

Once a name of a commodity is successfully recognized, the console displays the market rate of the item with current date. Save data button saves the same information in system's database for further analysis or future references. Open button is used to open the saved file from database.

(3) *Quit button*

Quit button helps in closing the application.

4   Database Preparation

Database is prepared for five agricultural commodities, annara (pomegranate), keela (banana), bhindxii (lady's finger), aaluu (potato), and tamaatar (tomato). Fifty samples of each of the commodities are prepared for training the system. MFCC

Table 1 Price of commodities

S. No.	Commodity	Price (Rs.)
1	Annara	80/kg
2	Keela	30/dozens
3	Bhindxii	40/kg
4	Aaluu	20/kg
5	Tamaatar	30/kg

values for each of these 250 utterances is calculated and first 13 values of each is saved in a training matrix file. This file is used to model a GMM which acts as a classifier. Once the classifier is successfully modeled, we apply it on real-time data for testing. Results are shown in confusion matrix.

Table 1 represents the price chart of the five commodities taken from a local market in Ranchi, Jharkhand.

5 Experimental Setup

Tool used for simulating the application is MATLAB R2014A. 2 channels, 16 bits resolution microphones were used for capturing voice samples with sampling rates of 44,100 samples per seconds. Figure 4 describes the front end and back end of the system. Once user presses the speak button, an audio-recorder object is created in the background and a temporary test wav file is generated containing the acoustic information. MFCC of the same is being evaluated and is fed to the GMM and Naïve Bayes classifier model for training and testing. Depending on the likelihood probability, decision is made and the classifier recognizes the appropriate word which is then passed onto the front console for display. Once a commodity is recognized, a table is queried for corresponding price of the commodities which is also sent to the front console for display. Current date is displayed using date function (Fig. 3). Block diagram of the front and back end is displayed in Fig. 4. Top section of the diagram displays the front end whereas the bottom half shows the back end. The extracted feature is fed to the trained GMM and Naive Bayes model to generate different output during testing.

6 Simulation and Results

Once the backend and front end of the GUI is designed, it is tested for different utterances.

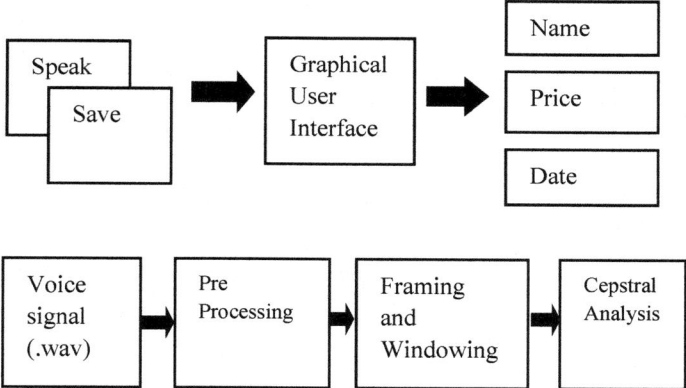

Fig. 4 Block diagram of front end and back end

6.1 The Flow of the Application Is as Follows

- An audiovisual message is played in Hindi language asking the user to speak the name of a product. This is shown in Fig. 5.
- The utterance is saved in a temporary wave file testing.wav.
- Feature extraction of the same is done and is being input to the GMM classifier.
- GMM outputs the product with maximum log-likelihood probability.
- After recognition of the said word, a query is made into the price table and corresponding price along with item name is fetched to the console for display along with current date.

Figure 6 shows the application in running mode where a user has spoken "Bhindxii" and corresponding audiovisual information of current price are generated.

Fig. 5 Prompt for user to speak

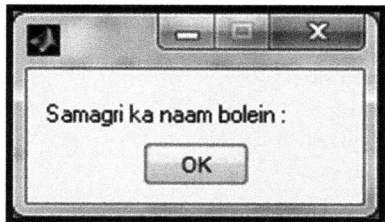

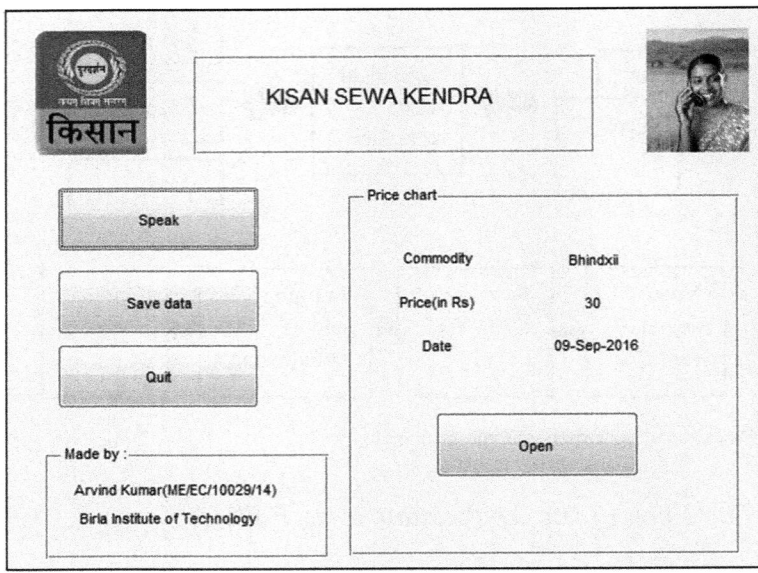

Fig. 6 Application in running mode

6.2 Confusion Matrix

Each of the model using GMM and Naïve Bayes classifier was tested with MFCC features of 50 utterances in a noise-free background.

Tables 2 and 3 present the confusion matrix for the recognition of each commodity for different classifiers. Figure 7 represents a comparative chart for the same.

Table 2 Confustion matrix for MFCC and GMM

Commodities	Commodities				
	Annara	Keela	Bhindxi	Aalu	Tamatar
Annara	**47**	0	0	3	0
Keelaa	4	**45**	1	0	0
Bhindxi	1	1	**48**	0	0
Aaloo	3	2	1	**41**	3
Tamatar	5	1	2	0	**42**

Table 3 Confustion matrix for MFCC and BAYES

Commodities	Commodities				
	Annara	Keela	Bhindxi	Aalu	Tamatar
Annara	**48**	0	0	2	0
Keelaa	0	**46**	0	4	0
Bhindxii	0	0	**48**	2	0
Aaloo	0	1	0	**49**	0
Tamatar	0	0	1	0	**49**

Fig. 7 Recognition efficiency for each commodity

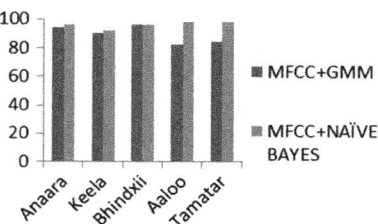

7   Conclusion and Future Scope

The proposed application for speech-based access to the price of agricultural commodity provides a simple alternative to confine the malpractice going around in transaction of Indian agricultural product. The recognition efficiency is found to be 89.2% in noise-free background for the combination of MFCC and GMM classifier whereas it is found to be 96% for the combination of MFCC and Naïve Bayes classifier. The proposed application is currently an offline system where "PRICE" table needs to be manually updated every day by the operator. Future work will focus on syncing the application with the agriculture website and automate the updating of price table by looking into the AGGMARK website. Model can be made more robust to handle noisy background by adding advance noise removal technique at preprocessing block during feature extraction methods.

References

1. Campbell, Joseph P. "Speaker Recognition: a tutorial." Proceedings of the IEEE 85.9 (1997): 1437–1462.
2. Meena, Kunjithapatham, Kulumani Subramaniam, and Muthusamy Gomathy. "Gender classification in speech recognition using fuzzy logic and neural network." Int. Arab J. Inf. Technol. 10.5 (2013), pp. 477–485.
3. Sahidullah M, Saha G. Design, analysis and experimental evaluation of block based transformation in MFCC computation for speaker recognition. Speech Communication. 2012 May 31;54(4), pp. 543–65.

4. Klapuri, Anssi, and Manuel Davy, eds. Signal processing methods for music transcription. Springer Science & Business Media, 2007.
5. Martinez J, Perez H, Escamilla E, Suzuki MM. Speaker recognition using Mel frequency Cepstral Coefficients (MFCC) and Vector quantization (VQ) techniques. Electrical Communications and Computers (CONIELECOMP), Feb 27, 2012, pp. 248–251. IEEE.
6. Al Marashli, A. and Al Dakkak, O., 2008, April. Automatic, Text-Independent, Speaker Identification and Verification System Using Mel Cepstrum and GMM. In Information and Communication Technologies: From Theory to Applications, 2008. ICTTA 2008. 3rd International Conference, pp. 1–6. IEEE.
7. Klapuri A. Multipitch analysis of polyphonic music and speech signals using an auditory model. IEEE Transactions on Audio, Speech, and Language Processing. 2008 Feb; 16(2), pp. 255–66.
8. Al Marashli A, Al Dakkak O. Automatic, Text-Independent, Speaker Identification and Verification System Using Mel Cepstrum and GMM. In Information and Communication Technologies: From Theory to Applications, 2008. ICTTA 2008. 3rd International Conference on 2008 Apr 7, pp. 1–6, IEEE.
9. Sinith MS, Salim A, Sankar KG, Narayanan KS, Soman V. A novel method for text-independent speaker identification using mfcc and gmm. In Audio Language and Image Processing (ICALIP), 2010 International Conference on 2010 Nov 23, pp. 292–296, IEEE.
10. Matlab R2014b Help- GUI building basics.

Dual Axis Solar Tracker for Solar Panel with Wireless Switching

Shahid Aziz and Mohammad Hassan

1 Introduction

Energy is the main driving force behind the development of any nation. Per capita energy consumption is considered an index of prosperity of a country. With the alarming rate of depletion of major conventional energy sources, nonconventional energy sources such as solar energy exists in abundance in our environment. It is ready to be harnessed, inexhaustible and more importantly a cleaner counterpart to fossil fuels.

Solar panels directly convert sun's radiation into electrical energy. The optimum efficiency of the majority of commercially available solar cells is in the range between 10 and 20% [1]. This indicates that there lies a scope for improvement. There are mainly three ways of increasing the efficiency of the solar panels, namely: (i) increase of cell efficiency (ii) maximizing the power output from the solar panel and (iii) use of a tracking system. Maximum power point tracking (MPPT) is the process of maximizing the power output from the solar panel, but this technology can only provide us the maximum power which can be received from stationary arrays of solar panels at a given time. In other words, this technology fails to increase the generation of power when the sun is not directly aligned with the system.

A solar tracker is a device used for orienting a photovoltaic solar array panel toward the sun for maximum power output. For regions lying along the equator, there are no significant changes in the apparent position of the sun during the various seasons; therefore use of single axis tracker having one degree of freedom will suffice. For regions lying far away from equator, the position of the sun keeps changing not only with the time of the day, but also around the year during various

S. Aziz (✉) · M. Hassan
Department of Electronics and Communication Engineering, Maulana Azad College of
Engineering and Technology, Magadh University, Patna 801113, India
e-mail: pune.shahid@gmail.com

© Springer Nature Singapore Pte Ltd. 2019
V. Nath and J. K. Mandal (eds.), *Proceeding of the Second International Conference
on Microelectronics, Computing & Communication Systems (MCCS 2017)*, Lecture Notes
in Electrical Engineering 476, https://doi.org/10.1007/978-981-10-8234-4_6

seasons, thereby dual axis trackers need to be employed. An additional circuitry and complexity are involved with dual axis trackers, but is worth the pain; as they prove to be more efficient in tracking the sun [2, 3]. The level to which efficiency can be improved depends on the efficiency of the tracking system and the weather conditions [3, 4]. More efficient a tracker is, more is the efficiency of the solar panel. Efficiency will also be considerably enhanced on a bright sunny day as opposed to a cloudy day [5].

There are broadly three stages involved in the design of a dual axis solar tracking system:

(i)   An input stage: It consists of light dependent resistors (LDRs) which are responsible for the conversion of sunlight into voltage. When the light source moves, the intensity of light falling on the four LDRs changes. This change is calibrated into voltage, using voltage dividers.

(ii) A control stage: It consists of a microcontroller. It receives the voltages from the LDRs, compares with the built-in comparators, and generates error. The microcontroller is programmed to ensure that it sends a required signal to the servomotor to act in accordance with the generated error.

(iii) Driver stage: It involves a servo motor. The servo motor has high enough torque to drive the panel. They are noise free and affordable [2]. The servo motor uses the error generated by the microcontroller and rotates through specific angles, so that the position of the solar panel is adjusted in such a way that the LDRs are at equal inclination.

2   Design and Implementation

2.1   Arduino IDE

The software design was done using Arduino IDE which was used for programming [6, 7]. The program was written using the C language. Autodesk 123D online circuit simulator was used for checking the code. The fritzing circuit editing software was used for drawing the circuit diagram.

2.2   Android Smartphone with Bluetooth App

To ease switching of solar tracker mechanism and battery connection with load point wireless switching is made use of. Wireless switching here is done using Bluetooth communication.

A smartphone having Android OS and with Bluetooth connectivity app here served as a remote or transmitter for this purpose. The Bluetooth connectivity app is made by using MIT app inventor site, this site is specialized in making app for Android OS. This app consists of five push-button each programmed for doing specific work.

- Button 1-For establishing connection between smartphone and Bluetooth receiver module
- Button 2-For controlling (on/off) relay 1
- Button 3-For controlling (on/off) relay 2
- Button 4-For turning on both relay at once
- Button 5-For turning off both relay at once
- The Bluetooth app is programmed to send signal in the form of alphabetical/numerical character each time a pushbutton is made ON or OFF.

2.3   Bluetooth Receiver Module

For receiving a wireless signal from smartphone, we have used a HC-05 Bluetooth module. The HC-05 Bluetooth module consists of six terminals out of which four are basically used namely V_{CC}, GND, TX, and RX.

The Bluetooth module establishes communication between smartphone and Arduino board. It manipulates the information as received by smartphone and transmits it to Arduino microcontroller in a suitable form.

2.4   Arduino Microcontroller for Wireless Switching (Microcontroller 1)

For manipulating or processing, the information as received by Bluetooth module a microcontroller is used. Here we have used an Arduino Uno microcontroller. The microcontroller is programmed to receive signal from smartphone via Bluetooth module in the form of alphabetical/numerical character and turn on/off the relay accordingly [6, 7]. The function performed by microcontroller is as follows:

- First of all, microcontroller checks whether there is any data in form of character available to be received.
- In case data is available, microcontroller saved it in char type variable.
- When pushbutton 1 is made on from Bluetooth app, microcontroller turns on the relay 1 by setting the pin IN1 to HIGH, similarly when pushbutton is made off, microcontroller turns off the relay 1 by setting the pin In1 LOW.

- When pushbutton 2 is made on from Bluetooth app, microcontroller turn on the relay 2 by setting the pin In2 HIGH, similarly when pushbutton is made off microcontroller turn off the relay2 by setting the pin In2 LOW.
- When pushbutton 3 is made on from Bluetooth app, microcontroller turn on both relay at once by setting both the pin In1 and In2 HIGH.
- When pushbutton 4 is made on from Bluetooth app, microcontroller turns off both the relay at once by setting both the pin In1 and In2 LOW.

2.5 Relay Module

Relay module consists of two separate relays here named as relay 1 and relay 2, these relays are controlled through microcontroller as per the instructions received from Bluetooth app.

Relay module has 4 input pin V_{CC}, GND, In1, and In2:

When In1 or In2 pin is set high, the corresponding relay closes its contact.

- Relay 1 is used to turn on/off the tracking mechanism by connecting/disconnecting the Arduino microcontroller responsible for tracking from power supply.
- Relay 2 can be used to connect/disconnect the load point from rechargeable battery which will be receiving power from solar panel.

2.6 Light Dependent Resistors (LDRs)

Four LDRs are used as a sensor, these LDRs detect any change in sunlight through a voltage divider circuit and generate output signal accordingly. The four LDRs are connected in parallel with 10k resistor to form a voltage divider circuit. Any change in light intensity falling on LDR changes the resistance of LDR which will correspondingly change the voltage output across 10k resistor

The voltage across 10k Ω resistor is fed to analog pin of microcontroller

- Voltage from voltage divider circuit of left top LDR (Lt) is fed to analog pin A0 of microcontroller 2
- Voltage from voltage divider circuit of right top LDR (Rt) is fed to analog pin A1 of microcontroller 2
- Voltage from voltage divider circuit of left down LDR (Ld) is fed to analog pin A2 of microcontroller 2
- Voltage from voltage divider circuit of right down LD R(Rd) is fed to analog pin A3 of microcontroller 2

These voltages are received in analog form which is then converted to digital form (0–1023) by microcontroller

2.7   Arduino Microcontroller for Tracking Mechanism (Microcontroller 2)

The microcontroller responsible for sunlight tracking receives the voltage from voltage divider circuit formed by LDR and generates output to two servo motor accordingly.

- First, Microcontroller 2 receives the voltages from voltage divider circuit formed by LDR using 4 analog pin (A0, A1, A2, A3).
- Second, it will compare the voltage received by circuit and generates output to two servos as follow:
- If the difference between (Lt + Rt)/2 and (Ld + Rd)/2 is found greater than tolerance level here set as 0.25 V (50 in digital) then the microcontroller tells the servo responsible for vertical motion to move in the direction of LDR's, whose voltage is found greater with compare to other. The servo will move in the direction of bottom LDR's if (Ld + Rd)/2 is greater than (Lt + Rt)/2, similarly if the (Ld + Rd)/2 voltage is found greater than (Lt + Rt)/2 then the microcontroller tells the servo responsible for vertical motion to move in the direction of bottom LDR's
- If the (Lt + Ld)/2 voltage is found greater than (Rt + Rd)/2 then the microcontroller tells the servo responsible for horizontal motion to move in the direction of left LDR's, similarly if the (Rt + Rd)/2
- Voltage is found greater than (Lt + Ld)/2 then the microcontroller tells the servo responsible for horizontal motion to move in the direction of right LDR's.

2.8   Servo Motor

For providing horizontal and vertical movement to solar tracker, we have put a high power servo motors (Horizontal servo and vertical servo). The servo motor receives error signal in the form of electrical pulse from microcontroller and acts accordingly.

2.9 Power Supply

A 12 V rechargeable battery is used to supply power to wireless switching and tracking mechanism. The battery receives power from solar panel.

- 12 V from rechargeable battery steps down to 9 V using LM2596 DC-DC adjustable converter to supply Arduino microcontroller.
- To supply two servos and relay module 12 V is stepped down to 5 V using 7805 voltage regulating IC. These ICs provide constant 5 V when connected with capacitor at both input and output.

3 Algorithm

The following are the general steps involved in tracking sunlight using the dual axis tracker (Fig. 1):

1. There is the input of the voltages from voltage divider circuit of four LDRs.
2. The inputs are analog. They are converted to digital values that range between 0 and 1023.
3. The four digital values are compared and the difference between them obtained.
4. The difference between the values obtained is the error proportional angle for the rotation of the servo motor.
5. If the difference is within tolerance no action has been taken.
6. If the difference is outside tolerance range servo will move until difference between voltage are within tolerance range.

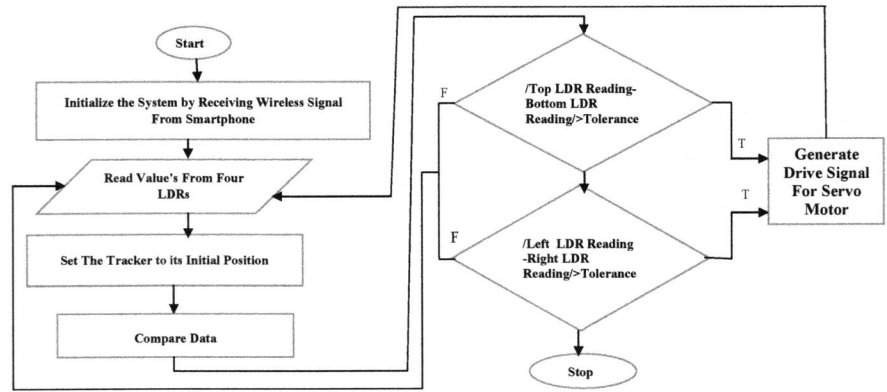

Fig. 1 Flowchart for the dual axis solar tracker

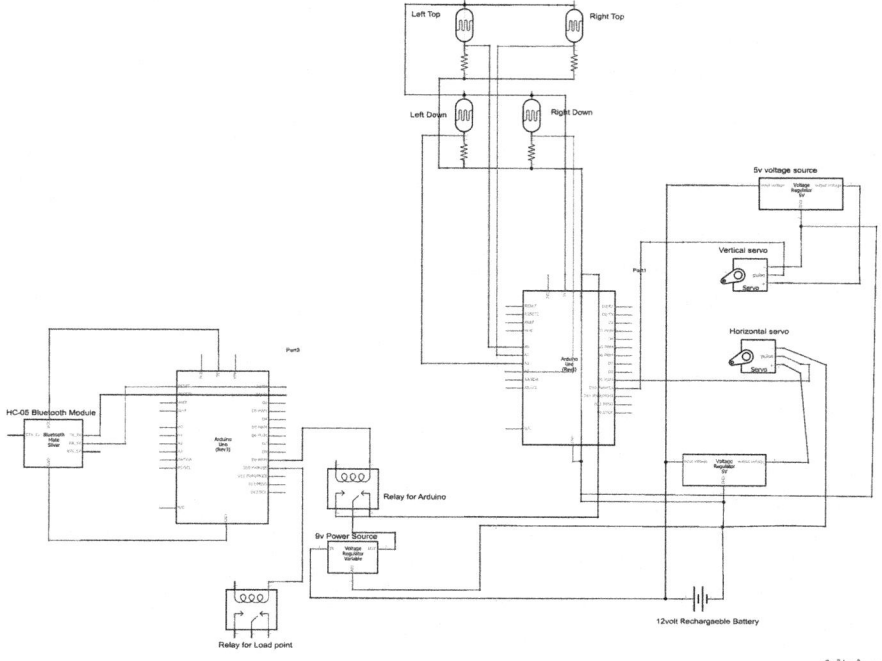

Fig. 2 Schematic of the dual axis solar tracker circuitry

4 Circuit Diagram

See Fig. 2.

5 Results and Analysis

5.1 Results

The results were taken from the LDRs for the solar tracking system and for the panel that has a fixed position. The result has been recorded by focusing 10 W LED floodlight at different angles ranging from 30° to 120°. Arduino has a serial that communicates on digital pins 0 (RX) and 1 (TX) as well as with the computer through a USB. If these functions are thus used, pins 0 and 1 can be used for digital input or output. The LDRs measure the intensity of light and therefor they are a valid indication of the power that gets to the surface of the solar panel. Arduino environment's built in serial monitor can be used to communicate with the Arduino board. To collect the results, a code was written that displays the reading taken from the four LDRs on serial monitor. Arduino board was used to connect the micro-controller to the computer. The code for reading the values that were recorded is loaded into the microcontroller [6, 7]. The various values are obtained and converted into volts. The VCC to the microcontroller and the LDRs is 5 V. The Atmega 328P has 1024 voltage steps and 5 V. When they are converted into digital values, the values will be in the range of 0–1023. The conversion is done using the relation

$$\text{LDR Output} = \frac{\text{Equivalent LDR output} * 5}{1023}$$

5.2 Analysis

From the readings and graph, it is clear that when we focus a 10 W flood light on the tracker with tracker mechanism ON, we get a higher reading of voltage across LDR. As voltage across LDR is measurement of light intensity falling on the tracker, it is clear evidence that efficiency has been improved [1]. The increase in efficiency can be calculated as a percentage and the two values (fixed panel/tracking panel) can be compared. It is to be noted that we have tested our solar tracking system by focusing artificial light and not sunlight. Artificial light is focused at different angles ranging from 30° to 130°, in step of 10°.

1. For Left Top LDR (Table 1 and Fig. 3)

   Percentage efficiency with fixed panel = 59.7818
   Percentage efficiency with tracking panel = 72.7455

Table 1 Left top reading for fixed and tracking panel

Angle in degrees	Left top LDR reading For fixed panel	Left top LDR reading for tracking panel
30	2.23	3.84
40	2.42	3.69
50	3.11	3.82
60	3.58	4.02
70	4.07	4.11
80	4.08	4.15
90	4.09	4.09
100	2.52	3.73
110	2.51	3.58
120	2.42	2.58
130	1.85	1.89

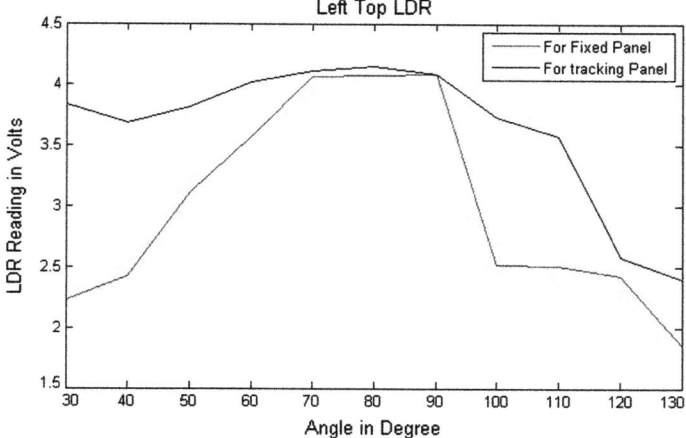

Fig. 3 Left top LDR reading as a function of angle of light incident

Table 2 Right top reading for fixed and tracking panel

Angle in degrees	Right top LDR reading for fixed panel	Right top LDR reading for tracking panel
30	2.34	3.82
40	2.83	3.92
50	3.25	4.09
60	3.92	4.12
70	3.93	4.18
80	4.01	4.13
90	4.07	4.17
100	3.06	3.58
110	2.22	3.36
120	2.20	2.63
130	1.56	1.75

2. For Right Top LDR (Table 2 and Fig. 4)

Percentage efficiency with fixed panel = 60.7091
Percentage efficiency with tracking panel = 72.2727

3. For Left Down LDR (Table 3 and Fig. 5)

Percentage efficiency with fixed panel = 56.9818
Percentage efficiency with tracking panel = 72.1818

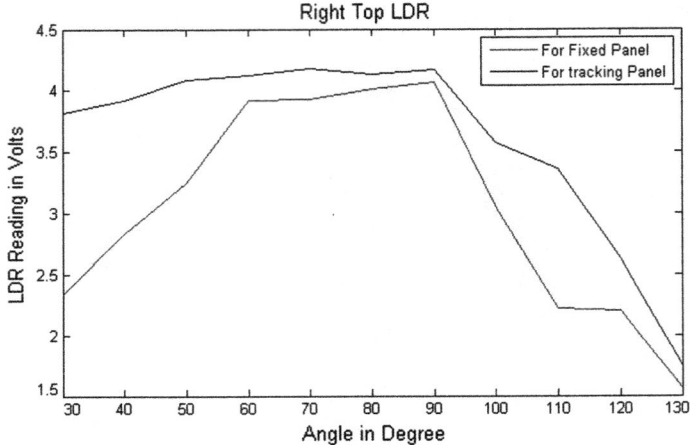

Fig. 4 Right top LDR reading as a function of angle of light incident

Table 3 Left down reading for fixed and tracking panel

Angle in degrees	Left down LDR reading for fixed panel	Left down LDR reading for tracking panel
30	1.43	2.93
40	1.47	3.71
50	1.80	3.76
60	3.79	3.99
70	3.93	4.09
80	3.95	4.09
90	4.03	4.1
100	3.37	3.69
110	2.89	3.48
120	2.62	3.04
130	2.06	2.82

4. For Left Down LDR (Table 4 and Fig. 6)

Percentage efficiency with fixed panel = 51.6727
Percentage efficiency with tracking panel = 67.2545
Average efficiency for fixed panel = 57.2863%
Average efficiency for tracking panel = 71.1136%
Increase in efficiency = 13.2872%

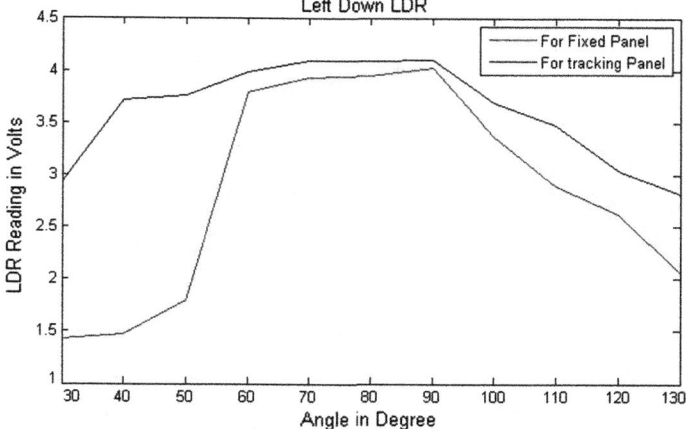

Fig. 5 Left down LDR reading as a function of angle of light incident

Table 4 Right down LDR reading for fixed and tracking panel

Angle in degree	Right down LDR reading for fixed panel	Right down LDR reading for tracking panel
30	1.26	2.53
40	1.43	3.44
50	1.67	3.45
60	3.59	3.83
70	3.67	3.93
80	3.71	3.81
90	3.94	3.98
100	2.98	3.83
110	2.43	3.16
120	2.13	2.45
130	1.61	2.58

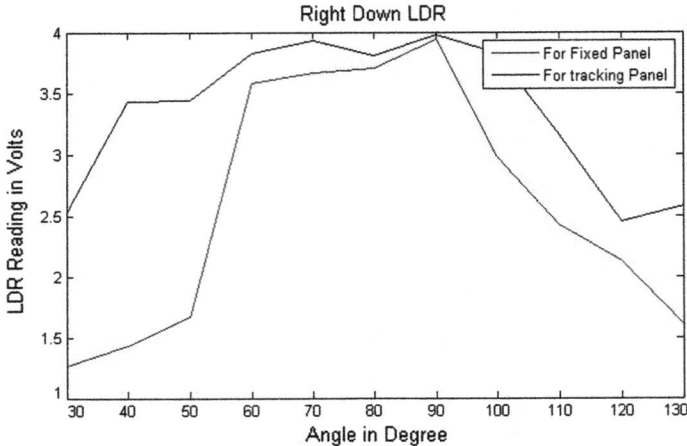

Fig. 6 Right down LDR reading as a function of angle of light incident

6 Conclusion

A solar panel that tracks the sunlight was designed and implemented. As a result, tracking was achieved. The system designed was a working model of dual axis tracker. Dual trackers are most suitable in regions where there is a change in the position of the sun. We have also seen that the dual axis trackers are more efficient than the single axis trackers [1]. The prototype was implemented with the minimum resources. The circuitry was kept simple, while ensuring that the efficiency is not affected. Further, wireless switching is added to ease the switching of the system, in case we want to shut down the system if we are moving out. Wireless switching for load point gives us a choice to either switch to battery mode (using stored energy from storage battery) or to switch AC mode (using energy directly from grid) (Table 5 and Fig. 7).

Table 5 Comparative study table

Parameter	Ref [2]	Ref [1]	Ref [5]	Proposed
Tracking mechanism	Single axis	Dual axis	Single axis	Dual axis
Key components	Stepper motor, μcontroller ATS9852, and OPAMP	PLC, DC motors, photo-sensors, encoders and power relays	Atmel 1 AT8C52 microcontroller, stepper motor	Arduino microcontroller, servo motor
Efficiency	–	17% in sunny day	–	13.2% tested by focusing 10 W lamp
Wireless switching	No	No	Graphical visual interface (GUI) with LCD display is used to interface the tracker with PC	An Android smartphone with Bluetooth connectivity served as a remote/ transmitter

Increase in efficiency =13.2872%

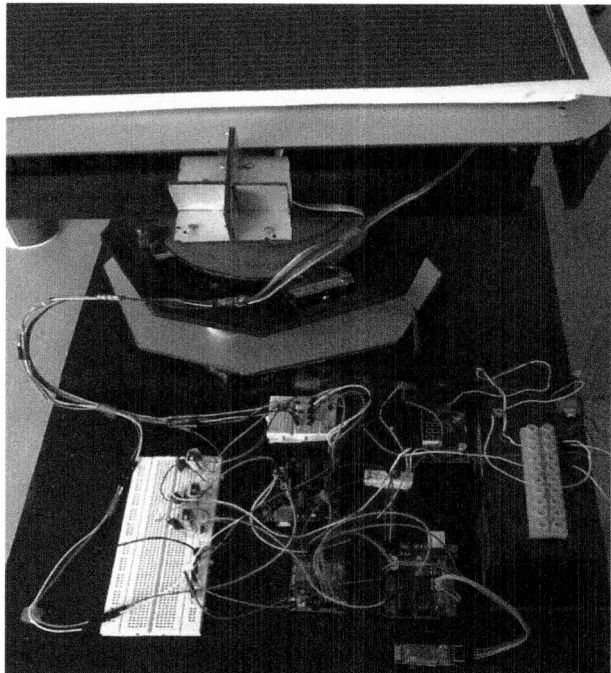

Fig. 7 Experimental set up for dual axis solar tracker

References

1. T.S. Zhan, W. Lin, M. Tsai, G. Wang. "Design and Implementation of the Dual-Axis Solar Tracking System." in Computer Software and Applications Conference (COMPSAC), 2013 IEEE 37th Annual.
2. S Chakrabory, N. Mukherjee, R. Biswas, T. Saha, A. Mohinta, N.K. Modi, D.P. Samajdar, "Microcontroller based Solar Tracker system using LDRs and Stepper Motor." in International Conference on Microelectronic Circuit and System (Micro-2015).
3. D.S. Ponni, A.R. Dhanbal, VIT University: Comparison of efficiency of Single axis tracker and dual axis tracker.
4. D. Cooke, " Single vs Dual Axis Solar Tracking", Alternate Energy eMagazine, April 2011.
5. S.O. Aliyu, M. Okwori and E.N. Onwuka. "A Prototype ASPC with Night-time Hibernation" in I.J. Intelligent Systems and Applications, 2016, 8, 18–25.
6. Michel Mc Roberts, "Beginning Arduino", Springer, Available: www.springer.com.
7. Michael Margolis, "Arduino Cookbook", O'Reilly Media.

A Node Stability Based Multi-metric Weighted Clustering Algorithm for Mobile Ad Hoc Networks

Naghma Khatoon and Amritanjali

1 Introduction

Mobile ad hoc networks (MANETs) in last decades have gained wide acceptance due to its inevitable characteristics of self-configuration, flexibility, low deployment cost, and distributed nature (Fig. 1). The achievements of theoretical contributions by many researchers in the area of MANET make it possible to be deployed for many practical applications like crisis management, battlefield applications, on the fly collaboration application, personal area network, commercial applications, etc.

Clustering is an important approach in order to achieve scalability in the presence of large network with large number of mobile nodes. Clustering is, in fact, not a routing protocol instead, it is a mechanism which provides a way of grouping mobile nodes into logical groups called clusters. Figure 2 illustrates that within a cluster, mobile nodes get different functions based on the status assigned to them. These are cluster heads (CHs), gateways (G1 and G2), and cluster members (CMs). A CH normally serves as a local coordinator for its cluster. It performs an arrangement for an intracluster transmission. Cluster gateways are nodes that lie in the transmission range of more than one CHs. It performs an arrangement for an intercluster transmission, i.e., it can access neighboring clusters and forward information between clusters. Cluster members are the ordinary nodes of a cluster. Here, routing path setup is done between cluster heads and gateways that act as the coordinator for its cluster and make control decisions for its members [1–4]. Each CH maintains two tables which are updated by periodic "HELLO" messages:

- Routing table—store routing information.
- Connectivity table—keep records of its cluster members and neighboring cluster heads and gateways.

N. Khatoon (✉) · Amritanjali
Department of Computer Science & Engineering, BIT Mesra, Ranchi 835215, India
e-mail: naghma.bit@gmail.com

© Springer Nature Singapore Pte Ltd. 2019
V. Nath and J. K. Mandal (eds.), *Proceeding of the Second International Conference on Microelectronics, Computing & Communication Systems (MCCS 2017)*, Lecture Notes in Electrical Engineering 476, https://doi.org/10.1007/978-981-10-8234-4_7

Fig. 1 A mobile ad hoc
network

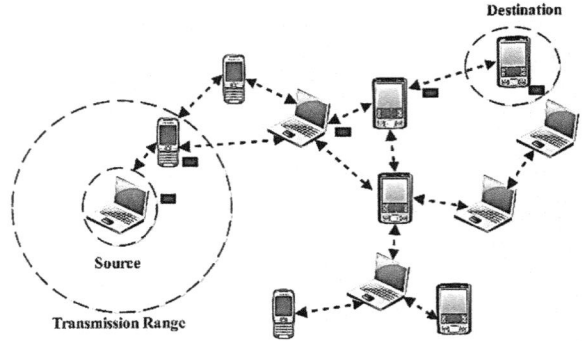

Fig. 2 Illustration of clusters
in MANET

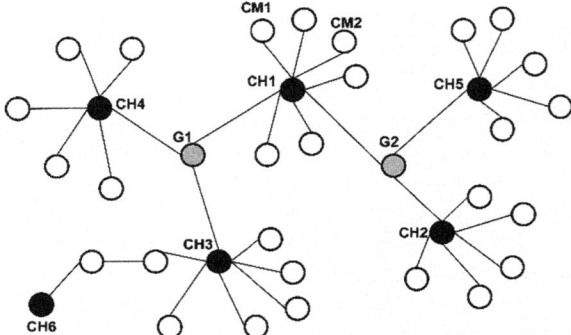

Thus, the obtained cluster structure provides the basis for a hierarchical network organization. The study of various clustering approaches reveals that clustering in MANET is a challenging task. This is due to the fact that a CH has to perform extra responsibilities and will consume more battery power than CMs, resulting in the frequent death of CHs which ultimately results in network partitioning and failure of communication links.

So, one of the most demanding tasks here is to elect the highly appropriate node to act as CH for each cluster which retains its neighborhood for longer time duration and maintains cluster stability. Thus, the CH selection should depend on the quality of nodes which can be achieved by combining various node's parameters such as the degree of connectivity, battery power, mobility, and the distance of a node from all its neighbors [1, 5–7]. The main benefaction of our proposed work is modifying the existing weight based clustering algorithm and presenting an improved method for clustering wireless mobile ad hoc networks based on node's eligibility and stability criteria.

The rest of the paper is organized as follows. A brief survey of related works on clustering algorithms proposed for ad hoc networks especially those developed for MANETs is presented in Sect. 2. The proposed stability based multi-metric weighted clustering algorithm (SM-WCA) analytical model and different CH selection parameters are explained in Sect. 3. An explanatory example is given in Sect. 4. Section 5 presents simulation results, which show the effectiveness of the proposed algorithm. Finally in Sect. 6, we conclude the paper and present some directions for future work.

2 Related Works

We have reviewed several MANET routing protocols. Flat structured routing protocols work well as long as the size of the network is small. As the network size increases, route establishment between a source–destination pair leads to blindly broadcasting of route request packets which leads to very high routing overhead, i.e., a congestive scenario is developed due to these routing mechanisms. Thus, for better performance of MANET, a good approach is to use hierarchical routing structure which aggregates existing routing protocols with clustering schemes for minimizing control overheads and improving throughput, spatial reuse, scalability, and power consumption.

In this section, we present a literature review of various clustering schemes proposed for MANETs and wireless sensor networks (WSNs) [6–12]. Amin et al. proposed energy efficient cluster-based routing protocol for MANETs [10]. Cluster-based routing protocol (CBRP), which is one of the powerful and scalable routing protocols for ad hoc networks, has been modified for energy consideration and is used with the basic MANET routing protocol AODV, thus generating less overhead and increasing throughput. Out of many resources, the energy of nodes is one of the critical factors in the operation of MANET. The performance of such a network can eventually be increased by increasing the lifetime of nodes, which in turn can be achieved by decreasing the energy consumption. Sleep mode is introduced for all idle nodes in the network except the CH and gateway nodes. This strategy helps to maximize the lifetime of nodes, their stability, connectivity, and saving battery in ad hoc networks. The work in this paper makes it feasible to conserve the energy of nodes, however, calculating energy consumption for sending and receiving packets each time leads to overhead and take time.

Several approaches have been proposed to cope with the limitations of mobility and battery dependency of mobile nodes in MANETs. Bokhari et al. in their paper [4] presented a classification of clustering techniques applied in MANETs that minimize the energy consumption by nodes which is one of the major concerns for wireless ad hoc networks and overall improves the power consumption and battery usage.

Although single metric CH selection algorithms are simple with less overhead, they are entirely dependent on single parameter without taking into consideration the other eligibility qualifications of nodes to become a CH. As a result, such clustering approaches increase the rate of re-affiliation and re-clustering. Chatterjee et al. have proposed a novel weight based clustering algorithm (WCA) for MANETs [5]. They considered the combined weight of more than one parameters, i.e., degree of connectivity, mobility, time till a node acts as a CH and the sum of distances to a node's neighbors for electing CHs. This algorithm is the basis for all weight based clustering approaches that consider many parameters for CH election. However, there are certain factors which reduce the efficiency of WCA like each time execution of the setup phase on a node's movement unnecessary dissolve a well-formed cluster and lack of node's stability. Choi and Woo [6] have presented a distributed weighted clustering algorithm (DWCA) for MANET which restrains configuration and reconfiguration of clusters and constraints on CHs in terms of power requirement. However, there are no parameters which define the stability of nodes while deciding CH.

In [3], a new enhanced version of weighted clustering algorithm has been proposed called enhancement on weighted clustering algorithm (EWCA). The parameters used for the selection of CHs are transmission power, transmission range, mobility, and battery energy. Here, the average number of cluster formation is reduced by avoiding the dynamic change of CH. EWCA takes care of load balancing and improve the stability of clusters in MANET.

3   SM-WCA Analytical Model

In this section, we illustrate the design philosophy and the basics of our proposed algorithm before going into details. The suitability of a node to become a CH is decided by a combination of parameters like stability factor, degree deviation, sum of distances with all the neighbors, and energy depletion. Our proposed SM-WCA is an improvement on WCA [5]. Like WCA, it effectively combines various parameters with certain weighing factors. These factors are chosen according to the mobility of nodes. As in [1], we are using transmission zone range for calculating stability factor. The flexibility of changing the weight factors helps us apply our algorithm to various networks.

A mobile ad hoc network can be modeled as an undirected graph $G = (V, E)$, where V is the set of mobile nodes and E is the set of links that exist between the nodes. There exists a bidirectional link E_{ij} between the nodes i and j when the distance between them, $dist(v_i, v_j) < t_{range}$ (transmission range) of the nodes. The cardinality of V ($|V|$) may change due to the death of existing nodes or entry of new nodes but the cardinality of $E(|E|)$ always changes. This is due to the creation of new links or deletion of existing links such as nodes that are mobile in the network.

The degree of a node v_i where $v_i \in V(G)$ is represented as the cardinality of the set of neighbors of v_i, $N(v_i)$.

$$\text{Deg}(v_i) = |N(v_i)|$$

The set of vertices $S \subseteq V(G)$, such that the union of $N(v_i)$, where $v_i \in S$, forms $V(G)$. The set S is called a dominating set such that every vertex of G belongs to S or has a neighbor in S.

3.1 Basics of the Proposed SM-WCA

The following features are considered in our clustering algorithm:

- The procedure of CH reelection is nonperiodic and is invoked only when the CHs in the dominating set did not cover all nodes in the network. Also, in our proposed SM-WCA, a node not satisfying the eligibility criteria is not going to calculate its combined weight and only becomes member of some elected CHs. This reduces system updates and also computation and communication cost.
- Each CH can ideally support only δ nodes. It is a constraint on the number of nodes that a cluster can handle ideally.
- A CH consumes more battery power than an ordinary node since it has extra responsibilities to carry out for its members. The battery power of a node can be used efficiently if it needs to communicate with its members that are quite closer to it.
- The degree of connectivity of the node changes frequently with the mobility of nodes. It is computed as the running average speed for every node till current time.

3.2 CH Selection Parameters in SM-WCA

Mobility leads to more number of CH reelection as well as link updation which results in poor cluster stability. For stable cluster formation, it is essential to consider the mobility of the nodes. The transmission range of a node v_i forms a circle, the radius of which is represented by r as shown in Fig. 3. Inside the circle, the neighboring nodes of v_i can exist either closer to v_i or far away, but within the transmission range of v_i. Accordingly, we can divide the transmission zone of v_i as trusted zone or risked zone. The inner circle with radius $\alpha_1 r$ forms the trusted zone of v_i. The circle with radius $\alpha_2 r$ (outer circle) represents the transmission range (t_{range}) of v_i. The coefficients α_1 and α_2 are suitably selected based on the mobility of the nodes in the network.

The zone having width $r(\alpha_2 - \alpha_1)$ forms the risked zone. All those nodes situated in the trusted zone are favorable nodes as they continue to hold the neighborhood for longer time period. On the other hand, the nodes situated in the risked zone are unfavorable as they are likely to leave the zone earlier compared to more centralized nodes in the trusted zone.

3.2.1 Stability Factor (STF)

Based on Fig. 3, mathematically, we are formulating zone factor (ZF), which determines where in the transmission range a node is lying. Accordingly, we are

Fig. 3 Transmission range
zones

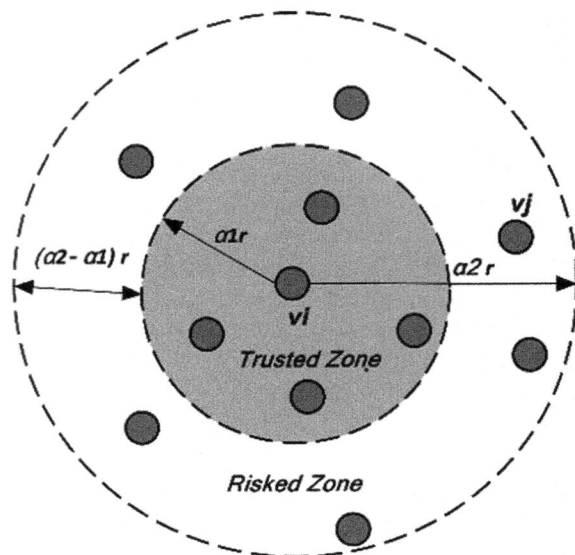

giving higher priority to favorable nodes (nodes belonging to trusted zone) and less
priority to unfavorable nodes (nodes belonging to risked zone). The priority of the
unfavorable nodes reduces as its distance increases. Zone factor (ZF) is calculated
as follows:

$$ZF(v_i, v_j) = \begin{cases} 1, & \text{if } \text{dist}(v_i, v_j) \leq \alpha_1 r \text{ for trusted zone} \\ 1 + \left[(\text{dist}(v_i, v_j) - \alpha_1 r)/(\alpha_2 - \alpha_1)r \right], & \text{if } \alpha_1 r < \text{dist}(v_i, v_j) < \alpha_2 r \text{ for risked zone} \end{cases}$$

where $\text{dist}(v_i, v_j)$ is the distance between the nodes v_i and v_j and $(\alpha_2 - \alpha_1)r$ is the
width of the risked zone.

Now we calculate the effective distance (ED) as follows:

$$ED(v_i, v_j) = ZF(v_i, v_j) \times \text{dist}(v_i, v_j)$$

Then, we calculate the cumulative effective distance (CED) from the node v_i to
all its neighboring nodes that are directly linked to it, i.e., within its transmission
range zone as follows:

$$\text{CED}(v_i) = \sum_{j=1}^{n} \text{ED}(v_i, v_j)$$

On the basis of above equations, Stability Factor (STF) is calculated as:

$$STF(v_i) = CED\,(v_i)/N(v_i) \qquad (1)$$

3.2.2 Degree Deviation (Δv)

The degree deviation of a node is measured as the relative deviation of its degree from that of ideal degree (δ).

$$\Delta v_i = |\delta - N(v_i)|/N(v_i) \qquad (2)$$

3.2.3 Sum of Distances (D)

This is an important parameter that affects the overhead on a CH and the quality of communication.

$$D(v_i) = \sum_{j=1}^{n} \text{dist}(v_i, v_j) \qquad (3)$$

3.2.4 Energy Depletion (ED)

In order to decide how much a node is capable of becoming a CH, it is a better choice to use the amount of energy consumed by it. If a node's depleted energy is high, it means its remaining battery energy is low which ultimately is not a better candidate to become a CH. In order to increase the network lifetime, a node with low depleted energy is chosen to act as a CH.

3.3 CH Election Procedure in SM-WCA

In our proposed SM-WCA, we considered two-phase clustering. Before the execution of clustering algorithm, network setup is performed called bootstrapping. In this, each node broadcasts its ID using CSMA/CA MAC layer protocol. Thereby every node collects their neighbor's information. In the setup phase of SM-WCA, we assumed that the depleted energy is constant and mobility is zero for all nodes as we considered a homogeneous network. So, in the first phase, i.e., initialization phase, the selection of CH is based on two parameters, i.e., degree deviation and the sum of distances to a node's neighbors, i.e., power consumption. As more power is required to communicate with nodes at larger distances. Therefore, a node with degree close to that of ideal degree and low requirement of power consumption is a suitable candidate to serve as CH. Using these two parameters, we calculate the combined weight of all nodes and the nodes having minimum combined weight are elected as CH. The initialization phase of SM-WCA consist of the following steps:

(1) Each node v_i calculates its degree deviation using Eq. 2 if it satisfies the check that it is not an isolated node, i.e., its degree of connectivity should not be NULL.
(2) Calculate the sum of distances with all neighbors for each node using Eq. 3.
(3) Calculate the combined weight $W(v_i)$ for each node v_i: $W(v_i) = W_1(\Delta v_i) + W_2 D(v_i)$ and share it with its neighbors.
(4) Select a node with minimum combined weight as CH.
(5) Delete node v_i and all its neighbors from G.
(6) Repeat Steps 1–5 for the remaining nodes not yet selected as a CH or assigned to a CH.

In the second phase (re-clustering phase), two additional parameters are considered, i.e., stability factor (STF) and energy depletion (ED). A node having low mobility and remaining battery level more than the threshold will participate in this phase. It is not periodic and invoked only on demand. If the number of nodes went into a region not covered by any CH become more than the predefined threshold, i.e., the number of clusters in the network becomes greater than the maximum number of clusters which depends upon the network topology, then the second phase of SM-WCA is invoked and the new dominant set is obtained. In this phase, we check eligibility criteria of each node before calculating its combined weight which ultimately reduces the computational and communicational overhead on nodes. Given below are the steps for re-clustering phase:

(1) A node is eligible if mobility is less than a predefined threshold and remaining battery energy is greater than a predefined threshold.
(2) For eligible nodes,

    (a) Calculate STF using Eq. 1.
    (b) Calculate the combined weight $W(v_i) = W_1 \text{STF}(v_i) + W_2 \text{ED}(v_i) + W_3(\Delta v_i) + W_4 D(v_i)$ and share it with its neighbors.

(3) Repeat the process like in initialization phase until all nodes are either selected as CH or assigned to a CH.

4 Explanatory Example of SM-WCA

The working of the proposed SM-WCA is illustrated with the help of a network scenario consisting of 21 mobile nodes with different mobility factor, degree of nodes, and remaining battery energy. The execution of the first phase, i.e., initialization phase, is shown in Table 1. It may be possible that a node with higher degree has neighbors belonging to the risked zone (unfavorable nodes), while some other node has less number of neighbors but they belong to the trusted zone

Table 1 Phase-I of SM-WCA (initialization phase)

Node #	Deg(v_i)	Δv_i	$D(v_i)$	$W(v_i)$
1	2	1	7.6	4.3
2	1	3	4	3.5
3	2	1	8	4.5
4	1	3	2.5	2.75
5	0	–	–	Isolated node
6	3	0.33	14	7.17
7	0	–	–	Isolated node
8	3	0.33	6.3	3.31
9	3	0.33	14	7.17
10	4	0	8.5	4.25
11	3	0.33	12.7	6.52
12	2	1	7.2	4.1
13	1	3	12	7.5
14	1	3	8.33	5.67
15	1	3	3.8	3.4
16	3	0.33	15	7.67
17	0	–	–	Isolated node
18	4	0	15	7.5
19	2	1	10	5.5
20	3	0.33	5	2.67
21	2	1	13	7

(favorable nodes). The clusters formed after initialization are as shown in Fig. 4. Details of execution of the re-clustering phase are shown in Table 2 and accordingly cluster formation is shown in Fig. 5. In our example re-clustering is invoked as nodes 5, 7 and 17 comes in a position when it is not covered by any of the CHs in the network and so they declare themselves as isolated clusters (Fig. 4) and the total number of clusters thus formed (nine in this case) become more than the maximum number of clusters allowed for this network scenario (seven in this case). The weighing factors are chosen arbitrarily such that $w_1 + w_2 + w_3 + w_4 = 1$. In this example, $w_1 = 0.5$ and $w_2 = 0.5$ for the initialization phase and $w_1 = 0.2$, $w_2 = 0.3$, $w_3 = 0.2$ and $w_4 = 0.3$ for the re-clustering phase and the value of δ is 4. These weighing factors are responsible for prioritizing the contribution of each and every component for selection of CHs. In Table 2, nodes 4, 12, 18, and 21 are not eligible for becoming CH as they have remaining battery energy lesser than the threshold. They are simply eliminated from taking part in the CH selection procedure and can only become cluster members of some cluster, thus reducing unnecessary computation and communication overhead. We can notice here in this example that some node although have the same degree of connectivity but have different combined weight. This is because it maybe possible that neighbors of a node belong closer to it or maybe far away, i.e., depending on whether they belong to trusted zone or risked zone stability factor changes.

Fig. 4 Cluster formation

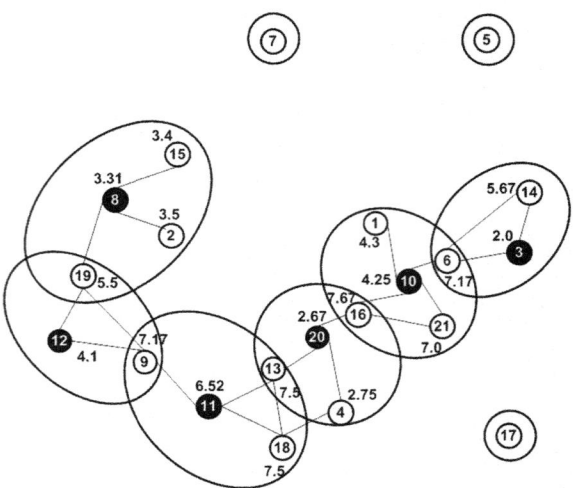

5   Simulation Results

In this section, we describe the simulation results of the proposed SM-WCA for
MANET. To evaluate our proposed work, we have performed pervasive simulation
experiments. We have used NS2 network simulator [13] for the execution of our
simulation work. To assess our proposed algorithm, the number of nodes varies
between 20 and 100. Nodes are deployed in the simulation area of 750 m × 750 m.
The transmission range of nodes varies between 10 m and 180 m and nodes are
initialized with the energy of 45,000 NJ. For the implementation of our proposed
work, we have summarized the simulation parameter setting in Table 3. The simulation results show the efficiency of SM-WCA over existing WCA and DWCA.

The performance of the proposed SM-WCA is measured in terms of number of
clusters, number of re-affiliations, and the choice of SM-WCA with AODV or
DSDV. The weighing factors used in our experiment are arbitrary values which can
be tuned as per system requirements. We also experimented the effect of changing
the density of network and compare our proposed algorithm with already existing
WCA and DWCA in terms of average number of clusters thus formed with change
in transmission range of nodes. For this, we have first implemented both WCA and
DWCA and then compared these with SM-WCA.

Figure 6 illustrates the average number of clusters formed with increase in
transmission range for number of nodes N, between 100 and 500. Transmission
range and average number of clusters formed are inversely proportional to each
other. This is due to the fact that as the transmission range of a CH increases, it will
cover a large area.

Figure 7 presents the average number of clusters formed with change in the total
number of nodes present in the network. From the figure, we can see that average

Table 2 Phase-II of SM-WCA (re-clustering phase)

Node #	Deg(v_i)	Δv_i	STF(v_i)	ED(v_i)	RBE(v_i)	$D(v_i)$	$W(v_i)$
1	3	0.33	1	2	8	7	2.97
2	1	3	3.5	3	7	3	3.2
3	2	1	2.8	4	6	5	3.34
4	1	3	1	8	2	2	NE
5	3	0.33	3.2	6	4	11	5.53
6	2	1	1.6	7	3	7	4.18
7	2	1	2.2	6	4	5	3.56
8	2	1	1.4	6	4	4	3.58
9	4	0	1	2	8	13	4.6
10	3	0.33	1	2	8	7	2.97
11	2	1	1.25	5	5	6.8	3.62
12	2	1	2.9	9	1	2	NE
13	2	1	1.4	4	6	10	4.78
14	3	0.33	1.1	3	7	8	3.29
15	2	1	1.3	2	8	5.4	2.61
16	4	0	3.0	1	9	8	3.5
17	2	1	1.4	4	6	10	4.78
18	1	3	2.1	8.5	1.5	3	NE
19	3	0.33	4.2	4.9	5.1	15	4.86
20	3	0.33	3.3	5	5	12	5.66
21	1	3	1.3	8.4	1.6	4	NE

Fig. 5 Cluster formation after re-clustering phase

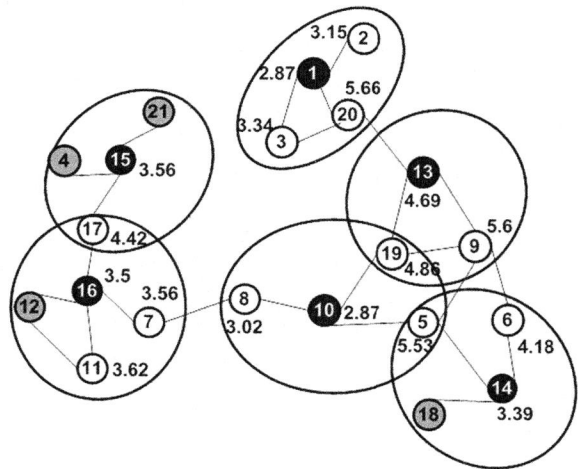

Table 3 Simulation
parameters setting

Parameter	Value
Number of nodes	20–100
Simulation area	750 m × 750 m
Simulation time	50 s
Initial energy	45,000 NJ
Packet size	512 bytes
Transmission range of nodes	10–180 m
Routing protocol	AODV
Movement model	Random-way point
Radio propagation model	Two ray ground

Fig. 6 Average number of
clusters versus transmission
range (R)

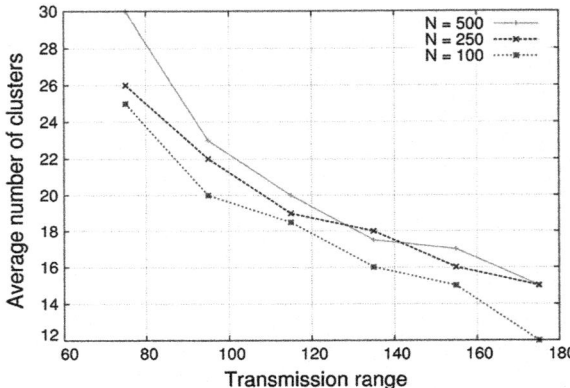

Fig. 7 Average number of
clusters versus number of
nodes (N) for SM-WCA,
WCA, and DWCA

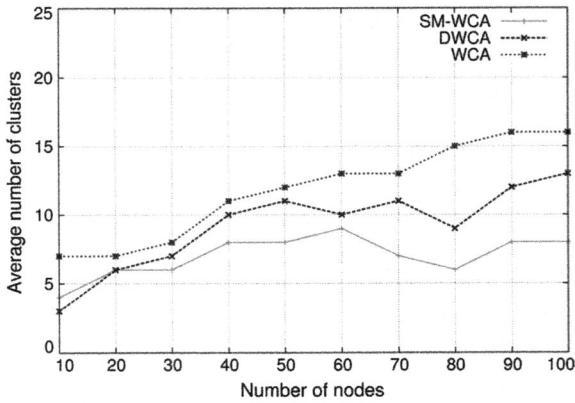

Fig. 8 Average number of clusters versus transmission range for SM-WCA and WCA

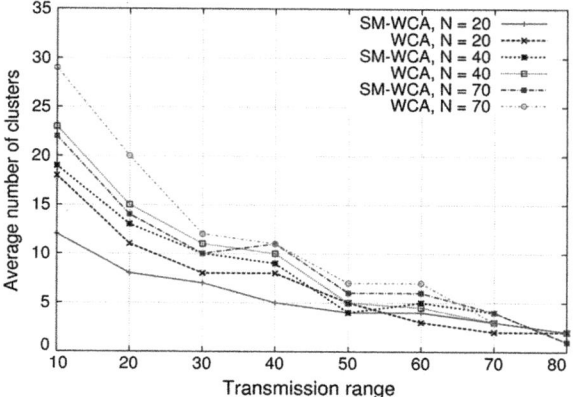

number of clusters for SM-WCA in comparison to WCA and DWCA is less as the number of nodes in the network increases.

Figure 8 shows the deflection of average number of clusters with change in transmission range for the proposed SM-WCA with WCA and DWCA for varying number of nodes between 20 and 70. From the figure, we noticed that there is an inverse relationship between the average number of clusters and the transmission range. SM-WCA constantly produced less number of clusters as compared to WCA. It shows that our algorithm gave a better performance than WCA in both the situations of high node density and high transmission range.

Figure 9 interprets the average number of re-affiliations with respect to the total number of nodes. As SM-WCA selects the most eligible nodes to act as CH by eliminating nodes with high mobility and lesser remaining battery energy at the initial stage of clustering and also selects nodes in the favorable zone to become a CH, it shows the better performance in terms of re-affiliations as compared to WCA and DWCA.

Figure 10 presents the result of applying the proposed SM-WCA with AODV and DSDV in terms of remaining battery energy of each node. As noticeably from the figure, it is clear that each node in SM-WCA if coupled with AODV has more remaining battery energy than that of SM-WCA coupled with DSDV. While applying SM-WCA with AODV, it consumes about 20–60% less energy than it is applied in DSDV. We also observed from the figure that there are four nodes lost due to energy depletion in case of DSDV while only one node is lost when the same is used with AODV. This is because there are more routing overheads with DSDV in maintaining routing tables and disseminating controls packets than AODV. Hence, our proposed algorithm gives better performance when it is coupled with AODV in terms of saving energy of nodes.

Fig. 9 Average re-affiliations
with respect to number of
nodes for SM-WCA, WCA,
and DWCA

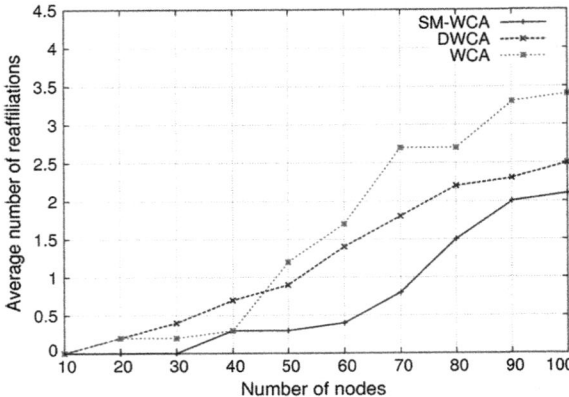

Fig. 10 Remaining energy
per node of SM-WCA using
AODV and DSDV

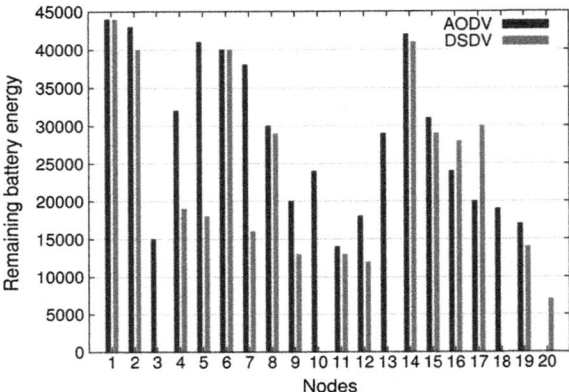

6   Conclusion

In our proposed SM-WCA, we are giving emphasis on individual parameters as
well as on the combined weight for selecting a CH. At the time of initialization,
energy consumption and degree deviation are considered to form balanced and
stable clusters. The re-clustering procedure of SM-WCA is nonperiodic and is
invoked only when the current dominance set of CHs is not covering all the nodes
in the network. Use of eligibility criteria in this phase helps to prevent weak nodes
to participate in the CH election process. It also reduces unnecessary computation
and communication overhead and increases network lifetime. The additional
parameters used in this phase helps to improve the cluster stability while selecting
appropriate CHs. In the future work, we will improve our proposed work to handle
load balance in a more effective way and also tackle the problem of trustworthiness
of nodes.

References

1. Aissa, M. & Belghith, A., 'A node quality based clustering algorithm in wireless mobile Ad Hoc networks', *Procedia Computer Science*, 32, pp. 174–181. Available at: http://dx.doi.org/10.1016/j.procs.2014.05.412, (2014).
2. Seno, S. A. H., Wan, T. C. & Budiarto R., 'Energy Efficient Cluster based Routing Protocols for MANETs', International Conference on Computer Engineering and Applications IPCSIT, vol. 2, pp. 380–384, (2009).
3. Bednarczyk, W. & Gajewski, P., 'An Enhanced Algorithm for MANET Clustering Based on Weighted Parameters', Universal Journal of Communications and Network 1(3): pp. 88–94, (2013).
4. Bokhari, D.M., Hamatta, H.S.A. & Siddigui, S.T., 'A Review of Clustering Algorithms as Applied in MANETs', International Journal of Advanced Research in Computer Science and Software Engineering Research, vol. 2, pp. 364–369, (2012).
5. Chatterjee, M., Das, S.K. & Turgut, D., 'An on-demand weighted clustering algorithm (WCA) for ad hoc networks', IEEE Global Telecommunications Conference (GLOBECOM '00), 3, pp. 1697–1701, (2000).
6. Choi, W. & Woo, M., 'A Distributed Weighted Clustering Algorithm for Mobile Ad Hoc Networks', Proceedings of the Advanced International Conference on Telecommunications and International Conference on Internet and Web Applications and Services (AICT/ICIW 2006), pp. 1–6, (2006).
7. Kaushik, G. & Goyal, S., 'An Clustering based AODV approach for MANET', Proc. of Int. Conf. on Emerging Trends in Engineering and Technology, pp. 901–904, (2013).
8. Li, C., Wang Y., Huang F. & Yang D., 'A Novel Enhanced Weighted Clustering Algorithm for Mobile Networks', 978-1-4244-3693-4/09/$25.00 ©2009 IEEE, pp. 1–4, (2009).
9. Mylsamy, R. & Sankaranarayanan, S., 'A new approach to trust evaluation for cluster based MANETs', Journal of Theoretical and Applied
10. Amine, D., Nasr-Eddine, B. & Abdelhamid, L., 'Energy efficient and safe weighted clustering algorithm for mobile wireless sensor networks', *Procedia Computer Science*, 52, pp. 641–646. Available at: http://www.sciencedirect.com/science/article/pii/S1877050914008953, (2014)
11. Schmidt, R.D.O., Antônio, M. & Trentin, S., 'MANETs Routing Protocols Evaluation in a Scenario with High Mobility', 978-1-4244-2066-7/08/$25.00 ©2008 IEEE, pp. 883–886, (2008).
12. Sucasas, V., Radwan A., Marques H., Rodriguez J., Vahid S. & Tafazolli R., 'Ad Hoc Networks A survey on clustering techniques for cooperative wireless networks' Ad Hoc Networks, 47, pp. 53–81. Available at: http://dx.doi.org/10.1016/j.adhoc.2016.04.008, (2016).
13. Project, T.V. et al. The ns Manual (formerly ns Notes and Documentation) 1, (3), (2011).

Design and Development of Microstrip Patch Antenna for Millimeter-Wave Application

Kumari Mamta, Raj Kumar Singh, Navin Kumar Sinha and Ritesh Kumar Keshri

1 Introduction

Antennas are the backbone of wireless communications. In today's world of fourth generation communication and advancing fast to the fifth generation (5G) stage, mobile and wireless communication without antenna cannot be thought of.

With ever-increasing usage and penetration to every part of the world, wireless communication technology has to develop rapidly to meet the demand. The 5G technology requires higher frequency bands to for supporting multi-Gbps data rates. Also, the idea is to gather infinite data broadcast within newest mobile technology [1]. Millimeter-wave communication systems using narrow beams suppress the interference of neighboring beams. By beam shifting, pointing toward the base station can be achieved effectively [2]. Low profile, high gain, lightweight, and simple structure antennas are the key requirement of modern wireless communication antenna. This is to take care of reliability, mobility, and high efficiency [3]. Microstrip patch antenna (MPA) meets all these standards. MPA is suited to planar as well as nonplanar surfaces. Their design considerations are simple. However, their relatively narrow bandwidth limits their usage in particular systems. Many

K. Mamta (✉) · N. K. Sinha
Department of Applied Physics, Cambridge Institute of Technology, Ranchi, India
e-mail: mamta.singh548@gmail.com

N. K. Sinha
e-mail: nksinha69@yahoo.com

R. K. Singh
Department of Physics, R. L. S. Y. College, Ranchi, India
e-mail: subrajon@yahoo.com

R. K. Keshri
Department of Electrical Engineering, VNIT, Nagpur, India
e-mail: riteshkeshri@gmail.com

© Springer Nature Singapore Pte Ltd. 2019
V. Nath and J. K. Mandal (eds.), *Proceeding of the Second International Conference on Microelectronics, Computing & Communication Systems (MCCS 2017)*, Lecture Notes in Electrical Engineering 476, https://doi.org/10.1007/978-981-10-8234-4_8

research works done across the globe on wireless communication have utilized a printed antenna for millimeter-wave band [4–6].

The narrow bandwidth of MPA can be broadened by various possibilities; like optimizing impedance matching, reducing substrate effective permittivity, increasing the substrate thickness, and incorporating multiple resonances.

Rectangular and circular configurations of the patch mostly worked on configurations. With these patch configurations, feed line flexibility, beam scanning, multifrequency operation, and broad bandwidth are easily obtainable.

Use of air as dielectric has considerably reduced dielectric constant and hence the antenna gain and efficiency have increased. This is different from some works reported on DRA and other antennas [7, 8]. By use of air as dielectric for substrate, the metallic and surface wave loss in the proposed antenna does not stand in way of gain and efficiency of antenna. With decrease in dielectric constant, the bandwidth has increased.

2 Antenna Design Configuration

In this paper, proposed antenna, based on microstrip technology, has been designed to perform at Ka band. The chosen substrate is 50 μm thick, RT5880 Rogers Corporation with dielectric constant 2.2 and relative permittivity of 0.009 loss tangent. Although there are various substrates available RT5880 Rogers Corporation has been chosen because of their suitability for high-frequency broadband application which demands minimal dispersion and losses. Further, their low water absorption ability makes them more suitable candidate for antenna application in high moisture environment, particularly in India during the monsoon and the sea coastal low land areas.

Design of antenna for millimeter-wave range is a challenging task before researchers because at this wavelength, the size of antenna is too small to handle. First, through MATLAB, antenna design and antenna parameters were worked out so that it performs at the said frequency with optimum features. Then the electronic behaviour of antenna is simulated in HFSS 13 to establish the desired features. Substrate size is taken exactly same as the size of the upper metallic surface of the antenna shown by blue color in Fig. 1. The left out space is due to air dielectric. By this, we have reduced the cost and material input of the substrate which in turn will cut down the cost of antenna fabrication. Probably, this has happened for the first time to the knowledge of authors.

The antenna is designed with a central frequency of 28 GHz. Antenna has an inset-fed transmission line. The feed line starts as a 50 Ω port, where the antenna is fed from. The main parameter determining the resonant frequency is the radiating element length, i.e., L_R. The bottom metallization, i.e., the ground plane, is a sheet of copper of a thickness of 18.5 μm. The copper sheet is fixed to the RT5880 Rogers Corporation. The total effective area occupied by the antenna is extremely small at 3.2 mm × 11.64 mm. Figure 2 shows the proposed antenna which

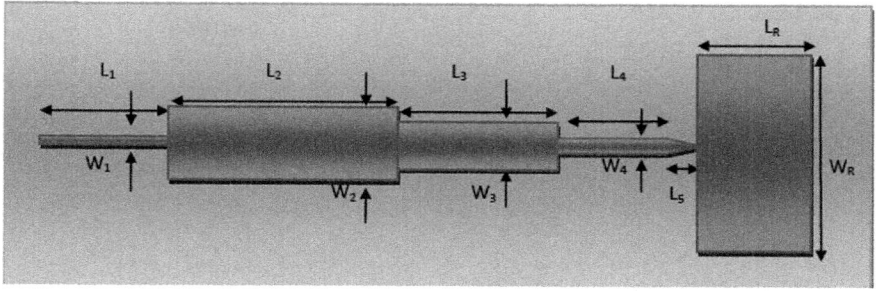

Fig. 1 Top view of antenna

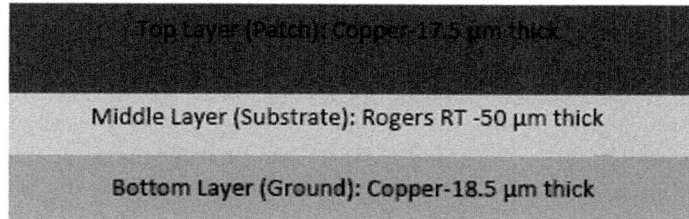

Fig. 2 Cross-sectional view of antenna

Table 1 Value of antenna dimensions and parameters

Parameter	Value (mm)	Parameter	Value (mm)
W_1	0.51	L_1	1.8
W_2	2.5	L_2	2.5
W_3	1.8	L_3	2.3
W_4	0.51	L_4	1.4
W_R	3.2	L_5	0.60
Wg/Lg	4.2/12.85	L_R	3.04

consists of three layers. The ground plane lower layer is of copper, substrate is middle layer, and top layer is patch of copper (Table 1).

3   Results and Discussion

Figure 3 presents measured and simulated reflection coefficient results $|S_{11}|$. The designed antenna can operate over an approximate BW of 1.4 GHz. By virtue of design, there exists air gap in the substrate which reduces the dielectric constant. This results in an increase in bandwidth [9]. The simulated and measured results

Fig. 3 Return loss S_{11} for
designed antenna

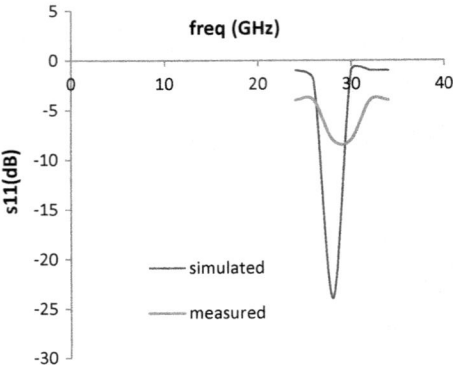

differ slightly. This difference can be mainly attributed to the observed shift in the resonant frequencies [10]. The shift in frequency can be reasoned to be originating from the dielectric constant of the substrate [11].

Figure 4 depicts the H-plane of radiation pattern at 28 GHz Similarly E-plane of radiation pattern at 28 shown in Fig. 5. Total efficiency and radiation efficiency are shown in Fig. 6 and simulated directivity of the designed antenna is shown in

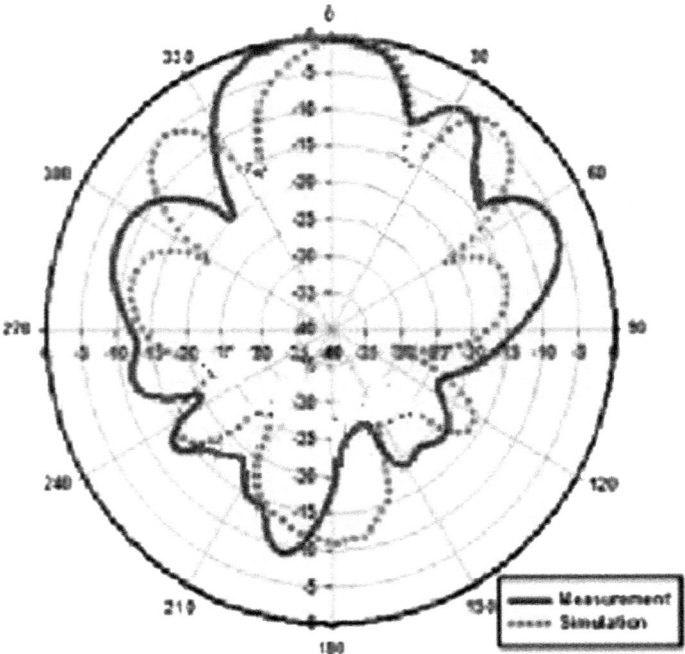

Fig. 4 H plane radiation pattern at 28 GHz

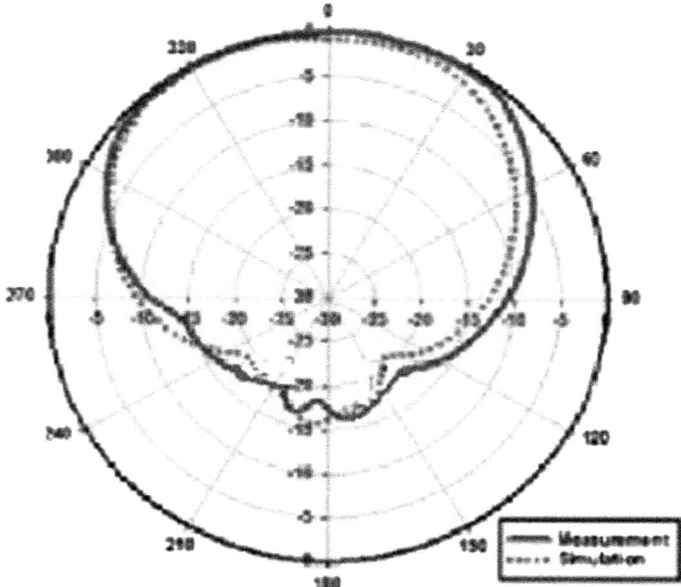

Fig. 5 E plane radiation pattern at 28 GHz

Fig. 6 Radiation efficiency and total efficiency characteristics of the proposed antenna for different scanning angles by simulation

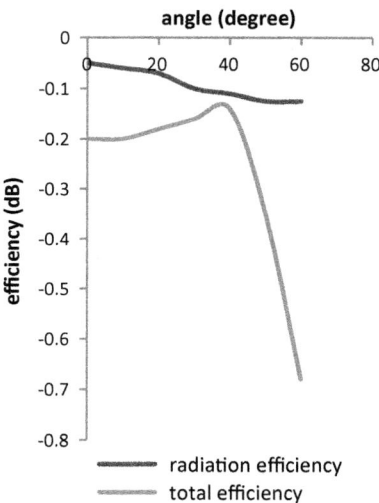

Fig. 7. For the scanning range of 0° to 40°, the antenna radiation is more than −0.2 dB. Similarly, total efficiency for the same scanning range is more than −0.3 dB. For different scanning angles, the antenna has high efficiencies. Further, the proposed antenna has around 13 dB directivity characteristic.

Fig. 7 Directivity of the proposed antenna for different scanning angles by simulation

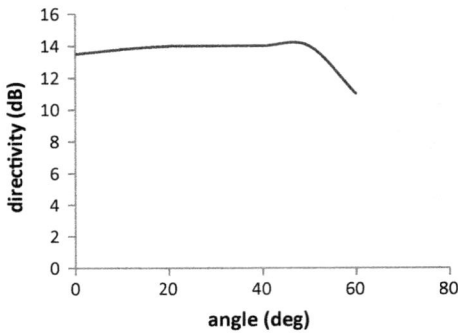

4 Conclusions

A novel microstrip patch antenna based on air-filled substrate aiming at 5G mobile communication in the millimeter-wave application has been reported. Low-cost substrate RT5880 Rogers Corporation is used to operate at Ka band frequency range. Fabrication of antenna and hence the measurement are due to take place. This will follow experimental verification on vector network analyzer.

References

1. R. G. S. Rao and R. Sai, "5G – Introduction & Future of Mobile Broadband Communication Redefined," Int. J. Electron. Commun. Instrum. Eng. Res. Dev., 2013, vol. 3, no. 4, pp. 119–124.
2. Z. Pi and F. Khan, "An introduction to millimeter-wave mobile broadband systems," Commun. Mag. IEEE, no. June 2011,, pp. 101–107.
3. Y. S. H. Khraisat, M. M. Olaimat, and S. N. Abdel-Razeq, "Comparison between Rectangular and Triangular Patch Antennas Arrays," Appl. Phys. Res., 2012, vol. 4, no. 2, pp. 75–81.
4. Chin, K. S., H. T. Chang, J. A. Liu, et al., "28-GHz patch antenna arrays with PCB and LTCC substrates," *Cross Strait Quad-Regional Radio Science and Wireless Technology Conference*, 2011, Vol. 1, pp. 355–358.
5. Tong, K. F., K. Li, and T. Matsui, "Performance of millimeter-wave coplanar patch antennas on low-k materials," *PIERS Online*, 2005, Vol. 1, No. 1, pp. 46–47.
6. Wang, D., H. Wong, K. B. Ng, and C. H. Chan, "Wideband shorted higher-order mode millimeter- wave patch antenna," *IEEE Antennas and Propagation Society International Symposium*, 2012, pp. 5–6.
7. N. H Shahadan, M. R. Kamarudin, N. A. Zainal,., "Investigation on feeding techniques for rectangular dielectric resonator antenna in higher-order mode for 5G applications," *Applied Mechanics and Materials*., 2015, Vol. 781.
8. Y. M Pan., K. W. Leung, and K. M. Luk, "Design of the millimeter-wave rectangular dielectric resonator antenna using a higher-order mode", *IEEE Trans. Antennas and Propag.*, 2011Vol. 59.

9. W. Bengal, "Effect of Dielectric Permittivity and Height on a Microstrip-Fed Rectangular Patch Antenna," IJECT, 2014, vol. 7109, no. 2, pp. 129–130.

10. C. Slot, "An Integrated Diversity Antenna Based on Dual-Feed," IEEE Antennas Wirel. Propag. Lett., 2014, vol. 13, pp. 301–304.

11. E. Kemppinen, Determination of The Permittivity of Some Dielectrics in The Microwave and Millimetre Wave Region. Oulun Yliopisto, Oulu, 1999.

A Review Article on Fault-Tolerant Control (FTC) and Fault Detection Isolation (FDI) Schemes of Wind Turbine

Chitrita Saha and Amit Kumar Singh

1 Introduction

Wind energy nowadays is the most cost-effective, environmental friendly and fastest growing source of energy around the globe. Investment in the wind power all over the world has expanded from $18 billion in 2006 to $60 billion in 2016 [1].

Though wind turbines nowadays contribute to be a large part in the production of power all over the world, the size of the wind turbines also tends to increase. Mainly vertical axis wind turbine is used. Megawatt size turbines are higher in cost and so their reliability is usually expected to be higher to generate more energy with shorter downtimes. In order to obtain this, advanced FDI and accommodation system should be introduced in the wind turbine. Normally fault detection and accommodation schemes are simpler. During simple faults, turbines are turned off. In order to improve the on-time of the turbine, there is a need to use advanced fault detection and isolation and accommodation schemes [2]. FDI algorithms are used for detection of irregularities in the sensor, actuator, and system of the wind turbines. Generally, three types of faults exist in wind turbine sensor faults, actuator faults, and system faults [3]. FDI procedure is implemented for assuming the severity of faults and for giving information about the faulty equipments whenever fault occurs in the turbine and accommodate these faults whenever possible [4]. Reduction of these faults at right times may increase the efficiency of the turbine as well as reduce the downtime and also reduce the repairing cost of the turbine by increasing its lifetime. If faults are not detected at right time this may lead to some major failures, which may result in higher cost breakdown. FDI is an important part

C. Saha (✉) · A. K. Singh
Department of EEE, Amity University, Noida, Uttar Pradesh, India
e-mail: chitrita.saha@yahoo.in

A. K. Singh
e-mail: aksingh20@amity.edu

© Springer Nature Singapore Pte Ltd. 2019
V. Nath and J. K. Mandal (eds.), *Proceeding of the Second International Conference on Microelectronics, Computing & Communication Systems (MCCS 2017)*, Lecture Notes in Electrical Engineering 476, https://doi.org/10.1007/978-981-10-8234-4_9

of FTC. FTC is a technique that prevent faults thus reducing the failures by taking proper action at proper time [5]. This paper is a literature review of several FDI and FTC schemes. This paper also includes few FTC and FDI techniques for horizontal axis wind turbine.

2 Literature Review

2.1 Fault-Tolerant Control (FTC) Schemes

In [6], a variable speed wind energy conversion system (WECS) is taken into consideration for modeling and simulation. The control strategy in this paper consists of two steps: wind turbine modeling and control system designing. The control system here consists of two loops: for regulation of torque and voltage, inner exciter loop is considered and a rotor loop in the outer position for maximizing the power captured and also for improving the reliability of the system. First, for an inner loop, a simple PI controller using a first-order system is implemented, but when the operating points changes or a wind turbine of variable speed is used, then the PI controller cannot provide acceptable performance also results in insufficient damping. At this stage, fuzzy logic controller (FLC) is used for regulation of voltage and power. Adaptive pitch controller is also used to control the pitch angle of the blades of the rotor for maximizing energy and to reduce loads. As an alternative pitch controller ANN is also designed. The performance of all the controllers is compared. It is concluded that ANN and FLC can be used for controlling WECS. Simulation results obtained are robust and satisfactory and give superior performance.

However, in [7] designing of an active and passive fault, tolerant linear parameter varying (LPV) controllers are taken into account. A wind turbine is considered that is working under full load condition and with pitch system faulty. The controllers are designed to handle the variations in both parameters, one during operating in normal conditions and other when the fault is introduced in the pitch system. This fault mainly occurs due to the presence of high air in the oil. Active fault-tolerant controller (AFTC) gives much better performance than passive fault-tolerant controller (PFTC), during the presence of faults as AFTC has the property of adapting with the faulty pitch system. PFTC does not depend on the algorithm for diagnosis of fault; therefore its structure is simpler and has lesser risk for false detection of faults. Problem of optimization for PFTC is solved using bilinear matrix inequalities (BMIs). This method is very difficult to solve because it uses an approach which is nonconvex that does not give convergence at the global minimum. AFTC is solved using convex approach of optimization and is simpler to solve and provide optimum results. Normally, AFTC should be implemented on system where designing of fault diagnosis system is required to be faster with lesser false detection risks. A PFTC should be preferred when diagnoses of faults are difficult or where the tolerance of false decisions in the system is zero. It concludes that AFTC is a better approach for control of faults in wind turbine (Fig. 1).

Better than AFTC and PFTC, a model predictive controller (MPC) is proposed for solving fault-tolerant control problem of a benchmark model of wind turbine [8]. In this paper, a hierarchal controller has been proposed with model predictive pre compensators, a global MPC, and supervisory controller. In model predictive, pre-compensators extended Kalman filter is designed for estimation of states in the system and parameters of fault. Based on estimates, a group of MPCs is designed for compensating the effects of faults for each component in the wind turbine. The performance of each component is optimized by global MPC. Management and

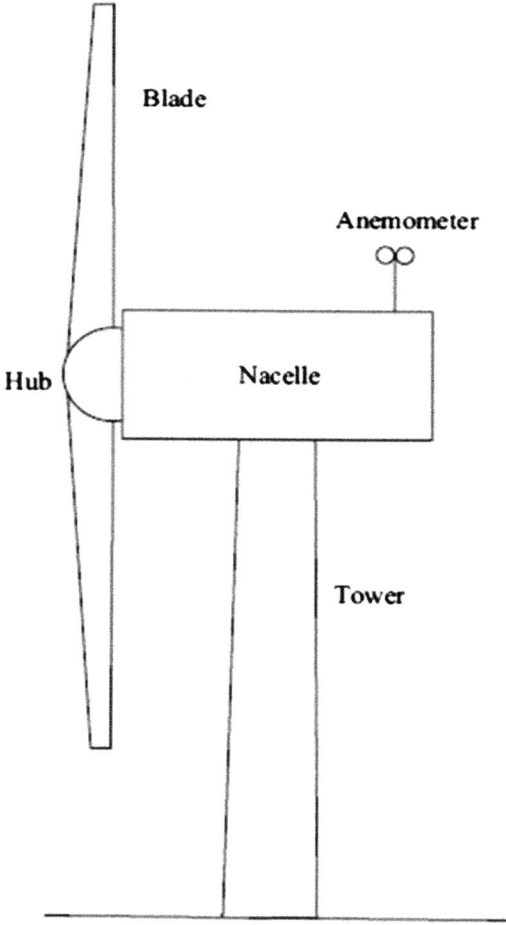

General outline of the wind turbine seen from the outside.

Fig. 1 Wind turbine outline

decision are a part of supervisory controller. Simulation results obtained are satisfactory which shows that the proposed controller can reduce actuator, system, and sensor faults and can improve the performance of the turbine. It shows that usage of MPC is a better option for fault-tolerant controllers.

A new and better method of FDI that is implemented on a benchmark model of wind turbine is marginalized likelihood ratio (MLR) [9]. This method is used for online detection of fault and isolating the fault. This technique helps to reduce the number of optimization problems that are needed for isolation of faults by using easy integrating methods. Moreover, MLR detects the time of the occurrence of faults and isolates that position of faults. MLR is efficient enough for detection and isolation of fault and also leads in reduction of effort required for computation. For critical processes, MLR is mostly used and also the results obtained are very reliable.

In [10], the use of virtual sensor and actuator approaches are mentioned for controlling sensor and actuator faults respectively using model matching principle. Hence, the FTC module is designed by taking information through the FDI that was formed earlier using set membership approach. A fault estimation scheme that was based on batch least square approach has been proposed. Simulation results and the performance of the FTC scheme are very much satisfactory.

In [11], a method for FTC is developed for wind turbines that are operated at a low wind speed by implementing the idea of estimation and compensation. In this strategy, PMIO is used as an intermediate between system and controller. This method uses PMIO for compensating the effect of rotor sensor and generator fault from TSDOFC. The challenges that are normally faced by wind turbine control system are nonlinearity in aerodynamic system, control of wind turbine due to variable wind speeds, and accurate effective wind speed (EWS) estimation without faults. In this paper, Lyapunov stability condition and proof for PMIO are being derived. Stability of TSDOFC is achieved with performance of horizontal axis wind turbine and D-stability constraints are developed using linear matrix inequality (LMI) problem. The benchmark model of wind turbine that is considered here is a nonlinear model. Implementation of this strategy can simplify the drivetrain fault estimation thus leading to maximization of power. It concludes that this strategy is best suited for control of faults in horizontal axis nonlinear wind turbine that operates at low wind speed.

In [12], an adaptive fault-tolerant control strategy is implemented on a benchmark model of wind turbine. The approach used here is based on the scheme that relies on the adaptive controllers that are designed using the online identified parameters of the system that is under diagnosis. This method has an advantage, i.e., preserving the original structure of logic-based switching digital controller scheme which was already implemented on the wind turbine benchmark model. Thus this strategy is implemented on the benchmark model with nominal operating conditions, different fault scenarios and also in the existence of disturbances and errors in measurements. The results that are obtained in the faulty condition and fault-free condition show effective control performance of wind turbine and also control of faults in a much better way.

Another better method of FTC using control allocation technique is implemented in the benchmark model [13] for compensating the effect of faulty actuators by using the law of reconfiguration. Fault in the actuator blade can lead to an error in the position of pitch of the respective blade. This leads to insufficient torque thus minimizing the output power. Control allocation is a generally used FTC scheme especially for actuator faults. The idea behind implementing this method is redistribution of the lost torque in actuator blades in order to get maximum power efficiency in spite of faulty actuator. The method is simulated using PI controller in the benchmark model. Effective simulation result is obtained which shows that the proposed methodology can control the fault in actuator blade as well as can improve the power efficiency.

The best technique for improvement of damping as well as the stability of a DFIG-based offshore wind farm (OWF) which is fed to a multi-machine system is by using a STATCOM [14]. For supplying adequate reactive power, PID damping controller is implemented. Moreover, for improvement in the performance of the system, a hybrid of PID and FLC is used. Implementation of static compensator (STACOM) along with hybrid FLC and PID results in the suppression of inherent oscillation thus improving the system stability and thus giving the best damping characteristics.

2.2 Fault Detection and Isolation Schemes (FDI)

A wind turbine benchmark model is considered here where GLR (generalized likelihood ratio test) is used for detection of the faults and isolating them [15]. Detecting a fault and isolating a fault in a wind turbine are difficult within given time constraints. Also due to practical implementation issues, computational effort should be low. By using GLR, accurate wind speed measurement can be achieved. Noise measurements as well as nonlinearities in drive train of wind turbine and other critical issues can be calculated using GLR. Here, GLR method along with FDT and FCT tests were used for detection and isolation of the faults. For pitch system, the fault detection and isolation module are divided into three divisions. The FDI module of pitch system and converter system both were designed using simple and linear Kalman filter and the simulation result for both pitch and converter system is acceptable. FDI module of a converter system is independent. However, the FDI module of drivetrain model is a function of pitch angle as well as wind speed therefore it is a dependent system.

In [16], a method of FDI for wind turbine is presented using a general method that is presented in Zhang et al. (2002, 2008). By using the benchmark model of the wind turbine, designing of fault detection estimator (FDE) and fault isolation estimator (FIE) are done without estimating the unknown wind speed and aerodynamic rotor torque. Here, the main focus is on designing of FDE and FIE for estimation.

A better method of FDI for a three-bladed horizontal axis wind turbine is using support vector machines (SVM) [17]. SVM are a data-based approach, therefore, it is robust. Additionally, it is also based on minimizing structural risk and also uses flexible and nonlinear kernels. In this work, for Kernel radial basis function was used. SVM is a better method for detection of sensor faults. Drivetrain fault due to friction and converter torque faults are also detected using SVM.

Usage of hybrid prototypes [18] is another better method for FDI for a benchmark wind turbine model that is estimated using input and output measurements of the turbine that are usually uncertain. The investigation process is assumed to be not linear and the measurements taken here are approximate measurements due to variation in wind speed. Hybrid modeling used here is a collection of some local affine model and each of them describing different operating points. This approach provides a wind turbine model, which was used for generation of residuals during diagnosis of faults. This strategy is effective enough as it was tested upon the data that was obtained from the simulated model of wind turbine. By using SVM detection and isolation of pitch angle, sensor fault is achieved.

In [5], an FDI system benchmark model of wind turbine is designed using generic automated design method, where minimum human decisions and assumptions are considered. This method is divided into three steps: generation of potential residuals generators, selecting residual generators, and finally constructing a diagnostic test. The second step is achieved using greedy selection algorithm and the third step is achieved by implementing a method based on K-L divergence. The performance by using generic automated design method is satisfactory enough and all the faults are detected within the feasible time limit. There are no false or missed detections.

In [3], model-based technique and up-down counters are used for detection of fault and isolation of faults in a wind turbine. Up and down counters are used for deciding on the fault residuals. Fault residuals are normally generated through parity equations and commonly used filtering methods. Up and down counters are used in aerospace industry for improvement of detection rules in a mission. By implementing this technique, fault can be easily detected and isolated. The benchmark model of wind turbine considered here is of 4.8 MW size. By implementing this technique faults in sensors, actuators and drivetrains can be detected easily without any false alarms and in lesser time without using any complex filtering and detection techniques. So this method is a simpler method for FDI of wind turbine.

In [19], a scheme for fault detection and isolation of a benchmark model of wind turbine is being proposed named as observer based technique. Pitch system, drivetrain, and converter all the three subsystems are taken into consideration. Kalman filter and diagnostic observer-based methods are used for generating residuals and for evaluating of residuals GLR and cumulative variance index are used. Using this strategy fault in drivetrain is eliminated band by using Kalman filter fault in the system and sensor is detected and isolated. Especially for the purpose of isolation of fault, residual generators are connected in a bank based on dual sensor redundancy. The parameters are designed in such a way so that false alarm is avoided and also faults are detected quickly. The FDI scheme performance is evaluated systematically by using Monte Carlo techniques.

Another method for detection and identification of fault is set theoretic method [20]. By using this method, fault is detected exactly when suitable assumptions are taken. This technique is an effective and efficient technique for detection of sensor, actuator, and composite faults.

In [9], fault in the benchmark model of wind turbine is diagnosed by applying set membership approach. This fault detection method is depended on usage of parity equations and unknown factors but noise description is bound along with the modeling errors. This method is used for checking the consistency of the model. The method detects the fault occurring on the line and theoretical ones are matched with them that are obtained by structural analysis. The approach that is proposed is tested on a wind turbine model and gives a consistent result.

Dat-driven algorithm is a more appropriate method for investing the fault and isolating the fault of wind turbine [21]. The benchmark model in this instance is set up using Simulink containing nonlinear lookup tables and wind disturbances that are not known. For designing a classical filter, linearization of Simulink model as well as errors are necessary. In order to avoid this difficulty, a data-driven design method is used which can design an FDI filter directly from the data available from the Simulink model. Hence, here only sensor faults with fixed values are targeted. Moreover, a new hardware redundancy based fault isolation technique is implemented here. Based on this, a bank of data-driven filters is designed for the benchmark model for fault detection and isolation and is used in parallel. The simulation result obtained is satisfactory and it is one of the better methods for FDI of sensor faults.

A data-driven modeling approach [22] is implemented on a benchmark model of wind turbine for detection and isolation of faults where hybrid prototypes are used which are identified from input and output measurements that are uncertain. The model of wind turbine used here is a nonlinear model and due to variable wind speed, the measurements are not exact. By applying this method, in spite of unknown measurements sensor and actuator faults can be detected and isolated easily.

In [23], fuzzy modeling and an identification approach are used for designing a PI fuzzy controller for the regulation of pitch angle as well as the generator torque of a wind turbine benchmark model. The idea behind this is to enhance the design of the regulator that is an alternative to switching controllers. For designing of fuzzy controller, knowledge of fuzzy modeling and identification scheme is required. Moreover, this controller is easier to design and implement than any other strategies and is also much more reliable. When result of PI fuzzy controller is compared with other controllers, the result obtained from PI fuzzy controller is much more satisfactory.

A more reliable and robust method for fault detection and isolation is set value observer (SVO) [4]. Here, SVO is tested using Monte Carlo simulation runs on a benchmark model of wind turbine. A FTC was designed on top of the proposed FDI algorithm, which enables the reconfiguration of the structure of the controller during the faults. To increase the performance and the operational availability of wind turbine, a robust controller was designed additionally.

3 Conclusions

This paper describes few techniques and methods of detection and control of fault in wind turbines as they are the major contributor of renewable energy source. Wind turbines are considerably preferred nowadays because of the desired levels of availability, efficiency, and reliability of source, i.e., wind. There are few key technologies adopted to ensure their reliability and efficiency of modern wind turbines operation which are diagnosis and advanced health monitoring, along with fault-tolerant control (FTC) and optimal control.

In nutshell, this paper is a review of several such discussed technologies on wind turbine. Among them, the importance is given on FTC and FDI of wind turbine using several methods. This paper as such creates a wholistic study on different methods of fault reduction, control, and detection of fault as well as isolation of faults and their results on the performance of wind turbine.

References

1. http://www.gwec.net/"Global Wind Energy Council". (2009, Jan. 24). [Online]. Available.
2. Peter Fogh Odgaard, member IEEE, Jakob Stoustrup & Michel Kinnaert, "Fault tolerant control of wind turbines: a benchmark model".
3. Ahmet Arda Ozdemir, Peter Seiler, and Gary J. Balas "Wind Turbine Fault Detection Using Counter-Based Residual Thresholding".
4. P. Casau, P. Rosa, S. M. Tabatabaeipour, C. Silvestre, "Fault Detection and Isolation and Fault Tolerant Control of Wind Turbines using Set-Valued Observers" 8th IFAC Symposium on Fault Detection, Supervision and Safety of Technical Processes (SAFEPROCESS) August 29-31, 2012. Mexico City, Mexico.
5. Carl Svärd, Mattias Nyberg "Automated Design of an FDI-System for the Wind Turbine Benchmark".
6. R. Chedid, Member IEEE, F. Mrad, Member IEEE, M. Basma, "Intelligent control of a class of wind energy conversion systems", IEEE transaction on energy conversion, Vol. 14, No. 4, December 1999.
7. Christoffer Sloth, Thomas Esbensen and Jakob Stoustrup, "Active and passive fault tolerant LPV control of wind turbines", 2010 American control conference, June 30-July 02, 2010.
8. X. Yang, J.M. Maciejowski, "Fault-tolerant model predictive control of a wind turbine benchmark", Cambridge University Engineering Dept., Cambridge CB2 1PZ, UK.
9. Fariborz Kiasi, Jagdeesan Prakash, Jong Min Lee and Sirish L. Shah, "Model based fault tolerant control using the marginalized likelihood ratio test", 2010 conference on control and fault tolerant systems, Nice, France, October 6-8, 2010.
10. Damiano Rotondo, Fatiha, Nejjari, Vicenç Puig, Joaquim Blesa. 8th IFAC Symposium on Fault Detection, August 29-31, 2012. Mexico City, Mexico "Fault Tolerant Control of the Wind Turbine Benchmark using Virtual Sensors/Actuators".
11. Montadher Sami* and Ron J. Patton, "An FTC Approach to Wind Turbine Power Maximisation via T-S Fuzzy Modelling and Control", 8th IFAC Symposium on Fault Detection, Supervision and Safety of Technical Processes (SAFEPROCESS) August 29-31, 2012. Mexico City, Mexico.

12. Silvio Simani, Paolo Castaldi "Adaptive Fault–Tolerant Control Design Approach for a Wind Turbine Benchmark",, 8th IFAC Symposium on Fault Detection, Supervision and Safety of Technical Processes (SAFEPROCESS) August 29-31, 2012. Mexico City, Mexico.
13. Jiyeon Kim, Inseok Yang, and Dongik Lee, "Control Allocation based Compensation for Faulty Blade Actuator of Wind Turbine", 8th IFAC Symposium on Fault Detection, Supervision and Safety of Technical Processes (SAFEPROCESS), August 29-31, 2012. Mexico City, Mexico.
14. Li Wang, Senior Member, IEEE and Dinh-Nhon Truong, "Stability enhancement of DFIG-based off shore wind farm fed to a multi machine system using a STATCOM", IEEE transactions on power systems, 2013.
15. F. Kiasi, J. Prakash, S. L. Shah, and J. M. Lee "Fault Detection and Isolation of a Benchmark Wind Turbine using the Likelihood Ratio Test".
16. Xiaodong Zhang, Qi Zhang, *Songling* Zhao, *Riccardo* Ferrari, Marios M. Polycarpou, and Thomas Parisini "Fault Detection and Isolation of the Wind Turbine Benchmark: an Estimation-based Approach".
17. Nassim Laouti, Nida Sheibat-Othman "Support Vector Machines for Fault Detection in Wind Turbines". Preprints of the 18th IFAC World Congress Milano (Italy) August 28 - September 2, 2011.
18. Silvio Simani, Paolo Castaldi, Marcello Bonfe "Hybrid Model–Based Fault Detection of Wind Turbine Sensors", Preprints of the 18th IFAC World Congress Milano (Italy) August 28 - September 2, 2011.
19. W. Chen, S. X. Ding, A. Haghani, A. Naik, A. Q. Khan, S. Yin. "Observer-based FDI Schemes for Wind Turbine Benchmark" Preprints of the 18th IFAC World Congress Milano (Italy) August 28 - September 2, 2011.
20. Florin Stoican, Catalin Florentin Raduinea, Sorin Olaru. "Adaptation of set theoretic methods to the fault detection of a wind turbine Benchmark" Preprints of the 18th IFAC World Congress Milano (Italy) August 28 - September 2, 2011.
21. Pierluigi Pisu and Beshah Ayalew, "Data Driven Fault Detection and Isolation of a Wind Turbine Benchmark".
22. Silvio Simani1, Paolo Castaldi, and Andrea Tilli. "Data–driven Modelling of a Wind Turbine Benchmark for Fault Diagnosis Application". Transaction on Control and Mechanical Systems, Vol. 1, NO. 7, PP. 278-289, NOV., 2012.
23. Silvio Simani "Data–Driven Design of a PI Fuzzy Controller for a Wind Turbine Simulated Model". IFAC Conference on Advances in PID Control PID'12 Brescia (Italy), March 28-30, 2012.

A Comprehensive Survey on Computational Grid Resource Management

Ankita and Sudip Kumar Sahana

1 Introduction

Grid computing has emerged as a new standard to solve complex scientific and engineering problems [1, 2]. A single machine with its processing capability is not able to solve such difficult problems. Eventually, the focus has drifted toward geographically distributed resources which have tremendous processing potential to solve a composite problem. Not only the processing power but the bandwidth and memory all are resources that an accredited user can use for a specific problem. Due to the recent advancements in grid computing, its domain is no longer restricted to scientific areas, both in the government sector, business areas, and industrial regions as well. Resource scheduling [3] in general is a process of mapping resources over different jobs with an aim to minimize job completion time and maximize resource utilization. Resource management is a difficult task due to the reason resources are widely distributed and heterogeneous in nature.

2 Resource Scheduling

2.1 Resource Detection

This is the very starting phase of resource scheduling [4]. In this phase, all the authorized resources presented in the grid are identified and listed down. Since grid

Ankita (✉) · S. K. Sahana
Computer Science and Engineering, Birla Institute of Technology, Mesra, Ranchi, India
e-mail: sharma.ankita0211@gmail.com

S. K. Sahana
e-mail: sudipsahana@gmail.com

© Springer Nature Singapore Pte Ltd. 2019
V. Nath and J. K. Mandal (eds.), *Proceeding of the Second International Conference on Microelectronics, Computing & Communication Systems (MCCS 2017)*, Lecture Notes in Electrical Engineering 476, https://doi.org/10.1007/978-981-10-8234-4_10

works in a dynamic environment, the scheduler must follow the same approach to have a continuous look into the changing state of information about resources because resource may be added or they may fail to work correctly at any time. This will help the scheduler at the time of resource selection.

2.2   Resource Selection

The scheduler starts analyzing the job queue's characteristic, i.e., the required CPU cycles needed by the job or its memory requirements. Depending on the job's constraints and its previous stage usage, the scheduler makes a list of appropriate resources from the available resources.

2.3   Schedule Generation

The scheduler starts analyzing the job queue's characteristic, i.e., the required CPU cycles needed by the job or its memory requirements. Depending on the job's constraints and its previous stage usage, the scheduler makes a list of appropriate resources from the available resources.

3   Classification of Resource Allocation Strategies

There are four classes of resource allocation strategies. The detailed description of each of these strategies is discussed in the next section (Fig. 1).

Fig. 1 Resource allocation strategies

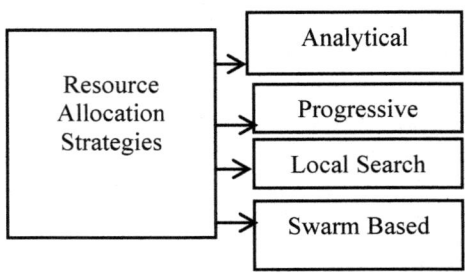

Fig. 2 Analytical approaches
(FCFS—first come first serve,
MET—minimum execution
time, MCT—minimum
completion time, OLB—
opportunistic load balancing)

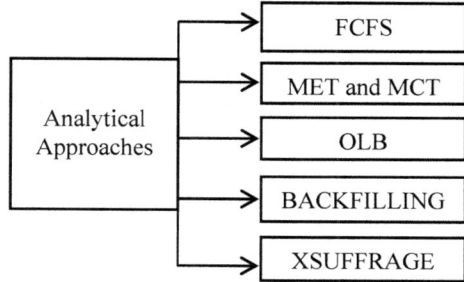

3.1 Analytical Approaches

Analytical or heuristic algorithms are basically approximate algorithms, i.e., these algorithms find out the best solution out of the possible ones. These algorithms are fast but not accurate (Fig. 2).

3.1.1 First Come First Serve (FCFS)

In FCFS, the scheduler assigns resources to the job in the sequence they arrive in the ready queue. Once the job is allocated to a particular resource, it will not leave the resource until the job is finished. It is a non-preemptive approach of scheduling [5].

3.1.2 Minimum Execution Time and Minimum Completion Time (MET and MCT)

The MET considers only the execution time of the job without having any knowledge about the current load on the grid. The intention is to allocate the job to the best resource. The MCT not only considers the execution time of the job but also the total resource length.

3.1.3 Opportunistic Load Balancing (OLB)

The thought behind OLB approach is to keep all the available resources busy at all time. The jobs are allocated to each resource without considering the execution time of the particular job on the allocated machine [6]. The approach results in long makespan which is its biggest drawback.

3.1.4 Backfilling Scheduling (BS)

The backfilling scheduling algorithm tries to eradicate the problem of FCFS. In this case, the other jobs with low resource requirement need not wait longer even if a job with high resource requirement is present in the queue. The shorter jobs will be executed until the particular resource becomes free for the longest job [7].

3.1.5 XSuffrage

The motivation behind this approach is very simple. The sufferer (if a job does not allocate a resource) allocates the resource first and has higher priority than a newly arrived job [8].

3.2 Progressive Approaches

Progressive or evolutionary approaches follow the classic Darwin's evolution theory which can be stated as "survival of the fittest". The basic idea is quite simple. In a population of individuals, the individuals who meet certain selection condition are allowed to reproduce. Progressive approaches include parameters like selection, mutation, and crossover. Each individual in the population represents a candidate solution. The best solution is determined as the population converges to the best individuals who satisfy certain conditions (Fig. 3).

3.2.1 Genetic Algorithm (GA)

GA is designed to mimic the processes in natural systems that are necessary for evolution. The concept of GA is inspired by the Darwinian's theory of "Survival of Fittest". The approach discussed by [9] uses genetic algorithms for grid scheduling. The author tried to minimize makespan time using a workflow allocation strategy. The author has used genetic algorithm and minimum completion time to schedule the grid [10].

Fig. 3 Progressive approaches (GA—genetic algorithm, EP—evolutionary programming, ES—evolution strategy, DE—differential evolution, GP—genetic programming)

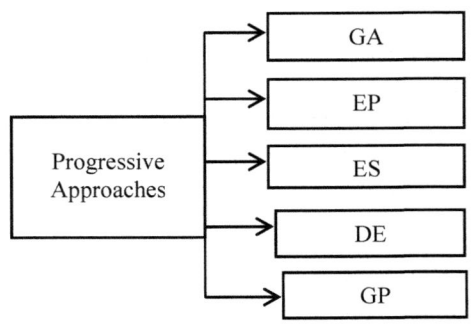

3.2.2 Evolutionary Programming (EP)

The EP was originally developed by Lawrence in 1960 and is very much similar to the GA in its concept. The underlying difference between the two approaches is that the EP does not use any operation like crossover. EP focuses emphasis on the behavioral connection between parents and their offspring. Rather than selecting the best individuals to reproduce as in case of GA, here each individual in the population is replicated into a new population. The offspring is then mutated and evaluated on the basis of fitness [11].

3.2.3 Evolution Strategy (ES)

The EP was originally developed by Lawrence in 1960 and is very much similar to the GA in its concept. The underlying difference between the two approaches is that the EP does not use any operation like crossover. EP focuses emphasis on the behavioral connection between parents and their offspring. Rather than selecting the best individuals to reproduce as in case of GA, here each individual in the population is replicated into a new population. The offspring is then mutated and evaluated on the basis of fitness [11].

3.2.4 Differential Evolution (DE)

Differential evolution is an approach that attempts to improve each and every candidate solution (chromosome) in the population. The population consists of a set of randomly generated integer values.

The multi-objective algorithm developed by Falco attempts to schedule jobs on the grid with an aim to maximize the performance of grid with minimum resource usage. Later on, similar algorithms were developed by Selvi and Talukder which aimed at minimizing the job completion time as well as total cost incurred in the process of job execution. The author has used differential evolution algorithm to solve a scheduling problem in grid computing with an aim to minimize the total execution of the jobs [12].

3.2.5 Genetic Programming (GP)

Genetic programming automatically creates a working computer program from a given problem. Genetic programming is based on Darwin's principle of natural selection and other bio-inspired operations (reproduction, crossover, mutation).

Implementation—Recently, progressive algorithms have been applied by NASA researchers to successfully evolve antennas for use on satellite, permutation flow-shop scheduling, and underwater image restoration.

Fig. 4 Local search
approaches (SA—simulated
annealing, TS—Tabu search,
HC—hill climbing)

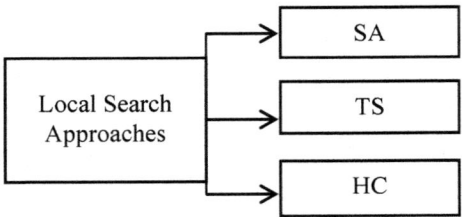

3.3   Local Search Approaches

Local search algorithms are iterative in nature and follow a sequential fashion (i.e., It moves from the current solution to a nearby one). It is iterative because it keeps on searching for good solutions on its neighbor relations. The best part of the local search algorithm is that it always returns a good solution even if the algorithm is interrupted at any moment of time (Fig. 4).

3.3.1   Simulated Annealing (SA)

Simulated annealing was originally proposed by Metropolis in 1953 to simulate the evolution of a solid in a hot bath until it reaches its equilibrium. In the annealing process, the temperature of the metal is first increased to a limit where its atoms can move freely. Then the temperature is gradually decreased so that the atoms can rearrange themselves and form a crystal. The cooling schedule is important here because if the crystal is cooled down too slowly or too quickly, then the proper crystal structure is not formed. The author proposed the use of simulated annealing which discusses how to search for best tasks during scheduling in grid environment [13].

3.3.2   Tabu Search (TS)

Tabu search is a metaheuristic which was proposed by Fred Glover in the mid-1980s. It is a local search procedure which explores the solution space beyond local optimality. Tabu search has been applied to many problems such as resource allocation, scheduling, routing, forecasting, data mining, etc. The main aim of tabu search mechanism was to inhibit some solution elements (or moves) and they are regarded as tabu, such that they cannot be used further in solution construction. The author has applied Tabu Search to a job-shop scheduling problem [14].

3.3.3 Hill Climbing (HC)

Hill climbing is one of the local search procedures to solve computer hard problems. Russel and Norvig [12] states that Hill climbing is best used in problems with "the property that the state description itself contains all the information needed for a solution". The algorithm suffers from the problem of local maxima [15].

Implementation—Extended double row layout problem, multistage distribution system planning.

3.4 Swarm-based Approaches

The swarm approaches are inspired by the behavior of living organisms that interact with large groups in nature. It could be a flock of birds, schools of fishes, a swarm of ants, or a swarm of bees. The best part of using a swarm algorithm is its speed of convergence (Fig. 5).

3.4.1 Ant Colony Optimization (ACO)

The concept of ACO was given by Marco Dorigo in 1992 where the ant's behavior in nature inspired the whole algorithm. In general, ants move randomly from one place to another in search of food. As soon as they found the food source, they take a load of it and return to their nest. During their return, they deposit a chemical substance called a pheromone on its path of traversal. This pheromone is sensed by the other ants and they tend to follow that path laid by the first ant that deposited pheromone. Eventually, this pheromone evaporates and hence it no longer attracts other ants on that path. Pheromone evaporation is important to avoid the convergence to a locally optimal solution. The search space would not have been fully explored if pheromone is not evaporated. ACO is a probabilistic approach in a manner that higher the pheromone concentration on the path more is the probability of that path being followed.

In the past few years, ant colony optimization technique has been used by various researchers for scheduling jobs [16] in the grid environment [17–19]. In [20], the author has shown how ants have been the agents in the ACO algorithms.

Fig. 5 Swarm-based approaches (ACO—ant colony optimization, PSO—particle swarm optimization, ABC—artificial bee colony)

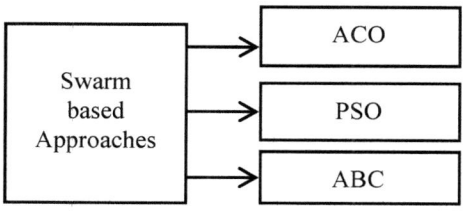

3.4.2 Artificial Bee Colony (ABC)

ABC stands for artificial bee colony and was given by Dusan Teodorovic in 2009 which motivated the action of bees in nature [21, 22]. Like other insects in nature, bees also have their own ability to show their potential in performing complex tasks. There are three categories of bees demonstrated by Dusan Teodorvic. The first bee is the employed bee that has already discovered a food source and exploiting it. The second bee is an onlooker bee that is still in the beehive and has not started the search for food and the third bee is called scout bee that is looking for a food source. The employed bee becomes a scout bee once its food source is relinquished. The author has used artificial bee colony algorithm and cluster-based HEFT for grid scheduling. With this approach, the grid was first partitioned into clusters for better utilization of resources [23]. The algorithm used in [24] is flexible and can easily be combined with other existing algorithms. Undoubtedly, this method gives better performance with minimum turnaround time, but it is not applicable for dynamic jobs.

3.4.3 Particle Swarm Optimization (PSO)

PSO stands for particle swarm optimization which was originally introduced by Kennedy, Eberhart, and Shi in 1995 which discusses the behavior of living organisms (fish schooling or bird flock) in nature. A set of random particles called population is first initialized in PSO. These particles traverse through the search space on the basis of some mathematical formulae and are evaluated after each iteration (at each time step). There are two best values which are considered for updating each and every particle. The first value is the value obtained by the particle itself during its course of traversal and that is called pbest (particle best). The other solution or value is the value given by the particle swarm optimizer and that is called gbest (global best). There is also a value called best (local best) which comes into the picture when the particle not as a part of the entire population but particates considering its neighbor as population. Then the second value is called lbest. The author has proposed a multi-objective PSO for job scheduling. There were three main objectives which consist of minimizing makespan, machine idle time, and total tardiness time to be achieved simultaneously.

Implementation—Virtual machine placement in cloud computing, Vehicle routing problems.

4 Discussion

Though all the approaches are applicable for scheduling in grid computing, they differ in terms of their performance. Evolutionary approaches work well for NP complete problems. For NP complete problems, though solution exists, one tries to

Table 1 Comparison of resource management strategies (NP—nonprobabilistic, P—probabilistic)

Approach	Parameters				
	Mode	Scalability	Resource utilization	Fault tolerance	Execution
Analytical	NP	Low	High	High	Serial
Progressive	P	High	Medium	Medium	Serial/parallel
Local search	P	Medium	High	Medium	Serial/parallel
Swarm	P	High	Low	Medium	Parallel

find the best in terms of complexity. Such problems are computationally intensive to solve through analytical approaches. Table 1 presents a comparison of these algorithms.

Table 1 briefly represents a comparison of various resource allocation strategies discussed so far. This comparison has been made on the basis of some standard parameters—Mode, scalability, resource utilization, fault tolerance, and execution.

The mode signifies if the approach is probable or not. Ant colony optimization is a probabilistic method since the probability of the path selected by the ants depends on the concentration of pheromone. The higher the concentration of pheromone, the higher is the probability of that path being selected by ants. The second parameter is scalability. Progressive and swarm-based approaches are highly scalable. This is one of the utmost reasons why researchers are focusing on these algorithms. Combining these algorithms give a substantial rise in performance as is proven by researchers.

Resource utilization is considerably high in analytical approaches as compared to swarm-based and progressive approaches.

The second last parameter is fault tolerance. Again analytical approaches are fault tolerant as compared to other approaches. Swarm algorithms generally do not have fault tolerance mechanisms. The last parameter is the type of execution that these algorithms follow. Analytical approaches provide serial execution of jobs while swarm approaches provide parallel execution and are highly scalable. For local search approaches, both serial and parallel modes of execution are possible.

So depending on the type of application, one can go for any of the approaches discussed above.

The next section discusses about the application of these algorithms in grid computing. In the past few years, the interests of researchers have been drifted to hybridized algorithms as these algorithms use the positive synergy of all the candidate algorithms. For example, in [25] the author has used a hybridized approach by combining genetic algorithm and artificial bee colony for grid scheduling. Simulation results showed that genetic algorithm based ABC performs better as compared to genetic algorithm and ABC when they were not hybridized.

Table 2 shows a report on the implementation of the approaches discussed in this paper. Two or more approaches can be combined to form a new approach which is called as a hybrid approach.

Table 2 An example of resource management algorithm in grid computing

Type	Paper	Approach	Remark
Analytical	An adaptive scoring job scheduling algorithm for grid computing [26]	Using scores for resource allocation in grid computing	The correct implementation of this approach depends on the value of cluster score. Not applicable for real-time grid applications
Progressive	An adaptive genetic algorithm for the grid scheduling problem [11]	Genetic algorithm	Simulation results showed that AGA performs better as compared to simple GA and max-min algorithm
Local search	Hill climbing-based decentralized job scheduling on computational grids [27]	Hill climbing	It is quite difficult for two nodes to be neighboring in a large-scale grid
Swarm based	Memory constrained ant colony system for task scheduling in grid computing [28]	Ant colony optimization	With this approach, the CPU idle time was minimized. This approach is not suited for dynamic jobs
Swarm based + progressive (hybrid)	A hybrid ant colony optimization algorithm for job scheduling in computational grids [29]	Ant colony optimization and 3 opt mutation operation	Mutation helps in diversification of the search space and hence this method less vulnerable to the problem of falling in local optima
Progressive + local search (hybrid)	A GA + TS hybrid algorithm for independent batch scheduling in computational grids [30]	Genetic algorithm and Tabu search	Here GA is applied first and its best output is given as input to the Tabu search in the next step. Its primary objective is to reduce makespan value

5 Conclusion

So far in this paper, we have discussed different types of resource allocation strategies that can be applied in grid computing. The research trend shows that the use of evolutionary techniques and swarm-based algorithms has gained popularity in recent years. The reason behind the popularity of these algorithms is their flexible nature. And with this appealing property, these algorithms can be combined with other approaches to gain substantial performance. Evolutionary and swarm-based approaches have already proven their potentiality. So researchers can make a noble attempt by forming a hybridized resource management algorithm in grid environment.

References

1. Foster, I., Kesselmen, C., Tuecke, S. (2001). The Anatomy of the Grid: Enabling Scalable Virtual Organisations. *International Journal of High Performance Computing Applications*, pp. 200–222.
2. Foster, I., Kesselmen, I. (1999). The Grid: Blueprint for a Future Computing Infrastructure. *Morgan Kaufmann Publishers*, pp. 1–593.
3. Nagariya, S., Mishra, M. (2013). Resource Scheduling in grid computing: A Survey. *International Journal of Advanced Research in Computer Science and software engineering*, 3(10), 735–739.
4. Buyya, R., Abramson, D., Giddy, J. (2000). Grid Resource Management, Scheduling and Computational Economy. In *Proc. Of the 2nd International Workshop on Global and Cluster Computing*, pp. 1–2.
5. Jiang, H., Ni, T. (2009). PB-FCFS–A Task Scheduling Algorithm Based on FCFS and Backfilling Strategy for Grid Computing, In *Pervasive Computing (JCPC)*, pp. 507–510.
6. Alharbi, F. (2012). Simple Scheduling Algorithm with Load Balancing for Grid Computing. *Asian Transactions on Computers*, 2 (2), 8–15.
7. Lokhande, S.F., Chavhan S.D., Jadhao, S.R. (2015). Grid Computing Scheduling Jobs Based on Priority Using Backfilling. *International Journal of Electrical, Electronics & Computer Science, Engineering*, pp. 68–72.
8. Ghazipour, F., Mirabedini, S.J., Harounabadi, A. (2016). Proposing a new Job Scheduling Algorithm in Grid Environment Using a Combination of Ant Colony Optimization Algorithm (ACO) and Suffrage. *International Journal of Computer Applications Technology and Research*, 5 (1), 20–25.
9. Joshua, R., Raj, S., Vasudevan, V. (2011). Grid Scheduling with Smart Genetic algorithm. *International Journal of Grid Computing and Multi Agent Systems*, 2(1), 1–10.
10. Carretero, J., Xhafa, F. (2007). Genetic Algorithm based schedulers for Grid Computing Systems. *International Journal of Innovative Computing, Information and Control*, 3(6), 1–19.
11. Wei, Z., Yang-Ping, B. (2012). An Adaptive Genetic Algorithm for the Grid Scheduling Problem. In *24th Chinese Control and Decision Conference*, pp. 730–734.
12. Russell, S. J., & Norvig, P. (2004). Artificial Intelligence: A Modern Approach. *Upper Saddle River, NJ: Prentice Hall.*
13. Fidanova, S. (2006). Simulated Annealing for Grid Scheduling Problem. In: *Modern Computing. In IEEE John Vincent Atanasoff International Symposium*, pp. 41–45.
14. Dell'Amico, M., Trubian, M. (1993). Applying Tabu Search to a job-shop scheduling problem. In *Annals of Operational Research*, 41 (3), 231–252.
15. Krishnamoorthy, N., Asokan, R. (2014). Optimal Resource Selection to promote Grid scheduling using Hill Climbing Algorithm. *International Journal of Computer Science and telecommunications*, 5 (2), 14–19.
16. Oshin, Chhabra, A. (2016). Job Scheduling using Ant Colony Optimization in Grid environment. In: *International Conference on Electrical, Electronics, and Optimization Techniques (ICEEOT)*, IEEE, pp. 2845–2850.
17. Mathiyalagan, P., Dhepthie, U.R., Sivanandam, S.N. (2010). Grid Scheduling using enhanced ant colony algorithm. *ICTACT journal on soft computing*, Volume 2, 85–87.
18. Ruhana, K., Mahamud, K. (2010). Ant colony algorithm for job scheduling in grid computing. In: *Fourth Asia International Conference on Mathematical/Analytical Modelling and Computer Simulation*, pp. 40–45.
19. Wei, L., Zhang, X., Li, Y., Li, Y. (2012). An Improved Ant Algorithm for Grid Task Scheduling Strategy, In International Conference on Applied Physics and Industrial Engineering, Volume 24, 1974–1981.
20. Blum, C. (2005). Ant colony optimization: Introduction and recent trends. In *Physics of Life Reviews*, Elsevier, pp. 353–373.

21. Alyaseri, S., Ruhana, K., Mahamud, K. (2013). Bee foraging behavior techniques for Grid Scheduling. *International Referred Journal of Engineering and Science*, 2 (4), 39–45.
22. Sha, D.Y., Lin, H.H. (2010). A multi-objective PSO for job-shop scheduling problems. *Expert System with Applications,* 37 (2), 1065–1070.
23. Teodorovic, D. (2009). Bee Colony Optimization. In *Innovations in Swarm intelligence 248*, 39–60.
24. Qureshi, M.B., Dehnavi M.M., Alla, N.M., Qureshi M.S., Hussain H., Rentifis I., Tziritas N., Loukopoulos T., Khan S.U., Xu C-Z., Zomaya A Y. (2014).Survey on Grid Resource Allocation Mechanisms. *Journal of Grid Computing*, 12 (2), 399–441.
25. Kaladevi, A.C, Srinath, M.V, Prabhakar, A. (2013). Reserved Bee Colony Optimization Based Grid Scheduling. In *International Conference on Computer Communication and Informatics*, pp. 1–6.
26. Chang, R.S., Lin, C.Y, Lin, C.F. (2012). An Adaptive Scoring Job Scheduling algorithm for grid computing. In *Information Sciences, Elsevier,* 207, 79–89.
27. Wang, Q., Gao, Y., Liu, P. (2006). Hill Climbing-Based Decentralized Job Scheduling on Computational Grids. In *Proc. of the First International Multi-Symp. on Computer and Computational Sciences*, IEEE, pp. 705–708.
28. Kokilavani, T., Amalarethinam, D.I.G. (2012). Memory Constrained ant colony system for task scheduling in grid computing. *International Journal of Grid Computing & Applications*, 3(3), 11–20.
29. Kumar, E.S., Sumanthi, A., Zubar, H.A. (2015). A hybrid Ant Colony Optimization algorithm for job scheduling in Computational grids. *Journal of Scientific and Industrial Research*, 74 (7), 377–380.
30. Xhafa, F., Kołodziej, J., Barolli, L., Fundo, A. (2011). A GA + TS Hybrid Algorithm for Independent Batch Scheduling in Computational Grids. In: *International Conference on Network-Based Information Systems*, pp. 229–235.

Thunderstorm Prediction Using Soft Computing and Wavelet

Kanchan Bala, Sanchita Paul and Mili Ghosh

1 Introduction

Three ingredients are needed in the formation of thunderstorm such as rising unstable air, moisture, and lifting mechanism. Earth is heated due to sun which warms the surface air. This warm air rises to mountain, hills, or areas where wet/dry or cold/warm bump together as result in rising the air motion. As the warm and humid air rises, heat is transferred from earth surface to upper layer of atmosphere. It contains water vapor which starts to cool, condenses, and forms a cloud. The cloud further grows upward area where the temperature is below the freezing point and creates different types of ice particles. These particles further grow due to condensing vapor. When two ice particles collide within cloud, they acquire some electric charge on them. A hues amount of electric charge is developed due to these collisions. The development of electric charges causes a bolt of lightning which creates thunder in the form of sound wave. The developing stage, mature stage, and dissipating stage are well-known life cycle of thunderstorm. In developing stage, a cumulus cloud is pushed upward due to updraft and form cumulus cloud look like a tower. This stage has no little rain but sometime associated the lightning. When the thunderstorm in mature stage in mature stage, updraft to feed the storm and precipitation starts to fall out from storm, creating a downdraft. As rain cooled air and downdraft spread out along the ground, it forms a line of gusty wind or gust front. The mature stage is generally associated with heavy rain, hail, tornadoes, strong

K. Bala (✉) · S. Paul
CSE, BIT, Mesra, Ranchi, India
e-mail: kanchanbala237@gmail.com

S. Paul
e-mail: sanchita07@gmail.com

M. Ghosh
Remote Sensing, BIT, Mesra, Ranchi, India
e-mail: milighosh2001@rediffmail.com

© Springer Nature Singapore Pte Ltd. 2019   109
V. Nath and J. K. Mandal (eds.), *Proceeding of the Second International Conference on Microelectronics, Computing & Communication Systems (MCCS 2017)*, Lecture Notes in Electrical Engineering 476, https://doi.org/10.1007/978-981-10-8234-4_11

wind, and lightning [1]. Hues amount of precipitation is produced and updraft is overcome due to downdraft and enter in dissipating stage. The gust front at ground cuts off the warm air which causes the thunderstorm. Even rainfall decreases in intensity but thunderstorm and lightning remain a danger. There is a continuous spectrum of thunderstorm type but thunderstorm is broadly classified into four categories—single-cell storm, multicell cluster storm, multicell line storms, and supercell storm. Single-cell thunderstorm is not strong and quite rare with lifetimes of 20–30 min. Most of the single-cell storms are non-severe but some of them may produce small severe weather event. It is poorly organized due to its location and random time. Single cell is difficult to forecast. Generally, multicell cluster is the common type of thunderstorm and consists of a group of cells. In different phases of thunderstorm life cycle, it moves as one unit along each cell. Multicell cluster may persist for several hour but each cell within cluster lasts only 20 min. Multicell cluster is generally stronger than single cell but less intense than supercell storm. It can produce heavy rainfall, moderate-size hail, and occasional tornadoes. Multicell line (squall line) storm is a group of storm with arranged in often accompanied by continuous, well-developed gust front. Squall line can be solid or there may be gaps and breaks in the line. Squall line can be hundreds of miles long but are generally only 10 or 20 miles wide. Squall lines can produce hail, heavy rain, and weak tornadoes. The supercell is a very organized, rare, strongest, and most deadly of thunderstorm. They associate with high threat to life and property. Supercell produces high severe weather events such as large hail, heavy rain and strong to violent tornadoes. Both thunder and lightning are produced from thundercloud. Every thunderstorm produces lightning.

2 Proposed Methodology

The ideal goal of proposed methodology is to predict whether the satellite image poses thunderstorm or no thunderstorm. The overall method for prediction of thunderstorm is presented in Fig. 1. In Fig. 1, original satellite image is taken as the input image for prediction. This input image may contain distinct types of noises like blur speckle, noise, and striping noise which should be removed. Individual textures like forest, water bodies, grass, asphalt, barren land, and concrete occur in input satellite image. All abovementioned textures are isolated to acquire the interest image and thus the other textures do not affect the accurate prediction of thunderstorm.

Clustering is used for segmentation of the input image into distinct clusters. The formations of clusters are the basis of similarity measure which uses the Euclidian distance. The segmented areas are based on pixel values of different textures which are further to compute a wavelength range. The pixel values for various textures used in segmentation method are presented in Table 1 [2]. Region of interest from satellite image is obtained using segmentation into distinct clusters established on pixel value [3]. In this paper, a clustering technique, Fuzzy C-means clustering is

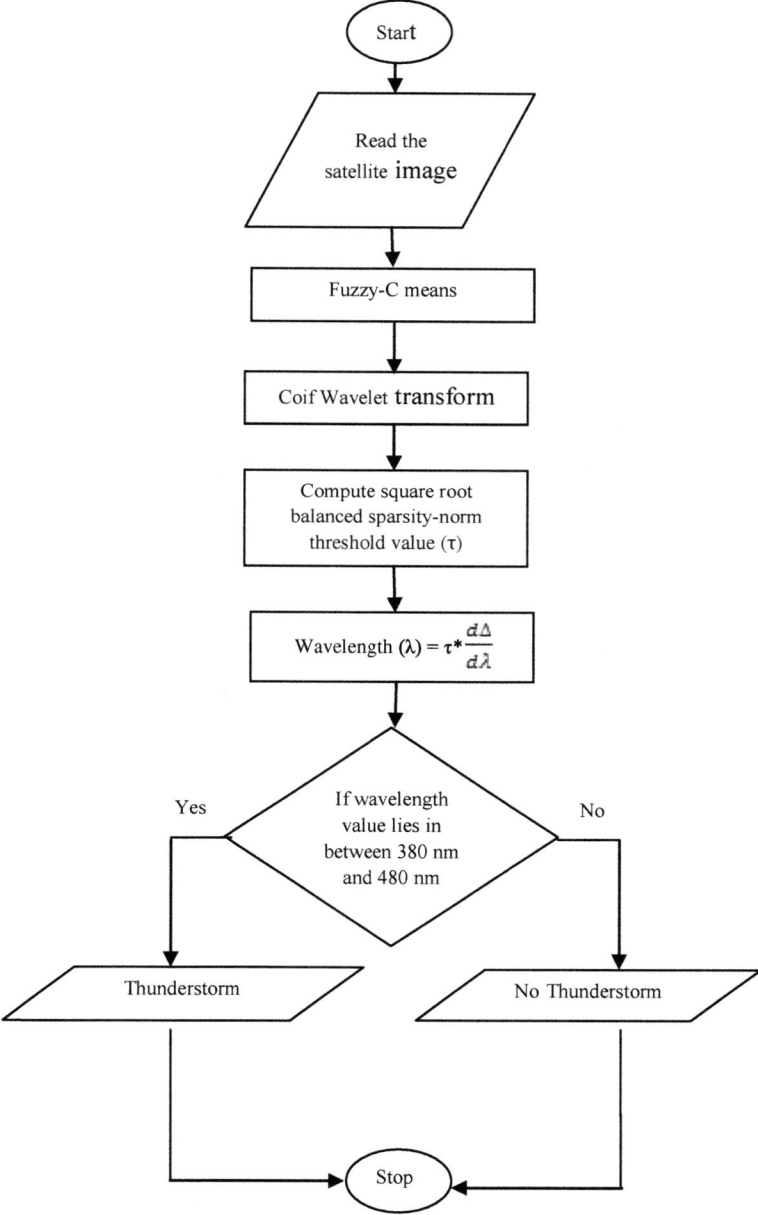

Fig. 1 Proposed method for prediction of thunderstorm

used for segmentation of image. Fuzzy C-means clustering segmented the image such that textures in a cluster have approximately same pixel value as compare to the other clusters formed [4–6]. Fuzz C-means is adopted to cluster the input image into six segments to escape the local minima. Whole procedure is performed in MATLAB tool. The wavelet transform is implemented on the feature extracted image. Here coif wavelet is chosen for transform. Coiflets has minimum signal disruption having scaling function and vanishing moments. It provides better regularity and enhanced steadiness than the efficient technique for image and signal processing [7, 8]. As an input satellite image is in RGB. Further, the coif wavelet naturally converts the RGB image into grayscale image and denoises the image. Denoised, clustered image is further adopted to compute the value of square root balance sparsity norm threshold.

A wavelength of satellite image lies in the range of 380–750 nm. Prediction of thunderstorm based on the estimation of wavelength range which lies in the visible spectrum between 380–480 nm. Whenever the thunderstorm occurs in the clustered, denoised image in frequency domain, the threshold value of square root balance sparsity norm should lie in between 9 and 11 [9]. Soft threshold is chosen to estimate the value of square root balance sparsity norm at $c(K) = \pm\tau$. The notations used in the paper are present in Table 2.

Square root balance sparsity norm threshold (τ), wavelength factor $\left(\frac{d\lambda}{d\lambda}\right)$ and wavelengths (λ) are computed using 2, 3, and 4 equations.

The soft threshold expression is presented in Eq. (1)

$$\overline{c_s}(k) = \begin{cases} \mathrm{sign}c(k)(c(k) - \tau), & c(k) \geq \tau \\ 0, & c(k) < \tau \end{cases} \tag{1}$$

For the evaluation of threshold values, Eq. (2) is used.

$$\tau = \sqrt{\frac{2\sigma^2 \log(n)}{n}} \tag{2}$$

Table 1 Different textures with pixel values

Textures	Pixel values
Waterbody	0–20
Forests	21–33
Grass	34–81
Asphalt	82–140
Barren lands	141–199
Concrete	200–224
Clouds	225–255

Symbol	Notation
c_s^-	Soft threshold value
σ^2	Variance of wavelength
τ_k	Fixed threshold value
n	Number of pixels
$\frac{d\Delta}{d\lambda}$	Wavelength factor
τ	Square root balance sparsity norm threshold

Table 2 Symbols and notation

Using Eq. (3), fixed threshold value is computed

$$\tau_k = \int_0^{255} \frac{d\Delta}{dx} \left[\sqrt{\frac{2\sigma^2 \log(n)}{n}} \right].dn \tag{3}$$

Wavelength is computed using Eq. (4)

$$\lambda = \tau * \frac{d\Delta}{d\lambda} \tag{4}$$

3 Experimental Result

Image data can be obtained from website of National Oceanic and Atmospheric Administration (NOAA) [10]. A threshold value is computed using the analysis of these images. Wavelength range of satellite image lies between 380 and 750 nm [11]. The results are presented in Tables 3 and 4.

Consider an image (I37.jpg). The computed wavelength of I37.jpg is 389.45 nm, which depicts the existence of thunderstorm but established historical outcome shows absent thunderstorm. Thus false prediction is obtained. Another image (I16. jpg) is considered and its wavelength is computed which is 381.29 nm. Computed wavelength lies within range of thunderstorm. Established historical result also indicates the presence of thunderstorm. Thus the prediction is true. The objective of proposed method is to predict the thunderstorm as accurate as possible. True positive (TP), true negative (TN), false positive (FP), and false negative (FN) values are computed in order to obtain the accuracy. TP indicates the positive tuples which are correctly labeled. TN indicates the negative tuple which are correctly labeled. FP specifies the negative tuples which are incorrectly labeled. FN indicates positive tuples which are incorrectly labeled. Sensitivity, accuracy, specificity, and precision are four basic performances which computed to test the working of the proposed system using equation. All these performance measures are defined in terms of percentage for ease of interpretation. Sensitivity, accuracy, specificity, and precision are based on Bayes' theorem

Table 3 Wavelength range computation for thunderstorm (TS) of image (I1...I50)

Image number	Variance (σ^2)	Square root balance sparsity norm threshold (τ)	Wavelength factor $\left(\frac{d\Delta}{d\lambda}\right)$	Wavelength $(\lambda) = \tau^*$ $\left(\frac{d\Delta}{d\lambda}\right)$	Historically established result
I1.jpg	5.90	9.65	40.36	389.52	Thunderstorm
I2.jpg	4.08	9.67	40.45	391.24	Thunderstorm
I3.jpg	6.87	10.39	40.43	420.10	Thunderstorm
I4.jpg	6.25	9.91	40.34	399.78	Thunderstorm
I5.jpg	6.11	9.80	40.30	395.03	Thunderstorm
I6.jpg	6.27	9.93	41.10	408.20	Thunderstorm
I7.jpg	7.66	10.92	41.57	456.10	Thunderstorm
I8.jpg	6.85	10.25	40.47	414.83	Thunderstorm
I9.jpg	5.90	9.63	40.43	389.42	Thunderstorm
I10.jpg	6.04	9.74	40.49	394.42	Thunderstorm
I11.jpg	6.26	9.92	40.56	402.39	Thunderstorm
I12.jpg	6.17	9.85	40.41	398.13	Thunderstorm
I13.jpg	5.86	9.60	39.83	382.45	Thunderstorm
I14.jpg	7.60	10.93	41.17	450.03	Thunderstorm
I15.jpg	7.66	10.97	42.84	470.03	Thunderstorm

$$\text{Sensitivity} = \frac{\text{TP}}{\text{TP} + \text{FN}}$$

$$\text{Specificity} = \frac{\text{TN}}{\text{FP} + \text{TN}}$$

$$\text{Accuracy} = \frac{\text{TP} + \text{TN}}{\text{TP} + \text{FP} + \text{FN} + \text{TN}}$$

$$\text{Precision} = \frac{\text{TP}}{\text{TP} + \text{FP}}$$

From Tables 3 and 4, performance measures are computed using TP, TN, FP, and FN and shown in Table 5.

The proposed method has less accuracy as compared to K-Means with haar wavelet [9] but more accurate than other methods which are shown in Table 6.

Table 4 Wavelength range computation for thunderstorm

Image number	Variance (σ^2)	Square root balance sparsity norm threshold (τ)	Wavelength factor ($\frac{d\Delta}{d\lambda}$)	Wavelength (λ) = $\tau * \frac{d\Delta}{d\lambda}$	Experimentally obtained result	Historically established result	Prediction
I16.jpg	5.74	9.5	40.13	381.29	TS	TS	True
I17.jpg	6.25	9.91	41.21	409.72	TS	TS	True
I18.jpg	6.29	9.94	41.03	387.54	TS	TS	True
I19.jpg	5.85	9.56	40.41	387.54	TS	TS	True
I20.jpg	6.20	9.87	36.09	356.28	No TS	No TS	True
I21.jpg	6.06	9.76	35.99	351.33	No TS	No TS	True
I22.jpg	6.00	9.71	40.40	392.36	TS	TS	True
I23.jpg	7.63	10.95	42.53	465.81	TS	TS	True
I24.jpg	12.69	14.12	37.06	523.36	No TS	TS	False
I25.jpg	5.16	9.01	34.98	315.25	No TS	TS	False
I26.jpg	7.64	10.96	42.85	469.72	TS	TS	True
I27.jpg	7.60	10.93	41.74	456.23	TS	TS	True
I28.jpg	6.11	9.80	36.04	353.21	No TS	No TS	True
I29.jpg	5.40	9.21	35.54	327.36	No TS	No TS	True
I30.jpg	6.12	9.81	40.53	396.91	TS	TS	True
I31.jpg	5.77	9.52	40.26	383.29	TS	TS	True
I32.jpg	6.01	9.72	40.37	392.42	TS	TS	True
I33.jpg	6.78	10.32	42.27	436.32	TS	TS	True
I34.jpg	5.90	9.63	40.93	394.24	TS	TS	True
I35.jpg	14.28	14.98	43.40	650.24	No TS	TS	False
I36.jpg	6.32	9.97	42.72	425.93	TS	TS	True
I37.jpg	5.16	9.01	43.22	389.45	TS	No TS	False

(continued)

Table 4 (continued)

Image number	Variance (σ^2)	Square root balance sparsity norm threshold (τ)	Wavelength factor $\left(\frac{d\lambda}{dj}\right)$	Wavelength $(\lambda) = \tau * \frac{d\lambda}{dj}$	Experimentally obtained result	Historically established result	Prediction
138.jpg	9.98	12.52	48.04	601.54	No TS	TS	False
139.jpg	5.55	9.34	42.33	395.42	TS	TS	True
140.jpg	6.12	9.81	40.28	395.23	TS	TS	True
141.jpg	7.67	10.98	43.64	479.25	TS	TS	True
142.jpg	5.61	11.25	30.01	337.67	No TS	No TS	True
143.jpg	6.26	8.62	38.38	330.85	No TS	No TS	True
144.jpg	4.60	8.50	37.10	315.40	No TS	No TS	True
145.jpg	5.51	12.01	27.75	333.36	No TS	No TS	True
146.jpg	14.55	15.12	32.75	495.32	No TS	No TS	True
147.jpg	6.25	9.91	40.45	400.95	TS	TS	True
148.jpg	5.62	9.40	40.53	380.99	TS	TS	True
149.jpg	13.07	14.33	39.56	486.23	No TS	No TS	True
150.jpg	5.42	8.91	35.94	320.25	No TS	No TS	True

Table 5 Performance measure for thunderstorm

Performance measure	Percentage (%)
Sensitivity	86
Accuracy	85
Specificity	84
Precision	90

Table 6 Performance measure for thunderstorm

Paper No.	Technique	Percentage (%)
12	ANN with LM algorithm	76
13	RS with SVM	71
14	Naïve Bayesian	65
15	KNN	72
Proposed method FCM with coif wavelet		85

4 Conclusion

The aim of proposed method is to predict the thunderstorm as accurately as possible. Fuzzy C-means clustering technique, coif wavelet transforms, and statistical techniques are used for the prediction of thunderstorms. Threshold value and wavelength range are statistical analysis used for real-time analysis of satellite image. This paper demonstrates the performance comparison of the proposed method with different techniques such as KNN, RS with SVM, Naïve Bayesian, ANN with LM algorithm in prediction of thunderstorm. The proposed system predicts 85% accuracy in prediction of thunderstorm.

References

1. Sutapa Chaudhari, "A probe for consistency in CAPE and CINE during the prevalence of severe thunderstorms: Statistical-Fuzzy Coupled approach," Atmospheric and Climate Science, pp. 197–205, 2011.
2. Murali Prasad Raja, Naraimha Prasad L V and Vasavi Krishna Yacham, "Real time detection of hurricanes from satellite cloud imagery using haar wavelet", Proceedings of the World Congress on Engineering and Computer Science, vol. 1, 22–24 October 2014.
3. C. Fraley and A. Raftery, "How many clusters? Which clustering method? Answers via model-based cluster analysis," The Computer Journal, 1998, pp. 578–588.
4. Nidhi Grover, "A study of various fuzzy clustering algorithms," International Journal of Engineering Research, vol. 3, issueno. 3, pp. 177–181, 01 march 2014.
5. Tara Silkumar, P. Yugander, P. Sreenivasa Murthy and B. Smitha, "Colour based image segmentation using Fuzzy C-Means clustering," International Conference on Computer and Software Modeling 20111.
6. A. M. Khan and Ravi. S, "International Journal of Soft Computing and Engineering," Vol. 3 (4), September 2013.

7. Abhinav Dixit and Swatilekha Majumdar, "Comparative analysis of coiflet and Daubechies wavelets using global threshold for image denoising," International Journal of Advance in Engineering & Technology, Nov. 2013.

8. Viswanath Kambhampaty, Rohith Gali and Narasimha Prasad, "A short term tornado prediction model using satellite imagery," First International Conference on System Informatics, Modeling and Simulation, 2014.

9. Kishore Kumar Reddy C, Anisha P R and Narsimha Prasad L V, "Detection of thunderstorms using data mining and image processing", pp. 226–231.

10. Rasika P. Gawande and Namrata D. Ghuse, "An efficient approach towards thunderstorm detection using saliency map" International journal of Computer Science and Information Technology, vol. 6(3), pp. 2429–2434, 2015.

11. http://en.wikipedia.org/wiki/visible_spectrum.

Design of RFID Tag Antenna with Impedance Matching Techniques at UHF Band

K. Rama Devi, A. Mallikarjuna Prasad and A. Jhansi Rani

1 Introduction

The microstrip antenna plays a vital role in communication world for four decades. Microstrip antenna is suitable for different applications like RFID tag, wireless communication, space communication, and biomedical, etc., due to their less weight, low-cost, and narrow bandwidth.

RFID tag is a device having transceiver, used for identification of objects which are made with metal based from few feet to hundreds of feet in direct contact. RFID tags are working in narrowband characteristics. These are classified into active RFIDs, passive RFIDs, and semi-active RFIDs [1–3]. RFID tags are working at different frequency ranges from low frequency to UHF. An antenna plays important role in RFID tag to transmit the signal and receive the reflected signal from the object.

The antenna to be designed is in the range 865–880 MHz which covers different spectral bands [4] used in several countries like India (865–867 MHz), Iran (865–868 MHz), South Africa (865.6–867.6 MHz), Singapore (866–869 MHz), etc. The basic RFID working principle is shown in Fig. 1. Designing of different antennas, enhancement techniques of inductive impedance of an antenna, and read range calculations with different commercially available chips are discussed in the following sections.

K. Rama Devi (✉) · A. Mallikarjuna Prasad
University College of Engineering, JNTUK, Kakinada, Andhra Pradesh, India
e-mail: kolisettyramadevi@gmail.com

A. Mallikarjuna Prasad
e-mail: a_malli65@yahoo.com

A. Jhansi Rani
V.R. Siddhartha College of Engineering, Vijayawada, Andhra Pradesh, India
e-mail: jhansi9rani@gmail.com

© Springer Nature Singapore Pte Ltd. 2019
V. Nath and J. K. Mandal (eds.), *Proceeding of the Second International Conference on Microelectronics, Computing & Communication Systems (MCCS 2017)*, Lecture Notes in Electrical Engineering 476, https://doi.org/10.1007/978-981-10-8234-4_12

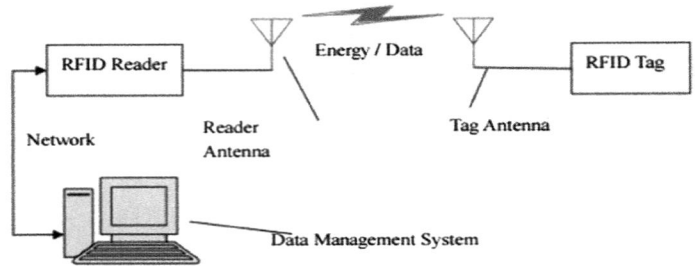

Fig. 1 Basic RFID design

2 RFID Tag Antenna Design Approach

To design a suitable RFID tag antenna, the important parameter to be considered is

(i) Read Range Equation

The most important characteristic of a passive RFID tag is the read range. The maximum distance of RFID reader is able to detect the modulated backscattered signal and successfully identify a tag. The read range may be roughly calculated using the FRIIS transmission formula [5–8]

$$R_{max} = \frac{\lambda}{4\pi} \sqrt{\frac{P_t G_t G_r \left(1 - |\Gamma|^2\right) \rho}{P_{th}}} \tag{1}$$

where

λ	length
P_t	Effective radiated power (i.e., 0.5, 2, and 4 W)
G_t	Gain of reader antenna (transmitter)
G_r	Gain of tag antenna (receiver)
P_{th}	Minimum threshold power which can activate the circuit device (i.e., -10, -15, -18, and -20 dBm)
ρ	Polarization mismatch between the reader and tag antenna
Γ	Voltage reflection coefficient
$|\Gamma|^2$	Power reflection coefficient
$\left(1 - |\Gamma|^2\right)$	Transmission coefficient

Fig. 2 Connection between
antenna and chip

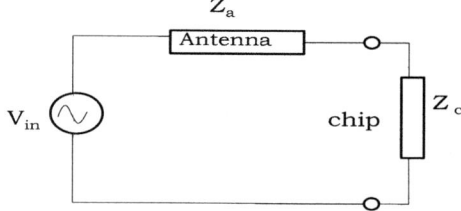

(ii) Impedance matching condition

According to the maximum power transfer theorem, tag antenna impedance is a
complex conjugate of a chip impedance [8] and the circuit is given in Fig. 2. The
voltage reflection coefficient is calculated from Eq. 2 [7, 9],

$$\Gamma = \frac{Z_c - Z_a^*}{Z_c + Z_a} \qquad (2)$$

Γ Voltage reflection coefficient (0–1)
$|\Gamma|^2$ Power reflection coefficient (0–1)
V_{in} Input voltage at antenna terminals

where Z_c and Z_a are the impedances of chip and tag antennas

3 Proposed Tag Antenna Designs and Analysis

This section presents the design aspects of conventional planar radiating system that
exhibits excellent impedance matching with commercially available RFID Chips. In
the current scenario, the major requirement of general purpose RFID tags is the
capability of reading even from distances greater than 1 m. It is evident from the
literature survey that long range detection is possible with proper impedance
matching, gain and threshold power.

It is a known fact that the radiating systems are resistive at resonant condition. It
is also possible to enhance the reactance of the radiating system and accordingly
many techniques are proposed. Among these are offset feed, parasitic patch, T-stub
matching, probe stub matching (shorting pin), and curved surface patch etc., to
name a few.

An attempt has been made to implement some of these techniques on the proposed hexagonal shape patch. The rectangle-shaped structure has a better efficiency
and bandwidth characteristics which are better than circular disk even though the
directivity is the same for both [10–12].

The aim of the antenna design is to match well with the frequently used though
limited to a few commercial chips. The complex impedance of a few chips is
tabulated and shown in Table 1 [4, 9, 13]. The chip impedance is almost constant

Table 1 Chip number and their impedance

S. No.	Commercial name of a chip	Complex impedance (Z_{IC}) Ω
1	IMPINJ-Monza-Series-5	$33 - j113$
2	NXP UCODE G2XM	$13.6 - j122$
3	AD-220 A Very Dennison	$8 - j98$
4	NXP UCODE G2XL	$20 - j150$
5	Alien Class-1 Chip	$6 - j125$
6	Alien H3	$16 - j160$

throughout the band of frequencies [7]. The work is extended to design a tag antenna well matched with chip 1 of impedance $33 - j113\ \Omega$ [4, 13].

The simulation results of an antenna are observed in terms of return loss (S_{11}), impedance plot, and radiation pattern (gain) at narrowband of operation 865.6–867.6 MHz. The designed antenna is thoroughly studied considering that these are bands recognized by India, America, Africa, Belgium, Denmark, etc. [14].

4   Design and Analysis of HPMSA

Moreover, the circular patch antenna has the advantage of less area occupancy than rectangular patch [10]. Better than the above two cases of rectangle and circular shape patches, incorporating the merits in both, a hexagonal structure is proposed here. The proposed structure area is 10% less than circular and 20% less than rectangular [10, 11]. An attempt has been made to achieve the required impedance to match the proposed structure.

4.1   Design of HPMSA

The proposed [10] HPMSA considers a patch of hexagonal shape on a RT-duroid substrate of high ε_r which is 10.0. Typical dimensions of the antenna substrate are 9 cm × 9 cm with a height of 0.156 cm simple probe feed (50 Ω). The dimension of the side arm of a hexagonal is calculated from basic circular patch dimensions given in [10], and the obtained value of 's' is 3.516 cm at the desired operating frequency or 870 MHz. The simulation step is followed by fabrication-based validation in which the arm size is further optimized to 3.36 cm. The photograph of fabricated antenna top view (Patch) and bottom view (ground and feed) is shown in Fig. 3a, b.

The return loss (Fig. 4) shows that antenna is resonated at 854 MHz with RL −4.978 dB. However, the patch performance is to be improved for the required application. The performance of antenna will be improved with slot technique and will be discussed in the next section.

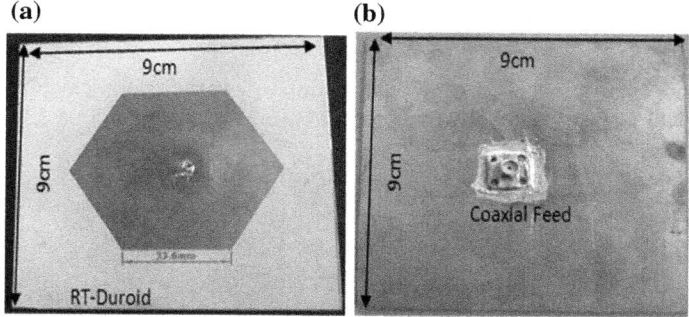

Fig. 3 HPMSA designed. **a** Patch shape. **b** Ground plane (back side)

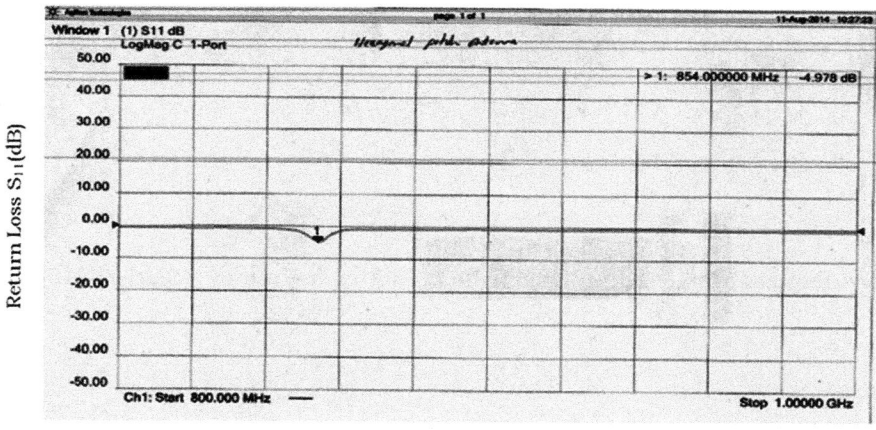

Frequency (MHz)

Fig. 4 HPMSA return loss S_{11} (dB) measured in network analyzer (NA)

4.2   Design of AHMSA

The HPMSA performance (Sect. 4.1) does not match the required frequency range. Hence, the technique of implementing a slot is considered and a similar hexagonal slot is made on the basic hexagonal patch. This is named as annular slot and hence the name AHMSA.

The merits of shells (i.e., slots) design are that it requires a smaller size than an orthodox hexagonal shape [11]. The slot models can be analyzed by using a cavity model [10].

4.3 Analysis of AHMSA

Let 'a' and 'b' be the inner and outer radius of an annular hexagonal and the resonance frequency can be given as [10, 11]

$$f_{\mathrm{m}} = \frac{K_{nm}c}{2\pi a \sqrt{\varepsilon_{\mathrm{reff}}}} \tag{3}$$

where k_{nm} are the roots of the characteristic equation
The modes are calculated from Eq. 4,

$$J_n'(kb)Y_n'(ka) - J_n'(ka)Y_n'(kb) = 0 \tag{4}$$

where Y_n' and J_n' are derivative Bessel function of second and first kind of order n. Equation 4 can be represented in another way as

$$J_n'(CX_{nm})Y_n'(X_{nm}) - J_n'(X_{nm})Y_n'(CX_{nm}) = 0 \tag{5}$$

where $C = b/a$ and

$$X_{nm} = K_{nm}a \tag{6}$$

ε_r is replaced with effective permittivity ε_e. The ε_e is given by Schneider [12] as

$$\varepsilon_e = \frac{1}{2}(\varepsilon_r + 1) + \frac{1}{2}(\varepsilon_r - 1)\left(1 + \frac{10t}{W}\right)^{\frac{-1}{2}} \tag{7}$$

where W = Microstrip of width = $b - a$.
According to the parallel plate model of circular ring, the microstrip line size is replaced with a parallel plate waveguide with identical ε_{re} and Z_o. The modified inner and outer radii are

$$a_e = a - \frac{3h}{4} \tag{8}$$

$$b_e = b + \frac{3h}{4} \tag{9}$$

For the given values 'a' and 'b', effective a_e and b_e are calculated. Then the characteristic equation is solved by replacing 'a' and 'b' by a_e and b_e, respectively.

After solving the characteristic equation for k_{nm}, the resonant frequency is determined from

$$f_m = \frac{k_{nm}c}{2\pi\sqrt{\varepsilon_{re}}} \tag{10}$$

Characteristics of Proposed AHMSA:

The proposed antenna with an annular hexagonal ring is designed according to the rules specified in Sects. 4.2 and 4.3. The optimal design dimensions (from Eqs. 8 to 9) are finally obtained and simulated in the HFSS simulator (Fig. 5). The outer hexagon radius is 3.156 cm. The arm length corresponds to 3.09 cm. The inner hexagon is a slot etched at the center with a radius of 0.8 cm and arm length 0.69 cm (Fig. 6). The feed position is discussed below.

Feed Position:

The output impedance of offset feed for the dominant mode at any distance from the center of the patch is given as [7]

$$R_{in}(\rho' = \rho_o) = \frac{1}{G_t}\frac{J_1^2(K\rho_o)}{J_1^2(Ka_e)} \tag{11}$$

where G_t(Total Conductance) $= G_{rad} + G_c + G_d$ Radiation conductance

$$G_{rad} = \frac{(K_0a_e)^2}{480}\int_0^{\frac{\pi}{2}}\left[J_{02}'^2 + \cos^2\theta J_{02}^2\right]\sin\theta d\theta \tag{12}$$

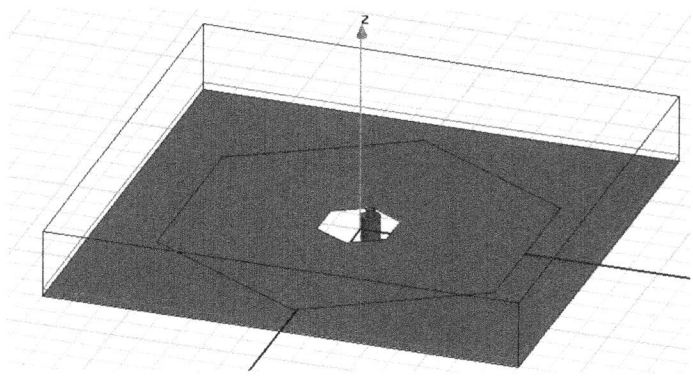

Fig. 5 Annular hexagonal microstrip antenna (AHMSA) in HFSS

Conduction due to ohmic loss

$$G_c = \frac{\varepsilon_{mo}\pi(\pi\mu_0 f_r)^{-3/2}}{4h^2\sqrt{\sigma}}\left[(Ka_e)^2 - m^2\right] \tag{13}$$

Conduction due to dielectric loss

$$G_d = \frac{\varepsilon_{mo}\tan\delta}{4\mu_0 h f_r}\left[(ka_e)^2 - m^2\right] \tag{14}$$

where a_e = effective radius.

Feed reactance

$$X_f = -\frac{\eta kh}{2\pi}\left[\ln\left(\frac{kd}{4}\right) + 0.577\right] \tag{15}$$

where d is the diameter of the feed probe.

The maximum reactance is present when the probe is nearer to the corner, and it is reduced when the probe is moving toward the center.

The probe is connected at $(-0.5\text{ cm}, 0, 0.156\text{ cm})$ for better impedance matching called offset feeding [15]. Simulated and fabricated antennas are shown in Figs. 5 and 6, respectively.

The return loss (both simulated and measured) of AHMSA are shown in Fig. 7a, b. It is evident from the plot that the resultant geometry offers resonant frequency at 868.54 MHz (simulated) and 869.90 MHz (measured) at RL −10 dB. The

Fig. 6 Fabricated antenna AHMSA, top view and ground plane

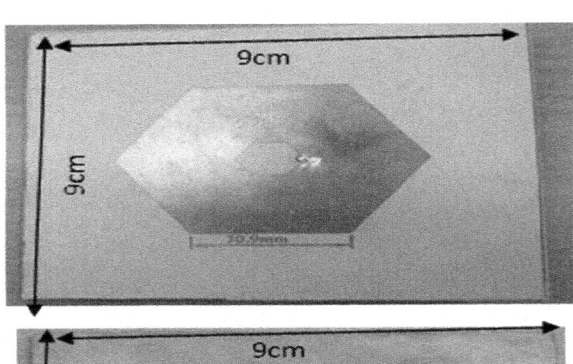

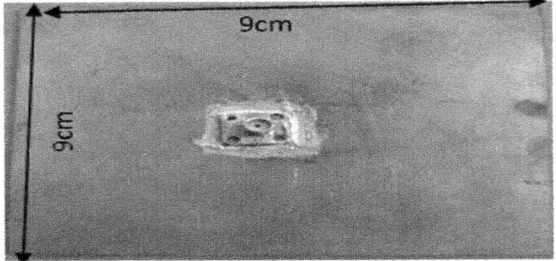

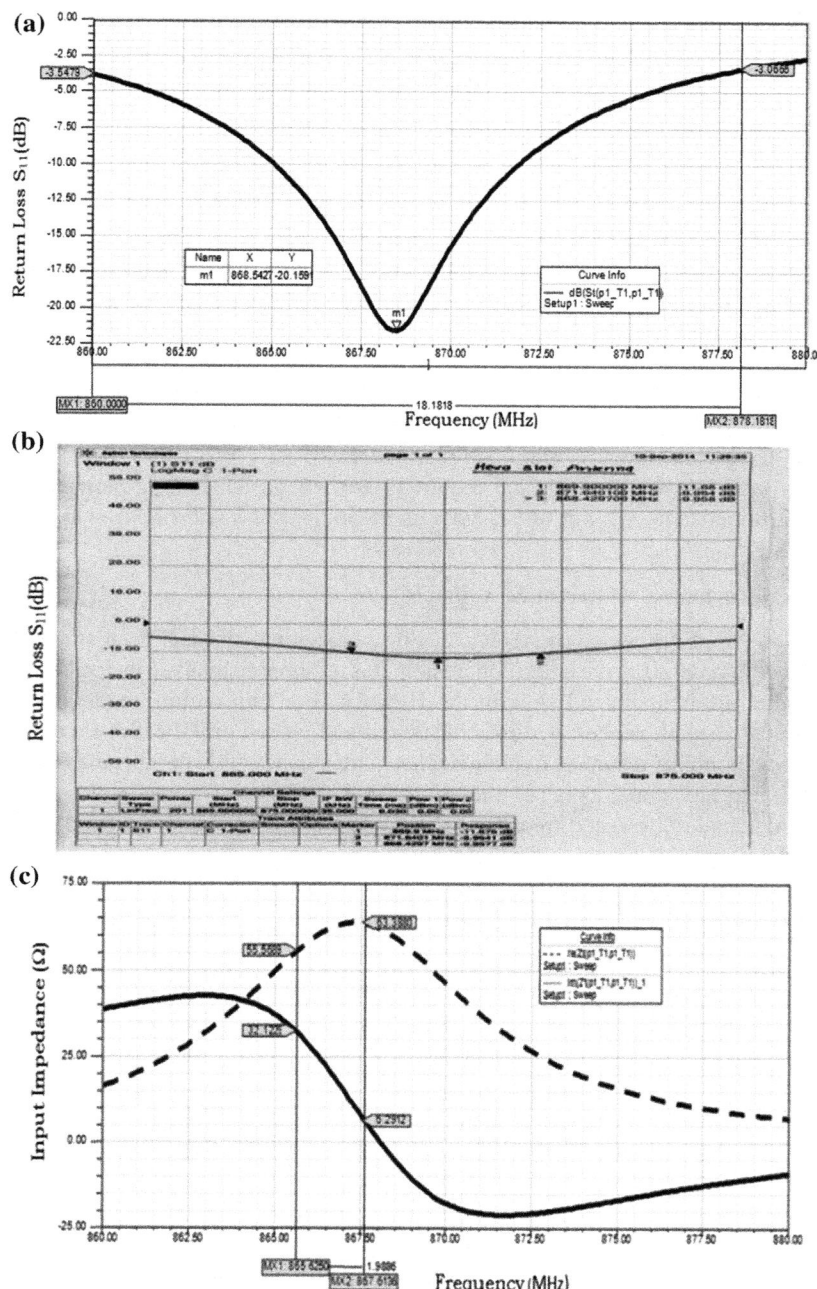

Fig. 7 Return loss of AHMSA. **a** Simulated. **b** Measured with NA. **c** Simulated impedance curve

Table 2 Response of HPMSA with different slot shapes (simulated)

Slot shape on HPMSA	Resonant frequency (MHz)	Frequency range (MHz) $S_{11} = -10$ dB
Rectangular	873.26	870.25–876.48
Pentagonal	873.87	864.02–883.90
Hexagonal	868.54	865.30–871.60

bandwidth at 10 dB RL line is 865.30–871.30 MHz (simulated) and 868.42–871.64 MHz (measured), respectively. Simulated results match well with the measured results.

The impedance versus frequency of AHMSA is shown in Fig. 7c. The impedance range at required band is 55.57 + j32.12 and 63.39 + j6.29 Ω. The RL (S_{11} curve Fig. 7a) shows that the antenna is tuned at 868.54 MHz and 3-dB HP (half power) bandwidth range is 860.00–878.18 MHz.

The 3-dB HP range covers the required narrowband RFID tag operating band. The tag impedance (Z_{tag}) at 865.6 MHz is 55.56 + j32.12 and 63.38 + j6.29 Ω at 867.6 MHz.

(iv) Performance of Different Slot Shapes

Till now, the performance of AHMSA with a hexagonal slot has been studied and analyzed. In order to arrive at the conclusion, it is required to analyze the performance with other different shaped slots in the HPMSA. For thin region, rectangular and pentagonal shaped patches are also considered and etched on HPMSA for analysis. The corresponding resonant frequencies and their bandwidth for rectangular, pentagonal, and hexagonal slots are shown in Table 2. The performance of hexagonal slot is matched to designed frequency and also nearer to the required band.

5 Impedance Matching Techniques

The impedance of an antenna at the measured band (Fig. 6) is not a perfect match to the selected chips. The following sections discuss the different techniques like parasitic patch, T-matching feed, and shorting pin to enhance the positive reactance of patch. The impedance values match the required IC chip 1. In each section, emphasis is laid on an analytical treatment of the technique followed by impedance matching measurements. Each technique mentioned in the following subsection is referred to different cases.

(A) Parasitic Annular Hexa-MSA (PAHMSA)

In this method [4, 16, 17], a parasitic annular hexagonal structure is arranged around the actual annular hexagonal antenna. This parasitic element acts as an inductive element of the antenna (acts as reflector). The proposed antenna is shown in Fig. 8a. The top of the substrate is well used without increasing the dimension of the antenna.

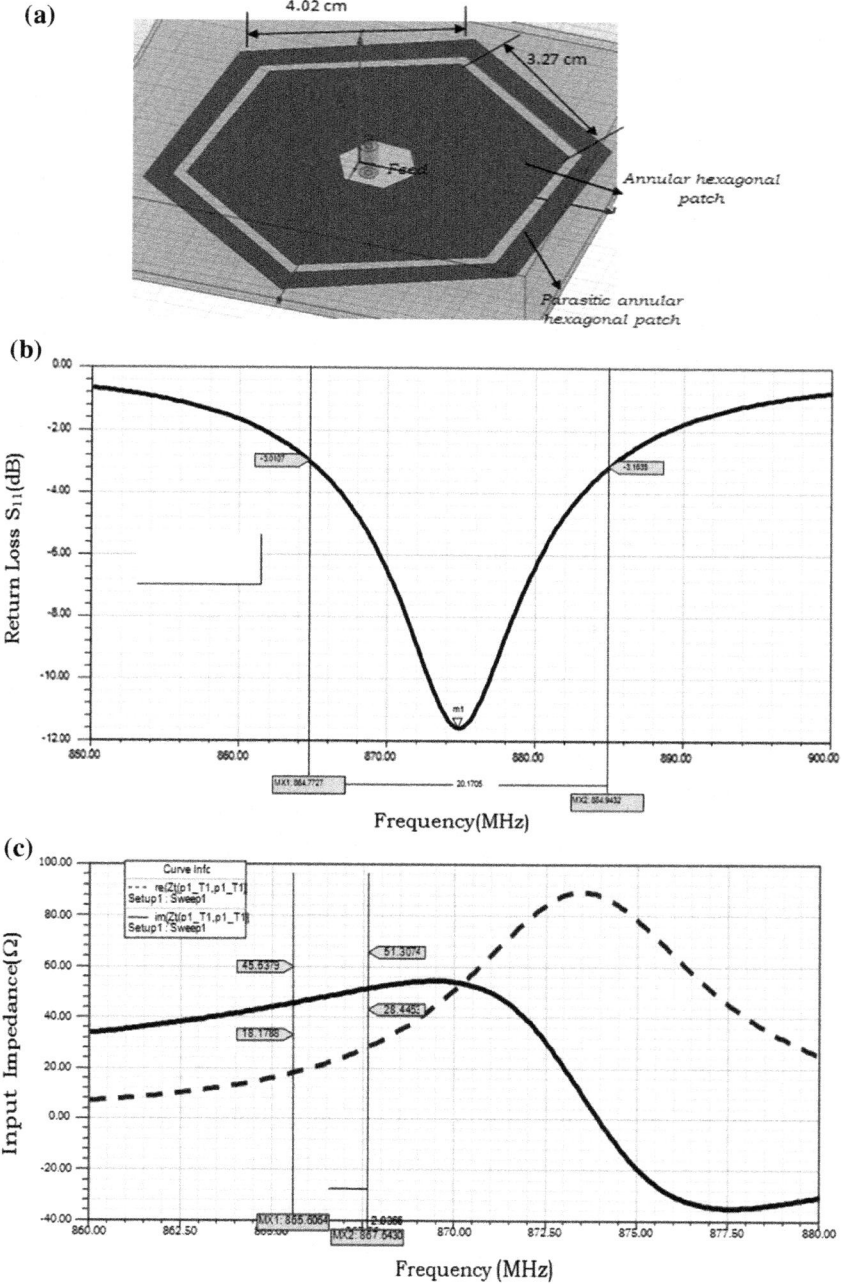

Fig. 8 a PAHMSA simulated geometry in HFSS. **b** PAHMSA return loss (S_{11}) (simulated). **c** PAHMSA impedance plot (simulated)

The RL (S_{11}) plot Fig. 8b shows that the antenna is tuned at 874.87 MHz and 3-dB HP frequency range is 864.77–884.94 MHz. The 3-dB HP range covers the required RFID Band [14].

The impedance is noted from Fig. 8c at 865.6 MHz is 45.63 + j18.17 and 28.44 + 51.31 Ω at 867.6 MHz. The inductive nature of the impedance is increased and real value decreases more than in the previous case.

(B) T-Stub Matching—AHMSA

The proposed antenna is shown in Fig. 9a. In this case, T-stub matching is proposed for better inductive matching of tag antenna to chip 1. It is a simple technique with no change in dimensions. The feed probe is located at the center of the patch and it is connected to the patch through T-stub matching as impedance converter [6, 18]. T-stub matching and their equivalent circuit are shown in Fig. 9b.

From Fig. 9b, two types of microstrip lines have different impedances; Step in width exists at the junction. This type of discontinuity capacitance C_s represents the fringing field capacitance at the junction. An expression for C_s is given by

$$C_s(\text{pF}) = \sqrt{W_1 W_2}\left[(10.1\log(\varepsilon_r) + 2.3)\frac{W_1}{W_2} - 12.6\log\varepsilon_r - 3.17\right] \quad (16)$$

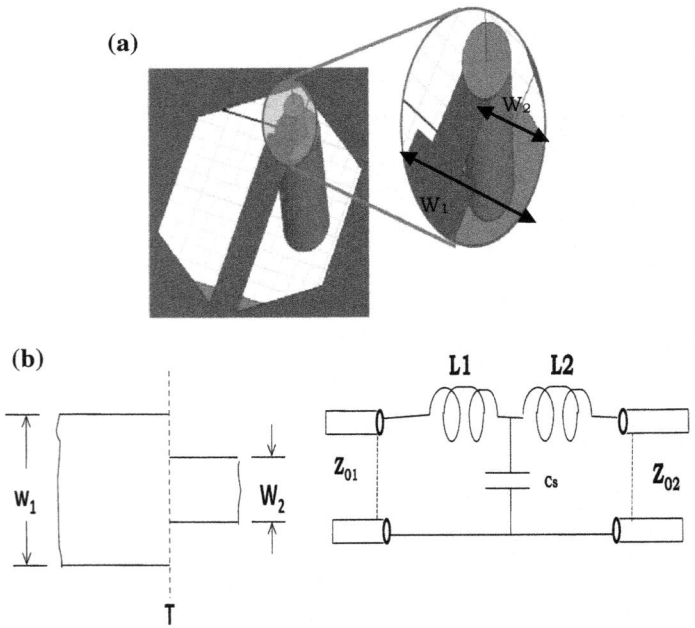

Fig. 9 T-stub matching AHMSA. **a** Proposed antenna. **b** T-matching (left), equivalent circuit (right)

The inductances L_1 and L_2 are obtained from the total discontinuity inductance L_s as

$$L_1 = \frac{L_{W1}}{L_{W1} + L_{W2}} L_s \qquad (17)$$

$$L_2 = \frac{L_{W2}}{L_{W1} + L_{W2}} L_s \qquad (18)$$

where L_W is inductance per unit length of microstrip of width W is given by

$$L_W = \frac{Z_0 \sqrt{\varepsilon_{re}}}{c} \qquad (19)$$

The expression for inductance L_s is also given by

$$L_s(\text{nH}) = h\left[40.5\left(\frac{W_1}{W_2} - 1.0\right) - 75\log\frac{W_1}{W_2} + 0.2\left(\frac{W_1}{W_2} - 1\right)\right] \qquad (20)$$

(or)

$$L_s(\text{nH}) = 0.000987h\left(1 - \frac{Z_{01}}{Z_{02}}\right)^2 \qquad (21)$$

where h is the thickness of the substrate (μm).

Based on Eqs. 16–21, W_1 and W_2 are selected as 2 and 1 mm, respectively. The T-stub matching antenna (Fig. 9a) is analyzed. The return loss (S_{11}) curve (simulated) is shown in Fig. 10a, the antenna is tuned at 873.11 MHz and 3-dB HP frequency range is 865.30–881.81 MHz. The 3-dB HP range covers the required RFID band. The impedance curve is shown in Fig. 10b.

The tag impedance (Z_{tag}) at 865.62 MHz is 41.18 + j85.04 and 64.83 + j76.95 Ω at 867.58 MHz.

(C) T-Matching with Shorting Pin (Probe or Inductive Load) AHMSA

The inductive nature of an antenna is improved with selection of feed position, parasitic patch, and T-stub matching. It can be further improved with a shorting pin or shorting probe or an inductive loading technique. The shorting pin [19, 20] is connected between ground and patch through substrate.

The value of ε_r is changed due to shorting pin and also it changes the resonant frequency [11]. One more advantage with shorting pin is that it reduces the effect on the results when placed on metallic objects. The post diameter is d (2 mm) and the strip width is W, for $d/W \ll 1$. The pin offers the series and shunt inductances. The series inductance (L_S) is neglected and shunt inductance (L_P) taken on the circular wire is given as

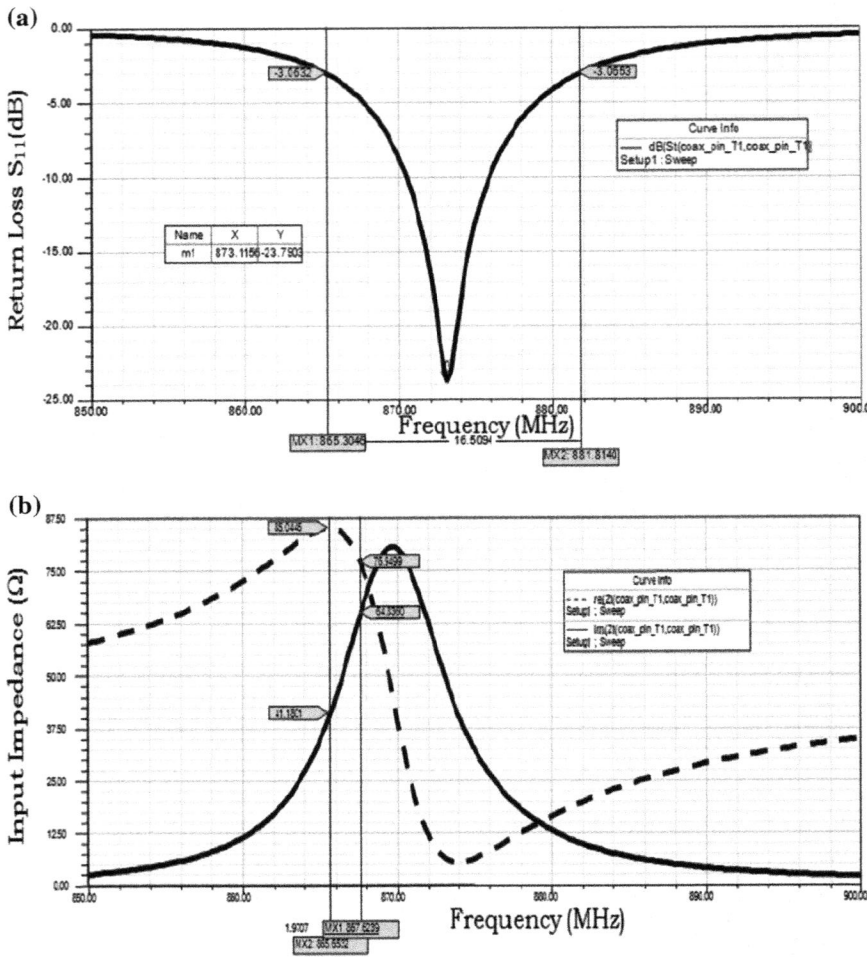

Fig. 10 **a** T-stub matching—AHMSA return loss plot (simulated). **b** T-Stub matching—AHMSA impedance plot (simulated)

$$L_P = 0.2h[\ln 4h/d + d/2h - 1] \tag{22}$$

The series resistance (R) due to the loss in the wire is given in Eq. 23

$$R = 2.63 * 10^{-3}h/d\sqrt{\rho f/\rho C_u} \tag{23}$$

where ρ is resistivity of the post metal and $\rho C_u = 1.72 * 10^{-4}$ Ω

The equivalent circuit of an antenna with a shorting pin is shown in Fig. 11. The shorting pin acts as parallel inductive load and it alters the current on the patch and

Fig. 11 Equivalent circuit of an antenna loading with shorting pin

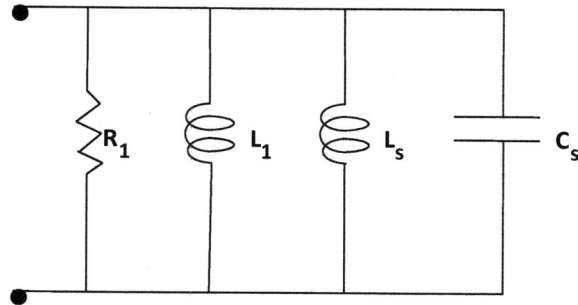

affects the input impedance of the patch. The input impedance is calculated from Eq. 24. The input impedance of an antenna is

$$Z_{\text{in}} = \frac{V_{\text{in}}}{I_0} \ \Omega \tag{24}$$

where

V_{in} is the input voltage at feed point
I_0 feed current

The resultant input impedance of an antenna at any location on the patch is calculated by Eq. 25,

$$z_{\text{in}} = -j\omega\mu_0 h \left\{ \begin{array}{l} \dfrac{1}{\pi a^2 k^2 (1-j\delta)} + \displaystyle\sum_{m=2}^{\infty} \dfrac{J_0^2(k_{0m}\rho_0)}{\pi a^2 J_0^2(k_{0m}a)\left\{k^2(1-j\delta)-k_{0m}^2\right\}} \\ + \dfrac{2}{\pi} \displaystyle\sum_{m=1}^{\infty}\sum_{n=1}^{\infty} \left(\dfrac{\sin n\Delta}{n\Delta}\right)^2 \dfrac{J_n^2(k_{nm}\rho_0)}{J_{0n}^2(k_{nm}a)} \dfrac{k_{0m}^2\cos^2 n\emptyset_0}{\left\{k^2(1-j\delta_{\text{eff}})-k_{nm}^2\right\}\left(k_{nm}^2 a^2 - n^2\right)} \end{array} \right\} \tag{25}$$

where δ_{eff} is called effective loss tangent of dielectric (including conductor loss, dielectric loss, and radiation loss).

When the shorting pin is placed at the corner of the patch, the normalized impedance (with 50 Ω) will increase w.r.to feed location when feed moves from edge of the patch to center and selects the best position of shorting pin. The proposed antenna in HFSS is shown in Fig. 12a and the designed antenna is shown in Fig. 12b, c. In this case, an inductive probe is added at an edge of the patch (green), along with T-stub (red) matching for better inductive matching of tag antenna to the chip 1.

It is a simple technique without altering the dimensions of an antenna. The antenna is analyzed and results are tabulated. Simulated return loss (Fig. 13a) shows that the antenna is tuned at 877.13 MHz and 3-dB HP range is 867.84–887.10 MHz. The simulated impedance response is shown in Fig. 13b. The tag impedances (Z_{tag}) at 865.6 and 867.6 MHz are 30.99 + j96.12, 49.08 + j101.57 Ω, respectively.

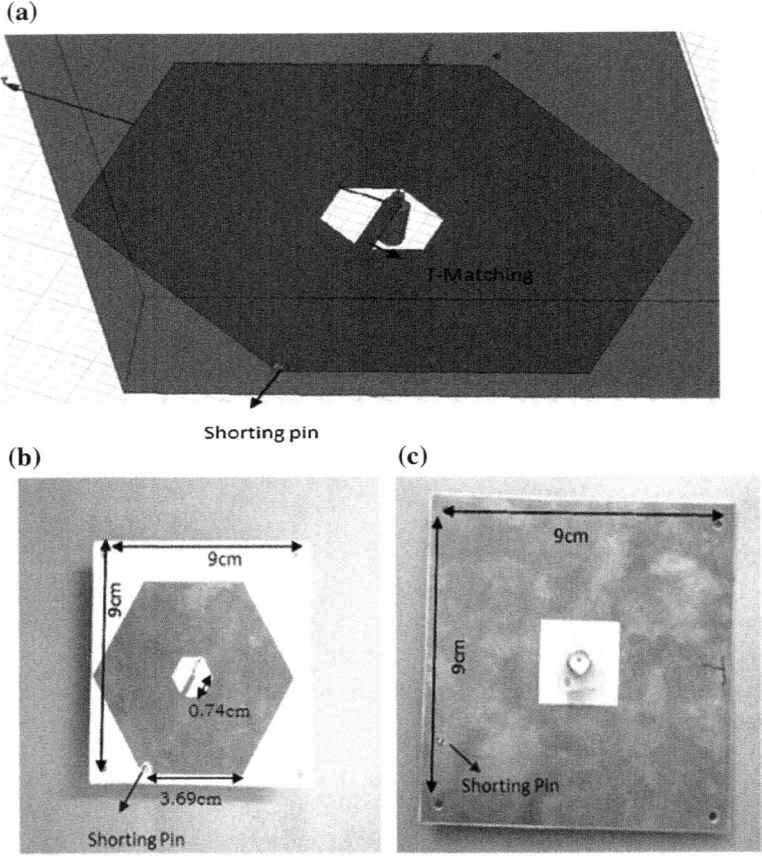

Fig. 12 **a** T-matching with shorting pin AHMSA design (HFSS). **b** Designed antenna—top view. **c** Designed antenna—bottom view

The designed antenna is tested with NA and response of return loss (Fig. 13c) shows −13.84 dB at the band of interest even though resonated at out of the range at 674 MHz with RL −46 dB. The corresponding VSWR variation is shown in Fig. 2.14d. It shows that, the performance characteristic of the Simulated and measured are in good agreement.

The gain of the antenna is calculated at both ends of the observation band as shown in Fig. 13e, f. The gain is almost the same during the band, −3.43 dB. The beam is radiated along 0° and HPBW is 120° on $\varphi = 0°$ plane.

The range performance of the antenna is calculated and tabulated in Tables 3 and 4. The range and PRC (power reflection coefficient) are calculated at chip threshold power −10 and −15 dBm at both band ends of frequencies 865.62 and 867.63 MHz. Tag impedance (Z_{tag}) matches to chip 1 more than the other chips. The minimum PRC at 865.62 MHz is 0.06 and read range is 5.78 m at −15 dBm

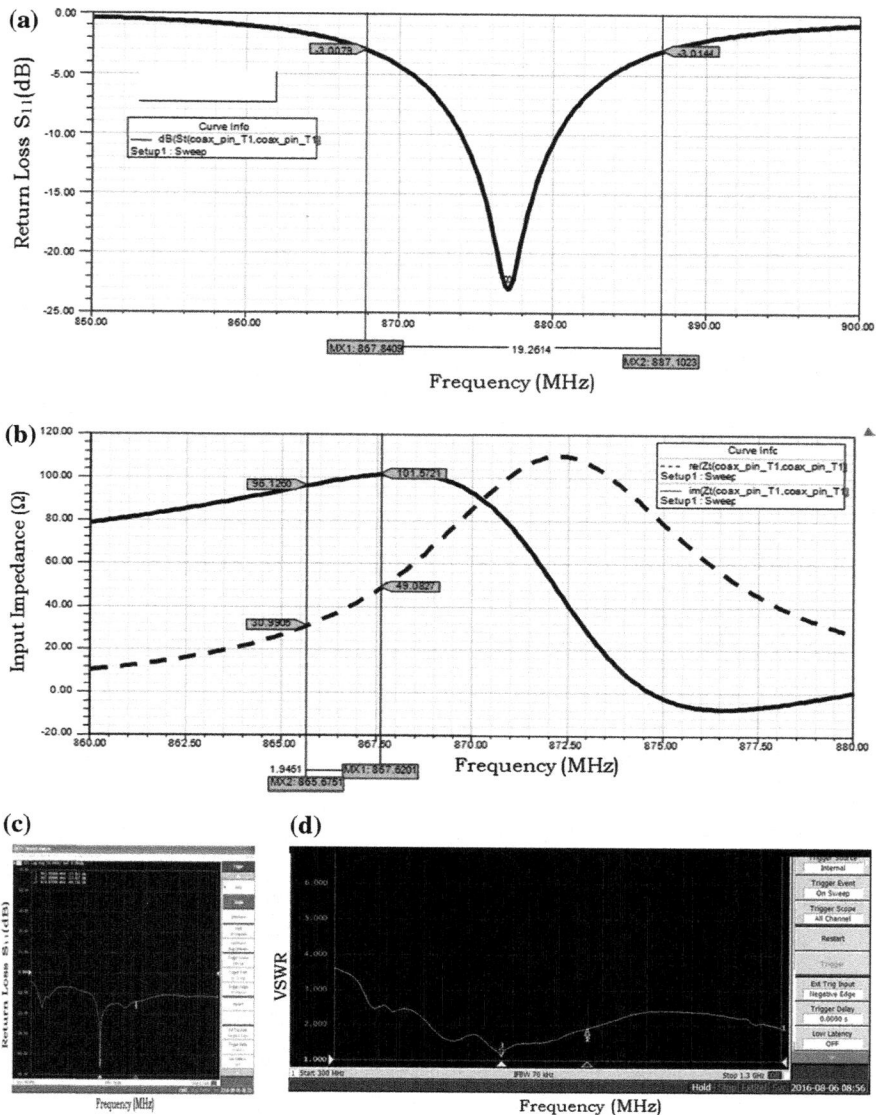

Fig. 13 Simulated. **a** Return loss. **b** Impedance plot. Measured with NA. **c** Return loss. **d** VSWR. **e, f** Radiation at frequencies at 865.62 and 867.63 MHz

threshold power. The maximum distance covered at 867.63 MHz is 5.81 m with a PRC 0.05 and the minimum distance with remaining chips is 3.18 m at 865.62 MHz.

(e)

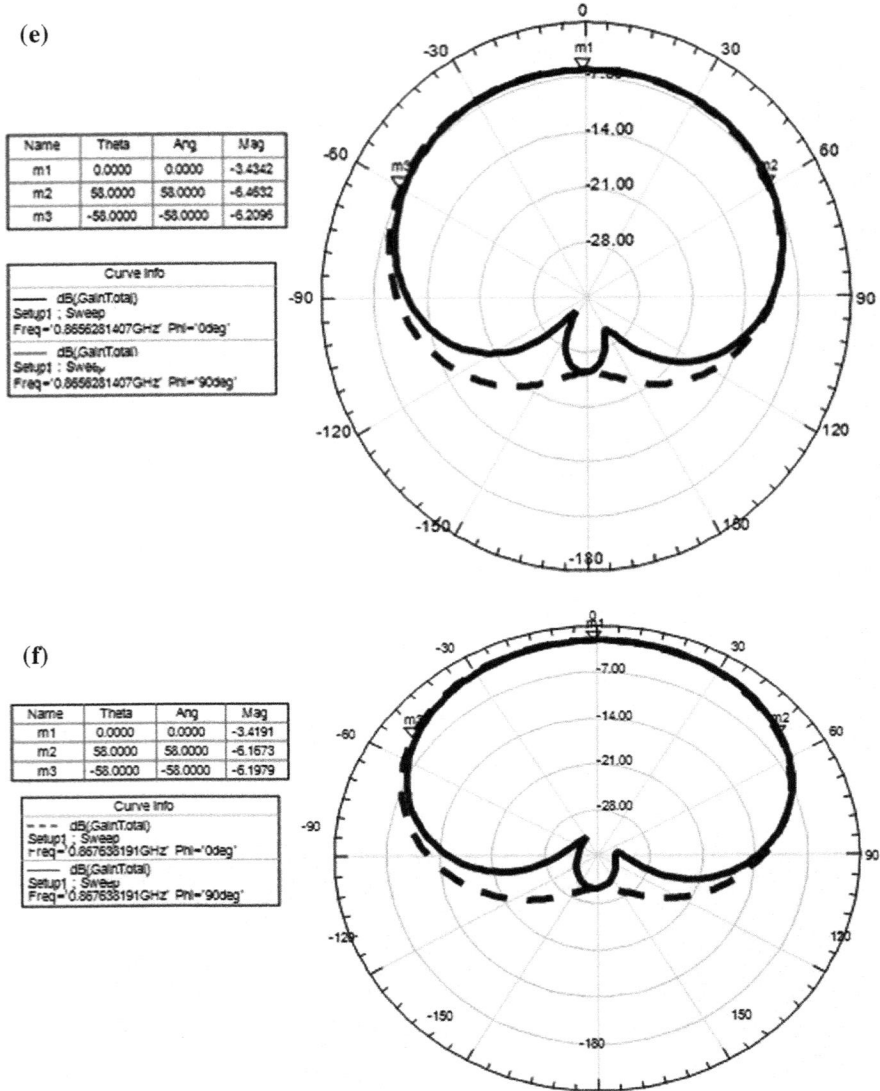

(f)

Fig. 13 (continued)

(D) Curved Shape Patch AHMSA

The proposed antenna is shown in Fig. 14a. In this case, the hexagonal patch edges are alternatively trimmed along with inductive probe and T-stub matching for better

Table 3 RFID tag read range measurement at frequency 865.62 MHz tag impedance (Z_{tag}) = 30.99 + j96.12 Ω, gain (G_{tag}) = −3.43 dB

S. No	Chip impedance Z_{IC} (Ω)	Power reflection coefficient (Γ^2)	Read range (m) at P_{th} = −10 dBm	Read Range (m) at P_{th} = −15 dBm
1	33 − j113	0.06	3.25	5.78
2	13.6 − j122	0.36	2.68	4.76
3	8 − j98	0.34	2.71	4.82
4	20 − j150	0.54	2.25	4.01
5	6 − j125	0.66	1.95	3.47
6	16 − j160	0.68	1.88	3.18

Table 4 Read range measurement at frequency 867.6 MHz tag impedance (Z_{tag}) = 49.08 + j101.5 Ω, gain (G_{tag}) = −3.42 dB

S. No	Chip impedance Z_{IC} (Ω)	Power reflection coefficient (Γ^2)	Read range (m) at P_{th} = −10 dBm	Read range (m) at P_{th} = −15 dBm
1	33 − j113	0.05	3.26	5.81
2	13.6 − j122	0.38	2.63	4.69
3	8 − j98	0.51	2.33	4.14
4	20 − j150	0.44	2.49	4.44
5	6 − j125	0.67	1.92	3.43
6	16 − j160	0.59	2.15	3.64

inductive matching of tag antenna to the chip and size reduction. It is a simple technique and the dimension of substrate is reduced from 9 to 7 cm (from 0.25λ to 0.2λ) length and width.

The return loss (S_{11}) shows (Fig. 14b) that the antenna is tuned at 877.89 MHz and 3-dB HP range is 867.84–888.63 MHz. The tag impedance (Z_{tag}) at 865.6 MHz is 26.75 + j86.88 Ω and 39.99 + j90.75 Ω at 867.6 MHz (Fig. 14c). This technique matches chip 1. The gain of the antenna is calculated at resonant frequency (f_r) (Fig. 2.15d)

(E) Performance Comparison of Proposed Techniques

The response of each technique is compared with remaining techniques in terms of read range versus chip number at threshold power of −15 dBm at frequencies 865.6 and 867.6 MHz shown in Fig. 15a, b. The technique T-matching with probe (black line) is better than the remaining techniques with all chips.

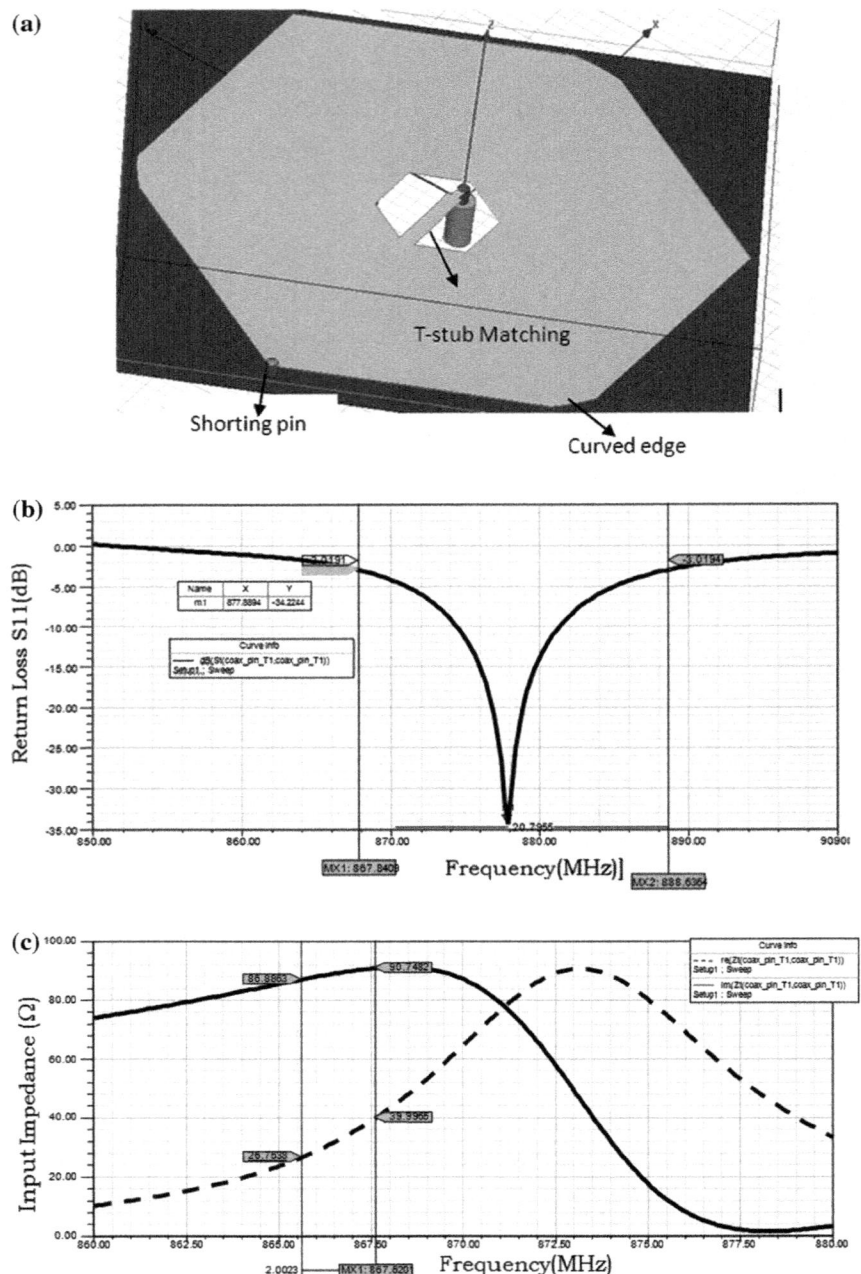

Fig. 14 **a** Simulated curved shape patch AHMSA. **b** Return loss. **c** Impedance matching

(a)

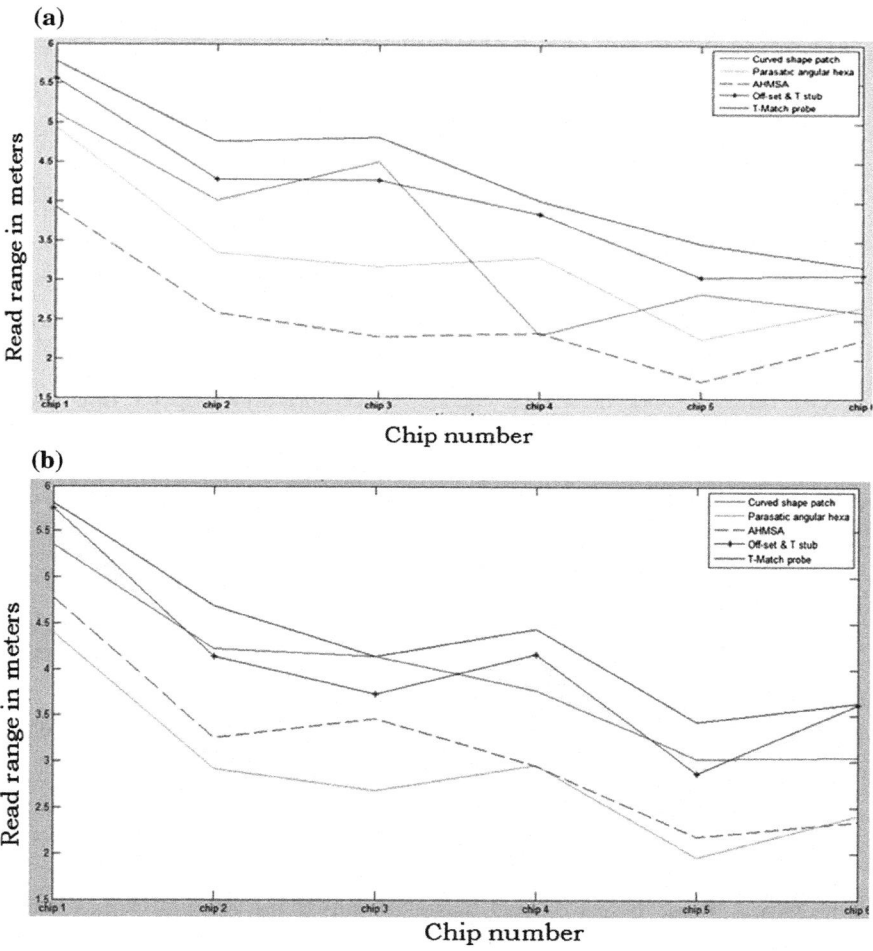

(b)

Chip number

Fig. 15 a, **b** Read range versus chip number at frequencies 865.6 and 867.6 MHz at $P_{th} = -15$ dBm

6 Conclusions

The system is designed for narrowband RFID Tag. The work started from HPMSA and achieved narrowband AHMSA. Its input impedance of antenna is well matched to chip 1. The minimum PRC 0.05 was achieved by introducing step-by-step inductive loading techniques to the basic HPMSA. The maximum read range 5.81 m with chip 1 and 3.64 m with chip 6 at −15 dBm threshold power was achieved. The proposed antenna (T-matching with stub AHMSA) was simple in design and analysis and well suited to the applications at selected frequency band.

References

1. K.B. Kumar, Domdouzis, "Radio-Frequency Identification (RFID) applications: A brief introduction", *Advanced Engineering Informatics* 21(4), 2007, pp. 350–355.
2. C.M. Roberts, "Radio frequency identification (RFID)", Computers & Security 25(1), 2006, pp. 18–26.
3. R. Weinstein, "RFID: A Technical Overview and Its Application to the Enterprise", *IEEE IT Professional* 7(3), 2005, pp. 6.
4. Zhang. J and Long. Y, "A Miniaturized via –Patch Loaded Dual –Layer RFID tag Antenna for Metallic Object Applications", *IEEE Antennas and wireless Propagation Letters*, Vol. 12, 2013, pp. 1184–1187.
5. Zhang. J and Long. Y, "A Dual-Layer Broadband Compact UHF RFID tag antenna for platform Tolerant Application", IEEE Transactions on Antennas and Propagation, Vol. 61, No. 9, Sep-2013, PP. 4447–4455.
6. A.A. Babar, S. Manzari, and U. Leena, "Passive UHF RFID tag for Heat sensing Applications", *IEEE Transactions on Antennas and propagation*, Vol. 60, No. 9, Sep-2012, pp. 4056–4064.
7. V.N. Pavel, K.V.S. Rao, F.L. Sandar, P. Vijay, R. Martinez, and H. Heinrich, "Power Reflection Coefficient Analysis for Complex Impedances in RFID tag Design", *IEEE Transactions on Microwave theory and Techniques*, Vol. 53, No. 9, Sep-2005, pp. 2721–2725.
8. H.D. Chen, S.H. Kuo, C.Y.S. Desmond, H.T. Ching, "Coupling-Feed Circularly Polarized RFID tag Antenna Mountable on Metallic Surfaces", *IEEE Transaction on Antennas and Propagation*, Vol. 60, No. 5, May- 2012, pp. 2166–2174.
9. J. Zhang and Y. Long, "A Dual-Layer Broadband Compact UHF RFID tag antenna for platform Tolerant Application", *IEEE Transactions on Antennas and propagation*, Vol. 61, No. 9, Sep-2013, PP. 4447–4455.
10. I.J. Bhal and P. Bhartia: Microstrip Antennas, Artech house.
11. J.R. James and P.S. Hall. Hand Book of Microstrip antennas, IEE Electromagnetic Waves Series 28, Peter Peregrinus Ltd., London, United Kingdom, 1989.
12. R. Garg, P. Bhartia, I. Bahl, A. Ittipiboon. Microstrip Antenna Design Handbook, Artech House, Inc. 2001.
13. J.P. Chen and P. Hsu, "A compact Strip Dipole Coupled Split-Ring Resonator Antenna for RFID tags", *IEEE Transactions on antennas and propagation*, Vol. 61, No. 11, Nov-2013, pp. 5372–5376.
14. http://www.gs1.org, Regulatory status for using RFID in the EPC Gen 2 band (860 to 960 MHz) of the UHF spectrum, 31 Oct-2014.
15. J.J. Tiang, M.T. Islam, N. Misran and J.S. Mandeep, "Circular Microstrip Slot Antenna for Dual Frequency RFID Application", *Progress in Electro Magnetics Research*, Vol.12, 2011, pp. 499–512.
16. I.J. Bahl, S.S. Stuchly and M.A. Stuchly, "A new microstrip radiator for medical applications", *IEEE Transactions on microwave theory and techniques*, vol., MTT-28, No. 12, Dec-1980, pp. 1464–1468.
17. A.E. Abdulhadi and A. Ramesh, "Design and Experimental Evaluation of Miniaturized Monopole UHF RFID tag Antennas", *IEEE Antennas and Wireless Propagation Letters*, Vol. 11, 2012, pp. 248–251.
18. H.T. Huy, X.T. Son, and P. Ikmo, "A Compact Circularly Polarized Crossed-Dipole Antenna for an RFID tag", *IEEE Antennas and wireless Propagation Letters*, Vol. 14, 2015, pp. 674–677.
19. M. Sanad, "Effect of the Shorting Posts on Short Circuit Microstrip Antennas", *IEEE*, 1994, pp. 794–797.
20. Schaubert D.H. Farrar F.G, Sindoris. A, and Hayes. S.T, "Microstrip Antennas with Frequency Agility and Polarization Diversity", *IEEE Transactions on Antennas and Propagation,* Vol. AP-29, No. 1, Jan-1981, pp. 118–123.

Decision Tree and Genetic Algorithm Based Intrusion Detection System

Chandrashekhar Azad and Vijay Kumar Jha

1 Introduction

Nowadays, Internet is universal and most of the organizations, either private or public, hosted their most valuable assets online in the digital form for easy accessibility, availability, sharing, etc. Today, Internet is the basic need for the human being like the food, shelter, and house, without Internet, the individuals and the corporate users are helpless in smooth running the day-by-day activity of the business. Individual and corporate users use web applications and services not only for the purpose to generate return and stay connected with clients or friends or competitor but applications are often used to store and share valuable information and services to the intended users [1]. Heavy access of the Internet motivates the service provider to give the security of the data, and information on web and also the transaction they carried out in the day-by-day activity. CERT reports [2] that the number of attack on computer network attack rises gradually and also the volume of records on the Internet also rises exponentially [3]. Due to attacks on computer network web service, the organizations and individuals face monetary losses, it also violates the integrity of the data and services of individuals or organizations in the web. Today there are access control, email security, firewalls, etc., security system that exist, but futile to deliver the security on WWW [4]. Literature in these [5–9] day by day domain of intrusion detection system (IDS) is gaining attention. The main reason behind the popularity of IDS is the size of network traffic log, which

C. Azad (✉)
Department of Computer Applications, National Institute of Technology, Jamshedpur,
Jamshedpur 831014, India
e-mail: csazad.ca@nitjsr.ac.in

V. K. Jha
Department of Computer Science and Engineering, Birla Institute of Technology,
Mesra, Ranchi 835215, India
e-mail: vkjha@bitmesra.ac.in

© Springer Nature Singapore Pte Ltd. 2019   141
V. Nath and J. K. Mandal (eds.), *Proceeding of the Second International Conference
on Microelectronics, Computing & Communication Systems (MCCS 2017)*, Lecture Notes
in Electrical Engineering 476, https://doi.org/10.1007/978-981-10-8234-4_13

increases exponentially and manual analysis of it is practically impossible. Therefore, there is a need for robust security system which may automatically extract valuable hidden facts from huge volume of data. Such, capabilities are present in machine learning, data mining, and pattern recognition [10, 11]. The machine learning, data mining, and pattern recognition can do prediction [12, 13], classification [14–16], clustering [17–19], and association rule mining [20–22]. Consequently, we can say that the field of intrusion detection system based on the data mining is the best encapsulation of the technological framework to overcome the effect of the intrusion in the WWW. Data classification in intrusion detection used to label the network traffic whether it is normal or anomalous or may further anomalous labeled data may categorized into various attack like DDOS, Trojan, malware, eavesdrop, etc. Data mining methods like outlier detection, feature selection, feature reduction, and data preprocessing methods are useful for intrusion detection in the host-based system and the network-based system [23].

2 Related Work

An IDS is hardware, software or both that attempt to monitor network activity. IDS may alert if any suspicious activity is found which may violate the integrity of the data and services in the WWW. Data mining is the set of process to find the hidden information from the large volume of the data. With the help of data mining, the data scientist may find the useful information which may be financially valuable as well as which is previously hidden. Anderson [24] addressed about IDS in his paper Computer Security Threat Monitoring and Surveillance, after that till date many of the innovations are done in the field of the intrusion detection by the researchers but still the data and services are not 100% secured in the web [25, 26]. Denning [27] proposed an intrusion detection model that is intrusion detection expert system (IDES), this is the first model of the IDS. The aim of the denning is to analyze the audit trail. Today, the field of IDS is fascination to the researchers because the popularity of Internet and IDS stands as a digital guardian in the web. Sindhu et al. [4] developed decision tree based intrusion detection. The developed system has been evaluated using detection percentage and the error percentage. Prema et al. have presented active rule-based IDS, which is an extension of the C4.5 method. The performance of the system is evaluated using the KDD CUP dataset. Mulay et al. [28] have presented a decision tree and SVM-based IDS, which is called tree structured multiclass SVM. Senthilnayaki et al. [29] have projected genetic algorithm and J48 based IDS. Muniyandi et al. [30] have proposed an anomaly detection system using the cascade K-Means and C4.5 algorithm. Panda et al. [31] have proposed an IDS using the multi-classifier system to detect the intrusion. Selvi et al. [32] have presented an intelligent decision-based intrusion detection system. This is based on the concept of Manhattan distance. Jiang et al. [33] have presented an incremental decision tree based IDS which is based on the concept of rough set theory. Koshal et al. [34] have projected a cascading of C4.5 and SVM to build a rule-based IDS. The decision tree C4.5 is proposed by Quinlan [35] and this is widely used classifier. The C4.5 based classifier gives better result in terms of detection rate. Many of the researchers use decision tree C4.5, REP tree, etc., to build intrusion detection, the main reason behind popularity of decision tree based IDS is, it is easy to implement and it is not a parametric method. Rule extracted through C4.5 is considered in two groups: small disjunct and large disjunct. Rule set which covers a small number of training data is called small disjunct and the large disjunct which covers a large number of training data [36–38]. C4.5 is unfair toward the large disjunct against the small disjunct [39–41]. The IDS developed using the C4.5's performance may affect the availability of the of small disjunct [42]. The small disjuncts in the decision tree based IDS are important as much as the large disjuncts, therefore the small disjuncts cannot be ignored. If the small disjuncts are ignored the system can be error-prone and the availability of small disjunct may affect the accuracy of the system and may also increase the false alarm rate.

3 Materials and Methods

3.1 Decision Tree

Decision tree is a collection of nodes and edges where edges may be empty but nodes never are empty. Nodes represent the condition and the edges contain the outcome of the test. In decision tree, leaf node contains the class labels and the non-leaf contains the features and the associated value of instance. In data mining, decision tree is treated as a classification system or model.

3.2 Attribute Selection Measures

3.2.1 Information Gain

$$\text{Info}(D) = -\sum_{i=1}^{n} P_i \log_2 p_i \tag{1}$$

$$\text{info}_A(D) = -\sum_{j=1}^{v} \frac{|D_j|}{|D|} \text{info}(D_j) \tag{2}$$

$$\text{Gain}(A) = \text{Info}(D) - \text{info}_A(D) \tag{3}$$

3.2.2 Gain Ratio

$$\text{Split Info}_A(D) = -\sum_{j=1}^{v} \frac{|D_j|}{|D|} \log_2\left(\frac{|D_j|}{|D|}\right) \tag{4}$$

$$\text{Gain Ratio} = \frac{\text{Gain}(A)}{\text{Split Info}_A(D)} \tag{5}$$

3.3 Genetic Algorithm

The genetic algorithm (GA) is used to optimize the initial solutions through natural evaluation process. GA takes chromosomes as an input and generates offspring from given chromosomes through genetic operations. The new offspring's fitness is evaluated and the best individuals are kept for the next generation [43]. In this paper, the GA is used to handle the problem of small disjunct in decision tree. Two ways to use the GA in handling small disjunct are as follows: (i) Instances covered by each leaf nodes individually as a training set, i.e., GA small (ii) All instances covered by all the leaf as a training set, i.e., GA large. In this, we focused on the second situation. The concept of the GA large is shown in Fig. 1.

Fig. 1 Decision tree with
GA large representation

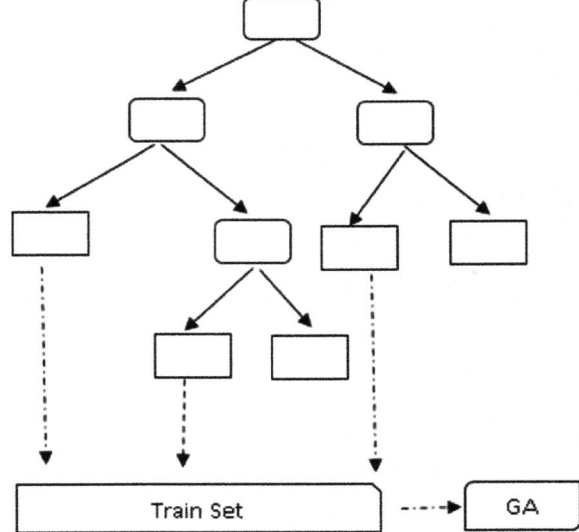

3.3.1 Individual Encoding

A chromosome in the genetic algorithm is the base to the operations and each consists of the n genes and each gene represents the one attribute. The first gene corresponds to the first attribute, the second gene corresponds to the second attribute, and so on. Each gene is divided into the three parts attribute, operator, and the value. Gene structure is shown in Fig. 2.

A_i is the ith attribute, O_i is the operator associated and V_{ij} is the jth value of the A_i. In between genes, an extra field is added that is B_i to show whether gene is active or not. The general structure of the genome is shown in Fig. 3.

Selection: In the genetic algorithm, the selection operation plays an important role in further evolution of the individuals. If the worse fitness value individual is chosen it may not provide the optimal solution. Roulette wheel selection, ranking based, tournament selection, etc., are the different selection methods.

Crossover: The crossover operation is used to produce the new individuals from the parents. This is the reproduction scheme in the genetic algorithm. The general crossover is shown in Fig. 4.

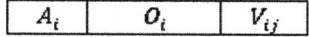

Fig. 2 Gene structure

| $A_1\ OP_1\ \{V_{1j}....\}$ | B_1 | .... | $A_i\ OP_i\ \{V_{ij}....\}$ | B_i | ...... | $A_n\ OP_n\ \{V_{nj}....\}$ | B_n |

Fig. 3 Structure of the genome

| dst_bytes < 3.5(1) | service = telnet(1) | srv_diff_host_rate < 0.28(1) | flag = SF(1) |

| dst_bytes > 3.5(0) | service= private(1) | srv_diff_host_rate < 0.3(1) | flag = SH(1) |

(a)　Parent (before crossover)

| dst_bytes < 3.5(1) | service = telnet(1) | srv_diff_host_rate < 0.3(1) | flag = SH(1) |

| dst_bytes >3.5(0) | service = private(1) | srv_diff_host_rate < 0.28(1) | flag = SF(1) |

(b)　Offspring or child (after crossover)

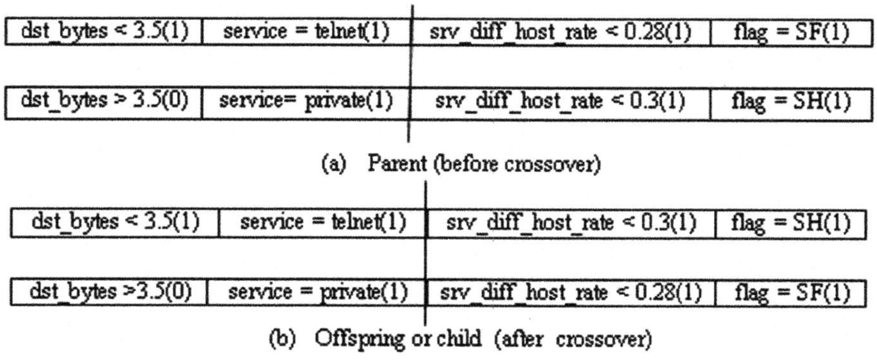

Fig. 4 Crossover (one point) of individual

(a)

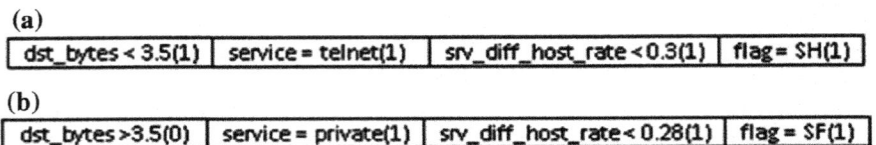

(b)

Fig. 5 Genome (before mutation)

(a)

(b)

Fig. 6 Genome (after mutation)

3.3.2 Mutation

In the previous step, crossover operation was used to produce offspring from the parents. The mutation task to maintain genetic diversity among the individuals of the different generations. The basic method is shown in Figs. 5 and 6.

3.3.3 Fitness Function

$$Fitness = Sensitivity * Specificity$$

4 Algorithm for Proposed IDS

Step 1. Load data set.

Step 2. Create a node N, instances in training set are belongs to same class then return N by labeling leaf node labeled label C;

Step 3. For each attribute, compute information gain ratio.

Step 4. Let attribute A has the highest information, create a node with best splitting node A.

Step 5. Remove the splitting attribute from the attribute list.

Step 6. For each outcome of the splitting criterion on A recursively, repeat Step 2–5.

Step 7. Find the decision rules from C4.5 decision tree.

Step 8. Find the small disjunct rule.

Step 9. Select parents.

Step 10. Perform crossover and mutation operation.

Step 11. Fitness value calculation and keep the best Offspring's.

Step 12. If maximum iteration or stopping criteria are fulfilled then go to Step 13 else go to 10.

Step 13. Return optimized rules and C4.5 decision rules which are not small disjunct rules.

Step 14. Stop.

5 Experimental Result and Discussion

5.1 Dataset

The KDD Cup data set is used.

5.2 Performance Parameters

$$\text{Accuracy} = \frac{\text{Correctly Classified Tuples}}{\text{Total Population}} \times 100$$

$$\text{Error} = \frac{\text{Incorrectly Classified Tuples}}{\text{Total Population}} \times 100$$

5.3 Results

See Tables 1, 2, 3 and 4.

Table 1 Results on data sample 1

Sample 1	Accuracy	Error
$T = 5$	99.29	0.71
$T = 10$	99.07	0.93
$T = 15$	98.84	1.16
$T = 20$	93.59	1.41
DT	96.93	3.07

Table 2 Results on data
sample 2

Sample 2	Accuracy	Error
T = 5	97.80	2.20
T = 10	98.09	1.91
T = 15	97.29	2.71
T = 20	96.87	3.13
DT	94.82	5.18

Table 3 Results on data
sample 3

Sample 3	Accuracy	Error
T= 5	99.94	0.06
T = 10	99.99	0.01
T = 15	98.77	1.23
T = 20	99.73	0.27
DT	98.66	1.34

Table 4 Comparative study
of proposed IDS with other
systems

System	Accuracy	Error
Proposed (T = 5)*	99.94	0.06
Proposed (T = 10)*	99.99	0.01
Proposed (T = 15)*	98.84	1.16
Proposed (T = 20)*	99.73	0.27
C4.5 (sample 1)	96.93	3.07
C4.5 (sample 2)	94.82	5.18
C4.5 (sample 3)	98.66	1.34
Random tree [4]	88.98	11.02
Random forest [4]	89.21	10.79
DTLW IDS [4]	98.38	01.62
Naive Bayes [4]	92.27	7.73
Rep tree [4]	89.11	10.89

*Best result on threshold

5.4 Discussion

This segment shows the comprehensive discussion of the outcome of the projected
IDS. The proposed system is tested on Windows 8.1 platform with the environment
JDK 1.8, 4 GB RAM and Keel. To test the proposed system, three samples of the
data sets are taken. Each sample is preprocessed to handle the outliers instance in
the data samples. The interquartile range of the weka is used to handle the outlier. In
the experimental process, training set is applied on the C4.5 and the system is
trained, then genetic algorithm is used to optimize the system. For the genetic
operations, we used the 1% mutation probability, 80% crossover probability. The
threshold levels 5, 10, 15, and 20 are taken to test the coverage of small disjuncts in

the decision tree C4.5 rules. The C4.5 decision tree is trained using default setting and the parameter values. The genetic algorithm is evaluated on the 100, 150, and 200 and 250 different generations. The result shown in the result section delivers the best results at the 200 generations. After learning of the system applied the testing data on the trained intrusion detection system based on the C4.5 and genetic algorithm. The effectiveness of the system is measured in terms of classification accuracy and classification error. In this work, results are expressed in percent. The results of the proposed method are compared with the C4.5 decision tree, Naïve Bayes, etc. From Table 4, it is clear that the proposed IDS delivers the superior accuracy and also has low error rate. High classification accuracy and the low classification error rate show the effectiveness of the system.

Table 1 presents the results of the proposed system on data sample 1, in which the system gives best results at the threshold level 5. The best results at the sample 1 are accuracy 99.29 and the classification error 0.71 at the threshold level 5. The second best results at the sample 1 are classification accuracy 99.07 and classification error 0.93 at threshold 10. The C4.5 decision tree is when demonstrated on the same sample it provides accuracy 96.93 and classification error 3.07 in percentage.

Table 2 presents the results of the proposed system on data sample 2, in which the system gives best results at the threshold level 10. The best results at the sample 2 are accuracy 98.09 and the classification error 1.91 at the threshold level 10. The second best results at the sample 2 are classification accuracy 97.80 and classification error 2.20 at threshold 5. The C4.5 decision tree is when demonstrated on the same sample it provide the accuracy 94.82 and classification error 5.18.

Table 3 presents the results of the proposed system on data sample 3, in which the system gives best results at the threshold level 10. The best results at the sample 2 are accuracy 99.99 and the classification error 0.01 at the threshold level 10. The second best results at the sample 2 are classification accuracy 99.94 and classification error 0.06 at threshold 5. The C4.5 decision tree is when demonstrated on the same sample it provide the accuracy 98.66 and classification error 1.34.

Table 4 shows the comparative study of the proposed system with the C4.5 decision tree, random tree, random forest, DTLW IDS, Naïve Bayes, and rep tree. Table 4 expresses comparative study of projected IDS. It is worth mentioning that the projected IDS is better in comparison to the other IDS. The projected IDS performed well at threshold level 10 and gave second-best outcomes at threshold level 5.

6 Conclusion

In this research paper, an IDS has been projected which is based on decision tree and genetic algorithm. The projected IDS may optimize the small disjunct rules in of decision tree C4.5 classifiers. In the proposed IDS, first training samples are pre-preprocessed to remove the outlier records, then C4.5 is used to train the IDS.

The C4.5 generates the decision rules, decision rules those which are small disjunct are optimized using genetic algorithm. It is worth to mention that the projected IDS gives the improved result in comparison to the methods such as the C4.5 decision tree, Naïve Bayes, etc.

References

1. Shon T, Moon J. A hybrid machine learning approach to network anomaly detection. Information Sciences. 2007 Sep 15; 177(18): 3799–821.
2. Ragsdale DJ, Carver Jr CA, Humphries JW, Pooch UW. Adaptation techniques for intrusion detection and intrusion response systems. In Systems, Man, and Cybernetics, 2000 IEEE International Conference on 2000 (Vol. 4, pp. 2344–2349). IEEE.
3. http://en.wikipedia.org/wiki/Internet_traffic [Accessed on February 18th, 2015].
4. Sindhu SS, Geetha S, Kannan A. Decision tree based light weight intrusion detection using a wrapper approach. Expert Systems with applications. 2012 Jan 31; 39(1): 129–41.
5. Azad C, Jha VK. Data mining in intrusion detection: a comparative study of methods, types and data sets. International Journal of Information Technology and Computer Science (IJITCS). 2013 Jul 1; 5(8): 75.
6. Kim J, Bentley PJ, Aickelin U, Greensmith J, Tedesco G, Twycross J. Immune system approaches to intrusion detection–a review. Natural computing. 2007 Dec 1; 6(4): 413–66.
7. Kumar G, Kumar K, Sachdeva M. The use of artificial intelligence based techniques for intrusion detection: a review. Artificial Intelligence Review. 2010 Dec 1; 34(4): 369–87.
8. Liao HJ, Lin CH, Lin YC, Tung KY. Intrusion detection system: A comprehensive review. Journal of Network and Computer Applications. 2013 Jan 31; 36(1): 16–24.
9. Ramakrishnan S, Srinivasan S. Intelligent agent based artificial immune system for computer security—a review. Artificial Intelligence Review. 2009 Dec 1; 32(1–4): 13–43.
10. Julisch K. Data mining for intrusion detection. In Applications of data mining in computer security 2002 (pp. 33–62). Springer US.
11. Liao SH, Chu PH, Hsiao PY. Data mining techniques and applications–A decade review from 2000 to 2011. Expert Systems with Applications. 2012 Sep 15; 39(12): 11303–11.
12. Fan W, Bifet A. Mining big data: current status, and forecast to the future. ACM sIGKDD Explorations Newsletter. 2013 Apr 30; 14(2): 1–5.
13. Han G, Jiang J, Shen W, Shu L, Rodrigues J. IDSEP: a novel intrusion detection scheme based on energy prediction in cluster-based wireless sensor networks. Information Security, IET. 2013 Jun; 7(2): 97–105.
14. Altwaijry H. Bayesian based intrusion detection system. In IAENG Transactions on Engineering Technologies 2013 (pp. 29–44). Springer Netherlands.
15. Azad C, Jha VK. Data mining based hybrid intrusion detection system. Indian Journal of Science and Technology. 2014 Jun 30; 7(6): 781–9.
16. Kim G, Lee S, Kim S. A novel hybrid intrusion detection method integrating anomaly detection with misuse detection. Expert Systems with Applications. 2014 Mar 31; 41(4): 1690–700.
17. Khan L, Awad M, Thuraisingham B. A new intrusion detection system using support vector machines and hierarchical clustering. The VLDB Journal—The International Journal on Very Large Data Bases. 2007 Oct 1; 16(4): 507–21.
18. Portnoy L. Intrusion detection with unlabeled data using clustering. 2000.
19. Rajeswari LP, Arputharaj K. An active rule approach for network intrusion detection with enhanced C4. 5 Algorithm. International Journal of Communications, Network and System Sciences. 2008 Nov 1; 1(4): 314.

20. Agrawal R, Srikant R. Fast algorithms for mining association rules. In Proc. 20th int. conf. very large data bases, VLDB 1994 Sep 12 (Vol. 1215, pp. 487–499).
21. El-Semary A, Edmonds J, Gonzalez-Pino J, Papa M. Applying data mining of fuzzy association rules to network intrusion detection. In Information Assurance Workshop, 2006 IEEE 2006 Jun 21 (pp. 100–107). IEEE.
22. Mabu S, Chen C, Lu N, Shimada K, Hirasawa K. An intrusion-detection model based on fuzzy class-association-rule mining using genetic network programming. Systems, Man, and Cybernetics, Part C: Applications and Reviews, IEEE Transactions on. 2011 Jan; 41(1): 130–9.
23. Davis JJ, Clark AJ. Data preprocessing for anomaly based network intrusion detection: A review. Computers & Security. 2011 Oct 31; 30(6): 353–75.
24. Anderson JP. Computer security threat monitoring and surveillance. Technical report, James P. Anderson Company, Fort Washington, Pennsylvania; 1980 Apr 15.
25. Azad C, Jha VK. Fuzzy min–max neural network and particle swarm optimization based intrusion detection system. Microsystem Technologies. 2016: 1–2.
26. Azad C, Jha VK. A Novel Fuzzy Min-Max Neural Network and Genetic Algorithm-Based Intrusion Detection System. In Proceedings of the Second International Conference on Computer and Communication Technologies 2016 (pp. 429–439). Springer India.
27. Denning DE. An intrusion-detection model. Software Engineering, IEEE Transactions on. 1987 Feb(2): 222–32.
28. Mulay SA, Devale PR, Garje GV. Intrusion detection system using support vector machine and decision tree. International Journal of Computer Applications. 2010 Jun; 3(3): 40–3.
29. Senthilnayaki B, Venkatalakshmi K, Kannan A. An intelligent intrusion detection system using genetic based feature selection and Modified J48 decision tree classifier. In Advanced Computing (ICoAC), 2013 Fifth International Conference on 2013 Dec 18 (pp. 1–7). IEEE.
30. Muniyandi AP, Rajeswari R, Rajaram R. Network anomaly detection by cascading k-Means clustering and C4. 5 decision tree algorithm. Procedia Engineering. 2012 Dec 31; 30: 174–82.
31. Panda M, Abraham A, Patra MR. A hybrid intelligent approach for network intrusion detection. Procedia Engineering. 2012 Dec 31; 30: 1–9.
32. Selvi R, Kumar SS, Suresh A. An Intelligent Intrusion Detection System Using Average Manhattan Distance-based Decision Tree. In Artificial Intelligence and Evolutionary Algorithms in Engineering Systems 2015 (pp. 205–212). Springer India.
33. Jiang F, Sui Y, Cao C. An incremental decision tree algorithm based on rough sets and its application in intrusion detection. Artificial Intelligence Review. 2013 Dec 1; 40(4): 517–30.
34. Koshal J, Bag M. Cascading of C4. 5 decision tree and support vector machine for rule based intrusion detection system. International Journal of Computer Network and Information Security. 2012 Aug 1; 4(8): 8.
35. Quinlan JR. C4. 5: programs for machine learning. Elsevier; 2014 Jun 28.
36. Carvalho DR, Freitas AA. A genetic algorithm-based solution for the problem of small disjuncts. In Principles of Data Mining and Knowledge Discovery 2000 Sep 13 (pp. 345–352). Springer Berlin Heidelberg.
37. Carvalho DR, Freitas AA. A hybrid decision tree/genetic algorithm method for data mining. Information Sciences. 2004 Jun 14; 163(1): 13–35.
38. Holte RC, Acker L, Porter BW. Concept Learning and the Problem of Small Disjuncts. In IJCAI 1989 Aug 20 (Vol. 89, pp. 813–818).
39. Carvalho DR, Freitas AA. A hybrid decision tree/genetic algorithm for coping with the problem of small disjuncts in data mining. In GECCO 2000 Jul (pp. 1061–1068).
40. Carvalho DR, Freitas AA. A genetic-algorithm for discovering small-disjunct rules in data mining. Applied Soft Computing. 2002 Dec 31; 2(2): 75–88.

41. Carvalho DR, Freitas AA. A Genetic Algorithm With Sequential Niching For Discovering Small-disjunct Rules. In GECCO 2002 Jul 9 (pp. 1035–1042).

42. Alcala-Fdez J, Sanchez L, Garcia S, del Jesus MJ, Ventura S, Garrell JM, Otero J, Romero C, Bacardit J, Rivas VM, Fernandez JC. KEEL: a software tool to assess evolutionary algorithms for data mining problems. Soft Computing. 2009 Feb 1; 13(3): 307–18.

43. Azad C, Jha VK. Genetic Algorithm to Solve the Problem of Small Disjunct In the Decision Tree Based Intrusion Detection System. International Journal of Computer Network and Information Security (IJCNIS). 2015 Jul 8; 7(8): 56.

A Novel Method of Image Denoising Using Nonlocal Algorithm

B. Jayalakshmi, G. Indraja and Saka Harshavardhan

1 Introduction

Advanced pictures assume a huge part in our everyday life like satellite television, intelligent traffic monitoring, signature validation, and astronomy. In advanced imaging, the procurement strategies and frameworks present different sorts of commotions and antiquities [1]. Removing the noise from the first image and storing the fine elements of the picture is the objective of picture denoising. It brings down the detectable quality of low complexity objects. Hence, commotion expulsion is fundamental in computerized imaging applications to upgrade the fine particulars of picture.

A computerized picture is generally determined as a matrix of gray level. The noise is mainly added due to the degradation, acquisition process, and data transformation process [2]. We know that the Gaussian noise is very compactable to add the original image. A normal way to get rid of the noise from the original image is to employ spatial filters.

A mean filter is the optional filter to evacuate the commotion. Linear filters cannot take away the noise from the image, the image is effected to blur sharp edges. Denoising algorithms can be divided into two types: Spatial filtering and frequency filtering. The spatial filtering is again divided into (a) linear filter (b) nonlinear filter. Spatial filtering refers to use like a low pass filter [3]. In frequency flattening filters, the amputation of noise is achieved by using the cutoff

B. Jayalakshmi (✉) · G. Indraja · S. Harshavardhan
Department of Electronics Engineering, VNITSW, Guntur, India
e-mail: jayabandarupalli2015@gmail.com

G. Indraja
e-mail: indraja246@gmail.com

S. Harshavardhan
e-mail: sakaharshavardhan@gmail.com

© Springer Nature Singapore Pte Ltd. 2019
V. Nath and J. K. Mandal (eds.), *Proceeding of the Second International Conference on Microelectronics, Computing & Communication Systems (MCCS 2017)*, Lecture Notes in Electrical Engineering 476, https://doi.org/10.1007/978-981-10-8234-4_14

frequency. These methods are time consuming and depend on the cutoff frequency. Filtering can be performed using wavelet domain in which noise is not completely removed and there is time complexity. The two principal limits in picture accuracy are blur and commotion. In the first rough approximation, the commotion value is

$$v(i) = u(i) + n(i)$$

where $u(i)$ is the practical value of original image at pixel i and $n(i)$ is the noise added. Hence $v(i)$ is corrupted image with noise. The most deal approach to the test of noise on a modernized picture is to incorporate Gaussian clamor [4]. All denoising calculations rely on a filtering parameter h. This parameter measures the level of filtering connected to the picture. The greater part of the strategies, the parameter h relies on guess of the commotion difference.

$$v = D_h v + n(D_h, v)$$

where $D_h v$ is more smooth than v.

We had to select the denoising methods for comparison. Here we have confusion arises which denoising technique had mostly used to remove the noise. In this paper, we discussed local smoothing filters like Gaussian filtering, anisotropic filtering, and neighborhood filtering. Gaussian smoothing administrator is a 2-D convolution administrator that is utilized to blur images and removes the fine points in the picture [5]. This is identical to the mean filter. But mean filter reduces the intensity of the pixel. Using Gaussian filter, the commotion is smoothed out in the meanwhile the signal is also contorted [6]. Gaussian convolution is optimum in level parts of picture. In this system, edges and surfaces are blurred. To avoid the drawbacks in Gaussian filtering, we use anisotropic filtering. The word anisotropic is originated as "an" means not, "iso" means same, and "tropic" means direction. That means it does not filter the same amount in all direction [7]. In computer graphics, this method is used to enhance the quality of the texture. In case of bilinear and trilinear filtering techniques, the image is affected by aliasing effects. The anisotropic filtering is used to reduce the noise present at the edges. This idea of a filter is Perona and Malik. In this technique, the limitations are edges are restored while the level and surfaces are degraded. Another system to evacuate the commotion is neighborhood filter. Any filter which gives back a pixel by taking a normal of the estimations of every single neighboring pixel [8]. There are different types of neighborhood filters: (a) 4-neighborhood (b) D-neighborhood (c) 8-neighborhood. The main limitation of this technique is it does not retrieve the original image.

2 Prepocessing Techniques

There are different types of preprocessing techniques: (a) DCT (b) DFT (c) FFT (d) KL techniques. A discrete cosine transform expresses the sum of cosine functions at different frequencies. These techniques are to compress the image. The FFT transformation technique computes Fourier transform of a sequence. The main limitation in FFT is that computations are more and has time complexity. In this paper, we discussed KL transform to compress the image in a more compact form. The KL transform is closely related to principal component analysis technique used in image processing. The KL transform is calculated using mean of random variables. The expectation is defined as

$$E(x_i x_j) = \begin{cases} 0 & i \neq j, \\ \sigma_i^2 & i = j, \end{cases}$$

The mean of the matrix is

$$X = [x_1 x_2 \ldots x_N]^T$$

The covariance of the matrix is

$$[C]_x = E\left[(x - m)(x - m)^T\right]$$

Let us consider an example for KL transform. The original matrix U is defined as

$$U = \begin{bmatrix} 2 & 4 & 5 & 5 & 3 & 2 \\ 2 & 3 & 4 & 5 & 4 & 3 \end{bmatrix}$$

Its vertical mean vector is

$$m = (1/N) \sum_{i=0}^{N-1} \rho_{u_i}$$

$$m = (1/6) \begin{bmatrix} 2+4+5+5+3+2 \\ 2+3+4+5+4+3 \end{bmatrix} = \begin{bmatrix} 3.5 \\ 3.5 \end{bmatrix}$$

And its covariance matrix

$$R_u = \frac{1}{(N-1)} \sum_{i=0}^{N-1} (\rho_{u_i} - \rho_\mu)(\rho_{u_i} - \rho_\mu)^T = \begin{bmatrix} \sigma_0^2 & c_{01} & c_{02} & \cdot \\ c_{01} & \sigma_1^2 & \cdot & \cdot \\ c_{02} & \cdot & \sigma_2^2 & \cdot \\ \cdot & \cdot & \cdot & \cdot \end{bmatrix}$$

$$R_u = \frac{1}{5}\left\{ \begin{bmatrix} 2.25 & 2.25 \\ 2.25 & 2.25 \end{bmatrix} + \begin{bmatrix} 0.25 & -0.25 \\ -0.25 & 0.25 \end{bmatrix} + \begin{bmatrix} 2.25 & 0.75 \\ 0.75 & 0.25 \end{bmatrix} \right.$$
$$\left. + \begin{bmatrix} 2.25 & 2.25 \\ 2.25 & 2.25 \end{bmatrix} + \begin{bmatrix} 0.25 & -0.25 \\ -0.25 & 0.25 \end{bmatrix} + \begin{bmatrix} 2.25 & 0.75 \\ 0.75 & 0.25 \end{bmatrix} \right\}$$

$$R_u = \begin{bmatrix} 1.9 & 1.1 \\ 1.1 & 1.1 \end{bmatrix}$$

$$\begin{vmatrix} 1.9 - \lambda & 1.1 \\ 1.1 & 1.1 - \lambda \end{vmatrix} = 0$$

$$\lambda^2 - 3\lambda + 0.88 = 0$$

$$\lambda_1 = 2.67\lambda_2 = 0.33$$

$$[R_u - \lambda_1 I]\phi_2 = 0$$

$$\phi = \left\{ \begin{matrix} \rho & \rho \\ \phi_1 & \phi_2 \end{matrix} \right\} = \left\{ \begin{bmatrix} \phi_{11} \\ \phi_{21} \end{bmatrix} \begin{bmatrix} \phi_{12} \\ \phi_{22} \end{bmatrix} \right\}$$

Equations for the eigenvector F_1

$$-0.77\phi_{11} + 1.1\phi_{21} = 0$$

$$1.1\phi_{11} - 1.57\phi_{21} = 0$$

$$\phi_{11} = 1.43\phi_{21}$$

Equations for the eigenvector F_2

$$1.57\phi_{12} + 1.1\phi_{12} = 0$$

$$1.1\phi_{12} + 0.77\phi_{22} = 0$$

$$\phi_{12} = -0.7\phi_{22}$$

In addition to Eqs. 1 and 2, if F_1 and F_2 are orthogonal eigenvectors, taking after conditions can be composed ,

$$\phi_{11}^2 + \phi_{21}^2 = 1$$

$$\phi_{12}^2 + \phi_{22}^2 = 1$$

The solution of these four equations is

$$\phi = \left\{ \begin{matrix} \rho & \rho \\ \phi_1 & \phi_2 \end{matrix} \right\} = \left\{ \begin{bmatrix} \phi_{11} \\ \phi_{21} \end{bmatrix} \begin{bmatrix} \phi_{12} \\ \phi_{22} \end{bmatrix} \right\} = \left\{ \begin{bmatrix} 0.82 \\ 0.57 \end{bmatrix} \begin{bmatrix} -0.57 \\ 0.82 \end{bmatrix} \right\}$$

Now, let us transform the image U

$$v_n = \phi^\tau u_n \quad 0 \in n \in N - 1$$

$$v = \begin{bmatrix} 0.82 & 0.57 \\ -0.57 & 0.82 \end{bmatrix} \begin{bmatrix} 2 & 4 & 5 & 5 & 3 & 2 \\ 2 & 3 & 4 & 5 & 4 & 3 \end{bmatrix}$$

$$v = \begin{bmatrix} 2.78 & 4.99 & 6.38 & 6.95 & 4.74 & 3.35 \\ 0.5 & 0.18 & 0.43 & 1.25 & 1.57 & 1.32 \end{bmatrix}$$

The covariance matrix for the transformed image is

$$R_v = \begin{bmatrix} 2.67 & 0 \\ 0 & 0.33 \end{bmatrix}$$

As observed over, the changed picture V is uncorrelated. 89% energy was packed in the first-row vector. For this situation, if we discard second row vector and inverse change it; let us see what we will have,

$$U' = \begin{bmatrix} 0.82 & -0.57 \\ 0.57 & 0.82 \end{bmatrix} \begin{bmatrix} 2.78 & 4.99 & 6.38 & 6.95 & 4.74 & 3.35 \\ 0 & 0 & 0 & 0 & 0 & 0 \end{bmatrix}$$

$$U' = \begin{bmatrix} 2.28 & 4.1 & 5.23 & 5.7 & 3.89 & 2.75 \\ 1.58 & 2.84 & 3.64 & 3.96 & 2.7 & 1.91 \end{bmatrix}$$

3 Nonlocal Means Algorithm

It is one of the techniques to remove the noise in digital image processing. We realize that the nearby mean filters take the average value as gathering of pixels encompassing a target pixel in the picture. Nonlocal means method takes the mean of all pixels in the picture by contrasting of these pixels within a objective pixel. This method first computes the similarity between two images [9]. This function computes the weighting function of pixels by taking the variance of the two pixels. If the variance of the two images is equal to zero, then the image is affected by less noise. In a similar way, we compare the pixels surrounded by the image. The weights are depending upon the closeness between the areas of the pixel. To find the resemblance among the pixels, we use weighted sum of the squares difference between the neighborhoods. The below figure shows the self-similarity in the image.

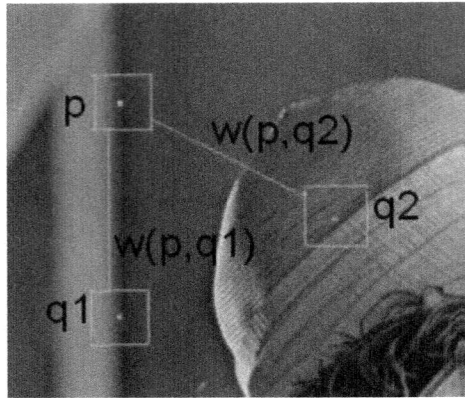

The above figure demonstrates three pixels in the image p, q1 and q2. P and q1 pixels have similarities in the neighborhood, but the pixel's p and q2 do not have the similar neighborhood. That is why q1 pixel will have stronger effect denoise value than p and q2. In the image, head-to-head pixels are mostly having comparable neighborhoods. On the off chance that there is a structure in the picture, non-adjoining pixels also have similar neighborhoods [10]. The pixel contains weighted normal of all the pixels in the picture. In the overhead figure, the heaviness of p and q1 is much superior than p and q2 on the ground that pixels p and q1 have same neighborhoods as pixels p and q2.

Each pixel p in the nonlocal means technique is computed by utilizing the formula

$$NL[v][i] = \sum_{j \in I} w(i,j)v(j)$$

where v is the noisy image and the range of $w(i,j)$ is given by $0 \leq w(i,j) \leq 1$ and

$$\sum_{j} w(i,j) = 1.$$

To figure the similarity between two neighborhoods, take the weighted amount of squares of neighborhood by utilizing the formula.

$$d(i,j) = \left\| V(N_i) - V(N_j) \right\|_{2,F}^{2}$$

For squared difference of neighborhoods, F is applied as neighborhood filter. The weights can be figured by using the formula

$$w(i,j) = \frac{1}{Z(i)} e^{\frac{-d(i,j)}{h}}$$

The standardizing constant can be given as

$$Z(p) = \sum_q e^{\frac{-d(p,q)}{h}}$$

4 Algorithm

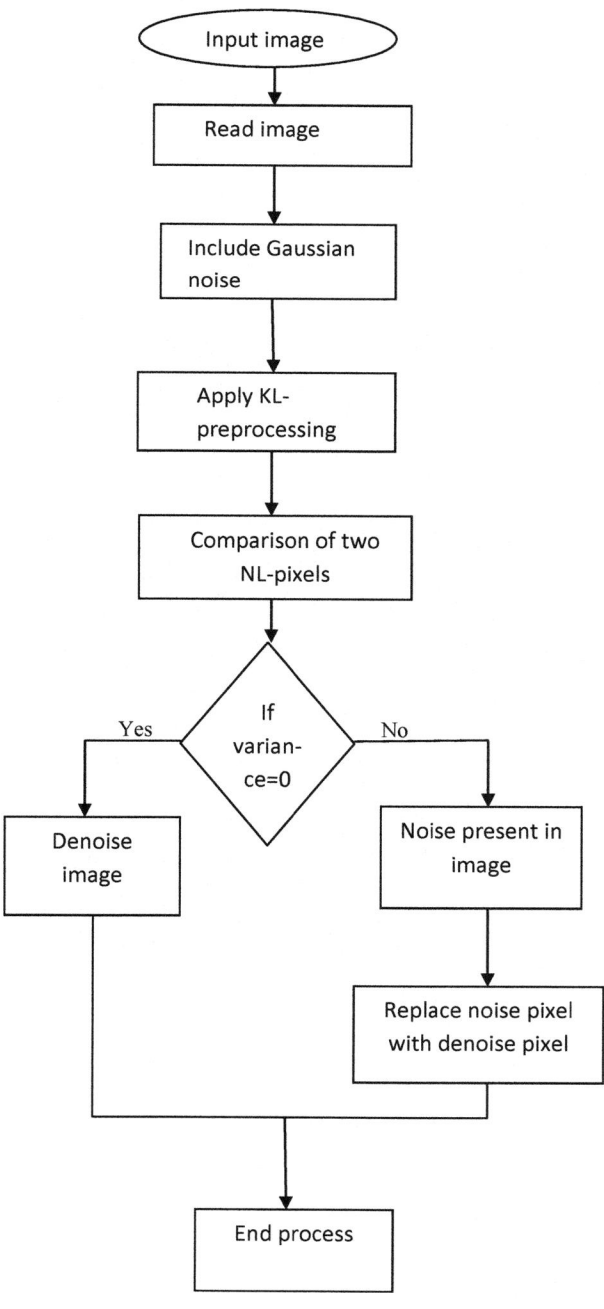

5 Result Analysis

5.1 Subjective Analysis

Gaussian filtering:

Anisotropic filtering:

Neighborhood filtering:

NL means filtering:

5.2 Objective Analysis

Types of images	Gaussian filter			Anisotropic filter			Neighborhood filter			NL means filter		
	PSNR	SSIM	MSE	PSNR	SSIM	MSE	PSNR	SSIM	MSE	PSNR	SSIM	MSE
Lenna	28.96	0.77	121.8	27.55	0.79	190.2	27.38	0.77	111.6	29.78	0.85	68.28
Coins	28.71	0.79	118.0	27.47	0.83	177.7	27.43	0.77	110.6	31.41	0.87	46.94
Cameraman	28.78	0.70	226.8	27.30	0.77	191.5	27.31	0.68	175.0	28.51	0.82	91.44
Boat	28.63	0.69	181.2	27.37	0.69	159.4	27.31	0.68	161.2	28.89	0.76	83.88
Texture	27.40	0.46	157.4	27.01	0.54	142.1	28.96	0.45	245.4	23.74	0.77	74.26

6 Conclusion

This idea speaks to a productive picture denoising based on the nonlocal means algorithm. As we talked about sometime recently, Gaussian convolution is optimum in the smooth parts of the picture but boundaries are blurred. Anisotropic filter is utilized to evacuate the distorting effect at the edges. It acts like a high pass filter, but the image is degraded. In denoising process, we include a noisy picture of which the commotion deviation is equivalent to that of the first noisy picture. It can progress the exactness of discovering comparative blocks by utilizing the nonlocal property of two pictures. In this segment, to check the qualities and performance of NL calculation, a simulation is carried out on the 512 * 512 bit dark scale. All simulations are done in the MATLAB7.0. Experimental results demonstrate the proposed NL means algorithm is effective when compared to local smoothing filters. The work of the picture acquisition and the processing speed are restricted and need to be further improved.

7 Future Scope

To minimize the time complexity, further investigation should be made. In future, the MATLAB can be changed over to VHDL code and actualized on FPGA kit keeping in mind the end goal to create ASIC for picture change.

References

1. An Adaptive Gaussian Filter For Noise Reduction and Edge Detection G. Deng and L. W. Cahill Department of Electronic Engineering, La Trobe University Bandoura Victoria 3083 Australia.
2. P. Perona and J. Malik. Scale space and edge detection using anisotropic diffusion. IEEE Trans. Patt. Anal. Mach. Intell., 12:629–639, 1990.
3. L. Yaroslavsky. Digital Picture Processing - An Introduction. Springer Verlag, 1985.
4. Image denoising using non-local means algorithm Tamanu Thakur 1, Sanjay Bhardwaj 2. Vol. 5, Issue 9, September 2016.
5. Comparative Study of Non-Local Means and Fast Non–Local Means Algorithm for Image Denoising. Pankaj Jain 3 and Mohit Mathura Department of Digital Communication, RGPV University, Bhopal, M.P., India. Communication, RGPV University, Bhopal, M.P., India.

A Novel Multimedia Encryption and Decryption Technique Using Binary Tree Traversal

Annu Priya, Keshav Sinha, Manu Priya Darshani and Sudip Kumar Sahana

1 Introduction

For the past several years, cellular network has increased exponentially and the number of customers connected to those communication networks needs more secrecy while transferring their data through those networks. In recent year, India is becoming a higher subscriber of the telecom networks. By the end of 2016, there are 10 Crore people connected to the telecom industry in the form of "2G", "3G", and "4G" system. Due to the progressive growth in GSM [1] network, it became a most omnipresent cellular network standards, by presently having more than 80% of the world's population. GSM was designed on the basis of multi-frame time division multiple access (TDMA) structure, where the control channels are allotted in fixed time and frequency domain. These channels should be scanned by different subscribers frequently to attaining the essential data to perform a network subscription. Though many people are using GSM as a primary network channel, huge amount of multimedia data has been communicated within it also. However, the distribution of the multimedia content to the open channels is highly insured and vulnerable to those attacks and it is inappropriate for communicating sensitive and valuable multimedia content such as (military, financial or personal) data and so it is

A. Priya (✉) · K. Sinha · S. K. Sahana
Department of Computer Science & Engineering, Birla Institute of Technology,
Mesra, Ranchi, India
e-mail: annu.priya12@yahoo.com

K. Sinha
e-mail: keshav.sinha@yandex.com

S. K. Sahana
e-mail: sudipsahana@bitmesra.ac.in

M. P. Darshani
Department of Physics, Birla Institute of Technology, Mesra, Ranchi, India
e-mail: manupriyadarshani05@gmail.com

© Springer Nature Singapore Pte Ltd. 2019
V. Nath and J. K. Mandal (eds.), *Proceeding of the Second International Conference on Microelectronics, Computing & Communication Systems (MCCS 2017)*, Lecture Notes in Electrical Engineering 476, https://doi.org/10.1007/978-981-10-8234-4_15

needed to be protected by the intruder and an eavesdropper. So data protection is a very high level priority while sending the multimedia contents through any communication channels to do this encryption algorithm [2, 3] which are introduced for multimedia data. A cryptography is one of the important mechanisms for protecting the multimedia data on intranets, extranets, and the Internet. A multimedia encryption is very different than the textual encryption because the size of the data is very large and more time consuming. A encryption technique like AES while taking 2 h for encrypting the 1 h video (or approx. 1 GB as in MPEG format) and similar time is consumed at decryption. The computational overhead is very vast while using a encryption algorithm so that is not suitable for the real-time multimedia communication. Figure 1 shows that how the multimedia content is being encrypted and decrypted using a classical encryption algorithm.

Figure 1 shows that original content is converted into encrypted form by using encryption key and encryption algorithm. And in the same way, the encrypted content is deciphered and converted into the original form using the same key and decryption algorithm. The paper will be primarily concerned with discussing and comparing the efficiency of binary tree based block cipher [4] with the standard symmetric algorithms such as AES, RC5, and RC6 by minimizing the time on compression and CPU overhead.

The motivation for multimedia encryption [5] is the primary concern for any individual user. It is very vital to protect the sensitive data before transmission or distribution. In past years, an access right control technique is used to validate users. The viewer can view the content on pay per view in this the user can provide their username and password to the server for downloading. While doing this process, it will undergo several attacks during transmission.

The organization of the paper is as follows: Sect. 2 describes the literature survey followed by an introduction in Sect. 1. Section 3 discusses research methodology.

Fig. 1 Encryption and decryption of multimedia data

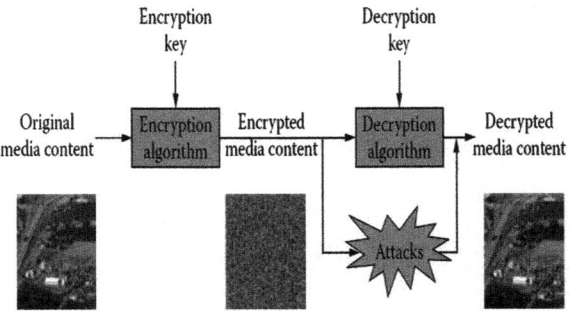

2   Literature Review

There are various methods used for encryption and decryption. Among those, the state-of-the art methods are DES algorithm [6], where it uses the stream header, DC and AC coefficient, frames and blocks. The algorithm is encrypting the 100% video stream. The variation in time may be captured while using DES or RSA. Spanos et al. [7] introduced a selective encryption technique which encrypts the intra-frames, video header, and MPEG format [8] code using the DES algorithm. The selective encryption [9] is given the high bit rate with full encryption. The only disadvantage is that it is not suitable for the distributed environment. In [10], the author proposed a video encryption technique where it splits the whole stream into two halves of random values odds and evens. So we get the ciphertext from two ends, it will reduce the encryption cost and time. Shi et al. [11] introduced the video encryption algorithm (VEA) for encryption of the video files. The algorithm will use the XORed [12] one-time secret key to exchange the sigh bit of DC coefficient. A lot of quality work has been performed by different researchers on multimedia encryption. Some of the methods are cited in Table 1.

Table 1 Chronological summary of research development in the field of multimedia encryption

Technique	Methodology	Method used	Advantage	Disadvantages
Agi and Gong [13]	Encryption offering using frames	Aegis algorithm	Less complexity	Low security
Jakimoski and Kocarev [14]	Use byte operations for linear and differential probabilities	Chaotic block cipher	Less time taken	Reduce the compression
Parzarci et al. [15]	Encrypt the video in the RGB color space, the four secret key transforms	Scrambling technique	Fast and secure	Reduce in compression performance
Zeng and Lie [16]	Selective bit scrambling, and/or rotation	Permutation-based algorithm	Highly compressible	Low entropy
Tang's [17]	Random permutation to remove the order of DCT coefficient	Transformation-based compression algorithm	Less time taken	Less secure
Shin et al. [18]	Using XOR operation for the grayscale image encryption	Binary and XOR	Less time	Low security
Sufyan et al. [19]	Encrypt the file by rabbit algorithm	Special Huffman tree (SHT)	Highly secure	Low entropy

3   Research Methodology

The objective of this work is to design a block cipher which will secure the multimedia contents over communication channels using complete encryption approach. It also reduces the computational time and strengthens the security, and it also overcomes with the transmission delay and overhead during the implementation of conventional algorithm. After study of various research papers, the methodology is developed based on complete encryption approach. During encryption process, a complete multimedia content is compressed and it is represented as a stream of bytes. The binary tree represents the block of stream into tree. The encryption and decryption process is true for any format either video or image.

3.1   Encryption Algorithm

The encryption algorithm consists of four phases: (i) Binary tree construction of plaintext block, (ii) Permutation-based random substitution, (iii) Row-wise matrix representation and, (iv) Column shifting of ciphertext. At first, the multimedia content is divided into blocks of sizes 16, 128, and 512 bits. The encryption algorithm is run for each block of plaintext for 'r' rounds. The framework for encryption is given in Fig. 2. The step-by-step procedure for proposed encryption strategy is discussed as below:

(i)   The first step is to get an original multimedia file. The file is in the form of stream of bytes so combined them to form blocks. Where each block has the size of $n = 2^{2m}$ bits, whereas $m = [2, 4, 6, 8]$ and the blocks are represented as $P_1, P_2, P_3, P_4 \ldots P_n$.

(ii) The next step is a key-based permutation where random permutation process is generated as $\pi = \{p_1, p_2, p_3 \ldots p_n\}$ of $\{1, 2 \ldots n\}$.

(iii) Arrange the generated block ($P_1, P_2, P_3, P_4 \ldots P_n$) as a complete binary tree. Whereas, in root node, the MSB of plaintext block is present and its consecutive bits are added as a left and right child. The LSB of a tree is attached as leaf node. By denoting the resulting tree using $\{B(j, x)$, whereas: 'j' = $\{0, 1, \ldots, l\}$ and l denotes the levels of the binary tree and 'x' denotes the position of a node on each level j.

(iv) Next step is substitution while choosing the node 'x' at the position of p_i whereas ($i = 1, 2, 3 \ldots n$) of the binary tree. The value of node 'x' is shown as Eq. 1.

$$Z = \sum_{j=0}^{i} B(j, x) + \sum_{k} B(k, x) \tag{1}$$

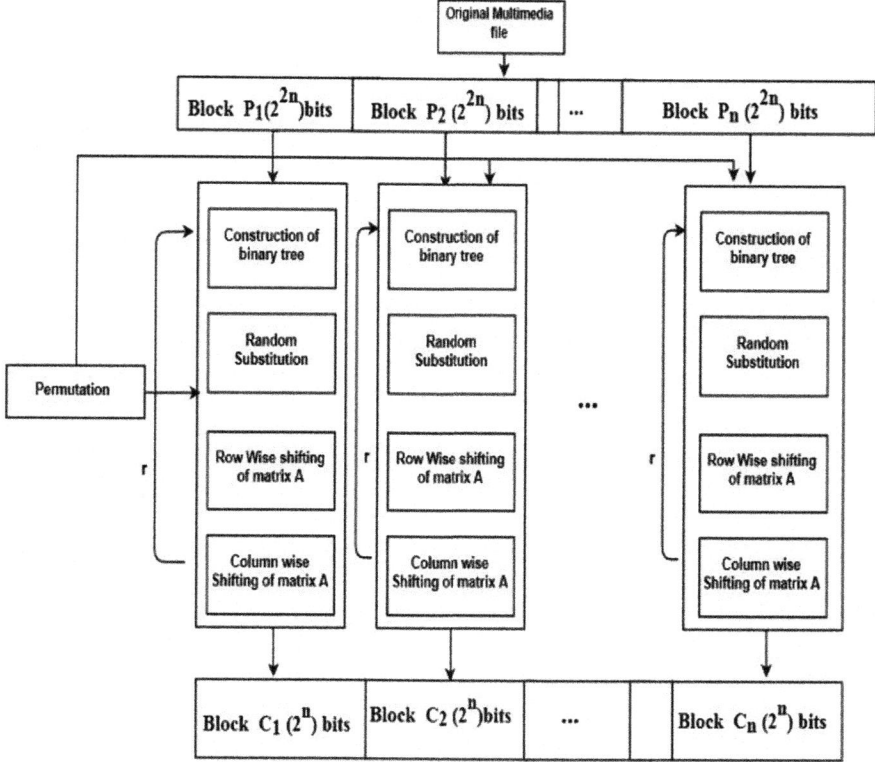

Fig. 2 Framework for the encryption scheme

whereas

$B(j, x)$ Value of node at level 'j' along with root path 'x'.
$B(k, x)$ Value of node 'k' in subtree rooted at 'x'.

If the value of 'Z' is even it is replaced by 0 and if the value of Z is odd it is replaced by 1 at node 'x'. The process is repeated for all the nodes of (P_1, P_2, P_3, P_4 … P_n) and the resultant bit is saved as a pseudo ciphertext C_p. The sum of the all node bit value is denoted by the 'Z'. So it traverses all the subnode of the tree until it reaches at 'x'.

(v) Next step is to construct the matrix A of size.

$$A = \log \sqrt{n} \times \log \sqrt{n}$$

And matrix R of size

$$R = \log \sqrt{n} \times \log \sqrt{n}$$

Arrange the bits of the ciphertext C_p and permutation position 'π' in row and column of 'A', respectively. Calculate the sum of each permutation for 'i' row of R. Whereas

$$S = p_{i1} + p_{i2} + p_{i3} + \ldots + p_i \sqrt{n}$$

While taking 'H' as the hamming distance of each row of A. If the permutation sum is odd then swift row 'i' to the left by H times otherwise shift right by H times. Do this process for all rows of A. After completion of row-wise permutation, the column-wise downward shift starts. For every H time, the sum of the permutation is calculated for column 'R' if the values are even then perform upward shift of 'A', otherwise move downward by H times. Do this process for all columns of 'A' matrix. And the resultant matrix is denoted by B.

(vi) Repeat the process of (iii) and (v) for 'r' times and store the ciphertext block in the C_1.

(vii) For every plaintext block, we will get the ciphertext while repeating the steps of (iii) through (vi) so that every plain block $P_2, P_3, \ldots, P_n$ will get the cipher blocks of $C_2, \ldots, C_n$.

3.2 Decryption Algorithm

Decryption is the reverse process of encryption where the cipher file is converted into original file. The framework for the decryption process is shown in Fig. 3.

Step-by-step process for Decryption of multimedia content is given as below.

(i) The first step is to get encrypted multimedia content and partitioned into a stream of bytes. Where each byte is combined to form a block, each block has $n = 2^{2m}$ bits, whereas $m = [2, 4, 6, 8]$. The encrypted file is presented as $C_1, C_2, C_3, \ldots C_n$.

(ii) The next step is the reverse permutation whereas

$$\pi \text{ as } \{p_n, p_{n-1}, \ldots, p_2, p_1\}$$

(iii) The next phase is to construct the encrypted data matrix of size.

$$B = \log \sqrt{n} \times \log \sqrt{n}$$

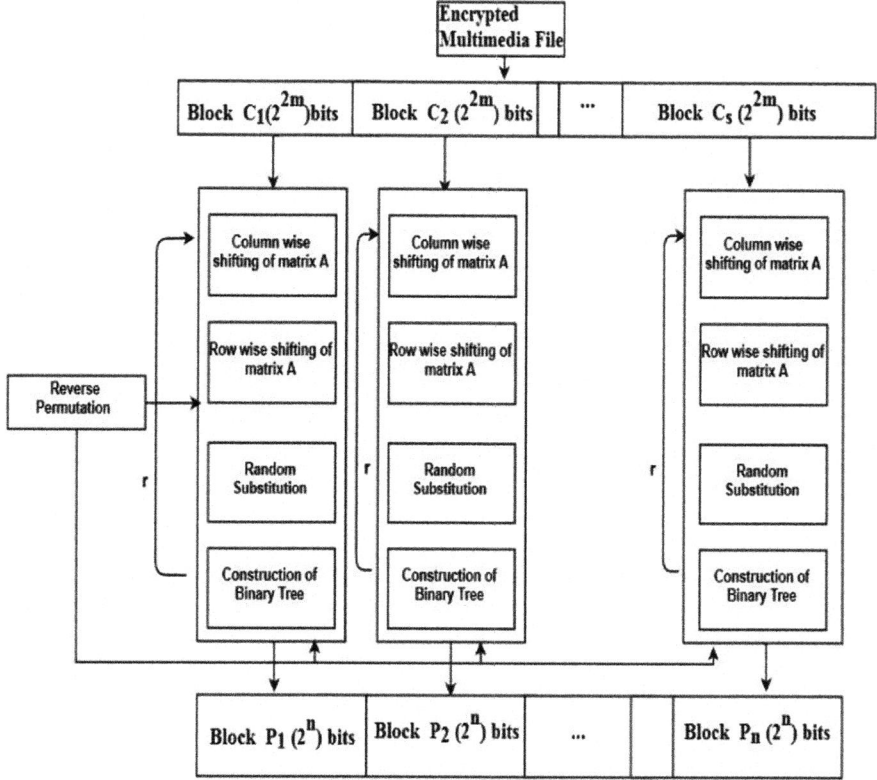

Fig. 3 Framework or the decryption scheme

And matrix R of size

$$R = \log \sqrt{n} \times \log \sqrt{n}$$

By ordering the bits of C_1 and permutation position 'π', respectively. The column- and row-wise shifting of B is performed by calculating the sum for each 'j' column of R.
whereas

$$S = p_{j1} + p_{j2} + p_{j3} + \ldots + p_j \sqrt{n}$$

where H is the humming distance of jth column of B. If the permutation sum is odd then shift the column to be downwards or else shift the jth column of B upwards by H times. Do the same process for all the columns of B. After completion of column-wise shift the ith row of the B is taken. If the sum of the permutation is even then shift the ith row of B to the left by H times or else shift right by H times. Do this for all the rows of B. and the resultant matrix is created as 'A'.

(iv)   Then the next step is to construct the binary tree of the matrix 'A' which is same as the encryption scheme.

(v)   For the binary tree of the matrix 'A', we apply the random substitution by selecting the node 'x' at position p_i. Whereas, $i = \{n, n-1...1\}$ of the tree. Equation 2 shows that how the value is replaced from the node 'x'.

$$Z' = \sum_{j=0}^{i} B(j, x) + \sum_{k} B(k, x) \tag{2}$$

whereas

$B(j, x)$ Node on level 'j' and the path of root node to 'x'.
$B(k, x)$ Value at node 'k' in the subtree at 'x'.

If the value of Z' is even then node x is replaced by '0' and if it is odd then it is replaced by '1'. Repeat this process for all the nodes at positions $p_n, p_{n-1}...p_2, p_1$ the resultant bits are formed as a pseudo plaintext P_c.

(vi)   Repeat the process of (iii) to (v) for 'r' number of times to get the original text block P_1.

(vii)   And repeat the step (iii) to (vi) for generating the remaining ciphertext block of the original multimedia file $C_2, C_3..., C_n$.

4   Experimental Setup and Measurement Model

Simulation work has been performed using Matlab toolbox version R2015a on Windows 8.1 platform having hardware specification: Processor Intel core i5 CPU frequency 1.70 GHz single core processor 4 GB RAM on ×64 bit operating system for benchmark image Lenna. The complexity of proposed encryption and decryption algorithms take $O(n)$ time. For creating every block of bits, it will take $O(n)$ time and for generating the pseudo permutation it will take $O(1)$ time. For substitution of each n elements, it will take $O(1)$ time. For the row- and column-wise computation, it will take $\sqrt{n}$ times.

The correlation coefficient is setup according to the accuracy and the performance of the proposed method. If the correlation coefficient for original image is of high value and for the encrypted image it would be low then we consider that system is strong one. Measurement model of the encrypted multimedia file is framed and shown in Eq. 3:

$$r_{xy} = \frac{\text{Cov }(x,y)}{\sigma_x \sigma_y} \qquad (3)$$

whereas

x and y Intensity value of two adjacent pixels of the image.
r_{xy} correlation coefficient.

The cov (x, y) and $D(x)$ are given as Eq. 4:

$$\text{Cov}(x,y) = \frac{1}{y} \sum_{i=1}^{n} (x_i - E(x))(y_i - E(y)) \qquad (4)$$

whereas

$$E(x) = \frac{1}{n} \sum_{i=1}^{n} (x_i)$$

$$\sigma_x = \sqrt{\frac{1}{n} \sum_{i=1}^{n} (x_i - E(x))^2}$$

whereas

$$E(y) = \frac{1}{n} \sum_{i=1}^{n} (y_i)$$

$$\sigma_y = \sqrt{\frac{1}{n} \sum_{i=1}^{n} (y_i - E(y))^2}$$

whereas

E is an expectation.
σ_x Standard Deviation of x.
σ_y Standard Deviation of y.

5 Result and Discussion

The proposed algorithm is used for all types of the multimedia files such as video and image. But the performance is measured separately for image and video files. Because of their features, the video file is stored in the form of frames and each frame has moving pictures. Whereas the image file is stored in a pixel format. The performance metrics for the video file are (i) Execution time, (ii) CPU utilization and (iii) Memory Consumption. And for the image file, the performance is measured by statistical analysis on pixel values.

Test Case 1: Video Files

Encryption and decryption of video file requires to satisfy the certain parameters such as: (i) Memory usage, (ii) Encryption and decryption time consumption, and (iii) CPU utilization.

1. Performance with Encryption and Decryption Time

For encryption and decryption, the 128-bit block size is considered. The input files are of different sizes such as (6.8, 8.3, 10.6, 15.6, 20.4) MB. The proposed algorithm will run for 5 rounds and the result is compared with the AES, RC5, and RC6. The encryption value is cited in Table 2 whereas the decryption value is cited in Table 3.

For the different sizes of the video format, a simulation test is performed and it is observed that the proposed algorithm exhibits better result than other existing algorithms. The graphical representation encryption/decryption time comparison is shown in Figs. 4 and 5.

Table 2 Encryption time comparison

Encryption time in milliseconds				
File size	AES	RC5	RC6	Proposed method
6.2	1486	1725	1948	685
8.3	1512	1826	2012	764
10.8	1695	1927	2156	978
15.6	1782	2018	2258	1024
20.4	1823	2123	2321	1122

Table 3 Decryption time comparison

Decryption time in milliseconds				
File size	AES	RC5	RC6	Proposed method
6.2	1502	1786	1979	695
8.3	1532	1879	2052	784
10.8	1701	1958	2178	986
15.6	1798	2058	2274	1068
20.4	1857	2148	2389	1162

Fig. 4 Graphical representation of encryption time comparison with AES, RC5, RC6, and proposed algorithm

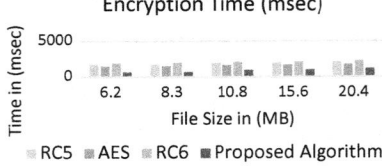

Fig. 5 Graphical representation of decryption time comparison of the proposed algorithm with AES, RC5, and RC6

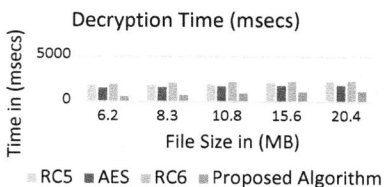

2. Performance with Memory Usage

During the execution of proposed algorithm used the physical memory to store the input data, intermediate value, and the ciphertext. So to calculate the memory usage we follow some steps such as

(i)   Initialization of the memory location.
(ii) Execute the algorithm.
(iii) Read the memory and find the difference between free and available space.
(iv) Read the garbage memory.
(v)   Subtract the memory consumed by the encryption code with the garbage space.

Here, we show the memory consumption by the proposed algorithm and compare the result with the existing algorithm. For simulation, the video file size has been taken as (6.2, 8.3, 10.8, 15.6 and 20.4) MB for the block size of 128 bits. The graphical representation is shown in Fig. 6.

The bar graph shows that the proposed algorithm consumes less memory usage than the AES, RC5, and RC6.

3. Performance with CPU Utilization

To measured the performance of the encryption algorithm, first we have to calculate that how much cpu memory is used by algorithm for reading and writting process. To do so, we calculate the CPU utilization of the existing algorithm AES, RC5, RC6, and the proposed algorithm for the input files of size (6.2, 8.3, 10.8, 15.6 and 20.4) MB for the block size of 128 bits. The graphical representation of CPU utilization is shown in Fig. 7.

The bar graph shows that proposed algorithm loads very low data to the CPU compare to the existing algorithm AES, RC5, and RC6.

Fig. 6 Graphical representation of memory usage

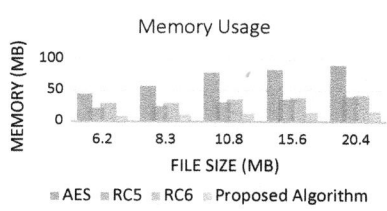

Fig. 7 Graphical
representation of the CPU
utilization

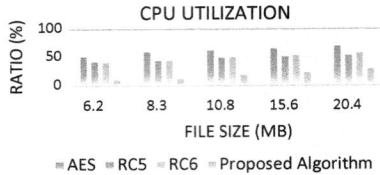

Test Case 2: Image File

(i) **Performance Analysis of Image Files**

During encryption of image file, the values are stored in the pixels form. In this paper, we use the benchmark image of Lenna having the 512×512 pixel.

Figures 8 and 9 show that the encrypted image is visually different than the original image. Whereas to measure the performance of the proposed algorithm, the correlation coefficient is used. It calculates the position of the pixel in the encrypted image. We have calculated the coefficient in horizontal, vertical, and diagonal adjacent pixels for both original and encrypted image is shown in Table 4.

Fig. 8 Original image

Fig. 9 Encrypted image

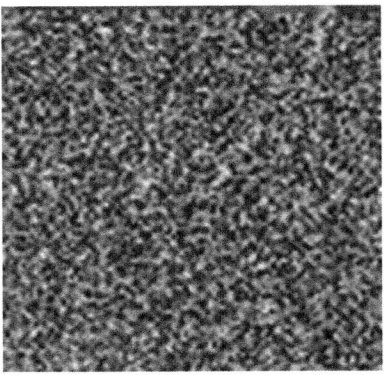

Table 4 Correlation coefficient of image

Image	Horizontal pixel	Vertical pixel	Diagonal adjacent pixels
Original image	0.9816	0.9858	0.9712
Encrypted image	0.01776	0.04912	0.00348

The results cited in Table 4 show that the correlation coefficients of the encrypted images are low than the original image. There are 10,000 randomly chosen pairs of pixels from (vertical, horizontal and diagonal). So we use to plot the graph which is shown in Figs. 10, 11 and 12. It is evidence that the correlation value of original image is high and concentrated in certain direction whereas in the case of encrypted image, the correlation value is low so it is scattered in almost all the regions. Figures 13, 14, and 15 show the correlation pixel value of the encrypted image.

Fig. 10 Correlation value of horizontal adjacent pixel of original image

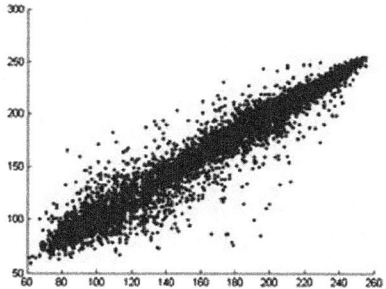

Fig. 11 Correlation value of vertical adjacent pixel of original image

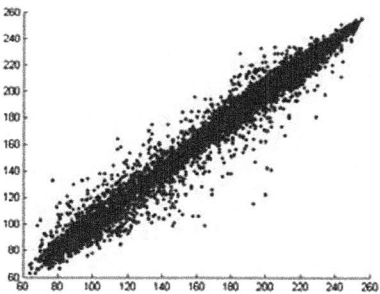

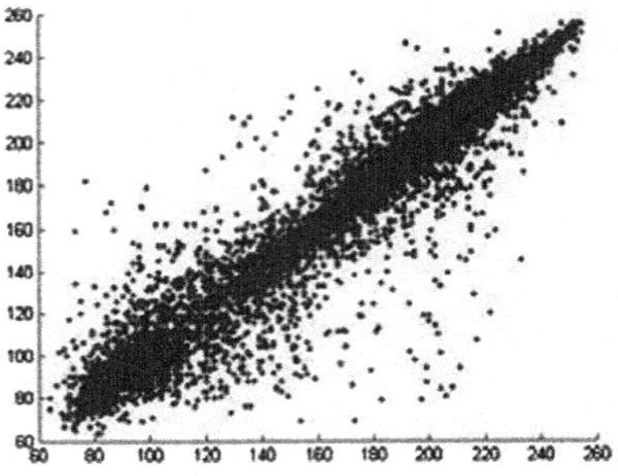

Fig. 12 Correlation value of
diagonal adjacent pixel of
original image

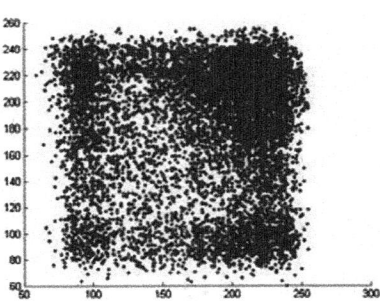

Fig. 13 Correlation value of
horizontal adjacent pixel of
encrypted image

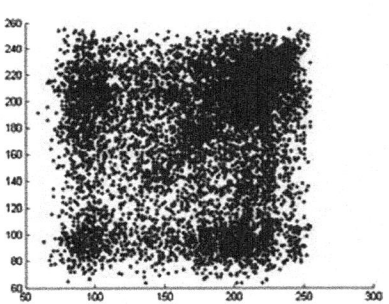

Fig. 14 Correlation value of
vertical adjacent pixel of
encrypted image

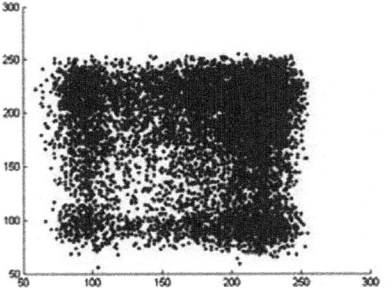

Fig. 15 Correlation value of
diagonal adjacent pixel of
encrypted image

6 Conclusions

The novelty of this paper is the implementation of binary tree based encryption and decryption algorithm for both image and video file. From the result section, it is evident that the proposed algorithm has a better performance than existing AES, RC5, and RC6 algorithms. The correlation coefficient exhibits better results while matching original image with the encrypted image for proposed algorithm. The implementation using binary tree be possible for large file problems and the trust area of the computer vision as per the proposed algorithm and result can be analyzed to find out the scope of the binary tree based encryption for the aforesaid application areas.

References

1. Keshav Sinha, Manu Priya Darshani, Shweta Kumari: "Enhanced, Efficient End-to-End Voice Encryption Using A5/3RC6 over GSM Network", Imperial Journal of Interdisciplinary Research (IJIR), 2017, vol-3, Issue-1.
2. Ali Saman Tosun, W-C Feng Tosun, "On Error Preserving Encryption Algorithms for Wireless Video Transmission", Proceedings of ACM International Conference on Multimedia, Ottawa, Canada, 2001, pp. 302–308.
3. E Alvarez, A Fernández, P García, J Jiménez, A. Marcano, "New approach to chaotic encryption", A Physics Letters, 1999, vol 263, no. 4–6, pp. 373–375.
4. Bruce Schneier, John Kelsey, Doug Whiting, David Wagner, Chris Hall, Niels Ferguson, "Twofish: A 128-Bit Block Cipher", June 1998.
5. Borko Furht, Daniel Socek, "Multimedia Security: Encryption Techniques," IEC Comprehensive Report on Information Security, International Engineering Consortium, Chicago, IL, 2003.
6. J. Meyer and F. Gadegast, "Security mechanisms for multimedia data with the example MPEG _ 1 video", Project Description of SECMPEG, Technical University of Berlin, Germany (1995).
7. G. A. Spanos and T. B Maples, "Security for Real-Time MPEG Compressed Video in Distributed Multimedia Applications", in Conference on Computers and Communications, 1995, pp. 72–78.
8. Changgui Shi, Bharat K. Bhargava, "A fast MPEG video encryption algorithm," in Proceedings of the 6th ACM International Conference on Multimedia, Bristol, UK, 1998, pp. 12–16.
9. Alattar and G. Al-Regib, "Evaluation of selective encryption techniques for secure transmission of MPEG video bit-streams", Proceedings of the IEEE International Symposium on Circuits and Systems, 1999, vol. 4, pp. 340–343.
10. Lintian Qiao, Klara Nahrstedt, "A new algorithm for MPEG video encryption," in Proceedings of the 1st International Conference on Imaging Science, Systems, and Technology (CISST 97), Las Vegas, USA), 1997, pp. 21–29.
11. Chang Gui Shi, Bharat K. Bhargava, "A fast MPEG video encryption algorithm", Proceedings of the 6th ACM International Conference on Multimedia, Bristol, UK, 1998, pp. 12–16.
12. Amitava Nag, Jyoti Prakash Singh, Srabani Khan, Saswati Ghosh, Sushanta Biswas, D. Sarkar Partha Pratim Sarkar, "Image Encryption Using Affine Transform and XOR Operation", International Conference on Signal Processing, Communication, Computing and Networking Technologies (ICSCCN 2011), pp. 309–312.

13. Iskender Agi and Li Gong, "An empirical study of secure MPEG video transmission," in Internet Society Symposium on Network and Distributed System Security, San Diego, Calif, USA, 1996, pp. 137–144.
14. G. Jakimoski, L. Kocarev, "Chaos and cryptography: block encryption ciphers based on chaotic maps", IEEE Transactions on Circuits and Systems Fundamental Theory and Applications, 2001, vol. 48, no. 2, pp. 163–169.
15. M. Pazarci, V Dipcin, "A MPEG2-Transparent Scrambling Technique", IEEE Transactions on Consumer Electronics", 2002, vol. 48, no. 2, pp. 345–355.
16. Wenjun Zeng, Shawmin Lei, "Efficient frequency domain selective scrambling of digital video," IEEE Transactions on Multimedia, 2003, vol. 5, no. 1, pp. 118–129.
17. Lei Tang, "Methods for encrypting and decrypting MPEG video data efficiently", Proceedings in the 4th ACM International Multimedia Conference & Exhibition, Boston, Mass, USA, 2009, pp. 219–229.
18. Chang-Mok Shin, Soo-Joong Kim, "Phase-Only Encryption and Single Path Decryption System Using Phase-Encoded Exclusive-OR Rules in Fourier Domain", Optical Review, 2006, vol 13, no. 2, pp 49–52.
19. Sufyan T. Faraj Al-Janabi, Khalida Shaaban Rijab, Ali Makki Sagheer, "Video Encryption Based on Special Huffman Coding and Rabbit Stream Cipher", Developments in E- systems Engineering, 2011, pp. 413–418.

ZC-CDTA Based Integrator Circuit Using Single Passive Component

Malladi Lakshmi Lavanya, Avireni Srinivasulu and V. Venkata Reddy

1 Introduction

Integrator using an operational amplifier (Op-amp) is a circuit that performs the integration operation with respect to its time, i.e., the output and the input voltages are directly proportional to each other. Most of the applications like analog to digital converters and wave shaping circuits are designed by incorporating the integrator circuits in their design. It is true that in analog signal dealing processes, the integrator is widely used as a building block. Nowadays integrator is being used in some current form and voltage form applications. If used in current form applications, promoting it helps in reduction of circuit complexity by high speed, slew rate, and bandwidth in the circuit. In most of the applications like communication and instrumentation systems and also in wave processing, active RC integrators are comprehensively used [1–3]. Some integrators that are designed with switched capacitors [4], so that the design area will be reduced. For a sampling of such circuits, extra clock circuit is needed. In general, switched capacitors are not helpful in high-frequency applications. In addition to switched capacitors, the switched resistors are also employed in the design of integrators [5]. The major disadvantage is that it using resistors makes the design more complicated. The fabrication cost will also increase. Voltage mode circuits are used in applications where linearity and less complexity are required. Prior to the ZC-CDTA active block, there are some active building blocks already designed. A some of them are operational transconductance amplifier (OTA) [6, 7], current conveyors (CC) [8],

M. L. Lavanya · V. Venkata Reddy
Electronics and Communication Engineering, VFSTR University (Vignan's University),
Vadlamudi, Guntur 522213, Andhra Pradesh, India

A. Srinivasulu (✉)
Department of Electronics and Communication Engineering, JECRC University,
Jaipur 303905, Rajasthan, India
e-mail: avireni_s@yahoo.com

© Springer Nature Singapore Pte Ltd. 2019
V. Nath and J. K. Mandal (eds.), *Proceeding of the Second International Conference on Microelectronics, Computing & Communication Systems (MCCS 2017)*, Lecture Notes in Electrical Engineering 476, https://doi.org/10.1007/978-981-10-8234-4_16

second generation current conveyors (CCII) [9–17], operational transresistance amplifier (OTRA) [18–21], current differencing buffered amplifier (CDBA) [22], [23], current differencing transconductance amplifier (CDTA) [24, 25], and Z-copy current differencing transconductance amplifier (ZC-CDTA). Majority of the filter applications is introduced by using these CDTA [26–28] and ZC-CDTA [29] devices.

Current mode applications are also built into an integrator process like CCII, OTA, and CDTA. However, the combination of Op-amp and OTA integrators is also designed [30, 31].

CDTA is a neoteric current form active block device propounded in 2003 by Biolek [32]. The designing of CDTA includes CDBA [33, 34] along with OTA. CDBA and OTA are the superior building blocks of Op-Amps. CDTA is the unique active element, which does not depend on parasitic input capacitances. Because of its nature of operation, it can be used in an extended frequency range. CDTA is helpful in some low current and voltage mode [35] applications, particularly in the region of filtering circuits. In the case of the CDTA, the impedance is low at the input and high at the output terminals. The current difference of the input terminals moves toward the output terminal into an exterior load. The main difference between the CDTA and ZC-CDTA is its Z-copy terminal. This terminal helps to copy the current through the floating branches. This ZC-CDTA is extensively used in filters because it is the most advanced current mode active building block. In this paper, the integrator is designed based on ZC-CDTA.

Z-Copy CDTA is the most recent current form active element. This type of circuits can be used in current mode as well as in voltage mode. It is the personalized version of CDTA. The circuit construction has two stages. The current differencing unit is the preliminary stage and the second stage involves dual output transconductance amplifier.

Monte Carlo is used for the execution of geometric analysis to generate the information about the performance of the circuits. In this paper, Monte Carlo analysis is performed to know the information about the mismatch and process variations. For each iteration, it uses different constraint values in each device used in the circuit. This simulation will also give the results that will be known how the manufacturing tolerances affect the overall production yield of the design. The Monte Carlo analysis is simulated by considering the appropriate number of bins in a frequency distribution. The number of bins taken depends on the number of points and the starting run up the number.

2 ZC-CDTA Fundamentals

Figure 1 is the symbolic representation of the ZC-CDTA. The internal transistor structure of Fig. 1 is revealed in Fig. 2. The ZC-CDTA is characterized by the matrix as given below. From this correlation between the input and output terminals can be determined.

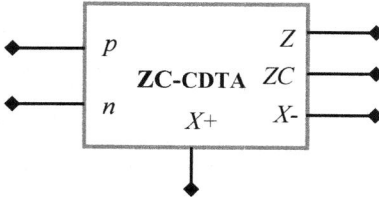

Fig. 1 Symbol for ZC-CDTA

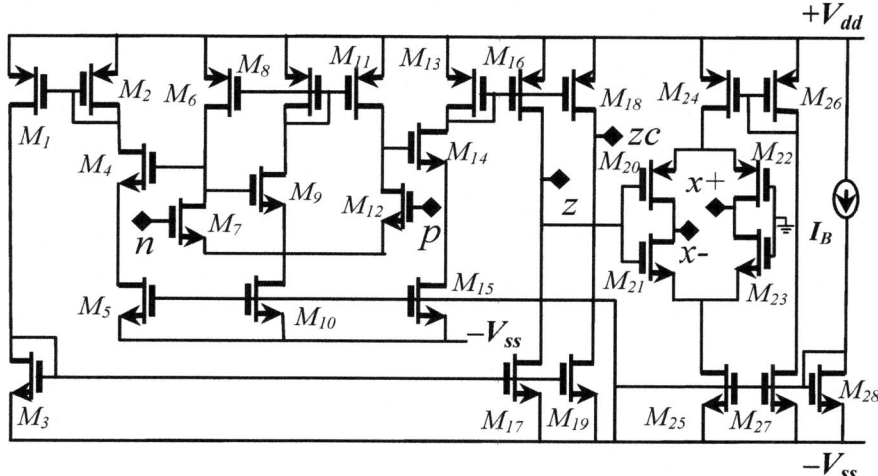

Fig. 2 Transistor level circuit diagram

$$\begin{bmatrix} V_p \\ V_n \\ I_Z \\ I_{ZC} \\ I_{X+} \\ I_{X-} \end{bmatrix} = \begin{bmatrix} 0 & 0 & 0 & 0 & 0 & 0 \\ 0 & 0 & 0 & 0 & 0 & 0 \\ 1 & -1 & 0 & 0 & 0 & 0 \\ 1 & -1 & 0 & 0 & 0 & 0 \\ 0 & 0 & g_m & 0 & 0 & 0 \\ 0 & 0 & -g_m & 0 & 0 & 0 \end{bmatrix} \begin{bmatrix} I_p \\ I_n \\ V_z \\ V_{ZC} \\ V_{X+} \\ V_{X-} \end{bmatrix} \qquad (1)$$

In the above matrix, the transconductance of the ZC-CDTA is represented by g_m. The ideal conditions of ZC-CDTA are

$$V_p = V_n = 0, \, I_z = I_{ZC} = I_p - I_n \text{ and } I_{X+} = -I_{X-} = g_m V_Z \qquad (2)$$

By using, the external bias current I_b and the aspect ratios of the transistors M_{21} and M_{23}, the transconductance of the ZC-CDTA can be controlled, which can be specified as

$$g_m = \sqrt{2\mu \, C_{ox} \left(\frac{W}{L}\right) I_b} \qquad (3)$$

The p and n terminals of Fig. 1 are having low impedance, the assisting terminal is Z, the high impedance output terminals are treated as $X+$ and $X-$ terminals. Moreover, the transconductance of ZC-CDTA is represented by g_m. ZC-CDTA has an added terminal ZC, identified as Z-copy and the difference of input terminals current is its outflow current. The current magnitudes are identical but varied in their current flow and also, the Z terminal voltage and the transconductance (g_m) multiplication gives the magnitudes. The matrix represented above in Eq. (1) gives the perfect conditions of the ZC-CDTA. The copy of the current at Z terminal makes available flexibility to the circuit designer.

The aspect ratios of the transistors used in Fig. 2 are specified in Table 1.

3   Proposed Integrator Circuit

The proposed integrator circuit having single ZC-CDTA is publicized in Fig. 3 and its transfer characteristics in Fig. 4. In this integrator circuit, the input is given to p terminal and considered its $X-$ terminal as the output, for which the capacitor is connected. Feedback path is affixed from the Z terminal toward the input terminal p. The ZC terminal, n, and the $X+$ terminals are linked to the ground.

Table 1 ZC-CDTA aspect ratios

Transistor numbers	W (μm)	L (μm)
M_1, M_4, M_5	3.0	0.6
M_2, M_6, M_8	10	1.0
M_3	25	2.0
M_7, M_9	3.0	0.6
M_{10}, M_{15}	2.0	0.18
M_{11}	16	1.0
M_{12}	3.0	0.6
M_{13}, M_{20}, M_{21}	16	1.0
M_{14}	40	0.5
M_{16}, M_{18}	6.0	1.0
M_{17}, M_{19}	10	1.0
M_{22}, M_{23}, M_{24}	6.0	1.0
M_{25}	12	2.0
M_{26}	6.0	1.0
M_{27}	20	2.0
M_{28}	10	2.0

Fig. 3 Proposed ZC-CDTA
integrator circuit

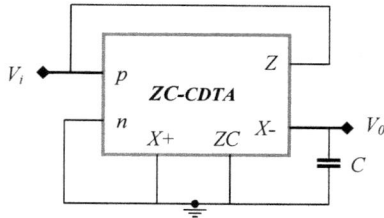

Fig. 4 Input and output
waveforms of the proposed
circuit

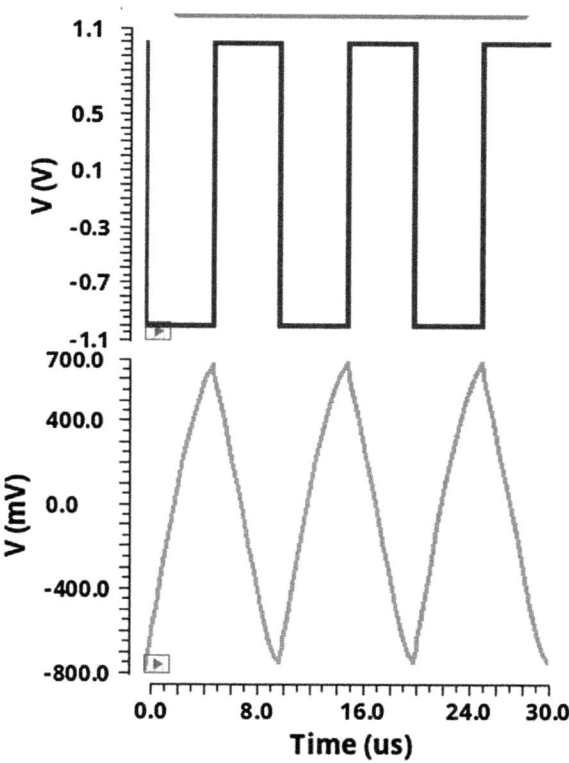

Input square waveform is applied at the p terminal. The current at the Z terminal
and p terminal are equal. One capacitor is connected in parallel to output terminal
X_-. The output amplitudes are linearly adjustable by bias current I_b. When the
square input is negative, the capacitor goes on charging and whenever the square
input is positive the capacitor acts as discharging agent.

From the idyllic conditions, which are specified in Eq. (2), the proposed circuit
transfer function can be written as

$$\frac{V_{o}}{V_{in}} = -\frac{g_m}{sC} \tag{4}$$

4 Simulation Results

By using Cadence and the model parameters of the 180 nm CMOS process, the results of the projected circuit are simulated. The supply rail voltage is ± 0.9 V and the biasing current has been taken as $I_b = 100$ μA. The capacitor value is considered as 400 pF. Square wave input is applied with ± 1 V_{pp} at a frequency of 100 kHz. The output is obtained as a triangular wave by giving a square wave as its input, which is revealed in Fig. 4.

The Monte Carlo evaluation is performed to achieve numerical data on the process variation, mismatch variation and both inclusive. The Monte Carlo analysis of the proposed circuit is performed by means of the combination of both process and mismatch variations for 200 samples with a supply voltage of ± 0.9 V at the output. The Monte Carlo transient analysis result in Fig. 3 is shown in Fig. 5. The Z-Copy output noise measurement of the proposed CDTA-based integrator is depicted in Fig. 6 [36].

Table 2 provides the comparative analysis of the candidate design and it is scrutinized to facilitate the proposed integrator circuit that requires only one active and one passive components.

Fig. 5 Monte Carlo result for the transient analysis of the proposed integrator output

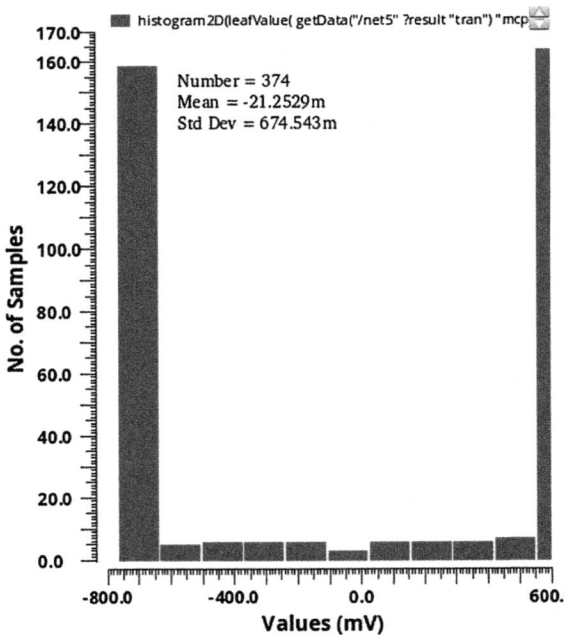

Fig. 6 Noise analysis of the proposed integrator circuit

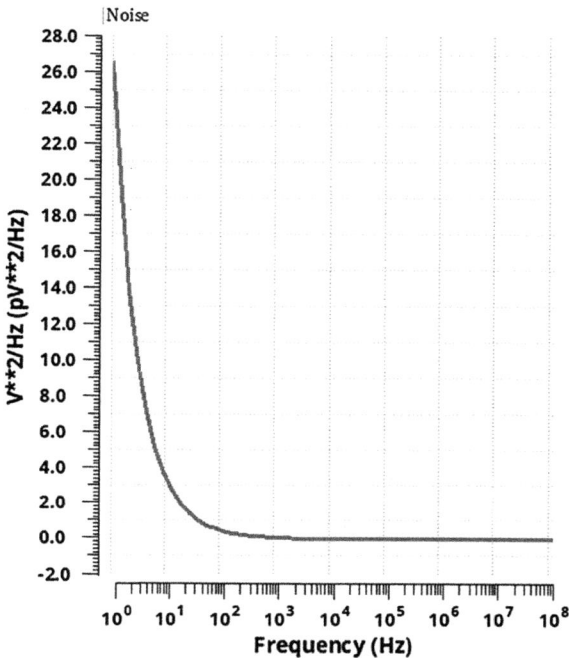

Table 2 Comparative analysis of candidate designs

Reference	Active element type	No. of active element(s)	No. of passive components
Sedra and Smith [2]	Op-amp	1	2
Lee and Tsao Fig. 5 [9]	CCII	2	4
Lee and Tsao Fig. 6 [9]	CCII	1	2
Patranabis and Ghosh Fig. 1(b) [10]	CCII	1	2
Chiu et al. [18]	OTRA	1	1
Proposed circuit of Fig. 3	ZC-CDTA	1	1

5 Conclusion

A new integrator circuit is implemented by using ZC-CDTA as an active block and a single capacitor. The circuits are implemented with 180 nm technology at the supply voltage ±0.9 V. The transient response and the Monte Carlo simulation results of the proposed integrator circuit are provided in this paper. The proposed Fig. 3 circuit is more appropriate in IC design applications, and look forward to be useful in current mode approach analog applications, mechanical systems, waveform generator applications, oscillators and other circuits.

References

1. J. Miliman and H. Taub, Pulse, Digital and Switching Waveforms. New York: MC Graw-Hill. 1976.
2. A. Sedra and K. C. Smith, Microelectronic Circuits, 5th ed. London, U.K.: Oxford Univ. Press, 1998, pp. 105–112.
3. R. L. Geiger and G. R. Bailey "Integrator design for high frequency active filter applications", *IEEE Trans. Circuits and Systems*, vol. CAS-29, no. 9, pp. 595–603, Sep 1982.
4. A. A. Tutyshkin, A. S. Korotkov "Current conveyor based switched-capacitor integrator with reduced parasitic sensitivity", in *Proc. IEEE International Conference on Circuits and Systems for Communications*, pp. 78–81, 2002.
5. A. Jiraseree-amornkun, A. Worapishett, E. A. M. Klumperink, B. Nauta, W. Surakampontorn, "Slew rate induced distortion in switched-resistor integrators", in *Proc. IEEE International Symposium on Circuits and Systems*, pp. 2485–2488, May 2006.
6. Won-Sup Chung, Hoon Kim, Hyeong-Woo Cha, Hee-Jun Kim, "Triangular/Square-Wave generator with independently controllable frequency and amplitude", *IEEE Transactions on Instrumentation and Measurement*, vol. 54, No. 1 pp. 105–109, Feb 2005.
7. M. Kumngerrn, "Realization of electronically tunable first-order all-pass filter using single-ended OTAs", in *Proc. of the IEEE Symposium on Industrial Electronics and Applications* (ISIEA), Bandung, Indonesia, pp. 100–103, Sep 23–26, 2012.
8. A. S. Sedra, K. C. Smith, "A second generation current conveyor its application", *IEEE Transactions on Circuit Theory*, vol. 17, no. 1, pp. 132–134, 1970.
9. Jiunn-Yih Lee and Hen-Wai Tsao, "True RC integrators based on current conveyors with tunable time constants using active control and modified loop technique", *IEEE Transactions on Instrumentation and Measurement*, vol. 41, no. 5, pp. 709–714, 1992.
10. D. Patranabis and D. K. Ghosh, "Integrators and differentiators with current conveyors," *IEEE Transactions on Circuits and Systems*, vol. CAS-31, no. 6, pp. 567–569, 1984.
11. A. Srinivasulu and D. Pal, "CCII+ based novel waveform generator with grounded resistance/capacitor for tuning", 21st *IEEE Applied Electronics International Conferences* (SEIC-2016), Pilsen, Czech Republic, pp. 247–252.
12. Fathi. A. Farag, "CMOS current mode integrator and differentiator for low voltage and low power applications", *Journal of Engineering sciences*, vol 42, no 1, pp. 149–164, Jan 2014.
13. Md. H. Maghami, A. M. Sodagar, "Fully-integrated, large-time-constant, low-pass, Gm-C filter based on current conveyors", in *proc. of the IEEE International Conference on Electronics, Circuits and Systems* (ICECS), Beirut, pp. 281–284, Dec 11–14, 2011.
14. D. Pal, A. Srinivasulu, M. Goswami, "Novel current mode waveform generator with independent frequency and amplitude control", in *Proc. IEEE Internation Symposium on Systems and Circuits*, ISCAS-2009, pp. 2946–2949.
15. M. T. Abuelmaatti, M. Ahmad. Al-Absi, "A current conveyor based relaxation oscillator as versatile electronic interface for capacitive and resistive sensors", *International Journal of Electronics*, vol. 92, issue 8, pp. 473–477, 2005.
16. Avireni Srinivasulu, "Current conveyor based relaxation oscillator with tunable grounded resistor/capacitor", *International Journal of Design, Analysis and Tools for Circuits and Systems* (Hong-Kong), vol. 3, no. 2, ISSN: 2071-2987, pp. 1–7, 2012.
17. A. Srinivasulu, "Current conveyor based square-wave generator with tunable grounded resistor/capacitor", in *Proc. of the IEEE Applied Electronics International Conference* (IEEE AEIC-09), Pilsen, Czech Republic, Sep 9–10, 2009, pp. 233–236, ISSN: 1803-7232.
18. Wenwei Chiu, Jiann-Horng Tsay, Shen-Iuan Liu, Hen-Wai Tsao and Jiann-Jong Chen, "Single-capacitor MOSFET-C integrator using OTRA", Electronics Letters, vol. 31, no. 21, pp. 1796–1797, 1995.
19. Lo. Yu-Kang, Hung-Chun Chien, "Switch controllable OTRA based square/triangular waveform generator", *IEEE Transactions on Circuits and Systems-II*, vol. 54, no. 12, pp. 1110–1114, Dec 2007.

20. P. Chandra Shaker, Avireni Srinivasulu, "Quadrature oscillator using operational transresistance amplifier", *in Proc. of the IEEE Applied Electronics 2014 International Conference* (IEEE AEIC-14), Pilsen, Czech Republic, ISSN: 1803–7232 pp. 117–120, Sep 9–10, 2014.

21. A. Srinivasulu and P. Ch. Shaker, "Grounded resistance/capacitance-controlled sinusoidal oscillators using operational transresistance amplifier", *WSEAS Transactions on Circuits and Systems*, vol. 13, pp. 145–152, 2014.

22. R. Nandi, Mousiki Kar, Sagarika Das, "Electronically tunable dual input integrator employing a single CDBA and multiplier: voltage controlled quadrature oscillator design", *Active And Passive Electronic Components*, vol. 2009, Article ID 835789, 5 pages.

23. S. Pisitchalermpong, Tangsrirat, W. Surakampontorn, "CDBA based multiphase sinusoidal oscillator using grounded capacitor", *in Proc. of the International Joint Conference* in Bexco, Busan, Korea, pp. 5762–5765, Oct. 2006.

24. A. U. Keskin, D. Biolek, "Current mode quadrature oscillator using current differencing transconductance amplifiers (CDTA)", *IEEE Proceedings on Circuits Devices Systems*, vol. 153, no. 3, pp. 214–218, 2006.

25. D. Dheer, Sagar Paliwal, Naved Ali, and Mohd. Samar Ansari, "A current-mode biquad filter for Zigbee applications using 45 nm ±0.75 V CMOS CDTA", *in Proc. IEEE International Symposium on Signal Processing and Information Technology* (ISSPIT), Noida, India, 2014, pp. 000131–000136, 15–17 Dec. 2014.

26. T. Bumrongchoke, D. Duangmalai, W. Jaikla, "Current differencing trans-conductance amplifier based current-mode quadrature oscillator using grounded capacitors", *in Proc. International Symposium on Communication and Informatics Technologies* (ISCIT), Tokyo, pp. 192–195, Oct. 2010.

27. D. Biolek, V. Biolkova, Z. Kolka, "Single-CDTA (Current Differencing Transconductance Amplifier) current-mode biquad revisited", *WSEAS Transactions on Electronics*, vol. 5, no. 6, pp. 250–256, 2008.

28. E. Alaybeyoglu and Hakan Kuntman, "A new reconfigurable filter structure employing CDTA for positioning systems", *in Proc. of the 9th International Conference on Electrical and Electronics Engineering* (ELECO), Bursa, pp. 37–41, 26–28 Nov. 2015.

29. N. Pandey and S. K. Paul, "Single CDTA-Based Current Mode All-Pass Filter and Its Applications", *Journal of Electrical and Computer Engineering*, vol 2011, Article ID 897631, 5 pages.

30. H. Kaabi, M. R.J. Motlagh, A. Ayatollahi "A novel current-conveyor-based switched-capacitor integrator", *in Proc IEEE International Symposium on Circuits and Systems*, vol. 2, pp. 1406–1408, May 2005.

31. Dattaguru. V. Kamath, "Overview of Op-amp and Ota based integrators", *International Journal Of Innovative Research In Electrical, Electronics, Instrumentation And Control Engineering*, vol. 3, issue 9, pp. 74–81, Sept 2015.

32. D. Biolek, "CDTA–building block for current-mode analog signal processing" *in Proc. of the European Conference on Circuit Theory and Design*, pp. 397–400, 2003.

33. E. Alaybeyoglu, H. H. Kuntman, A. Guney, M. Altun, "Design of positive feedback driven current-mode amplifiers Z-Copy CDBA and CDTA and its filter Analog applications", *Analog Integrated Circuits and Signal Processing*, vol. 81, issue 1, pp. 109–120, Oct 2014.

34. L. Tan, K. Liu, Yu Bai and J. Teng, "Construction of CDBA and CDTA behavioral models and the applications in symbolic circuits analysis", *Analog Integrated Circuits and Signal Processing*, vol. 75, issue 3, pp. 517–523, June 2013.

35. S. K. Rai and M. Gupta, "Current differencing transconductance amplifier (CDTA) with enhanced performance and its application", *Analog Integrated Circuits and Signal Processing*, vol. 86, issue 2, pp. 307–319, Feb 2016.

36. S. Koziel, "Noise analysis and optimization of continuous-time active-RC filters", in *Proc. IEEE International Conference on Electronics, Circuits and Systems*, ICECS 2004, Tel Aviv, Israel, 14–15 Dec. 2004, pp. 45–48.

Road and Traffic Sign Detection Using Colour Segmentation

Saka Harshavardhan, Vadlamudi Madhavi and Sajja Tejaswi

1 Introduction

Road and traffic signs are characterized as a visual dialect that can be comprehended by drivers. These speak to the present traffic circumstance on the road, indicate risk and challenges around the drivers, give them notices and help them with their directing by giving valuable data that assemble safe driving and convenience [1, 2]. The human visual observation depends on physical and mental conditions. In some conditions, they are affected by factors such as exhaustion and observatory skills. This information given to the drivers in correct time can prevent accidents, save lives of people, increase driving performance of vehicles and reduce the pollution caused by vehicles [3–5]. Colour representation plays an important role. That provides information to the driver to ensure the objectives of the road sign. The colours of the road and traffic signs are completely different from nature. Detection of road and traffic signs provide safety for the driver. Intelligent transport systems are closely related to emerging technologies: the technologies are Internet, mobile data services, smart sensors, artificial intelligent. Road and traffic sign detection is the one of the important fields in the intelligence transport system. The road and traffic signs play an important role in daily life. These signs define a visual language that can be understand by the drivers [1]. The field of road sign detection is not very old; the first paper appeared in Japan in 1984. Many segmentation methods for the detection of road and traffic signs. Many research teams and companies are con-

S. Harshavardhan (✉) · V. Madhavi · S. Tejaswi
Department of Electronics and Communication Engineering, VNITSW, Guntur, India
e-mail: sakaharshavardhan@gmail.com

V. Madhavi
e-mail: vadlamudimadhavi456@gmail.com

S. Tejaswi
e-mail: tejaswisajja20@gmail.com

© Springer Nature Singapore Pte Ltd. 2019
V. Nath and J. K. Mandal (eds.), *Proceeding of the Second International Conference on Microelectronics, Computing & Communication Systems (MCCS 2017)*, Lecture Notes in Electrical Engineering 476, https://doi.org/10.1007/978-981-10-8234-4_17

ducted research in this field, and colossal measure of work has been finished. Different types of techniques have been used, and big improvements have been attain during the last decade. The detection system can be realized by either colour information, shape information, and both of them. If both are present its give better results. Many reviews represent that the location can be accomplished regardless of the possibility that either shading or shape is absent. Road and traffic sign detection for many segmentation methods are present. They are histogram-based method, clustering method, edge detection method, dual growing method and dual clustering method thresholding method. But we used thresholding method. This is the best method for detecting road and traffic signs in terms of speed and efficiency. The identification of the road and traffic signs is attained by one main stage that is detection. In detection phase, the colour image is taken as preprocessed, enhanced and cropped image. The size of cropped image size will be fixed as 512 * 512. The competence and high speed of the detection play an important role in road and traffic detection process, because it reduces the search space and indicate only potential regions. The sign will be detected based on the number of white pixels present in the image. In this paper, we use resize, interpolator, decimator, RGB thresholding.

2   Overview of Imaging Model and Algorithm

2.1   Learning Process

In learning process, calculate the threshold value of different types of road and traffic signs. In that threshold value means to count the number of white pixels in different signs. This process is done before taking in the input image. In that, first take the input sign after that perform either interpolation operation or decimation operation depending on the size of input sign. If the size of input sign is greater than 512 * 512 size of image, perform decimation operation on input sign. If the size of input sign is less than 512 * 512 size of image, perform interpolation operation input image. After getting the 512 * 512 size of image, perform RGB thresholding on the image. In that image remove small or large regions. Small regions that means noise, vehicle tail light cover, road pavement, etc. Large regions mean large red background (building wall). Remove large axis ratio. Axis ratio means the ratio of major axis to minor axis of sign. After that calculate the threshold values of different signs (Fig. 1).

Fig. 1 Image observation
model

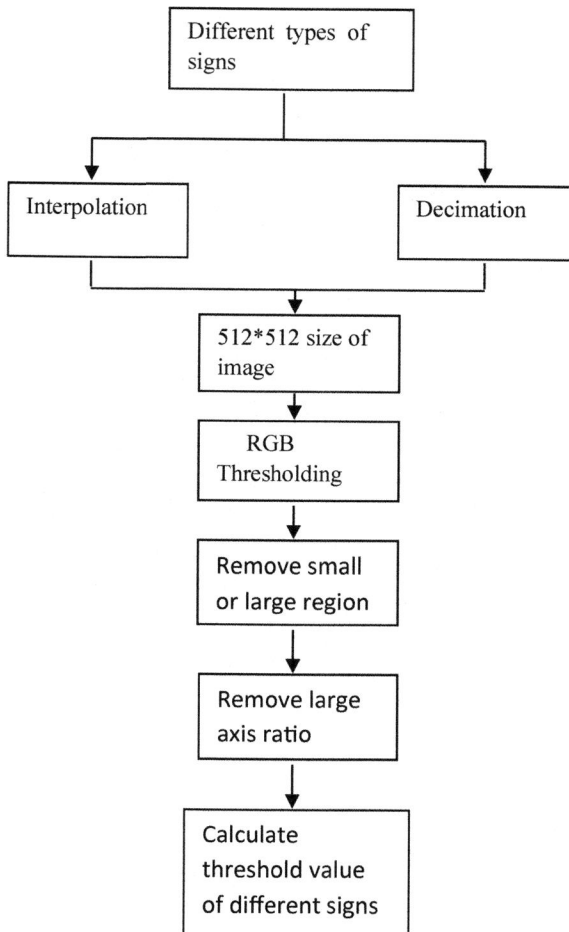

2.2 Algorithm

The overall procedures of this algorithm is

1. Calculate the threshold values of different signs, before choosing the input image.
2. Taking the input image.
3. Either interpolation or decimation operation done based on size of image for getting 512 * 512 size of image.
4. R, G, B domain thresholding.
5. Remove small and large regions.

6. Dilation to connect fragmented part of the signs.
7. Remove regions whose major/minor axis ratio is too large.
8. Count the no of white pixels in the image.
9. If count equal to threshold value to detect the sign.

Since both stop sign and yield sign are red, they can be separated from background by thresholding in RGB domain using the following scheme, as mentioned in [3]:

$$g(x,y) = k1 \begin{Bmatrix} Ra \leq fr(x,y) \leq Rb \\ Ga \leq fg(x,y) \leq Gb \\ Ba \leq fb(x,y) \leq Bb \end{Bmatrix} \qquad (1)$$

where $fr(x; y)$; $fg(x; y)$; $fb(x; y)$ is the components in red, green, and blue domain, respectively, and Ra; Rb; Ga; Gb; Ba; Bb is the corresponding lower and upper thresholds in those domains, respectively. However, this scheme has a disadvantage that the thresholds are affected by the lighting condition of the image thus making it difficult to find a universal threshold set that is applicable to images with varying lighting conditions. As suggested by de la Escalera and Salichs [6], a slightly modified thresholding scheme is used in this work. Instead of directly thresholding on blue and green domain, the thresholding on these two components is applied on the ratio of corresponding component to red component that is

$$g(x,y) = \begin{Bmatrix} Ra \leq fr(x,y) \leq Rb \\ Ga \leq \dfrac{fg(x,y)}{fr(x,y)} \leq gb \\ Ba \leq \dfrac{fg(x,y)}{fr(x,y)} \leq Bb \end{Bmatrix} \qquad (2)$$

This scheme turns out to be less affected by lighting condition than original scheme and provide better segmentation results (Fig. 2).

Fig. 2 Different types of input signs

2.3 Segmentation of Stop Sign

See Fig. 3.

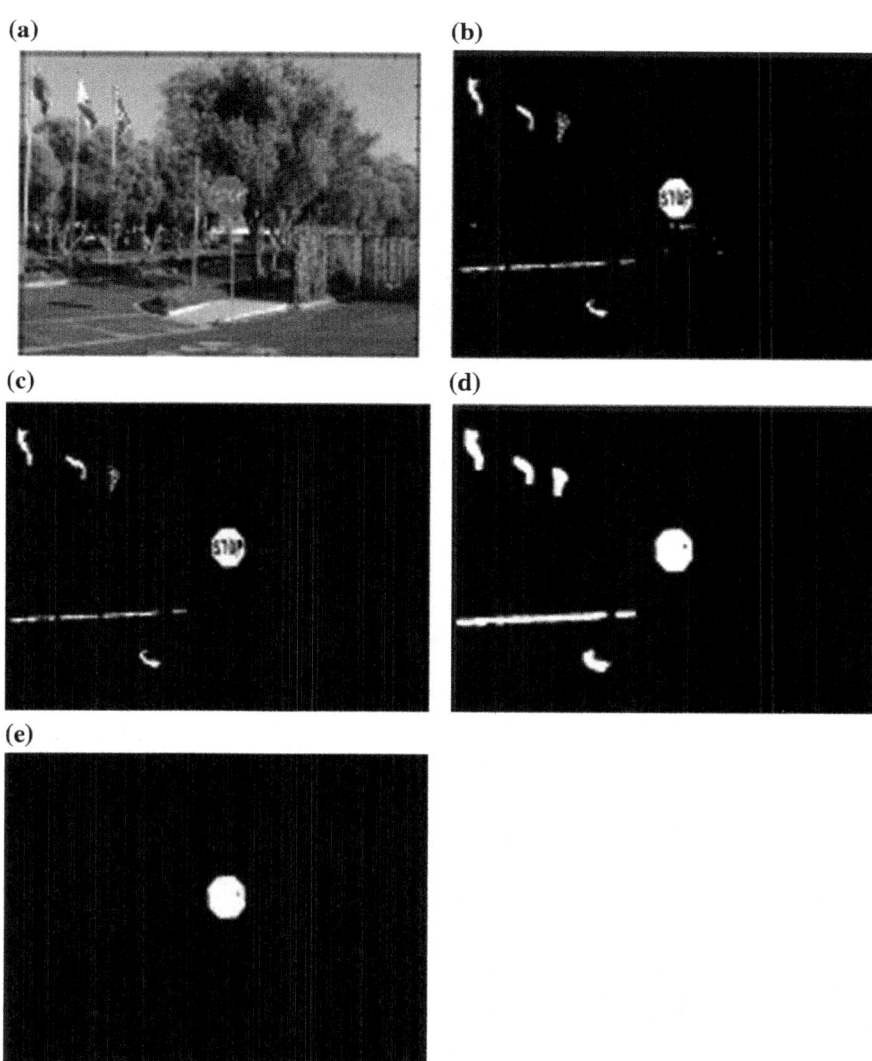

(a)

(b)

(c)

(d)

(e)

Fig. 3 **a** Original image, **b** image after RGB thresholding, **c** after removing small region, **d** after dilation, **e** after removing region whose axis ratio is large

2.4 Segmentation of Yield Sign

See Fig. 4.

2.5 Thresholding Choosing

All the thresholds are chosen based on their performance on training images. For each of the image in the testing group, a small area in the traffic sign is manually selected and its red component and the ratio of green to red, and blue to red in this area $(f_r(x,y), \frac{f_g(x,y)}{f_r(x,y)}$ and $\frac{f_b(x,y)}{f_r(x,y)}$ in Eq. 2) are computed. Then, the thresholds are selected based on their distributions across all the images. The thresholds for size limit and depending on the number of white pixels in sign detect which type of sign. The chosen thresholds are as follows: RGB domain threshold, as those defined in Eq. 2: $Ra = 65$, $Rb = 255$, $Ga = 0$, $Gb = 0.7$, $Ba = 0$, $Bb = 0.7$; size threshold: lower limit is 40 pixels and upper limit is 400,000 pixels. In that, only detect the red

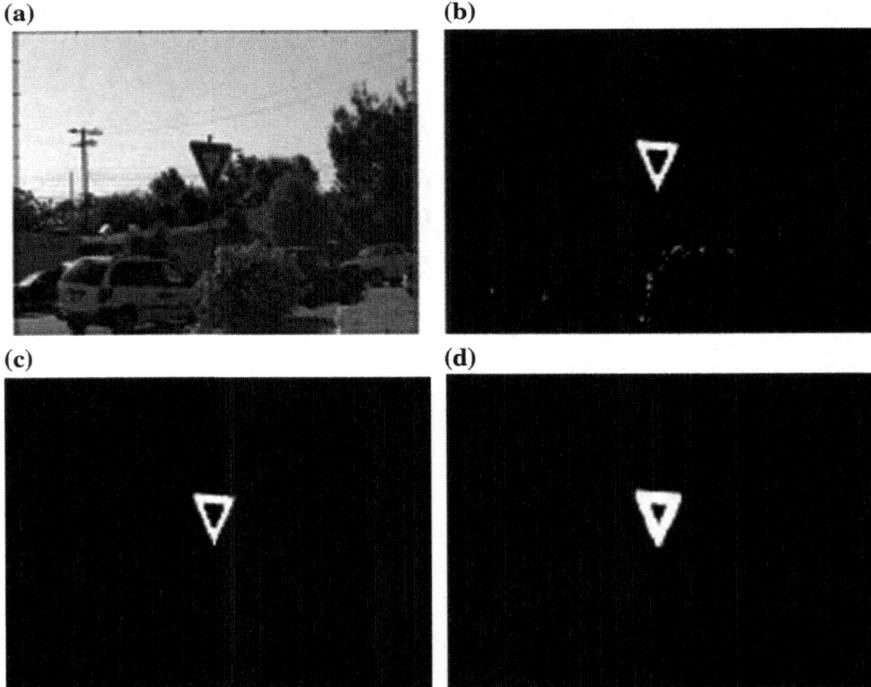

Fig. 4 **a** Original image, **b** image after RGB thresholding, **c** after removing small region, **d** after dilation and removing region whose axis ratio is large

colour road and traffic signs. The main reason behind this is most of the signs are red in colour. But blue and green signs will also be detected. Message will be detected after detecting the sign.

3 Proposed Algorithm

3.1 Learning Process

In that, first calculate threshold value of different signs before taken into input image. In that, threshold value means to count the number of white pixels in different signs. In this, process take the input image. After that perform either interpolation operation or decimation operation depending on the size of input sign. If the size of input sign is greater than 512 * 512 size of image, perform decimation operation on input sign. If the size of input sign is less than 512 * 512 size of image, perform interpolation operation input image. After getting the 512 * 512 size of image, perform RGB thresholding on the image. In that image remove small or large regions. Small regions mean noise, vehicle tail light cover, road pavement, etc. Large regions mean large red background (building wall). Remove large axis ratio. Axis ratio means the ratio of major axis to minor axis of sign (Fig. 5).

Then count the number of white pixels in input sign. After that compare the count value of input sign to threshold value of different signs. If count value is equal to threshold value then detect sign. The chosen threshold values (no of white pixels) of different signs are as follows: Stop Sign: 28,428, Yield Sign: 1,48,026, Wrong Sign: 89,679, Closed Sign: 7960.

4 Performance Evaluation

In this section, we assess experimentally the performance of the proposed algorithm. Two traffic and road sign images are considered through which we demonstrate the feasibility of the proposed approach for detection of road and traffic signs. In this paper, the simulations are done with two number of road and traffic sign images. In that, sign detection can be done based on the number of white pixels (Figs. 6 and 7).

Stop Sign:

(A) Input sign
(B) Red thresholding
(C) Green thresholding
(D) Blue thresholding

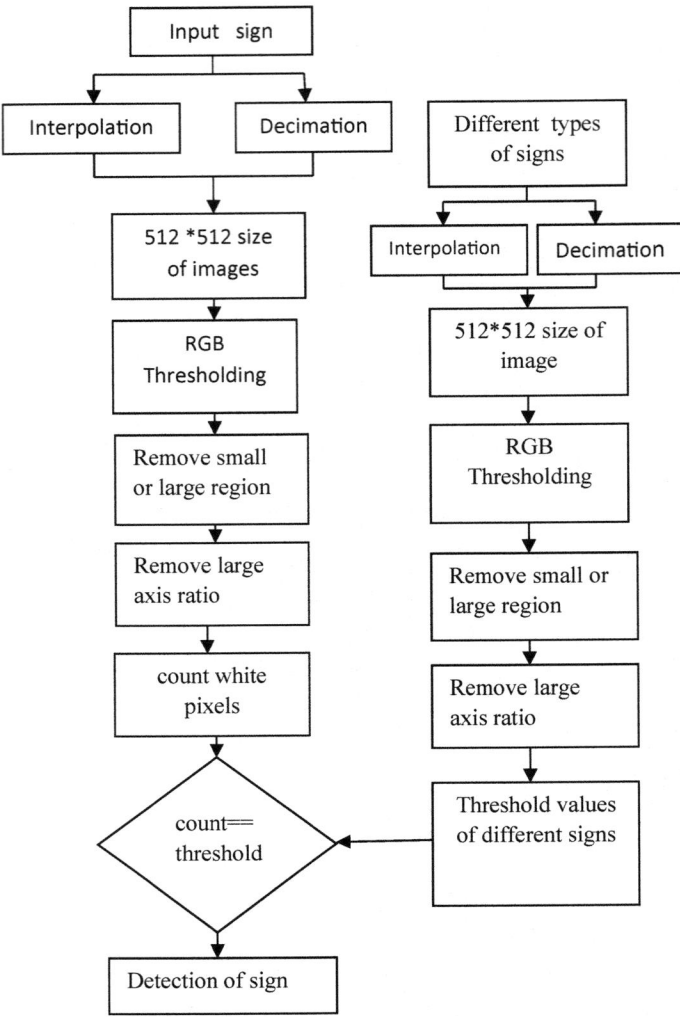

Fig. 5 Flow chart for the proposed algorithm

(E) Original image
(F) Selected area
(G) Remove small region
(H) Threshold area
(I) Output image
(J) Message of stop sign

Fig. 6 Resulted stop sign images using colour segmentation

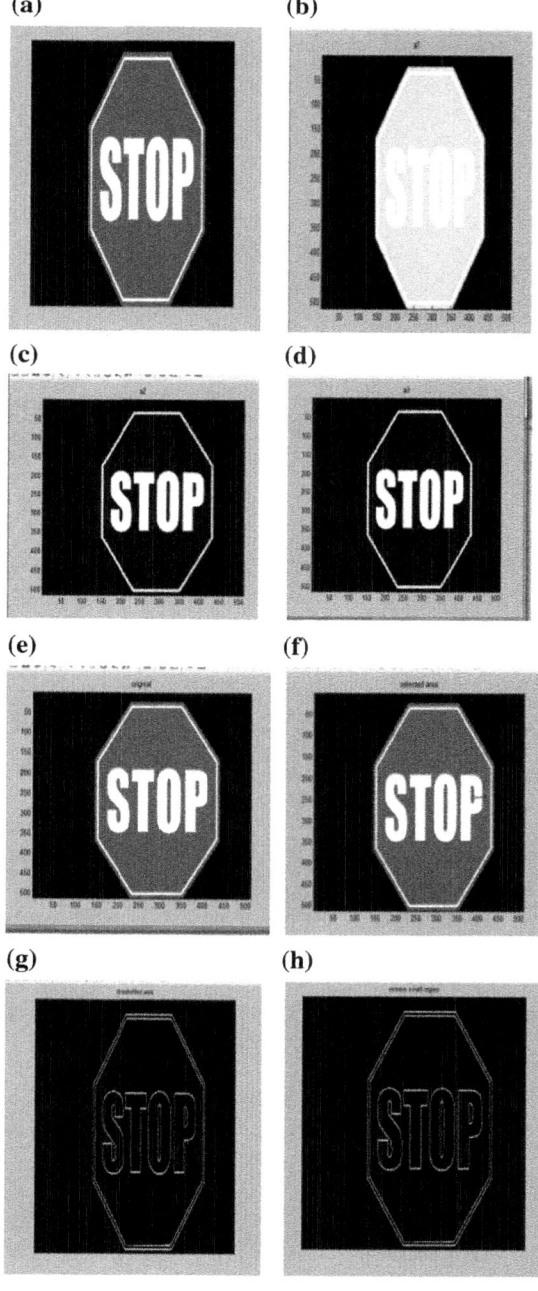

S. Harshavardhan et al.

Fig. 7 Resulted yield sign images using colour segmentation

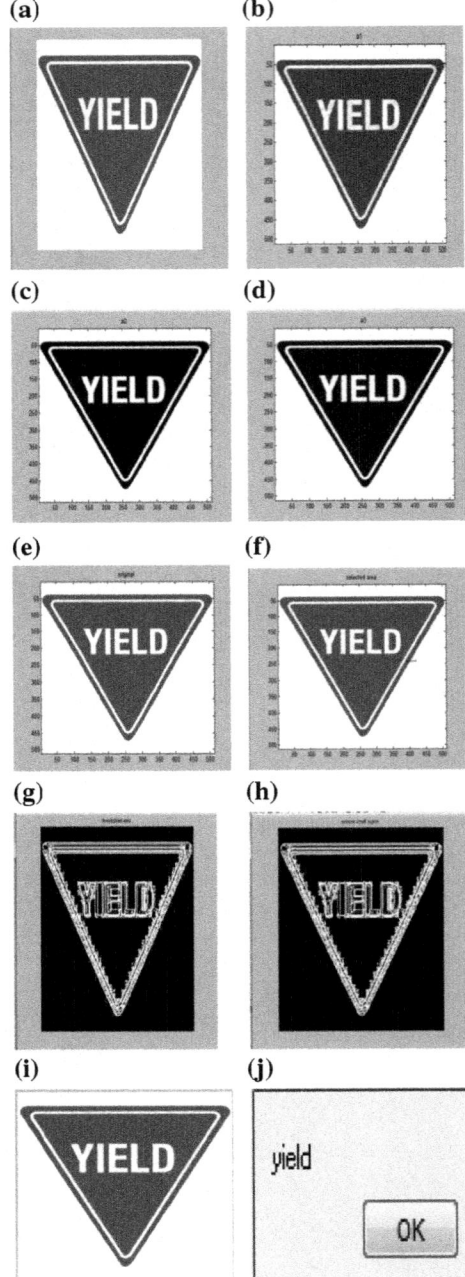

Table 1 Detected, missed and false detected traffic signs

	Zheng method			Proposed method		
	Total	Stop sign	Yield sign	Total	Stop sign	Yield sign
	20	15	5	20	14	6
Detected	16	13	3	18	14	4
Missed	4	2	2	2	0	2
False detected	2	2	0	1	0	1

Yield Sign:

(A) Input sign
(B) Red thresholding
(C) Green thresholding
(D) Blue thresholding
(E) Original image
(F) Selected area
(G) Remove small region
(H) Threshold area
(I) Output image
(J) Message of yield sign

5 Detection Rate

See Table 1.

6 Conclusion

This proposed algorithm has the field for improvement to improve the detection rate. This algorithm achieved about 90% correct detection rate in all the images.

Future Scope Future scope of this project is implementing this algorithm for video signals.

References

1. C. Fang, C. Fuh, S. Chen, and P. Yen, "A road sign recognition system based on dynamic visual model," presented at Proc. 2003 IEEE Computer Society Conf. Computer Vision and Pattern Recognition, Madison, Wisconsin, 2003.
2. C. Fang, S. Chen, and C. Fuh, "Road-sign detection and tracking," *IEEE Trans. on Vehicular Technology*, vol. 52, pp. 1329–1341, 2003.

3. L. Estevez, and N. Kehtarnavaz, "A real-time histographic approach to road sign recognition," presented at Proc. IEEE Southwest Symposium on Image Analysis and Interpretation, San Antonio, Texas, 1996.
4. C. Fang, S. Chen, and C. Fuh, "Road-sign detection and tracking," IEEE Trans. on Vehicular Technology, vol. 52, pp. 1329–1341, 2003.
5. A. de la Escalera, J. Armingol, and M. Mata, "Traffic sign recognition and analysis for intelligent vehicles," Image and Vision Computer, vol. 21, pp. 247–258, 2003.
6. Arturo de la Escalera and Miguel Angel Salichs, "Road Traffic Sign Detection and Classification", IEEE Transactions on Industrial Electronics, vol. 44, No. 6, December 1997.

Trimmed Median Filter for Removal of Noise from Medical Image

Pawan Kumar, Mahesh Chandra and Sanjeev Kumar

1 Introduction

During acquisition, transmission and storing medical images are generally affected by impulse noise or by speckle noise. Impulse noise is also known as salt-and-pepper noise. In this type of noise, the pixel values are changed either 0 or 255. Medical images are very sensitive in nature. If mammogram images are corrupted by noise then it will affect the performance in finding microcalcification present in the images. Median filter removes salt-and-pepper noise present in the images. Median filter works well only for low noise density. Filters such as central weighted filter, adaptive median filter, progressive switching median filter, decision-based algorithm [1–5] are very popular methods for removing the noise present in the images. These filters work well for digital images but the performances of these filters are not good for medical images. Hong et al. [6] developed a method based on genetic algorithm for removing the noise present in the images. This method uses operators such as crossover and mutation for filtering the filter value assigned to each combination based on a fitness function. Elaiyaraja et al. [7] proposed a method based on quaternion model proposed by Hamilton in 1843 and vector median filter (VMF). VMF is used to remove the noise present in the medical images. Alpha-trimmed mean filter [8] is a nonlinear filter. This filter is used to

P. Kumar · S. Kumar (✉)
Electronics and Communication Engineering Department, Cambridge Institute
of Technology, Ranchi, India
e-mail: sanjeevsingh.ece@gmail.com

P. Kumar
e-mail: pawan_aloysius1@yahoo.com

M. Chandra
Electronics and Communication Engineering Department, Birla Institute of Technology,
Mesra, Ranchi, India
e-mail: shrotriya69@rediffmail.com

© Springer Nature Singapore Pte Ltd. 2019
V. Nath and J. K. Mandal (eds.), *Proceeding of the Second International Conference
on Microelectronics, Computing & Communication Systems (MCCS 2017)*, Lecture Notes
in Electrical Engineering 476, https://doi.org/10.1007/978-981-10-8234-4_18

remove the salt-and-pepper noise present in the image. This filter uses a parameter "α" called trimmed factor. This parameter is used to control the number of values to be trimmed. This filter works well up to 50% noise density but blurring takes place when noise density is high. In decision-based algorithm [4] for low noise density, median value is used for replacing the noisy pixels but at high noise density, median value of neighborhood pixels is used. Further improvement is done by Decision-Based Unsymmetrical Trimmed Variant Filter (DBUTVF) [9]. In this filter detection of noisy pixels are done for 3×3 windows. After that, counting of noisy pixels is done, and further processing is done on the basis of number of noisy pixels. If the number of noisy pixels is ≤ 3, then noisy pixels are replaced by median of unsymmetrical trimmed output. If the number of noisy pixels is ≥ 3, then noisy pixels are replaced with mean of the rank order unsymmetrical trimmed output. The proposed method used the concept of alpha-trimmed mean filter, decision-based algorithm, and decision-based unsymmetric trimmed median filter. The proposed method shows better result for both low and high noise density. The paper is organized into four sections. Section 2 describes the proposed algorithm. Section 3 gives the detail description of simulation and results of proposed algorithm. Section 4 concludes the paper.

2 Proposed Algorithm

The medical images are generally affected by impulse and speckle noises. In this paper, a method is proposed for removing the salt-and-pepper noise present in medical images. The flowchart of proposed filter is shown in Fig. 1. The proposed technique uses a non-overlapping window of size 2×2. If the selected window contains either 0 or 255, then replace all pixels by median value of the previous window. If in the selected window all pixels are not noises, then find median value of noise-free pixels and replace the noisy pixels with this median value. The proposed method is tested with MIAS database images. The MIAS mammograms have been carefully selected from United Kingdom national breast screening program. This database contains 322 images of 161 different patients. Medico-lateral oblique view is used for preparing the mammogram database. All the images were digitalized at a spatial resolution of 0.05 mm pixel size with 8-bit density resolution.

3 Simulation Result

The proposed method is tested with MIAS database images. From MIAS database, images mdb004, mdb016, and mdb022 are used for testing of proposed method. This method is used to remove the noise present in the mammogram images. The database images are of size 1024×1024. From the noisy images, a Region of Interest (ROI) of size 256×256 is cropped. The proposed algorithm is applied to

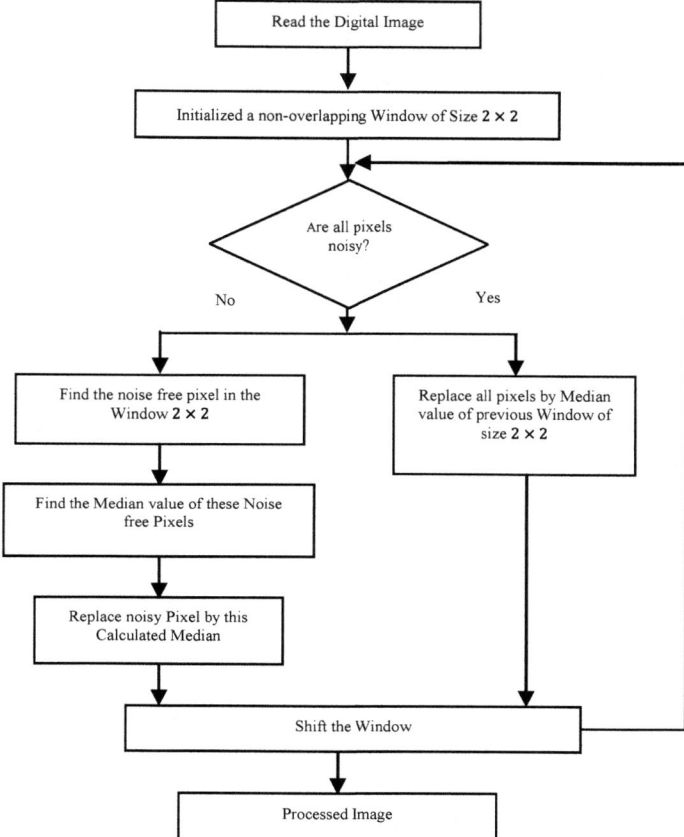

Fig. 1 Block diagram of proposed algorithm

this ROI. Figures 2a, 3a and 4a show the original database image. Figures 2b, 3b and 4b show the noisy images, and Figs. 2c, 3c and 4c show the ROI. Figures 2d, 3d and 4d show processed image. The STMF is compared with MF, AMF, DBA, Decision-Based Unsymmetrical Trimmed Variant Filter (DBUTVF), and modified quaternion vector filter (MQVF) [10]. The comparison of different methods is done in terms of MSE and PSNR. These are mathematically defined as

$$\text{MSE} = \frac{1}{M \times N} \sum_{i=1}^{M} \sum_{j=1}^{N} (x(i,j) - \hat{x}(i,j))^2 \tag{1}$$

$$\text{PSNR} = 10\log_{10}\left(\frac{255^2}{\text{MSE}}\right) \tag{2}$$

(a)   (b)

(c)   (d)

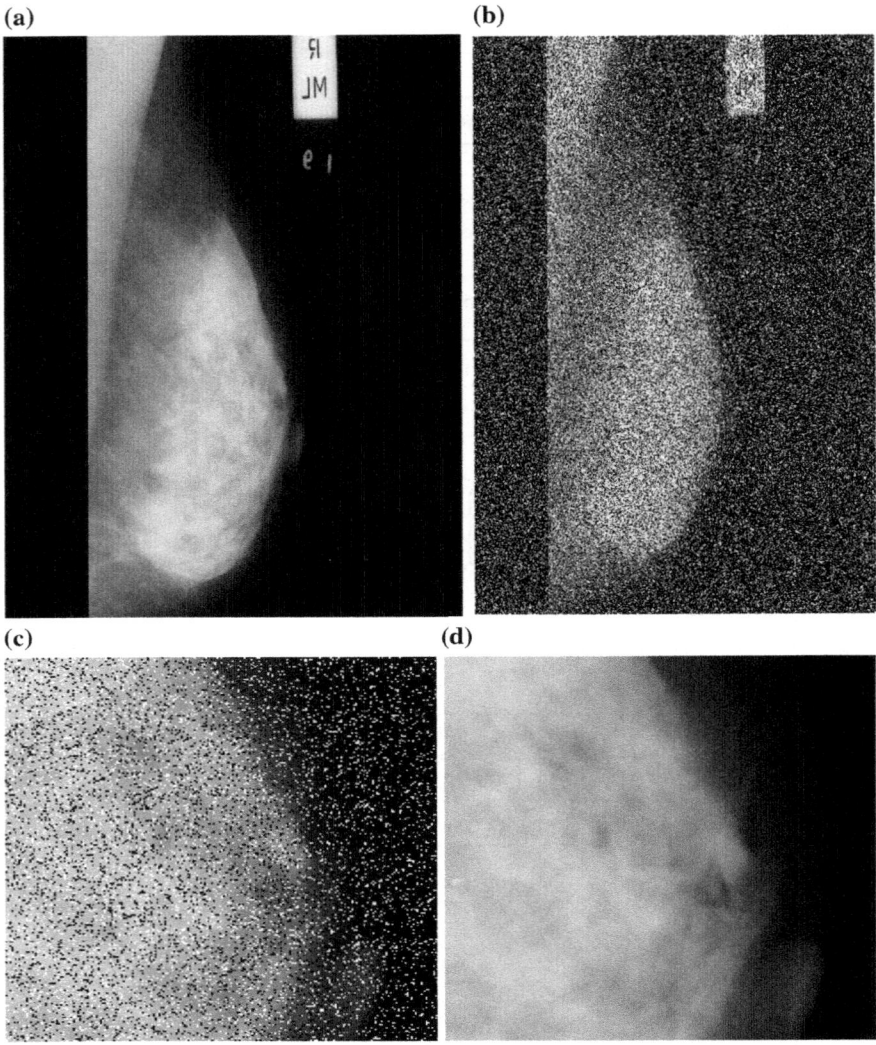

Fig. 2 **a** Original mdb004 image. **b** Noisy image. **c** ROI selected from noisy image. **d** Processed image

where $M \times N$ shows the size of the image. All experiments are performed on MATLAB2014a and window 8. The Intel(R) Core (TM) i5 CPU M480 @ 2.67 GHz and 8 GB RAM Laptop are used for all experiments. The STMF shows better result when compared with MF, AMF, DBA, MQVF, and DBUTVF. The results for ROI selected from mdb004 are shown in Tables 1 and 2. Tables 1 and 2 show the MSE versus noise density and PSNR versus noise density ranging from 10 to 90% for the same image, respectively. Tables 1 and 2 shows that for low to

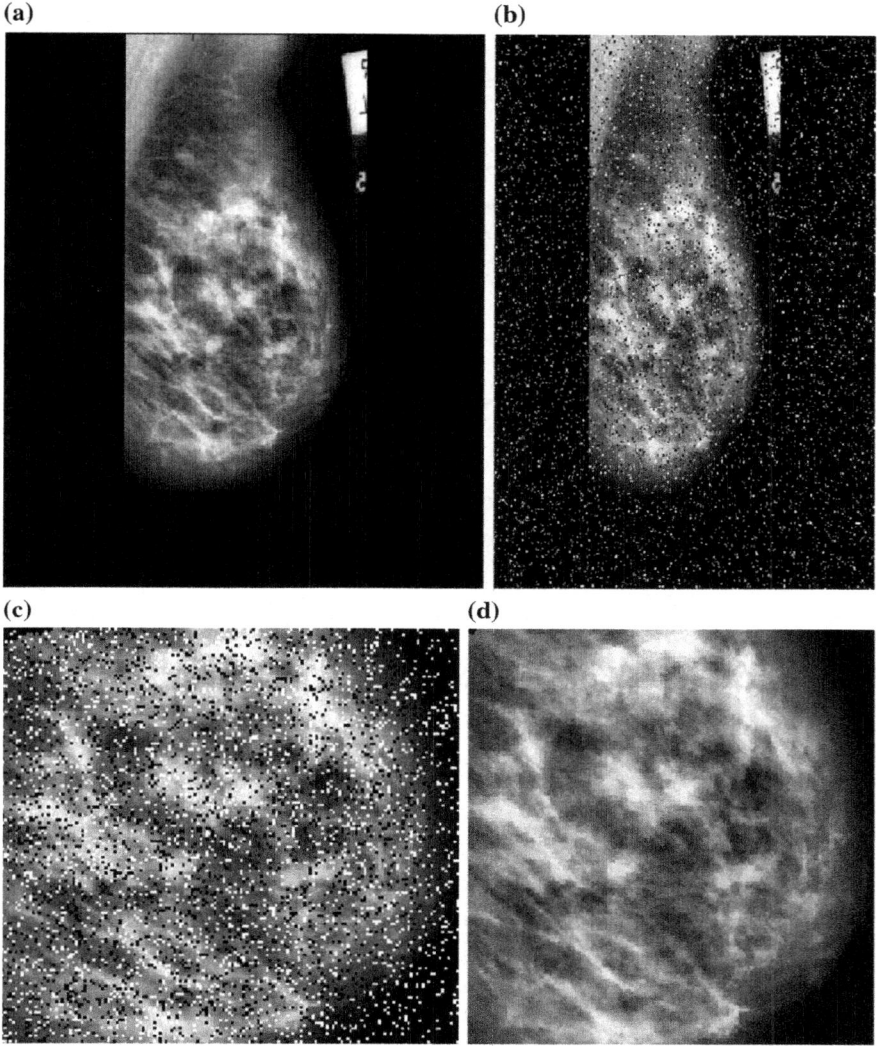

Fig. 3 **a** Original mdb016 image. **b** Noisy image. **c** ROI selected from noisy image. **d** Processed image

high noise density the MSE value and PSNR value is low and high, respectively. Figures 5 and 6 show graphical results. From these results, it is found that the performance of STMF is better than existing algorithms. Even at 90% noise density, the STMF gives better result in terms of MSE and PSNR value.

(a)   (b)

(c)   (d)

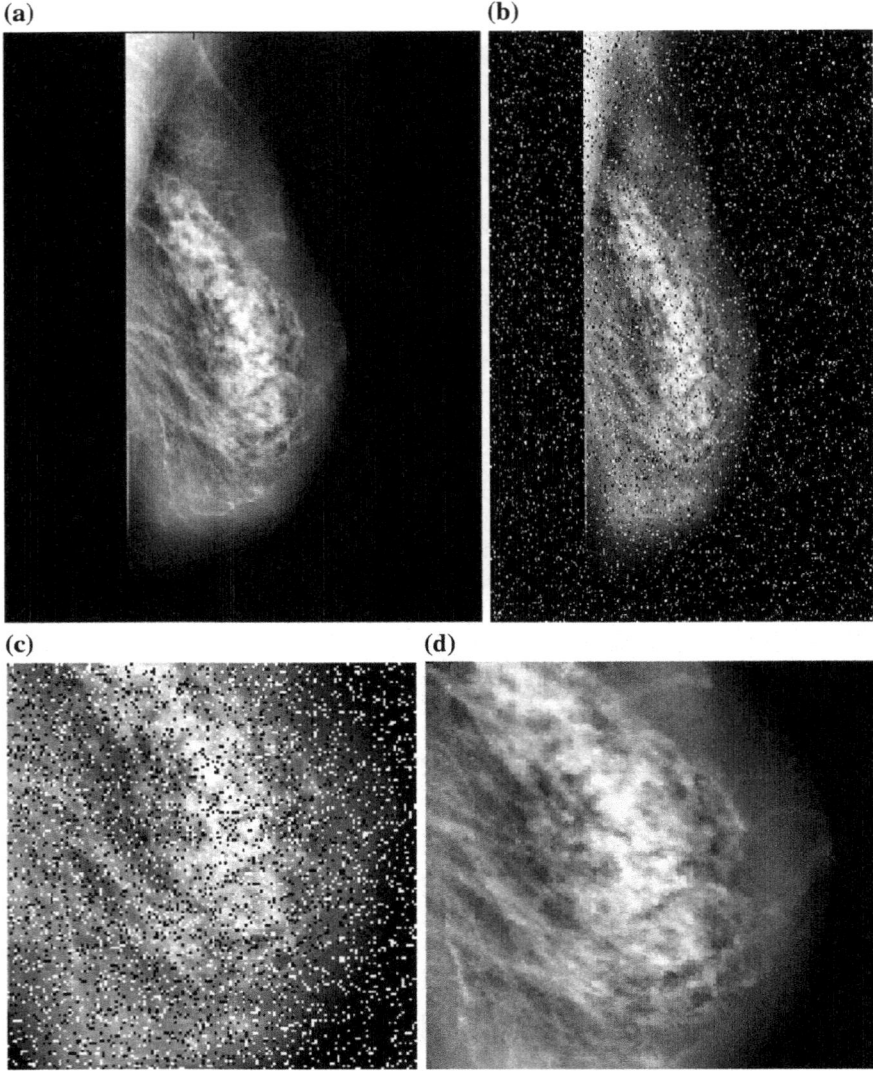

Fig. 4 **a** Original mdb022 image. **b** Noisy image. **c** ROI selected from noisy image. **d** Processed image

Table 1 MSE value for ROI of mdb004

Noise (in %)	ME	AMF	DBA	DBUTVF	MQVF	Proposed
10	38.99	0.69	37.48	1.21	0.695	0.41
20	72.60	1.03	69.42	1.88	1.55	0.83
30	115.63	1.53	102.56	6.61	3.98	2.33
40	182.05	2.30	161.49	13.71	8.93	5.23
50	308.93	7.73	174.57	40.00	23.66	7.08
60	663.79	43.61	225.48	95.95	48.31	10.16
70	1527.30	297.20	282.86	138.38	72.79	11.14
80	3683.03	1880.30	355.47	237.18	94.64	25.89
90	9468.70	6971.71	528.73	390.91	166.37	69.37

Table 2 PSNR value for ROI of mdb004

Noise (in %)	ME	AMF	DBA	DBUTVF	MQVF	Proposed
10	32.22	49.73	32.39	47.32	49.71	52.07
20	29.52	47.96	29.71	45.39	46.23	48.91
30	27.49	46.27	28.02	39.93	42.13	44.46
40	25.53	44.51	26.04	36.76	38.62	40.94
50	23.23	39.24	25.71	32.11	34.39	39.63
60	19.91	31.73	24.59	28.31	31.29	38.06
70	16.29	23.41	23.61	26.72	29.51	37.66
80	12.47	15.38	22.62	24.38	28.37	33.99
90	8.368	9.69	20.89	22.21	25.92	29.78

Fig. 5 MSE versus noise density for mdb004 image

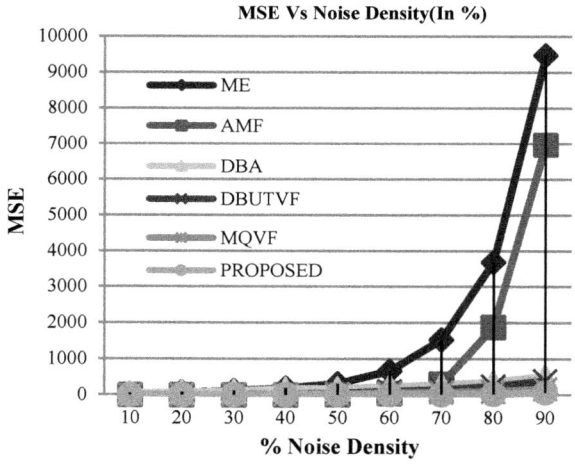

Fig. 6 PSNR (in dB) versus
noise density for mdb004
image

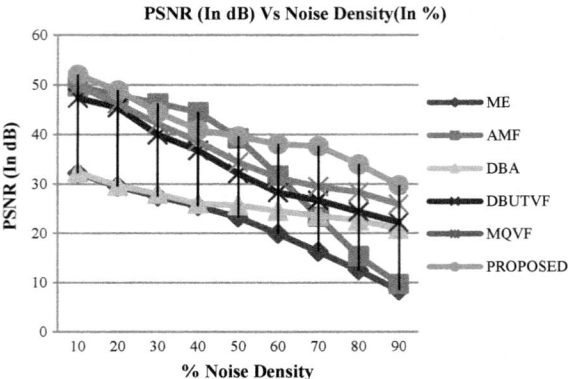

4 Conclusion

The proposed method introduces a new technique for removing the impulse noise present in the mammogram images. The performance of this filter is better than all other filters for all range of noise density. Even at high noise density, the values of MSE and PSNR are better in comparison to the existing algorithms. Further improvement can be done in terms of simulation time. The performance of STMF is better in comparison to MF, AMF, DBA, MQVF, and DBUTVF. This filter works well for low to high noise density ranging from 10 to 90%.

References

1. H. Hwang and R. A. Hadded, "Adaptive median filter: new algorithms and results", *IEEE Transactions on Image Processing,* vol. 4, Issue 4, pp. 499–502, 1995.
2. Z. Wang and D. Zhang, "Progressive switching median filter for the removal of impulse noise from highly corrupted images", *IEEE Transactions on Circuits and Systems-II,* vol. 46, pp. 78–80, 1999.
3. K. K. V. Toh and N. A. M. Isa, "Noise adaptive fuzzy switching median filter for salt-and-pepper noise reduction", *IEEE Signal Processing Letters,* vol. 17, Issue 3, pp. 281–284, 2010.
4. K. S. Srinivasan and D. Ebenezer, "A new fast and efficient decision based algorithm for removal of high density impulse noise", *IEEE Signal Processing Letters,* vol. 14, Issue 3, pp. 189–192, 2007.
5. S. Esakkirajan, T. Veerakumar, A. N. Subramanyam and C. H. Prem Chand, "Removal of high density salt and pepper noise through modified decision based unsymmetric trimmed median filter", *IEEE Signal Processing Letters,* vol. 18, Issue 5, pp. 287–290, 2011.
6. Jin Hyuk Hong, S. Bae Cho and Ung Keun Cho, "A Novel Evolutionary Method to Image Enhancement Filter Design: Method and Applications", *IEEE Transactions on Systems,* vol. 39, Issue 6, pp. 1446–1457, 2009.

7. G. Elaiyaraja, N. Kumaratharan and C. R. Prapau, "Modified Quaternion Based Impulse Noise Removal with Adaptive Threshold from Color Video Sequences and Medical Images", *Middle-East Journal of Scientific Research*, vol. 23, Issue 7, pp. 1382–1389, 2015.

8. J. B. Bednar and T. L. Watt, "Alpha-Trimmed Means and their Relationship to Median Filters", *IEEE Transactions On Acoustics, Speech and Signal Processing*, vol. 32, Issue 1, pp. 145–153, February 1984.

9. V. Vedanarayanan, "Content Based Noise Removal using non Linear Filter for the Elimination of outliers in mammogram images for effective identification of lesion", *Research Journal of Pharmaceutical, Biological and chemecal science*, vol. 6, Issue 3, pp. 237–244, 2015.

10. Elaiyaraja, N. Kumaratharan and C. Rama Prapau, "Modified Quaternion Based Impulse Noise Removal with Adaptive Threshold from Color Video Sequences and Medical Images", *Middle-East Journal of Scientific Research*, vol. 23, Issue 7, pp. 1382–1389, 2015.

Characterization of Interfering Signal in S Band of IRNSS

Surya Gupta, Darshna D. Jagiwala and Shweta N. Shah

1 Introduction

IRNSS is an emerging satellite-based navigation system offering an independent positioning and timing services over India and neighboring regions being launched by the Indian government. The objective of the system is to provide an indigenous and independent navigation system for national growth. The system is expected to provide accuracy similar to the Global Positioning System (GPS), and also to provide accurate real-time position, velocity, and time to all users on various platforms in the Indian region. The proposed IRNSS system consists of seven satellites to complete its constellation and covering 24 MHz of bandwidth in L5 band (1164–1189 MHz) and 16 MHz in S band (2483.5–2500 MHz) of frequency spectrum to transmit its signal from satellite to ground station. The number of systems that make use of radio frequency spectrum is increasing day by day which results in crowding frequency spectrum significantly. The Signal power level at IRNSS receiver antenna is very low as it is coming after traveling a long distance, making it vulnerable to Radio Frequency Interference (RFI) whose sources are ground-based. The nominal separation and provided guard band are often not sufficient to avoid the problem.

RFI is the major and devastating error of every GNSS system. There are different types of RF interfering signal including Chirp, Continuous Wave (CW), and Pulse. There are various methods to detect interference mainly pre-correlation and

S. Gupta (✉) · D. D. Jagiwala · S. N. Shah
Electronic Engineering Department, Sardar Vallabhbhai NIT, Surat, India
e-mail: guptasurya92@gmail.com

D. D. Jagiwala
e-mail: darshna12@gmail.com

S. N. Shah
e-mail: shahshweta13@gmail.com

© Springer Nature Singapore Pte Ltd. 2019
V. Nath and J. K. Mandal (eds.), *Proceeding of the Second International Conference on Microelectronics, Computing & Communication Systems (MCCS 2017)*, Lecture Notes in Electrical Engineering 476, https://doi.org/10.1007/978-981-10-8234-4_19

post-correlation [1], most of them are based on some property of received signal and proposed to deal with particular types of interference, but a method applicable to all types of interference is still a big problem. This paper provides an overview of the vulnerability of IRNSS receiver to Chirp, CW, and Pulse type interfering signal in S band of IRNSS and focused on some pre-correlation methods where the signal is modeled as random data and used its statistical characteristic, AGC, and Analog-to-Digital Converter (ADC) values for detecting and characterizing interference. Organization of paper is as follows: parameters considered for detection and characterizing are discussed in Sect. 2. In Sect. 3, experimental setup and flow of experiment are explained, and results are shown in Sect. 4. Section 5 concludes this paper.

2 Detection Techniques

2.1 Chi-Square Goodness of Fit

The chi-square test GoF belongs to the class of methods known as *tests concerning goodness of fit*. This method is applied by formulating the hypothesis testing problem:

$$\longrightarrow \quad H_0 \qquad RFI \text{ Absent}$$

$$\longrightarrow \quad H_1 \qquad RFI \text{ present}$$

Method based on the statistical characteristic of GNSS signal and interfering signal. The only requirement of this method is the process distribution of un-interfered signal. It takes a set of samples of the interfered and un-interfered signals for each bin, and the number of measurements is counted and represented by vector $E = \{E_1, E_2, \ldots, E_k\}$, this step serves to build the Probability Density Function (PDF) in the form of histogram. At this point, two histograms are available: reference histogram (E) and observed histogram (O). Test statics $T_x(x_m)$ is evaluated in order to discriminate two hypotheses. χ^2-test evaluates the probability, designated as *p-value* [2]:

$$P_m = P_r\{T_x(x) > T_x(x_m)\} \tag{1}$$

where $T_x(x)$ is approximated to χ^2 distributed with $k-1$ degree-of-freedom. It is observed that if $P_m \approx 1$ mean $T_x(x_m) \approx 0$ and no interference is present while $P_m \approx 0$ indicates that interference is present.

2.2 Parameters for Interference Detection

2.2.1 Histogram of ADC Values

ADC is the first active component in receiver chain, after antenna and all analog signal processing stage signal entered into ADC. Basic implementation of ADC uses only one bit. For multi-bit ADC signal, power has to be maintained at a certain level to make ADC work in its active range. AGC is used in GNSS receiver's front end to adjust the power level of incoming signal at the input of ADC to minimize quantization losses. According to the type of interference, AGC will either amplify or attenuate the signal level in order to exploit the ADC input range; hence, it is a valuable tool to detect and characterize interference [3]. The strength of signal at receiver end is very low or under noise floor, so the histogram of ADC sample will always be in Gaussian shape. Variation in AGC value can be better identified by observing the histogram of un-interfered and interfered signal. AGC value is mainly driven by noise and interfered signal rather than GNSS signal so if interference is under noise floor it cannot be detected by AGC. Power spectral density is another parameter to detect the presence of interference.

2.2.2 Power Spectral Density

The most common method in the frequency domain is based on power spectral estimation. It is based on the Fourier transform and implemented by FFT block. It can be defined as [4]:

$$P(k_{\max}) = |X(k_{\max})|^2 \tag{2}$$

where $X(k)$ is kth FFT equivalent defined as

$$X(k) = \sum_{n=0}^{N-1} x(n) \exp\left(-j\frac{2\pi nk}{N}\right) \tag{3}$$

where N is the number of samples of $x(n)$.

In many applications, this technique is simple and an efficient tool for inter-ference monitoring. The received GNSS signal is completely under the noise floor, and in the frequency domain, the contribution of thermal noise is constant, so PSD helps in identifying the existence of interference [5]. The power level of received signal increased or decreased due to constructive and destructive interference of the signals having the same frequency band, so average energy of the received signal is more when interference is present.

2.2.3 Standard Deviation

Interference detection and characterization utilizes statistical behavior (mean, standard deviation) of incoming GNSS signal, where the received signal is treated as random data. These parameters vary uniquely according to the type of interference and applicable to all types of interferences as no prior information of the signal is needed. The standard deviation of interfered signal frequency can be used to classify the type of interference [6].

3   Experiment Setup and Work Flow

In Fig. 1a, the hardware setup to measure the actual C/N0 is shown, the GUI IR-DAS (Data Analysis Software) version 3.0 is used to capture the IF data from IRNSS ACCORD receiver, and the experiment flow diagram is shown in, respectively, Fig. 1b. The entire setup for this experiment is done at the communication research lab, SVNIT, Surat, India.

To evaluate the performance of the method discussed above, CW, Chirp, and Pulse type interference signals generated using GNU SDR kit. These signals were targeted toward IRNSS antenna in an interval of 4 h in order to corrupt the received signal at the user end. At RF front end digitized IF samples of interfering and un-interfered signals are taken from the USB port of the ACCORD IRNSS receiver. The further simulation is done on IF data in MATLAB software R2014a to generate the simulation results. Simulation parameter of IRNSS receiver front end is discussed in Table 1.

4   Simulation Results

To check the potentiality of the above-discussed method in interfered and un-interfered environment, simulation results of digital IF data are shown here. The received signal of IRNSS was corrupted by transmitting CW, Chirp, and Pulse type interference in S band during a day, and the respective graph for *p-value* is plotted. The results of *p-value* in [6] for GPS receiver are considered as a reference, and the same is being calculated and plotted for IRNSS receiver.

p-value of the sample window goes to zero where interference is present. Figures 2, 3 and 4 compare the *p-value* when CW, Chirp, and Pulse type interference were targeted toward IRNSS antenna at different time instances.

p-value can be used to characterize the type of interference as well, observing Fig. 2; Pulse interference can be properly identified as *p-value* is going to zero during Pulse intervals only. By observing Fig. 3, it can be said that CW interference is affecting the entire spectrum of IRNSS. While the presence of Chirp interference

(a)

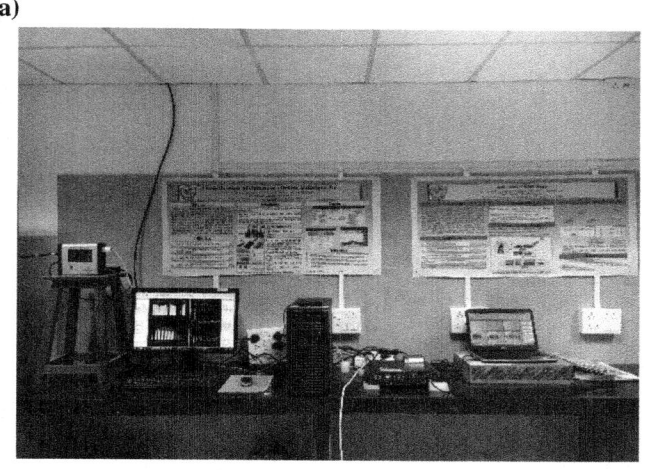

(b)

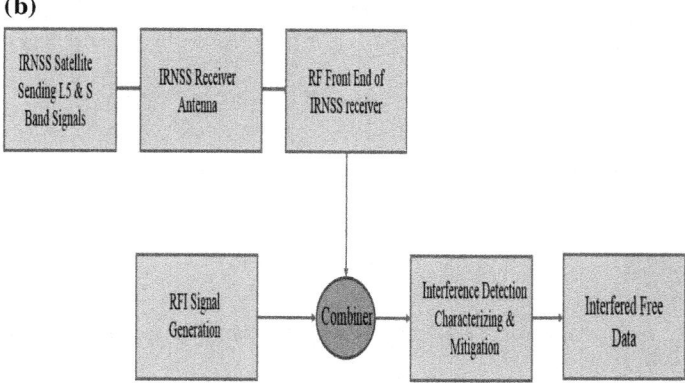

Fig. 1 **a** Hardware setup of experiment. **b** Experiment flow diagram

Table 1 Front end parameters of IRNSS receiver [7]

Parameter of IRNSS-UR	Value
Frequency of S band (MHz)	2492.028 ± 8.25
IF frequency (MHz)	72.221
Sampling frequency [F_s (MHz)]	56
Antenna polarization	RHCP
Bits/sample	4
LNA gain	>20 dB
LNA noise figure	<2 dB

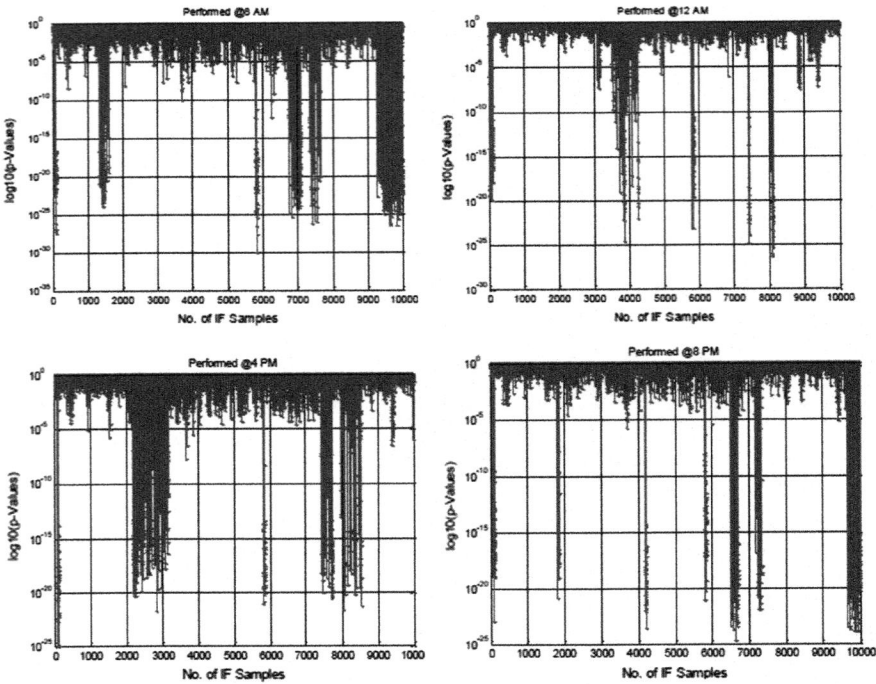

Fig. 2 Comparison of *p-value* evaluated at the different time instants when S band is targeted with Pulse interference

cannot be confirmed because of continuously changing behavior of *p-value,* as shown in Fig. 4.

The value of AGC is changing according to the type and intensity of the interference present. The IRNSS signal is under the noise floor; hence, the shape of the histogram must be Gaussian.

Taking reference from [8] histogram of samples of ADC is plotted. Changes in AGC value at the front end of an IRNSS receiver in the presence of CW, Chirp, and Pulse type interference signal in the S band can be clearly observed by histogram in Fig. 5.

The presence of interference can also be reported by observing the power spectrum [6].

The power spectrum of the signal with interfere signal changes significantly with time and several spurious peaks always appear over the frequency range with respect to the type of interfering signal in comparison to the un-interfered signal, and the same can be observed in Fig. 6.

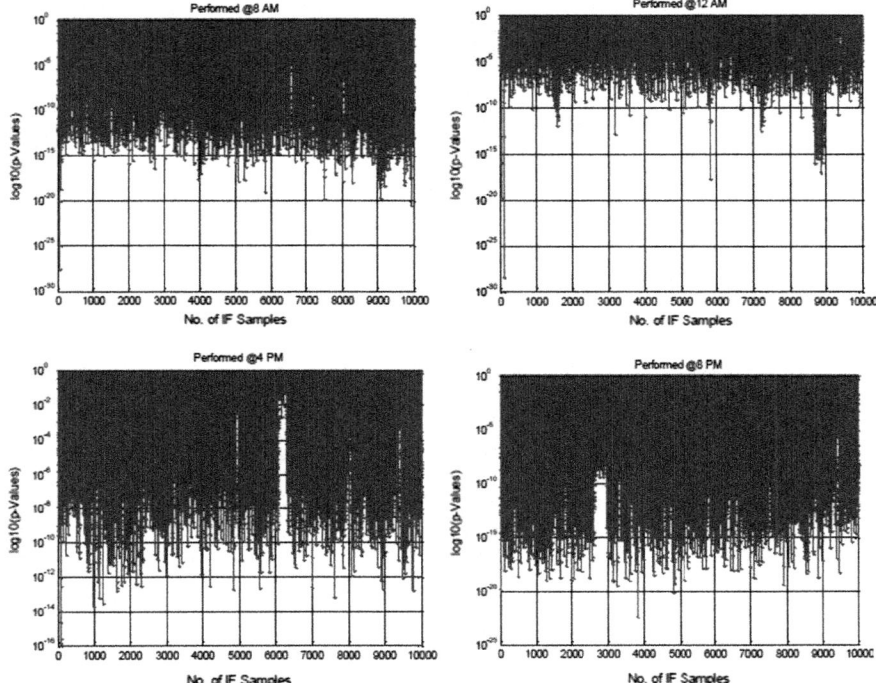

Fig. 3 Comparison of *p-value* evaluated at the different time instants when S band is targeted with CW interference

In addition, statistical characteristic, i.e., SD is used to characterize the type of interference by calculating variations in their estimated frequency. Taking the reference from [9], SD for the interfered signal is plotted. From Fig. 7, it can be inferred that CW is affecting the entire IRNSS frequency spectrum hence the value of SD is higher for CW. The frequency of the Chirp signal is continuously varying and when it matches with received signal frequency, interference is generated, so SD for Chirp is less than CW. While for the Pulse type signal, interference is present at only at Pulse intervals so the value of SD is minimum.

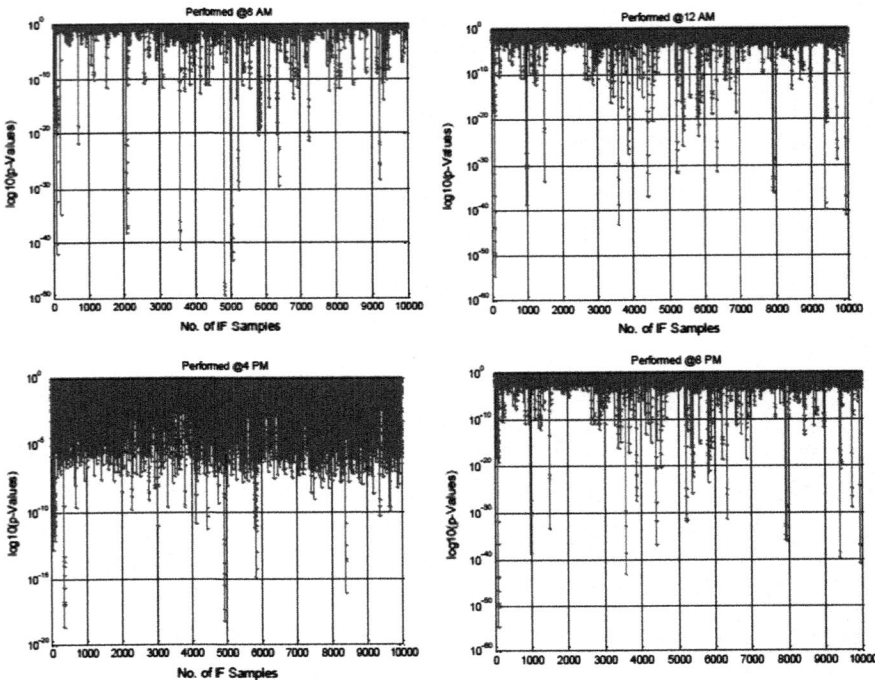

Fig. 4 Comparison of *p-value* evaluated at the different time instants when S band is targeted with Chirp interference

Fig. 5 Histogram of received IF data with Pulse, CW, and Chirp interference in S band

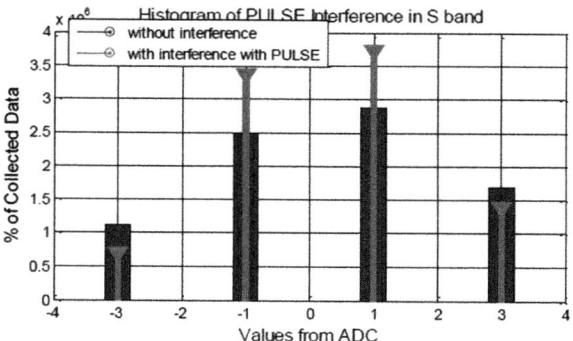

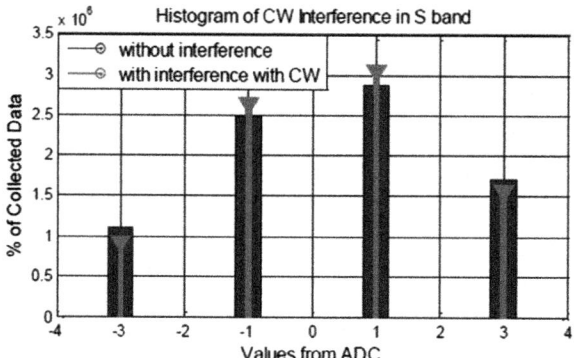

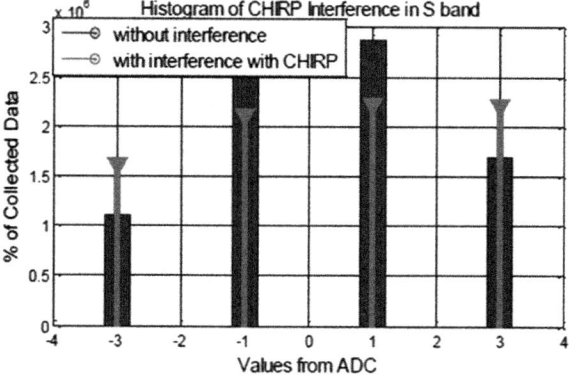

Fig. 6 PSD at front end of
IRNSS with Pulse, CW, and
Chirp interference

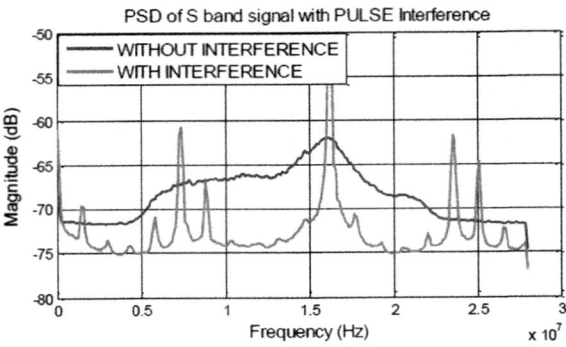

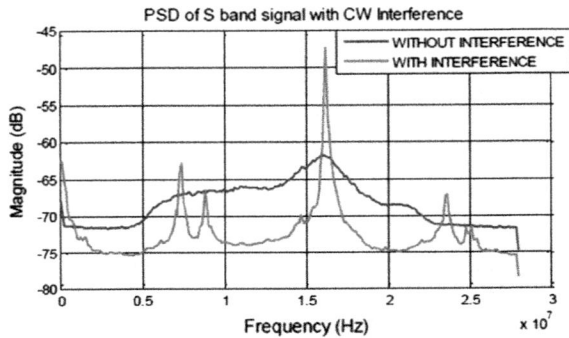

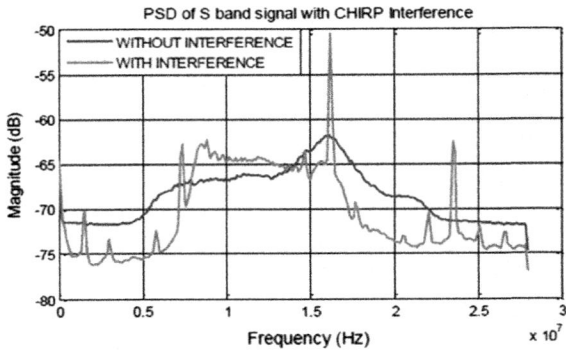

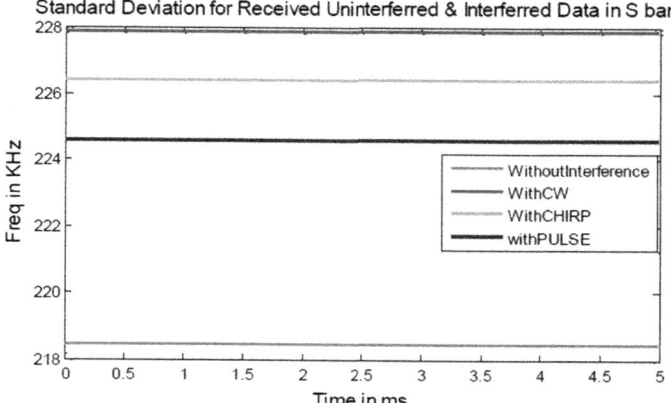

Fig. 7 Characteristic of SD for the different types of interferences

5 Conclusion

Interference monitoring has been considered as one of the major issues of all GNSS systems and investigation of effective algorithms for detection and characterization is a major challenge. This paper focuses on some parameters for the detection and characterization of interference in S band of IRNSS and also the presence of interference is confirmed by observing *p-value* of the chi-square method. The result of histogram shows how AGC value is changing in the presence of interference, and also PSD of interfered signal is having various peaks over the frequency range. The simulation results demonstrate that CW and Pulse can be better identified by observing the *p-value* of chi-square method while SD of estimated frequency is suitable for characterizing CW, Chirp, and Pulse interference in S band. In future, proper interference mitigation technique will be investigated for these types of interference signals to give interference-free environment to IRNSS users.

Acknowledgements We are very grateful to the Director, Space Application Center (SAC), **Shri Tapan Mishra** and special thanks to researchers of SAC, **Shri Atul Shukla** and **Shri Yagnesh Patel,** for providing the necessary guidance to do this type of analysis on IRNSS receiver.

References

1. Sheetal M. Ruparelia, Priyanka L. Lineswala, Darshana D. Jagiwala, Mr. Mehul V. Desai, Shweta N. Shah Upena D. Dalal "Study of L5 Interferences in IRNSS", Proc. Of GNSS user meet pp. 45, 2015.
2. Motella, Beatrice, Marco Pini, and Letizia Lo Presti. "GNSS interference detector based on Chi-square Goodness-of-fit test." *Satellite Navigation Technologies and European Workshop on GNSS Signals and Signal Processing, (NAVITEC), 6th ESA Workshop on* IEEE, pp. 1–6, 2012.

3. Bastide, Frederic, Dennis Akos, Christophe Macabiau, and Benoit Roturier. "Automatic Gain Control (AGC) as an interference assessment tool." In *ION GPS/GNSS, 16th International Technical Meeting of the Satellite Division of The Institute of Navigation*, pp. 2042, 2003.
4. Motella, Beatrice, and Letizia Lo Presti. "Methods of goodness of fit for GNSS interference detection." *IEEE Transactions on Aerospace an Electronic Systems* vol. 50, no. 3, pp. 1690–1700, 2014.
5. Fadaei, Nahal. "Detection, Characterization and Mitigation of GNSS Jamming Interference Using Pre-Correlation Methods." *IEEE Transaction on Aerosapce and Electronic System,* vol 3, pp 1690–1699, July 2014.
6. Balaei, Asghar Tabatabaei, and Andrew G. Dempster. "A statistical inference technique for GPS interference detection." *IEEE Transactions on Aerospace and Electronic Systems vol* 45, no. 4, pp 1499–1510, oct 2009.
7. Indian Regional Navigation Satellite System, Signal in Space ICD, For Standard Positioning Service, Version 1.0, ISRO, June-2014.
8. Isoz, Oscar. "*Interference detection and localization in the GPS L1 frequency band.*" PhD diss., Luleå tekniska universitet, 2012.
9. Yang, Jeong Hwan, et al. "Intentional GNSS interference detection and characterization algorithm using AGC and adaptive IIR notch filter." *International Journal Aeronautical and Space Sciences* 13.4, pp 491–498, 2012.

Enhancing Multibiometric System Security Using ECC Based on Score Level Fusion

Sandip Kumar Singh Modak and Vijay Kumar Jha

1 Introduction

Multibiometric system offers several advantages compared to the unimodal biometric system. In multibiometric system, two or more individual modalities are combined together to form a secure and better performance system, since the unimodal system suffers from lots of problem like spoof attack, intra-class variation, noise in senses data and matching problem [1]. E-commerce plays an important role in information and communication technology including online trading, supply chain management, and online transaction [2]. There are lots of researchs done in the area of e-banking to provide the security during the transaction. Biometric-based system can be used as a tool to provide the security during the online transaction. But unimodal-based biometric system suffers from problem like intra-class variation, noisy-sensor data, non-universality, and spoofing attacks [3]. To overcome this problem, we propose a multibiometric-based system using elliptic curve cryptography based on score level fusion. The fusion level in multibiometric system can be classified as sensor level, feature level, score level, and decision level. Further, the overall fusion technique in multibiometric system can we divided as pre-matching (sensor, feature) and post-matching (score, decision). It is very difficult to deal with the feature level fusion because at this level the feature sets.

Generated from multiple modalities are in different in nature as we can see in the fusion of fingerprint and face. Researcher preferred to use score level fusion due to ease of combining matching score. Due to lack of information, it is very difficult to

S. K. S. Modak (✉) · V. K. Jha
Department of Computer Science and Engineering, Birla Institute of Technology,
Mesra, Ranchi 835215, India
e-mail: modaknit@gmail.com

V. K. Jha
e-mail: vkjha@bitmesra.ac.in

© Springer Nature Singapore Pte Ltd. 2019
V. Nath and J. K. Mandal (eds.), *Proceeding of the Second International Conference on Microelectronics, Computing & Communication Systems (MCCS 2017)*, Lecture Notes in Electrical Engineering 476, https://doi.org/10.1007/978-981-10-8234-4_20

select decision level fusion as a fusion strategy in a multibiometric system [4]. One of the most popular cryptographic techniques used in current era is elliptic curve cryptography (ECC) which is one of the forms of public key cryptography. In the public key cryptography, each and every users participating in communication channel has a pair of private and public key. The private key is kept by the receiver, and public key is announced to the public [5]. The key size used in ECC is relatively smaller than other key size used in public key cryptography algorithm like RSA. In general, 160-bit key size used in ECC provides the similar type of security as in 1024-bit RSA [6]. The mathematical equation of ECC over Prime field is $y^2 = (x^3 + ax + b) \bmod p$, where $4a^3 + 27b^2 \neq 0$. a and b are two non-negative integers less than p. For each value of a and b, we got a point in elliptic curve. In this paper, we propose the concept of biometric signature based on ECC. Digital signature is a cryptographic concept which is used for the authentication purpose. It is basically used to maintain the integrity of the message and ensure that the origin of the message is correctly identified. At first Cryptographic Hash function is used to get the message digest of a message. In the process of signature creation, sender uses his/her private key to encrypt the message digest and at the receiver end to verify the signature the receiver can use sender public key and same Hash function [7].

The remainder of this paper is organizing as follows. Section 2 discusses the score level of fusion strategy. Section 3 provides explanation regarding Elliptic curve cryptography based biometric signature. Section 4 gives details about the proposed work. Section 5 concludes the paper.

2 Score Level Fusion

In this fusion method, the different score produce by different or same modalities are combined into a single score, and then, the decisions are taken as genuine or imposter according the resulting score value is greater than or less than threshold value. The score level fusion is also known as a measurement level fusion. There are two popular score level fusion approaches, and these are classification-based and combinational-based approach [3]. There are lots of research work have been done in the area of score level fusion including logistic transform method to integrate the output score from three different fingerprint matching algorithm [8]. Jain et al. in [4] explain the performance of different normalization technique based on score level fusion. They use face, fingerprint, and palm print as biometric modalities. Fusion at score level by combining the score from multiple biometric using triangular norms (t-norms) [9]. Mohi-ud-Din and Salah-ud-din et al. [10] proposed a score level multimodal fusion based on sum and product rule. They use hand geometry and fingerprint as a biometric modality. The matching scores generated by the individual matcher may not be homogenous due to a different scale or in the different probability distribution. In order to overcome this limitation, three fusion schemes have been introduced, i.e., density-based scheme; transformation-based scheme; and classifier-based scheme.

3 Biometric Signature Based on ECC

In the process of signature creation, sender uses his/her private key to encrypt the message digest, and private key is generated by the user biometric trait. On the other side, receiver uses sender public key to verify the signature. The message digest is nothing but Hash value of the Original message [7]. In this paper, we use the concept of biometric signature based on ECC which is applied on OTP generated by the bank transaction server.

Generation of Keys
In this paper, we use $y^2 = (x^3 + x + 13) \bmod 31$ as an elliptic curve with parameter $a = 1$, $b = 13$ and $p = 31$. $G = (9, 10)$ is the first point on this elliptic curve known as generating point. The private key denoted by E_p is a Hash value of 512 bytes iris code of genuine user and public key E_q is generated with the help of private key.
$E_q = E_p * G$, where E_p is private key, and G is generating point.

Signature Generated by Sender: To sign the OTP (one-time password), the following algorithms are used:

a. Evaluate $e_{\text{otp}} = \text{Hash}(otp)$ where Hash() is a cryptographic Hash function.
b. Evaluate $r_{\text{otp}} = x_{\text{otp}} (\bmod n)$
   where $(x_{\text{otp}}, y_{\text{otp}}) = k * G$, $k \in [1, n]$, and n is order of the point G.
c. $s_{\text{otp}} = k^{-1} (e_{\text{otp}} + E_p r_{\text{otp}}) \bmod n$, k^{-1} is the multiplicative inverse of k.
d. The pair $(r_{\text{otp}}, s_{\text{otp}})$ is the resultant signature denoted by S.

OTP (one-time password) and S are share to the receiver. E_q and G are used for the signature verification purpose.

Verification of Signature by the Receiver: The following algorithms are used for verification purpose:

a. The signature is only valid if the received value of r_{otp} and s_{otp} is in the range $[1, n-1]$.
b. Evaluate $e_{\text{otp}} = \text{Hash}(otp)$, Hash () is the same function used in signature generation.
c. Calculate $w_{\text{otp}} = s^{-1} (\bmod n)$.
d. Calculate $u_{\text{otp}} = e_{\text{otp}} w_{\text{otp}} (\bmod n)$ and $v_{\text{otp}} = r_{\text{otp}} w_{\text{otp}} (\bmod n)$.
e. $(m_{\text{otp}}, n_{\text{otp}}) = u_{\text{otp}} * G + v_{\text{otp}} * E_q$.

Finally, the signature is considered as a valid if $m_{\text{otp}} = r_{\text{otp}} (\bmod n)$.

4 Proposed Work

In the enrolment phase, the fingerprint feature extraction is carried out using Minutia Extractor algorithm, and the extracted feature is stored in Fingerprint database, and face feature extraction is carried out using LBP Extractor algorithm and extracted feature are stored in Face database and iris feature extraction is carried out using Log Gabor extractor algorithm, and finally, the extracted features are stored in Iris database shown in Fig. 1. The verification phase is divided into two sections. In the first level of verification, the resultant score after the score level fusion is compared with the system threshold value, and if it is found to be more than threshold value, then the user is considered as a genuine user and can access the second level of verification phase; otherwise, the user is considered as an imposter. In the second level of verification phase, we use the concept of biometric signature based on elliptic curve cryptography (ECC), where the signature is generated with the help of user iris code to provide security to the OTP, which is generated by the bank server and send to the user mobile. So in this paper, we use biometric signature as a tool to provide security. And finally, genuine user verifies the signature and accesses the original OTP for making transaction. The verification phase is shown in Fig. 2.

Fingerprint Feature Extraction
In this paper, we use the minutiae based algorithm for feature extraction. Each feature vector consisting of minutia point along with following parameter: (a) x-coordinate, (b) y-coordinate, and (c) orientation i.e. (x, y, φ).

Fingerprint Preprocessing: Image enhancement is the first step in preprocessing and for this Histogram equalization and Fast Fourier Transform (**FFT**) are used [11]. The main role of Histogram equalization is to organize the pixel value distribution of an image. **FFT** are applied to joins the false broken point of ridge and improve the contrast between ridge and furrows. Binarization is the process to convert the grayscale image into a binary image with 0-value for ridge and 1-value for furrows [12].

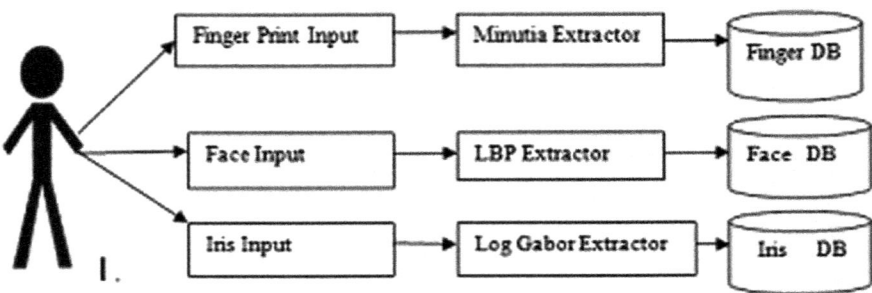

Fig. 1 Enrollment phase

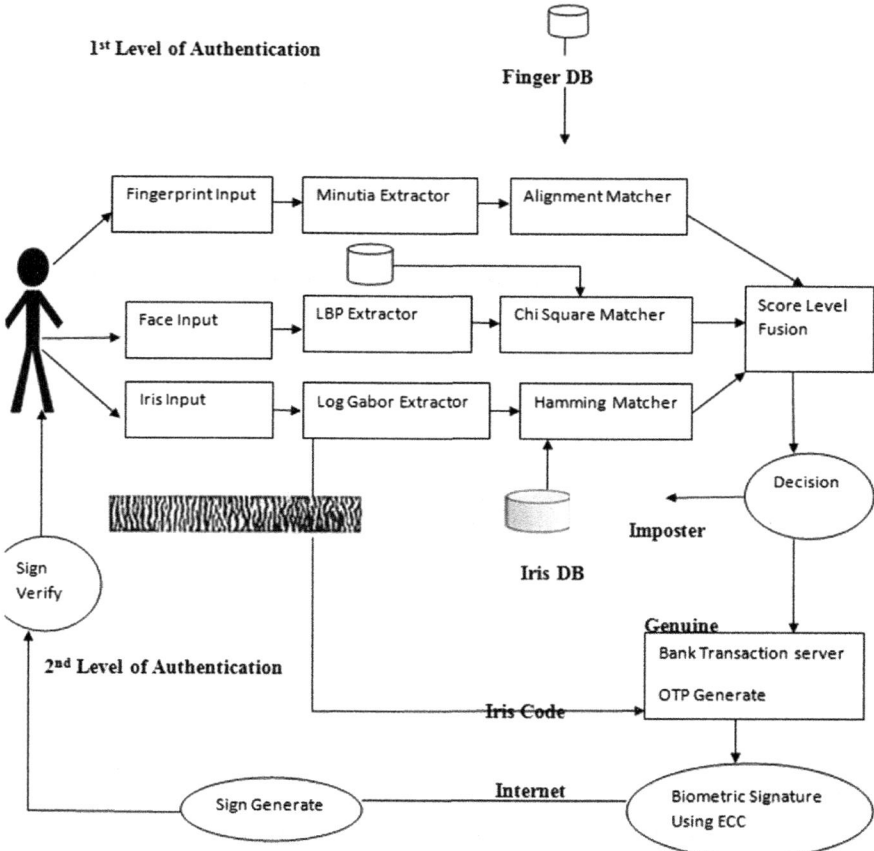

Fig. 2 Verification phase

Fingerprint Minutiae Extraction: The enhanced fingerprint image is then used for minutia point extraction. The whole minutia extraction process carried out with the help of bifurcation and termination.

a. **Bifurcation**: In this process, the ridge is treated as a bifurcate if it has three ridge pixels as a neighbor.
b. **Termination**: In this process, the ridge is treated as a terminated if it has two ridge pixels as a neighbor.

After the minutia extraction, the next step is selection of region of interest (ROI) [13].

Iris Feature Extraction: The following are the steps for iris feature extraction.

Iris Segmentation: The iris segmentation is the first step of iris feature extraction. The main goal of segmentation is to extract the required iris region from the inner and outer boundary of eye [12]. The two well-known methods of iris segmentation are canny edge detection [14] and circular Hough transform [14]. The basic purpose of both these methods is to detect the iris boundaries and find their radius and center. One of the main features of canny operator is that it can handle the noisy image very well. Canny edge detection algorithm is used to get the exact boundaries of the iris image. [15].

Iris Normalization: The main purpose of normalization is to transform the segmented iris region into fixed dimension. The iris matching is affected by some problem like pupil dilation and stretching of iris [12]. The most common normalization method proposed by Daugman is Daugman's rubber sheet model [16] which state that the remapping $(x(r, \varnothing), y(r, \varnothing)) \to I(r, \varnothing)$ of iris image from Cartesian coordinate to polar coordinate system can be represented as

$$x(r, \theta) = (1 - r)x_p(\theta) + r\, x_s(\theta),$$
$$y(r, \theta) = (1 - r)y_p(\theta) + r\, y_s(\theta),$$

where $(x_p(\varnothing), y_p(\varnothing))$ and $(x_s(\emptyset), y_s(\emptyset))$ are respectively the coordinate nearest to the papillary boundaries and r is radius in the interval [0, 1].

Face Recognition with LBP

In this paper, we use local binary pattern (LBP) for feature extraction of face. LBP is one of the most successful feature extraction method for face introduced by Ojala et al. [17].

The working procedures of LBP are as follow:

a. If any 8 neighbor pixels around the center pixel have a higher or equal gray value than the center pixel then center pixel has a value "one".
b. Otherwise, the center pixel has a value "zero".

Finally, the LBP code for center pixel is calculated by combining the eight one or zero to a binary form which is shown in Fig. 3. The next step of feature extraction is to find the different size of neighborhood points of center pixel. For this, a circle with radius R_{face} is formed with reference to center pixel. Some x sample points are selected from the surface of the circle, and value of each pixel around the circle is compared with center pixel. For this purpose, bilinear interpolation is used which is shown in Fig. 4 where we show three circles with different values of P and R. LBP is one of the simple methods of feature extraction used in many application of facial recognition system. Different versions of LBP are completed LBP (CLBP), dominant LBP (DLBP), and LBP histogram Fourier (LBP-HF) which outperforms the original LBP [18].

Normalization: The main task of normalization in multibiometric is to convert the matching score generated by different modalities into a common format. The most

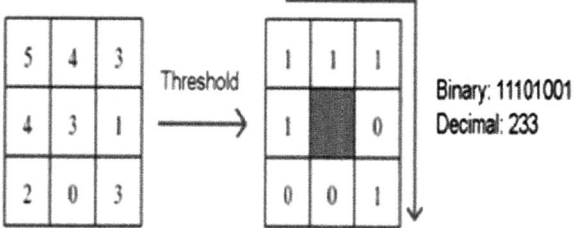

Fig. 3 LBP face

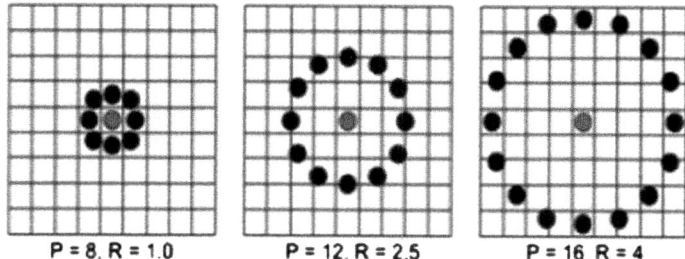

Fig. 4 Bilinear interpolation

popular normalization technique is Min-max normalization [4]. Min-max normalization is given by the following form with respect to Iris:

$$s_{\text{iris}} = \frac{M_{s_{\text{iris}-\min_{\text{iris}}}}}{\max_{\text{iris}} - \min_{\text{iris}}},$$

where max and min are the maximum and minimum score value and M_S is the generated matching score.

Weighted Sum Rule-Based Score Fusion: After we get normalized score $(x_1, x_2, \ldots, x_n)$ from a particular person, the fused score f_s is calculated using the formula $f_s = w_1x_1 + w_2x_2 + \cdots + w_nx_n$, where $n = 1, \ldots, m$ indicate the biometric matcher and w_i represent the weight assigned to the matcher i [19]. In this paper, we use the weighted sum rule-based score fusion and experimentally select the weight for fingerprint ($w_1 = 0.2$), face ($w_2 = 0.45$) and iris ($w_3 = 0.35$) such that

$$\sum w_i = 1.$$

Decision Module: The main task of this module is to decide whether the claimed user is valid or not. For this purpose, we use system threshold value as comparison parameter. The user is considered as a genuine if the resultant fused score value is above the threshold value; otherwise, it considered as an imposter. Only the genuine user can access the bank server and move to the second level of authentication phase, where an OTP (one-time password) is generated by the bank server, and it is forward to the user mobile for making the transaction.

Generation of Ecc Private and Public Key: The next step in second level of authentication is generation of private and public key using elliptic curve cryptography. In this paper, we use $y^2 = (x^3 + x + 13) \bmod 31$ as an elliptic curve, where $a = 1$, $b = 13$ and $p = 31$. $G = (9, 10)$ is the first point on elliptic curve, which is also known as a generator point denoted by G. The private key is the hash value of 512 bytes iris code denoted by d_a. And the public key $Q_a = G * d_a$ is used for the purpose of verifying the biometric signature. In this paper, we use biometric signature based on Ecc to protect our OTP from the attacker. This OTP is sent to the user mobile and user verifies the signature with the help of private and public key. The key attraction of Elliptic curve cryptography over RSA is that it offers small key size with higher security, this is one of the reasons of popularity of ECC.

Experimental Result: To evaluate the performance of the proposed system we use the database of fingerprint, iris, and face. The first fingerprint database is **CASIA-Fingerprint V5** which contains 20,000 fingerprint images for 500 subjects. Randomly we select fingerprint image of 100 subjects and each of them contains 10 samples. Among the 10 samples, 5 samples are used for training purpose and rest 5 samples are used for testing purpose. The 2nd database is **UBIRIS** iris database [20] which is a database of iris image. We randomly select iris image of 100 subjects and each of them contains 10 samples. Among the 10 samples, 5 samples are used for training purpose and rest 5 samples are used for testing purpose. The third most important database is **ORL** face database [21] consist of 2500 colored facial image of 250 subjects. We randomly select face image of 100 subjects and each of them contain 10 samples. FAR (false acceptance rate) and FRR (false rejection rate) are two important performance parameter of biometric. The performance of the system directly depends on FAR and FRR. To improve the performance of the system, we always try to minimize the value of FAR and FRR. False acceptance rate (FAR) is the probability of incorrectly accepted a person as a genuine user, due to incorrectly matching the template with the existing template at verification stage. False rejection rate (FRR) is the probability of incorrectly rejecting access to a genuine user. In this paper, we use score level fusion of fingerprint, iris, and face based on weighted sum rule. Figures 5, 6 and 7 show the

Fig. 5 Performance of score level fusion based on sum rule

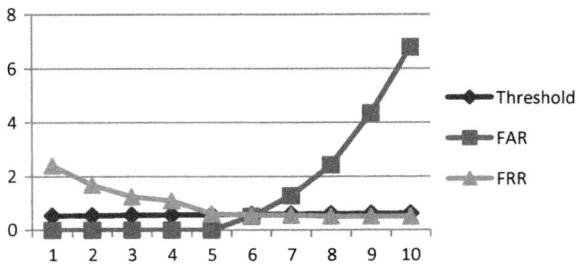

Fig. 6 Performance of score level fusion based on product rule

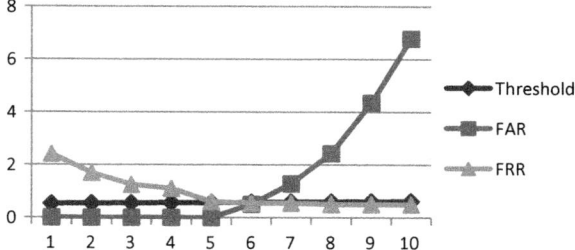

Fig. 7 Performance of score level fusion based on weighted sum rule

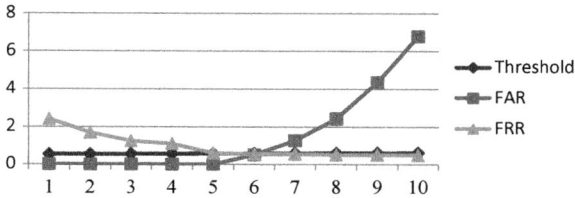

performance evaluation of score level fusion based on sum rule, product rule, and weighted sum rule. And the experiment result shows that the performance of weighted sum rule (99.8%) is far better than the performance of sum rule (98.3%) and performance of product rule (99.2%).

5 Conclusion

In this paper, we propose a multibiometric system which is based on score level fusion. We use the concept of biometric signature based on ECC to enhance the security of the system. Score level fusion plays an important role in multibiometric verification system because at this fusion level, there has sufficient information to make distinguish between genuine and imposter. Biometric play an important role to maintain the high level of security in the e-banking transaction. But unimodal-based biometric system have suffered from some problem like intra-class variation, noise in senses data, spoofing attack, and non-universality. One of the solutions of this problem is multibiometric. Basically, in verification phase, we divide the system into two modules. In the first level of authentication after the score level fusion of three modalities, the resultant score is compared with the system threshold value and if it is found to be above, then the user is considered as a genuine user and can move to the second level of authentication; otherwise, user considered as an imposter. The system uses biometric signature based on ECC to provide the security.

References

1. C. Prathipa and L. Latha, "A survey of multimodal biometric fusion and template security techniques." IJARCET volume 3, pp 3511–3516, 2014.
2. Mahto, Dindayal, and Dilip Kumar Yadav. "Enhancing security of one-time password using elliptic curve cryptography with biometrics for e-commerce applications." *Computer, Communication, Control and Information Technology (C3IT), 2015 Third International Conference on.* IEEE, 2015.
3. Oloyede, Muhtahir O., and Gerhard P. Hancke. "Unimodal and Multimodal Biometric Sensing Systems: A Review." *IEEE Access* 4 (2016): 7532–7555.
4. Anil Jain, Karthik Nandakumar, Arun Ross, "Score Normalization in Multimodal Biometric System", Journal of Pattern Recognition, Vol. 38, 2005, pp. 2270.
5. Anoop, M. S. "Elliptic curve cryptography." An Implementation Guide (2007).
6. Rajam, S. Thiraviya Regina, and S. Britto Ramesh Kumar. "Enhanced elliptic curve cryptography." *Indian Journal of Science and Technology* 8.26 (2015).
7. Mohammadi, Shahriar, and Sanaz Abedi. "ECC-based biometric signature: A new approach in electronic banking security." *Electronic Commerce and Security, 2008 International Symposium on.* IEEE, 2008.
8. Jain, Anil K., Salil Prabhakar, and Shaoyun Chen. "Combining multiple matchers for a high security fingerprint verification system." *Pattern Recognition Letters* 20.11 (1999): 1371–1379.
9. Hanmandlu, Madasu, et al. "Score level fusion of multimodal biometrics using triangular norms." *Pattern recognition letters* 32.14 (2011): 1843–1850.
10. Ghulam Mohi-ud-Din, Salah-ud-din, et al. "Personal identification using feature and score level fusion of palm-and fingerprints." *Signal, Image and Video Processing* 5.4 (2011): 477–483.
11. Sherlock, B. G., D. M. Monro, and K. Millard. "Fingerprint enhancement by directional Fourier filtering." *IEE Proceedings-Vision, Image and Signal Processing* 141.2 (1994): 87–94.
12. Aboshosha, Ashraf, et al. "Score Level Fusion for Fingerprint, Iris and Face Biometrics." *International Journal of Computer Applications* 111.4 (2015).
13. Brindha, V. Evelyn, and A. M. Natarajan. "Multi-modal biometric template security: Fingerprint and palmprint based fuzzy vault." *Journal of Biometrics and Biostatistics* 3.6 (2012): 1–6.
14. Daugman J. (2004) IEEE Journals on Circuits and Systems for Video Technology, 14(1), 21–39.
15. Gawande, Ujwalla, Mukesh Zaveri, and Avichal Kapur. "Fingerprint and iris fusion based recognition using RBF neural network." *Journal of Signal and Image Processing* 4.1 (2013): 142.
16. Altun, A. Alpaslan, H. Erdinc Kocer, and Novruz Allahverdi. "Genetic algorithm based feature selection level fusion using fingerprint and iris biometrics." *International Journal of Pattern Recognition and Artificial Intelligence* 22.03 (2008): 585–600.
17. T. Ojala, M. Pietik¨ainen and D. Harwood, "A comparative study of texture measures with classification based on feature distributions" Pattern Recognition vol. 29, 1996.
18. Ahonen, Timo, Abdenour Hadid, and Matti Pietikäinen. "Face recognition with local binary patterns." *European conference on computer vision.* Springer Berlin Heidelberg, 2004.
19. He, Mingxing, et al. "Performance evaluation of score level fusion in multimodal biometric systems." *Pattern Recognition* 43.5 (2010): 1789–1800.
20. H. Proenca and A. Alexandre, UBIRIS Iris Image Database: http://iris.di.ubi.pt.
21. AT&T Laboratories Cambridge, The ORL Database of Faces: http://www.cam-orl.co.uk/facedatabase.html.

Generation of Photographic Mosaic Using Apache Spark and Scalding for Image Processing

Santhoshi Rupa Gayatri Neralla

1 Introduction

In recent years, big data has played a vital role in the industry. Some technologies are developed to enable streamline processing of enormous amount and a wide range of data.

The amount of data created every day worldwide is growing rapidly. With the introduction of electronic innovation and smart gadgets, a huge quantity of digital information is produced every day. This huge data is hard to process making use of traditional modern technologies. Capability to assess this massive quantity of data is bringing a brand-new era of performance growth and advancement in technology. Big data is a collection of datasets so large and facility that it comes to be difficult to process it utilizing standard data source management tools or data handling applications.

Big Data Attributes: The velocity, volume, as well as variety are typically utilized to explain various facets of big data. The

Three features make it simple to specify the nature of the information and the software platforms offered to examine.

Volume: Volume is the important element of big data because it enforces a requirement for scalable storage. Huge ventures currently have a large quantity of information collected as well as archived throughout the years using system logs, record maintaining, etc. The quantity of this information conveniently specifies where traditional database management systems could not have the ability to manage it. Big data innovations provide a remedy to develop value from this substantial as well as previously unused/tough to process information.

S. R. G. Neralla (✉)
Electrical and Computer Engineering, Sungkyunkwan University, Suwon-Si, South Korea
e-mail: rupamay20@skku.edu

© Springer Nature Singapore Pte Ltd. 2019
V. Nath and J. K. Mandal (eds.), *Proceeding of the Second International Conference on Microelectronics, Computing & Communication Systems (MCCS 2017)*, Lecture Notes in Electrical Engineering 476, https://doi.org/10.1007/978-981-10-8234-4_21

Velocity: Information is moving into companies at a tremendous rate. The Internet, modern mobile innovations have allowed to produce a data recede to the customers. On the Internet purchasing has reinvented between consumer and company.

Communications: The development of the mobile phone period, there is also additionally better location-based data produced as well as it becomes vital to be able to make use of this the massive quantity of information.

Variety: Big Data is the tendency to maintain the data.

Considering that the majority of this is created as soon as and also read many times. Big Data thinks that there could be insights concealed in all information. The frequently used active open-source software framework amongst them is Apache Hadoop [1–4]. The benefit of Spark that gives the advantage over Hadoop is its light fastening speed for large-scale data processing [3].

Apache Spark exposes high-level API for, data pipelines, and one that is available in Java and Scala [5]. It is a more powerful tool than Hadoop [1].

Scalding is a Scala library that makes it simple to write MapReduce jobs in Hadoop (by eliminating the need to write the raw map and reduce functions).

This article emphasizes how Apache Spark may be used to construct an image-processing pipeline for generating photographic mosaic that consists of two dependent tasks: a map-only task and a map-reduce task. Then operate it using Spark, on datasets ranging from small to an extensive collection of library images. And hence compare the execution time of Spark with Scalding.

Photographic Mosaic: Photographic Mosaics are usually formed from a collection of still images. The effect is to recreate same source image (e.g., a face) by substituting small portions of the source image with many other images (which we will call a tile) which has the same average color in each grid of the original picture. At a distance, the mosaic will seem to look like the original picture, while up close, the individual tiles can be viewed embedded in the image. When we look as a whole, it seems to be one image, but in reality, the image consists of hundreds of smaller images. Practically, it is a problem of image search for selecting the best match for a number of input images in a set of library images and as the latter one increases it becomes a big data problem.

"We may want to know the impact of disturbance—harvesting, thinning, fires, storms—things that lead to changes in forests," said Goward.

In 1974, researchers launched the very first photomosaic of 48 adjoining Unites States. The map is made up of images from 595 cloud free black and white images gathered by Landsat in 1972 between July 25 and October 31.

The web-enabled landsat data (WELD) project currently develops mosaics of the United States yearly making use of computer system automation as well as 40 years of calibrated data.

Mosaic has actually been primarily done by artists. Nowadays, such mosaic job is taken care as one unrealistic making strategy by researchers of computer system graphics. The pictures made by this method improve the artistic property of the image. In the past, several materials need to be gathered for a very period of time for this mosaic to be created. Nevertheless, this issue is substantially resolved by low-cost digital devices that are released in wide range and variety.

As a result of current growth of modern digital technology, these photomosaic innovations are most widely used in television promotions, poster, magazines, installation music video, etc. It takes very long time to make photomosaic pictures even by specialized visual graphic developers because mosaic pictures are selected manually by them. Therefore, it is tough for general individuals to make photomosaic pictures. This paper proposes a fast automatic generation of photomosaic algorithms through block matching in Spark and scalding using Best Match selection.

There are many applications, such as medical imaging confidential enterprise archives, storage systems, systems, album of personal photograph, and also military image databases. Such kind of pictures generally contains private or confidential data so they must be protected from the loss of this private data.

A good study by Battiato classified Mosaic into four types: (i) crystallization mosaic, (ii) ancient mosaic, (iii) photomosaic, and (iv) puzzle image mosaic. First and second are acquired by breaking down a source image into sections (with various colors, appearance, dimensions, and rotations) and get back the picture by appropriately painting the tiles, therefore they both are called tile mosaics. We get third and fourth types of mosaics by putting images from a huge database to cover a chosen source image, and these both are called multi-picture mosaics. It is used for the application of hidden communication of secret images. Hence, we propose a fast and new method in this paper that generates mosaic images with large database from google.usercontent.com by using spark.mllib in Hadoop MapReduce and showed that it is 17 times faster than scalding.

The remaining part of the article is presented as follows: Sect. 2 discusses the literature survey, in Sect. 3, we discussed the tools and technology used, and in Sect. 4 we propose a methodology, two distinct ways of creating mosaics and their algorithms are presented. Section 5 elucidates the experimental results, compared graphs are shown in this section. Section 6 discusses the mathematical modeling, and Sect. 7 presents comparison work, and in the final sections, we will discuss the conclusion and the future work.

2 Literature Review

Many frameworks have been developed to facilitate processing of complex data pipelines and to widen the productivity margins of evolving MapReduce applications. Some of the productivity frameworks include Apache pig, Crunch and Cascading [6]. Many frameworks like Clojure-based Cascalog, Scala-based Scoobi, Scalding, and Scrunch have been developed using programming languages which support functional-like method for performing Hadoop computations. Due to high adaptability and adoption, Scala (Crunch) frameworks are more in number.

Szul and Bednarz [7] illustrate how Scalding framework can be used to create an effective solution to a big data image-processing problem of creating photographic mosaics and had compared it to a Hadoop API (MapReduce) based implementation.

An automatic photomosaic algorithm through block matching and intensity adjustment has been proposed by Lee [2]. A framework called E-scalding that creates photographic mosaic using cloud storage over big data and Microsoft Azure acting as a storage repository for the image dataset is proposed in [3]. In this image fusion technique is applied for optimizing to improve the resultant mosaic cell and compared performance in Hadoop (MapReduce), scalding and E-scalding with image fusion.

We would like to determine the effect of disruption- towards, fires, harvesting, thinning factors that cause changes in forests, "stated Goward". In 1974, researchers launched the very first photomosaic of 48 adjoining Unites States. The map is made up of images from 595 cloud free black and white images gathered by Landsat in 1972 between July 25 and October 31. The web-enabled landsat data (WELD) project currently develops mosaics of the United States yearly making use of computer system automation as well as 40 years of calibrated data [8].

By developing mosaic picture a brand-new strategy for safe and secure image transmission is suggested in [9], which converts a secret image into mosaic images which resemble just like source image.

A method that creates mosaic images with huge databases is proposed in [10].

3   Tools and Technology Used

Hadoop: It is difficult to leave out Hadoop while discussing about big data. Hadoop is the open resource software program system managed by the Apache Software. Hadoop was inspired by papers released by Google, explaining its strategy to handle an avalanche of information, and has given that become the criterion for saving, processing and also analyzing numerous terabytes, and even petabytes of data. Hadoop has attracted the inspiration from Google File System (GFS). Hadoop have three vital features (a) storage, (b) processing, and resource administration. It is currently making use of Yahoo, eBay, LinkedIn, and Facebook.

Hadoop Attributes: (a) Fault-tolerant, (b) built-in redundancy, (c) automatic scale up and down, and (d) move computation to data.

HDFS: HDFS is a distributed Java-based file system created to operate on commodity hardware. HDFS has name node and data node. The Name Node handles all HDFS metadata. It maintains the namespace and the file system and provides access to files by the user. The file system namespace operations like closing, opening, and renaming the directories are done by the name node. Name node is very important to HDFS.

Data Node: There are many data nodes, generally one per cluster that manages the storage which is attached to the nodes. A file is fragmented into one or more blocks internally and these blocks are stored in data nodes. The job of data node is to serve read and write requests. Additionally, on the instruction of the name node block creation, deletion, and replication are done by the data node. It also determines the mapping of blocks to Data Nodes. The Data Nodes are responsible for serving read and write requests from the file system's clients. The Data Nodes also perform block creation, deletion, and replication upon instruction from the Name Node.

MapReduce: MapReduce is a software application structure presented by Google to do parallel processing on large datasets assuming that large dataset storage space is dispersed over a huge number of machines. The data can be processed either on filesystem (Unstructured) or in a database (Structured). In order to reduce communication overhead MapReduce takes the advantage of locality of data Distributed processing of Map and Reduce are allowed by MapReduce paradigm. But each mapping operation must not be dependent on others, all the maps can be processed in parallel. However, in practice, it is restricted by the number of independent data sources and/or the number of CPUs near each source. In the same way, a group of 'reducers' can do the reduction phase, provided that all outputs of the map task which share the same key are given to the same reducer at the same time.

Spark: Apache Spark, is a framework for processing large data that can be run in distributed mode on a cluster. Spark applications are run as independent sets of processes on a cluster, all coordinated by a central coordinator. This central coordinator can connect with three different cluster managers, Spark's Stand-alone, Apache Mesos, and Hadoop YARN (Yet Another Resource Negotiator). Spark has a programming model which is similar to MapReduce that expands itself with a property of data sharing abstraction called resilient distributed datasets. With capabilities like real-time processing and inbuilt data storage of memory Spark takes MapReduce paradigm to the next level of hierarchy with minimum shuffles in the data processing, thus achieving performance which is several times faster than many other big data technologies Spark has mainly three interfaces:

Data Storage: Spark works fine with any Hadoop compatible sources of data including HDFS, HBase, Cassandra, etc. For data storage, Spark uses HDFS file system.
API: The API provides the application developers to create applications based on Spark using a standard API interface. Spark provides API for Scala, Java, Python programming languages.
Resource Management: Spark can be used as a Stand-alone server or in contrast can be used on a distributed computing framework like Mesos or YARN.

The Spark architecture with its interfaces and storage system is shown in Fig. 1. Spark is a framework which is often run on Hadoop YARN.

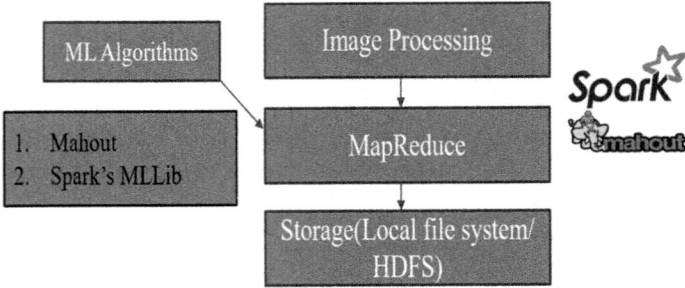

Fig. 1 Configure Spark MLlib in Hadoop MapReduce

4 Methodology

Fast Data Processing with Spark tells how to write distributed map-reduce style programs with Spark. There are many algorithms like clustering, regression, and collaborative filtering, etc., available. Spark.mllib package model in a machine learning algorithm supports many models like K-means, Gaussian mixture, etc. It solves the problem of unsupervised learning which aims to group subsets of the entities with one another depending on the likeliness and sameness. The scheduling and the job flow is managed by the driver and it is available for all the time the application is running. Though we run on yarn the driver and the client process is same to initiate the job. In interactive mode, the shell itself is the driver process.

In contrast, to Hadoop, the Spark framework constructs a Directed Acyclic Graph (DAG) of execution before it schedules the task. The information about transformations that is to be performed on data and Spark is provided by DAGs which is able to intelligently combine many of the transformations into a single stage and then execute transformations all at once—an idea initially pioneered by Microsoft Research in ProjectDryad.Spark Context driver is very important [11].

In this work, we have collected the initial source image from flicker and configured spark using MLlib library (apache.spark.mllib) in Hadoop MapReduce. First, the source image is stored in the local disk and the library set of images to be compared are from user.googlecontent.com. Transformations for generating keys are performed and actions for values are performed for image processing using flatmap, groupBy, reduce, etc., steps for reducing the computation time by dividing the images in the library to be compared into batches. The architecture used in this work to configure apache.spark.mllib in Hadoop MapReduce is depicted in Fig. 2. MLlib in Spark is an in build machine learning library which has algorithms like Clustering, regression, Collaborative filtering, etc. These algorithms work with streaming data like linear regression which used K-means clustering or ordinary least squares.

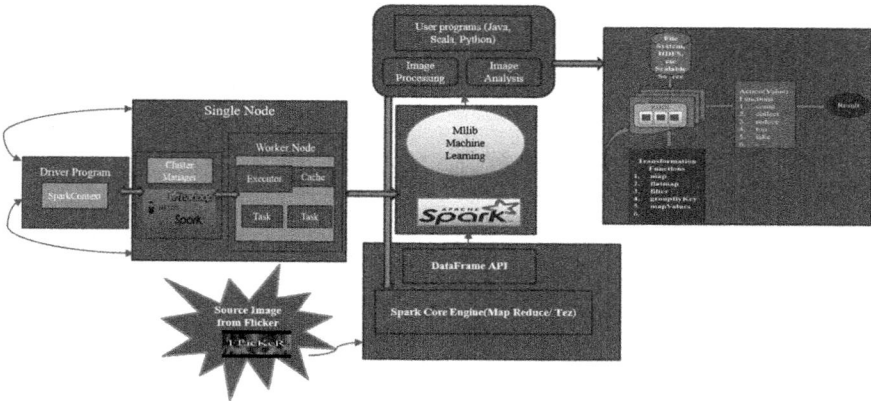

Fig. 2 Spark-based photographic image-processing architecture

4.1 Approach of the MapReduce Using MLlib to Configure Spark

Generation of mosaic could be simply viewed by a MapReduce model. During the map stage, we compute the minimum distance for the image which is most suitable to fit in the grid of the original image that has the average color when compared to many other images in the library set.

The output of the map stage is seen as the heat map showing most appropriate and feasible match (there is a mapping to one heat map for every reference image).

Reduce Phase: We merge all the heat maps in reduce stage as shown in Fig. 3. Therefore, find the image which best suits for each and every grid cell (here, for example, skin color) and choose it as the replacement.

The flowchart is shown in Fig. 4.

In Fig. 4 the first step in the flowchart is that it reads the source image here stored in local disk taken from flicker. Next, we collect the library images and choose dimensions for the mosaic grid. Depending on the tile dimensions, the

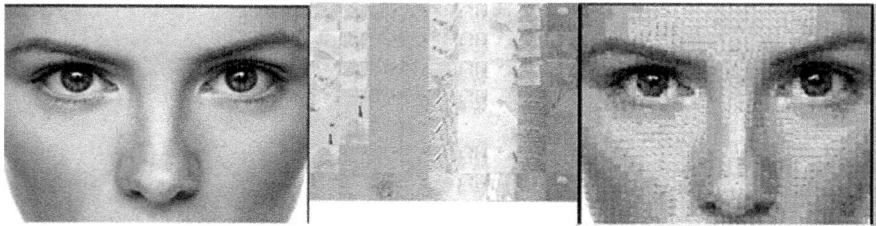

Fig. 3 Heat maps produced in map phase and choosing best match in reduce phase to create mosaic

Fig. 4 Process for generating
photographic mosaic

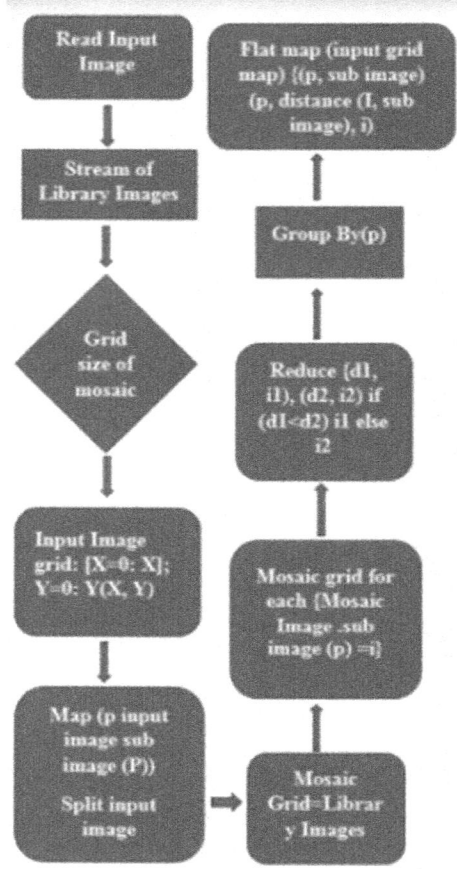

source image is split into grids in the next step, and then selects suitable images
from the library images and then in the reduce phase I selects the most suitable
image at a minimum distance which has same average color of the grid. Finally, it
embeds the tiles on the grids of the source image thus creating a mosaic.

The pseudo-code (Scala) for the scalding is represented by an algorithm below:

Algorithm 1: Proposed Approach

1. Read source image
2. Read (id, image) // Read Library images
3. Filter RGB images (id, image)
4. Distance calculation (pos, dis, id) // Calculate the distance to cells
5. Group By position (pos, [dis, id)]) // Groups by position
6. Minimum distance (pos, min_id) // calculate the min distance

7. Join By 'id='min_id // if image is at a minimum distance place it
8. Write (pos, img) // Write Mosaic Grid

Scalding reduce tasks is the best Matches Selection step and the Join step [3, 7]. The mosaic stream does not have any particular perception of MapReduce phases. But still, it is changed to a streaming pipeline of two Maps. The conceptual flow for creating photographic mosaics is shown in Algorithm 1.

First, we read the source image from flicker.

The second step is to create input stream of tuples (id, image) that is the library images.

And then the third step is to filter those images based on RGB pixel values.

The fourth step is to calculate the distance.

And in the fifth step we Group By the position all the selected images in the library.

In the sixth step, we will calculate the minimum distance and check for the likeliness or sameness with the grid cell in the source image for same average color.

In the seventh step, we transform the pipeline in order to produce (pos, min-id) tuples, which are then combined with the original stream by Join By step.

And in the final step, the image is written as (pos, image) to the output.

5 Experimental Results

The generation of mosaic building algorithm has been carried out in two ways

1. Scalding: Scalding is a Scala library which helps to define MapReduce tasks. Scalding flow was operated on the datasets. It has fewer lines of code by using Group By and Join steps. Scalding is a Scala library which is built on top of cascading using "import cascading.flow.hadoop.HadoopFlowprocess". Scalding performs data analytics on datasets with a Scala-based API.
2. Apache Spark: In MapReduce used Apache.spark.Mllib library to configure the Spark Configuration. Maven is a dependency to configure the spark and it is a tool used for building Java projects [12].

Map side of Scalding is optimized with Partial Aggregation though it does not use combiners. In MapReduce used apache.spark.mllib library to configure the Spark Configuration write the code as shown in Fig. 5.

MLlib is inbuilt Spark's adaptable machine learning library comprising of many algorithms and utilities like regression, clustering, collaborative filtering, classification, dimensionality reduction [12].

Here the keys are the Image URLs and values are the binaries of the pixel values.

These are combined with Hadoop files of format [Text, Bytes Writable]. The URLs of the images and their binary values are utilized as [key, value] pairs, respectively. Moreover, the final photographic mosaic is scaled to 8. The spark is configured with bindings as shown in Figs. 6 and 7.

```
History    
import org.apache.hadoop.mapreduce.lib.output.FileOutputFormat;
import org.apache.hadoop.mapreduce.lib.output.SequenceFileOutputFormat;
import org.apache.hadoop.util.GenericOptionsParser;
import org.apache.spark.SparkConf;
import org.apache.spark.api.java.JavaSparkContext;
import org.apache.spark.mllib.linalg.Vector;
import org.apache.spark.mllib.linalg.Vectors;
import org.jfree.ui.RefineryUtilities;
public class CreateMosaicMapReduce {
    private static final String CONFIG_NAME_PSZUL_MOSAIC_SOURCE_FILE =
    private final static Path tempPath = new Path("target/mosaic-temp");
    public static int c=0;
    public static class CreateMosiacMapper extends Mapper<Text, BytesWri
        private int cellSize = 30;
        private int computeSize = 10;
        private Map<Integer, BufferedImage> gridCells = null;
```

Fig. 5 Sample code of configuring Spark

The minimum distance is calculated and the image which best fits in the grid as tile with dimensions 30 * 10 in the original image is selected as shown in Fig. 8.

Even for the biggest datasets, this configuration enables all the map jobs to run in parallel. Every dataset comprises of 30×10 JPG embedded images, which are collected from Flicker [13].

After the selection of the images from the library, the mosaic was created in the original image. The original and Photographic mosaic were shown in Fig. 9.

It checks the extension of the image for (png, jpg, jpeg, etc.). It compresses the library image into thumbnail to fit into the grid. Then calculates distance. The equation for calculating minimum distance between two points is given by Euclidian distance is given by Eq. (1) and is discussed in detail [2].

$$d(p,q) = \sqrt{(p_1 - q_2)^2 + \cdots + (p_n - q_n)^2} = \sqrt{\sum_{i=1}^{n} (p_i - q_i)^2} \qquad (1)$$

In order to measure the likeness of RGB pixel values, the above Eq. (1) is modified as Eq. (2).

```
43      private final static Path tempPath = new Path("target/mosaic-temp");
44   public static int c=0;
45   public static class CreateMosiacMapper extends Mapper<Text, BytesWritable, IntWritable, Text> {
46      private int cellSize = 30;
47      private int computeSize = 10;
48      private Map<Integer, BufferedImage> gridCells = null;
```
```
Output - Apache_Spark_and_Map_Reduce (run)
  run:
  SLF4J: Class path contains multiple SLF4J bindings.
  SLF4J: Found binding in [jar:file:/media/4AD2B41F02B41165/2015-16hadoopprojects/Apachespark/code/Apache_Spark_and_Map_Reduce/spark-assembly-1.4.0-hadoop2.4.0.jar!/
  SLF4J: Found binding in [jar:file:/media/4AD2B41F02B41165/2015-16hadoopprojects/Apachespark/code/Apache_Spark_and_Map_Reduce/spark-examples-1.4.0-hadoop2.4.0.jar!/
  SLF4J: Found binding in [jar:file:/media/4AD2B41F02B41165/2015-16hadoopprojects/Apachespark/code/Apache_Spark_and_Map_Reduce/MRStreamer-assembly-0.9.2.jar!/org/slf
  SLF4J: See http://www.slf4j.org/codes.html#multiple_bindings for an explanation.
  SLF4J: Actual binding is of type [org.slf4j.impl.Log4jLoggerFactory]
  WARNING: org.apache.hadoop.metrics.jvm.EventCounter is deprecated. Please use org.apache.hadoop.log.metrics.EventCounter in all the log4j.properties files.
  fdmfdmfdfd
  16/11/10 17:52:56 INFO spark.SparkContext: Running Spark version 1.4.0
  16/11/10 17:52:59 WARN util.NativeCodeLoader: Unable to load native-hadoop library for your platform... using builtin-java classes where applicable
  16/11/10 17:52:59 INFO spark.SecurityManager: Changing view acls to: projects
  16/11/10 17:52:59 INFO spark.SecurityManager: Changing modify acls to: projects
  16/11/10 17:52:59 INFO spark.SecurityManager: SecurityManager: authentication disabled; ui acls disabled; users with view permissions: Set(projects); users with mo
  16/11/10 17:53:00 INFO slf4j.Slf4jLogger: Slf4jLogger started
  16/11/10 17:53:00 INFO Remoting: Starting remoting
```

Fig. 6 Output showing that the Spark is configured

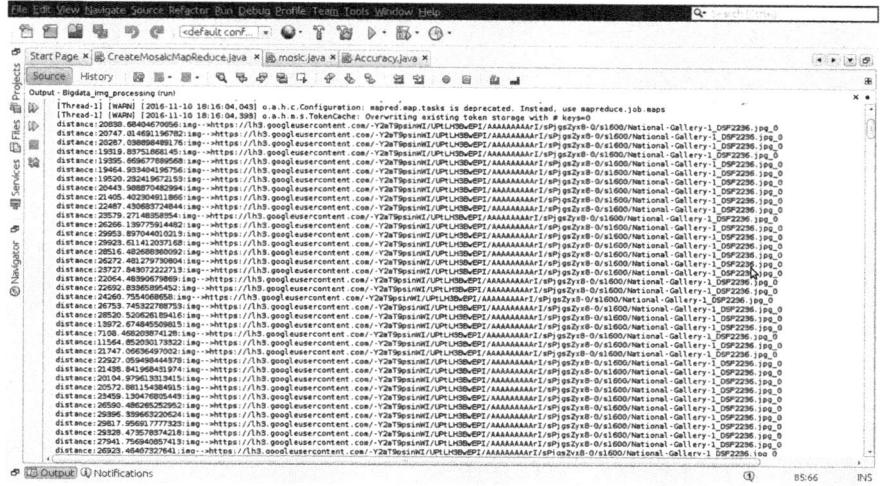

Fig. 7 Best match selection and calculating the distance

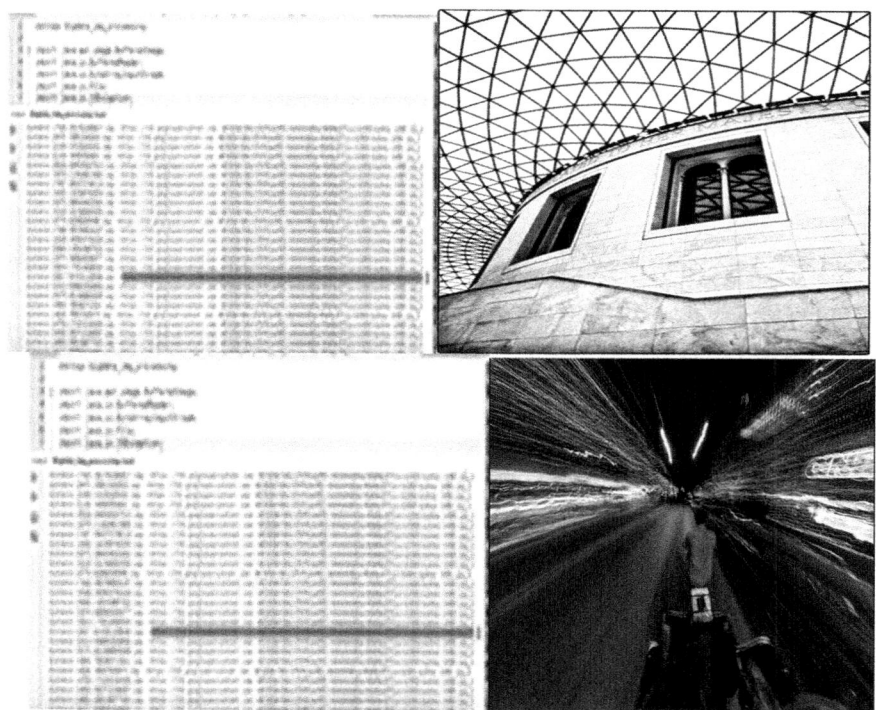

Fig. 8 Streaming images from online for BestMatchSelection

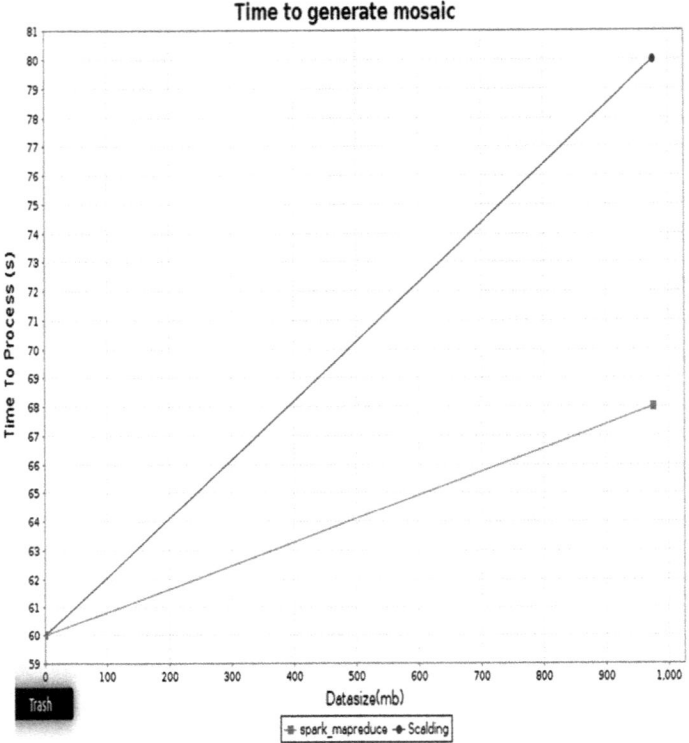

Fig. 9 The comparison graph of spark and scalding execution times for generating mosaic

$$D = \sqrt{(r_1 - r_2)^2 + (g_1 - g_2)^2 + (b_1 - b_2)^2} \tag{2}$$

where D is the difference between the Euclidean distances between compared pixels of the RGB values. The compared pixels seem to be more similar when the D value is low.

The images are compared with the grid of the original image continuously from the library set for BestMatchSelection with dimensions as shown as sample in Fig. 8.

And finally, at a minimum distance, the image which has dimensions of 30 * 10 and almost same average color of the block is selected as the tile to fit in the block.

The results indicate that Spark and scalding API's perform very well for large datasets, but the spark is 17% faster than scalding as shown in Fig. 9. Spark and Scalding platforms are used to create mosaics for datasets that are varying between 0 MB to 1 GB on a single-node Hadoop cluster which is 200 GB HDD, Intel(R) CPU i7- 5500 2.4 GHz CPU and RAM of 12 GB, Spark and Hadoop running on

Table 1 Original, small, and large dataset mosaic sizes

Time(s)	Spark	Scalding
Big-image set	69	80
Small-image set	60	62

Ubuntu 14.04. We varied the datasets sizes and compared the execution time for spark_mapreduce and scalding.

The comparison of times between Spark and Scalding was compared in Table 1.

The results show that Spark takes less time than scalding for large image set when compared to small-image set.

Therefore, for large image datasets, Spark is more effective.

The pixel resolution of the original image is 1152 * 921 and the pixel resolution of mosaic image is 1850 * 1450. The details are presented in Table 2.

The sizes of the original and the photographic mosaic image were shown in Table 2.

The original image is embedded with small rectangular blocks which are selected according to BestMatchSelection algorithm. The blocks all are of equal size of 30 * 10. And the size of small dataset is 507.9 kB and that of large dataset is 397.7 kB. The size of original image is 312.74 kB.

The heat map which is most suitable for many generated heat maps in map phase is selected at a minimum distance in the reduce phase. The minimum distance is calculated by the Euclidean distance formula.

The original face image and the tiles embedded mosaic image are shown below in Fig. 10.

Table 2 Comparison of time for generating mosaics in Spark and Scalding

Tile size	30 * 10
Original image size	312.74 kB
Size of small dataset mosaic	507.9 kB
Size of large dataset mosaic	397.7 kB
Pixel resolution of the original image	1152 * 921
Pixel resolution of the mosaic	1850 * 1450

Fig. 10 The original and
photographic mosaic images

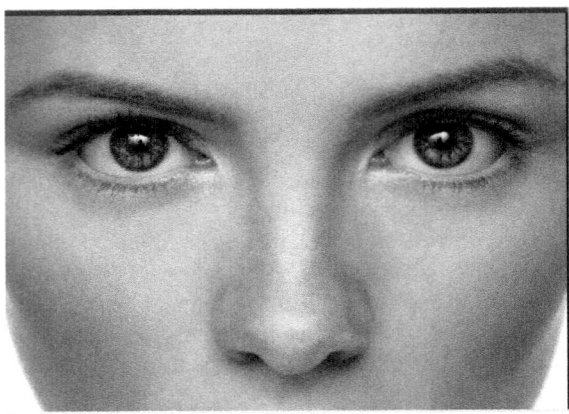

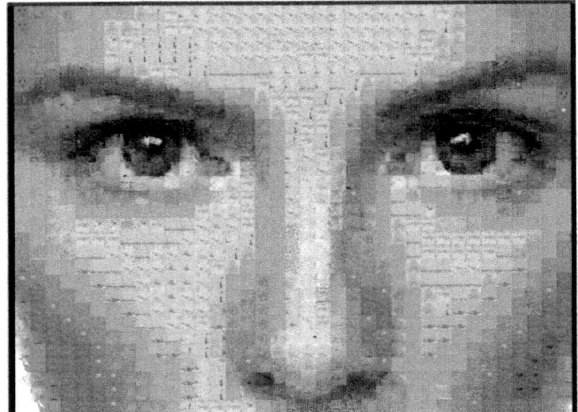

6 Mathematical Modeling

Time taken for a generation of the photographic mosaic was calculated from the
graph as the difference between the time taken to process in scalding (t_s) and spark
to the time taken to process in spark (t_k).

$$((t_s - t_k)/t_k) * 100 = \text{percentage spark is faster than scalding.}$$

7 Comparison with Previous Work

In this work, Hadoop MapReduce using spark.mllib is 17 times faster than
Scalding. Also, the lines of code are less for scalding. In [7], they have used 6-node
cluster for analyzing data of 9 MB to 6 GB in 64 GB of RAM and proved that

scalding is 13 times faster than Hadoop MapReduce. Also, in [3] authors mentions E-scalding technique improved version of traditional scalding methods and proved that E-scalding is better than Hadoop MapReduce and Scalding in a 6 node cluster in 64 GB RAM for data ranging from 10 MB to 64 GB.

But in this work, we used only single node and achieved 17 times better performance than scalding.

Apache Spark implementation for image processing Photographic Mosaic is nowhere reported in the literature. Hence, it is a novel way to conclude that this work is new and Apache Spark is faster than any other framework for large datasets.

Software Used This work is carried out in Ubuntu 14.04 version on VMware workstation and has used Java netbeans ide.

1. Hadoop
2. Spark
3. JDK-1.8
4. Netbeans-8.0
5. ubuntu-14.04 LTS

We run this on a single-node cluster in 4 GB RAM.

8   Conclusion and Future Work

This paper presents the implementation of photographic mosaic using the Apache Spark framework for improving the speed when compared to Scalding and Hadoop to the problem of image processing for large datasets for developing a photographic mosaic. The purpose of this work done within this paper was to show the applicability of the Apache Spark framework for construction of image-processing pipelines also to enhance the operation of the execution when compared to Hadoop API and Scalding. Although the mosaic production is easy, it is substantially beneficial from the high-tech performance offered by scalding (e.g., group By or join) and many intricate pipelines might be constructed. Scala has a less code complexity, Apache Spark framework is much faster for processing large datasets. Further, work will concentrate on the photomosaic image through block matching and image adjustment and execution on a Hadoop cluster with GPU service.

Acknowledgements Our sincere thanks to Prof. Dong Ryeol Shin for their continuous support and inspiration. This work was supported financially by the MISP (Ministry of Science, ICT and Future Planning), Korea under the National Program for Excellence in SW(2015-0-00914) supervised by the IITP (The Institute for information and communications Technology Promotion).

References

1. Cartwright, Angela (2007) *Mixed Emulsions: Altered Art Techniques for Photographic Imagery.*
2. Lee, H. "Generation of Photo-Mosaic Images through Block Matching and Color Adjustment". World Academy of Science, Engineering and Technology, International Science Index 87, International Journal of Computer, Electrical, Automation, Control and Information Engineering (2014), 8(3), 457–460.
3. U. Mehraj Ali, A. John Sanjeev Kumar and Anantha Kunar, 2016. Hybrid Big Data Image Processing for Creating Photographic Mosaics Using Hadoop and Scalding. Asian Journal of Information Technology, 15: 418–423.
4. K. Wiley, A. Connolly, S. Krughoff, J. Gardner, M. Balazinska, B. Howe, Y. Kwon and Y. Bu, "Astronomical Image Processing with Hadoop," in *Astronomical Data Analysis Software and Systems XX. ASP Conference Proceedings*, 2011.
5. Zaharia, M.: An Architecture for Fast and General Data Processing on Large Clusters. Association for Computing Machinery, New York, NY, USA (2016).
6. Almeer, Mohamed. (2012). Cloud Hadoop Map Reduce For Remote Sensing Image Analysis. J. Emerg. Trends Comput. Inf. Sci. 3.
7. Szul, Piotr, and Tomasz Bednarz. "Productivity Frameworks in Big Data Image Processing Computations - Creating Photographic Mosaics with Hadoop and Scalding, Procedia Computer Science, 2014.
8. https://earthobservatory.nasa.gov/Features/LandsatBigData/.
9. http://inpressco.com/wp-content/uploads/2015/01/Paper32195-198.pdf.
10. http://www.ijcttjournal.org/Volume4/issue-5/IJCTT-V4I5P86.pdf.
11. https://spark.apache.org/docs/1.6.0/programming-guide.html#resilient-distributed-datasets-rdds.
12. Meng, X., Bradley, J., Yavuz, B., Sparks, E., Venkataraman, S., Liu, D., Freeman, J., Tsai, D., Amde, M., Owen, S., Xin, D., Xin, R., Franklin, M.J., Zadeh, R., Zaharia, M., Talwalkar, A.: Mllib: Machine learning in apache spark arXiv:1505.06807 (2015).
13. Twitter, "Scalding," 2011. [Online]. Available: https://github.com/twitter/scalding.

Mitigation of Congestion in Transmission Line Using Series Smart Wire

Nibha Rani, Pallavi Choudekar, Divya Asija and P. Vishnu Astick

1 Introduction

In the multi-buyer/multi-seller system, congestion of transmission lines has become a severe problem due to relatively extension of load. The ever-increasing load demand has further necessitated the deployment of power generated by renewable energy sources in the market. This leads to distributed generation which has further intensified the congestion of lines. Out of the various problems in modern deregulated power system network, Definition of congestion is as the inefficacy of conducting line to carry power optimally. Due to increase in demand, some transmission lines are overloaded and that leads to congestion in transmission network. This further effects the performance of transmission line and results in reliability issue.

To overcome this congestion, various new technologies have been employed like construction of new line, various FACTS devices, phase shifting transformer and many others. But all these technologies suffer from pricing or reliability issues for adoption. Out of all the alternatives, D-facts are considered to be more cost-effective.

In this paper SW module in series with conducting line is used for management of congestion. Smart wire module in series introduces extra reactance in line which limits the flow during congestion and hence shifts the load to underutilised line.

N. Rani (✉) · P. Choudekar · D. Asija · P. Vishnu Astick
Amity University, Noida, Uttar Pradesh, India
e-mail: nibharani279@gmail.com

P. Choudekar
e-mail: pallaveech@gmail.com

D. Asija
e-mail: dasija@amity.edu

P. Vishnu Astick
e-mail: astick.vishnu@gmail.com

© Springer Nature Singapore Pte Ltd. 2019   249
V. Nath and J. K. Mandal (eds.), *Proceeding of the Second International Conference
on Microelectronics, Computing & Communication Systems (MCCS 2017)*, Lecture Notes
in Electrical Engineering 476, https://doi.org/10.1007/978-981-10-8234-4_22

This SW module activates whenever the congestion is detected during normal operating condition. It bypasses the circuit leaving passive assets in circuit. Hence, works on fail normal operation. Since smart wire module is connected in series it will add introduction to line but not able to reduce it, thus during it shifts power to another line reliving it further without construction of any new line.

The proposed method has been validated on IEEE 15 bus system using MATLAB. The IEEE 15 network has total active power generation as $P = 1.29$ MW, reactive power generation as $Q = 1.31$ MVAR. Total active power demand $P = 1.23$ MW and reactive power demand $Q = 1.25$MVAR. It has 15 loads across 15 buses. The centre of bus carries **11 kV**.

In this paper, Sect. 1 deals with introduction of the congestion management and smart wire. Section 2 comprises literature survey. Section 3 deals with working of smart wire connected in series. Section 4 simulation and results obtained from the proposed model. Section 5 deals with conclusion and further scope of the model.

2 Literature Survey

Congestion management can be done by using FACTS control devices. The optimal location of FACTS control devices can be optimised by using various intelligent techniques. For optimally locating FACTS for minimisation of losses, load at each bus and voltage stability is considered [1].

FACTS devices location can be optimised for controlling the congestion by developing simple and effective model for its correct location. 5 and 75 bus system were used for testing [2].

Adoption of Facts devices is widely hampered by its cost and reliability issue. Hence, DFACTS devices became an alternative approach for cost-effective power flow control [3, 4].

The optimal location of DFACTS devices is done on the basis of sensitiveness of conducting line losses w.r.t. the line impedances [5].

A new distributed approach for congestion management by increasing system reliability can be directly achieved by using current limiting diodes [6].

Power flow control in line can be achieved by DSS compensation. Research shows that DFACTS devices provide cost-effective congestion management. DSSC directly connected to heavy voltage lines do not essentially need high voltage insulation, thus have lower cost [7].

Minimisation of losses by rearranging the network shows that losses have been maintained low. Smart wire module can be considered for overloaded lines for congestion management [1].

3   Operation of Smart Series Wire

The power in transmission line flows from generating bus to meet the load demand at load bus. The transmitted power is inversely related to the reactance of the line and directly to the sine of angle between generating bus and load bus.

$$P_{ij} = \frac{V_i V_j \sin \delta}{X_{ij}} \tag{1}$$

where V_i is voltage ith bus V_j is jth bus voltage.

From the above equation, real power flow control of the line can be done by either changing angle between bus voltages or by changing line impedance.

A phase shifting transformer can change the angle between bus voltages but it would be expensive to implement phase shifting transformer hence it limits dynamic control. Otherwise, a smart wire module can control the reactance of the line by allowing to control the real power flow. Reactance is inversely proportional to power hence as the reactance increases, the power of the line reduces.

Smart wire was developed by Georgia tech it introduces control of reactance in the transmission line to control the transmission line power hence manages congestion problem.

The idea of the smart wire module is to have a system with a single turn transformer and a switch (static) with single thyristor connection. The air gap of smart wire is responsible for production of required magnetising inductance, thus eliminating the use of additional inductor. The STT turn ratio is chosen such that it would always guarantee small secondary current even during faulty conditions for proper working of switch (static). For switching current liming diode is used (Fig. 1).

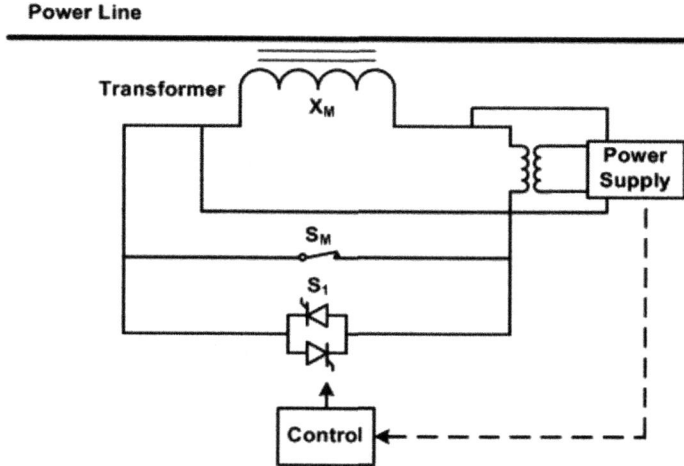

Fig. 1 Proposed smart wire module

Current limiting diode operates up to a defined limit. If the current exceeds the threshold value, diode behaves as open switch, and transformer magnetising inductance is tuned as per the expected value by setting the air gap and reactance is introduced to the line which enhances the overall reactance of transmission line, reducing power flow limit. Hence shifts the extra load to another underutilised line. The firing angle of thyristor controls the current division.

$$P_{ij} = \frac{V_i V_j \sin \delta}{X_{ij} + X} \tag{2}$$

Congestion management can be achieved by connecting smart wire in series. During normal operating condition, smart wire module is short-circuited through current liming diode. As the load increases the current raises the certain value and diode operates as open switch. Then through thyristor control, desired reactance is added to the line, limiting the flow of current in line which reduces power flow in the line shifting the extra load demand to another underutilised line.

4 Simulation and Results

For demonstration, IEEE 15 bus system is considered which is modelled in MATLAB/Simulink.

It consists of substation as its generating unit and 15 loads connected to it at each bus. Its generation capacity $P = 1.29$ MW, $Q = 1.31$ MVAR. Total load connected to it is $P = 1.23$ MW and $Q = 1.25$ MVAR. The system has been modelled at voltage 11 kV.

For implication of the proposed model, load at bus 7 is varied and line connecting bus 6 to 7 is considered congested. The results are observed in terms of line current (Fig. 2).

In Fig. 3, the model consists of a current limiting diode that bypasses the current during normal operation, and a thyristor pair that operates during congestion. STT has been used to ensure small current in secondary side during faulty condition.

Figure 4 signifies that under normal condition current in line 6–7 is 14.49 A. But as the load raises to 75%, the load current of the changes to 23.67 A hence creates congestion. As the congestion appears current limiting diode opens and extra reactance of is added in series. Thus, the overall reactance of the line increases limiting the power transfer capability. Hence shifts the extra added load to some other underutilised line.

From Fig. 5, it is observed that to increase the range of control we can further connect more number of switches in series. The connection of more no of smart wire modules makes power line behave as current limiting conductor itself, to remove congestion.

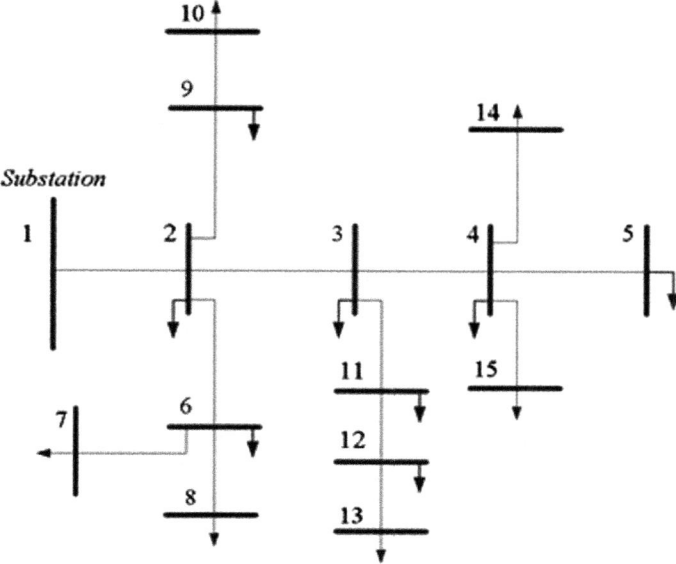

Fig. 2 Single-line diagram of IEEE 15 bus system

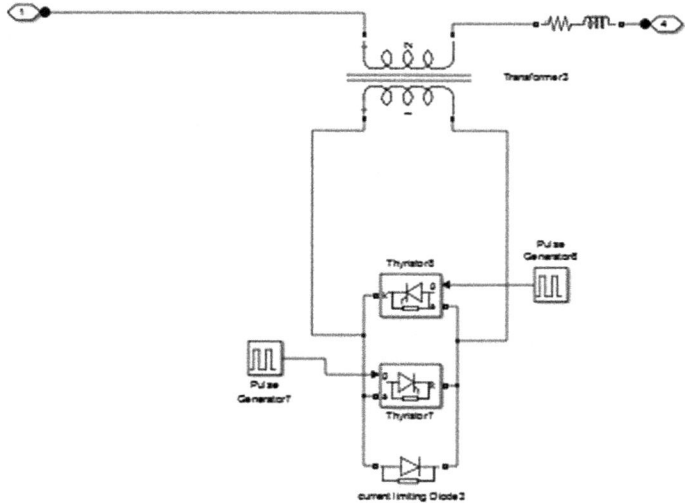

Fig. 3 Simulink model of single smart wire module

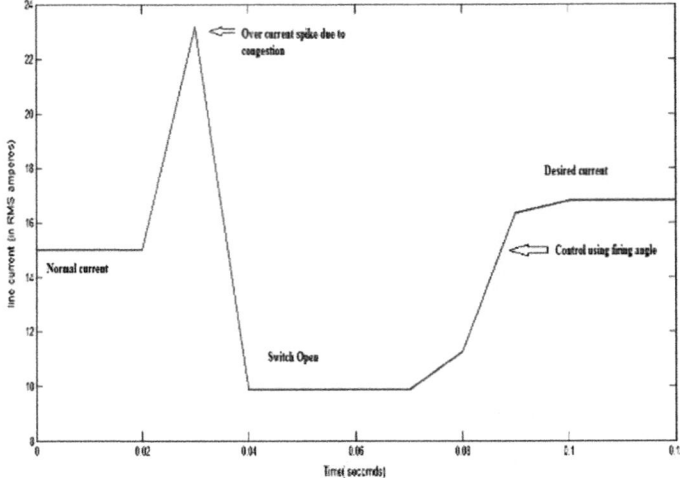

Fig. 4 Variation of line current with single smart wire module connected in series

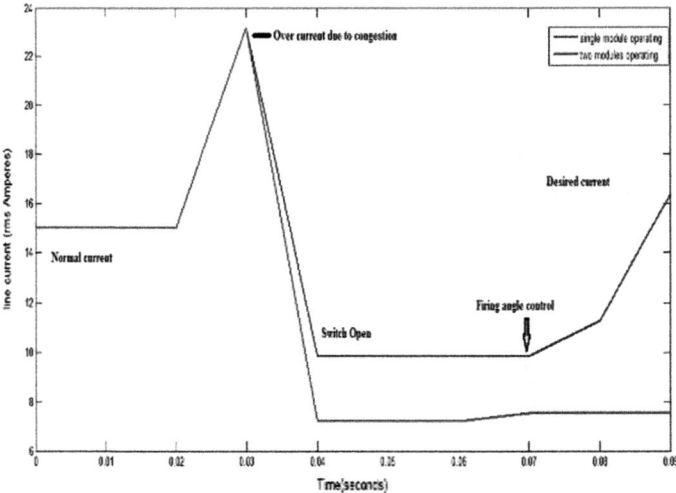

Fig. 5 Effect of adding no. of smart wire module in series

5 Conclusion

Increase in load demand, results in congestion of transmission lines. In the proposed work, the idea of using smart wire in series with power line that itself enables the conducting line to act like a current limiting conductor thus reduces overloading of line. It is very cost-effective and reliable way of congestion management.

References

1. Sananda Pal, A. Neogi, S. Biswas, M. Bandyopadhyay and S. Sengupta Loss Minimization and Congestion Management of a Power Distribution Network through its Reconfiguration-International Journal of Electrical, Electronics ISSN No. (Online): 2277–2626 and Computer Engineering 2(2): 95-99(2013) Special Edition for Best Papers of Michael Faraday IET India Summit-2013, MFIIS-13.
2. D. Venugopal, A. Jayalaxmi Blue Eyes Intelligence Engineering & Sciences Publication Pvt. Ltd. Congestion Management by Optimal Choice and Allocation of FACTS Controllers using Genetic Algorithm- International Journal of Soft Computing and Engineering (IJSCE) ISSN: 2231–2307, Volume-4 Issue-3, July 2014.
3. Electrical Transmission System & Smart Grid, Selected Entries from the Encyclopedia of Sustainability Science and Technology by Begovic, Miroslav M. (Ed.), Springer, 2013.
4. D. Divan and Harjeet Johal, "Distributed FACTS- A New Concept for Realizing Grid Power Flow Control", IEEE Power Electronics Specialist Conference 2005.
5. Frank Kreikebaum, Student Member, IEEE, Debrup Das, Student Member, Yi Yang, Member, IEEE, Frank Lambert, Senior Member, IEEE, Prof. Deepak Divan, Fellow, IEEE. Smart Wires – A Distributed, Low-Cost Solution for Controlling Power Flows and Monitoring Transmission Lines- international conference on Innovative Smart Grid Technologies Conference Europe (ISGT Europe), 2010 IEEE PES.
6. S. N. Singh and A. K. David Department of Electrical Engineering Hong Kong Polytechnic University, Kowloon, Hongkong Congestion Management by Optimizing FACTS Device Location- International Conference on Electric Utility Deregulation and Restructuring and Power Technologies, 2000. Proceedings. DRPT 2000.
7. Deepak M. Divan, Fellow, IEEE, William E. Brumsickle, Senior Member, IEEE, Robert S. Schneider, Member, IEEE, Bill Kranz, Randal W. Gascoigne, Dale T. Bradshaw, Member, IEEE, Michael R. Ingram, Senior Member, IEEE, and Ian S. Grant, Fellow, IEEE -A Distributed Static Series Compensator System for Realizing Active Power Flow Control on Existing Power Lines- IEEE transactions on power delivery, vol. 22, no. 1, january 2007.

Mobility Aware Distributed Clustering and Routing Algorithm Based on $A*$ Search for Mobile Ad Hoc Networks

Naghma Khatoon and Amritanjali

1 Introduction

The rapid advancement in wireless technology, hardware and the growth of small low powered devices make it possible to deploy infrastructure-less networks anywhere and anytime. MANETs are such a network that consists of devices on move and are deployed for some specific applications (Fig. 1). Such networks are best suitable in situations where infrastructure is not available or deployment of infrastructure is not cost effective like search and rescue operations, disaster recovery, military services and communication between vehicles and roadside equipment as vehicular ad hoc networks (Fig. 2). However, the management of such network is difficult due to its self-configuring nature and absence of any central authority. Furthermore, scalability is another issue which in last decades becomes one of the most focused areas of communication research.

Clustering is an important approach which solves many problems of MANET and provides network scalability and increases its lifetime. Here nodes are divided into virtual groups called clusters with a cluster head (CH) in each cluster which serves as a local coordinator for its cluster (Fig. 3). Communication from source to destination is done via CHs and gateway nodes which are within the transmission range of more than one CHs and thus, conserve energy of other nodes. Also, cluster-based MANET improves network management as route setup is localized with clusters and so reduces the routing table of other nodes.

However, CHs bear extra workload of intracluster and intercluster transmission. This results in early depletion of energy and death of CHs which ultimately

N. Khatoon (✉) · Amritanjali
Department of Computer Science & Engineering, BIT, Mesra, Ranchi 835215, India
e-mail: naghma.bit@gmail.com

© Springer Nature Singapore Pte Ltd. 2019
V. Nath and J. K. Mandal (eds.), *Proceeding of the Second International Conference on Microelectronics, Computing & Communication Systems (MCCS 2017)*, Lecture Notes in Electrical Engineering 476, https://doi.org/10.1007/978-981-10-8234-4_23

Fig. 1 A mobile ad hoc network

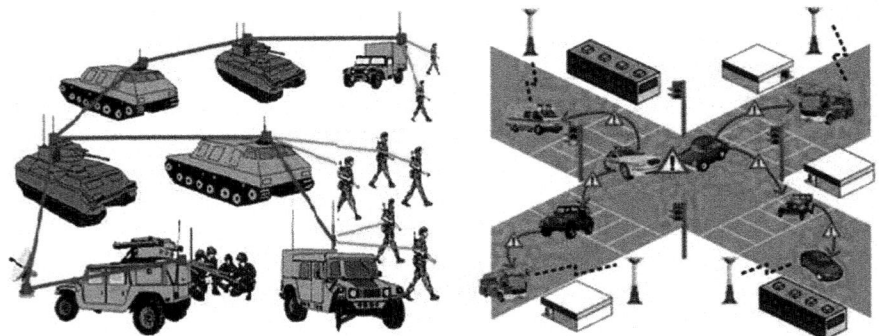

Fig. 2 MANET applications

partition the network and degrade network lifetime. Thus, designing an efficient clustering algorithm with optimal intercluster path from source to destination becomes a good approach to increase the network lifetime. Many researchers have proposed various clustering algorithms [1–6, 7] where CHs are selected prudently. However, the focus should also be given on optimal intercluster routing so as to find the best path from source to destination with cheapest cost. This ultimately will increase the lifetime of CHs by transmitting packets on optimized path. In this paper, we propose a mobility aware clustering and routing algorithm that addresses the problem of both clustering and routing in MANET.

In order to increase network lifetime of such networks, clusters are created on the basis of node's stability and depleted energy so as to select a node to serve as a CH for longer duration of time. Once CHs are decided, routing from source to

Fig. 3 An illustration of
clusters in MANET

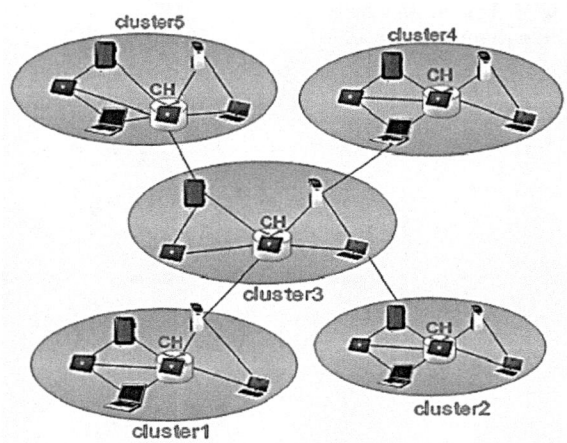

destination via CHs needs to be judiciously taken. For this, we use a heuristic approach to find the optimal path from source to destination and use the *A** graph search technique to find the cheapest route from source to destination.

2 The Proposed Algorithm

Our main objective of the proposed method is to prolong network lifetime of MANET by combining both clustering and intercluster multi-hop routing problems. The network setup is performed in three phases: CH selection, cluster formation and intercluster multi-hop routing. The details of these steps are given in the below subsections.

2.1 Cluster Head Selection and Cluster Formation

To select a most eligible node to act as CH from ordinary nodes, a combined weight of different parameters are used. Mobility leads to more number of CH reelection as well as link updating which results in poor cluster stability. Thus, for making stable clusters it is essential to consider the mobility of the nodes. The transmission range of a node say n_i forms a circle with radius r consisting of say k nodes. Inside the circle, the neighboring nodes of n_i can exist either closer to n_i or far away, but within the transmission range of n_i ($T_{\text{range}}(n_i)$). Accordingly, we can divide the transmission zone of n_i as trusted zone or risked zone [1]. The inner circle with radius $\alpha_1 r$ forms the trusted zone and the zone having width $r(\alpha_2 - \alpha_1)$ forms the risked zone (Fig. 4). The coefficients α_1 and α_2 are suitably selected based on the mobility of the nodes in the network.

Fig. 4 Transmission range
zones with mobile nodes

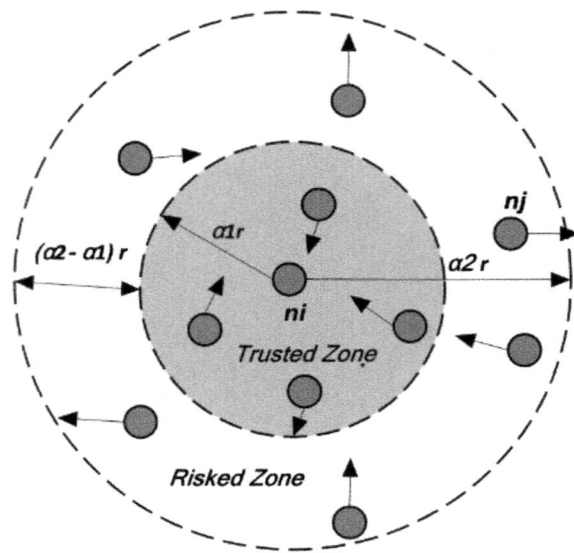

In order to decide how well suited a node is to behave as a CH, we take its
relative mobility, distances with neighbors and the number of neighbors in its direct
communication range. In our proposed algorithm, we calculate relative mobility
based on received signal strength between two successive "HELLO" packets which
is inversely proportional to the distance between the sender and the receiver [8].
The relative mobility at node n_i with respect to node n_j, $M_{n_i}^{\text{Rel}}(n_j)$, is calculated as

$$M_{n_i}^{\text{Rel}}(n_j) = 10 \ \log_{10} \frac{RxPr_{n_j \rightarrow n_i}^{\text{new}}}{RxPr_{n_j \rightarrow n_i}^{\text{old}}} \tag{1}$$

where, $RxPr_{n_j \rightarrow n_i}^{\text{new}}$ is the new and $RxPr_{n_j \rightarrow n_i}^{\text{old}}$ is the old receiving power of the
"HELLO" packet from node n_j to node n_i. In case when $RxPr_{n_j \rightarrow n_i}^{\text{new}} < RxPr_{n_j \rightarrow n_i}^{\text{old}}$,
$M_{n_i}^{\text{Rel}}(n_j)$ is negative, it means n_j is moving away from n_i as shown in Fig. 4. When
$RxPr_{n_j \rightarrow n_i}^{\text{new}} > RxPr_{n_j \rightarrow n_i}^{\text{old}}$, $M_{n_i}^{\text{Rel}}(n_j)$ is positive, i.e., n_j is coming closer to n_i.

For each neighboring node, n_j, of node n_i, we find its range indicator $R_{\text{Ind}}(n_j, n_i)$
with respect to n_i. On the basis of distance between the nodes and its relative
mobility, we categorize it as below,

$$R_{\text{Ind}}(n_j, n_i) = \begin{cases} 0, & \text{if } \alpha_1 r < \text{dist}(n_i, n_j) \leq \alpha_2 r \wedge M_{n_i}^{\text{Rel}}(n_j) < 0 \\ 1, & \text{if } \text{dist}(n_i, n_j) \leq r \wedge M_{n_i}^{\text{Rel}}(n_j) > 0 \\ 1 + \frac{\text{dist}(n_i, n_j) - \alpha_1 r}{(\alpha_2 - \alpha_1) r}, & \text{if } \text{dist}(n_i, n_j) \leq \alpha_1 r \wedge M_{n_i}^{\text{Rel}}(n_j) < 0 \end{cases} \tag{2}$$

$$\text{STD}(n_i) = \left(\sum_{j=1}^{k} R_{\text{Ind}}(n_j, n_i) \times \text{dist}(n_i, n_j) \right) / N(n_i) \tag{3}$$

If a node is in risked zone and its mobility is negative, then it is moving away from the node which is calculating its weight to be selected as CH. Such neighboring node is discarded as it will get away from the transmission range of the concerned node. If a node is anywhere within the transmission range and its relative mobility is also positive, it means that node is coming closer to the node calculating its weight. In the third situation when a node is in trusted zone but is moving away, then its contribution in finding range indicator depends on how far it is moving away from the concerned node as shown in Eq. (2). Finally we calculate stability deviation of node n_i($STD(n_i)$) as the summation of range indicator multiplied with distance $(\text{dist}(n_i, n_j))$ and the whole divided by its cardinality of neighboring nodes $(N(n_i))$ as given in Eq. (3).

In order to decide how much a node is capable of becoming a CH, it is a better choice to use the amount of energy consumed by it. If a node's depleted energy is higher than a threshold, it means its remaining battery energy is low, means it is not a better candidate to become a CH. In order to increase the network lifetime, a node with low depleted energy is chosen to act as a CH.On the basis of the stability deviation and energy depletion, the combined weight is calculated in Eq. (4). Here W_1 and W_2 are weighing factors such that $W_1 + W_2 = 1$, the values of which are selected depending upon the network scenario.

$$W(n_i) = W_1 \times \text{STD}(n_i) + W_2 \times \text{ED}(n_i) \qquad (4)$$

A node which is less deviated from stability and depleted less energy is a good candidate to become a CH. Thus, a node with minimum weight among its neighbors will declare itself as a CH and broadcast CH advertisement message. A neighboring node willing to join this CH will send join request message which will be accepted by it. In case, if a neighboring node got CH advertisement message from more than one CH, it will join that CH which has less weight, and in case of a tie, it will join to any CH randomly (Fig. 5).

2.2 Intercluster Multi-hop Routing

We now describe here the proposed scheme for intercluster multi-hop routing using a heuristic technique called as the A^* pathfinding method. It achieves better performance by using heuristics to guide its search and find the cheapest cost to reach from source to destination. A source node sends its packets for desired destination to its own CH using its TDMA schedule. Now, it is the duty of the source CH to find the optimal path to the destination CH via different CHs which act as relay nodes to the destination. The source CH initializes an OPEN and CLOSED list and put its node id in the OPEN list as the start node which acts as a parent for its successor nodes. Now it calculates a partial path for all its successor nodes using function $f(n)$ as below:

Input: A set of nodes $N = \{n_1, n_2, ..., n_m\}$.

Output: cluster formation with elected set of CHs $\Psi = \{ch_1, ch_2, ..., ch_k\}$.

Step 1: Each node (say n_i) get aware of its neighbors by broadcasting "HELLO" messages.

Step 2: n_i calculate its combined weight, $W(n_i)$ using equation (4) and broadcast it to all its neighbors.

Step 3: Set flag =1

Step 4: *While* (flag $= = 1 \wedge n_i$ is receiving $W(n_j)$
                                $\forall n_j \in T_{range}(n_i)$) *do*

    4.1: *if* $W(n_j) < W(n_i)$ *then*

        4.1.1: n_i gives up CH candidature and set flag=0

    endif

endwhile

    4.2: *If* (flag == 1) And (no incoming $W(n_j)$ to n_i) *then*

        4.2.1: n_i declare itself as CH & broadcast CH_Adv msg to all its neighbors.

    end

        4.2.2: n_i accept join_Req msg from its neighbors & form a cluster.

    else n_i receives CH_Adv msg from other node and send join_Req msg to join it.

    If n_i receives CH_Adv msg from more than one node *then* it will send join_Req msg to a node with minimum weight.

    end

    end

    end

Stop.

Fig. 5 CH election and cluster formation algorithm

$$f(n) = g(n) + h(n) \tag{5}$$

Here, $g(n)$ is the actual cost (i.e., Euclidean distance) from the source node to its one-hop neighbor and $h(n)$ is the estimated cost (heuristic value) from this one hop neighbor to the desired destination. From the set of paths so found, it will select the path having least cost to the destination, i.e., least value of $f(n)$ and put it in the CLOSED list. The process is repeated until OPEN list is null. For each successor of a node, there are three situations. First, if the successor node is a new node which is neither in OPEN or CLOSED list, we simply calculate its heuristic value, adjust the parent, and put it in the CLOSED list. The second situation is when the successor node is already in the OPEN list, i.e., it already has $f(n)$ value. In that case, we create the parent pointer, compute the g value and h value. If this new path is better than the old one, readjust the parent pointer and g value as shown in the algorithm in Fig. 6. The third case is when the successor belongs to the CLOSED list, i.e., we already visited it earlier. Here we do readjustment of pointers, if got better path like

Input: (1) A set of nodes $N = \{n_1, n_2, ..., n_m\}$ and a set of elected CHs $\Psi = \{ch_1, ch_2, ..., ch_k\}$. (2) A source node and the goal node i.e. the destination.

Output: optimal path from source to destination

Step 1: $OPEN \leftarrow (S);\ Parent\ (S) \leftarrow Nil;\ CLOSED \leftarrow NULL$ # *initialize OPEN and CLOSED list.*

Step 2:   **while** $OPEN \neq NULL$ **do**
      2.1: Pick best node n_a having least value of f(n)
      2.2: Delete the best selected node from OPEN and add
            it to CLOSED list.
    end

Step 3:   *For each n_m in Successors, there are three cases*
      case I : $n_m \notin$ OPEN and $n_m \notin$ CLOSED (for new nodes)
         3.1: Compute h(n_m)
         3.2: Parent(n_m) $\leftarrow n_a$
         3.3: g(n_m) = g(n_a) + k(n_a, n_m) # k(n_a, n_m) is the
                                cost of edge going
                                from n_a to n_m
         3.4: f(n_m) $\leftarrow$ g(n_m) + h(n_m)
         3.5: Add n_m to CLOSED
      case II : $n_m \in$ OPEN **then** check
                If g(n_a) + k(n_a, n_m) < g(n_m) **then**
                  2.1: Parent(n_m) $\leftarrow n_a$
                  2.2: g(n_m) = g(n_a) + k(n_a, n_m)
                  2.3: f(n_m) $\leftarrow$ g(n_m) + h(n_m)
          end
      case III : $n_m \in$ CLOSED
         3.1: Do like in case II.
         3.2: **If** better path found **then**
         3.3: Re-adjust the Parent pointer and g value as
                in case II and propagate the improved cost
                to sub-tree below m.
          end

Stop.

Fig. 6 Intercluster multi-hop routing algorithm

in second case, propagate the improved cost to sub-tree below the successor. In this way, we get an optimal path from source to destination. The routing is not on the basis of number of hops but on the basis of cost of the path. So, if two or more paths having the same number of hops from source to destination, the path with minimum cost, i.e., minimum distance will be selected for transmitting data packets. Also, as the path setup is among CHs, and these CHs are prudently chosen which are more stable, so the chances of path breakage will be less. The algorithm gives better performance in situations when the number of mobile nodes is high especially used in vehicular ad hoc networks or military purposes.

3 Conclusion

In this paper, we have proposed mobility aware distributed clustering and routing scheme based on the $A*$ search algorithm. The basic idea behind this work is developing less overhead technique to route data from source to destination in those real applications where nodes are highly mobile and routes from source to destinations needs to be heuristically developed like large-scale military applications or vehicular ad hoc networks. Both clustering and routing have been solved to meet the challenges. Nodes are clustered based on its mobility, energy consumed, distances with neighbors and its connectivity with neighbors. The three parameters, i.e., mobility, distances, and connectivity have been combined to develop a single parameter called as stability deviation. A node which is less deviated from stability and depleted less energy will serve as CH for longer duration of time. For routing between CHs, an important $A*$ search technique is applied. In this, heuristically, we are calculating path from source CH to destination CH, and the one which is of cheapest cost, based on distance, is selected. Thus, the overall stability is improved with less overhead. In future, we work forward to implement it with more realistic scenarios.

References

1. M. Aissa and A. Belghith, "A node quality based clustering algorithm in wireless mobile Ad Hoc networks", 5th International Conference on Ambient Systems, Networks and Technologies (ANT-2014), Procedia Computer Science vol. 32, pp. 174–181 (2014).
2. A. Z. Ahwazi and M. R. NooriMehr, "MOSIC: Mobility-Aware Single-Hop Clustering Scheme for Vehicular Ad hoc Networks on Highways", (IJACSA) International Journal of Advanced Computer Science and Applications, Vol. 7, No. 9, pp. 424–431, (2016).
3. M. Chatterjee, S. K. Das, and D. Turgut, "An on-demand weighted clustering algorithm (WCA) for ad hoc networks," 2000 IEEE Glob. Telecommun. Conf. (GLOBECOM '00), vol. 3, pp. 1697–1701, (2000).
4. S. Chandra, I. Saha, P. Mitra, B. Saha and S. Roy, "A Brief Overview of Clustering Schemes Applied on Mobile Ad-hoc Networks", Vol. 5, Issue 2, pp. 667–675, (February 2015).

5. A. A. Abbasi and M. Younis, "A survey on clustering algorithms for wireless sensor networks", computer communications, ScienceDirect, vol. 30, pp. 2826–2841, (2007).
6. W. Choi and M. Woo, "A Distributed Weighted Clustering Algorithm for Mobile Ad Hoc Networks", Proceedings of the Advanced International Conference on Telecommunications and International Conference on Internet and Web Applications and Services (AICT/ICIW 2006), pp. 1–7, (2006).
7. M. Bheemalingaiah, M. M. Naidu and D. S. Rao, "Energy Aware Clustered Based Multipath Routing in Mobile Ad Hoc Networks", I. J. Communications, Network and System Sciences, vol 2, pp. 91–168 (2009).
8. P. Basu, N. Khan, and T.D.C. Little, "A Mobility Based Metric for Clustering in Mobile Ad Hoc-Networks", Proc. IEEE ICDCS 2001 Workshop on Wireless Networks and Mobile Computing, Phoenix, AZ, pp. 1–19, (April 2001).

Floating Admittance Matrix Approach to Model Development of Active Devices and Circuits

Meena Singh and B. P. Singh

1 Introduction

As stated in the abstract, the modeling suggests guidelines to minimize time and cost of the process. In other words, modeling enhances the process of redesign, a relatively new and easier field in engineering education. The main objective of the designer is to stay competitive in the global market by constantly adjusting their approach to the art of design with continuous improvement and modification of the process and product. This can be achieved using some scientific approach. The scientific approach that enhances time, cost and energy is nothing but called the modeling [1–18].

The engineering education must address the problems faced properly to infuse the attitudinal change toward the acceptance level of mathematics and modeling.

As technology approaches its limit, fundamental changes do not occur at the same pace as it used to happen previously. For instance, the sudden announcement of BJT as the replacement of vacuum tubes gave shock and surprise to both engineers and industrialists. The surprise was in the form of perception to all of us including the designer, whether the device of miniaturized size will be able to hold the current, voltage and power dissipation required for industries supported by the gigantic size of the vacuum tubes. The shock, on the other hand, was for the industrialist who had made huge investments in procuring large inventory of different types of tubes and accessories and foundries used in the industries. But medium scale technology of mm size proved its might by producing equivalent transistor that handled such currents, voltage, and power dissipations required by the industries.

M. Singh (✉) · B. P. Singh
Department of ECE, University Polytechnic, BIT, Mesra, Ranchi, India
e-mail: meena71_singh@rediffmail.com

B. P. Singh
e-mail: bpsinghgkp@gmail.com

© Springer Nature Singapore Pte Ltd. 2019
V. Nath and J. K. Mandal (eds.), *Proceeding of the Second International Conference on Microelectronics, Computing & Communication Systems (MCCS 2017)*, Lecture Notes in Electrical Engineering 476, https://doi.org/10.1007/978-981-10-8234-4_24

The change in analog technology to the digital and then multiple chip packed in a capsule with its relevant terminal connection as pins out were used for the considerable amount of time. Slowly, vertical technology allowed multiple chip design to single chip design, hardware approach to software approach for testing and finalizing the system design. Ultimately, computers came as the mainstay for all aspects of design to production. This suggests that the concepts of engineering process cannot change drastically, but constant revision in its modes and applications is possible. As an instance, we can think that the concept of lint filter in the washing machine remained the same even in the electrical circuit. However, its mode and applications went on continuous revision from physical to electrical, passive to active, hardware to software and finally to digital type.

1.1 Mathematics

Mathematics helps in the thinking process of the human being for formulating the ideas to optimize the functionality of the process and phenomena. Having this in mind, it becomes narrative that a successful and innovative engineer must equip himself or herself with adequate knowledge of mathematics to engage in abstract thinking about the system.

1.2 Modeling

Model developed in one physical environment must replicate its behavior under observations in other environment. Modeling is essential for the preparation of detailed course curriculum of any discipline, whether it may be science, arts, engineering, fine arts, economics, etc. For instance, to investigate the effectiveness of an engineering curriculum, engineering colleges/institutes in Fig. 1 should:

- Be able to predict the program outcome of the proposed curriculum.
- Evaluate and assess the proposed curriculum.
- Identify the points of shortcomings.
- Act timely to overcome the shortcomings.

Thus, the modeling becomes a powerful tool to perform task as per the assignments. For instance, an effective engineering program curriculum can be modeled as in Fig. 1. This model helps engineering colleges/institutions/universities to engage in *continuous evaluation of their proposed program, find the pitfalls, and get suggestions to improve and modify it in a timely manner.*

Any educational program starts with some objective and outcome. Based on the objectives and outcome, the syllabi are prepared. The development of working

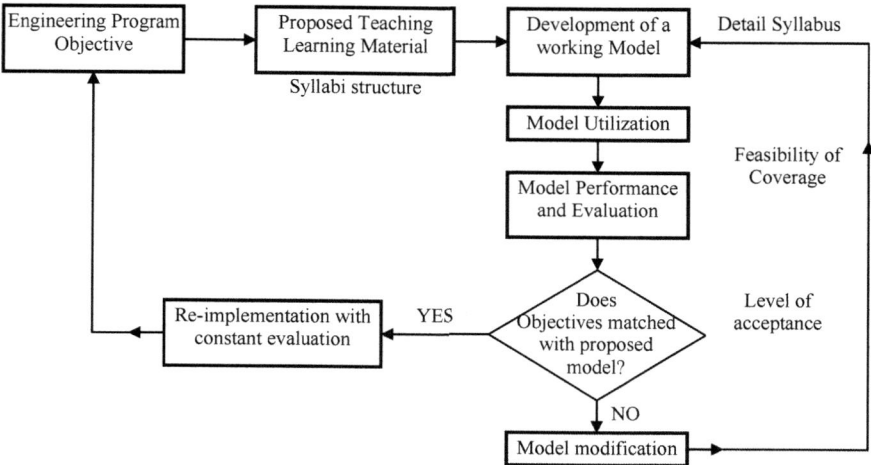

Fig. 1 Model for an effective curriculum development

model embeds the methodology and teaching aids to be used to convey to the learners in the best manner. Model utilization considers the matching of theoretical aspects with practical setups based on understanding of transfer of the knowledge. The model performance utilizes the rubrics for evaluation going down to the fraction of marks. Lastly, if the result deviates much from the set objectives and outcomes, a relook has to be given after each complete cycle of the program; sometimes even before it completes the 1st cycle.

The electronic process and objects are generally used as system model because electronic systems have unusual combination of properties and characteristics. A mathematician, Verlan [2], developed the mathematical model for integral equation. This essentially provides a procedure to find out the mathematical relationships between the known input source and unknown electrical network parameters. The following brief description of Verlan's work shows how mathematical modeling helps engineers to study the behavior of a system and invokes suggestions for further development of a more accurate system.

The RC circuit can act as a simple integrator or a first-order low-pass filter. The response of the RC circuit of Fig. 2 to a Step input of series interconnected resistor and capacitor is described by the nonhomogeneous differential equation. If the step input voltage (V_{in}) is applied at $t = 0$, what will be the form of the output? The same current passes through resistor and capacitor and hence

$$\frac{V_R}{R} = i = C\frac{dV_C}{dt} \tag{1}$$

Fig. 2 RC integrator

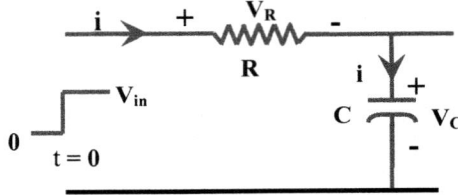

Solving it with the Initial conditions: $t = 0$, $V_C = 0$, and, as $t \rightarrow \infty$, $V_C \rightarrow V_{in}$

$$V_C = V_{in}\left[1 - \exp\left[\frac{-t}{RC}\right]\right] \tag{2}$$

A plot of this response is given in Fig. 3. Reducing the value of the time constant $\tau = RC$ (i.e., reducing either R or C) changes the output faster and at any given time the expected output voltage will reach sooner.

1.3 Background

Most of the times engineers try to analyze a system to control or optimize it using a mathematical model. During the process of analysis, modeling helps in estimating how an unforeseen circumstance can affect the system. The mathematical model usually describes a system by a set of mathematical equations that define relationships among its variables.

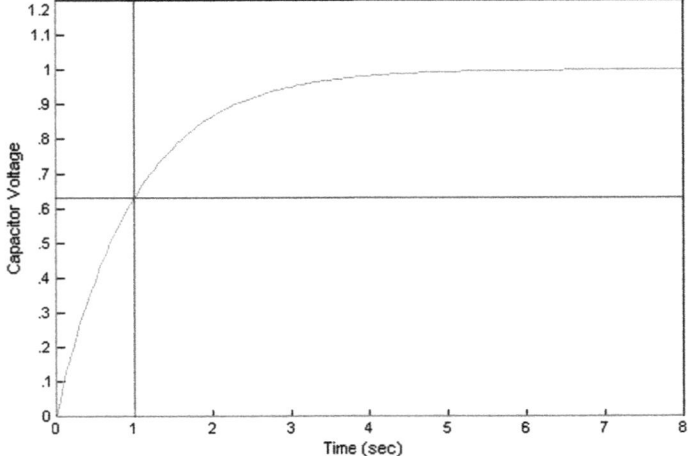

Fig. 3 Low pass filter response for step input

1.4   Prior Information

Basically, there are two types of modeling techniques: *black box* or *white box*. A system is called black box if no prior information is known. On the other hand, a white-box or clear (glass) box model has all necessary information known. Practically none of these two models: white box or black box does exist in reality. All systems are somewhere between the black-box and white-box models. So such concept provides just a guideline for approaching the modeling problems.

1.5   Model Evaluation

Evaluation is the crucial part of modeling to guess whether the developed model represents the actual system or not. This part is not so simple as there are different types of evaluation schemes. One scheme may give better match w.r.t. the other scheme(s).

The cross validation, the most general approach, gives the extent of the experimental fits in. The cross validation assumes two subsets: training data and verification data. The training data estimates the model parameters; whereas the verification data gives the information to extent of the closeness to developed model.

1.6   Scope of Model

The interpolation and extrapolation are two techniques normally used in matching the system model. The interpolation gives an idea about how well the properties of the system describe the model behavior between observed data points. On the contrary, the extrapolation is used to predict the behavior of the system model beyond observed data set.

All physical systems such as graph, scattered plots, structures, diagrams, circuits, etc., can be represented by mathematical modeling to predict its behavior and functionality under different conditions and excitations. The model gives an overall idea of the internal combination and compositions of different parameters involved in the system formation.

Here in this article to follow, our main interest lies in describing the mathematical modeling of the active and passive components and circuits using these components using Floating Admittance Matrix (FAM). There are a number of tools such as MATLAB/ MATHEMATICA available that can be used to evaluate simulated mathematical model for complicated circuits.

2 FAM of FET

The floating admittance matrix of FET [19–23] assuming $g_g = 0$ is expressed as

$$\begin{bmatrix} i_1/i_g \\ i_2/i_d \\ i_3/i_s \end{bmatrix} \begin{bmatrix} 0 & 0 & 0 \\ g_m & g_d & -g_m - g_d \\ -g_m & g_d & g_m + g_d \end{bmatrix} \begin{bmatrix} v_1/v_g \\ v_2/v_d \\ v_3/v_s \end{bmatrix} \tag{3}$$

Equation (3) is the basic mathematical model of an active device, i.e., MOSFET/FET in the form of a floating admittance matrix. This equation will be used in all circuits containing MOSFET/FET to design and analyze the behavior of any amplifier. The major advantages of floating admittance matrix approach to circuit analysis in this article will become evident w.r.t. the conventional small signal equivalent circuit method soon we start analyzing any circuit. For the purpose, we would consider common source amplifier first.

2.1 Common Source Amplifier (CS) [22, 23]

If we try to compare the topology of a CS with that of the corresponding counterpart in BJT technology, then we find the Common Emitter (CE) amplifier is topologically identical to that of the CS amplifier. For the small signal analysis of the amplifier, the coupling and bypass capacitors are replaced by short circuits. The dc power supplies (V_{GG} and V_{DD}) are also replaced by the short circuit. After these replacements, the circuit is called ac circuit as shown in Fig. 4.

Floating admittance matrix for the circuit of Fig. 4 is written as

$$\begin{bmatrix} g_s + G_G & 0 & -g_s - G_G \\ g_m & g_d + G_D + G_L & -g_m - g_d - G_D - G_L \\ -g_m - g_s - G_G & -g_d - G_D - G_L & g_s + G_G + g_m + g_d + G_D + G_L \end{bmatrix} \tag{4}$$

The voltage ratio between output port and input port of Fig. 4 is expressed as

Fig. 4 AC circuit of CS amplifier

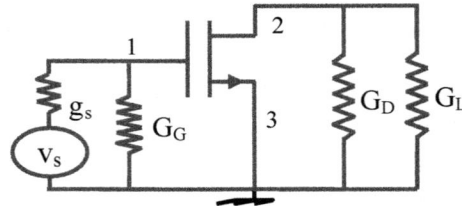

$$A_v\Big|_{13}^{23} = \text{sgn}(2-3)\,\text{sgn}(1-3)(-1)^{2+3+1+3}\frac{\left|Y_{23}^{13}\right|}{\left|Y_{13}^{13}\right|} \tag{5}$$

$$\left|Y_{23}^{13}\right| = g_m, \quad \left|Y_{13}^{13}\right| = g_d + G_D + G_L$$

$$A_v\Big|_{13}^{23} = -\frac{g_m}{g_d + G_L + G_D} = -g_m(r_d\|R_D\|R_L) \tag{6}$$

As $r_d \gg R_D$ and R_L, Eq. (6) reduces to

$$A_v = -\frac{g_m}{g_d + G_L + G_D} = -g_m\,(R_D\|\,R_L) \tag{7}$$

The effective resistance seen across the input port of Fig. 4 is expressed as

$$Z_i = Z_{13} = \frac{\left|Y_{13}^{13}\right|}{\left|Y_3^3\right|_{g_s=0}} \tag{8}$$

$$\left|Y_3^3\right|_{g_s=0} = (g_s + G_G)(g_d + G_D + G_L) = G_G\,(g_d + G_D + G_L) \tag{9}$$

$$Z_i = Z_{13} = \frac{g_d + G_D + G_L}{G_G(g_d + G_D + G_L)} = R_G \tag{10}$$

The effective resistance across the output port in Fig. 4 is expressed as

$$Z_O = Z_{23} = \frac{\left|Y_{23}^{23}\right|}{\left|Y_3^3\right|_{G_L=0}}$$
$$\left|Y_{23}^{23}\right| = g_s + G_G$$
$$\left|Y_3^3\right|_{G_L=0} = (g_s + G_G)(g_d + G_D + G_L)$$
$$= (g_s + G_G)(g_d + G_D) \tag{11}$$

$$Z_O = Z_{23} = \frac{\left|Y_{23}^{23}\right|}{\left|Y_3^3\right|_{G_L=0}} = \frac{g_s + G_G}{(g_s + G_G)(g_d + G_D)}$$
$$= \frac{1}{(g_d + G_D)} = r_d\|R_D \tag{12}$$

3 FAM of BJT

The FAM of BJT is written as

$$
\begin{bmatrix} i_1 \\ i_2 \\ i_3 \end{bmatrix}
\begin{bmatrix}
g_i & -g_i h_r & -g_i(1-h_r) \\
g_m & g_0 - g_m h_r & -g_0 - g_m(1-h_r) \\
-g_m - g_i & -g_o + g_m h_r + g_i h_r & g_o + g_m(1-h_r) + g_i(1-h_r)
\end{bmatrix}
\begin{bmatrix} v_1 \\ v_2 \\ v_3 \end{bmatrix}
\quad (13)
$$

3.1 Common Emitter Amplifier [19–21, 23]

The next circuit for analysis here is the ac circuit of CE amplifier as shown in Fig. 5.

Now, the FAM including active device and passive components in Fig. 5 is written as

$$
\begin{bmatrix}
g_i + g_s & -g_i h_r & -g_i(1-h_r) - g_s \\
g_m & g_0 - g_m h_r + G_L & -g_o - g_m(1-h_r) - G_L \\
-g_m - g_i - g_s & -g_o + (g_m + g_i)h_r & g_o + g_m(1-h_r) + g_i(1-h_r) \\
& -G_L & -g_s - G_L
\end{bmatrix}
\quad (14)
$$

Some of the first-order and second-order cofactors that will be used; are evaluated from Eq. (14) as

$$
\left. |Y_3^3| \right|_{g_s=0} = \frac{1 + h_{oe}R_L}{h_{ie}R_L}, \quad |Y_{13}^{13}| == \frac{h_{ie} + \Delta h_e R_L}{h_{ie}R_L}
$$

where, $\Delta h_e = h_{ie}h_{oe} - h_{fe}h_{re}$.

$$
|Y_3^3| = \frac{1 + h_{oe}R_L + (\Delta h_e R_L + h_{ie})g_s}{h_{ie}R_L}
$$

$$
\left. |Y_3^3| \right|_{G_L=0} = \frac{h_{oe} + \Delta h_e g_s}{h_{ie}}, \quad |Y_{23}^{23}| = g_s + g_i = \frac{r_s + h_{ie}}{r_s h_{ie}}
$$

Fig. 5 CE amplifier

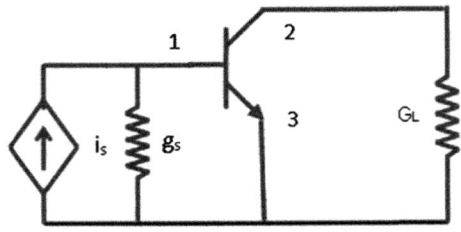

The voltage ratio between output port and input port of Fig. 5 is expressed as

$$A_v\Big|_{13}^{23} = \frac{v_{23}}{v_{13}} = \text{sgn}(2-3)\text{sgn}(1-3)(-1)^{2+3+1+3}\frac{|Y_{23}^{13}|}{|Y_{13}^{13}|}$$

$$= \frac{-h_{fe}R_L}{h_{ie}+\Delta h_e R_L} \tag{15}$$

The effective resistance across input port of Fig. 5 is expressed as

$$Z_i = Z_{13} = \frac{|Y_{13}^{13}|}{|Y_3^3|}\Big|_{g_s=0} = \frac{h_{ie}+\Delta h_e R_L}{1+h_{oe}R_L} \tag{16}$$

The effective resistance across output port of Fig. 5 is expressed as

$$Z_o = Z_{23} = \frac{|Y_{23}^{23}|}{|Y_3^3|}\Big|_{G_L=0} = \frac{r_s+h_{ie}}{h_{oe}r_s+\Delta h_e} \tag{17}$$

The current ratio in Fig. 5 is expressed as

$$A_i\Big|_{13}^{23} = \frac{i_2}{i_1} = \frac{i_c}{i_b}$$

$$= \text{sgn}(2-3)\text{sgn}(1-3)(-1)^{2+3+1+3}\frac{|Y_{23}^{13}|}{|Y_3^3|}G_L \tag{18}$$

$$= \frac{-h_{fe}}{1+h_{oe}R_L+(\Delta h_e R_L+h_{ie})g_s}$$

$$= \frac{-h_{fe}}{1+h_{oe}R_L} \quad \text{for } g_s \to 0 \tag{19}$$

Equation (19) is available in the book which is the approximate ones after assuming $g_s \to 0$. Our method is simple and yields exact ones.

3.2 Common Collector Amplifier [19–21, [23]

The simplest ac circuit of a CC amplifier shown in Fig. 6 is called emitter follower also. The FAM for this circuit is written as

$$\begin{bmatrix} g_i+g_s & -g_s & -g_i \\ g_m-g_s & g_0+g_s+G_E & -g_o-g_m-G_E \\ -g_m-g_i & -g_o-G_E & g_o+g_m+g_i+G_E \end{bmatrix} \tag{20}$$

Fig. 6 Common collector
amplifier

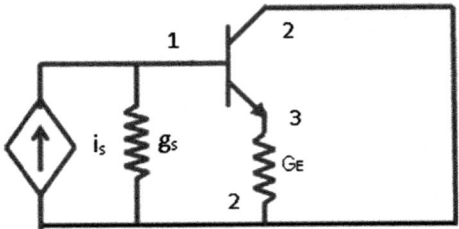

The ratio of voltage across the output port and the input port of Fig. 6 is expressed as

$$A_v\big|_{12}^{32}= \text{sgn}(3-2)\text{sgn}(1-2)(-1)^{3+2+1+2}\frac{\left|Y_{32}^{12}\right|}{\left|Y_{12}^{12}\right|} \tag{21}$$

$$\begin{aligned} A_v &= -\frac{-(g_i+g_m)}{g_i+g_m+g_o+G_E} = \frac{g_i(1+h_{fe})}{g_i(1+h_{fe})+h_{oe}+G_E} \\ &= \frac{(1+h_{fe})R_E}{h_{ie}+(1+h_{fe})R_E+h_{oe}h_{ie}R_E} \end{aligned} \tag{22}$$

The effective resistance across the input port of Fig. 6 is expressed as

$$Z_i = Z_{12} = \frac{\left|Y_{12}^{12}\right|}{\left|Y_2^2\right|}\bigg|_{g_s=0}, \quad \left|Y_2^2\right|_{g_s=0}= (g_0+G_E)g_i$$

$$Z_i = Z_{12} = \frac{g_0+g_m+g_i+G_E}{(g_0+G_E)g_i} = \frac{h_i+h_oh_iR_E+(1+h_f)R_E}{1+h_oR_E} \tag{23}$$

The effective resistance across the output port of Fig. 6 is expressed as

$$Z_o = Z_{32} = \frac{\left|Y_{32}^{32}\right|}{\left|Y_2^2\right|}\bigg|_{G_L=0}, \quad \left|Y_{32}^{32}\right| = g_ig_s(h_i+r_s)$$

$$\left|Y_2^2\right|_{G_L=0}= g_ig_s\{h_o(h_i+r_s)+(1+h_f)\}$$

$$Z_o = Z_{32} = \frac{(h_i+r_s)}{h_o(h_i+r_s)+(1+h_f)} \tag{24}$$

$$Y_o = \frac{1}{Z_o} = h_o + \frac{h_i+r_s}{1+h_f}$$

3.3 BJT Phase Splitter [20, 21, 23]

Figure 7 is the basic configuration of phase splitter circuit in which the emitter resistance and collector resistance are used to split the phase of 180°. Figure 7 is simplified ac circuit of the phase splitter.

The floating admittance matrix for the phase splitter amplifier shown in Fig. 7 is written by inspection as

$$
\begin{bmatrix}
g_i + g_s & -g_i h_r & -g_i(1 - h_r) & -g_s \\
g_m & g_0 - g_m h_r & -g_o - g_m(1 - h_r) & -G_L \\
-g_m - g_i & -g_o + g_m h_r + g_i h_r & g_o + g_m(1 - h_r) + g_i(1 - h_r) & -G_E \\
-g_s & -G_L & -G_E & g_s + G_L + G_E
\end{bmatrix}
$$

(25)

The voltage ratio of output port and the input port in Fig. 7 is expressed as

$$
A_v\big|_{14}^{24} = \frac{v_{24}}{v_{14}} = \text{sgn}(2 - 4)\,\text{sgn}(1 - 4)(-1)^{2+4+1+4}\frac{\left|Y_{24}^{14}\right|}{\left|Y_{14}^{14}\right|}
$$

$$
\left|Y_{24}^{14}\right| = \begin{vmatrix} g_m & -g_o - g_m(1 - h_r) \\ -g_i - g_m & g_o + (g_i + g_m)(1 - h_r) + G_E \end{vmatrix}
$$

$$
= -g_m g_i h_{re} + g_m G_E - g_i g_o + g_m g_i h_{re} = \frac{h_{fe} - h_{oe} R_E}{h_{ie} R_E}
$$

$$
\left|Y_{14}^{14}\right| = \frac{\Delta h_{re}(R_E + R_L) + h_{ie} + (1 + h_{fe})R_E + R_E(R_L h_{oe} - h_{re})}{h_{ie} R_E R_L}
$$

Hence,

$$
\cong \frac{-h_{fe} R_L}{h_{ie} + (1 + h_{fe})R_E + R_E R_L h_{oe}} \cong \frac{-h_{fe} R_L}{h_{ie} + (1 + h_{fe})R_E}
$$

(26)

Fig. 7 BJT phase splitter

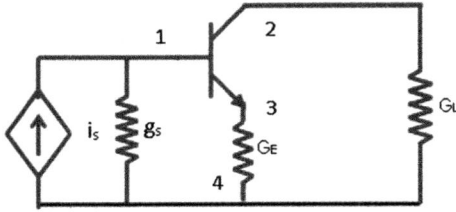

$$A_v\big|_{14}^{34} = \frac{v_{34}}{v_{14}} = \text{sgn}(3-4)\text{sgn}(1-4)(-1)^{3+4+1+4}\frac{\left|Y_{34}^{14}\right|}{\left|Y_{14}^{14}\right|}$$

$$\left|Y_{34}^{14}\right| = \frac{1+h_{fe}+h_{oe}R_L}{h_{ie}R_L}$$

(27)

$$A_v\big|_{14}^{34} = \frac{v_{34}}{v_{14}} = \frac{\left|Y_{34}^{14}\right|}{\left|Y_{14}^{14}\right|}$$

$$= \frac{(1+h_{fe}+h_{oe}R_L)R_E}{\Delta h_{re}(R_E+R_L)+h_{ie}+(1+h_{fe})R_E+R_E(R_L h_{oe}-h_{re})}$$

$$= \frac{(1+h_{fe})R_E}{h_{ie}+(1+h_{fe})R_E}$$

(28)

$$\text{If } R_L = R_E, \text{ then } A_v\big|_{14}^{34} = -A_v\big|_{14}^{24}$$

(29)

Hence, it splits the phase between v_{24} and v_{34} by $180°$.
The effective input resistance across the input port of Fig. 7 is expressed as

$$Z_i = Z_{14} = \frac{\left|Y_{14}^{14}\right|}{\left|Y_4^4\right|_{g_s=0}}$$

$$\left|Y_4^4\right|_{g_s=0} = g_i\{h_{oe}(R_E+R_L)+1\}G_E G_L$$

Hence, $Z_i = Z_{14}$

$$= \frac{\Delta h_{re}(R_E+R_L)+h_{ie}+(1+h_{fe})R_E+R_E(R_L h_{oe}-h_{re})}{h_{ie}R_E R_L g_i\{h_{oe}(R_E+R_L)+1\}G_E G_L}$$

(30)

$$= \frac{h_{ie}+(1+h_{fe})R_E}{1} = h_{ie}+(1+h_{fe})R_E$$

(31)

The current transfer ratio in Fig. 7 is expressed as

$$A_i\big|_{14}^{24} = \text{sgn}(2-4)\text{sgn}(1-4)(-1)^{2+4+1+4}\frac{\left|Y_{24}^{14}\right|}{\left|Y_4^4\right|}G_L$$

$$\left|Y_4^4\right| = G_L G_E g_i g_s\{(h_{ie}+r_s)+(1+h_{fe})R_E\}$$

$$A_i = \frac{-g_m G_E G_L}{g_i g_s G_L G_E\{h_{ie}+(1+h_{fe})R_E+r_s\}}$$

(32)

$$= \frac{-h_{fe}r_s}{h_{ie}+(1+h_{fe})R_E+r_s}$$

The resistance across the output port of Fig. 7 is expressed as

$$Z_O = Z_{24} = \frac{|Y_{24}^{24}|}{|Y_4^4|}\bigg|_{G_L=0} \tag{33}$$

$$|Y_4^4|_{G_L=0} = g_o g_s g_i G_E (r_s + r_i + R_E)$$

$$|Y_{24}^{24}| = g_i G_E g_s \{h_{ie} + h_{ie} h_{oe} R_E + (1 + h_{fe}) R_E + (1 + h_{oe} R_L) r_s\}$$

$$\begin{aligned}
Z_O = Z_{24} &= \frac{g_i G_E g_s \{h_{ie} + (1 + h_{fe}) R_E + (1 + h_{oe} R_E) r_s + h_{oe} h_{ie} R_E\}}{g_i g_o G_E g_s \{h_{ie} + r_s + R_E\}} \\
&= \frac{1}{h_{oe}} + \frac{(h_{ie} + r_s) R_E}{h_{ie} + r_s + R_E} + \frac{h_{fe} R_E}{h_{oe}(h_{ie} + r_s + R_E)}
\end{aligned} \tag{34}$$

4 Conclusions

The proposed method of analysis using FAM approach is simple and without any approximation results in the exact expressions. The method is, more or less, mathematical in nature. For more than four or five node networks, computer can be used to evaluate the cofactors of the FAM of the circuit developed using all types of components, i.e., active devices and passive both.

References

1. Pieter Eykhoff, System Identification, Springer (Hardcover), 1974.
2. Verlan, V.V, Network Analysis by the Method of Integral Equation, Journal of Electron Modeling, 5, 5,1021–1031, 1985.
3. Wai-Kai Chen, On second-order cofactors and null return difference in feedback amplifier theory, International Journal of Circuit Theory and Applications, Volume 6 Issue 3, Pages 305–312, Dec 2006.
4. Balbanian, N. and Bickart, T.A., Electrical Network Theory, John Wiley.
5. Friedman and R. Gulliver, Mathematical modeling for instructors, Technical Report 1254, Institute for Mathematics and its Applications, 1994.
6. E. Edwards and M. Hamson, Guide to Mathematical Modeling, CRC, 1990.
7. J. G. Andrews and R. R. McLone, Mathematical Modeling, Butterworths, 1976.
8. D. Burghes, P. Galbraith, N. Price, and A. Sherlock, Mathematical Modeling, Prentice Hall, 1996.
9. R. R. Clements, Mathematical Modeling: A Case Study Approach, Cambridge Press University, 1989.
10. M. Cross and A. O. Moscardini, Learning the Art of Mathematical Modeling, Ellis Horwood, 1985.
11. C. L. Dym and E. S. Ivey, Principles of Mathematical Modeling, 58 Academic Press, 1980.
12. Bender, E.A., An Introduction to mathematical Modeling, Wiley, 1978. 40. P. Doucet and P.B. Sloep, Mathematical Modeling in the life Sciences, Ellis Horwood, 1992.

13. Aris, Rutherford [1978] (1994), Mathematical Modeling Techniques, New York: Dover. ISBN 0-486-68131-9.
14. Bender, E. A. [1978] (2000), An Introduction to Mathematical Modeling, New York : Dover. ISBN 0-486-41180-X.
15. Lin, C. C. & Segel, L. A. (1988), Mathematics Applied to Deterministic Problems in the Natural Sciences, Philadelphia: SIAM. ISBN 0-89871-229-7.
16. Gershenfeld, N. (1998), The Nature of Mathematical Modeling, Cambridge University Press ISBN 0-521-57095-6.
17. Baehr, Jason. (2006), "A Priori and A Posteriori," Internet Encyclopedia of Philosophy.
18. I. Getreu, Modeling the Bipolar Transistor, Beaverton, Ore: 98 Tektronix, Inc., 1976.
19. Otso Juntunen (1998); A Two-Port S-Parameter Data Transformation; Circuit Theory Laboratory Report Series, CT-35, Helsinki University of Technology, Finland, Espoo.
20. B. P. Singh, Unified Approach to Electronics Circuit Analysis: IJEEE, Vol, pp. 276–285, July 1978.
21. B. P. Singh, Active Bridge for Measurement of the Admittance Parameters of the Transistor; Indian Journal of Pure & Applied Physics, Vol.15, pp. 783–786, Nov. 1976.
22. B. P. Singh, A New Active Bridge for Measuring FET Parameters; J Phys. E Scientific, Instruments, Vol. 11, pp. 667–670, 1978.
23. B.P. Singh, Meena Singh, Sanjay Kumar Roy, and S.N. Shukla, "Mathematical Modeling of Electronic Devices and its integration," Proceedings of National Seminar on Recent Advances on Information Technology, pp. 494–502, Feb. 6–7, 2009, Indian School of Mines Dhanbad University, Published by Allied Publishers Pvt. Ltd.

An Efficient Approach for Monuments Image Retrieval Using Multi-visual Descriptors

Ravi Devesh and Jaimala Jha

1 Introduction

Nowadays, numerous amount of images are being captured by different image acquisition techniques; these images database are used in many day-to-day applications, for example, military, remote sensing, engineering, medical field, crime prevention, environmental, astronomy, multimedia, and many other applications [1].

For handling the gigantic amount of stored and exchanged image information, some automatic image retrieval techniques are required. In case of database having less number of images, it is practicable to discover a required image simply by browsing while in case a database containing thousands of images it is not practicable to browse the desired image thus some more effective techniques are needed. Image retrieval is the procedure of finding the images having similar kind of content as query image given by the user based upon some similarity functions [2, 3].

Image retrieval can be implemented based on the following two methods:

(a) Text-Based Image Retrieval
(b) Content-Based Image Retrieval

2 Related Work

The word CBIR is used for explaining the retrieval of images based on visual descriptors that are extracted from image automatically in CBIR system [4]. The descriptor used for retrieval system can be either local descriptor or global

R. Devesh (✉) · J. Jha
Department of CSE & IT, MITS, Gwalior, Madhya Pradesh, India
e-mail: scholar.mits@gmail.com

J. Jha
e-mail: jaimala.jha@gmail.com

© Springer Nature Singapore Pte Ltd. 2019
V. Nath and J. K. Mandal (eds.), *Proceeding of the Second International Conference on Microelectronics, Computing & Communication Systems (MCCS 2017)*, Lecture Notes in Electrical Engineering 476, https://doi.org/10.1007/978-981-10-8234-4_25

descriptor. But the process used for extraction of descriptors must be automatic. The set of descriptors used has an important role during retrieval of similar images. A robust descriptor set is made by combining shape, texture, and color features. Shape feature is broadly used in multimedia information service system (MISS).

Padmashree Desai et al. in 2013 proposed a CBIR system which classifies the archaeological monuments based on local features [5]. In this paper, morphological gradients with invariant moments for shape feature and gray level co-occurrence matrix (GLCM) for texture feature extraction are used. For comparing the results of proposed system, Sobel and Canny operators are used.

Shilpa Yaligar et al. in 2013 proposed a system for recognition and fetching of archaeological monuments utilizing visual descriptors [6]. Database used consists of 500 images with five categories. For shape, feature morphological gradients were calculated, and then invariant moments were applied on the gradients; GLCM is used for the extraction of texture feature.

Fuxiang Lu et al. in 2016 proposed improved local binary pattern (ILBP) that describes texture feature of an image more efficiently [7]. A significant group of basic primitives is discovered like lines, cross-intersections, and T-junctions; these are generally unnoticed by uniform LBP method.

Malay S. Bhatt et al. in 2015 proposed a system which extracts genetic programming evolved spatial descriptors and on the basis of linear support vector machine (SVM), classification of Indian monuments visited by tourists is performed [8].

Not much work is done in this field. So, it is still a dynamic research field. We need sophisticated systems which could correctly identify, classify, and retrieve monuments images using local or global features.

3 Proposed System

In this paper, it is proposed a CBIR system which is used to retrieve the images of monuments based on low-level features. It forms the proficient combination of shape, texture, and color features of an image. Here, the dataset consists of images of monuments of different categories. The images presented in database are loaded.

The subsequent images are given as input to feature extraction techniques to extract desired features and form a set of extracted features. These features are chosen proficiently so that retrieval accuracy and recall rate is improved. In the last step, the similar images from database are retrieved. This method produces noble outcomes for the monument images in comparison to other methods available.

In the proposed methodology, it used morphological operators along with invariant moment in extracting the shape features [5], improved local binary pattern (ILBP) in extracting the texture features and RGB color histogram in extracting the color feature of images [7, 9]. The proposed CBIR system is depicted in Fig. 1. Similarity metrics are used for evaluating the distance among the query image, and N similar images present in the database. Precision and recall values are evaluated for the purpose of performance evaluation of the proposed system.

Fig. 1 Image retrieval
system

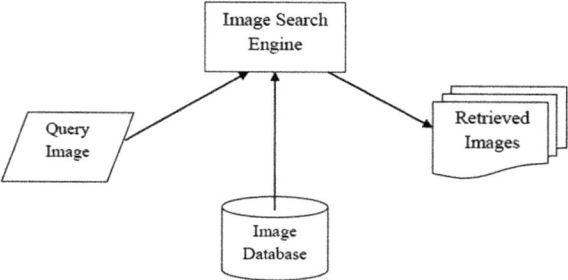

4 Feature Extraction

Mathematical morphology has a sole benefit in detecting boundary of an image, as it is based on set operation and nonlinear in nature. Morphological methods can be applied on image with distinct structuring element. Such operations are Erosion, Dilation, Open, and Close [5, 6].

5 Overview of Mathematical Morphology

The elementary notion of mathematical morphology explains use of structuring element of particular type to determine and elicit the relative shape in image to attain the intents of image analysis and identification [6]. In morphology, structuring element is applied on given image, and output image is obtained. Distinct morphological operations such as Dilation, Erosion, Open, and Close can be enforced on image with distinct structuring element [6].

Erosion is given in Eq. (1)

$$(f\theta b)(x,y) = \min\{f(x+i,y+j)-b(i,j)|(x+i,y+j)\}. \tag{1}$$

Dilation is given in Eq. (2)

$$(f \oplus b)(x,y) = \max\{f(x-i,y-j)+b(i,j)|(x-i,y-j)\} \tag{2}$$

$$\text{Basic gradient} = \text{Image Dilated} - \text{Image Eroded}$$
$$\text{Internal gradient} = \text{Image Original} - \text{Image Eroded}$$
$$\text{External gradient} = \text{Image Dilated} - \text{Image Original}.$$

Five edge maps will be acquired for given image. Shape descriptors are acquired for each of the boundary images. Out of five edge maps obtained, only internal gradient, external gradient, horizontal gradient, and vertical gradients edge maps are used.

Figure 3 depicts edge maps acquired after applying morphological gradients on input image (Fig. 2).

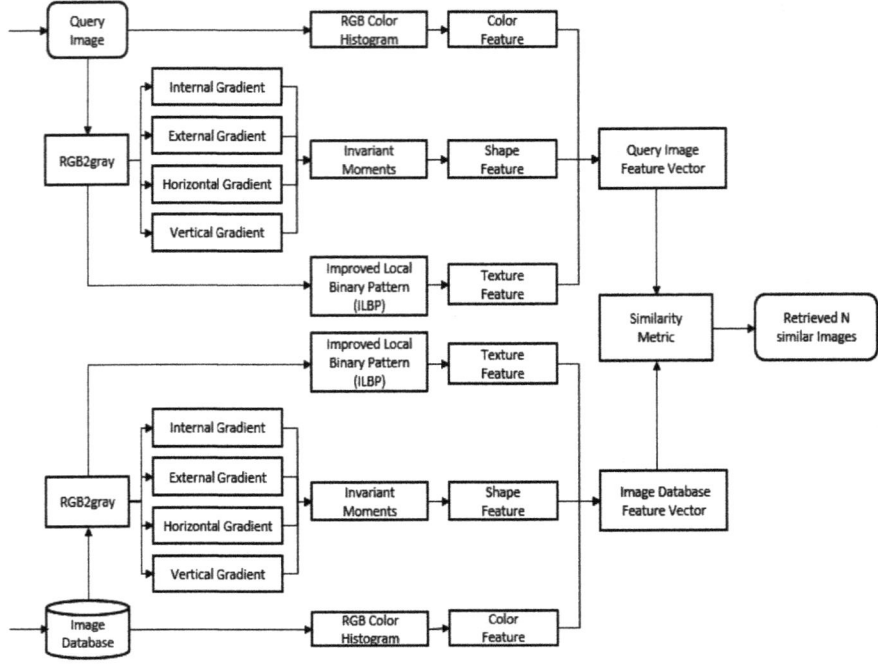

Fig. 2 Architecture of proposed CBIR system

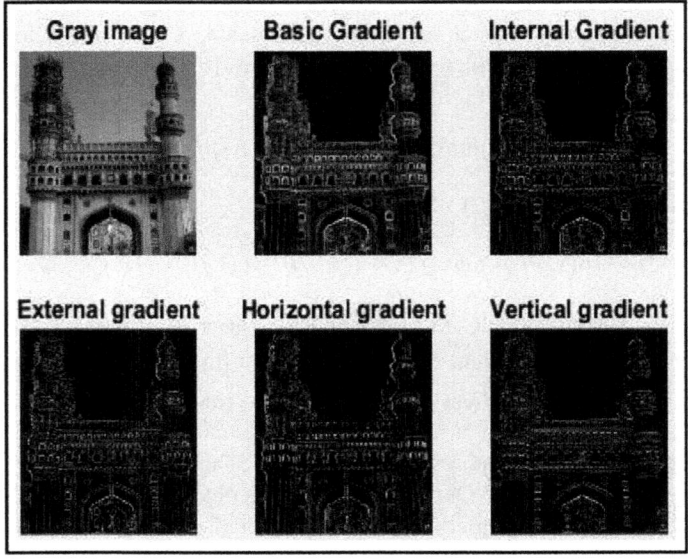

Fig. 3 Boundary maps acquired for the given image

6 Shape Feature

Seven (2-D) moment invariants that are unaffected by rotation, mirroring, translation, and change in scaling are known by Eq. (3). Shape descriptor vector comprises seven descriptor values [5, 10].

$$
\begin{aligned}
I1 &= \eta_{20} + \eta_{02} \\
I2 &= (\eta_{20} - \eta_{02})^2 + 4\eta_{11}^2 \\
I3 &= (\eta_{30} - 3\eta_{12})^2 + (3\eta_{21} - \eta_{03})^2 \\
I4 &= (\eta_{30} + \eta_{12})^2 + (\eta_{21} + \eta_{03})^2 \\
I5 &= (\eta_{30} - 3\eta_{12})(\eta_{30} + \eta_{12})\left[(\eta_{30} + \eta_{12})^2 \right. \\
&\quad \left. - 3(\eta_{21} - \eta_{03})^2\right] + (3\eta_{21} - \eta_{03}) \\
&\quad (\eta_{21} + \eta_{03})\left[3(\eta_{30} + \eta_{12})^2 - (\eta_{21} + \eta_{03})^2\right] \\
I6 &= (\eta_{20} - \eta_{02})\left[(\eta_{30} + \eta_{12})^2 - (\eta_{21} + \eta_{03})^2\right] \\
&\quad + 4\eta_{11}(\eta_{30} + \eta_{12})(\eta_{21} + \eta_{03}) \\
I7 &= (3\eta_{21} - \eta_{03})(\eta_{30} + \eta_{12}) \\
&\quad \left[(\eta_{30} + \eta_{12})^2 - 3(\eta_{21} + \eta_{03})^2\right] - (\eta_{03} + 3\eta_{12}) \\
&\quad (\eta_{21} + \eta_{03})\left[3(\eta_{30} + \eta_{12})^2 - (\eta_{21} + \eta_{03})^2\right]
\end{aligned}
\tag{3}
$$

7 Texture Feature

Local binary pattern (LBP) feature for extraction of texture information was introduced by Ojala et al. [11]. This operator is stout against change in illumination and is characterized by a computational simplicity and capability to encode texture details. Every pixel in LBP feature is described by a binary code. Each pixel's (central) gray level is tested with its eight neighborhood of size (size 3 × 3) [12]. If neighborhood pixels' value is greater than the central pixel, then the result is fixed to one else to zero. For producing binary code, multiply the results with weights given by 2's powers. For the central pixel (x, y) LBP code is given as follows [7]:

$$
\text{LBP}_{P,R} = \sum_{p=0}^{p-1} s(g_p - g_c)2^p,
\tag{4}
$$

$$s(z) = \begin{cases} 1, & \text{if } z \geq 0, \\ 0, & \text{otherwise,} \end{cases} \tag{5}$$

where g_c is the value of intensity for the pixel in center (x, y), and pth neighbor value is g_p. When a neighbor does not fall at integer coordinates, bilinear interpolation is used to determine its intensity value.

An extension of original LBP is called uniform LBP, denoted as $\text{LBP}_{P,R}^{u2}$.

$$U\left(\text{LBP}_{P,R}\right) = \sum_{p=0}^{p-1} \left| s\left(g_{p+1} - g_c\right) - s\left(g_p - g_c\right) \right|, \tag{6}$$

where g_p is equal to g_0. If $U(\text{LBP}_{P,R}) \leq 2$, then LBP will be called as uniform (Fig. 4).

ILBP can be defined as

$$\text{ILBP}_{P,R}^{ri} = \begin{cases} \sum_{P=0}^{P-1} s\left(g_p - g_c\right), & \text{if } U\left(\text{LBP}_{P,R}\right) \leq 2, \\ P - 1 + \frac{U}{2}, & \text{otherwise,} \end{cases} \tag{7}$$

where U is defined in (6). By doing so, the $\text{ILBP}_{P,R}^{ri}$ operator has $3/2P$ different output labels.

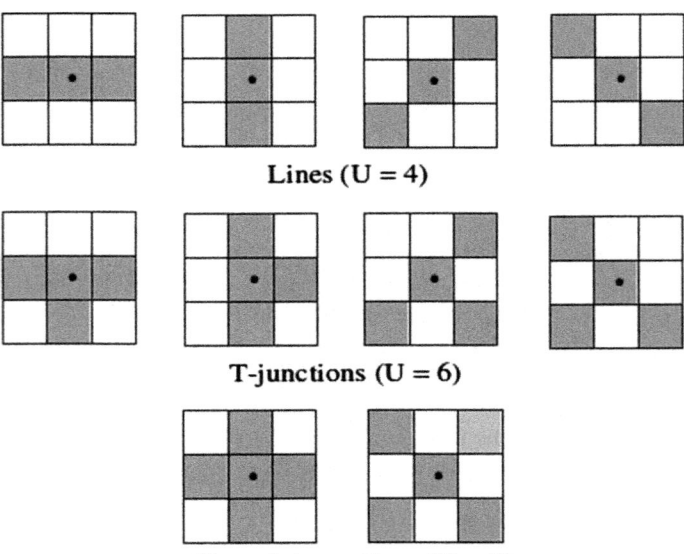

Lines (U = 4)

T-junctions (U = 6)

Cross intersections (U = 8)

Fig. 4 Basic primitives detected by gray and white rectangles relate to bit values 1 and 0 in binary forms [7]

8 Color Feature

The distribution of colors in an image can be represented with color histogram. Color descriptor is a subtle and comprehensible descriptor of an image, and generally, histogram-based techniques are preferred to show it [9, 13]. The basic benefit of color histogram is its high speed, require less memory and not subject to change in size of the image, and other factors.

Let there are M rows and N columns in an image. Each element of color space is separated into three bins such as B_{red}, B_{green}, and B_{blue}. Also let $R(i, j)$, $G(i, j)$, $B(i, j)$ be the intensity values in each portion of color space, and Ir, Ig, Ib be the bin index values of the color image's histogram using three factors. By dividing $H($Ir,Ig, Ib$)$ by the size of image, normalized histogram is obtained. The pseudocode for RGB color histogram is given as shown below:

```
1. Initialization:
for I_r = 0 to B_red - 1
    for I_g = 0 to B_green - 1
        for I_b = 0 to B_blue - 1
            H(I_r, I_g, I_b) = 0
        end for
    end for
end for
```

```
2. Updating Histogram:
for I = 0 to M - 1
    for j = 0 to N - 1
        Ir = LR(i,j) × B_red/256J
        I_g = LG(i,j) × B_green/256J
        I_b = LB(i,j) × B_blue/256J
        H(I_r, I_g, I_b) = H(I_r, I_g, I_b + 1)
    end for
end for.
```

Here, $\lfloor x \rfloor$ is a floor function which returns largest integer smaller than x.

9 Analysis and Outcomes of Proposed System

Database of 360 images of six different classes is used to verify the performance of the proposed system. For the purpose of measuring the retrieval effectiveness of the system, precision and recall values are used.

10 Image Database

The most vital task is the collection of data, no database is directly available of monument images for experimental purpose. The database presented in this paper consists of six different categories. Each image used here is rotated left, rotated right, rotated 180°, flipped horizontal, and flipped vertical. Database is collected by visiting the archaeological monuments individually and images are captured, each image is of 72 dpi. Camera used for taking the images is of 13 megapixels, a feasible distance from the monument while capturing the monument is maintained so that image captured is of good quality and consists of more details. Preprocessing is done to make the database suitable for experimental purpose.

Six randomly selected images from all categories are used as query images, some images used as query image are shown in Fig. 5.

Fig. 5 Sample images used as query

11 Similarity Metrics and Performance Evaluation

For calculating similarity between two images, six similarity metrics are used, namely: Euclidean (L2), Cityblock, Chebyshev, Minkowski, Cosine, and Normalized Euclidean. All these six similarity metrics are used one by one to calculate distance between the query image feature vector and feature vector of each image from database. Comparison table is also built on the basis of retrieval accuracy for different metrics.

Performance evaluation is done by using precision and recall. Average precision as well as recall is computed based on the number of retrieved relevant images. Basic methods of evaluating precision and recall are given by Eqs. (8) and (9)

$$\text{Precision} = \frac{|\{\text{Relevant images}\} \cap \{\text{Retrieved images}\}|}{|\{\text{Retrieved images}\}|} \tag{8}$$

$$\text{Recall} = \frac{|\{\text{Relevant images}\} \cap \{\text{Retrieved images}\}|}{|\{\text{Relevent images}\}|}. \tag{9}$$

Table 1 gives the average precision (%) of six different similarity metrics for six different categories of images. Six images of each category are selected randomly from the database. These selected images are then provided as queries and precision (%) for, respectively, selected image is calculated, and then average of the resulted precision is taken.

Figure 6 illustrates the plot of precision (%) of six different similarity metrics for six different classes. Six random images are chosen from the image database. Chosen images are then given as queries and result is attained.

Table 2 gives the average recall (%) of six different similarity metrics for six different categories of images. Six images of each category are selected randomly from the database. These chosen images are then provided as queries and recall (%) for each selected image is calculated, and then average of the resulted precision is taken.

Table 1 Average precision (%) using six similarity metrics

| Image category | Similarity metrics (Nr = 50) | | | | | |
	Euclidean	City block	Minkowski	Chebyshev	Cosine	Normalized Euclidean
Category 1	94.66	96.00	94.66	93.00	95.66	94.66
Category 2	87.00	84.00	86.66	83.66	81.00	85.00
Category 3	90.66	89.33	90.66	87.00	83.66	90.33
Category 4	76.66	87.00	76.66	68.00	74.66	76.66
Category 5	98.66	94.66	98.66	92.66	95.66	98.66
Category 6	75.66	72.66	75.66	74.66	69.33	77.00

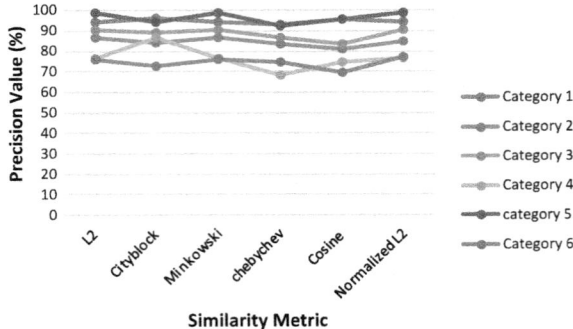

Fig. 6 Plot of average precision (%) of different similarity metrics for six different categories

Table 2 Average recall (%) using six similarity metrics

Image category	Similarity metrics (Nr = 50)					
	Euclidean	City block	Minkowski	Chebyshev	Cosine	Normalized Euclidean
Category 1	78.56	79.68	78.56	77.19	79.39	78.56
Category 2	72.21	69.72	71.92	69.43	67.23	70.55
Category 3	75.19	74.11	75.19	72.21	69.38	74.94
Category 4	63.62	72.21	63.62	56.44	61.96	93.91
Category 5	81.34	78.56	81.88	76.90	79.39	81.88
Category 6	62.79	60.30	62.79	61.96	57.54	63.91

Figure 7 shows the plot of recall (%) of six different similarity metrics for six different classes. Six random images are chosen from the image database. These images are then given as queries and result is obtained.

Fig. 7 Plot of average recall (%) of different similarity metrics for six different categories

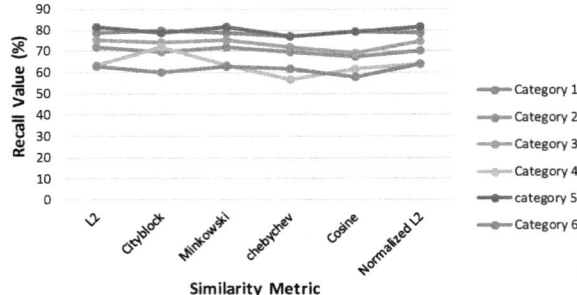

12 Results

The results generated by proposed system are good; it can be perceived in the graphs of precision and recall, where average precision and average recall of six categories of images are shown in Figs. 5 and 6. 50 images are retrieved for the input query image. A snapshot of retrieved similar images is depicted in Fig. 8.

Results for the implemented system show that the system is efficient enough to retrieve similar kind of images for monument images database. The database used for experimental purpose is synthetic database and may vary from the database used in other papers. The system presented in this paper includes all the three descriptors so it is able to differentiate and retrieve images more accurately and can be applied on any database of monument images.

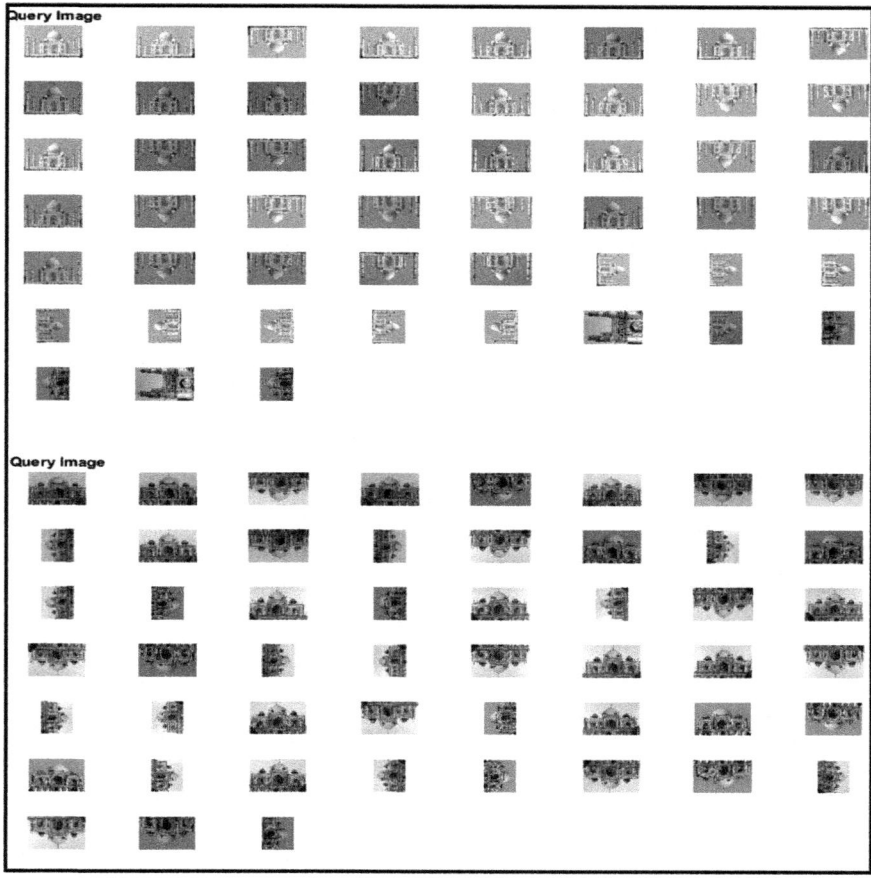

Fig. 8 Retrieved N similar images

13 Conclusion

The CBIR system proposed in this paper is an efficient approach to retrieve monument images. Shape feature of an image is extracted by using mathematical morphology, texture feature is extracted using ILBP, and color feature is extracted using RGB color histogram. The retrieval technique present in this paper for monument image retrieval is better than other retrieval techniques, because color feature is also considered in this proposed system which produces more precise results.

References

1. Ruchi Jayaswal and Jaimala Jha, "A hybrid approach for image retrieval using visual descriptor," in ICCCA 2017 unpublished.
2. Ravi Devesh, Jaimala Jha and Ruchi Jayaswal, "Retrieval of Monuments Images Through ACO Optimization Approach," International Research Journal of Engineering and Technology (IRJET), Vol. 4, Issue 7, pp. 279–285, 2017.
3. Ruchi Jayaswal, Jiamala Jha and Ravi Devesh, "An Efficient Method of Image Mining using k-Medoid Clustering Technique," International Journal of Computer Science and Engineering (IJCSE), Vol. 5, Issue 3, pp. 206–214, 2017.
4. Manish K. Shriwas and V. R. Raut, "Content Based Image Retrieval: A Past, Present and New Feature Descriptor," in International Conference on Circuit, Power and Computing Technologies [ICCPCT], 2015.
5. Padmashree Desai, Jagadeesh Pujari, N.H. Ayachit and V. Kamakshi Prasad, "Classification of Archaeological Monuments for Different Art forms with an Application to CBIR," in International Conference on Advances in Computing, Communications and Informatics (ICACCI), ISBN: 978-1-4673-6217-7/13/ 2015.
6. Shilpa Yaligar, Sanjeev Sannakki and Nagaratna Yaligar, "Identification and Retrieval of Archaeological Monuments Using Visual Features," in Proceedings of International Conference on Emerging Research in Computing, Information, Communication and Applications (ERCICA) 2013.
7. Fuxiang Lu, Jun Huang, "An Improved Local Binary Pattern operator for texture classification," ICASSP 2016 pp. 1308-1311, ISBN: 978-1-4799-9988-0/16.
8. Malay S. Bhatt and Tejas P. Patalia, "Genetic Programming Evolved Spatial Descriptor for Indian Monuments Classification," in IEEE International Conference on Computer Graphics, Vision and Information Security (CGVIS) 2015 pp. 131–136.
9. C. Singh and K. Preet Kaur, "A fast and efficient image retrieval system based on color and texture features," J. Vis. Commun. (2016), https://doi.org/10.1016/j.jvcir.2016.10.002.
10. Jaimala Jha and Dr. Sarita Sign Bhaduaria "Review of Various Shape Measures for Image Content Based Retrieval," International Journal of Computer & Communication Engineering Research Nov. 2015.
11. T. Ojala, M. Pietikainen and D. Harwood, "A comparative study of texture measures with classification based on featured distributions," Pattern Recognition, Vol. 42, pp. 425–436, 2009.

12. Leila Kabbai, Mehrez Abdellaoui and Ali Douik, "Content Based Image Retrieval Using Local and Global Features Descriptor," in 2nd International Conference on Advances Technologies for Signal and Image Processing –ATSIP'2016, March 21–24, 2016, Monastir, Tunisia, pp. 151-154, 2016.
13. Kanwal Preet Kaur, "On Comparative Performance Analysis of Color, Edge and Texture Based Histogram for Content Based Color Image Retrieval," ISBN: 978-1-4799-6896-1/14, 2014.

Comparison of ANN-Based MPPT Controller and Incremental Conductance for Photovoltaic System

Ruchira, Ram N. Patel and Sanjay Kumar Sinha

1 Introduction

Power generation through renewable energy sources is gaining large attention due to the fast rate of reduction of fossil fuel reserves and increasing pollution level. Out of all the renewable energy sources, the use of solar energy provides the following advantages: (1) It is freely available in abundance. (2) It is a clean source of energy and does not produce any greenhouse gases, and there is no problem of waste decomposition as in case of nuclear waste. (3) Its maintenance cost is very low. (4) It does not produce any noise [1, 2].

However, solar power has certain drawbacks also. The power supply obtained through the *PV* panel is not reliable as the *PV* array characteristics are dependent on the solar irradiation and temperature. Since the environmental conditions do not remain constant throughout the day, so power obtained is not constant. To resolve this problem, MPPT controllers have been made to increase the output power of the *PV* module [3, 4].

The literature survey reveals that there are different algorithms available to put into practice the maximum power point tracking [5, 6]. The comparison between various techniques can be done on the basis of the cost involved, circuit complexity, convergence speed and sensor dependence. The various algorithms can be divided into following categories: (1) conventional algorithms, namely, perturb and observe

Ruchira (✉) · S. K. Sinha
Amity University Uttar Pradesh, Noida, Uttar Pradesh, India
e-mail: er.ruchiragarg@gmail.com

S. K. Sinha
e-mail: sinha.sanjay66@gmail.com

R. N. Patel
SSCT, Bhilai, India
e-mail: ramnpatel@gmail.com

© Springer Nature Singapore Pte Ltd. 2019
V. Nath and J. K. Mandal (eds.), *Proceeding of the Second International Conference on Microelectronics, Computing & Communication Systems (MCCS 2017)*, Lecture Notes in Electrical Engineering 476, https://doi.org/10.1007/978-981-10-8234-4_26

(P&O) and incremental conductance (IC); (2) Hill climbing techniques; (3) ripple correlation current; and (4) intelligent algorithms based on fuzzy logic control, artificial intelligence techniques.

Perturb and observe (P&O) is a simple and straightforward technique but its performance gets degraded when step size is selected due to the trade-off between accuracy and speed [7]. Incremental conductance (Inc. Cond) technique performs better than the P&O for variable step increment. However, it has a drawback of showing the power oscillations close to the optimum power point. This algorithm is affected by the change in weather conditions [8]. Intelligent MPPT techniques, like fuzzy logic control and artificial neural network (ANN), are getting preference because of their robust, fast and accurate performance. The use of intelligent algorithms accurately tracks the MPP under changing weather conditions [9–15].

This paper presents an MPPT controller design based on ANN. In the proposed technique, two inputs signals are given: (i) *PV* voltage and (ii) *PV* current. The present paper is organized into following sections. Section 1 presents the introduction followed by *PV* modelling and *PV* array non-linear characteristics presented in Sect. 2. In Sect. 3, the system configuration has been discussed. Further MPPT techniques have been discussed in Sect. 4. Simulation and results have been discussed in Sect. 5. At last, the paper is concluded in Sect. 6.

2 Non-linear Behaviour of *PV* Array

Figure 1 shows the electrical equivalent circuit of a *PV* cell. It is basically a PN junction diode which gives DC output power when it is exposed to sunlight. Ideally, a solar panel can be shown as a current source in parallel with a diode. To show the practical behaviour, the series and shunt resistances are added in the circuit. I_{ph} is the photon generated current, I_D is the diode current and I_{sh} and R_{sh} are the shunt current and resistance, respectively. A solar cell is basically the main component of the solar generation. When the number of cells is joined in series, they form a *PV* module. Generally, 36 or 72 cells are connected in series.

Fig. 1 Model of PV cell

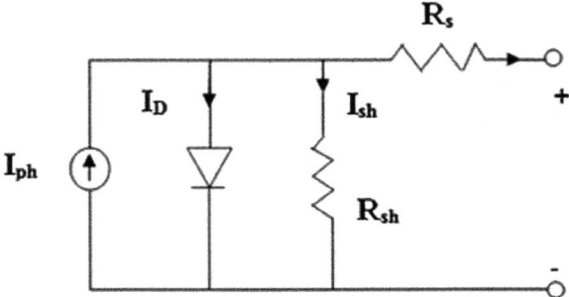

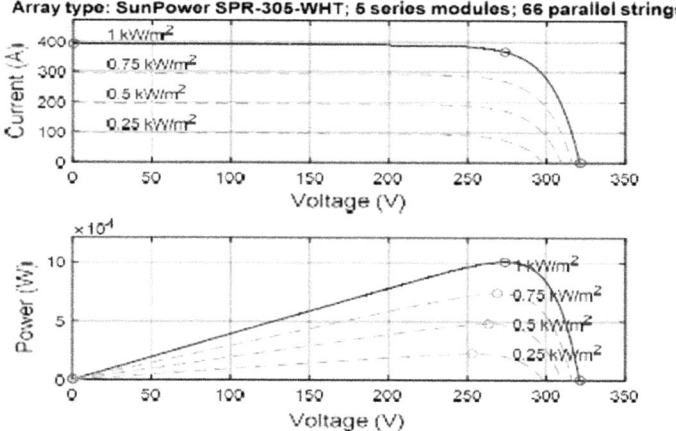

Fig. 2 *I-V* and *P-V* characteristics of Sun Power SPR-305WHT

Figure 2 shows the electrical characteristics of the *PV* cell. The characteristics of the solar cell changes with change in temperature, irradiation, etc. Maximum power point (MPP) is the point at which the *PV* module can produce maximum power. The operation of *PV* module(s) at the MPP (operating condition) must be guaranteed by the inverters and the same is accomplished with MPPT. The MPP power generation is approximately 160 W with voltage in the range of 23–38 V with open-circuit voltage below 45 V.

3 System Configuration

In the present work, 330 Sun Power (SPR305) *PV* array of 100 kW has been considered. Each module consists of parallel connection of 66 strings and further, there is a series connection of five such modules. In Table 1, the specifications of the *PV* array have been mentioned. The complete system consists of five main components. The first main component is the *PV* array which is the energy conversion device. Second is the boost converter which raises the voltage level of the output of the *PV* array. Then, the next component is the three-phase inverter used to get the ac output. Another two important components are the MPPT controller and transformer connected to the grid. The *PV* array is connected to a 25 kW grid through DC-DC converter or chopper and three-phase inverter. At 1000 W/m² solar irradiance, the *PV* array can deliver a maximum power of 100 kW. A boost converter with frequency 5 kHz has been designed to step up the voltage from 272 to 500 V. The duty cycle of the converter is controlled by the MPPT algorithm, namely, incremental conductance and integral regulator technique, and the results have been compared with the proposed ANN-based MPPT method.

Table 1 Specifications of each *PV* module

Number of series-connected cells in one module: 96
Open-circuit voltage: V_{oc}: 64.2 V
Short-circuit current: I_{sc}: 5.96 A
Voltage at maximum power point, V_{mpp}: 54.7 V
Current at maximum power point, I_{mpp}: 5.58 A
PV panel power: 305.2 W
No. of panels in series: 5
No. of panels in parallel: 66
Total power: 66 * 5 * 305.2 W = 100.7 kW
Output voltage of boost converter: 500 V
Switching frequency of boost converter: 5 kHz

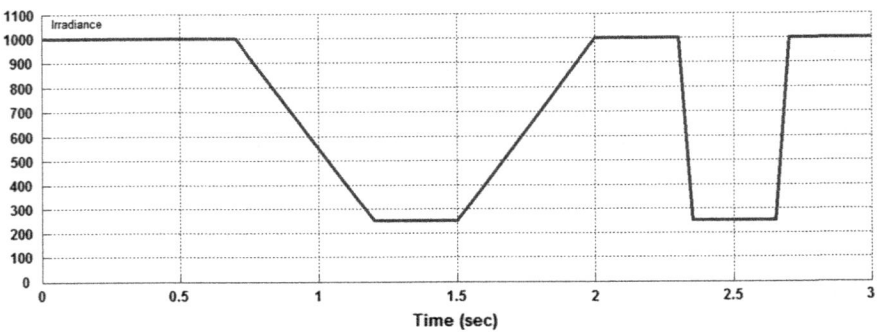

Fig. 3 Variable irradiation as input to the PV array

The temperature and irradiation are given as input to the *PV* array. The changing irradiation is considered here as shown in Fig. 3.

4   MPPT Algorithm

In this paper, first, the conventional algorithm, namely, incremental conductance has been applied to the system. The output of the *PV* panel depends on the physical conditions like solar irradiation and temperature. At maximum power point (MPP), the *PV* system can supply the maximum power to the load and the system works with maximum efficiency. So MPPT methods are employed to maximize the output of *PV* array by continuously tracking the maximum power point. In this paper, incremental conductance algorithm and ANN-based MPPT controller has been simulated to obtain the MPPT and the results have been discussed.

Generally, a solar panel gives an output of 30–40% of the input incident solar irradiation. In case of mechanical tracking, mechanical rotation of *PV* modules is done so that maximum power can be generated under a given condition. In case of electrical tracking, electronic circuitry is used to ensure that maximum power transfer takes place from source to the load. For electrical tracking, an algorithm is used which is based on the principle of impedance matching between load and *PV* array. Hence, the problem of tracking the maximum power point can also be stated as an impedance matching problem. This impedance matching is done by using a DC-DC boost converter, this converter used for the impedance matched by changing the duty cycle (*D*) of the switch.

We can calculate the power of the solar module by measuring the voltage and current through voltage and current measuring devices and then using the multiplier to obtain the power signal. This power is an input to the algorithm which adjusts the duty cycle of the switch, resulting in the adjustment of the reflected load impedance as per the power output of *PV* array. In this paper, the effectiveness of an important control algorithm has been thoroughly investigated via mathematical simulation.

5 Incremental Conductance Controller Method

The main advantage of the IC algorithm is that it is able to track the MPP under varying physical conditions, which is the major drawback of perturb and observe algorithm. If the *PV* array impedance matches with the effective impedance of the converter reflected across the array terminals, then the MPP is tracked. Figure 4 shows the power–voltage (*P-V*) curve of the *PV* array. The slope of the curve is zero at MPP, i.e. $dP/dV = 0$ as shown in Fig. 4. If the change in IC is greater than the negative of instantaneous conductance, then this represents the condition of left side of the MPP and when the change in IC is less than the negative of instantaneous conductance, then this represents the condition of right side of the MPP.

Mathematically, the equations to know about the MPP have been reached can be written as

$$\text{At MPP}, /dV = -I/V \tag{1}$$

Fig. 4 Graph showing power versus voltage for PV array

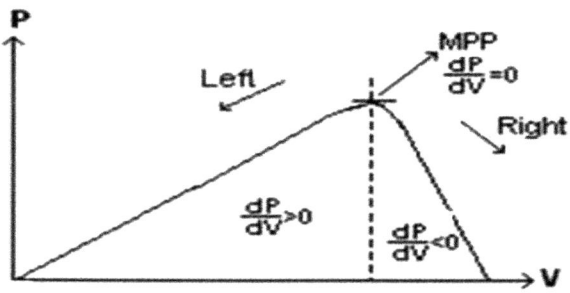

If the operating point is on to the left of MPP, then

$$dI/dV > -I/V \qquad (2)$$

If the operating point is on the right of the MPP, then

$$dI/dV < -I/V \qquad (3)$$

The flow chart of the IC algorithm has been shown in Fig. 5.

The IC algorithm further stops perturbing the operating point if the MPPT has reached the MPP. In any case, if the condition is not fulfilled, the relation between dP/dV and −I/V tells the direction of the MPPT operating point perturbed [16, 17].

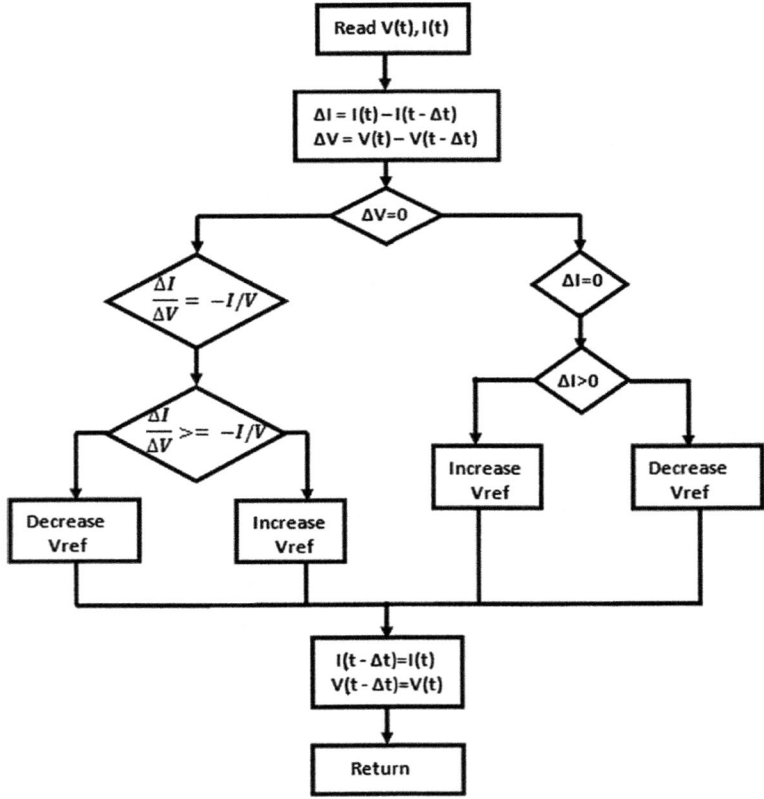

Fig. 5 Flow chart of incremental conductance algorithm

6 Artificial Neural Network (ANN) based MPPT

The concept of ANN is inspired by the working of human brains. The functioning of ANN revolves around the processing elements called neurons interconnected by links and having some adjustable weights assigned to them. The ANN is trained with a set of data and has the ability of comprehension. The major advantage of ANN is its self-learning and self-organized nature. By self-organized it is meant that ANN represents the information which it receives during learning time [8, 18–20].

ANN is used in the applications that require pattern recognition or data classification. Any artificial neural network basically consists of three layers, i.e. input layer, hidden layer and output layer. The input layer receives the information or data from some file or directly from the sensors in case of real-time application. This data is now processed based on the weight associated with a link between input layer and hidden layer. Similarly, weight is also assigned between hidden and the output layer. These weights are adjustable and affect the output. Depending upon the number of hidden layers used, the structure is named as single layer or multilayer.

The working of ANN follows either of the two topologies. These are feedforward and feedbackward approach. In feedforward approach, the signal travels in one direction from input to output. The feedback concept is not used. This type of approach is also called top-down or bottom-up approach. However, in feedbackward network, signals can travel in both the directions, and the output of one layer forms a loop by connecting to the previous layer. These networks are powerful and possess memory.

In the present work, multilayer feedforward network has been used as shown in Fig. 6. The *PV* voltage and current are given as the inputs to the proposed network. The output is taken as the duty cycle of the converter to track the MPP.

The hidden layer consists of 30 neurons and uses tangent sigmoid activation function to give the output while linear activation function is used to obtain the output of output layer. The net is obtained by training (supervised) with trainlm function-Levenberg–Marquardt algorithm.

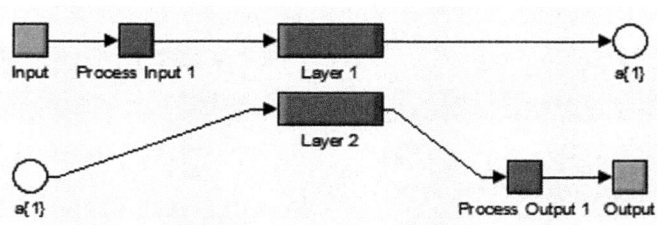

Fig. 6 Multilayer feedforward network

7   Simulation and Results

Figure 7 shows the SIMULINK model of incremental conductance algorithm-based MPPT for grid-connected *PV* system. The simulation results have shown that voltage obtained is 274.42 V and power is 80.88 kW. This voltage is boosted to 499.96 V with the help of Boost Converter.

Figure 8 shows that variation of different parameters with time. These parameters are power, voltage, duty cycle and input irradiation to the *PV* array.

Figure 9 shows the variation of voltage obtained with incremental conductance-based MPPT and ANN-based MPPT. The graph clearly indicates that the voltage curve obtained with ANN controller gives a better response than the incremental conductance-based MPPT.

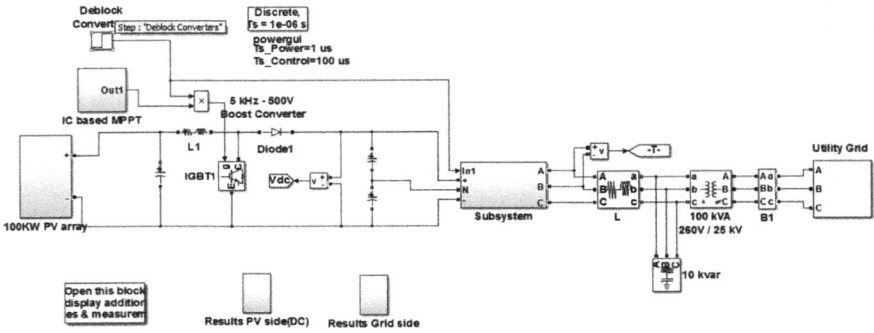

Fig. 7   SIMULINK model of IC-based MPPT system for grid-connected PV system

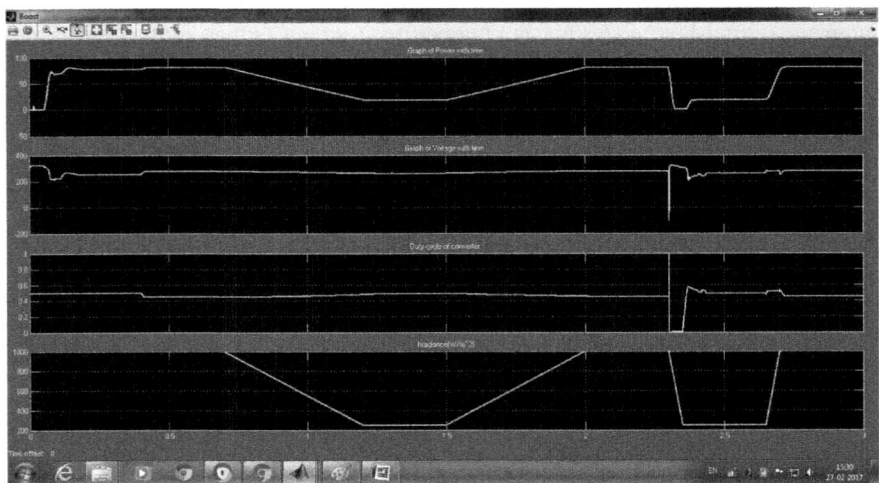

Fig. 8   Graph showing the variation of power, voltage, duty cycle and irradiation

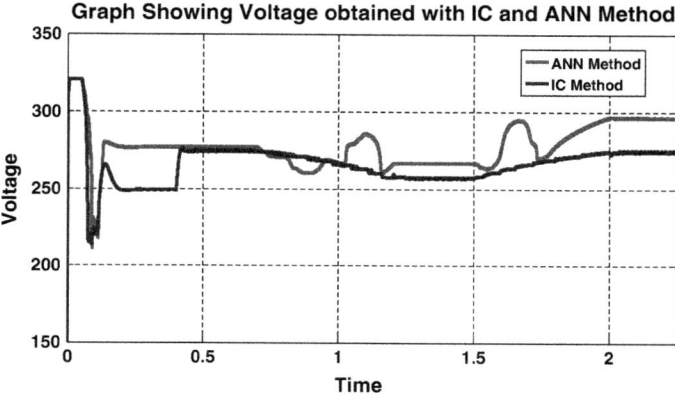

Fig. 9 Graph showing comparison of voltage obtained with IC algorithm and ANN controller

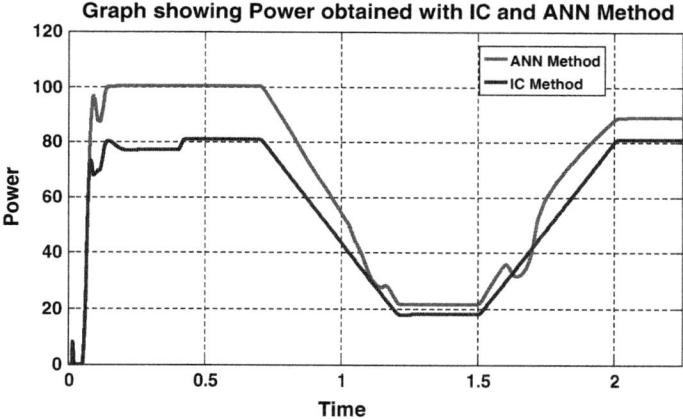

Fig. 10 Graph showing comparison of power obtained with IC algorithm and ANN controller

Figure 10 shows the comparison graph of power output obtained with incremental conductance-based MPPT and ANN-based MPPT. The graph clearly indicates that the power curve obtained with ANN controller gives a higher value than the IC-based MPPT.

8   Conclusion

From the simulation results of ANN-based MPPT controller and incremental conductance algorithm for MPPT, it can be concluded that better response is found for the proposed ANN controller than the conventional incremental conductance algorithm. The output power obtained is higher in case of ANN controller than the IC algorithm. The comparison curve has been shown in Fig. 9.

References

1. R. Khanaki, M.A.M. Radzi and M. H. Marhaban, "Comparison of ANN and P&O MPPT methods for PV applications under changing solar irradiation," IEEE Conference on Clean Energy and Technology (CEAT), pp. 284–287, 2013.
2. Y. Yongchang and Y. Chuanan, "Implementation of a MPPT controller based on AVR Mega16 for photovoltaic systems," International Conference on Future Electrical Power and Energy Systems, vol. 17, pp. 241–248, 2012.
3. Bragard, M.; Soltau, N.; Thomas, S.; De Doncker, R.W.;, "The Balance of Renewable Sources and User Demands in Grids: Power Electronics for Modular Battery Energy Storage Systems," Power Electronics, IEEE Transactions on, vol.25, no.12, pp. 3049–3056, Dec. 2010.
4. Bialasiewicz, J.T.;, "Renewable Energy Systems With Photovoltaic Power Generators: Operation and Modeling," Industrial Electronics, IEEE Transactions on, vol. 55, no. 7, pp. 2752–2758, July 2008.
5. Esram, T.; Chapman, P.L.;, "Comparison of Photovoltaic Array Maximum Power Point Tracking Techniques" Energy Conversion, IEEE Transactions on, vol.22, no.2, pp. 439–449, June 2007.
6. Jain, S.; Agarwal, V.; "Comparison of the performance of maximum power point tracking schemes applied to single-stage grid-connected photovoltaic systems" Electric Power Applications, IET, vol. 1, no.5, pp. 753–762, Sept. 2007.
7. Elgendy, M.A.; Zahawi, B.; Atkinson, DJ. "Assessment of Perturb and Observe MPPT Algorithm Implementation Techniques for PV Pumping Applications" Sustainable Energy, IEEE Transactions on, vol. 3, no. 1, pp. 21–33, Jan. 2012.
8. Safari, A.; Mekhilef, S.; "Simulation and Hardware Implementation of Incremental Conductance MPPT With Direct Control Method Using Cuk Converter," IEEE Transactions on Industrial Electronics, vol. 58, no. 4, pp. 1154–1161, April 2011.
9. Shahroooz Hajighorbani, M.A.M. Radzi, M.Z.A. Ab Kadir, S. Shafie, Razieh khanaki, M.R. Maghami, "Evaluation of Fuzzy Logic Subsets Effects on Maximum Power Point Tracking for Photovoltaic System" International Journal of Photoenergy Volume 2014.
10. Giraud, F.; Salameh, Z.M. "Analysis of the effects of a passing cloud on a grid-interactive photovoltaic system with battery storage using neural networks," Energy Conversion, IEEE Transactions on, vol. 1 4, no. 4, pp. 1572–1577, Dec 1999.
11. Jung Woo Baek; Jae Sub Ko; Jung Sik Choi; Sung Jun Kang; Dong Hwa Chung;, "Maximum power point tracking control of photovoltaic system using neural network," Electrical Machines and Systems (ICEMS), 2010 International Conference on, vol., no., pp. 638–643, 10-13 Oct. 2010.
12. Zhang, L.; Yunfei Bai; AI-Amoudi, A.; "GA-RBF neural network based maximum power point tracking for grid-connected photovoltaic systems," International Conference on Power Electronics, Machines and Drives, pp. 18–23, 4–7 June 2002.

13. Ramaprabha, R.; Mathur, B.L.; Sharanya, M.;, "'Solar array modelling and simulation of MPPT using neural network,"' International Conference on Control, Automation, Communication and Energy Conservation, 2009. INCACEC 2009, pp. 1–5, 4–6 June 2009.

14. Di Vincenzo, M.e.; Infield, D.; "Artificial Neural Network for real time modelling of photovoltaic system under partial shading," IEEE International Conference on Sustainable Energy Technologies (ICSET),, vol., no., pp. I–5, 6–9 Dec. 2010.

15. Kaliamoorthy, M.; Sekar, R.M.; Raj, I.G.e.; "Solar powered single stage boost inverter with ANN based MPPT algorithm," IEEE International Conference on Communication Control and Computing Technologies (ICCCCT), pp. 165–170, 7–9 Oct. 2010.

16. Ankita Arora, and Prerna Gaur. "AI based MPPT methods for grid connected PV systems under non linear changing solar irradiation", 2015 International Conference on Advances in Computer Engineering and Applications, 2015.

17. Ahmed Saidi, Chellali Benachaiba, "Comparison of IC and P&O algorithms in MPPT for grid connected PV module"; International Conference on Modelling, Identification and Control, Nov. 15–17, 2016.

18. Nambiar, Nirupama, RoseMary S. Palackal, Greeshma K.V, and Chitra A. "PV fed MLI with ANN based MPPT", International Conference on Computation of Power Energy Information and Communication (ICCPEIC), 2015.

19. Arash Anzalchi, Arif Sarwat, "Artificial Neural Network Based Duty Cycle Estimation for Maximum Power Point Tracking in Photovoltaic Systems"; Proceedings of the IEEE South eastCon 2015, April 9–12, 2015.

20. D. S. Karanjkar, S. Chatterji and Shimi S. L, Amod Kumar, "Real Time Simulation and Analysis of Maximum Power Point Tracking (MPPT) Techniques for Solar Photo-Voltaic System"; International Conference RAECS 06–08 March, 2014.

Low Power Implementation of 32-Bit RISC Processor with Pipelining

Sneha Mangalwedhe, Roopa Kulkarni and S. Y. Kulkarni

1 Introduction

Variety types of processors exist in the market, among them few of the processors are designed using hardware description language (HDL) like VHDL (very high-speed integrated circuit HDL) and Verilog-HDL is used for writing a particular type of the processor. RISC (reduced instruction set computer) is one of efficient computer architecture which is used for the high-speed and low-power application. The processor uses reduced instruction set (RISC) to enable the pipelined execution of the instruction. This increases the throughput and performance.

In order to improve the RISC processor, performance pipelining method is used. CPU has to do many processes. When one process is under execution, instead of waiting for the process to get finish, pipelining helps to run the new process simultaneously without troubling the preceding process. To achieve this, each part of the process is separated into several pipeline stages. After each clock, the result of the process is stored into the next pipeline stage and allows other operation to begin without interrupting the prior process. By doing this, path of all the stages can be used. Hence, all the stages in the path can be used concurrently.

Pipeline hazards are of three types. They are control hazard, structural hazard and data hazard. In the proposed architecture, hazard detection unit is designed to reduce data hazard and control hazard.

S. Mangalwedhe (✉) · R. Kulkarni
Department of Electronics and Communication, KLS GIT, Belagavi, Karnataka, India
e-mail: snehamangalwedhe@gmail.com

R. Kulkarni
e-mail: rrkulkarni@git.edu

S. Y. Kulkarni
RITM, Bangalore, India
e-mail: sy_kul@yahoo.com

© Springer Nature Singapore Pte Ltd. 2019
V. Nath and J. K. Mandal (eds.), *Proceeding of the Second International Conference on Microelectronics, Computing & Communication Systems (MCCS 2017)*, Lecture Notes in Electrical Engineering 476, https://doi.org/10.1007/978-981-10-8234-4_27

The Framework of the proposed system contains MIPS-based RISC Core. RISC core is a microprocessor intended to do small set of instructions to speed up the processor. MIPS includes five stages of pipeline, they are instruction fetch (IF), instruction decode (ID), execution (EX), memory access (MEM) and write back (WB) stages. Different sub-blocks employed are instruction memory (IM), ALU, data memory (DM), register file and the rest. The objective is to reduce the power consumption of the processor using multiplexer-based clock gating technique and to include hazard detection unit for the effective functioning of the pipeline. RISC processor is used in all places including billions of embedded system. This paper will help to understand the design of the RISC processor in detail. Design implementation is done in Verilog-HDL.

Paper is organized as follows: Sect. 2 is a literature survey. Sections 3 and 4 cover low power technique and MIPS architecture with internal blocks explanation, respectively. While Sect. 5 deals with the instruction format and instruction set. Sections 6 and 7 provide implementation and simulation results with explanations.

2 Literature Survey

Few of the major low power techniques are clock gating, sub-clock power gating, advanced branch technique, power gating and HDL modification techniques. In this survey, the diverse papers referred are specified below.

One of the ways to improve the system performance is by employing dynamic power management (DPM) [8], which dynamically alters system power consumption by turning ON the only required number of blocks for the requested service. And it deactivates the blocks when they are idle for a long time.

Joseph and Sabarinath [1] proposed front-end design process to accomplish low power by using clock gating technique. It is a power decreasing technique which stops the clock signal to arrive at different blocks of the RISC processor. The absence of the clock signal stops the flip-flop or register to take transition. Thus, the power consumption of the RISC processor can be decreased to great extent by using clock gating technique. But this technique increases gate count and design area.

Clock gating technique makes the block to sleep, in order to place the processor in further power reduction mode. This process continues until the external signal suspends it. Control block does op-code reading, decoding and finishes operation by controlling the flow of data. Clock gating can be applied to the levels such as system architecture [11], sub-blocks, gates and to the logic design. Clock gating method reduces power dissipation of high-performance circuit effectively.

Leakage power is foremost component of the overall power consumption. It is mainly due to the drain of the power when the circuit is in active mode. Power gating technique [10] is applied for reducing leakage power. To separate the circuit in the standby mode, a power switch with thick T_{ox} and high V_{th} is incorporated between ground and the logic. Power gating technique is based on predictive control. Large energy consumption may occur due to repeated misprediction of the control logic.

Another technique to reduce the leakage power in digital circuit is by using Sub-clock power gating [9]. This technique simultaneously scales the frequency and voltage. Within clock cycle, power reduction during active mode is done by using sub-clock power gating.

Another method [7] combines power gating and clock gating to accomplish leakage and dynamic power saving. This method also decreases the overhead related to inclusion of control logic because of its involvement. In order to estimate useful viability of the proposed approach, author [7] established an analysis tool to analyse the RTL design and decide whether to integrate power gating and clock gating or not. Here tool takes the decision to insert sleep transistor for accurate Power Gating approach to reduce total circuit power.

Islam et al. [2] designed 32-bit energy efficient RISC processor which is aimed at data-intensive and branch-related calculation. Design includes architecture and circuit technique to avoid the energy wastage. In branch specific application, advance branch technique is applied which further reduces wastage of energy by reducing the stalls. Low power data memory access strategy is also included in the architecture.

Kumar and Rattan [8] have used the HDL modification and state machine technique to reduce the power consumption at system level. They have applied dynamic power management to the RISC processor, here power management unit maintains the database of access to the IO devices, while processing. The authors have applied low power clock gating technique within the code for power optimization. Hence, this paper highlights on HDL modification combined with clock gating technique. Power reduction achieved in this technique is 13.33%.

Through different papers, it is seen that the work is concentrated on methods such as advance branch technique, HDL modification, clock gating and the rest to decrease the total power consumption.

3 Low Power Technique

In processor design, dissipation of the power is the main problem. Clock signal consumes a considerable fraction of total power dissipation. The main source of power dissipation is the clock signal power due to its high load and frequency. Clock signal neither performs computation nor carries any information. It is mainly used for synchronization. Adopting Clock gating technique, anyone can decrease the power consumption in a gated module by blocking unwanted switching activity of the clock signal.

Two main sources of power consumption are leakage and dynamic power [3, 4] consumption. Leakage power consumption exists when the device is idle or in standby mode. And dynamic power consumption is because of logic or switching transition in flip-flops or in logic gates. Power (P) consumption equation [6] is given by

$$P = \frac{1}{2}C_0 V_{in}^2 Nf \qquad (1)$$

where C_0 is capacitor, V_{in} is input voltage and f is frequency. Reducing the number of transition (N) reduces the dynamic power consumption.

One way to decrease the consumption of dynamic power is by adopting clock gating (CG) [3–5] technique. This reduces the clock power dissipation. In this technique, multiplexer is used. Control signal acts as a select line. And one of the inputs to the multiplexer is clock and another input is grounded. It clock gate the blocks which are not used. The output generated by the multiplexer acts as a gated clock signal to the block. This clock gating (CG) disables clock to a circuit that is not in use. In this paper, enable signal is used as the control signal. By doing this, it stops avoidable discharging or charging of capacitance and saves the clock power when the circuit is idle.

4 Proposed Low Power MIPs Architecture

The architecture of the 32-bit MIPS-based RISC processor is shown in Fig. 1. It consists of following stages:

1. *Instruction fetch (IF) stage*: This is the first stage of the processor, which fetches the instruction from the instruction memory (IM) based on the address which PC contains. The size of the PC and Instruction is 32-bit. Then, the fetched instruction is sent to the next stage.
2. *Instruction decode (ID) stage*: This is the second stage of the processor, which decodes the instruction and separates the op-code and operand from the instruction. This stage also accesses the register file to read the register.
3. *Execute (EX) stage*: This is the third stage of the processor, where actual computation takes place. Execute stage contains ALU which performs arithmetic, logical and shift related operation based on the op-code value.
4. *Memory access (MEM) stage*: This is the fourth stage of the processor and it contains data memory (DM). It performs loading and storing operations.
5. *Write back stage (WB) stage*: The fifth or last stage of the processor is write back. This stage writes or stores the result into the register.
6. *Hazard detection unit*: Hazard detection unit is designed to detect the control hazard and data hazard. It detects the hazard and tries to reduce it by introducing stall in order to get the correct result.

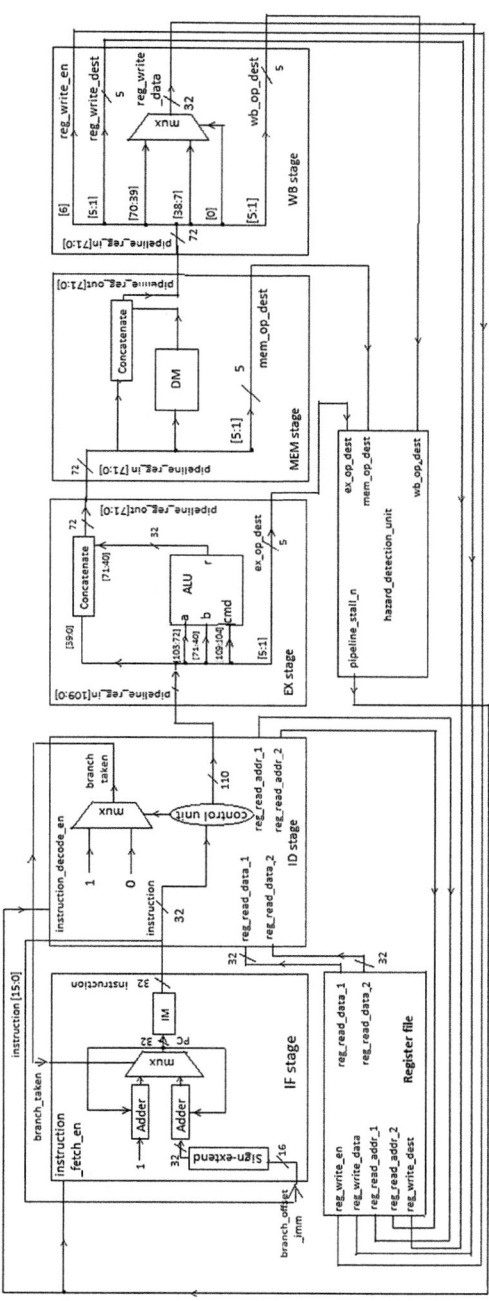

Fig. 1 Architecture of the 32-bit MIPS-based RISC processor

5   MIPS 32-Bit Instruction Set Description

The instruction set of MIPS in this paper is divided into two different types of instructions. They are as follows:

1. *Register type (R-type)*: In register type (R-type) instruction, both the operands are registered on which computation is carried out. Src1 and Src2 are the two source registers used for computation. Dest is the destination register used for storing the computed result. Table 1 shows the format for the R-type instruction. In R-Type instruction bits from 31 to 26 is used for Op-code, from bits 25 to 21 is used to indicate the address of destination register (Dest), bits from 20 to 16 and 15 to 11 corresponds to address of Source-1 and Source-2 (Src2) register, respectively. And the remaining 11-bits are reserved.

2. *Immediate Type (I-Type)*: It is used for immediate addressing mode and for branch operation. In immediate addressing mode, one of the operands is immediate value. In case of branch operation, immediate value acts as offset address and gets added with PC to branch to that address. Table 2 shows the format for the I-Type instruction.

   In I-type instruction, bits 31–26 correspond to Op-code, bits from 25 to 21 correspond to the address of the destination register (Dest), and bits from 20 to 16 are used to represent the address of the source-1 register (Src1). Remaining 16-bits are used as an immediate value. In case of branch operation, it is considered as offset address.

Table 3 shows the instruction used in the RISC processor. Op-code and its corresponding operation are shown.

Table 1 Register type (R-type)

Op-code	Dest	Src1	Src2	Reserved
31–26	25–21	20–16	15–11	10–0

Table 2 Immediate type (I-type)

Op-code	Dest	Src1	Immediate value
31–26	25–21	20–16	15–0

Table 3 MIPS instructions

Sl No.	OPCODE	Operation	Description
1	000000	OP_NOP	NO OPERATION
2	000001	OP_ADD	Result = A + B
3	000010	OP_SUB	Result = A − B
4	000011	OP_AND	Result = A & B
5	000100	OP_OR	Result = A|B
6	000101	OP_XOR	Result = A^B
7	000110	OP_NOT	Result = ∼B
8	000111	OP_SL	Result = A ≪ B
9	001000	OP_SR	Result = A ≫ B (sign ext)
10	001001	OP_SRU	Result = A ≫ B
11	001010	OP_LD	Rd = mem [Rs]
12	001011	OP_ST	mem[Rs] = Rt
13	001100	OP_BZ	Branch pc = pc + offset
14	100001	OP_ADDi	Result = A + imm
15	100010	OP_SUBi	Result = A − imm
16	100011	OP_ANDi	Result = A & imm
17	100100	OP_ORi	Result = A|imm
18	100101	OP_XORi	Result = A^imm
19	100110	OP_NOTi	Result = ∼imm
20	001101	MOV	Result = B
21	101101	MOVi	Result = imm

6   Implementation

Figure 2 shows how data flows in the RISC processor. First instruction fetch stage fetches the instruction from the instruction memory based on address contained in the program counter. Then decode stage separates the op-code and operands. Execute stage performs the computation with the help of the ALU. Memory access stage loads or stores the result. Write back stage stores the result back into the destination register.

Figure 3 shows the symbol of top RTL view. Internal RTL view of the processor with low power technique is shown in Fig. 4. Internal stages of the processor are shown in Figs. 5, 6, 7, 8, 9, 10 and 11.

Fig. 2 Flow chart for 32-bit
RISC processor

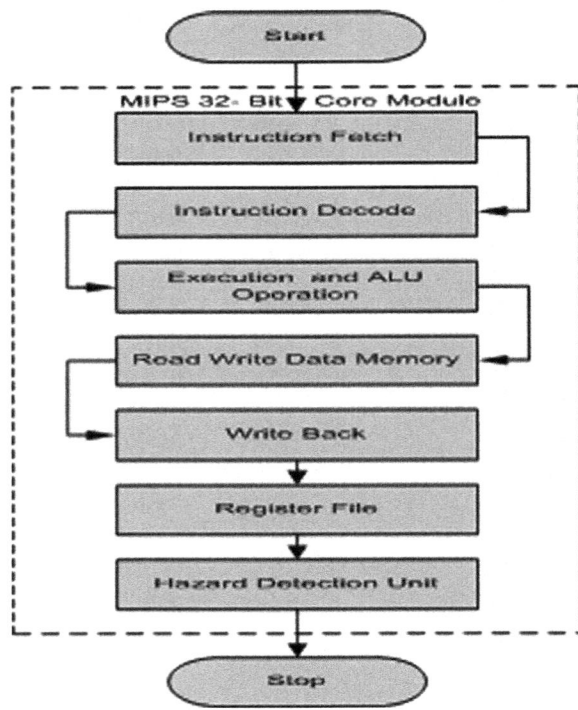

Fig. 3 Top RTL view
symbol of the RISC processor

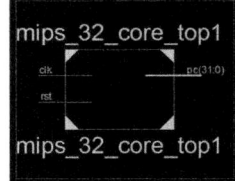

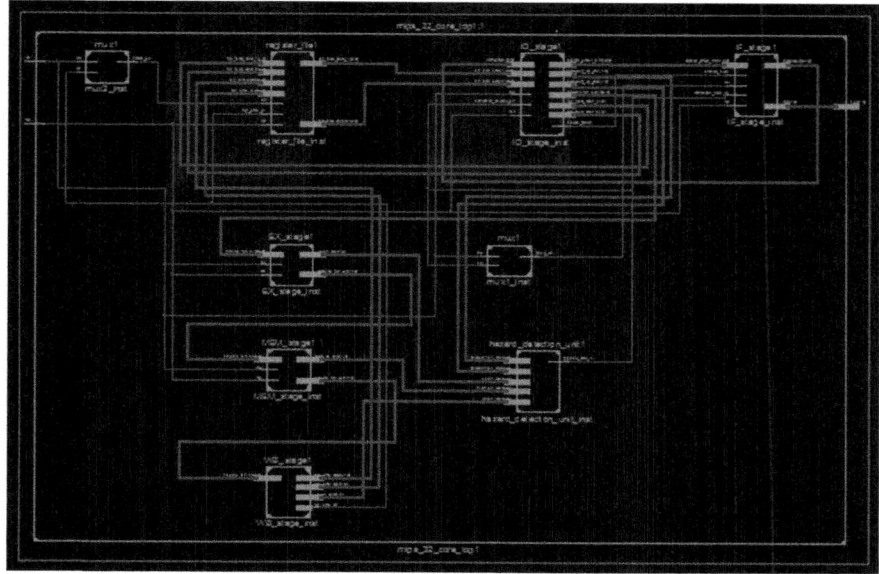

Fig. 4 Internal RTL view of the processor with low power technique

Fig. 5 Instruction fetch
(IF) stage symbol

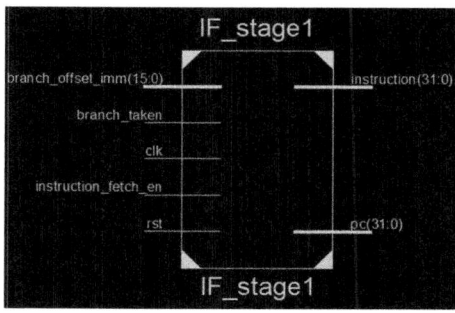

Fig. 6 Instruction decode
(ID) stage symbol

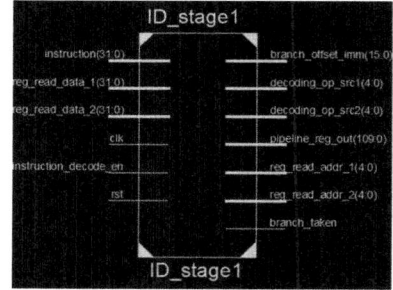

Fig. 7 Execute stage
(EX) stage symbol

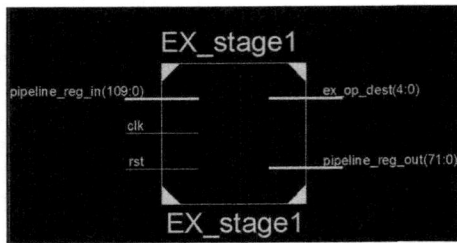

Fig. 8 Write back
(WB) stage symbol

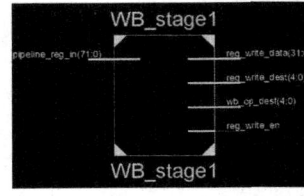

Fig. 9 Memory access
(MEM) stage symbol

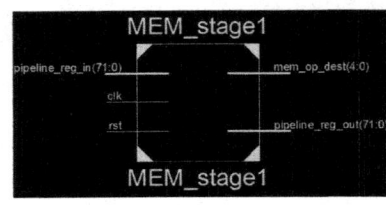

Fig. 10 Hazard detection
unit symbol

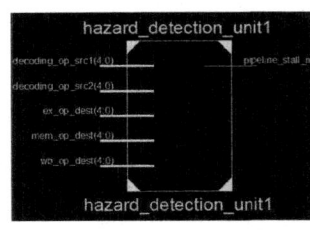

Fig. 11 Register file symbol

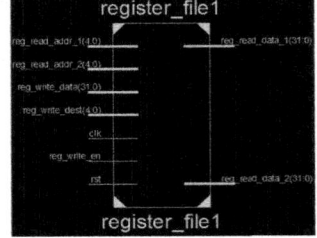

Table 4 Power of the processor without low power technique

Sl No.	Frequency (MHz)	Dynamic power (W)	Leakage power (W)	Total power (W)
1	100	0.112	0.015	0.127
2	250	0.270	0.017	0.287
3	500	0.551	0.021	0.572
4	1000	1.097	0.034	1.131

Table 5 Power of the processor with low power technique

Sl No.	Frequency (MHz)	Dynamic power (W)	Leakage power (W)	Total power (W)
1	100	0.099	0.015	0.114
2	250	0.214	0.016	0.230
3	500	0.427	0.019	0.446
4	1000	0.854	0.027	0.881

Table 6 Comparative power analysis of RISC Processor for varied technologies

Device	Process technology (nm)	Frequency (MHz)	Power [8][a] (W)	Power (W)	
				Without clock gating	With clock gating
SPARTAN3	90	98.09	0.144	0.121	0.110
VERTIX5	65	321.048	0.777	0.375	0.291
VERTIX6	40	401.881	1.44	0.463	0.363

[a]Architecture containing peripherals

8 Conclusion

RISC processor with five-stage pipeline is designed and verified using XILINX 14.7 ISE simulator and Spartan 6 family. Modules developed in this architecture are —Fetch, Decode, Execute, Memory access, Write back, hazard detection unit, register file, ALU, instruction memory and data memory. By employing multiplexer-based clock gating technique, power consumed by the processor is reduced. The proposed architecture is tested for different frequencies, namely: 100, 250, 500 and 1000 MHz, and the respective percentage of power reduction is as follows: 10.23, 19.86, 22.02 and 22.10%. The area of the processor is increased to 2–3% after applying low power technique.

9 Future Scope

The power optimization of the proposed architecture can further be reduced by creating power domains based on the functionality of the modules. The other possibility is to inhabit the low power technique such as dynamic clock gating and adaptive clock gating to obtain the maximum power reduction.

References

1. N. Joseph, Sabarinath, et al, "FPGA based implementation of high performance architectural level low power 32-bit RISC core", International conference on advances in recent technologies in communication and computing, Annual IEEE, pp. 53–57, 2009.
2. Shofiqul, Debanjan, et al, "Design of high speed pipelined execution unit of 32-bit RISC processor", Annual IEEE, pp. 1–5, 2006.
3. Roopa Kulkarni, S.Y. Kulkarni, "Power analysis and comparison of clock gated techniques implemented on a 16-bit ALU", Proceedings of International conference on circuits, communication, control and computing, pp. 416–420, November 2014.
4. Roopa Kulkarni, S.Y. Kulkarni, "Energy efficient implementation of 16-bit ALU using block enabled clock gating technique", Annual IEEE India conference (INDICON), 2014.
5. Aneesh Raveendran, Vinayak Baramu Patil, David Selvakumar, Vivian Desalphine, "A RISC-V Instruction Set Processor-Micro architecture Design and Analysis", International conference on VLSI system, Architectures, Technology and Application (VLSI-SATA), 2016 IEEE.
6. Mohit N. Topiwala, Saraswathi, "Implementation of a 32-bit MIPS based RISC processor using cadence", IEEE International Conference on Advanced communication control and Computing technology (ICACCCT), 2014.
7. Li Li, Ken Choi, Haiqing Nan, "Effective algorithm for integrating clock gating and power gating to reduce dynamic and active leakage power simultaneously", Quality Electronic Design (ISQED), 2011 12th International symposium, pp. 1–6, 2011.
8. Narender kumar, Munish Rattan, "Implementation of embedded RISC processor with Dynamic Power Management for low power embedded system on SOC", Proceedings of 2015 RAECS UIET Punjab university Chandigarh 21–22nd December 2015. IEEE.
9. Jattin N, Bashir M, et al, "Sub clock Power-gating technique for minimizing leakage power during active mode", Design Automation and Test in Europe Conference and Exhibition, pp. 1–6, 2011.
10. Hao Xu, Ranga Vemuri and Wen-Ben Jone, "Dynamic characteristics of power gating during mode tradition", VLSI system, IEEE Transactions on, VOL. 19, NO. 2, pp. 237–249, February 2011.
11. Shmuel Wimer and Israel Koren, "The optimal Fan-out of clock network for power minimization by adaptive gating", VLSI systems, IEEE Transaction on, NO. 99, pp. 1–9, 2011.

A Novel SINR-Based Cooperative Radio Resource Allocation Mechanism (SBC-RRAM) for LTE/Wi-Fi Radio Access System in Smart Home Environment

Bollampally Joy Persis and Sakuru K. L. V. Sai Prakash

1 Introduction

Internet of Things or IOT is one of the fastest growing technological industries. With the announcement of smart cities by Government of India, the growth rate of the IOT in India has taken a new dimension aiming at managing the services such as electricity, gas, petrol, foodstuff, health services, safety, transportation, etc. in a more efficient manner. Almost more than 1000 companies are developing IOT devices before the standards are evolving. This made the network engineers relook at the demand of wireless access. Therefore, the heterogeneous IOT devices need novel network architecture with non-deterministic channels. The wireless channels are mostly rogue in behaviour, vulnerable to channel characteristics, dependent on indoor environment, providing access based on SINR of the channel. As a result of the channel being non-deterministic, the capacity of the channel varies affecting the data throughput of the network. And also there are many applications in the network which require fixed bandwidth all the time. As an example, a biometric sensor which reports the health of a patient requires a fixed bandwidth channel over a large period of time. Apart from such nodes, there are also emergency service nodes like gas leak detector which are to be given high priority and hence require a reasonably high data rate. The network also contains nodes, which collaborate among themselves and complete the given task internally and remain passive most of the time. These nodes can do away with lower resource shares. Hence, a resource allocation mechanism based on SINR is proposed which takes into account the channel conditions while allocating the resources. A comparison is made between the resource allocation

B. Joy Persis (✉) · S. K. L. V. Sai Prakash
Department of Electronics and Communication Engineering, National Institute
of Technology, Warangal, Telangana 506004, India
e-mail: bollampellyjoy@gmail.com

S. K. L. V. Sai Prakash
e-mail: sai@nitw.ac.in

© Springer Nature Singapore Pte Ltd. 2019
V. Nath and J. K. Mandal (eds.), *Proceeding of the Second International Conference
on Microelectronics, Computing & Communication Systems (MCCS 2017)*, Lecture Notes
in Electrical Engineering 476, https://doi.org/10.1007/978-981-10-8234-4_28

mechanisms in which one is based on SINR and the other is Dandachi et al. [1] without considering the SINR.

Many communication technologies like Z-wave, NFC, Sigfox, Neul and LoRaWAN have emerged mainly for IOT coexisting with Wi-Fi, LTE, Bluetooth, etc. All the technologies have different communication capabilities. The maximum data rates of Z-wave, NFC, Sigfox, Neul, LoRaWAN, Ethernet, Wi-Fi, LTE and Bluetooth are 100 kbps, 420 kbps, 1000 bps, 100 kbps, 50 kbps, 54 Mbps, 50 Mbps and 1 Mbps, respectively. Hence, it is evident that managing the resources and allocating them efficiently is a crucial topic when such access IOT devices coexist in a network.

In this paper, we considered one LTE 5 Mbps channel and one 15 Mbps broadband connection for the study. It is assumed that at least minimal services to all the nodes are provided by these two channels. We also assumed that the 15 Mbps broadband connection provides non-deterministic capacity and can be accessed by means of three Wi-Fi access points (APs). The resource allocation controller will allocate the access to IOT devices based on their SINR and request demand.

The remaining part of the paper is organized in sections as follows. Section 2 reviews the previous work done related to the scope of our problem, the architecture of the system model proposed is described in Sect. 3. The proposed resource allocation mechanism is given in Sect. 4, and the modified resource allocation mechanism with SINR is discussed in Sect. 5. The results obtained through simulations are given in Sect. 6, and the conclusion of the work is given in Sect. 7.

2 Literature Review

Resource allocation and scheduling in wireless networks with multiple non-deterministic channels is a topic of much interest among the researchers in the current years. The allocation is with respect to many views: offloading data traffic between multiple access networks, the optimal distribution of resources to maximize throughput and to minimize congestion as well as load balancing.

Multihoming is considered in the paper [1], to offload the data traffic between 3G and WLAN networks to decrease the congestion and increase throughput. Local Proportional Fairness and Global Proportional Fairness schemes by considering a fast fading channel are studied and compared in this paper. A dynamic optimization scheme for load balancing is proposed in [2], which considers the load of the multiple access network and channel quality and then splits the packets between the access network. A radio resource allocation scheme for offloading mobile tasks is proposed in [3].

In [4], a game theoretic framework is used in CDMA networks to propose a distributed pricing-based algorithm which allocates data rate and power jointly. This distributed resource allocation scheme is also extended to an environment containing multiple cells. An optimization of utility function by using a softened

SIR requirement along with adding a penalty on consumption of power and then allocating power to the users is proposed in [5].

In [6], maximum throughput is achieved in a multi-hop wireless network by implementing a randomized power allocation algorithm.

In [7] the resource allocation is done based on the energy efficiencies of the user equipment and aims to guarantee proportional fairness. However, this scheme is restricted only to the users in small cell networks. A resource allocation approach which takes into account various communication parameters like PER, time delay, etc. and forms an optimization function is proposed in the paper [8].

Various opportunistic scheduling schemes which are channel aware are discussed in [9]. These algorithms map to the functionalities of different layers of the protocol suite. A hybrid MAC approach based on ALOHA is proposed for 5G networks which aim to decrease congestion in radio access and channel access delay [10].

Although there is much research on the resource allocation mechanisms, in this paper, we propose a novel resource allocation mechanism by exploring the true utility of the users based upon their priority and need. This mechanism maximizes the resource share among the users based on the SINR of the channel such that proportional fairness is maintained. The network which we considered has two access networks, i.e. 5 Mbps LTE link and 15 Mbps broadband connection. The availability of these two networks are random in nature and do not provide a deterministic resource.

3 System Model

The system in consideration is heterogenous environment comprising LTE 5 Mbps cellular link and 15 Mbps broadband link and both these links coexist. The network consists of a smart home controller (SHC) scheduling node connected to n number of Wi-Fi APs and an LTE link, where $n = 1, 2, 3$ in this case. The SHC scheduler implements a scheduling mechanism for a fair resource allocation. This SHC scheduler is implemented as a service provider to the IOT devices allocating resources based on network and resource availability as shown in Fig. 1.

Let w_i^n / l_i^n be the Wi-Fi/LTE IOT devices such that $\forall i = 1 \ldots k$ associated with an nth ($n = 1, 2, 3$) Wi-Fi cell or LTE link. The scheduling is done such that each user gets a share of the resources. Let s_l be the share of the LTE resources allocated to the l users and s_w^i be the resource share of the Wi-Fi users. We studied the resource allocation problem under two assumptions: (1) network consists of nodes with uniform resource requirement and (2) network of nodes with non-uniform resource requirements.

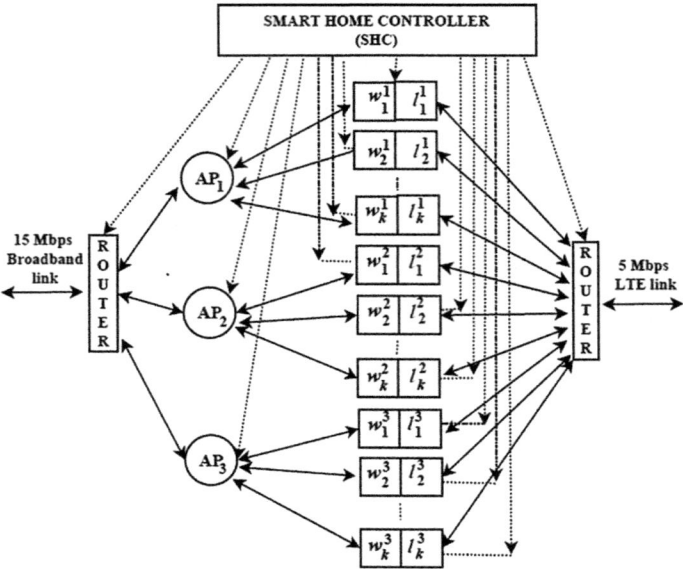

Fig. 1 System model

- Uniform Wi-Fi nodes: All the nodes in this network have equal priority and the resources are shared equally among them.
- Non-uniform Wi-Fi nodes: Some of the Wi-Fi nodes demand a bigger share of the resources and hence called aggressive nodes and some nodes require just a small share of the resources. Such nodes are called non-aggressive nodes. As a whole, the network is non-uniform as it has both aggressive and non-aggressive nodes. For example, in smart home application, we may have a node say, a temperature sensor which transmits the sensed data every given interval of time, a node used for streaming of videos. The same system may also contain a node which remains inactive and only becomes active when triggered, e.g. motion detector. Hence, the temperature sensing node and video streaming are aggressive nodes whereas the motion detecting node is non-aggressive.

Our aim is to optimize the resources given to the users in both the cases of uniform nodes and non-uniform nodes.

4 Resource Allocation Mechanism Without Considering SINR

The two mechanisms for resource allocation, i.e. uniform proportional fairness (UPF) and Weighted Proportional Fairness (WPF) are discussed in this section. UPF mechanism is applied to the network of uniform nodes whereas WPF is applied to the network of non-uniform nodes. The capacity is considered as the resource to be allocated here. Using these mechanisms, we allot an optimized capacity share to each user in both the network classes.

4.1 Uniform Proportional Fairness (UPF) [1]

In this mechanism, the resources are allocated to the users by considering both the access network as a whole. The share of the resources allocated to each user is obtained by maximizing the total utility of all the users in both the networks (considered as a whole). The utility function is the product of a number of users in an access network and the logarithmic function of the resource allocated to them. As we wish to maximize the resource shares of all the users by considering both the networks as a whole, the function which is to be optimized is a sum of the utility functions of both the access networks, i.e. LTE and Wi-Fi. The logarithmic function is used to ensure fairness to all the users.

The capacity of LTE channel is c_l and of Wi-Fi channel is c_w and the share allocated to each LTE user is s_l and for Wi-Fi users is s_w^i. The UPF mechanism is applied to the network composed of two access networks as a whole. The APs serve uniform nodes. The optimization problem maximizes the utility function of all the users. The problem can be stated as

Maximize:

$$l \log s_l c_l + \sum_{i=1}^{n} w_k^i \log \left(s_w^i c_w \right) \tag{1}$$

Subject to:

$$l s_l = 1 \tag{2}$$

$$w_k^i s_w^i = 1 \tag{3}$$

4.2 Weighted Proportional Fairness (WPF)

The WPF mechanism is almost similar to the UPF except for one thing. A weight is assigned to the resource shares in the logarithmic function to allot a priority for the aggressive and non-aggressive nodes. Let p be the fraction of Wi-Fi nodes which are aggressive implying that there are $p_i w_k^i$ aggressive nodes and $(1 - p_i) w_k^i$ non-aggressive nodes in the network. We assume that α is the weight (or priority) associated with aggressive nodes and β is the weight associated with non-aggressive nodes. The share of aggressive nodes is denoted by $s_{\alpha w}^i$ and of non-aggressive nodes is $s_{\beta w}^i$. The WPF resource allocation mechanism is implemented in this network of non-uniform nodes. The optimization problem can be stated as

Maximize:

$$
\begin{aligned}
l \log s_l c_l + \sum_{i=1}^{n} p_i w_k^i \log\left(\alpha_i s_{\alpha w}^i c_w\right) \\
+ \sum_{i=1}^{n} (1 - p_i) w_k^i \log\left(\beta_i s_{\beta w}^i c_w\right)
\end{aligned}
\tag{4}
$$

Subject to:

$$
l s_l = 1 \tag{5}
$$

$$
p_i w_k^i \alpha_i s_{\alpha w}^i + (1 - p_i) w_k^i \beta_i s_{\beta w}^i = 1 \tag{6}
$$

Both the mechanisms considered above give almost equal throughputs. Hence, any one of them can be used when the channel is deterministic. However, when the channel is non-deterministic, the Signal to Interference plus Noise ratio (SINR) of the channel is to be considered for fair resource allocation. In the next section, a resource allocation mechanism based on the channel conditions is proposed.

5 Resource Allocation Based on SINR

In this mechanism, the resource allocation is done based on the channel conditions. This ensures that all the nodes are given a fair share of resources and also the priority is kept in mind. This mechanism is based on the WPF mechanism.

To ensure the network is highly reliable and also throughput of every link is maximum, the share of resources allocated to all the users is to be maximized as well as the SINR should also be maximized.

If g_i, P_i is the gain and Power transmitted by the transmitter, respectively, then the SINR is given by

$$
\frac{g_i P_i}{N_0 + \sum_{j=1, j \neq i}^{k} g_j P_j} \tag{7}
$$

The process gain of the channel is the ratio of the capacity of the channel to the data rate provided. λ is the data rate, s_x (where x is either l or α or β) is the share of the resource.

So, the capacity of a link is given by $s_x c_y$ where y is either LTE or Wi-Fi. Hence, process gain is given by $\frac{s_x c_y}{\lambda}$.

The Utility function which is to be maximized is given as
Maximize

$$
l \log \left(1 + \frac{s^i_{\alpha w} c_w}{\lambda} \frac{g_i P_i}{N_0 + \sum_{j=1, j \neq i}^{k} g_j P_j} \right)
$$
$$
+ \sum_{i=1}^{n} p_i w^i_k \log \left(\alpha_i \left(1 + \frac{s^i_{\alpha w} c_w}{\lambda} \frac{g_i P_i}{N_0 + \sum_{j=1, j \neq i}^{k} g_j P_j} \right) \right) \qquad (8)
$$
$$
+ \sum_{i=1}^{n} (1 - p_i) w^i_k \log \left(\beta_i \left(1 + \frac{s^i_{\beta w} c_w}{\lambda} \frac{g_i P_i}{N_0 + \sum_{j=1, j \neq i}^{k} g_j P_j} \right) \right)
$$

Subject to:

$$
ls_l = 1 \qquad (9)
$$

$$
p_i w^i_k \alpha_i s^i_{\alpha w} + (1 - p_i) w^i_k \beta_i s^i_{\beta w} = 1 \qquad (10)
$$

The solution to the optimization function is obtained through Lagrange's method.

6 Simulation Results

For simulation, a network with one LTE eNodeB and three Wi-Fi access points is simulated. The capacity of LTE channel is 5 Mbps and the capacity of each Wi-Fi channel is 5 Mbps. The total of three Wi-Fi channels is 15 Mbps. All the simulations are done using Matlab.

In Fig. 2, a comparison is made between the weighted proportional fairness resource allocation mechanisms with and without SINR considerations of the channel. Uniform SINR is considered, i.e. both the Wi-Fi and LTE channels are assumed to have same SINR. The number of users for the simulation purpose is taken as 20 and offered traffic load is 5 Mbps per user. It is observed that the throughput in the case of resource allocation using channel SINR is much higher than that of the channel without SINR.

Non-uniform SINR conditions are considered for Wi-Fi and LTE channels, and a comparison is made between the weighted proportional fairness mechanisms with and without SINR considerations of the channel keeping all the other parameters same as in simulation shown in Fig. 2. It is clearly evident that there is an increase in throughput when we consider non-uniform SINR of the channels when compared to uniform SINR considerations of the channel. This simulation is shown in Fig. 3.

Fig. 2 Throughput of a link versus the SINR of the channel (with uniform SINR for Wi-Fi and LTE channels)

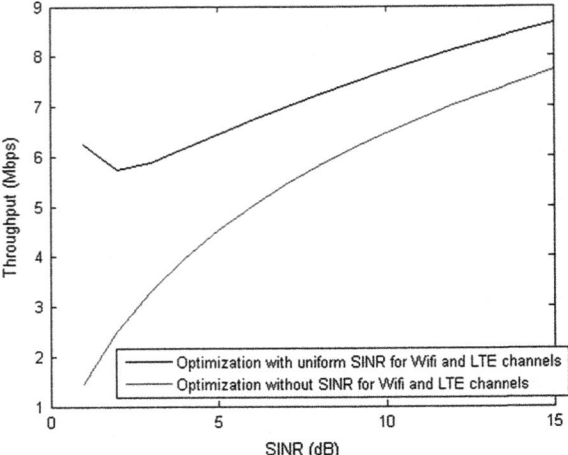

Fig. 3 Throughput of a link versus the SINR of the channel (with non-uniform SINR for Wi-Fi and LTE channels)

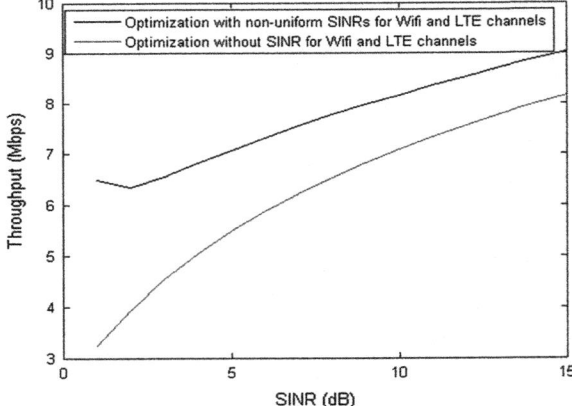

A comparison between the weighted proportional fairness resource allocation mechanisms without considering SINR and with considering SINR is done by varying the offered traffic load of the network is shown in Fig. 4. In this simulation, the SINR is fixed at 10 dB for both the channels. It is observed that the channel with SINR is having higher throughput than the channel without SINR.

In Fig. 5, the SINR of Wi-Fi channel is considered to be 10 dB and that of LTE fixed at 5 dB and then a comparison is made between the weighted proportional fairness resource allocation mechanisms without considering SINR and with considering SINR is done by varying the offered traffic load of the network.

Fig. 4 Throughput of a link versus the offered traffic load (with uniform SINR for Wi-Fi and LTE channels)

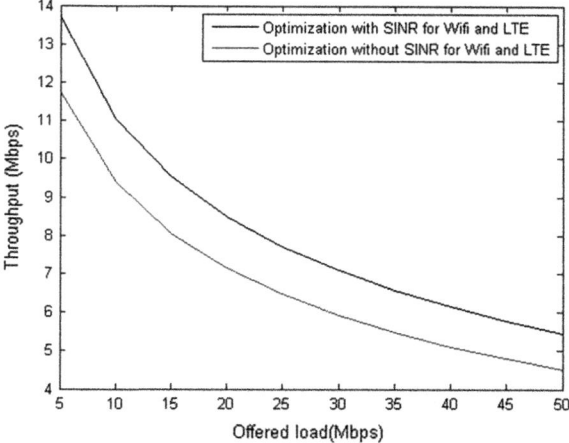

Fig. 5 Throughput of a link versus the offered traffic load (with non-uniform SINR for Wi-Fi and LTE channels)

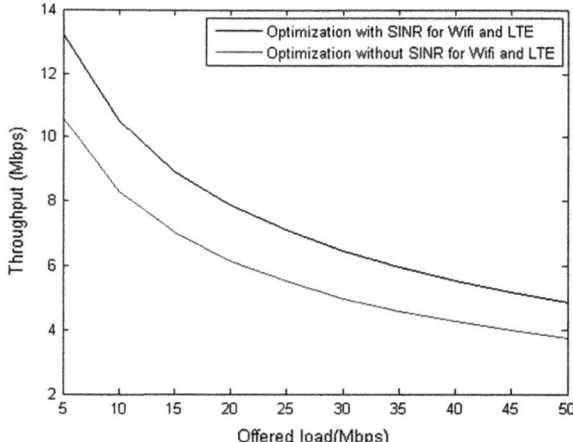

In Fig. 6, the throughput of a single user link is studied under both the mechanisms by varying the number of Wi-Fi users. It can be seen that both the networks degrade similarly for increasing number of users.

The resource allocation mechanism based on SINR is compared with the resource allocation mechanism proposed in paper [1]. Table 1 shows the comparison. The comparison is made for the two mechanisms for the network consisting of 5, 10 and 30 nodes.

Fig. 6 Throughput of a link versus the number of users

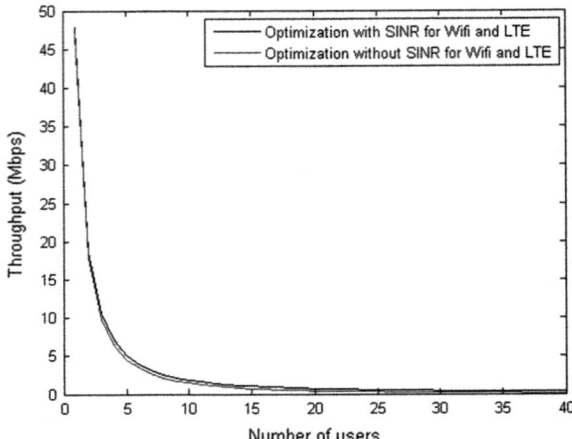

Table 1 Comparison between resource allocation mechanisms: one based on channel conditions and other without channel conditions

Parameter (for non-uniform SINR condition)	Resource allocation mechanism based on channel conditions (Mbps)	Resource allocation mechanism proposed in paper (Mbps) [1]
Throughput when SINR is 5 dB	7.072	5.496
Throughput when SINR is 10 dB	8.18	7.089
Throughput when SINR is 15 dB	9.033	8.175
Throughput when offered load is 15 Mbps	8.902	6.98
Throughput when offered load is 30 Mbps	6.458	5.001
Throughput when offered load is 45 Mbps	5.161	3.95
Throughput when number of nodes is 10	1.874	1.462

7 Conclusion

We have dealt with the problem of resource allocation in a heterogenous network based upon the channel conditions. As SINR is one of the best ways to determine the channel conditions, the share of the resource allocated to a given node is calculated with respect to SINR. The throughput is maximized with the help of a Utility function which is an optimization function dependent on SINR and share of the resources allocated. Though there is much research work on resource allocation based on SINR, this paper throws a different perspective by introducing priorities to the nodes. The comparison is made between the resource allocation without SINR and resource allocation based on SINR. Results show that resource allocation based on SINR gives more throughput than the technique without SINR.

References

1. G. Dandachi, S. E. Elayoubi, T. Chahed, N. Chendeb, and H. Jebalia, "Comparing resource allocation schemes in multi-homed lte/wifi access networks," in 2015 IEEE 82nd Vehicular Technology Conference (VTC2015-Fall), Sept 2015, pp. 1–6).
2. R. Yang, Y. Chang, J. Sun, and D. Yang, "Traffic split scheme based on common radio resource management in an integrated LTE and HSDPA networks," in 2012 IEEE Vehicular Technology Conference (VTC Fall), Sept 2012, pp. 1–5.
3. Y. Cao, T. Jiang, and C. Wang, "Optimal radio resource allocation for mobile task offloading in cellular networks," IEEE Network, vol. 28, no. 5, pp. 68–73, September 2014.
4. M. R. Javan and A. R. Sharafat, "Efficient and distributed sinr-based joint resource allocation and base station assignment in wireless cdma networks," IEEE Transactions on Communications, vol. 59, no. 12, pp. 3388–3399, December 2011.
5. M. Xiao, N. B. Shroff, and E. K. P. Chong, "A utility-based power-control scheme in wireless cellular systems," IEEE/ACM Transactions on Networking, vol. 11, no. 2, pp. 210–221, Apr 2003.
6. H. W. Lee, E. Modiano, and L. B. Le, "Distributed throughput maximization in wireless networks via random power allocation," IEEE Transactions on Mobile Computing, vol. 11, no. 4, pp. 577–590, April 2012.
7. W. Jing, X. Wen, Z. Lu, Z. Hu, and T. Lao, "Radio resource allocation with proportional-fair energy efficiency guarantee for smallcell networks," in 2016 IEEE 27th Annual International Symposium on Personal, Indoor, and Mobile Radio Communications (PIMRC), Sept 2016, pp. 1–6. M. Young,
8. C. Han, J. M. Jornet, E. Fadel, and I. F. Akyildiz, "A cross-layer communication module for the internet of things," Computer Networks, vol. 57, no. 3, pp. 622 – 633, 2013. [Online]. Available: http://www.sciencedirect.com/science/article/pii/S138912861200357X.
9. X. Lin, N. B. Shroff, and R. Srikant, "A tutorial on cross-layer optimization in wireless networks," IEEE Journal on Selected Areas in Communications, vol. 24, no. 8, pp. 1452–1463, Aug 2006.
10. D. Posnakides, C. X. Mavromoustakis, G. Skourletopoulos, G. Mastorakis, E. Pallis, and J. M. Batalla, "Performance analysis of a rate-adaptive bandwidth allocation scheme in 5 g mobile networks," in 2015 IEEE Symposium on Computers and Communication (ISCC), July 2015, pp. 1–7.

Optimal Location of Thyristor Controlled Series Compensator to Assuage Congestion in Transmission Network

Pallavi Choudekar, Shagun Kachwaha, Aaditya Jhunjhunwala and Anchal Dua

1 Introduction

Dependence on electrical energy is increasing with the expansion in industrialization. Due to this, the power sector is growing rapidly with few uncertainties. Growth in the load demand is likely to strain the existing power systems, in which the transmission systems are to work near to the operating limits and also causes congestion. Thus, the transmission system should be flexible enough that it can handle unpredictable power supply and other demand conditions. FACTS devices (especially TCSC) are one the ways to reduce the transmission congestion and allow better utilization. TCSC is to be connected in series with the transmission line to reduce the transmission losses. Also, TCSC is a variable impedance type compensator. The sensitivity factor is used to find the best location of TCSC for the enhancement of static performance of the system. In [1], Optimal Location of Series FACTS devices to achieve maximum system load ability with minimum installation cost. Load flow analysis is used for sensitivity analysis [2]. To overcome problems like voltage stability, voltage compensating devices like SVC are used to find optimal location. Weak bus locations are identified by FVSI. Loss reduction provided by SVC contributes in cost saving of power generation [3]. Optimization method inspired by law of gravity is used to reduce the reactive power losses in system by attaching/adding TCSC in series [4].

P. Choudekar (✉) · S. Kachwaha · A. Jhunjhunwala · A. Dua
Amity University, Noida, Uttar Pradesh, India
e-mail: pallaveech@gmail.com

S. Kachwaha
e-mail: shagunkachwaha@gmail.com

A. Jhunjhunwala
e-mail: aadityajhunjhunwala@gmail.com

A. Dua
e-mail: anchaldua1@gmail.com

© Springer Nature Singapore Pte Ltd. 2019
V. Nath and J. K. Mandal (eds.), *Proceeding of the Second International Conference on Microelectronics, Computing & Communication Systems (MCCS 2017)*, Lecture Notes in Electrical Engineering 476, https://doi.org/10.1007/978-981-10-8234-4_29

After placing TCSC, generator scheduling and investment expenditure are reduced using the method which is popularly known as particle swarm optimization. Optimal Location for TCSC is determined by calculating sensitivity index [5]. Congestion is managed by finding optimal location and size of Unified Power Flow Controller in transmission line network. Sizing of UPFC and Optimal Location can be identified by a fuzzy-based approach [6]. For least loss configuration where there is over-loading, reconfiguration of existing network is done with sectionalizing switches. For congestion management, a new technology of Smart Wire is used for high reliability and low cost [7]. Congestion mitigation through TCSC, Formulation includes maximization of social welfare function subject of real and reactive power balance and capability curve constraints TCSC location can also be decided by Mixed Integer Non-Linear Programming (MINLP) [8]. To enhance the load of transmission line, TCSC is used. For maximum power transfer capability, we can use IGSA. Maximum power loss is considered as fitness function [9]. Congestion occurs in the weakly controlled network. The bidding of higher price and withdrawing capacity are analysed. Location Marginal Price (LMP) gives the impact on transmission con-gestion [10]. TCSC is utilized for the issue of congestion management. Location of TCSC can be determined by LMP-based method. To reduce generation cost and manage congestion in deregulated electricity markets we set its parameters [11]. There are two new methods for placement of TCSC which is similar to the sensitivity factor. These new techniques are known as LMP difference methodology and con-gestion rent methodology. These two methods are profitable because LMP is an outgrowth of the security restrained OPF, and congestion rent is a part consequence of the LMP difference [12]. For the planning, operation and control, power flow studies are necessary. PSAT is one of the MATLAB software OPF is used practically with PSAT. It determines the setting of OPF having objective function for mini-mizing fuel cost [13]. For finding the solution to OPF problems when controllable FACTS devices are used, PSO also works. Two FACTS devices can be used, i.e. Thyristor Control Series Capacitor and Thyristor Control Phase Shifter. Location of both FACTS devices can be determined by sensitivity analysis [14]. Overloading of transmission line is called congestion. Congestion weakens the whole mechanism and invalidates optimization of the network. Congestion increases the cost and makes power less efficient. Congestion is needed to be eliminated through various means [15]. FACTS devices location can be determined by PSO in minimum cost of installation and improves system load ability. Voltage limits for the buses and thermal limits for the lines are main constraints for finding the optimal location [16]. PSAT is one of the easily accessible and open sources proprietary software it is included in MATLAB & GNU software kit for the design and analysis of different sizes of electric power systems. PSAT provides convenient graphical representations and Simulink planted editor for one line diagrams [17]. To solve OPF problems, efficient and reliable nature approach is used. These can be integration of fuzzy systems with genetic algorithm and PSO for OPF problem control variables [18]. Proper location of FACTS is important. To assuage the congestion of the system, we can use sensitivity factor to find the best and efficient location of TCSC [19]. TCSC is constructed in such a way that it includes a microcontroller. It has the ability to administer line impedance through the thyristor controlled capacitor in series with transmission line [20]. Voltage stability evaluation is experienced in safe operation of power system. Proximity of voltage instability is predicted. The effectiveness of LVSI is differentiated here [21]. Voltage stability is important and can lead to a condition of total collapse of system. LVSI, FVSI and LQP can be determined to voltage stability [22].

2 Series Compensation

The controllers which introduce voltage in course of the line are called as series controllers. They are usually devices having variable impedance. They basically feed or utilize variable reactive power. Their essential function is to reduce the inductance of the line.

Example: SSSC (Static Synchronous Series Compensator), TCSC (Thyristor Controlled Series Capacitor).

3 Theory Analysis of TCSC

3.1 Distinct Modes of TCSC

TCSC is required in power system because it extensively regulates the reactance of the transmission line so that satisfactory load to balance can be maintained. TCSC has several benefits. It can operate in different modes depending on the firing angle.

Thyristor Blocking Mode:

When thyristor valve is off, an inductive branch is opened. Then TCSC is said to be in Blocked Mode. This forces TCSC to operate as fixed series capacitor.

Thyristor Bypass Operating Mode:

In this mode, thyristor valve is always kept in ON position, i.e. the valve is triggered constantly, and thus, it keeps in conducting position all the time. In this mode, when thyristor is bypassed, then TCSC behaves similarly when series capacitor is connected in shunt to the inductor. In this way, the current through TCSC is reduced. Through this method, stress on capacitors is reduced so that during the faulty conditions the balance between load and supply can be maintained.

Vernier Operating Mode:

In this mode, the control over parameters is made by constantly regulating the firing angle. The firing angle stretches from 0 to 90° for every half cycle.

There are two modes in Vernier Operating Mode:

- Capacitive Boost Mode
- Inductive Boost Mode

Capacitive Boost Mode:

The thyristor valve with forward voltage is triggered ahead of capacitor voltage crossing zero, so that flow of current is through the inductive division of the thyristor. This leads to dynamically increasing the capacitance of the TCSC without the need of replacing it with the larger capacitor within the device itself.

Inductive Boost Mode:

It is not favourable for the steady-state operation. In this mode, the rotating current in the thyristor branch of TCSC is larger than the line current. Due to high thyristor current, the voltage waveform of capacitor is highly distorted in nature.

3.2 Modelling of TCSC

Thus, TCSC limits the line current during the faulty conditions and also allows increased compensation by using different modes of operation. TCSC damps the sub-synchronous resonance which is due to the torsional oscillations and inter-area oscillations. The reason behind this damping of oscillations is due to controller controlling the compensator. Thus, it has the ability of transferring more power over the long distances (Fig. 1).

Modelling of this device has become essential because it is used for different parameters.

1. Controlling the flow of power.
2. Damping of system oscillations.
3. For suppression of sub-synchronous resonance.
4. Fault current limitation.

TCSC is a device which is added in series to compensate the reactive losses in the transmission line. TCSC is supplemented and adjusted in series with the line

Fig. 1 Schematic diagram of TCSC

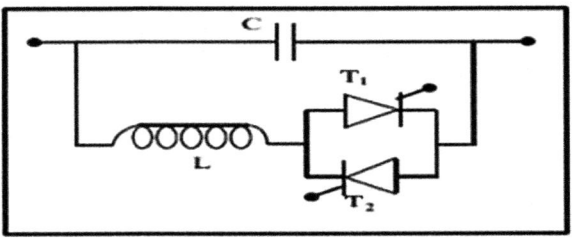

and function of series controller is to introduce voltage in the course of the line. Both types of reactance compensation are possible: reactive and inductive, with the help of appropriate values of capacitors and inductors. The appropriate values of capacitor and inductor can be realized with the help of reactance equations:

$$X_L(\alpha) = X_L \frac{\pi}{\pi - 2\alpha - \sin 2\alpha} \tag{1}$$

where

X_L is the inductive reactance.
α is the firing angle of the thyristor.
X_L varies with respect to α.

$$X_C = \frac{1}{2\pi f C} \tag{2}$$

where

X_C is the capacitive reactance.

$$\begin{aligned} X_{\text{TCSC}}(\alpha) = & -X_C + C_1(2(\pi - \alpha) + \sin(2(\pi - \alpha))) \\ & - C_2 \cos^2(\pi - \alpha)(\omega \tan(\omega(\pi - \alpha)) - \tan(\pi - \alpha) \end{aligned} \tag{3}$$

$$C_1 = \frac{X_C + X_{LC}}{\pi} \tag{4}$$

$$C_2 = \frac{4X_{LC}^2}{X_L \pi} \tag{...5}$$

$$X_{LC} = \frac{X_C X_L}{X_C - X_L} \tag{6}$$

$$\omega = \sqrt{\frac{X_C}{X_L}} \tag{7}$$

The efficient reactance of Thyristor Controlled Series Capacitor operates in three different regions:

1. Capacitive region ($\alpha_{\text{Clim}} \leq \alpha \leq 180°$).
2. Resonance region ($\alpha_{\text{Llim}} \leq \alpha \leq \alpha_{\text{Clim}}$).
3. Inductive region ($90° \leq \alpha \leq \alpha_{\text{Llim}}$) (Fig. 2).

Fig. 2 Operating modes of TCSC

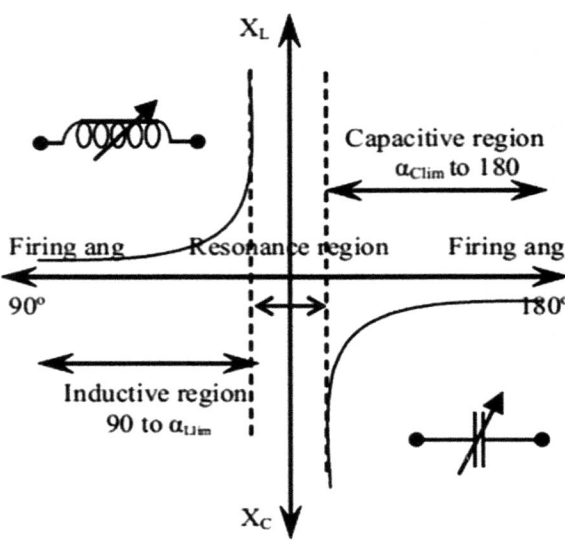

4 Optimal Location Methods

Large power systems usually undergo unstable voltage conditions and stability limits are usually violated. Due to decrease in voltage level system usually undergoes into a state of voltage instability and disturbance, power system stability is the capacity of the system to regulate voltage levels at all the buses having disturbances. By installing FACTS at most appropriate and correct location can lead to wonders because system stability increases and violations are reduced. When the index value is close to 1, then system represents unstable conditions and is on the edge of instability. If the index value crosses 1, i.e. when system has exceeded stability limit, then system comes into the state of voltage failure, which can also lead to total power failure.

FACTS are usually equipped on the most critical areas of the power system, i.e. on lines and buses which are determined by various Line Voltage Stability Indices. With the help of series or shunt controllers, we can enhance power transfer capacity and voltage profile of the system. TCSC is a series device which introduces series compensation in the transmission lines. The advantage of using TCSC is that it works in two modes: capacitive and inductive with the help of suitable choice of firing angle.

The optimal position of series FACTS controller, TCSC helps in increasing the voltage profiles of the buses where it is positioned.

Line Voltage Stability Index

Line Voltage Stability Indices play an essential role in controlling the system stability. It is most essential in ascertaining the weakest bus and the most critical line of the system. Voltage stability poses dynamic problems, and thus with the help of various indices, we can find the proximity to current operating point of steady stability limit and can stop the power system from getting collapsed. In this paper, we have made use of the following line indices to determine the system stability:

Fast Voltage Stability Index (FVSI)—It works on the principle of flow of power through a single line. For critical transmission line, we can calculate the index with the help of the following equation:

$$FVSI = \frac{4Z^2 Q_b}{V_a X} \tag{8}$$

where

Z symbolizes the impedance of the transmission line in the system.
X symbolizes transmission line reactance.
Q_b represents the flow of reactive power at the accepting end.
V_a symbolizes dispatching end voltage.

To ensure that the system is stable we have to make sure that the value of this index should be way below unity, preferably closer to zero.

Line Stability Factor (LQP)—The transmission lines which have indices values closer to unity determines that they are unstable or on the verge on instability, and are violating the stability conditions. If the values of this index go beyond the determined value 1, the bus connected to this line undergoes abrupt voltage drop and the system collapses. LQP can be determined by the following equation:

$$LQP = 4 \left(\frac{X}{V_a} \right) \left(\frac{X P_a^2}{V_a^2} + Q_b \right) \tag{9}$$

where

X symbolizes transmission line reactance.
V_a symbolizes the voltage at sending end.
P_a symbolizes flow of active power at the dispatching end.
Q_b symbolizes flow of reactive power at the accepting end.

Line stability index (L_{MN})—This index explains and determines the stability factor of each and every line which is linked or coupled among two distinct buses. The system is known to be stable until the values are extremely less than unity. As the L_{MN} value increases the stability decreases and hence chances for system collapse increase. L_{MN} index can be determined using the following equation:

$$L_{MN} = \frac{4 X_{ab} Q_b}{\{V_a \sin(\theta - \delta)\}^2} \tag{10}$$

where

X_{ab} symbolizes transmission line reactance.

Q_b is the symbol used for flow of reactive power at accepting end.

V_a is representation of the dispatching end voltage.

θ symbolizes impedance angle of the line.

δ Symbolize the angle difference between the dispatching and accepting end voltage.

5 Results

In this paper, we have managed to improve the bus voltages and line flows in a standard IEEE 30 bus system with the use of FACTS devices. We used different indices to find out the most critical line and the weakest bus. By using this, we can improve the voltage regulation and the stability of the system. Significant improvements are made in the voltage levels and line flows in any bus system by using Thyristor Controlled Series compensator. Since these devices are currently very expensive, hence their use is not very common. But by finding the appropriate location for and placing the FACTS devices, where the losses are most and the line seems to be critical according to the indices we can improve the system by placing them on the optimal location. Power flow is carried out on IEEE 30 bus system. Stability indices are found out using results of power flow. Optimal location is found out by these indices and TCSC is placed at optimal location. Some transmission lines are reaching their thermal stability limits. The critical lines are further identified by the parameters and then the congestion is managed by adding TCSC at the most optimal location (Fig. 3).

Table 1 shows LQP index calculations of the top 5 most critical lines in the network where we can manage congestion.

Fast voltage stability index calculations also represent the buses which are overloaded and need a corrective action (Tables 2 and 3).

LMN index shows that the values which are way closer to 1 are the buses with the critical conditions and are thus unstable.

Best location for TCSC is the line connected between bus 27 and bus 30 which is observed from Table 1, Table 2 and Table 3.

Improvement in percentage loading and reduction in losses can be observed leading to management of congestion of critical lines (Fig. 4).

Result with and without TCSC is shown in Table 4, here we can compare the difference between two. We were able to reduce the losses with the help of TCSC and by placing it at optimal location, we were able to mitigate congestion.

Fig. 3 IEEE 30 bus system

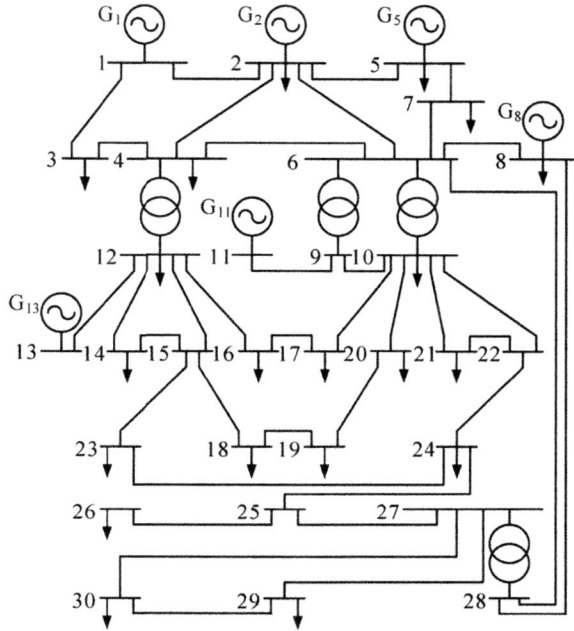

Table 1 LQP indices

From bus (a)	To bus (b)	LQP
27	*30*	*0.384298*
28	27	0.307876
27	29	0.260705
24	25	0.192488
9	10	0.12089

Table 2 FVSI indices

From bus (a)	To bus (b)	FVSI
27	*30*	*0.473507*
27	29	0.321586
28	27	0.292069
24	25	0.253314
29	30	0.150576

Table 3 LMN indices

From bus (a)	To bus (b)	LMN
27	*30*	*0.557042*
27	29	0.380349
24	25	0.265241
29	30	0.193842
25	27	0.142647

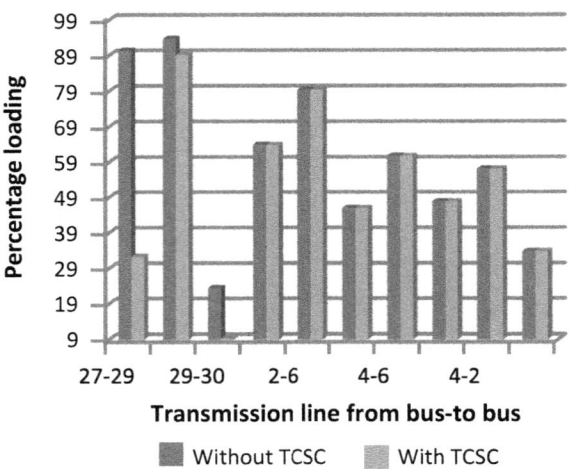

Fig. 4 Percentage reduction in line loading

Table 4 Generation and losses

	Total generation		Total losses	
	Without TCSC	With TCSC	Without TCSC	With TCSC
Real power (MW)	274.81	274.686	13.932	12.28
Reactive power (MVar)	178.41	178.489	53.076	50.48

6 Conclusion

By using TCSC for controlling congestion, we have improved reactive power losses and real power losses.

The reduction in real power losses was observed to be 11.81% and that of reactive power losses is 4.88%.

Line loading reduction of various lines was realized, which determined that all the transmission lines were under their thermal rating. Maximum reduction was inspected to be 64% in the line between buses 27 and 30.

References

1. "Optimal Location of TCSC with Minimum InstallationCost using PSO" K. Satyanarayana, B.K.V. Prasad, G. Devanand, N. Siva Prasad PPDCET, AP, India K.L. University, AP, India. Divisional Engineer, KTPS, AP, India
2. "Sensitivity Analysis for the IEEE 30 Bus System using Load-Flow Studies" Ibrahem Totonchi, Hussain Al Akash, Abdelhadi Al Akash and Ayman King Abdullah II School for Electrical Engineering, Princess Sumaya University for Technology, Amman-Jordan

3. "Voltage Profile Improvement and Loss Reduction Using Optimal Allocation of SVC" Sriparna Roy Ghatak Debarghya Basu Parimal Acharjee, Electrical Engineering Department, KIIT University, Electrical Engineering Department, NIT Durgapur, Electrical Engineering Department, NIT Durgapur, Bhubaneswar, India Durgapur, India Durgapur, India

4. "Optimal Placement Of TCSC For Reactive Power reserve Management With Reactive Power Loss minimization Using Hybrid Psogsa" L. N. Mrunalini Devi1 & A. Surya Prakash Rao, Department Of E.E. E, Sir C. R. Reddy College Of Engineering, Andhra Pradesh, India, Sr. Assistant Professor, Department of E.E.E, Sir C. R. REDDY College of Engineering, Andhra Pradesh, India

5. "Transmission Congestion Management with TCSC using Bacterial Foraging-Particle Swarm Optimization" Manasarani Mandala and C. P. Gupta, Member, IEEE

6. "Congestion Management in Deregulated Power System by Fuzzy Based Optimal Location and Sizing of UPFC" UMA V./P. LAKSHMI/ J. D. ANUNCIYA Department of Electrical & Electronics Engineering, Anna University College of Engineering, Guindy, Chennai

7. "Congestion Management of a Multi-bus Transmission System using Distributed Smart Wires" Sananda Pal, Samarjit Sengupta, Department of Applied Physics, University of Calcutta, Kolkata, India

8. "Congestion Management Using Optimal Placement of TCSC in Deregulated Power System", Kanwardeep Singh, Vinod K. Yadav, Arvind Dhingra, Guru Nanak Dev Engineering College, Ludhiana, India

9. "Enhancing the Power System Load ability Using TCSC: Improved Gravitational SearchAlgorithm" Anwar S. Siddiqui, Manisha Rani Professor, Department of Electrical Engineering, Jamia Millia Islamia, Jamia Nagar, New Delhi, India, PhD. Scholar, Department of Electrical Engineering, Jamia Millia Islamia, Jamia Nagar, New Delhi, India

10. "Impacts of Transmission Congestion on Market Power in Electricity Market" H. He, Z. Xu, Member, IEEE, and G. H. Cheng

11. "Locating and Parameters Setting of TCSC for Congestion Management in Deregulated Electricity Market" M. Joorabian, department of electrical engineering, Shahid Chamran University, Ahwaz, Iran, M. Saniei, department of electrical engineering, Shahid Chamran University, Ahwaz, Iran, H. Sepahv and, department of electrical engineering, Shahid Chamran University, Ahwaz, Iran

12. "Locating series FACTS devices for congestion management in deregulated electricity markets" Naresh Acharya, N. Mithulananthan, Electric Power System Management, Energy Field of Study, Asian Institute of Technology, P.O. Box 4, Klongluang, Pathumthani 12120, Thailand

13. "Optimal Power Flow Using PSAT" Dr. Amudha1 and V.J. Vijayalakshmi, Department of Electrical and Electronics Engineering, Professor and Head, Karpagam University, Coimbatore, Tamil Nadu, India. Assistant Professor (Senior Grade), KPR Institute of Engineering and Technology, Coimbatore, Tamil Nadu, India

14. "A New Method To Incorporate Facts Devices in Optimal Power Flow Using Particle Swarm optimization" Chandrasekaran, Arul Jeyaraj, Sahayasenthamil, M. Saravanan power system, Lecture, EEE Dept. P.S.N.A college of Engg. And Tech., Dindigul, India, power system, Lecture, EEE Dept. P.S.N.A college of Engg. And Tech., Dindigul, India, power electronic, Lecture, EEE Dept. P.S.N.A college of Engg. And Tech., Dindigul, India. Associate professor, EEE Dept. Thiyagarajar college of Engg., Madurai, India.

15. "Transmission Congestion and its Social Effects" H. He, Z. Xu, Member, IEEE

16. "Application of particle swarm optimization technique for optimal location of FACTS devices considering cost of installation and system load ability" M. Saravanan, S. Mary Raja Slochanal, P. Venkatesh, J. Prince Stephen Abraham, Electrical and Electronics Engineering Department, Thiagarajar College of Engineering, Madurai 625015, India, Received 13 April 2005; received in revised form 5 November 2005; accepted 8 March 2006

17. "An Open Source Power System Analysis Toolbox" F. Milano, Member, IEEE

18. "Optimal Power Flow Solution Using GA-Fuzzy and PSO-Fuzzy" S. Kumar • D. K. Chaturvedi, Received: 31 July 2012/ Accepted: 24 January 2014, The Institution of Engineers (India) 2014
19. "Optimal Placement of TCSC Based on A Sensitivity Approach for Congestion Management" Srinivasa Rao Pudi, S.C. Srivastava, Senior Member, IEEE
20. "Design of Thyristor Controlled Series Capacitor for High Voltage Controllability and Flexibility" Ravi Kant Kumar, Sanjeet Kumar, Santan Kumar and K.S.S. Prasad, Department of EEE, Bharath University, Chennai, India
21. "Comparative Study of Line Voltage Stability Indices for Voltage Collapse Forecasting in Power Transmission System" H. H. Goh, Q. S. Chua, S. W. Lee, B. C. Kok, K. C. Goh, K. T. K. Teo
22. "An Assessment of Voltage Stability based on Line Voltage Stability Indices and its Enhancement Using TCSC" Merlyn Mathew, Sudeshna Ghosh, D. Suresh Babu, Dr. A.A. Ansari (M.Tech Scholar, Department of Electrical & Electronics, LNCTE Bhopal (M.P)) (Assistant Professor, Department of Electrical & Electronics, LNCTE Bhopal (M.P)) (Assistant Professor, Department of Electrical & Electronics, Sree Vidya nikethan Engineering College, Tirupati (A.P)) (Professor & Head, Department of Electrical & Electronics, LNCTE Bhopal(M.P))

An Efficient Brightness Preserving Contrast Enhancement Technique Using Discrete Wavelet Transform and Singular Value Decomposition

Priyanka Gupta, Jamvant Singh Kumare, Uday Pratap Singh
and Rajeev Kumar Singh

1 Introduction

In current computing environment, image processing is very vital area of study. It is a process in which input an image and output might be either any image or some characteristics. It has several applications like in medical science, identification of fingerprints, remote sensing and earth science. Producing well contrast images is the strong necessity in various areas like fault detection and analysis of biomedical images.

Common methods that are used in enhancement of image contrast are linear contrast stretching as well as Histogram equalization (HE), but the images produced by this procedure have over lightning and unnatural contrast. To deal with the drawback of these techniques, so many enhancement strategies of image have been proposed [1–10]. Among all these techniques, HE is also a basic enhancement technique, to achieve overall better enhancement in images; it equalized the histogram of image by scattering and stretching the dynamic range of given image. HE produces visual artefacts and gives an abnormal look to the processed image and results having unnecessary contrast enhancement of image, various enhancement techniques have been presented like BBHE [1] and BPDHE [6] for removing these

P. Gupta (✉) · J. S. Kumare · R. K. Singh
Department of CSE/IT, M.I.T.S, Gwalior, India
e-mail: guptapriya071@gmail.com

J. S. Kumare
e-mail: jamvantsingh09@gmail.com

R. K. Singh
e-mail: rajiv.mits1@gmail.com

U. P. Singh
Department of Applied Science, M.I.T.S, Gwalior, India
e-mail: usinghiitg@gmail.com

© Springer Nature Singapore Pte Ltd. 2019
V. Nath and J. K. Mandal (eds.), *Proceeding of the Second International Conference on Microelectronics, Computing & Communication Systems (MCCS 2017)*, Lecture Notes in Electrical Engineering 476, https://doi.org/10.1007/978-981-10-8234-4_30

cons of traditional HE. In BBHE method, given image decomposes into segments according to the mean value. Then equalization performed on every single sub-image. This technique does not give better contrast enhancement of the input image. To overcome this problem, BPDHE technique has been presented. However, in this technique, each segment maps into to a new discrete range and then applies the equalization on histogram, since the mean brightness of image changes by changing the discrete range of image; at the final stage, the processed image needs to be normalized. Even though these strategies are visually brighter than HE, they cannot adjust the level of enhancement. Alternatively, techniques based on SVD proposed to improve the contrast of images and to overcome the limitations related to the HE methods [11–14]. In SVE technique [11] equalize the image pixel's singular value matrix, which holds the information related to intensity of given image, obtained by SVD. The HE, BPDHE, and SVE techniques are compared with DWT-SVD technique, and the results show that performance of DWT–SVD is superior to others. There are various DWT-based techniques proposed for improvement in resolution of image.

The technique of resolution enhancement is based on wavelet enhanced the resolution of an image as well as do not loss the edge information but it flops to produce the better contrast image. The SVD can also use for enhancement of image contrast. The leading cause of using SVD, it holds information of illumination [7].

Here, we attempted to develop an efficient image enhancement technique by using our proposed methodology and performed the experimental analysis by comparing it with other's existing methods. There are various modules in this paper. Part II explains related work of various researchers in image enhancement. Section III describes about DWT, BPDHE and SVD with our suggested work. The experimental work is done in part IV. Part V mentions conclusion with the future scope of this work followed by references.

2 Related Work

In 2010, Anbarjafari and Demirel [11] presented a method for improving resolution of images based on CWT. In this technique, given image decomposed into four sub-bands using DTCWT. Then we can get high-resolution image by the interpolation of given image and sub-bands with high frequency and finally by performing Inverse DT CWT. The qualitative metric and quantitative parameter, i.e. PSNR, show that proposed technique provides better enhancement over the traditional resolution enhancement techniques.

In 2010, Demirel et al. [15] presented a technique in which input image having low contrast decomposes into the frequency sub-bands by using DWT and after the decomposition, in next step, the matrix of SVD can be calculated for the low-low (LL) sub-band image. This method is called (DWT–SVD) reforms the enhanced output image by performing the IDWT.

In 2011, Demirel and Anbarjafari [16] proposed a new strategy for improving the resolution of satellite images based on DWT and through interpolated output of given image and sub-bands with high frequency. In this method, given image decomposes into sub-bands of different frequencies after that sub-bands having high frequency and given image having low-resolution interpolated and then perform IDWT for generating image of high resolution.

In 2013, Srinivas and Venkatesh [12] performed a study of SWT, DWT, DWT and SWT, and DTCWT-based resolution enhancement of satellite image and got better enhanced and sharper image using these wavelet-based enhancement techniques. As a result, the enhancement by DWT-SWT has high resolution than enhancement using DWT.

In 2014, Sharma and Verma [17] proposed a modified algorithm proposed in [12]. This technique based on gamma correction for enhancement in contrast of satellite images using SVD and DWT. In this method, intensity transformation done by gamma correction improves illumination by using SVD. This presented technique confirmed the effectiveness of its method by comparing with Demirel's [12] by calculating entropy, PSNR and EME (measure of enhancement).

In 2014, Priyadarshini et al. [18] presented a method based on DWT, SWT and BPDHE. Image having low contrast and low resolution is decomposed by using SWT and DWT then the sub-bands are interpolated and some intermediate process is performed to produce an image with better resolution image but having less contrast. The contrast of image is improved by using BPDHE, and the output image after applying BPDHE technique is decomposed by DWT, and the new LL sub-band is acquired by using SVD. Inverse wavelet transform is performed to generate output image. Given image contrast and resolution is enhanced by this technique. The quantitative and qualitative parameters are measured to depict that the presented technique gives a better result than the traditional method.

In 2015, Atta, Abdel-Kader and Farouk [19] presented a technique which is variation of the SVD-based contrast enhancement methods. SVD is used for preservation of the mean brightness of an input image. In this technique, the weighted aggregation of input image's singular matrices of the given image and its histogram equalization (HE) image is considered to attain the equalized image's matrix having singular values. Through results, it depicts that the presented method preserves the brightness of image more precisely with relatively minor pictorial artefacts.

In 2016, Sharma and Khunteta [20] presented a resolution and contrast enhancement technique based on bicubic interpolation for resolution enhancement which is applied on sub-bands having high frequency of given image and difference image obtained by subtraction of LL sub-band of given image and given image itself parallel the SVD deal with brightness enhancement which is applied on LL sub-band of histogram equalized and LL sub-band of input image. LL sub-band and components having high frequency of an image can be obtained by DWT transform. Whereas, resolution enhancement works on HL, LH, HH sub-bands and contrast enhancement works on LL sub-band.

In 2016, Jayashree et al. [21] presented a technique for mammographic images. In this, RMSHE technique is used for contrast enhancement and the SWT-DWT filtering applied for edge preservation and resolution enhancement. EME result depicts that proposed method gives a better result than HE and RMSHE technique.

3 Proposed System

In our suggested technique, we are using DWT, SVD and BPDHE. First, DWT applied on the captured image and contrast enhanced image by BPDHE technique. Through DWT, we get four frequency sub-bands and new LL component of image is obtained by performing SVD on lower component of both images as depicted in the figure.

3.1 Discrete Wavelet Transform

It decomposes four different frequency sub-bands: low-high low-low and high-high, high-low are obtained by the decomposition through DWT of the given image. DWT can overcome the restriction of Fourier transform [23, 24] which is unable to provide both analysis of frequency and time resolution at the same point. IDWT splits the information row-wise, column-wise and diagonally.

In DWT, decomposition image is generated by filtering the given 2D-image initially along with rows followed by filtering the resultant image along the column in decomposition at every level as shown in Figs. 1 and 2 [17].

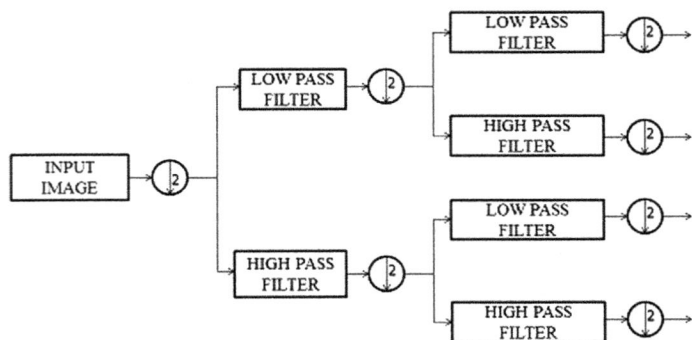

Fig. 1 The 2D-DWT decomposition [17]

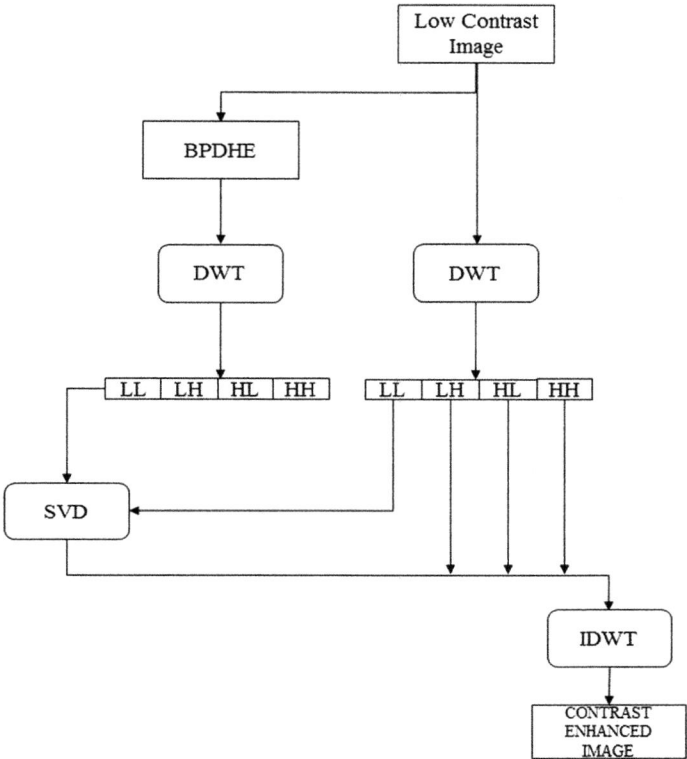

Fig. 2 Flowchart of presented technique

3.2 *Singular Value Decomposition*

SVD has various applications like for recognition of face, feature extraction and compression [22] and also for the enhancement of the images having less resolution and low contrast. SVD is a technique which preserves the maximum signal energy of an input image into numbers of coefficients as optimal as possible. Through scaling of singular value matrix, SVD-based methods expand the contrast of images by the decomposition of the image's matrix in a minimum square sense into an optimal matrix. The information of illumination represented by maximum signal energy. DWT decomposes the given image into four frequency bands, and SVD will work only on the LL frequency band component is considered for the purpose of illumination enhancement but we deal with only the LL band because the other frequency band edge information contained by other frequency band and under any condition we do not want to distorted image's edge information. The SVD of an image can be depicted in the following form:

$$I = U_i \overline{\sum}_I V_i \tag{1}$$

where U_i and V_i are matrices which deliberated as the eigenvectors and $\overline{\sum}_I$ is a diagonal matrix that has singular value and its diagonal has non-zero value's square root [1]

In our suggested method, given image contrast is enhanced by BPDHE and after this new LL, sub-band is generated by taking the SVD of given image and enhanced image. The coefficient ratio can be evaluated by using Eq. 1.

$$\beta = \frac{\max\left(\sum_{LL_A}\right)}{\max\left(\sum_{LL_A}\right)} \tag{2}$$

where the LL_A is high contrast image's SVM and LL is SVM of the given input image. After calculating the correction coefficient (β) composed a new LL sub-band by Eqs. 2 and 3

$$\overline{\sum}_{LL_A} = \beta \sum_{LL_A} \tag{3}$$

$$\overline{LL}_A = U_{LL_A} \overline{\sum}_{LL_I} V_{LL_A} \tag{4}$$

Then to generate output enhanced image recombine this generated new LL band and input image's LH, HL, HH sub-bands by IDWT

$$\overline{A} = IDWT\left(\overline{LL}_A, LH_A, HL_A, HH_A\right) \tag{5}$$

Through the IDWT output image having better contrast obtained. The results by vision point of view through the suggested method are depicted in Figs. 3, 4 and 5. By all these results we able to perceive the variance between original given image and output image having better contrast. Images captured by Satellite as well as simple low contrast images improved by the suggested method.

3.3 Brightness Preserving Dynamic Histogram Equalization

The BPDHE technique is one of the techniques which focus on the improvement in contrast of image. It involves four stages, in the initial stage involves computation of histogram based on fuzzy statistics. It is able to address the uncertainty of histogram values in a vastly improved manner compared to traditional crisp histograms consequently delivering a smooth histogram. Thus for this specific application, use of

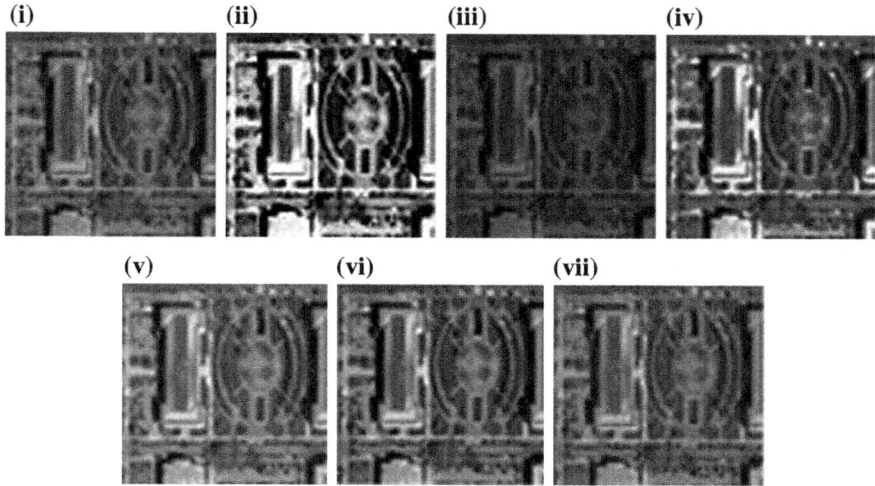

Fig. 3 **a** Original image, **b** HE, **c** SVE, **d** BPDHE, **e** Demirel's [15], **f** Nitin's [17] and **g** the proposed technique. *Source* [25]

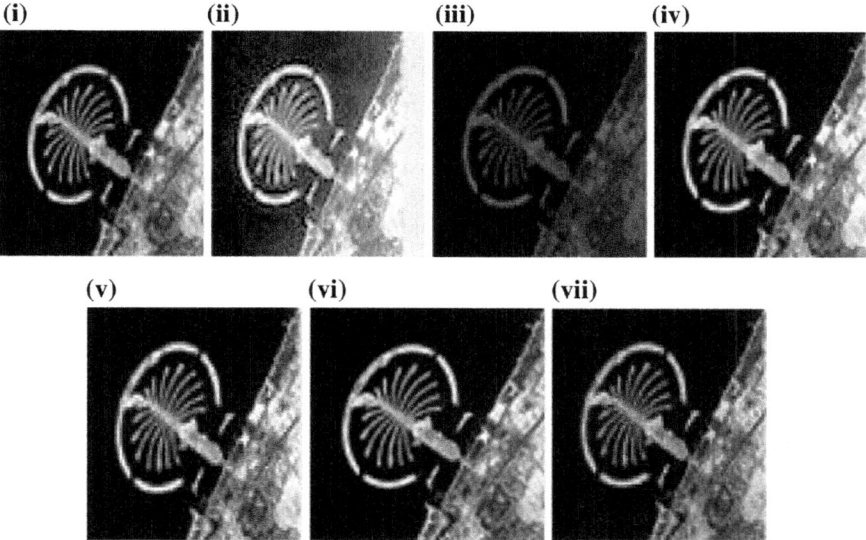

Fig. 4 **a** Original image, **b** HE, **c** SVE, **d** BPDHE, **e** Demirel's [15], **f** Nitin's[17] and **g** the proposed technique

histogram based on fuzzy is suitable. In the second stage dividing the histogram obtained in stage one based on local maxima and segments them to achieve numerous cluster of histogram. Along this way, each valley portion forms a segment

(i)   **(ii)**   **(iii)**   **(iv)**

(v)   **(vi)**   **(vii)**

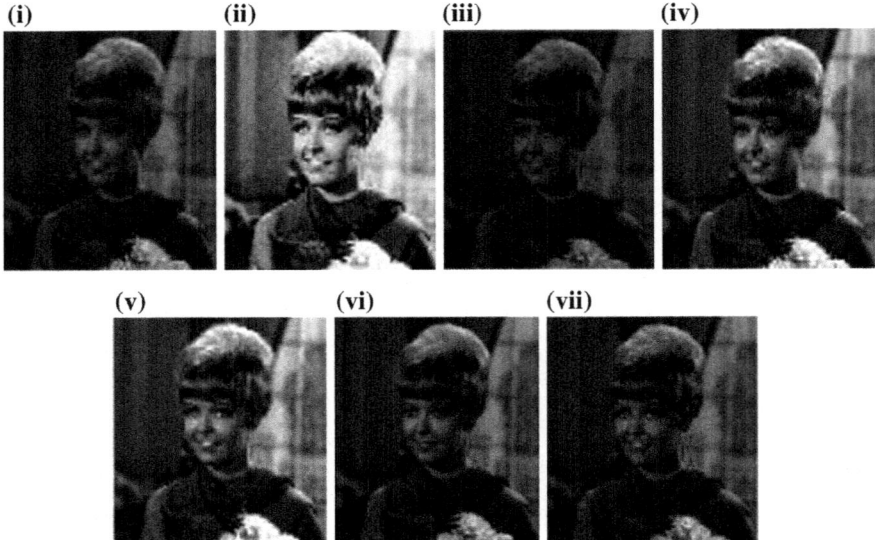

Fig. 5 **a** Original image, **b** HE, **c** SVE, **d** BPDHE, **e** Demirel's [15], **f** Nitin's[17] and **g** the proposed technique

between two straight local maxima. Then dynamic histogram equalization (DHE) method is performed on the histogram's peaks that do not get remapped, and this outcome gives better protection of image's mean brightness while enhancing the image's contrast. The third step is sub-histograms equalization by dynamic histogram equalization (DHE) method [8]. A function is used to achieve the equalization of image based on aggregate count of pixels in the segment. The fourth step is image brightness normalization. The image found after the DHE has the mean brightness that is somewhat better than the given image. The normalization process to eliminate this deviation is applied on the output image.

3.4 Proposed Algorithm

Step 1. Input an image having low contrast.
Step 2. Perform the Preprocessing step on the image. First, resize the given image to 256 × 256 and then transform it from RGB image to GRAY image.
Step 3. Apply BPDHE contrast enhancement technique.
Step 4. Apply wavelet transform on contrast enhanced image and image having low contrast.
Step 5. Apply SVD on LL component of images.
Step 6. Perform IDWT to generate the enhanced image.

4 Experimental Results

4.1 Implementation Details

A number of tests were accomplished to evaluate the effect the enhancement techniques in these low contrast input images of 256×256 pixel resolutions. In our paper, 25 images have been tested but few results are shown to reduce the size of paper. The performance of the presented mechanism compared with HE, BBHE, DWT-SVD and BPDHE technique. It is a difficult job of evaluating the degree of enhancement in images but, for comparing the result of several contrast enhancement methods, we required to have an unbiased tool. In this work our objective to focus on brightness preservation to obtain the consistent and meaningful results, for we use metrics based on mean and standard deviation. With these metrics, histogram equalization can attain the better result even though it may create artefacts in vision and possibly may produce an abnormal look.

All these result represents that the output image, i.e. contrast improved by presented method is brighter and sharper than the other traditional methods. Not only visual but also quantitative measurements for the better effect of proposed method using PSNR and MSE parameters. By using the following formula, we can calculate the PSNR value of an image:

$$ \text{PSNR} = 10 \, \log_{10} \left(\frac{R^2}{\text{MSE}} \right) \tag{6} $$

Here R is taken as 255 because of images are signified by 8 bit and by using the following formula the MSE of given image Ii and the produced image Io can be calculate:

$$ \text{MSE} = \frac{\sum_{i,j} (I_i(i,j) - I_o(i,j))^2}{\text{MAX}} \tag{7} $$

4.2 Subjective and Objective Assessments

Table 1 is showing the comparative study of the presented method and HE, SVE, BPDHE, Demirel's method [15] and Nitin's method [17] by means of calculating PSNR for various images.

Figures 3, 4, 5, 6 and 7 represent a low contrast input image (a) and in (b) image obtained after HE in (c) processed image by SVE and BPDHE technique in (d), (e) image after Demirel's method (f) image after Nitin's method and (g) shows the output image by the proposed technique. The HE image, i.e. satellite image sat1.jpg has a PSNR value 14.93 and MSE value 2.1034e+03. Figures 6 and 7 depict the graph representation of PSNR value and MSE value, respectively.

Table 1 Comparison of PSNR values of images

Images	Method					
	HE	SVE	BPDHE	Demirel's [15]	Nitin's [17]	Proposed method
Sat1	14.93	17.55	22.60	20.66	32.64	35.34
Sat2	11.07	19.37	24.12	11.93	32.67	35.89
Sat3	12.83	14.54	24.81	16.49	22.73	31.50
Couple	7.66	44.43	38.10	8.20	33.92	51.63
Girl	10.15	23.62	36.08	10.56	31.61	48.60

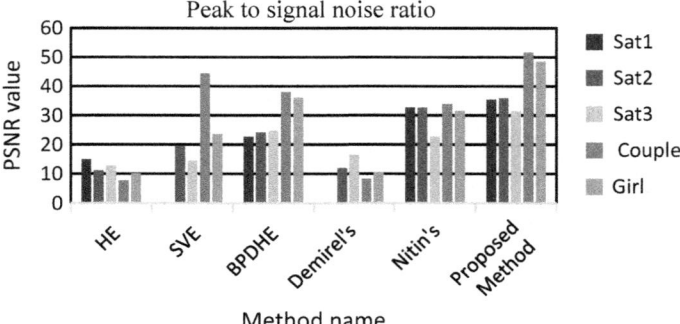

Fig. 6 Peak-to-signal noise ratio

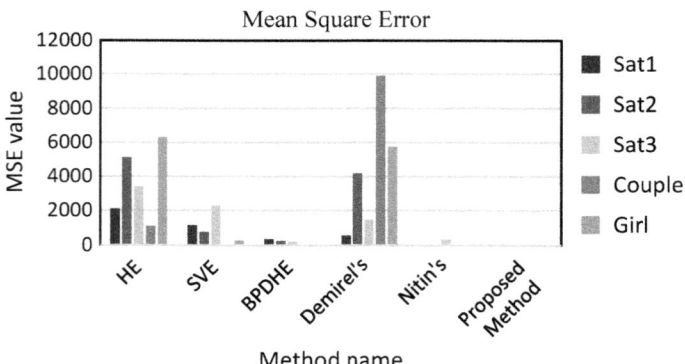

Fig. 7 Mean square error graph

Similarly, Table 2 also shows the comparison of all these techniques for same images by MSE value. Overall, the results in Tables 1 and 2 represent that our proposed method performs better over the mentioned and traditional techniques.

Table 2 Comparison of MSE values of images

Images	Method					
	HE	SVE	BPDHE	Demirel's [15]	Nitin's [17]	Proposed method
Sat1	2.1034e+03	1.1498e+03	359.42	562.33	35.66	19.55
Sat2	5.1129e+03	757.08	258.38	4.1977e+03	35.37	16.86
Sat3	3.4090e+03	2.3000e+03	216.38	1.4696e+03	349.25	46.34
Couple	1.1229e+04	2.36	10.12	9.8977e+03	26.56	0.45
Girl	6.3174e+03	284.45	16.13	5.7596e+03	45.15	0.90

This image is a grayscale image of 256×256 pixels and the proposed technique results the contrast enhanced image with a PSNR value 35.34 and MSE value 19.55, visibly depict the improvement on the image.

Hence by observing the output and corresponding values, we can say that brightness of image has been enhanced twice that of the given image as well as. This work finds applications in various fields mainly where the satellite images have been used. Images that are captured by satellite have numerous applications in geology, agriculture, landscape, meteorology regional, and conservation planning of biodiversity, education, intelligence, forestry and warfare. Contrast enhanced image by the presented technique leads to improvement in all of the above fields.

5 Conclusion and Future Scope

In this paper, we focus on modified DWT and SVD-based brightness preserving and contrast enhancement method. This proposed technique basically deals with images having low contrast, we apply proposed method on given image having low contrast and improved image by BPDHE. Also, apply SVD on both LL sub-band of BPDHE contrast improved image and after this illumination of image increases. Results envisage that proposed method have good results while comparison with other existing approaches. In future, we can use different contrast enhancement techniques, and gamma correction can also be used for image contrast enhancement and try to minimize the MSE, so that the resulting images have good appearance and high PSNR value than all traditional and proposed techniques.

References

1. Y.T. Kim, "Contrast enhancement using brightness preserving bi-histogram equalization", IEEE Trans. Consum. Electron. 43 (1997) 1–8.
2. S.D. Chen, A.R. Ramli, "Minimum mean brightness error bi-histogram equalization in contrast enhancement", IEEE Trans. Consum. Electron. 49 (2003) 1310–1319.

3. S. Chen, A. Ramli, "Preserving brightness in histogram equalization based contrast enhancement techniques, Digital Signal Process. 14 (2004) 413–428.
4. N.A.M. Isa, C.H. Ooi, "Adaptive contrast enhancement methods with brightness preserving", IEEE Trans. Consum. Electron. 56 (2010) 2543–2551.
5. J.Y. Kim, L.S. Kim, S. Hwang, An advanced contrast enhancement using partially overlapped sub-block histogram equalization, IEEE Trans. Circuits Syst. Video Technol. 11 (2001) 475–484.
6. H. Ibrahim, N.S.P. Kong, "Brightness preserving dynamic histogram equalization for image contrast enhancement", IEEE Trans. Consum. Electron. 53 (2007) 1752–1758.
7. T.K. Kim, J.K. Paik, B.S. Kang, Contrast enhancement system using spatially adaptive histogram equalization with temporal filtering, IEEE Trans. Consum. Electron. 44 (1998) 82–86.
8. C.C. Sun, S.J. Ruan, M.C. Shie, T.W. Pai, "Dynamic contrast enhancement based on histogram specification", IEEE Trans. Consum. Electron. 51 (2005) 1300–1305.
9. Y. Wan, Q. Chen, B.M. Zhang, Image enhancement based on equal area dualistic sub-image histogram equalization method, IEEE Trans. Consum. Electron. 45 (1999) 68–75.
10. M.A.A. Wadud, M.H. Kabir, M.A.A. Dewan, O. Chae, "A dynamic histogram equalization for image contrast enhancement", IEEE Trans. Consum. Electron. 53(2007) 593–600.
11. H. Demirel, G. Anbarjafari, M.N. Jahromi, "Image equalization based on singular value decomposition", in: Proc. 23rd IEEE Int. Symp. Comput. Inf. Sci., Istanbul, Turkey, (2008) pp. 1–5.
12. P. Bala srinivas B. Venkatesh, "Comparative Analysis of DWT SWT, DWT & SWT and DTCWT Based Satellite Image Resolution Enhancement", IJECT Vol. 5, Issue 4, (2014).
13. A.K. Bhandari, A. Kumar, P.K. Padhy, "Enhancement of low contrast satellite images using discrete cosine transform and singular value decomposition", World Acad. Sci. Eng. Technol. 55 (2011) 35–41.
14. R. Atta, M. Ghanbari, Low-contrast satellite images enhancement using discrete cosine transform pyramid and singular value decomposition, IET Image Proc. 7 (2013) 472–483.
15. H. Demirel, C. Ozcinar, G. Anbarjafari, "Satellite image contrast enhancement using discrete wavelet transform and singular value decomposition", IEEE Geosci. Remote Sens. Lett. 7 (2010) 333–337.
16. H. Demirel and G. Anbarjafari, "Discrete wavelet transform-based satellite image resolution enhancement," IEEE Trans. Geoscience and Remote Sensing, vol. 49, no. 6, (2011) pp. 1997–2004.
17. Nitin Shanna,Om Prakash Venna, "Gamma Correction Based Satellite Image Enhancement using Singular Value Decomposition and Discrete Wavelet Transform IEEE International Conference on Advanced Communication Control and Computing Technologies (ICACCCT) 2014 ISBN No. 978-1-4799-3914-5/14/2014 IEEE.
18. M. Priyadarshini, Ms. R. Sasikala, Dr. R. Meenakumari, "Novel Approach for Satellite Image Resolution and Contrast Enhancement Using Wavelet Transform and Brightness Preserving Dynamic Histogram Equalization", IEEE 2016.
19. Randa Atta, Rabab Farouk, Abdel-Kader, "Brightness Preserving Based on Singular Value Decomposition for Image Contrast Enhancement", Optik 126 (2015) 799–803.
20. Aditi Sharma, Ajay Khunteta, "Satellite Image Contrast and Resolution Enhancement using Discrete Wavelet Transform and Singular Value Decomposition", International Conference on Emerging Trends in Electrical, Electronics and Sustainable Energy Systems (ICETEESES–16) 2016.
21. K. Akila, L.S. Jayashree, A. Vasuki., "A Hybrid Image Enhancement Scheme for Mammographic Images", Advances in Natural and Applied Sciences. 10(6); Pages: 26–29 2016.
22. S.S. Agaian, B. Silver, K.A. Panetta, "Transform coefficient histogram-based image enhancement algorithms using contrast entropy", IEEE Trans. Image Process. 16 (2007) 741–758.

23. H. Demirel and G. Anbarjafari, (2011) "Image resolution enhancement by using discrete and stationary wavelet decomposition," IEEE Trans. Image Process., vol. 20, no. 5, pp. 1458–1460.
24. Turgay Celik, Tardi Tjahjadi, "Resolution Enhancement Using Dual Tree Complex Wavelet Transform", IEEE Geo Science and Remote Sensing Letters, Vol 7, No 3, (2010) pp. 554–557.
25. Satellite Image got from- http://www.satimagingcrop.com/

Computer-Assisted Valuation
of Descriptive Answers Using Weka
with RandomForest Classification

Ruhi Dubey and Rajni Ranjan Singh Makwana

1 Introduction

We are living in the century where Internet and computers are one of the basic requirements of the education system. Most of the exams like CAT, GATE and GMAT are performed online only, whereas exams such as AFCAT, UGC NET, etc. are performed using OMR techniques which are evaluated using digital valuation. But today also most of the university exams are illustrative in nature. They require human valuation which consumes a big amount of time as well as efforts. But many times it happens that marks vary from student to student due to lack of attention, concentration and mood swings of evaluator which further deals with revaluation of copies and results in unnecessary expenditure by students. It also affects them economically and also harms the reputation of university. There are many systems that are already implemented regarding auto-valuation of illustrative answers which uses the technologies like natural language processing, artificial intelligence, Java applications, etc. but they also need extra expenditure for developing such systems which indirectly effects the economy.

Here, a system is proposed which will help the evaluator in correct valuation of copies and which did not need any special training or any special tools. It can be developed using free tools available so that any institute or organization can easily use it and can perform an effective auto-valuation of illustrative answers.

R. Dubey (✉) · R. R. S. Makwana
Department of CSE & IT, Madhav Institute of Technology and Science, Gwalior, India
e-mail: ruhidubeygwalor@gmail.com

R. R. S. Makwana
e-mail: ranjansingh06@gmail.com

© Springer Nature Singapore Pte Ltd. 2019
V. Nath and J. K. Mandal (eds.), *Proceeding of the Second International Conference on Microelectronics, Computing & Communication Systems (MCCS 2017)*, Lecture Notes in Electrical Engineering 476, https://doi.org/10.1007/978-981-10-8234-4_31

2 Related Work Done

(1) Mohan et al. [1] proposed an algorithm for valuation of illustrative type examination. Their technique utilizing the parts of speech components like nouns, pronouns, verbs, adverbs and adjectives as the pre-announced clusters. SVM classifier is used to evaluate test samples. The authors claim that method is good for disquisition type answers only and not working accurately for formula-based and mathematical type questions.
(2) Kaur et al. [2] proposed an algorithm for the valuation of single sentence illustrative answers. Similarity measurement is performed between student answers and standard answer based on full or partial string match. The work does not contain sufficient examples to validate the system.
(3) Sunil Kumar et al. [3] presented a significant work which utilizes bagging classifier for the valuation of illustrative answers. Author claim that on an average 76% of accuracy is obtained when tested across five datasets using tenfold cross validation using naive Bayes, logistic regression, RandomForests, support vector machine (SVM), decision stump and decision trees. Nevertheless, this paper appears to suffer from two drawbacks. The applied dataset consists of student written essays, it could be better if the author would provide some specific questions and their answers to train the classifiers because essays valuation is totally different from illustrative answer valuation, second instead for tenfold validation, an unseen test dataset should be supplied to test the system.
(4) Mamčenko et al. [4] proposed an illustrative model to recognize concealed patterns in student's answers. Clustering techniques are applied so that we can group similar types of the objects together. The result includes total time spend, time spent to give incorrect answer and time spend to give correct answers; however, the proposed research is not directly related to the valuation of illustrative answers.

3 Description

In the proposed approach, the computer-assisted system is described by which descriptive answers can be evaluated. The classifier used for the experiment is RandomForest classifier.

RandomForest Classification: RandomForest classification [5] is the combination of a group which predicts the tree model of the data. In this classification, every tree pivots on the values of a random vector specimen such that they are independent of each other and every tree in the forest has similar allocation. RandomForest classification algorithm is an entity learning method, i.e. a group of items viewed as a whole rather than individually, for classification, regression and other tasks. RandomForest classifier uses the principle of average.

Dataset Collection: To build a required dataset, a form was developed with the help of Google docs and a descriptive test was performed consisting of eight questions and was conducted for B.E. first-year students of computer science branch involving 88 students by which we got 704 answers out of that we select 630 answers. Further, those answers are divided into 530 answers for training dataset and 159 answers for testing dataset. Answers are divided randomly. These answers got evaluated by human expert, and marks are allotted manually on the scale of 0–3 where 0 is considered as worst and 3 as best.

Test samples are first evaluated by manual evaluator, and it also was given to the classifiers for automated valuation. Classifier valuation results are compared to the manual valuation results with the objective that classifier valuation and manual valuation will produce the similar results.

4 Experimental Details

Experiment Setup: Experiment is carried out on the system having Windows 8 with 1.3 GHz Intel i3 processor and 3 GB Ram. Weka V 3.6.11 machine learning workbench is utilized for the classification of illustrative answers.

Experiment Steps: Fig. 1 show the experimental steps performed.

Tokenizer: It breaks the incoming answer string down into a stream of terms or tokens. A simple tokenizer divides the string up into terms wherever it encounters whitespace or punctuation [6]. For examples, four answers are given as follows.

Fig. 1 Experimental steps

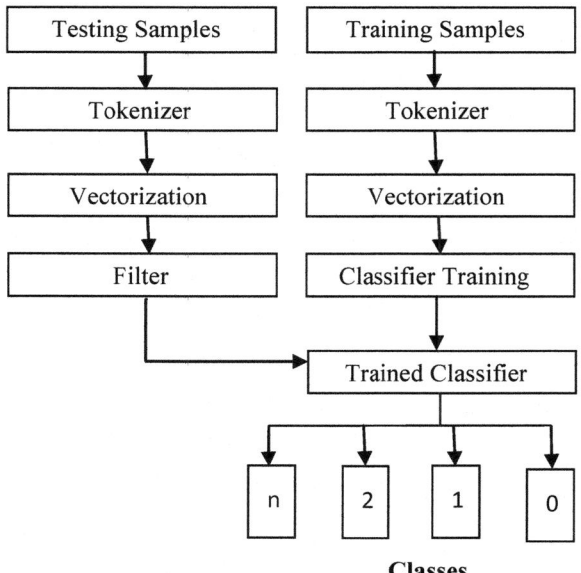

Fig. 2 Vectorization

$$\begin{array}{cccccccc}
\text{Great} & \text{work} & \text{excellent} & \text{worst} & \text{ever} & \text{no} & \text{comment} & \text{Class}
\end{array}$$

$$V1 = [1,\ 1,\ 0,\ 0,\ 0,\ 0,\ 0,\ 2]$$

$$V2 = [0,\ 1,\ 1,\ 0,\ 0,\ 0,\ 0,\ 2]$$

$$V3 = [0,\ 1,\ 0,\ 1,\ 1,\ 0,\ 0,\ 0]$$

$$V4 = [0,\ 0,\ 0,\ 0,\ 0,\ 1,\ 1,\ 1]$$

Ans. 1 = 'great work', Ans. 2 = 'excellent work', Ans. 3 = 'worst work ever' and Ans. 4 = 'no comment'. Identified tokens by tokenizer are as follows: great, work, excellent, worst, ever, no and comments.

Vectorization: Tokens obtained from the tokenizer are processed to transform into a column vector [6]. Each vector row is represented by following structure:

$$V = [V_1, V_2, V_3 \ldots V_n, \text{Class}] \tag{1}$$

Example: Identified tokens in the previous example are transformed into column vector, here Class attribute represents the marks (scale 0–3) given by manual evaluator, given in Fig. 2.

Here, $V1$, $V2$, $V3 \ldots Vn$ are the column vectors, and class is the final classified column vector.

Training: RandomForest classifier is used for training dataset.

Filter: For successful classification, both training and testing file should have the same name, type and equal number of attributes (column vector) [6]. However, in this work training and testing samples are having unequal column vectors. So it is required to make them compatible; therefore, test samples are preprocessed by the arbitrary filter to achieve vector dimension compatibility. The composition of the filter is based solely on the training data and test instances will be processed by the filter without changing their composition.

Classification: As mentioned the RandomForest algorithm is utilized to evaluate the test samples under the scale of 0–3 and produces the result. It is expected that train classifier would evaluate test samples as human evaluates them.

5   Observations and Results

The dataset is applied to the RandomForest classifier and observed result is measured by following factors. Below A, B, C and D are true positive, false negative, false positive and true negative, respectively.

Table 1 Confusion matrix

		Detected	
		Positive	Negative
Actual	Positive	A: True positive	B: False negative
	Negative	C: False positive	D: True negative

The classification of the answers is based on the following factors. By these factors, the classification is justified and accordingly, marks are allotted to each answer. These factors are liable for the classification and allotment of the marks (Table 1).

True Positive Rate (TP Rate)/Recall: It is the section of cases whose results are positive, which were accurately classified as positive and calculated by using the following equation:

$$\text{Recall} = A/A + B \tag{2}$$

False Positive Rate (FR Rate): It is the section of cases whose results should be negative but those were inaccurately classified as positive, as calculated by the following equation:

$$\text{FP Rate} = C/C + D \tag{3}$$

Precision: It represents the part of the predicted positive cases which were correct and is calculated using the following given equation:

$$\text{Precision} = A/A + C \tag{4}$$

F-Measure: The F-Measure evaluates some mean as follows:

$$F = 2 \cdot \frac{\text{Precision} * \text{Recall}}{\text{Precision} + \text{Recall}} \tag{5}$$

ROC Curve: This curve diagnoses the level of the experiment on the same graph, i.e. how excellent, good and worthless experiments are. A zone of 1 represents an accurate evaluation; a zone of 0.5 represents a test that has no or very little value.

Kappa Statics: If the result of statics is 1, then it specifies accurate accordance whereas if the result of statics is 0, then it specifies accordance equal to chance.

Classification %: It depends on the no. of samples correctly classified. Here, t is the no. of sample cases correctly classified, and n is the total no. of sample cases.

$$\text{Classification } \% = 100 * \frac{t}{n} \tag{6}$$

Table 2 displays the exact result that is classified by the classifier without any threshold value. But there may be some variations in human evaluated and computer evaluated answers therefore to minimize the error rate we have set a threshold

Table 2 Summary of implementation

Correctly evacuated	Incorrectly evacuated	FP rate	Precision	Recall	F-measure	ROC area	Kappa statics	Classification %
98	61	0.616	0.265	0.613	0.616	0.608	0.775	61.6352

value of ±1 mark. That is, if the difference in between human evaluated and computer evaluated marks is ±1 mark, then no notification will be generated but if the difference exceeds the threshold value, then an alert will be generated for rechecking of copy.

Table 2 represents the summary of the experiment conducted with the above-mentioned factors and has concluded with the accuracy of 61.63% without any threshold value. Afterwards, we have calculated the result with ±1 threshold value and achieved the accuracy of 96.22% as specified in Table 3. Figure 3 represents the similarity between human evaluated marks and computer evaluated marks. In the experiment, marks are evaluated in four classes, i.e. 0, 1, 2 and 3. The

Table 3 Result analysis

	Correctly classified	Incorrectly classified	Classification %
Classification including threshold value	153	6	96.22
Classification excluding threshold value	98	61	61.63

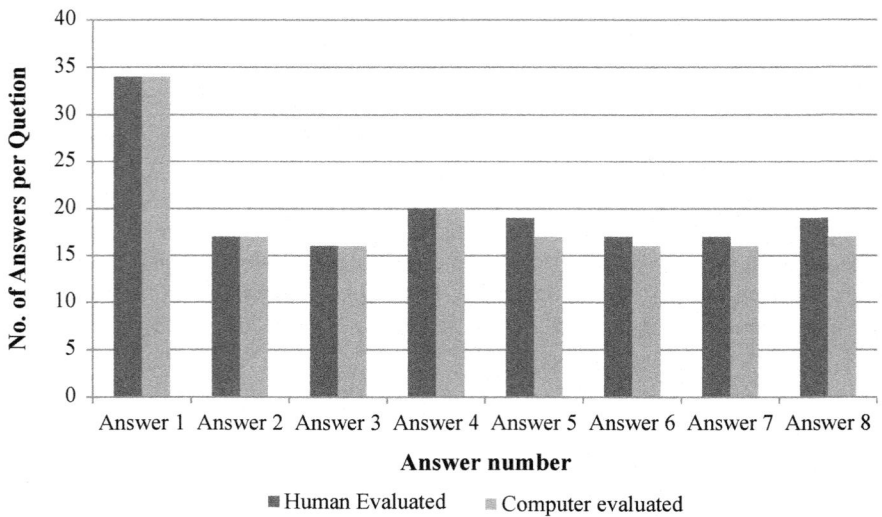

Fig. 3 Similarity between human evaluated and computer evaluated with threshold value

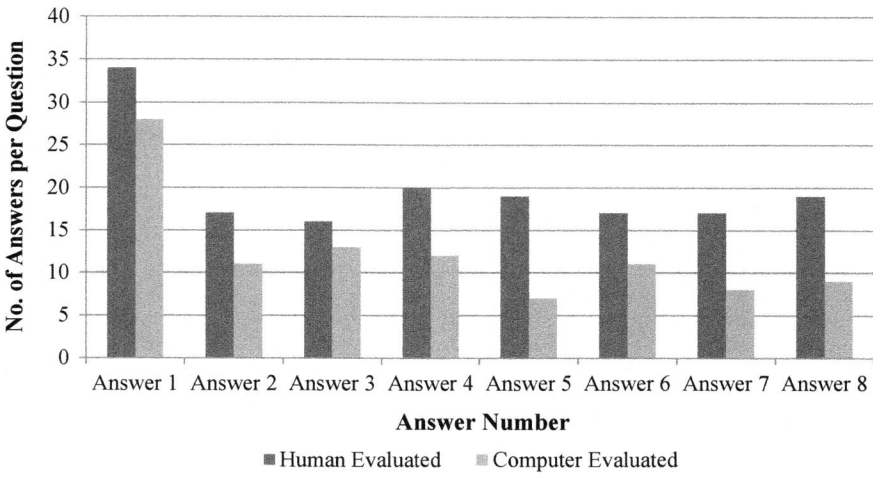

Fig. 4 Similarity between human evaluated and computer evaluated without threshold value

chart represents the similarity in human valuation and computer valuation, for example, for answer 1 we have evaluated 34 answers by both human and machine and both have marked exactly same marks for all the. Figure 4 represents the similarity between human evaluated and computer evaluated without threshold value. Here, both values are different.

6 Conclusion

The main purpose of this system is to reduce the revaluation of copies in the university. Details of how the approach can be used are briefly described in this paper. Moreover, details of the functions used in the approach are also explained. In descriptive examination system, there are a lot of human efforts required and more when revaluation of some copies adds up as an additional task. To reduce extra efforts in valuation, this approach is proposed. It can be used by compiling it with a framework so that non-technical person can also use it, and the approach can be used in real-life systems. However, the proposed method may also be used as 'Computer assisted Manual valuation of illustrative answers' where each answer was first evaluated by computers and then manual valuation; if the difference between manually allocated marks and auto-evaluated marks is higher than the predefined threshold, then an alert will be generated to the human evaluator in order to re-correct and evaluate it again.

References

1. A. K. Mohan, M. H. M. K. Prasad, "A Novel Feature Clustering Algorithm for Valuation of Descriptive Type Examination", Int. J. Comput. Appl, vol. 98, no. 9, July 2014, pp: 35–41.
2. A. Kaur, M. Sasikumar, S. Nema, S. Pawar, "Algorithm for Automatic Valuation of Single Sentence Descriptive Answer", Int. J. Inventive Engineering and Sciences, vol. 1, Issue-9, August 2013, pp: 6–9.
3. C. S. Kumar and R. J. R. Sree, "An Attempt to Improve Classification Accuracy through Implementation of Bootstrap Aggregation with Sequential Minimal Optimization during Automated Evaluation of Descriptive Answers," vol. 7, no. September, pp. 1369–1375, 2014.
4. J. Mamčenko, I. Šileikienė, J. Lieponienė, and R. Kulvietienė, "Evaluating the Data of an e-Examination System Using a Descriptive Model in Order to Identify Hidden Patterns in Students Answers," Online J. Comput. Sci. Inf. Technol., no. 12, pp. 45–49.
5. P. Sharma, "International Journal on Recent and Innovation Trends in Computing and Communication Comparative Analysis of Various Decision Tree Classification Algorithms using WEKA," pp. 684–690.
6. R. Dubey and R. R. S. Makwana, "Comparative Analysis of Computer Assisted Valuation of Descriptive Answers using WEKA with different classification algorithms," vol. 4, no. 6, pp. 5–10, 2017.
7. A. Dhokrat, "Automated Answering for Subjective Examination", Int. J. of Computer Appl, vol. 56, no. 14, pp. 14–17, 2012, ISSN: 0975 – 8887.
8. V. P. Parmar and C. K. Kumbharana, "Question Answer Formulation, Evaluation Techniques And Comparison Of MCQ Type With One Word Answer", Int. J. Scientific and Research Publications, vol. 6, no. 3, pp. 459–463, ISSN: 2250–3153, 2016.
9. S. Praveen, "An Approach to Evaluate Testability", Int. J. Innov. Res. Comput. Commun. Eng., vol. 2, no. 11, pp. 6410–6413, ISSN: 2320-9801, 2014.
10. A. Tüfekçi, H. Ekinci, and U. Köse, "Development of an internet-based exam system for mobile environments and evaluation of its usability", Mevlana Int. J. Educ., vol. 3, no. 4, pp. 57–74, 2013.
11. R. Divya and M. Kumar, "Enhanced Digital Assessment Of Examination With Secured Access", Int. J Advanced Studies in Computer Science and Engineering, vol. 3, no. 10, pp. 33–37,ISSN: 2278-7917, 2014.
12. S. J. Aboud, "Secure E-Exam Scheme", Int. J. Science and Research, vol. 3, no. 9, pp. 2200–2203, ISSN: 2319-7064, 2014.
13. R. K. Sungkur, I. Beekoo, D. L. Bhookhun, "An Enhanced Mechanism For The Authentication Of Students Taking Online Exams", IEEE AFRICON Conf., 2013.
14. L. Devasena, "Comparative Analysis of Random Forest, REP Tree and J48 Classifiers for Credit Risk Prediction", Int. J. Comput. Appl., pp. 975–8887, ISSN: 0975-8887, 2014.
15. J. Ali, R. Khan, N. Ahmad, I. Maqsood, "Random Forests and Decision Trees", Int. J. Computer Science Issues, vol. 9, issue 5, No 3, ISSN (Online): 1694-0814, 2012.
16. R. Mishra, N. Singh, M. Gusai, J. Patel, N. Modi "Ontology Based Algorithm And Life Cycle For Subjective Answer Assessment System", Int. J. Advance Research In Science And Engineering, vol. 4, Special Issue (01), ISSN(Online): 2319-8354, 2015.
17. A. Guruji, Mrunal, M. Pagnis, S. M. Pawar, P. J. Kulkarni, "Evaluation of Subjective Answers Using GLSA Enhanced with Contextual Synonymy" Int. J. Natural Language Computing, vol. 4, no. 1, 2015.
18. M. Syamala Devi, Himani "Subjective Evaluation using LSA Technique" Int. J. Computers and Distributed Systems, vol. 3, no. 1, pp. 15–19, ISSN: 2278-5183, 2015.

A 70.8 MW Wideband CMOS Low-Noise Amplifier for WiMAX Application

Sumit Singh, Deepak Prasad and Vijay Nath

1 Introduction

High data rate and high mobility have proved to be the backbone of modern wireless communication. On the other hand, the introduction to WiMAX in the wireless communication takes it to next level due to its low sensitivity and high dynamic range in the receiver system.

To provide larger area coverage, technology has moved from WPAN, WLAN to WMAN, which offers mobility in a larger area. IEEE standard for WMAN is IEEE 802.16 [1], established in the year 1999. Due to higher coverage and high throughput performance, it can be used for cellular backhaul support. It can effectively work in a disaster situation [2, 3]. IEEE 802.16a is an evolution of IEEE 802.16 which is known as WiMAX. It is an appropriate technology for last mile connection. It offers higher range, wider bandwidth and high data rate. However, system design in this frequency becomes more complex due to interference. The first released active standard of WiMAX, i.e. IEEE 802.16, addressed Line-of-Sight (LOS) environments at high-frequency bands operating in the 10–66 GHz range and Non-Line-of-Sight (NLOS) environments in the band between 2 and 11 GHz. Looking forward, IEEE 802.16e adds mobility and enables applications similar to cellular system and personal digital assistants in the frequency range of 2-11 GHz within three bands, namely, (1) 2.5–2.9 GHz, (2) 3.4–3.6 GHz and (3) 5.2–5.9 GHz. WiMAX offers a data rate of 70 Mbps and covers up to 50 km of range [3]. In recent years' technology become more advanced and its keep on emerging day by day. This emerging technology prominently used in our day-to-day live, which helps us to find new solutions for arising problems. Also, the current markets demand latest technology. As we know, technology has been moved from using vacuum tubes to transistors and then to

S. Singh · D. Prasad (✉) · V. Nath
VLSI Design Group, Department of Electronics and Communication Engineering,
Birla Institute of Technology, Mesra, Ranchi 835215, Jharkhand, India
e-mail: prasaddeepak007@gmail.com

© Springer Nature Singapore Pte Ltd. 2019
V. Nath and J. K. Mandal (eds.), *Proceeding of the Second International Conference on Microelectronics, Computing & Communication Systems (MCCS 2017)*, Lecture Notes in Electrical Engineering 476, https://doi.org/10.1007/978-981-10-8234-4_32

integrated circuits (IC), which allows a lot of electronics paraments to accumulate in a smaller area [4]. As days passed on, electronics device size becomes smaller day by day. To make the smaller device size, CMOS technology has been used from past few years. From last two decade, CMOS technology keeps on emerging and it will become more advanced in coming years [5].

In any communication systems transceivers are subdivided into two parts, one is transmitting part and another one is receiving part. At the receiver side, a filter is used to pick the desired signal, and the received RF signal from the transmitter is very weak to be utilised further. The surrounding noises and interference need to be removed from the desired received signal without adding any internal noise. In general, signal-to-noise ratio is almost negligible for the received RF signals which are around −100 dBm. So, a low-noise amplifier (LNA) is used to amplify the desired signal. LNA is a high gain and low noise figure device.

LNA is the first component in a communication receiver chain, which is being used for amplifying the desired received signals. LNA is a key component in a receiver side, because LNA itself creates very less noise and a high gain. The noise figure of the LNA is dominated the noise figure of remaining receive chain components. It creates a great impact on receiver [6]. Hence, it is advised to keep low noise figure of LNA to maintain the system's noise figure low. On the other hand, the gain of LNA should neither be too low which may increase the noise contribution nor be too high to degrade the linearity of the system [7].

In superheterodyne receiver, the RF filter and image rejection filter which are matched to be 50 Ω are placed before and after the LNA. Input/output impedance matching is highly required in LNA, where the discrete components are used, which are too expensive. Hence, an alternate less expensive architecture should be proposed. The introduction of VLSI reduces the power consumption up to milliwatt. Hence, a low power LNA has been proposed in this research paper. The features and specifications of an ideal LNA are listed in Table 1 [8].

The other portion of the paper is organised as follows: Sect. 2 includes the description of cascode LNA, and Sect. 3 demonstrates the result and discussion. Finally, Sect. 4 binds the conclusion on the proposed wideband low-noise amplifier (LNA).

Table 1 Characteristic and specification of wideband LNA	Features	Specification
	Low noise	NF < 3 dB
	Gain	A_v > 15 dB
	Linearity	IIP3 > −10 dB
	Power consumption	– <50 MW
	Frequency range	2–11 GHz
	Other characteristics	Power supply—1.5 V 50-Ω input impedance

2   Wideband LNA Circuit Details

Figure 1 shows the schematic diagram of our proposed wideband LNA. In general, there are two kinds of transistor frequently being used for any kind of amplifier design. Those are known as CS and CG transistor configuration. Both the transistor configurations hold their own advantage over others. CS amplifier configuration provides high gain with good noise performance [9], and CG amplifier configuration provides low power consumption circuit, robust against parasitic and has stable circuit [10]. But there is one limitation in CG amplifier configuration transistor, i.e. it has weak noise performance. So, CS amplifier configuration has been used for our proposed wideband LNA design. Because our goal is to get good gain and good noise performance. Also, inductive source degenerated technique has been adapted for our circuit design. By placing an inductor at the source of the input transistor this technique can be achieved. The purpose behind this technique is to give a good input impedance matching to the LNA without adding any noise.

This proposed LNA has been designed by making a cascoding combination of two transistors, because cascode LNA combination assures to provide low power consumption, high reverse isolation, good noise figure and high-power gain [11]. In lower band of microwave frequencies, the noise sources of upper cascode transistor are degenerated by the lower transistor output impedance. More importantly, cascode combination has superior noise performance.

In our proposed LNA, RF input is feed to the circuit CS amplifier transistor M_1. Before degenerated cascode pair M_1 there is a series combination of inductor (L_{i1}) and capacitor (C_{i1}) is added. As the LNA's input impedance has a reactive part, the

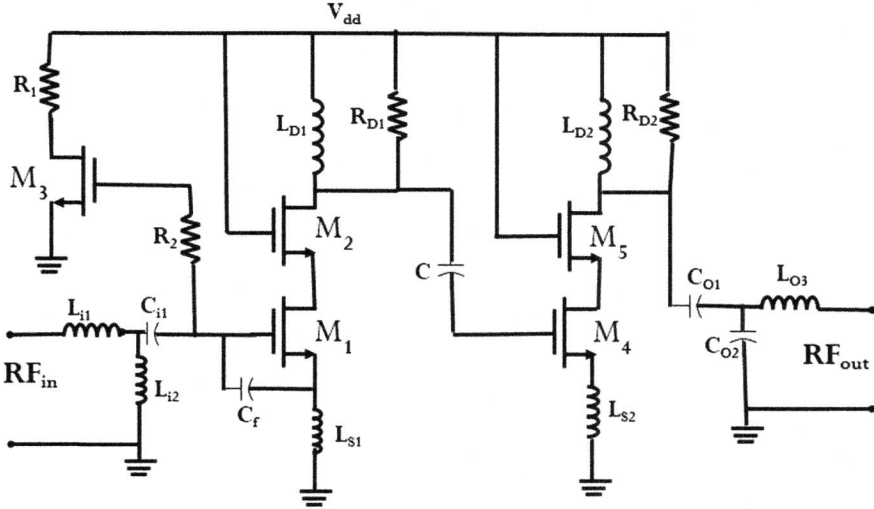

Fig. 1 Proposed wideband low-noise amplifier

combination of L_{i1}, C_{i1}, C_f and L_{s1} helps to resonate the reactive part over multiple frequencies within the band of interest; thus, it can able to provide a wider bandwidth. The capacitor C_f is parallel with C_{gs} of M_1, which helps further to increase the bandwidth. So, a rough calculation for LNA's input impedance can give Z_{in} as follows:

$$Z_{in} = S(L_{I1} + L_{S1}) + \frac{1}{S}\left(\frac{1}{C_{I1}} + \frac{1}{C_f}\right) + \frac{g_{m1}L_{S1}}{C_f} \tag{1}$$

The M_2 transistor is a cascode transistor for the M_1, which is being used for isolate the input to output matching of the LNA. To make the M_1 transistor to work under saturation region, it has to be biased. Biasing is done by the current mirror M_3 transistor and M_2 transistor is biased directly by the V_{dd} power supply. Inductor L_{D1} and L_{D2} are being used for providing exact power supply to the drain of the transistor M_2 and M_5. Resistor R_{D1} and R_{D2} are being used for stability of the circuit, and these resistors help in getting a higher gain. Besides, inductor L_{D1}, L_{D1} and resistor R_{D1} and R_{D2} being used as a resonator circuit. Resonator circuit is used to enable the device application to a particular frequency. In our LNA circuit, there are two resonators have been used to peak two frequencies, and the capacitor C is used for holding the frequency within two frequencies. So, these are the prime constraint to make a wideband. Also, two gain stages have been used to achieve a higher gain but the power consumption is still limited to first gain stage. These two gain stages are the two CS cascaded amplifiers M_1, M_2 and M_4, M_5 as shown in Fig. 1. Circuit overall gain is calculated by multiplying individual again as follows:

$$A_V = A_{V1} \times A_{V2} \tag{2}$$

where A_{V1} is the gain of first gain stage and A_{V2} is the gain of the second gain stage. Individual gain can be calculated as

$$A_{V1} = G_{m1} \times R_{out1} \tag{3}$$

$$A_{V2} = G_{m2} \times R_{out2} \tag{4}$$

G_m and R_{out} are transconductance and output impedance of the gain stages, respectively. Approximate calculation using the above mention methods renders in Eqs. (6)–(9).

It is a common practice to use a buffer at the output end of the LNA for matching the output of the LNA to the next stage. However, buffer degrades the LNA gain, and it draws a high power from the circuit. So, practicing to use a buffer is not convincible. Instead, we have used a t-shape matching circuit, which is series combination of inductor and capacitor and a parallel capacitor. Same matching circuit has been used at the input but instead of a parallel capacitor, an inductor has been used to achieve a 50-Ω input–output impedance matching.

Whereas the length of all the transistors is fixed for this LNA design, the width has been varied to achieve the desired parameters. Table 2 shows the transistor's width to length ratio.

Previously in noise theory, we have already studied that the noise performance for any cascade device highly depends on first few stages of that network, which is known as 'Friis formula' for noise [11]. Friis formula is described as

$$\mathrm{NFfront}_{\mathrm{end}} = \mathrm{NF}_{\mathrm{LNA}} + \left(\frac{\mathrm{NF}_{\mathrm{subsequent}} - 1}{\mathrm{G}_{\mathrm{LNA}}} \right) \tag{5}$$

$\mathrm{NF}_{\mathrm{subsequent}}$ Referred input noise figure.
$\mathrm{NF}_{\mathrm{LNA}}$ Core LNA's noise figure.
$\mathrm{G}_{\mathrm{LNA}}$ Gain of LNA.

The noise generated by subsequent stages can be reduced by gain of the LNA. But the noise which is generated by LNA itself is fed directly to the receiver frontend. So, it is advised to make LNA's gain high and keep low noise figure. Also, to keep power consumption low for low power wireless application.

(A) *Stability analysis*:

For any practical circuit, it is mandatory to make the circuit stable. If the circuit is not stable, then it would show unnecessary behaviours. LNA circuit's stability can be defined by two parameters; those are rollet's factor (K) and Δ_s, where K should be great than 1, and Δ_s should be less than 1.

$$G_{m1} = \frac{(g_{m2} + 1/r_{\mathrm{out2}})\left(\frac{g_{m1}r_{O1}}{sC_f} - sL_{s1}\right)}{\left[s(L_{i1}+L_{s1}) + \frac{1}{s}\left(\frac{1}{C_{i1}} + \frac{1}{C_f}\right)\right]\left[(g_{m2} + \frac{1}{r_{O2}})(r_{O1}+sL_{s1}) + 1\right] + sL_{s1}\left(g_{m2} + \frac{1}{r_{O2}}\right)\left(\frac{g_{m1}r_{O1}}{sC_f} - sL_{s1}\right)} \tag{6}$$

$$R_{\mathrm{out1}} = [r_{O1} + r_{O2} + sL_{s1} + g_{m2}r_{O2}(r_{O1}+sL_{s1})] \| R_{D1} \| sL_{D1} \tag{7}$$

$$G_{m2} = \frac{r_{O4}g_{m4}g_{m5} + r_{O4}g_{m4}}{r_{O4}g_{m5} + r_{O4}} \tag{8}$$

$$R_{\mathrm{out2}} = [r_{O4} + r_{O4}g_{m5}] \| R_{D2\|}sL_{D2} \| \left[\frac{1}{sC_{O1}} + sL_{O1}\right] \tag{9}$$

Table 2 Transistor ratio (W/L)

Transistor	Width (μm)	Length (nm)
M_1	100	45
M_2	90	45
M_3	3	45
M_4	100	45
M_5	90	45

where K and Δ_s can be defined as follows:

$$K = \frac{1 - |S_{22}|^2 - |S_{11}|^2 + |\Delta_s|}{2|S_{12}S_{21}|} \tag{10}$$

$$\Delta_s = S_{11}S_{22} - S_{12}S_{21} \tag{11}$$

where

S_{11} Input reflection coefficient with the output matched.
S_{21} Forward transmission gain or loss.
S_{12} Reverse transmission or isolation.
S_{22} Output reflection coefficient with the input matched.

(B) *Small signal analysis*:

Small signal model for proposed LNA circuit is shown in Fig. 2.

Where C_{gs} shows gate to source capacitance of the transistor $M1$, which is identify as C. And V_{gs} comprise as gate to source voltage of the transistor M1. g_m denotes as transconductance of transistor $M1$. After applying the Kirchhoff's voltage law (KVL) at the input, we can write as

$$V_{in} = I_{in}jwL_s + \frac{I_{in}}{jwC} + I_a jwL_s \tag{12}$$

At the output, it is found that

$$V_{gs} = \frac{I_{in}}{jwC} \tag{13}$$

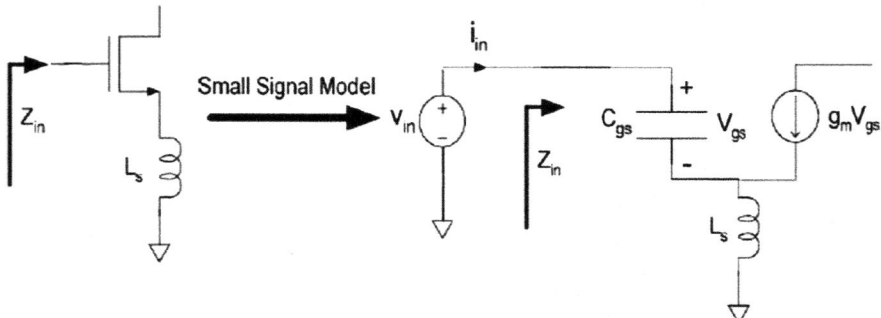

Fig. 2 Small signal model

$$I_o = g_m V_{gs} = \frac{g_m I_{in}}{jwC} \tag{14}$$

By substituting Eqs. (7) in (4), we get

$$V_{in} = I_{in} jwL_s + \frac{I_{in}}{jwC} + \frac{g_m I_{in}}{jwC} jwL_s \tag{15}$$

$$V_{in} = I_{in} \left[jwL_s + \frac{1}{jwC} + \frac{g_m L_s}{C} \right] \tag{16}$$

So, we get input impedance as

$$Z_{in} = \frac{V_{in}}{I_{in}} = \left[jwL_s + \frac{1}{jwC} + \frac{g_m L_s}{C} \right] \tag{17}$$

For impedance matching, there are certain condition must be certified:

$$\text{Re}[Z_{in}] = \frac{g_m L_s}{C} = R_s \tag{18}$$

$$\text{Im}[Z_{in}] = jwL_s + \frac{1}{jwC} = 0 \tag{19}$$

In Real part, it need to be equal with 50 Ω. In case that input impedance matching operated at centre frequency, then imaginary term has to be nullified. So, we can write as

$$W_c = 1/\sqrt{CL_s} \tag{20}$$

At lower significance values, if

$W < W_c$,   It will be capacitive in nature.
$W > W_c$,   It will be inductive in nature.

Calculation of NF is an important consideration for any LNA circuit. Calculation for noise performance of any circuit is called as Noise Factor (F). Generally, it exhibits in decibel and commonly cite as noise figure (NF).

$$NF = 10 \log_{10} F \tag{21}$$

where F is defined as:

$$F = \frac{SNR_{in}}{SNR_{out}}$$

Or

$$F = \frac{SNR_{\text{in}}}{SNR_{\text{out}}} = \frac{P_{sig}/P_{RS}}{SNR_{out}} \tag{22}$$

where

SNR_{in} Signal-to-noise ratios at i/p.
SNR_{out} Signal-to-noise ratios at o/p.
P_{sig} Input signal power.
P_{RS} Source resistance noise power.

It follows that

$$P_{\text{sig}} = P_{\text{RS}}.F.SNR_{\text{out}} \tag{23}$$

And

$$P_{\text{RS}} = \frac{4kTR_s}{4} \frac{1}{R_{\text{in}}} = kT \tag{24}$$

3   Result and Discussion

Our proposed LNA circuit has been simulated in Cadence Virtuoso gpdk 45 nm CMOS technology over a 1.5 V power supply. There are two important things has been considered before analysing the proposed LNA circuit's simulation. First one is, we have not considered any parasitic capacitance values for the MOSFETS. Rather we can consider parasitic values during the layout design. Second one is, instead of using an RF CMOS, we have considered a regular CMOS in our design, because RF CMOS design is the substantial variability of device temperature and circuit parameters with process [15, 16]. Also, it gives extremely small size to avoid parasitic elements [17, 18]. So, we have considered a regular COMS (proficiently work up to frequency 60 GHZ) for our design.

The LNA circuit becomes stable for the values of $K > 1$ and $\Delta_s < 1$, which is shown in the below figure. Figures 3 and 4 proves that the condition for K and Δ_s is satisfying to make the circuit stable.

Characteristics of the S-parameters have also shown below. Where S_{21} represents gain of the circuit, and S_{11} and S_{22} are represent the input and output reflection coefficient. S_{21}, S_{11}, S_{22} values have demonstrated in Figs. 5, 6 and 7, respectively.

From Fig. 5, we are able to show that out LNA circuit is achieved a high gain. Input/output reflection coefficient is also represented for matching of the input/output. In general, the value for reflection coefficient should be less than -10 dB [19]. Proper matching is required for maximum power delivery from input to output. We are able to achieve this desired value as shown in Figs. 6 and 7.

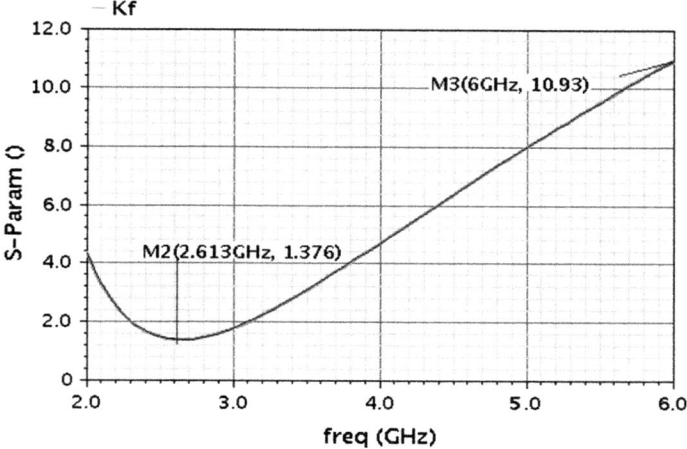

Fig. 3 K (rollet)-factor

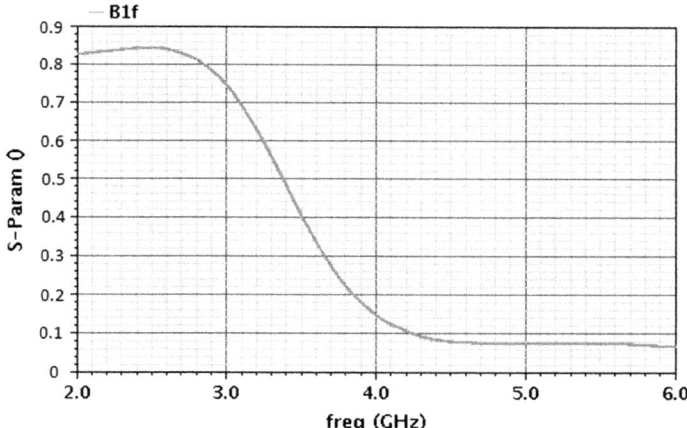

Fig. 4 Δ_s plot of LNA

As we have discussed before the Noise figure of any LNA circuit is very important parameter. NF should be less.As a result, we are able to achieve a very less NF figure for a wideband 2–6 GHz band. And we are getting an NF of less than 2 dB for the band of interest. Figure 8 shows the NF of LNA.

Figure 9. Shows that the gain compression because of nonlinearity in the system which causes the power gain to diverge from its ideal curve. At the point where power gain is drop below 1 dB from the idealised curve is referred as 1 dB compression point. To avoid nonlinearity a system it must work several decibels below this level. In our proposed LNA circuit, it is −12.12 dBm as shown in below figure.

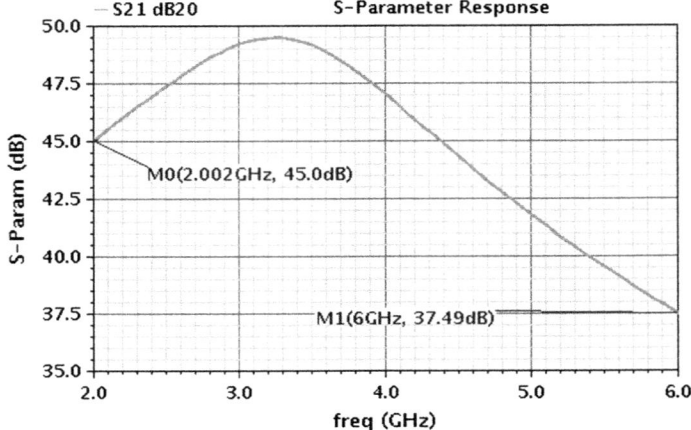

Fig. 5 S_{21} (gain) plot of LNA

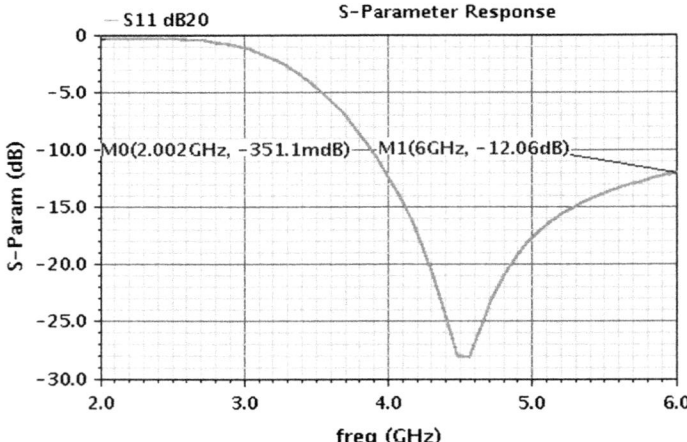

Fig. 6 S_{11} input reflection coefficient

Power consumption is another important concern for any LNA. Power consumption should be less for the low power device. But it is advised to keep a low power for wireless application devices. After simulation, we are able to get 70.88 MW power consumption which out proposed LNA drew from the 1.5 V power supply. Figure 10 shows the power consumption of the circuit.

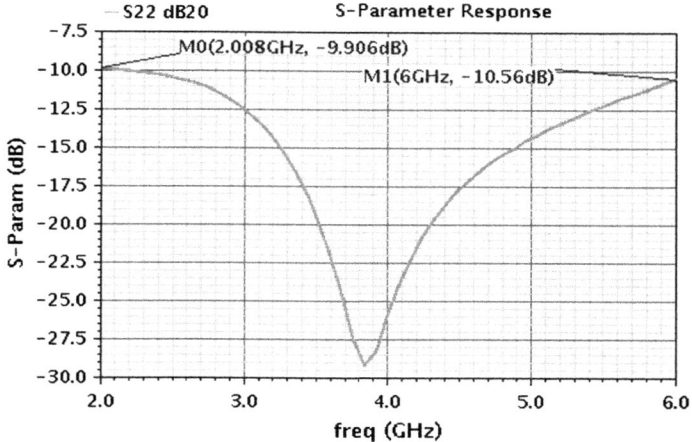

Fig. 7 Output reflection coefficient

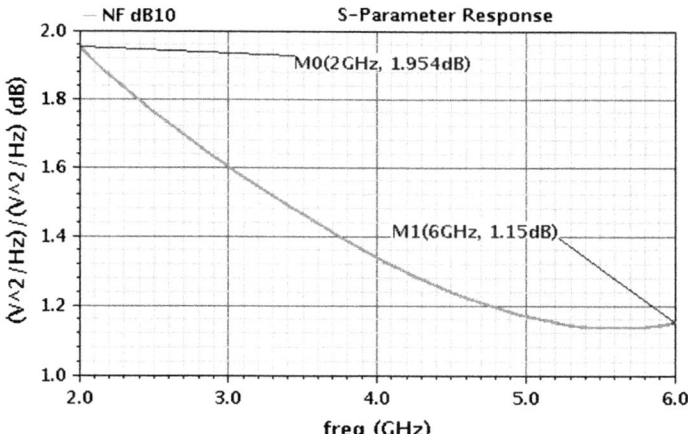

Fig. 8 NF of LNA

Process corner analysis tells about different effects and variations found in the gain and NF of a circuit when it comes into the real environment. Table 3 represents the values for gain and NF for different process corners.

Performance summary of LNA is sum up and represented in Table 4, which is compared with some previously published works.

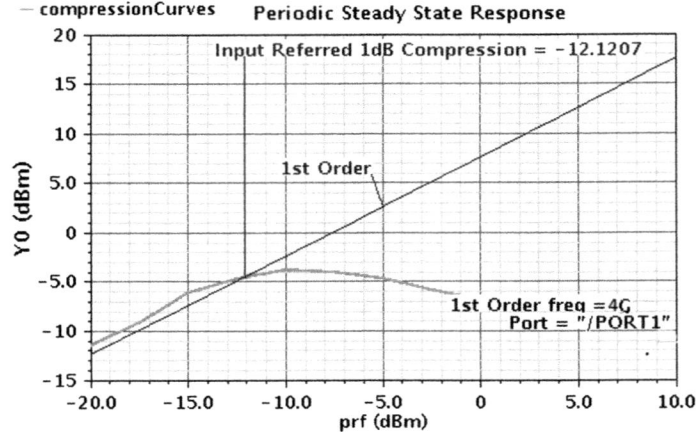

Fig. 9 1 dB compression point

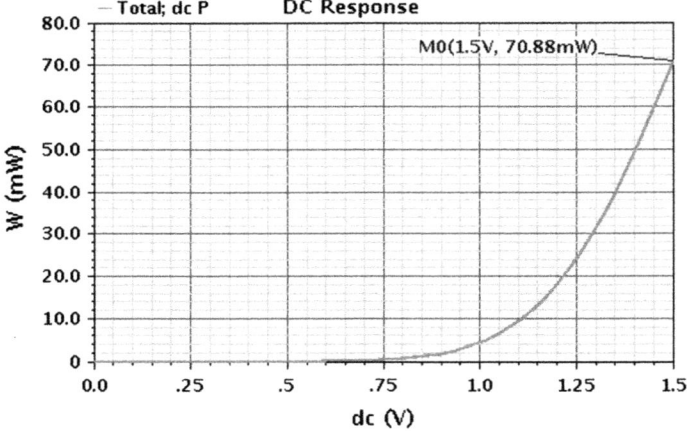

Fig. 10 Total power consumption of proposed LNA

Table 3 Process corner summary of proposed LNA

Process corners	S11 (dB)	S12 (dB)	S21 (dB)	S22 (dB)	NF (dB)
SS	−12.58	−69.88	46.48	−19.85	1.32
FF	−8.069	−65.07	45.72	−20.35	1.485
TT	−10.25	−66.015	46.37	−46.47	1.392
NN	−12.44	−66.96	47.03	−26.12	1.341

Table 4 Performance summary comparison with published works

Specification	Ref. [12]	Ref. [13]	Ref. [14]	This work
Technology (nm)	130	65	90	45
Operating frequency (GHz)	4–6	0.1–4.3	0.1–3	2–6
S11 (dB)	<−10	<−8	<−39.9	−27.95
S21 (dB) gain	25.5	21	17.007	45–37
Noise figure (dB)	1.9	2.8-4	2.73	1.1–1.9
Power supply (V)	1.5	1.2	2	1.5
Power consumption (MW)	20.9	2	2.56 mA	70.88

4 Conclusion

In this paper, a wideband CMOS LNA designed has been presented for the IEEE 802.16a(WiMAX) standard, which is studied and tested several times to accomplish an approach that can help to deliver a high gain, low NF, low power consumption and fully integrated LNA design. From the proposed LNA, we are able to get high gain, low noise figure and low power consumption circuit. As the gain is high, it reflects its effect on the linearity of the circuit. Gain and linearity trade-off can be done to get a better linearity in future. It is designed and simulated through the Cadence Virtuoso analogue design environment using gpdk 45 nm library. Since this proposed LNA shows suitable Gain and noise figure, it can be used for practical circuits.

Acknowledgements We are thankful to Prof. Dr. V. R. Gupta, HOD ECE, BIT Mesra, for his cooperation. We also extend our gratitude to our Vice-Chancellor, Dr. M. K. Mishra for his encouragement. We are also thankful to RESPOND ISRO Ahmadabad for funding this project.

References

1. M. A. G. Lorenzo and M. T. G. d. Leon, "Comparison of LNA Topologies for WiMAX Applications in a Standard 90-nm CMOS Process," *2010 12th International Conference on Computer Modelling and Simulation*, Cambridge, 2010, pp. 642–647.
2. R. Kundu, A. Pandey, S. Chakraborty and V. Nath," A CMOS low noise amplifier based on common source technique for ISM band application," Microsystem Technologies, Vol 22 (11) 2016
3. N. Yadav, A. Pandey and V. Nath," Design of CMOS low noise amplifier for 1.57 GHz," IEEE International Conference Microelectronics, Computing and Communications (MicroCom), 23-25 Jan 2016, Durgapur, India.
4. Tyagi. S, Saurav. S, Pandey. A, Priyadarshini. P, Ray. M, Pal. B. B and Nath. V, (2017) A 21nW CMOS Operational Amplifier for Biomedical applications. In: Nath V. (eds) Proceedings of the International Conference on Nano-electronics, Circuits & Communication Systems. Lecture Notes in Electrical Engineering, Vol 403, Springer.
5. Kumar. V, Singh. K.K, Pandey. A, Nath. V, (2017) Design of Comparator in Sigma-Delta ADC Using 45 nm CMOS Technology. In:Nath V. (eds) Proceedings of the International

Conference on Nano-electronics, Circuits & Communication Systems. Lecture Notes in Electrical Engineering, Vol 403, Springer.

6. S. Kumar, "Wireless Communication Fundamental and Advance Concepts," River publisher series in communication, vol. 1, pp. 117–121, 2015.

7. Telecom Regulatory Authority of India, "consultation Paper on Allocation and pricing of spectrum for 3G services and Broadband Wireless Access," consolations Paper no.9/2006.

8. T. H. Lee, "The Design of CMOS Radio-Frequency Integrated Circuit", ed. 1st, New York: Cambridge University Press, chap.11, p. 277–289,1998.

9. J. Le Ny, B. Thudi, J. McKenna, "A 1.9 GHz Low Noise Amplifier," EECS 522 Analog Integrated Circuits Project, Winter 2002.

10. S. Shekhar, J. S. Walling, S. Aniruddhan and D. J. Allstot, "CMOS VCO and LNA Using Tuned-Input Tuned-Output Circuits," in *IEEE Journal of Solid-State Circuits*, vol. 43, no. 5, pp. 1177–1186, May2008. https://doi.org/10.1109/jssc.2008.920360.

11. X. Fan, H. Zhang and E. SÁnchez-Sinencio, "A Noise Reduction and Linearity Improvement Technique for a Differential Cascode LNA," in *IEEE Journal of Solid-State Circuits*, vol.43, no.3, pp. 588–599, March 2008. https://doi.org/10.1109/jssc.2007.916584.

12. A. A. Abidi, "On the operation of cascode gain stages," in *IEEE Journal of Solid-State Circuits*, vol. 23, no. 6, pp. 1434–1437, Dec1988.https://doi.org/10.1109/4.90043.

13. Choong-Yul Cha and Sang-Gug Lee, "A 5.2-GHz LNA in 0.35-μm CMOS utilizing inter-stage series resonance and optimizing the substrate resistance," in *IEEE Journal of Solid-State Circuits*, vol. 38, no. 4, pp. 669–672, Apr 2003. https://doi.org/10.1109/jssc.2003.809523.

14. D. K. Shaeffer and T. H. Lee, "A 1.5-V, 1.5-GHz CMOS low noise amplifier," in *IEEE Journal of Solid-State Circuits*, vol.32,

15. B. Razavi, " RF Microelectronics": Prentice Hall PTR, chap. 2, 2006.

16. B. Razavi, "CMOS technology characterization for analog and RF design," in *IEEE Journal of Solid-State Circuits*, vol. 34, no.3, pp. 268–276, Mar1999. https://doi.org/10.1109/4.748177.

17. S. Arshad, F. Zafar and Q. u. Wahab, "Design of a 4–6 GHz wideband LNA in 0.13 μm CMOS technology," *2012 IEEE International Conference on Electronics Design, Systems and Applications (ICEDSA)*, Kuala Lumpur, 2012, pp. 125–129. https://doi.org/10.1109/icedsa.2012.6507780.

18. Z. Pan, C. Qin, Z. Ye and Y. Wang, "A Low Power Inductor Less Wideband LNA With G_m Enhancement and Noise Cancellation," in *IEEE Microwave and Wireless Components Letters*, vol. 27, no.1, pp. 58–60, Jan.2017. https://doi.org/10.1109/lmwc.2016.2629969.

19. A. Pandey, M. Pusalkar and P. Dwaramwar, "A 0.1–3 GHz, 90 nm CMOS wideband LNA employing positive negative feedback for gain, NF and linearity improvement," *2016 International Conference on Advanced Communication Control and Computing Technologies (ICACCCT)*, Ramanathapuram, 2016, pp. 147–152. https://doi.org/10.1109/icaccct.2016.7831618.

Computation of Discrete Fourier Transform (FFT): A Review Article

Shaik Qadeer, Mohammed Zafar Ali Khan
and Mohammed Yousuf khan

1 Introduction

The discrete Fourier transform (DFT) is one among the important fundamental signal processing operation needed. And its computer algorithm which can be efficiently computed is termed as the Fast Fourier transform (FFT). Today, its use is even more needed, as it is falling in top 10 algorithms which have its influence on the practice of engineering and development in the twentieth century [1]. A large number of mobile phones, tablets, and PCs [2, 3] compute the FFT on a large scale for many of sound, video, and image processing applications. The computer has been instrumental in studying the FFT, and numerous articles have been published about it over the past few decades [3, 4]. Few of these publications discuss the basic modification to make it more efficient [5, 6], whereas the other covers their implementation.

This article reviews the classification of FFT for radix-2 only. In this class, the FFT is derived by dividing the sequence either in frequency domain or time domain till it reaches two. The rest of the work is described as follows: The first part of Sect. 2 (i.e., 2.1) covers a classification of FFT, its restriction, the properties used in and the keywords. The second part of Sect. 2 (i.e., 2.2) covers introduction, and derivation of algorithm is defined in Sect. 2.2. The last part of Sect. II (i.e., 2.3) covers Matlab code for FFT algorithm with a few applications. The conclusion is given in Sect. 3.

S. Qadeer (✉)
Electrical Engineering Department, Muffakhamjah College
of Engineering and Technology, Hyderabad, India
e-mail: haqbei@gmail.com

M. Zafar Ali Khan
Department of Electrical Engineering, Indian Institute of Technology Hyderabad,
Sangareddy, India

M. Yousuf khan
Polytechnic, MANUU, Hyderabad, India

© Springer Nature Singapore Pte Ltd. 2019
V. Nath and J. K. Mandal (eds.), *Proceeding of the Second International Conference on Microelectronics, Computing & Communication Systems (MCCS 2017)*, Lecture Notes in Electrical Engineering 476, https://doi.org/10.1007/978-981-10-8234-4_33

2 The Fast Fourier Transform

This section describes the Radix-2 FFT, starting with a few fundamentals the detailed algorithm is covered.

2.1 The Fundamentals of FFT

The general equation which describes the classification of FFT is $N = r^{\vartheta}$, where N is the number of DFT points, r defines radix (example $r = 2$ is termed as radix-2, $r = 4$ as radix-4, etc.), and ϑ indicate the stages in FFT algorithm. The review article is restricted to radix-2 FFT. The derivation of this FFT either decimates time domain sequence (called as DIT-FFT) or frequency domain sequence (called as DIF FFT) repeatedly till their size reaches 2.

The possible restriction of radix-2 FFT can be observed from its computing size equation: $N = 2^{\vartheta}$, with integers values for ϑ we get sizes for FFT as 2, 4, 8, 16, 32, etc. However, the sizes 3, 5, 6, 9, etc., cannot be computed.

To proceed further for algorithm, let us discuss the properties responsible for the possible saving in computing arithmetic of DIT radix-2 FFT. The first property is from the multiplier of DFTs power series called as twiddle factor. It is a complex number and offer symmetry property over its half period as $W_N^{k+N/2} = -W_N^{k}$. The other one is the periodic property of DFT. As the algorithm is derived with decimation of sequence, we can take the advantage of this property in each stage.

Now, we discuss the keyword used in the algorithm. It is as follows:

- Butterfly diagram: This is the generic diagram used to draw the flowchart of the algorithms. Example will be covered in Sect. 2.2.
- Bits reverse operation: In both the method of radix-2 FFT, there is a need to shuffle the data. This shuffling operation is called as Bits reverse operation shown in Fig. 1.
- In-place computation: Computation of all stages at a single place/single array is called as "In-place computation". Example for $N = 8$ will be covered in Sect. 2.2.

2.2 Derivation of FFT Algorithm

This subsection discusses the detail FFT algorithm. The discussion is restricted to DIT-FFT, due to its simple structure and good signal to quantization noise ratio. DIT-FFT is a class of radix-2 FFT in which the decomposition is done in time domain. Let us decimate the N-point input $p(n)$ into two halves as $q_1(n)$ and $q_2(n)$,

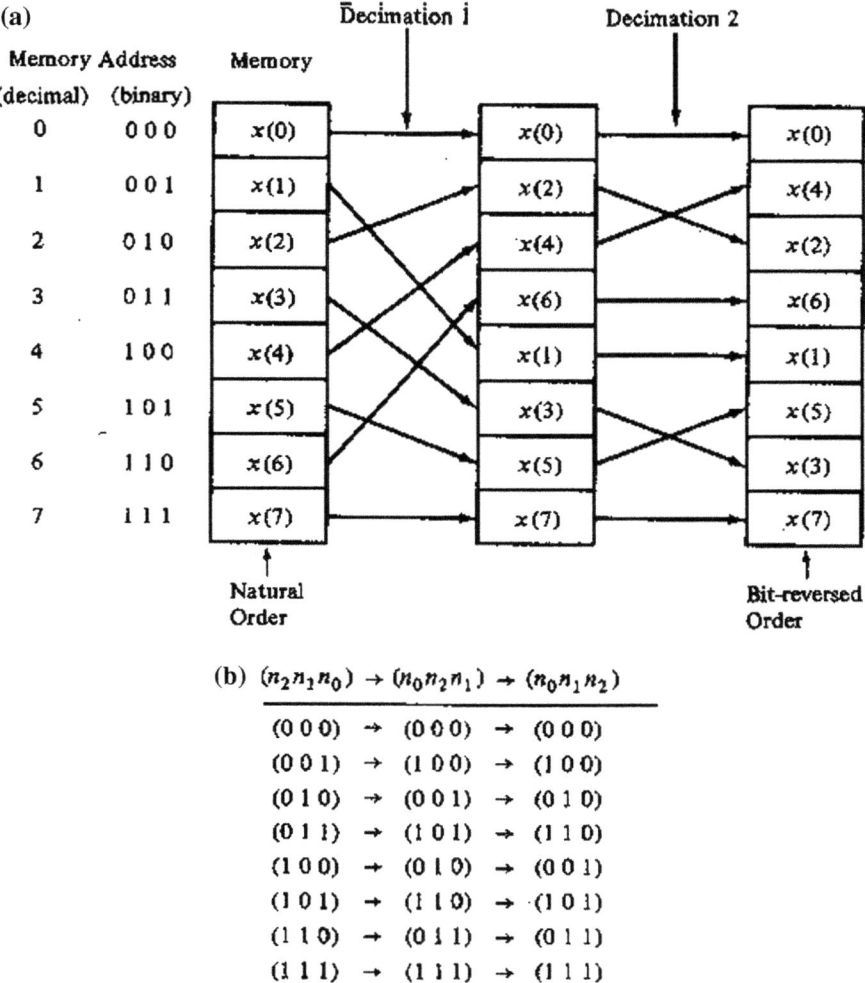

Fig. 1 Decimation process, i.e., division of $p(n)$ in each stage into two parts according to even and odd order as shown above—bit reverse operation

corresponding to the even-numbered and odd-numbered samples of $p(n)$, respectively, that is

$$q_1(n) = p(2n)$$

$$q_2(n) = p(2n+1), \quad n = 0 \dots \frac{N}{2} - 1$$

The derivation process for the N-point DFT is as follows [3, 4]:

$$
\begin{aligned}
P(k) &= \sum_{n=0}^{N-1} p(n) W_N^{kn}, \quad k = 0 \ldots N-1 \\
&= \sum_{n \text{ even}} p(n) W_N^{kn} + \sum_{n \text{ odd}} p(n) W_N^{kn} \\
&= \sum_{m=0}^{N/2-1} p(2m) W_N^{2km} + \sum_{n=0}^{N/2-1} p(2m+1) W_N^{k(2m+1)}
\end{aligned}
\tag{1}
$$

With $W_N^2 = W_{N/2}^1$ Eq. (1) will become

$$
\begin{aligned}
P(k) &= \sum_{m=0}^{N/2-1} q_1(m) W_{N/2}^{km} + W_N^k \sum_{m=0}^{N/2-1} q_2(m) W_{N/2}^{km}, \\
&= Q_1(k) + W_N^k Q_2(k), \quad k = 0 \ldots N-1
\end{aligned}
\tag{2}
$$

where $Q_1(k) = \mathrm{DFT}_{N/2}\{q_1(n)\}$ and $Q_2(k) = \mathrm{DFT}_{N/2}\{q_2(n)\}$. Using the periodic property of half size DFTs (i.e., $Q_1(k)$ and $Q_2(k)$) as $Q_1(k+N/2) = Q_1(k)$ and $Q_2(k+N/2) = Q_2(k)$ with twiddle factors property introduced in Sect. 2.1 ($W_N^{k+N/2} = -W_N^k$), Eq. (2) can now be modified as

$$
\begin{aligned}
P(k) &= Q_1(k) + Q_2(k) W_N^k, \quad k = 0 \ldots \frac{N}{2} - 1 \\
P\left(k + \frac{N}{2}\right) &= Q_1(k) - Q_2(k) W_N^k, \quad k = 0 \ldots \frac{N}{2} - 1
\end{aligned}
\tag{3}
$$

Equation (3) is called as the first step of decimation. From this, we can draw the butterfly diagram as shown in Fig. 2.

Extending the above decimation to the computation of $N/4$-point DFTs in terms of $Q_1(k)$ and $Q_2(k)$ relations:

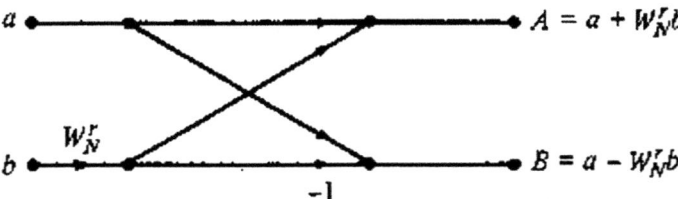

Fig. 2 Basic butterfly diagram for DIT-FFT algorithm

$$Q_1(k) = F[q_1(2n)] + W_{N/2}^k F[q_1(2n+1)],$$

$$Q_1\left(k + \frac{N}{4}\right) = F[q_1(2n)] - W_{N/2}^k F[q_1(2n+1)], \quad k = 0\ldots\frac{N}{4} - 1, \quad n = 0\ldots\frac{N}{4} - 1$$

$$Q_2(k) = F[q_2(2n)] + W_{N/2}^k F[q_2(2n+1)], \tag{4}$$

$$Q_2\left(k + \frac{N}{4}\right) = F[q_2(2n)] - W_{N/2}^k F[q_2(2n+1)], \quad k = 0\ldots\frac{N}{4} - 1, \quad n = 0\ldots\frac{N}{4} - 1$$

$F(*)$ represents fourier transform

This process is repeated till the size of the sequence is reduced to two. As an example for $N = 8$, the block diagram is shown in Fig. 3, whereas the in-place computation diagram is shown in Fig. 4. Figure 3 or Fig. 4 shows that the computation of DFT is distributed into ϑ stages. Figure 3 or Fig. 4 shows for $N = 8$, the number of stages which is three, and the size of DFT in stage-I, stage-II, and stage-III is 2, 4, and 8, respectively. The computation complexity can be calculated either using (3) or (4). Considering (3) for calculation. Let the following representation: $M(N) \rightarrow$ number of complex multiplication, $A(N) \rightarrow$ number of complex addition (or subtraction). Equation (3) reveals that each stage consuming $2N/2$ complex additions and $1N/2$ complex multiplications (as there are $N/2$ butterflies in this stage). As this process is repeated ϑ $((\log_2 N)$ times. Hence, N-point FFT takes $N \log_2 N$ complex additions and $N/2 \log_2 N$ complex multiplications. Note that each complex multiplication takes four real multiplications and two real additions, whereas each complex addition takes two real additions.

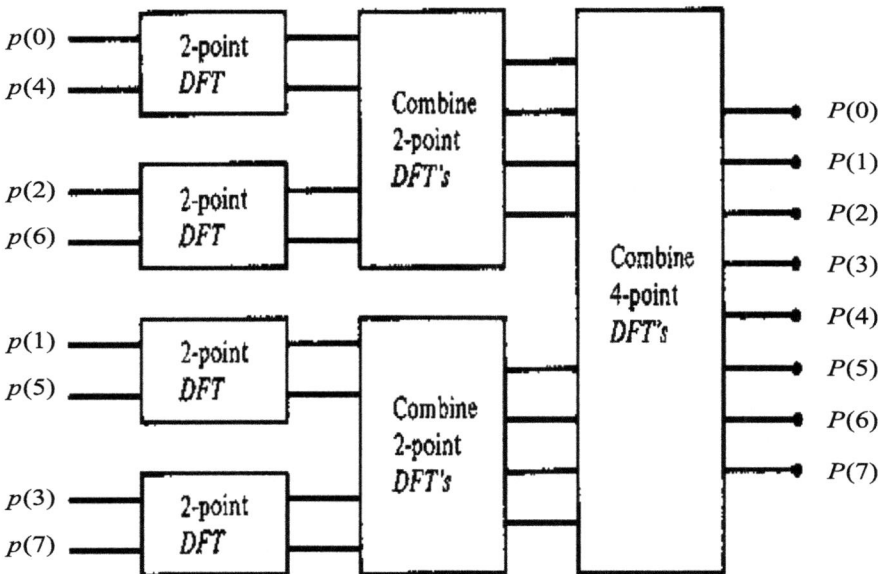

Fig. 3 Block diagram of 8 point DFT with DIT-FFT method

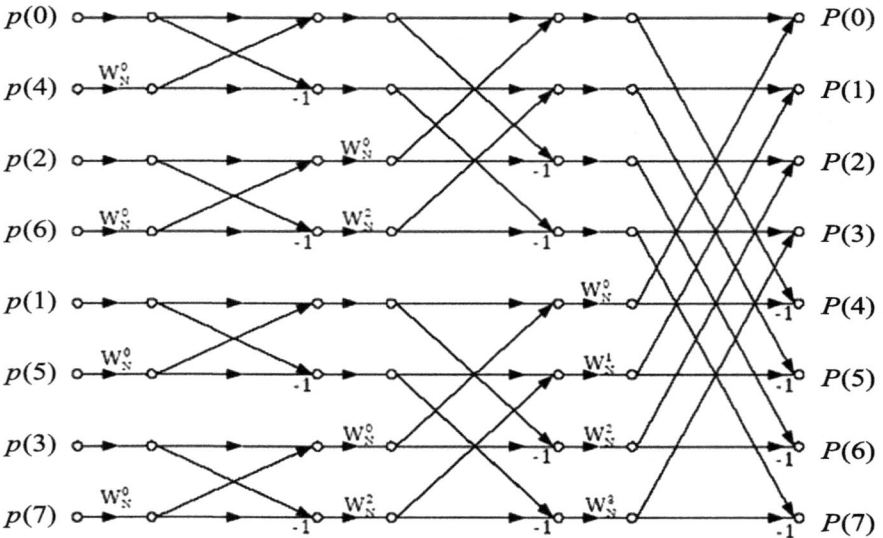

Fig. 4 In-place computation diagram for $N = 8$ point DFT with DIT-FFT

2.3 The Results Related to FFT Algorithm

This subsection gives Matlab simulation, a few applications, and comparison of computation of FFT. The self-explanatory Matlab function for the *N-point* FFT is shown in Program 1.

```
function y = fft_dit(x)

    N = length(x);
    x = x(bitrevorder(1:N));
    q = log2(N);


    for j = 1:q
        m = 2^(j-1);
        w = exp(-pi*1i/m).^(0:m-1);
        for k = 1:2^(q-j)
            s = 2*(k-1)*m + 1;%index-s
            e = 2*k*m;% index-e
            r = s + (e-s + 1)/2;
        %   index-m
            A     = x(s:r-1);
            y_bottom = x(r:e);
```

```
          B = w .* y_bottom;
          y = [A + B, A - B];
          x(s:e) = y;
     end
  end
```

Program 1: Matlab code for *N*-point FFT simulation

An application of FFT as spectrum analyser for the verification of sampling theorem is given in Program 2.

```
%verification of sampling theorem
clear all;close all;clc
k = 5;
fm = input('enter max freq of sin signal');
choice = menu('menu','equal to ST','less than ST','greater than ST');
   if(choice ==1)
                 fs = 2*fm;
         elseif(choice ==2)
                 fs = 1.2*fm;
         else
                     fs = 10*fm;
         end
n = 0.0001:1/fs:k/fm;
xn = sin(2*pi*n*fm);
xf = abs(fft(xn));
N = length(xf);
f = (0:N-1).*fs/N;
subplot(1,2,1);
stem(f,xf);
xlabel('freq');
ylabel('mag of |xf|');
title('spectrum analyser');
subplot(1,2,2);
plot(n,xn);
xlabel('discrete time');
ylabel('amplitude of xn');
title('cro');
```

Program 2: FFT as spectrum analyzer in verification of sampling theorem application

The cases obtained after the execution of Program 2 for fm = 200, is shown in Fig. 5a–c.

The other application of FFT cover here can be used for discrete system analysis as its use for linear and circular convolution operation shown in Program 3:

```
clc;clear all;
x = input('enter i/p x(n)');
h = input('enter i/p h(n)');
choice = menu('menu','Circular Convolution','Linear Circular Convolution');
if(choice ==1)
    if(length(x) ==length(h))
        xf = fft(x);
        hf = fft(h);
        yf = xf.*hf;
        y = ifft(yf);
    else
        display('for entered input circular convolution is not possible');
    end
else
    %linear using circular convolution
    xf = fft([x,zeroes(1,length(h)-1)]);
    hf = fft([h,zeroes(1,length(x)-1)]);
    yf = xf.*hf;
    y = ifft(yf);
end
display('Convolution of x and h is y = ');
y;
```

Program 3: Use of FFT for linear and circular convolution operation

The comparison of computational count of Radix-2 FFT with direct DFT is shown in Table 1. It is observed that DFT with FFT decreases computational count exponentially with increase in N value.

Fig. 5 **a** Frequency and time domain plot for equal to sampling frequency case; **b** frequency and time domain plot for less sampling case; **c** frequency and time domain plot for more than sampling case

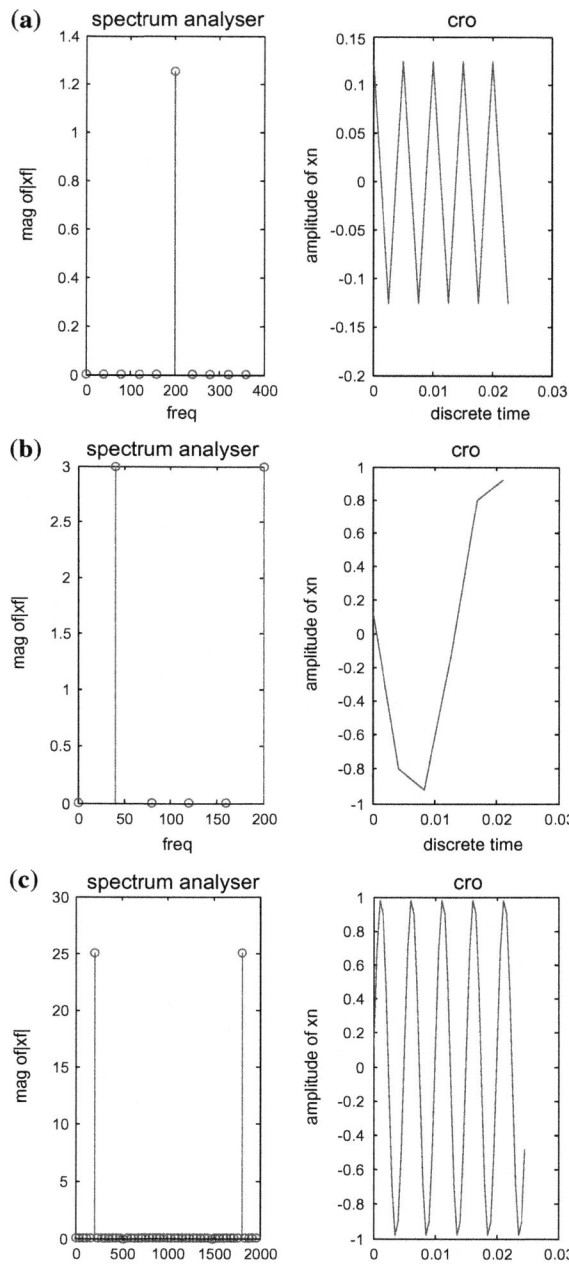

Table 1 Comparison of computational count of FFT with direct DFT

Size of FFT (N)	Computation with direct DFT method		Computation with Radix-2 FFT method	
N	C^* multiplication N^2	C additions $N(N-1)$	C multiplication $N/2\log_2 N$	C additions $N\log_2 N$
4	16	12	4	8
16	256	240	32	64
32	1024	992	80	160
64	4096	4032	192	384
128	16,384	16,256	448	896

C^* stands for complex

3 Conclusion

In this chapter, FFT algorithm is reviewed. The review covers an introduction to applications. In addition to this, the source codes in Matlab for FFT with frequently used DFT applications like spectrum analyzer, verification of sampling and convolution theorems are covered.

References

1. J. Dongarra and F. Sullivan. Guest editors' introduction: The top 10 algorithms. Computing in Science & Engineering, PP 22–23, 2000.
2. Jason Fitzpatrick. An interview with Steve Furber. Communications of the ACM, Vol. 54(5), PP. 34–39, 2011. Dylan McGrath. IDC cuts PC microprocessor forecast. EETimes, September 2011.
3. Rao, K. R., Kim, Do Nyeon, Hwang, Jae Jeong. Fast Fourier Transform - Algorithms and Applications, Springer Netherlands 2010.
4. Nussbaumer, Henri J. Fast Fourier Transform and Convolution Algorithms, Springer Series in Information Sciences, 1982.
5. M. Z. A. Khan and S. Qadeer, "Streamlined real-factor FFTs," *2010 18th European Signal Processing Conference*, Aalborg, 2010, pp. 567–571.
6. M. Z. A. Khan and S. Qadeer, "A new variant of Radix-4 FFT," *2016 Thirteenth International Conference on Wireless and Optical Communications Networks (WOCN)*, Hyderabad, 2016, pp. 1–4. https://doi.org/10.1109/wocn.2016.7759873.

Ramanujan Sums and Signal Processing: An Overview

Debaprasad De, K. Gaurav Kumar, Archisman Ghosh
and M. K. Naskar

1 Introduction

A lot has been glorified about the great Indian mathematician, Srinivasa Ramanujan. He was born in 1887 in Erode, Tamil Nadu. He grew up with ordinary education and never got any formal advanced teaching in mathematics. Still, he developed his mathematical ideas and made immense contributions to number theory, mathematical analysis, and infinite series [1]. His genius was noticed by some of the Indian mathematicians and suggested him to contact renowned Western mathematicians. Among them, the only one who took his correspondence seriously was the legendary Cambridge mathematician, G. H. Hardy. Hardy recognized the significance of the formulae derived by Ramanujan in his letter and made arrangements for Ramanujan to come to Cambridge and collaborate with him. This collaboration continued for 5 years until Ramanujan fell ill and returned back to India. In his phenomenal life of 32 years, Ramanujan wrote many more formulas which fascinate mathematicians even today [2, 3]. One can consult the detailed work of Ramanujan in his collection of papers compiled by Hardy himself [1].

D. De (✉)
Techno India, Salt Lake, Kolkata 700091, India
e-mail: dpdatju@gmail.com

K. Gaurav Kumar · A. Ghosh · M. K. Naskar
ADESLab, Jadavpur University, Kolkata 700032, India
e-mail: kgauravkumar35@gmail.com

A. Ghosh
e-mail: archismanghosh12@gmail.com

M. K. Naskar
e-mail: mrinaletce@gmail.com

© Springer Nature Singapore Pte Ltd. 2019   391
V. Nath and J. K. Mandal (eds.), *Proceeding of the Second International Conference
on Microelectronics, Computing & Communication Systems (MCCS 2017)*, Lecture Notes
in Electrical Engineering 476, https://doi.org/10.1007/978-981-10-8234-4_34

In 1918, Ramanujan introduced an exponential summation known as Ramanujan Sums (RS) in one of his papers. Because of inherent orthogonal property, RS can be used to obtain convergent finite-duration expressions for many number-theoretic arithmetic functions such as Mobius function $\mu(n)$, Euler's totient function (n), sum of divisors of a number $\sigma(n)$, etc. Besides, owing to the integer property, RS can be used to simplify the computations of Arithmetic Fourier Transform (AFT), Discrete Fourier Transform (DFT), and Discrete Cosine Transform (DCT) coefficients for special type of signals. In recent times, scientists have realized the importance of this concept and extended the domain of its applications from number theory to signal processing.

In 1950, Cohen [4, 5] investigated the even symmetric signals in number theory and observed that the DFT coefficients of this class of signals can be computed by forming integer-valued weighing coefficients of signals. It was later proved that these integer-valued coefficients are nothing but well-known Ramanujan Sums (RS). RS can be expanded to get Ramanujan Fourier Transform (RFT). It is observed in [6] that RFT can achieve faster implementations with less memory requirement than conventional DFT as only specific co-resonant frequency terms are used in its computation. Planat [7] used RFT as a method to analyze low-frequency noise in periodic, quasi-periodic, and complex time series as an alternative to DFT. An RFT-novel approach is presented in [8] to characterize the structural properties of protein in amino acid sequences. The usability of this transform has been expanded by Lagha and Bensebti [9] in estimation of the Doppler spectrum for weather radar signal. The utility of this transform in biomedical signals such as T-wave alternans has been reported by Mainardi et al. [10]. Sugavaneswaran et al. [11] introduced an extension to time–frequency analysis of signals from 1-D RFT into 2-D space. Chen et al. [12] have introduced matrix-based RFT in sparse signal analysis and concluded that RFT provides better signal energy conservation for analysis, similar to DFT. Yin et al. [13] have provided a novel RFT-based alignment-free computational method for comparative analysis of DNA sequences. The research studies showed that in presence of noise, RFT is more efficient in capturing signal periodicities than DFT. These studies provide deeper insights in understanding properties of RFT and opened newer dimensions in applying this concept in different research domains.

RS is a powerful promising technique, having a wide range of applications from number theory, signal processing to biomedical studies. It provides an alternative solution which is error-free and is more signal energy efficient than conventional techniques. It has been useful in finding out patterns and periodicities in signals in presence of noise. Thus, owing to several advantages discussed, RS can prove to be a tremendous boost to VLSI signal processing. The implementation of RS in signal processing can develop faster, less erroneous processors with less complex computations involved.

2 Ramanujan Sum

Ramanujan Sum (RS), a trigonometric summation has the form [14]

$$c_d(n) = \sum_{\substack{k=1 \\ (k,d)=1}}^{d} e^{\frac{j2\pi kn}{d}} = \sum_{\substack{k=1 \\ (k,d)=1}}^{d} W_d^{-kn} \tag{1}$$

Here, the notation (k, d) denotes the greatest common divisor (gcd) of k and d. Thus, $(k, d) = 1$ means that k and d are coprime. For example, for $d = 12$, the coprime values of k are 1, 5, 7, and 11.

$$c_{12}(n) = e^{\frac{j2\pi n}{12}} + e^{\frac{j10\pi n}{12}} + e^{\frac{j14\pi n}{12}} + e^{\frac{j22\pi n}{12}}$$

Ramanujan showed that on adding the complex exponentials $e^{\frac{j2\pi kn}{d}}$ over the coprime values of k and $d(1)$, the result $c_d(n)$ is always real and integer valued. The first few Ramanujan sequences are presented below for one period $0 \leq n \leq d - 1$.

$$
\begin{aligned}
c_1(n) &= 1 \\
c_2(n) &= 1, -1 \\
c_3(n) &= 2, -1, -1 \\
c_4(n) &= 2, 0, -2, 0 \\
c_5(n) &= 4, -1, -1, -1, -1 \\
c_6(n) &= 2, 1, -1, -2, -1, 1 \\
c_7(n) &= 6, -1, -1, -1, -1, -1, -1 \\
c_8(n) &= 4, 0, 0, 0, -4, 0, 0, 0 \\
c_9(n) &= 6, 0, 0, -3, 0, 0, -3, 0, 0 \\
c_{10}(n) &= 4, 1, -1, 1, -1, -4, -1, 1, -1, 1 \\
c_{11}(n) &= -1, -1, -1, -1, -1, -1, -1, -1, -1, -1, 10 \\
c_{12}(n) &= 0, 2, 0, -2, -4, 0, -2, 0, 2, 0, 4
\end{aligned}
\tag{2}
$$

Relation between RS and primitive roots
RS $c_d(n)$ can be related to primitive roots. It is defined as sum of nth powers of all the dth primitive roots of unity as W_d^{-k} is the primitive dth root of unity if and only if $(d, k) = 1$. It is known that α is the dth root of unity if $\alpha^d = 1$. α is called the dth primitive root of unity if $\alpha^d = 1$, but $\alpha^n \neq 1$ for positive integer $n < d$.

Relation between RS and Mobius Function

Ramanujan established following relation between $c_d(n)$ and Mobius function $\mu(n)$ to prove that $c_d(n)$ are integer valued. Ramanujan showed [14] that

$$c_d(n) = \sum_{q|(d,n)} \mu\left(\frac{d}{q}\right) q \tag{3}$$

Another important relation between $c_d(n)$ and $\mu(n)$, provided by Hardy [15] is

$$c_d(n) = \frac{\mu(m)\psi(d)}{\psi(m)}, \quad m = \frac{d}{(d,n)}, \quad \psi(.) \text{ is Euler.} \tag{4}$$

The Mobius function and Euler's totient function are explained in Appendix of the chapter. The above formulas are obtained by expressing RS in Dirichlet convolution form and performing Mobius inversion on them.

The motivation behind introducing RS was to show that several arithmetic functions in number theory can be expressed as linear combinations of $c_d(n)$, i.e.,

$$x(n) = \sum_{d=1}^{\infty} x_d c_d(n) \tag{5}$$

An arithmetic function is an infinite sequence defined for $1 \le d < \infty$ and is usually integer valued. x_d are nothing but the coefficients of Ramanujan Transform. For example, let us consider the function $\sigma(n)$, which calculates the sum of all the divisors of n, including 1 and n for any positive number n. As an example, $\sigma(6) = 1 + 2 + 3 + 6 = 12$ can be found. RS can be used to expand $\sigma(n)$ as infinite series [14]

$$\sigma(n) = \frac{n\pi^2}{6} \sum_{d=1}^{\infty} \frac{c_d(n)}{d^2}, \quad n \ge 1 \tag{6}$$

Many amazing expansions of number-theoretic arithmetical functions are present in [14].

3 Methods to Calculate RS in Signal Processing

RS has diverse applications in signal processing but there are few ways to compute RS values. Literature survey shows that mainly three methods exist for computing the values of RS. These include (i) calculating RS values by definition, (ii) using an expression comprising number-theoretic Mobius function and Euler's totient function [16, 17], and (iii) a Z-transform method [6]. These methodologies can be broadly classified into two types, i.e., recursive and non-recursive method. Recursive method calculates RS values recursively following a set of rules, such as calculating RS by definition. Non-recursive techniques use a straightforward expression to compute RS values, such as calculating RS by Mobius and Euler's totient expression. The Z-transform method is proposed by Samadi, where the RS values are used to calculate DFT coefficients of even symmetric periodic signals.

Recursive Method

$c_d(n)$ can be computed recursively in the following manner: if $c_d(n)$, for $d < D$ is known, a combination of these RS values can be used to compute $c_D(n)$. Starting with $c_1(n) = 1$, $c_d(n)$ for $d > 1$ can be recursively found by simple manipulations of previously computed RS values. Calculating RS values by definition is an example of recursive method.

The recursive expression for calculating RS from lower order RS is [15].

$$c_d(n) = d\delta((n))_d - \sum_{\substack{d_k|d \\ d_k < d}} c_{d_k}(n), \quad \text{where } d_k|d \text{ implies that } d_k \text{ is divisor of } d.$$

(7)

$$\delta((n))_d = 1 \quad \text{if} \quad n = 0 \mod d \quad \text{and} = 0 \quad \text{otherwise.}$$

Equation (7) gives us a recursive technique to compute $c_d(n)$ starting with $c_1(n) = 1$. For example, $c_{12}(n) = 12\delta((n))_{12} - c_1(n) - c_2(n) - c_3(n) - c_4(n) - c_6(n)$. $c_3(n)$, $c_4(n)$, and $c_6(n)$ can in turn be calculated recursively. Only simple additions and subtractions are involved in this technique.

Non-recursive Methods

In addition to recursive method, there exists a non-recursive method for obtaining the RS values. Non-recursive method uses straightforward expressions to directly compute RS values. Generally, this method is preferred over recursive one. Calculating RS using Mobius function and Euler's totient function and Z-transform methods fall into this category.

RS using Mobius function and Euler's totient function

This non-recursive method of calculating RS is widely used. This method [5] utilizes the relationship between RS and Mobius function to calculate RS values. Equation (3–4) shows the relation between the two. Mobius function $\mu(n)$ is a number-theoretic function, defined in Appendix of this chapter. For example, $\mu(10) = 1$; $\mu(5) = -1$. Euler's totient function $\psi(d)$ stores the number of positive integers which are coprime to d and less than or equal to d. It is defined in Appendix of the chapter. For example, (10) = 4; $\psi(5) = 4$. This method seems simple but Mobius and Euler's totient functions demand a priori knowledge of factorization of the period of the RS. It is difficult to estimate the complexity of factorization. Thus, this method is not simple enough for hardware programming applications.

The architecture for calculating RS using the above method is shown in Fig. 1. In this figure, the block-level architecture is presented. Instead of computing

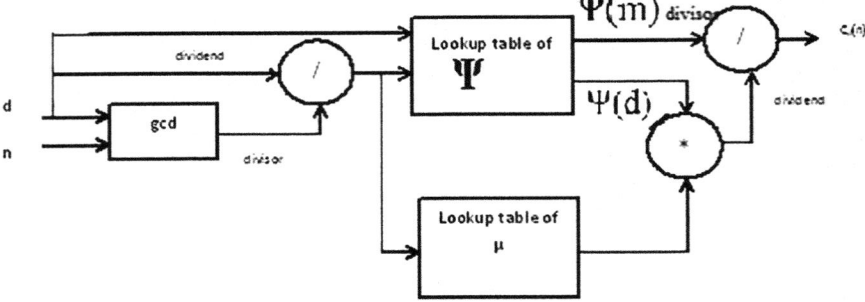

Fig. 1 Architecture for calculating RS

Mobius function and Euler's totient function, Look-Up Tables (LUTs) are used to store their values and access them when required.

RS using Z-transform method

Samadi and his colleagues proposed the Z-transform method [6] to calculate one-sided Z-transform of RS for even periodic signal, which is given by $C_d(z) = \sum_{n=0}^{\infty} c_d(n)z^{-n}$, where $C_d(z)$ is one-sided Z-transform of RS $c_d(n)$.

$$C_d(z) = \frac{z^{1-\psi(d)} \frac{d}{dz} F_d(z)}{F_d(z^{-1})} \tag{8}$$

Here, $\psi(d)$ is the Euler's totient function. $F_d(z)$ is the dth cyclotomic polynomial defined as

$$F_d(z) = \prod_{\substack{(U,d)=1 \\ 0 \le U \le d-1}} (z - W_d^{-U}) \tag{9}$$

W_d^{-k} is the primitive root of unity. The proof of these expressions is given in [6]. Using the above equations, $C_{10}(z)$ is calculated as

$$C_{10}(z) = \frac{4 - 3z^{-1} + 2z^{-2} - z^{-3}}{1 - z^{-1} + z^{-2} - z^{-3} + z^{-4}}$$

For computing different $C_d(z)$, different expressions are to be derived depending upon the cyclotomic polynomial $F_d(z)$. Thus, this method is not generic and is suitable for computing RS values with fixed period d.

4 FIR Representations of RS

The main idea behind RS is to express number-theoretic functions as a linear combination of the form shown in (5). This is called Ramanujan sum expansion of $x(n)$. The coefficients x_d can be calculated by the following formula [7, 18, 19]:

$$x_d = \frac{1}{\psi(d)} \left(\lim_{M \to \infty} \frac{1}{M} \sum_{n=1}^{M} x(n) c_d(n) \right) \tag{10}$$

Here, $\psi(d)$ is the Euler's totient function. The arithmetic functions defined by (5) are infinite-duration sequences and the coefficients can be calculated by the above formula.

For FIR representations of RS [20], the sequences need to be finite duration with $x(n) = 0$ for all n, except in the range $1 \le n \le N$, the limit becomes

$$\lim_{M \to \infty} \sum_{n=1}^{M} \frac{x(n) c_d(n)}{M} = \lim_{M \to \infty} \sum_{n=1}^{N} \frac{x(n) c_d(n)}{M} \to 0 \tag{11}$$

which shows that $x_d \to 0$ for each d. Thus, the conventional approach (10) fails to provide the correct expansion of (5).

Another mathematical approach has been considered to prove the inefficiency of conventional method. Let us consider an FIR sequence $x(n)$ with one period of periodic signal, i.e.,

$$x(n) = x(n+N) \tag{12}$$

For fixed d, the limit in (10) can be expressed as $\lim\limits_{k \to \infty} \sum\limits_{n=1}^{kdN} \frac{x(n) c_d(n)}{kdN}$. As $x(n)$ and $c_d(n)$ have been defined with periods N and d, respectively, $x(n) c_d(n)$ repeats after every Nd samples. So the above limit equals $\sum_{n=1}^{dN} \frac{x(n) c_d(n)}{dN}$, and

$$x_d = \frac{1}{dN\psi(d)} \sum_{n=1}^{N} x(n) c_d(n) \tag{13}$$

Using (12), the expression (1) for $c_d(n)$,

$$\sum_{n=1}^{N} x(n) c_d(n) = \sum_{n=1}^{N} x(n) \sum_{i=0}^{d-1} c_d(n + iN)$$

$$= \sum_{n=1}^{N} x(n) \sum_{\substack{k=1 \\ (k,d)=1}}^{d} W_d^{kn} \sum_{i=0}^{d-1} W_d^{ikn}$$

As $(k, d) = 1$, the inner sum is nonzero only when N is a multiple of d. Hence, $x_d = \frac{1}{N\psi(d)} \sum_{n=1}^{N} x(n)c_d(n)$, where d is divisor of N and is zero otherwise. The representation of N-point FIR sequence is thus, given by

$$x(n) = \sum_{d_i|N} x_{d_i} c_{d_i}(n), \quad 1 \le n \le N \tag{14}$$

Here, $d_i|N$ implies that d_i are divisors of N and where

$$x_{d_i} = \frac{1}{N\psi(d_i)} \sum_{n=1}^{N} x(n)c_{d_i}(n) \tag{15}$$

The matrix form of the above representation is

$$\mathbf{x} = \begin{bmatrix} c_{d_1} & c_{d_2} & \cdots & \cdots & c_{d_\emptyset} \end{bmatrix} \mathbf{q} \tag{16}$$

where $\emptyset$ is the number of divisors of N, and $\mathbf{q}$ is the column vector of elements x_{d_i}. For example, if $N = 12$, the set of divisors d_i is $\{1, 2, 3, 4, 6, 12\}$, so $x(n)$ can be expressed as $x(n) = x_1 c_1 + x_2 c_2 + x_3 c_3 + x_4 c_4 + x_6 c_6 + x_{12} c_{12}$. Thus, using this approach, any arbitrary FIR sequence of length N cannot be represented as the number of coefficients is less than number of samples in the signal.

Another way to successfully represent FIR sequence using RS by slightly changing our conventional approach is given in [20]. Let us consider the following expansion:

$$x(n) = \sum_{d=1}^{N} x_d c_d(n), \quad 0 \le n \le N - 1 \tag{17}$$

Here, the N sequences of $c_d(n)$ are used. In matrix vector form, Eq. (17) can be expressed as

$$\begin{bmatrix} x(0) \\ x(1) \\ \vdots \\ \vdots \\ x(N-1) \end{bmatrix} = C_N \begin{bmatrix} x_1 \\ x_2 \\ \vdots \\ \vdots \\ x_{N-1} \end{bmatrix} \tag{18}$$

Here, $\mathbf{C_N}$ denotes the vector comprising the RS values such that the dth column of $\mathbf{C_N}$ contains $c_d(n)$ repeated with period d unless N terms are obtained. For example, using Ramanujan sequences defined in (2), C_5 can be expressed as

$$C_5 = \begin{vmatrix} 1 & 1 & 2 & 2 & 4 \\ 1 & -1 & -1 & 0 & -1 \\ 1 & 1 & -1 & -2 & -1 \\ 1 & -1 & 2 & 0 & -1 \\ 1 & 1 & -1 & 2 & -1 \end{vmatrix}$$

The Ramanujan expansion, in Eq. (18), can be applied for any FIR sequence as $N \times N$ matrix (C_N) always have full rank. The proof of its full rank is given in [20]. Although the columns of C_N form integer basis for $\mathbb{C}^N$, many of the columns do not contain integer number of periods of $c_d(n)$. Thus, the columns of the matrix are not orthogonal. But any N-point FIR sequence can be represented by this form and is called *first Ramanujan FIR representation*. It can be shown that due to non-orthogonal nature of C_N, it is impossible to find out the period (periodic components) of unknown signal waveform using this above representation (17) and (18). The matrix C_N has elements $c_{d_i}(n)$ repeated with period d_i until there are N elements in the ith column. A new technique of representation is introduced which instead of having only one column for each d_i, considers using $\psi(d)$ circularly shifted versions, i.e., for each d_i, the vector G_{d_i} can be defined as

$$G_{d_i} = \begin{bmatrix} c_{d_i} & c_{d_i}^{(1)} & \cdots & \cdots & c_{d_i}^{(\psi(d)-1)} \end{bmatrix} \tag{19}$$

where $c_{d_i}^{(k)}$ denoted circularly shifting by k. This space spanned by $c_d(n)$ is called Ramanujan Subspace S_d, which contains period-d signals with dimension $\psi(d)$. This subspace captures much broader kind of period-d signals and has an integer basis comprising $c_d(n)$ and its shifted components. By combining finite number of such periodic subspaces (19) to form a composite matrix F_N, a finite-duration signal (FIR) can be represented.

$$F_N = \begin{bmatrix} G_{d_1} & G_{d_2} & \cdots & \cdots & G_{d_K} \end{bmatrix} \tag{20}$$

Here, K is the number of divisors of N. It is known [20] that the sum of the Euler's totients, over all the divisors of N, is exactly equal to N, i.e.,

$$\sum_{d_i|N} \psi(d_i) = N \tag{21}$$

The matrix vector representation of N-point FIR sequence using this method is shown below:

$$x = F_N g \tag{22}$$

where x is a $N \times 1$ vector, F_N is a $N \times N$ vector described above, and g are coefficients of expansion. This is called *Ramanujan Periodic Representation* (**RPR**).

The matrix vector form (22) can be further expressed as

$$x(n) = \sum_{d_i|N} \sum_{l=0}^{\psi(d_i)-1} \beta_{il} c_{d_i}(n-l) \tag{23}$$

where $c_{d_i}(n)$ are RS, and this representation is called *second Ramanujan FIR representation*. Using Eq. (24), the total number of terms in the double summation is equal to N. The inner sum is nothing but the Ramanujan subspace S_{di}, i.e., the subspace spanned by $\psi(d_i)$ columns of G_{d_i}. Since the column spaces of G_{d_i} are orthogonal for different d_i, the components x_{d_i} are orthogonal projections of "$x(n)$" onto "K" Ramanujan subspaces. Therefore, any FIR sequence can be decomposed into K Ramanujan subspaces. DFT also has a similar property but the coefficients of expansions in this form, i.e., matrices C_N and F_N, are real integer valued, which makes computation easy and facilitates quantization error-free calculations as simple additions and subtractions are involved. The corresponding transformation of RPR from $\mathbf{x}$ to $\mathbf{g}$ is called *Ramanujan Periodicity Transform* (**RPT**) and is used to find out the hidden periodicities in an unknown waveform effectively unlike the previous two representations.

$$\mathbf{g} = F_N^{-1}\mathbf{x} \tag{24}$$

5 Applications

5.1 Ramanujan Fourier Transform

Ramanujan Fourier Transform (RFT) is first introduced by Planat [7] in 2002. In this paper, a frequency domain signal, RFT, based on Ramanujan Sums was proposed. For an N-point input sequence $x(n)$, RFT is defined by

$$x(n) = \sum_{d=1}^{\infty} x_d c_d(n)$$

Inverse RFT is given by the formula

$$x_d = \frac{1}{\psi(d)} \lim_{M \to \infty} \frac{1}{M} \sum_{n=1}^{\infty} x(n) c_d(n)$$

$1/\psi(d)$ is used to normalize the norm of basis RS. It is quite similar to DFT and the frequency domain samples are called Ramanujan Spectrum. These equations are defined earlier as (5) and (10), but to have continuity of the topic they are discussed again.

Another way to calculate RFT coefficients is the Matrix method. In 2013, Chen et al. [12] introduced matrix method to calculate RS transform values where 1-D and 2-D forward and inverse RS transforms are represented by means of matrix multiplication. Let us define a matrix for a signal with period M

$$A(d,j) = \frac{1}{\psi(d)M} c_d(\mathrm{mod}(j-1,d)+1) \tag{25}$$

Here, d and j belong to $[1, M]$. The input signal $x(n)$ can be

$X = (x(1), x(2), \ldots, x(M))^{\mathrm{T}}$, where T denotes the transpose of vector. The forward 1-D RS of signal X can be given by

$$Y = AX \tag{26}$$

where $Y = (y(1), y(2), \ldots, y(M))^{\mathrm{T}}$ contains the RFT coefficients for signal X. The inverse 1-D RS can be realized by

$$X = A^{-1}Y \tag{27}$$

Similarly, the 2-D forward and inverse RS can be calculated by

$$Y = AXA^{\mathrm{T}} \quad \text{and} \quad X = A^{-1}Y(A^{-1})^{\mathrm{T}} \tag{28}$$

The mathematical proofs and descriptions are present in [12]. A number of functions $(x(n))$ defined in the Appendix and their corresponding RFTs x_d are tabulated in Table 1.

In [7], Planat et al. introduced RFT and applied it in time–frequency analysis. It was observed that RFT works better than DFT, which is discussed in later part of the chapter. In [21–26], some theoretical aspects of Ramanujan Sums are discussed. In [6, 27], a special class of signals, i.e., even symmetric and odd-symmetric signal, is introduced, and DFT coefficients of the same are computed using RS.

In [6], Samadi et al. introduced a special class of signal known as even symmetric signal. An even symmetric signal $x_r(n)$ with period "r" and time index "n" is defined as

Table 1 Number-theoretic functions and their corresponding RFT

Number-theoretic function $x(n)$	RFT x_d
$\frac{\sigma(n)}{n}$	$\frac{\pi^2}{6}\frac{1}{d^2}$
$\frac{\psi(n)}{n}$	$\frac{6}{\pi^2}\frac{\mu(d)}{\psi_2(d)}$
$b(n)$: $\frac{\psi(n)\wedge(n)}{n}$	$\frac{\mu(d)}{\psi(d)}$
$C(n)$	$\left(\frac{\mu(d)}{\psi(d)}\right)^2$

$$x_r(n) = x_r(\gcd(n,r)), \quad \text{for } \forall n \tag{29}$$

where $\gcd(p, q)$ denotes the greatest common divisor of p and q. Thus, for an integer "m",

$$\begin{aligned} x_r(n + mr) &= x_r(\gcd(n + mr, r)) = x_r(\gcd((n + mr)\bmod r,)) \\ &= x_r(\gcd((n \bmod r),)) = x_r(n) \end{aligned} \tag{30}$$

Further, it can be shown that a symmetric signal with period r follows

$$x_r(n) = x_r(r - n)$$

If even signal is symmetric with respect to mod r, then it must be symmetric with respect to mod mr, i.e., $x_r(n) = x_r((n + r)r)$ implies $x_r(n) = x_r((n + mr) \bmod r)$

Using these properties, an even symmetric signal can be represented. For example, with period $(r) = 10$, the signal $x_r(n)$ can be expressed as

$$\langle x_r(n), n = 0, \ldots, 9 \rangle = \langle a, b, c, b, c, d, c, b, c, b \rangle \tag{31}$$

Similarly, for $r = 20$

$$\langle x_r(n), n = 0, 1, 2, \ldots 19 \rangle = \langle a, b, c, b, d, e, c, b, d, b, f, b, d, b, c, e, d, b, c, b \rangle \tag{32}$$

In [6], the concept of DFT using RS for even symmetric signal is introduced. The DFT sample at Kth frequency for even symmetric signal $x_r(n)$ is given by

$$X_r(K) = x_r(0) + \sum_{\substack{d|r \\ d>1}} x_r\left(\frac{r}{d}\right) c_d(K), D > 1, D|r \tag{33}$$

Authors of [28] have called it RFT and proposed hardware architecture for the same. The proposed architecture is implemented in FPGA and is compared with conventional DFT. It is observed that this method is faster and consumes less hardware resources as compared to conventional DFT methods. Detailed methodologies and implementations have been presented in [28]. The introduction of RS in VLSI architectures will yield faster low power consuming processors which require lesser hardware resources. Figures 3 and 4 testify the claim. The block-level architecture of DFT using RS is shown in Fig. 2.

The proposed architecture for DFT using RS for even signals [6] has been coded in Verilog HDL and implemented on Vertix-5 FPGA kit. The simulator used is Xilinx ISE. The observations are noted for different periods of signals and are plotted as histogram in Figs. 3 and 4.

The proposed architecture has some disadvantages. The maximum possible number of distinct signal values is $\prod_{p^{\text{prime}}} (m_p + 1)$ if $r = \prod_{p^{\text{prime}}} p^{m_p}$. For $r = 20$, only

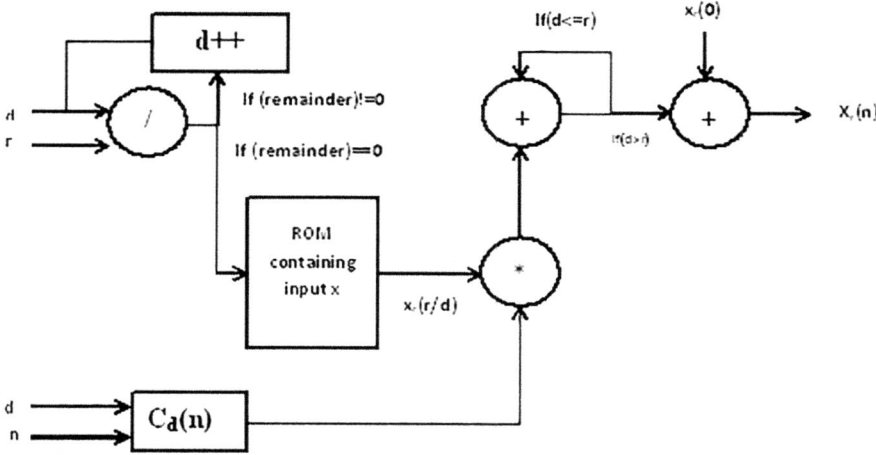

Fig. 2 Architecture for DFT using RS

Fig. 3 Computation time versus technique used

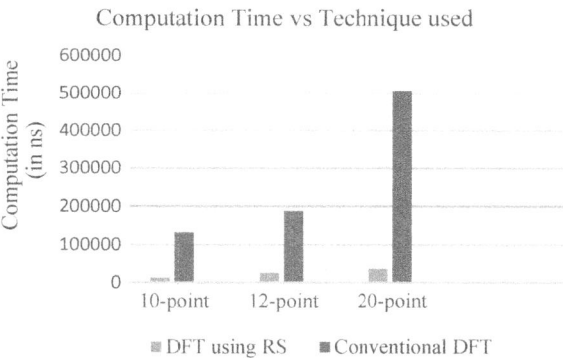

Fig. 4 Device resource utilization versus technique used

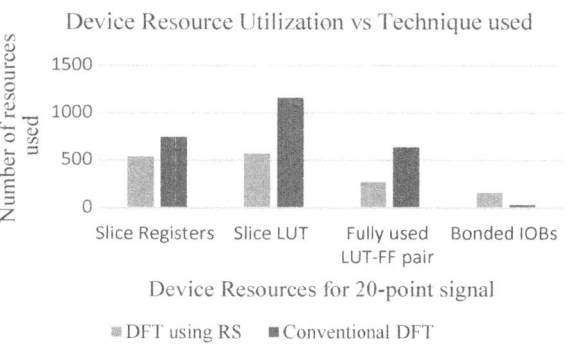

six distinct values are obtained. To increase the number of distinct values and to make this more powerful, Pei et al. [27] introduced a special class of odd-length $4N$ periodic symmetric signals and showed how the odd Ramanujan Sums can be used as weighting coefficients to compute their pure imaginary DFT integer-valued coefficients. A signal $x_r(n)$ is called an odd signal (mod r) if

$$
\begin{aligned}
x_r(n) &= x_r(\gcd(n,r)), \quad \text{if} \quad \frac{r}{\gcd(n,r)} = 0 \text{ mod } 4 \text{ and } \frac{n}{\gcd(n,r)} = 1 \text{ mod } 4 \\
&= -x_r(\gcd(n,r)), \quad \text{if} \quad \frac{r}{\gcd(n,r)} = 0 \text{ mod } 4 \text{ and } \frac{n}{\gcd(n,r)} = 3 \text{ mod } 4 \\
&= 0, \quad \text{elsewhere}
\end{aligned}
$$

(34)

From the above definition, it is clear that this type of signal is periodic with periodicity r, i.e., $x_r(n + kr) = x_r(n)$. Moreover, this type of signal possesses odd symmetry, i.e., $x_r(r - n) = -x_r(n)$. The proof is in [27].

Using all these properties, any sequence type of $4N$ number can be derived. The value of signal $x_8(n)$ in its main period may be represented as

$$\langle x_8(n), n = 0, 1, \ldots 7 \rangle = \langle 0, p, q, -p, 0, p, -q, -p \rangle \tag{35}$$

Similarly for $r = 12$ and 20, similar types of sequences are obtained in its main period.

$$\langle x_{12}(n), n = 0, 1, ..11 \rangle = \langle 0, p, 0, q, 0, p, 0, -p, 0, -q, 0, -p \rangle \tag{36}$$

$$\langle x_{20}(n), n = 0, 1, ..19 \rangle = \langle 0, p, q, -p, 0, r, 0, -p, 0, s, 0, -s, 0, p, 0, -r, 0, p, -q, -p \rangle$$

(37)

Generally for a given $r = 4 \prod_{p^{\text{prime}}} p^{m_p}$, the number of distinct values in input sequence is $\prod_{p^{\text{prime}}} (m_p + 1)$. Now, using this type of sequence, it can be shown that this Ramanujan Sums is of odd type. The proof of this statement is given in [27].

In [27], Pei et al. showed that how these even and odd signals can be combined to increase the number of distinct values. One method is used to get odd signal is circular shift method where a sequence is circularly shifted by $\frac{r}{4}$ times of even sequence of same dimension. By simply adding these, new sequences are obtained that contain more number of distinct integers. Pei et al. showed that instead of using this type of odd sequence, if the proposed new type of odd sequence is used, equal or more number of distinct integers are obtained than that found by adding the even and odd counterpart.

For example, adding (32) and (37), a new sequence having 12 distinct values can be represented as

$$\langle z_{12}(n), n = 0, 1, 2, \ldots 19 \rangle = \langle a, p+b, q+c, b-p, d, r+e, c, b-p, d, s+b, f, b$$
$$-s, d, p+b, c, e-r, d, b+p, c-q, b-p \rangle \tag{38}$$

Using circular shift method, the resulting odd signal is represented as

$$\langle y_r(n), n = 0, 1, 2, \ldots 19 \rangle = \langle a', b', c', b', d', e', c', b', d', b',$$
$$f', b', d', b', c', e', d', b', c', b', a', b', c', b', d' \rangle \tag{39}$$

Adding Eqs. (32) and (39), new sequence having eight distinct values can be obtained. The above example exemplifies the fact that the odd–even method of [27] has more distinct basis than circular shift method. A rigorous mathematical proof is given in [27]. In practice, it is convenient to decompose the original signal into odd–even signal than using circular shift method.

5.2 Ramanujan Sums in Determining Periodicities in Signals

The RPT, i.e., Ramanujan periodic transform, discussed above, has undue advantage over the conventional and *First Ramanujan FIR representation*. It results in sparse representation which further helps to identify the periodicity of $x(n)$ accurately as shall be demonstrated. The RPR and RPT assist in identifying the exact periodicity of the waveform. It is known that (i) each of the orthogonal projection $x_{d_i}(n)$ has period equal to d_i. (ii) If the nonzero projections are represented as $x_{d_{i_k}}(n)$ for $1 \leq k \leq L(L \leq K)$. Then, periodicity of $x(n)$ is the least common multiple (l.c. m.) of these d_{i_k}s. Hence, the second Ramanujan FIR representation enables us to find out the periodicity of the signal by simply identifying the nonzero projections and taking l.c.m. of these projection values. The second Ramanujan FIR representation can be applied to identify a single period in a signal, multiple hidden periodic components in a waveform, denoising a noisy periodic signal, etc. These applications, with examples, are illustrated in [20].

Example

Considering the second FIR representation, RS can be applied to find out the single period in FIR signal. For further examples, one can see [20]. Figure 5 shows an FIR signal $x(n)$ with $N = 256$ points and periodic with $d = 64$ points (a divisor of N). Considering the *second Ramanujan FIR representation* (25–26), the projections and the energies of projection can be computed and plotted for different N. Figure 6 shows the plot of the coefficients of expansion (gk), and Fig. 7 shows the energy of projections, i.e., $x_{d_i}(n)$. The l.c.m. of the divisors with projection peaks gives the required period of the signal. Figure 8 shows how the first Ramanujan

Fig. 5 FIR signal with
$N = 256$ points and
periodicity 64 points

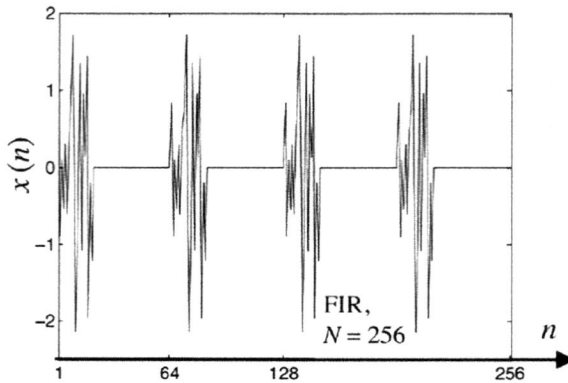

Fig. 6 Coefficient of
expansion ($\mathbf{g_k} = \mathbf{b_k}$) in second
Ramanujan FIR
representation

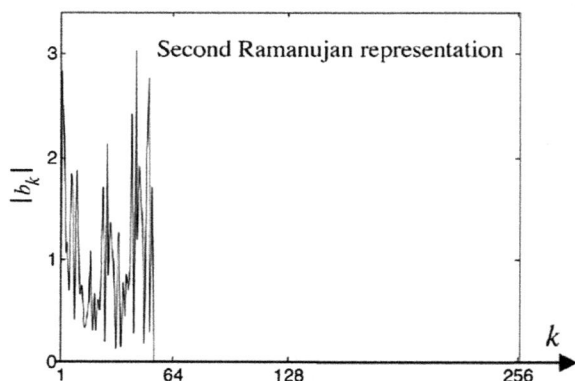

Fig. 7 Plot of projection
energies using second
Ramanujan representation

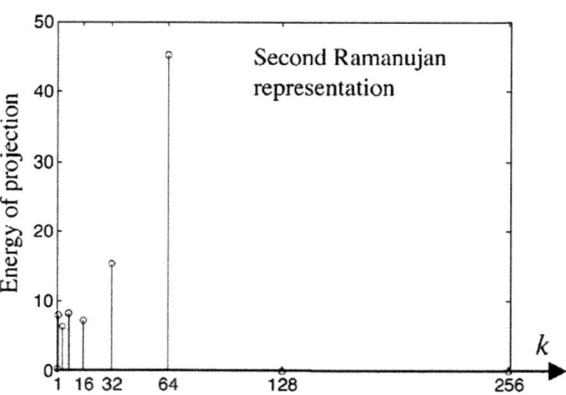

Fig. 8 Coefficient of
expansion in first
Ramanujan FIR
representation [20]

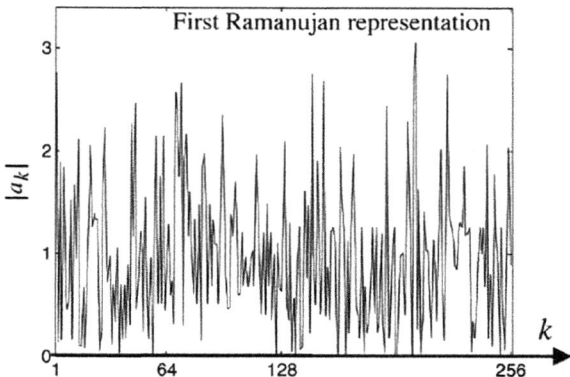

representation fails in this process. The DFT-based technique also results in the same period but the ease of computation in handling integer-based transformation gives RS-based solution an upper hand.

5.3 Ramanujan Sums in Time–Frequency Analysis

Any signal of interest can be expressed as a weighted sum of orthonormal basis functions. The selection of basis functions is done based on the signal type and the corresponding applications. Weights or coefficients are then calculated using the orthogonal relationship among the basis functions. Here comes the need of Ramanujan Sums. It is previously discussed in detail that Ramanujan Sums have orthogonal property for which it can be used as an element in time–frequency analysis. In [11], Sugavaneswaran et al. showed this clearly and introduced a novel class of RFT-based time–frequency transforms, constituted by Ramanujan Sums basis. Through the time–frequency analysis, simultaneous access of time and frequency information can be done which is not possible either by time domain signal or by its frequency spectrum where information can be obtained only with respect to time or with respect to the frequency. In this paper, the authors used Wigner–Ville distribution [29] function for the time–frequency analysis.

Figure 9 depicts the overall architecture for the time–frequency analysis through Ramanujan Sums [11]. Here, the analytic representation is obtained by first taking the Hilbert transform of the input signal $x(t)$. With the help of the Ambiguity Function (AF), the analytic function is then mapped onto the Ambiguity Domain (AD), where AF is a time-varying autocorrelation function which is defined as below:

$$A(\emptyset, \tau) = \int x\left(t - \frac{\tau}{2}\right)x^*\left(t - \frac{\tau}{2}\right)e^{-j\emptyset t}dt \qquad (40)$$

where $x^*(t)$ is the complex conjugate of the signal $x(t)$. Using the autocorrelation function as defined in (40), a signal can be mapped on to the AD. The reason of mapping the signal on AD is that, in this domain, the signal terms appear localized around the origin, and the cross-terms are located away from the origin [30]. In their work, they used a modified AF representation computed by using the RS (on the time-varying autocorrelation values) as

$$\grave{A}(q, \tau) = \int x\left(t - \frac{\tau}{2}\right)x^*\left(t - \frac{\tau}{2}\right)c_q(t)dt \qquad (41)$$

Such a computation results in an overall reduction in the number of AD coefficients. These AD-mapped coefficients can then be transformed into a sparser time–frequency representation, using the RFT computed from their characteristic RS.

There are many reasons for the growing interests in RS. The following points outline this [11]

1. RFT can be applied on the entire signal length where DFT can be used on signals of limited length.
2. TF-RFT has the capability to capture the dominant periodicity of a signal of interest.
3. From computational complexity perspective, RFT is better than AF and DFT as shown in Table 2.
4. Apart from these, RFT provides better noise immunity over AF and DFT.

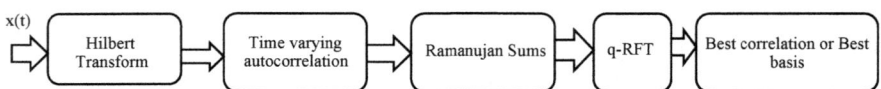

Fig. 9 Overall architecture for the proposed class of transforms [11]

Table 2 Computational complexity of various algorithms in time–frequency analysis [11]

Domain	Complexity	Comments
Input	$O(N)$	$N \rightarrow$ signal length
Ambiguity Domain (AD)	$O(N \times N)$	Can be reduced up to $O(N \times N/2)$ due to the symmetric nature
FFT	$O(N \log_2 N)$	On need basis Fourier computation window length can be reduced
RFT	$O(Q \log_2 Q)$	$Q = \gcd(q_r, q_c)$ and $Q < N$

5.4 Application of Ramanujan Sums in Low-Frequency Noise or 1/f Noise Analysis

The unavailability of nonstatistical model of 1/f noise due to its randomness has motivated M. Planat et al. to work on it, where he used an arithmetical approach [29]. In their work [31], it is shown that using RS, arithmetical functions can be analyzed better as RFT has the ability to extract quasi-periodic features and fine aperiodic features (low-frequency signal) that the DFT fails to show [7, 14, 15, 32]. As RS analysis uses the properties of irreducible functions, it is favorable to analyze rich time series signal showing a $1/f^{\alpha}$ $(0 < \alpha < 2)$ FFT dependence. The authors [31] established their claim by showing two examples: first on the data from the stock market (for which the price index FFT follows a $1/f^2$-law) and in the second one on the data obtained from solar cycle activity (for which the coronal index follows a $1/f$-law). Thus, RFT proved to be a magnifying glass for analyzing low-frequency signal and $1/f$ noise.

6 Conclusions

It has been observed how a deep, profound mathematical theory finds applications in engineering science. RS has been introduced as a number-theoretic concept to express arithmetical functions but it finds applications in diverse areas, which include signal processing, information theory, spectrum estimation, coding theory, and even biomedical areas. It is concluded that the concept of RFT and RPT using RS provide an efficient solution as compared to conventional DFT. Further, RPT identifies hidden periodicities in an unknown signal for which conventional methods like DFT fails. Another advantage of RFT and RPT is that they work on integer basis, making them easy to handle than the complex real coefficients of DFT. It is because of this integer basis of RS, the hardware implementation becomes easier and less error-prone as handling of integers requires less word length and involves negligible quantization error. Moreover, the hardware cost is less, and the time required for computation is less as real integers are dealt. The hardware realizations of architectures based on RS will be the topic of interest in the field of VLSI in future owing to the stated advantages of RS. In VLSI area, RS-based architectures provide better results with less power consumption as compared to conventional models. Since RS has diverse applications, the realized hardware can be reconfigured to be used in different applications. Besides these, being such a powerful mathematical technique, research on finding and applying the concept of RS in newer areas of interest is underway.

Appendix

1. **σ(n) {Sum of divisors}**
   The function $\sigma(n)$ is the sum of positive divisors of n, i.e., $\sigma(n) = \sum d$ if $d|n$.
2. **(n) {Euler's totient function}**
   The function (n) is defined as the number of positive integers which are less than and coprime with n. For example, $(6) = 2$ since $\{1, 5\}$ are the only two positive integers which are less than and coprime with 6.

$$\psi(n) = n \prod_i \left(1 - \frac{1}{n_i}\right) \text{ Such that } n = \prod_i n_i^{\alpha_i} \text{ where } (\alpha_i) \text{ is prime.}$$

$$\psi_2(n) = n^2 \prod_i \left(1 - \frac{1}{n_i^2}\right)$$

3. **μ(n) {Mobius function}**
   Mobius function $\mu(n)$ is a number-theoretic function and is defined as

$$\begin{aligned}
\mu(n) &= 1 && \text{if } n = 1 \\
&= (-1)^k && \text{if } n = p_1 p_2 \ldots p_k \\
&= 0 && \text{otherwise}
\end{aligned}$$

Here, p_i are distinct prime numbers.

4. **Λ(n) {von Mangoldt function}**

$$\begin{aligned}
\wedge(n) &= \{\ln p && \text{if } n = p^\beta, p \text{ is prime} \\
&= && \text{Otherwise}
\end{aligned}$$

5. **C(n)**
   The function $C(n)$ is defined as

$$\begin{aligned}
C(n) &= 2C_2 \prod_{p|n} \frac{p-1}{p-2}, && \text{if } n \text{ is odd} \\
&= 0, && \text{if } n \text{ is even}
\end{aligned}$$

Here, $p > 2$ is a prime and $p|n$ implies p divides n. The value of twin prime constant, C_2 is 0.660

References

1. Hardy G. H., Seshu Iyer P. V., and Wilson B. M., Collected papers of Srinivasa Ramanujan, Cambridge University Press, London, 1927.
2. Berndt B. C., Ramanujan's notebooks, Springer-Verlag, Inc., N. Y., 1991.

3. Andrews G. E., Berndt B. C., Ramanujan's lost notebook, Springer, N. Y., 2005.
4. Cohen E., "A class of arithmetical functions," in Proc. Nat. Acad. Sci. U.S.A., vol. 41, 1955, pp. 939–944.
5. Cohen E., "Representations of even functions (mod r). I. Arithmetic identities," Duke Math. J., vol. 25, pp. 401–421, 1958.
6. Samadi S., Ahmad M. O., Swamy M., Ramanujan sums and discrete fourier transforms, IEEE Signal Processing Letters 12 (4) (2005) pp. 293–296.
7. Planat M., Ramanujan sums for signal processing of low frequency noise, in: Frequency Control Symposium and PDA Exhibition, 2002. IEEE International, IEEE, 2002, pp. 715–720.
8. Mainardi L., Pattini L., Cerutti S., 2007. Application of the Ramanujan Fourier transform for the analysis of secondary structure content in amino acid sequences. Methods Inf Med 46, pp. 126–129.
9. Lagha M., Bensebti M., 2009. Doppler spectrum estimation by Ramanujan Fourier transform (RFT). Digital Signal Processing 19, pp. 843–851.
10. Mainardi L., Bertinelli M., Sassi R., 2008. Analysis of t-wave alternans using the Ramanujan transform, in: Computers in Cardiology, 2008, IEEE. pp. 605–608.
11. Sugavaneswaran L., Xie S., Umapathy K., and Krishnan S., "Time frequency analysis via Ramanujan sums," IEEE Signal Processing Letters, vol. 19, pp. 352–355, June 2012.
12. Chen G., Krishnan S., Bui T. D., "Matrix Based Ramanujan Sums Transforms", IEEE Signal Processing Letters, vol. 20, No. 10, pp. 941–944, October, 2013.
13. Yin C., Yin X.E., Wang J., " A Novel Method for comparative analysis of DNA sequences by Ramanujan Fourier transform", https://doi.org/10.1089/cmb.2014.0120.
14. Ramanujan S., "On certain trigonometrical sums and their applications in the theory of numbers," Trans. Cambridge Philosoph. Soc., vol. XXII, no. 13, pp. 259–276, 1918.
15. Hardy G. H. and Wright E. M., An Introduction to the Theory of Numbers. New York, NY, USA: Oxford Univ. Press, 2008.
16. Hardy G. H., "Note on Ramanujan's trigonometrical function, and certain series of arithmetical functions," in Proc. Cambridge Philosoph. Soc., 1921, vol. 20, pp. 263–271.
17. Vaidyanathan P. P., "Ramanujan sums in the context of signal processing Part I: Fundamentals," IEEE Trans. Signal Process., vol. 62, no. 16, pp. 4145–4157, 2014.
18. Maddox J., "Möbius and problems of inversion," Nature, vol. 344, no. 29, p. 377, Mar. 1990.
19. Carmichael R. D., "Expansions of arithmetical functions in infinite series," in Proc. London Math. Soc., 1932, pp. 1–26.
20. Vaidyanathan P. P., "Ramanujan sums in the context of signal processing: Part II: FIR representations and applications," IEEE Trans. on Signal Proc., vol. 62, no. 16, pp. 4158–4172, Aug., 2014.
21. Haukkanen P., "Discrete Ramanujan Fourier transform of even functions (mod r)", Indian J. Math. Math. Sci., vol. 3, no. 1, pp. 75–80, 2007.
22. Toth L. and Haukkanen P.,"The Discrete Fourier transform of r-even functions", Acta Univ. Sapientiae, Mathematica, vol. 3, no. 1, pp. 5–25, 2011.
23. Laohakosol V., Ruengsinsub P. and Pabhapote N.," Ramanujan Sums for signal processing of low frequency noise", Phys. Rev. E., 2006.
24. Anderson D.R., Apostol T.M., "The Evolution of Ramanujan Sums and Generalizations", Duke Mathematical Journal, vol. 20, no. 2, pp. 211–216.
25. Apostol T.M., "Arithmetical properties of generalized Ramanujan Sums", Pacific J. of Mathematics, vol 41, No. 2, 1972.
26. McCarthy P.J.," A generalization of Smith's determinant", Canad. Math. Bull,, vol. 29, no. 1, pp. 109–113, 1986.
27. Pei S.-C. and Chang K.-W., "Odd Ramanujan sums of complex roots of unity," IEEE Signal Process. Letters, vol. 14, pp. 20–23, Jan. 2007.
28. Debaprasad De, Archisman Ghosh, K Gaurav Kumar, Mrinal Kanti Naskar, "An efficient FPGA Implementation of Discrete Fourier and Inverse Discrete Fourier Transforms using Ramanujan Sums", Circuits, Systems, and Signal Processing, Springer (submitted).

29. H. G. Gadiyar and R. Padma, "Ramanujan-Fourier series, the Wiener-Khintchine formula, and the distribution of prime numbers," Physica A, vol. 269, pp. 503–510, 1999.
30. Cohen L., "Time-frequency distributions—A review," Proc. IEEE, vol. 77, no. 7, pp. 941–981, Jul. 1989.
31. Planat M., Minarovjech M.and Saniga M., "Ramanujan sums analysis of long-period sequences and 1/f noise", EPL J.; https://doi.org/10.1209/0295-5075/85/400052008.
32. Cohen E., "An extension of Ramanujan's sum. III. Connections with totient functions," Duke Math. J. 23, 1956, pp: 623–630

Analysis and Design of Bandgap Reference (BGR)

R. Akshaya and Siva Yellampalli

1 Introduction

Variation of performance of a bandgap reference circuit with temperature parameters like CTAT and PTAT is highlighted in this unit. Bandgap reference (BGR) circuit, its basic principles like temperature variations, supply variations, and process variations are demonstrated in this unit.

1.1 Introduction to Bandgap Reference (BGR)

Bandgap reference (BGR) is an analog integrated circuit which is built with basic analogy of canceling positive temperature coefficient which is PTAT and negative temperature coefficient which is CTAT that gives fixed voltage/current reference independent of TVP (temperature, supply voltage, and process) variations. In addition to canceling PTAT and CTAT for constant reference voltage, BGR circuit should be immune to variations in other parameters such as process variations and supply voltage variations. Figure 1 illustrates all the parameters of the designed BGR which should be stable to generate a constant output voltage.

The BGR is a vital component in many analog and mixed-signal circuits such as ADC/DAC, memory, oscillators, PLLs, high-resolution data converters like A/D and D/A converters, and battery-operated dynamic random access memories (DRAMs) for circuit biasing [1, 2]. The performance of many mixed analog/digital systems is limited by variation in reference voltage and power supply noise coupling errors introduced by integrated voltage references. To overcome this problem, precision voltage reference circuits have become a necessity for all integrated circuits.

R. Akshaya · S. Yellampalli (✉)
VTU Extension Centre, UTL Technologies Limited, Bangalore, India
e-mail: siva.yellampalli@gmail.com

© Springer Nature Singapore Pte Ltd. 2019
V. Nath and J. K. Mandal (eds.), *Proceeding of the Second International Conference on Microelectronics, Computing & Communication Systems (MCCS 2017)*, Lecture Notes in Electrical Engineering 476, https://doi.org/10.1007/978-981-10-8234-4_35

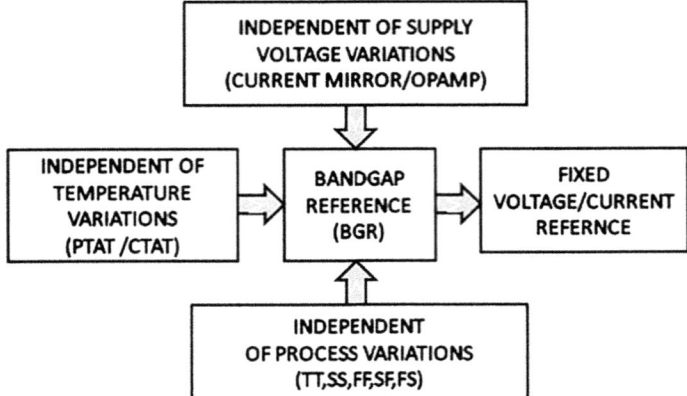

Fig. 1 Block diagram of BGR

1.2 Fundamentals of Bandgap Reference (BGR)

The term bandgap in the name of BGR circuit originates from the physical nature of the silicon. Figure 2 [3] shows a simplified energy band diagram of semiconductor.

The E_C line indicates bottom edge of the conduction band and E_V is top edge for valence band. The distance between these two bands is the bandgap energy E_g. This bandgap energy decreases as the temperature increases; see Eq. (1.1) [3, 4],

$$E_g(T) = E_g(0) - \frac{\alpha T^2}{T + \beta},\qquad(1.1)$$

where $E_g(0)$, α, β are fitting parameters for different materials (Ge, Si, GaAs). For silicon at 300 K room temperature, Eq. (1.1) can be rewritten as Eq. (1.2) [3] substituting all the constant values

Fig. 2 A simplified energy band diagram

$$E_g(300) = 1.166 - \frac{0.473 \times 10^{-3} \times 300^2}{300 + 636} \cong 1.12 \text{ eV} \qquad (1.2)$$

As result of Eq. (1.2), 1.12 eV is the amount of energy that electron needs to overcome gap between the valence and conduction band. The theoretical output voltage of BGR is also equal to 1.12 V [5], and therefore it is called as "Bandgap". Therefore, BGR is the significant voltage reference used due to its minimized temperature coefficient (TC).

1.3 Temperature Dependency of Bandgap Reference

Variation in reference voltage is due to variation in the supply voltage, temperature, and process parameters associated with voltage reference circuit. There are multiple techniques available to design a stable voltage reference [6, 7] which will overcome the variations in supply voltage, temperature, and process parameters. BGR technique is one which is most widely used in integrated circuits [3]. In the first-order BGR circuit, constant reference voltage is obtained by proper PVT compensation. The variation of reference voltage with respect to temperature is measured by the parameter temperature coefficient (TC). The variation of the output reference voltage with respect to temperature can be expressed using the Taylors polynomial shown in Eq. (1.3) [3].

$$\rho(T_0) = \rho(T_0)\left(1 + \alpha(\Delta T) + \beta(\Delta T)^2 + \gamma(\Delta T)^3 + \cdots\right) \qquad (1.3)$$

where $\Delta T \rightarrow$ Difference between temperature T and the nominal temperature (T_0), $\rho(T_0) \rightarrow$ Parameter ρ at temperature (T_0), and α, β, and γ are temperature coefficients (TC).

If TC of higher order (β, γ) is negligible, then we can consider the dependency to be linear, and Eq. (1.3) can then be written as

$$\rho(T_0) = \rho(T_0)(1 + \alpha(\Delta T)) \qquad (1.4)$$

From Eq. (1.2), it can be observed that if the parameter ρ is assumed to be the reference voltage, V_{ref}, and α is greater than zero, then the value of V_{ref} increases with temperature. This phenomenon is called PTAT (Proportional to Absolute Temperature). The relation between V_{ref} and α for PTAT condition is illustrated in Fig. 3a. Similarly, if α is smaller than zero, then the V_{ref} decreases with temperature and it is called CTAT and is illustrated in Fig. 3b [5, 8].

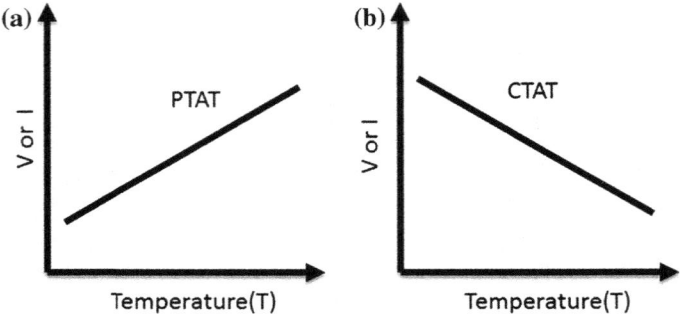

Fig. 3 **a** PTAT and **b** CTAT voltage reference

1.4   Bandgap Reference Basic Principle

Figure 4 [1, 5] shows the principle of ideal BGR reference circuit. The ideal BGR circuit has zero TC. From Fig. 3, we observe that BGR circuit exploits CTAT and PTAT to compensate temperature dependence. Figure 5 shows the zero TC due to PTAT and CTAT cancelation.

In practice, PTAT and CTAT vary to different scales, to match the variations, and to compensate each other PTAT is multiplied by constant K as shown in Eq. (1.5) [9]

$$V_{\mathrm{REF}} = V_{\mathrm{BE}} + K V_T \tag{1.5}$$

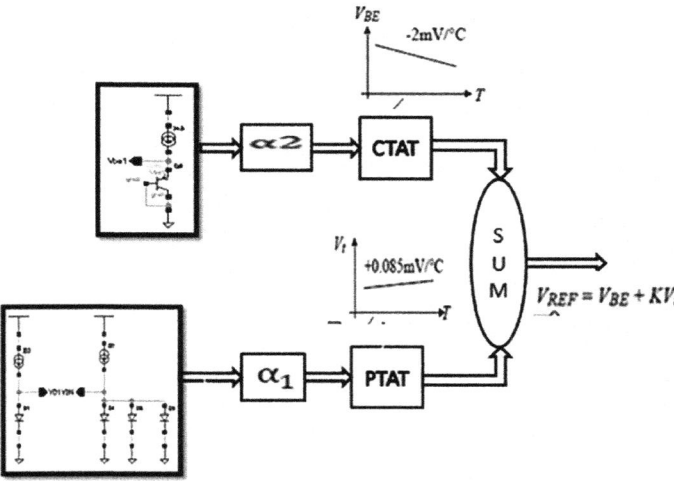

Fig. 4 Bandgap basic principle

Fig. 5 Zero TC due to PTAT
and CTAT cancelation

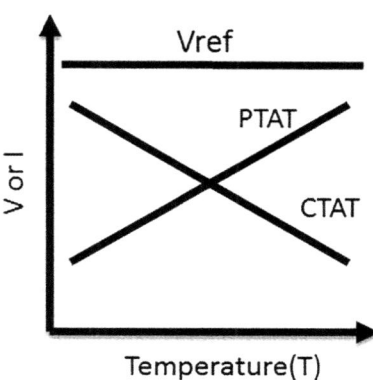

In addition to temperature dependence, the output reference voltage in BGR is dependent on other parameters such as variations in supply voltage and process variations. In the following section, the mathematical equations representing the effect of these parameters on output reference voltage is presented. There are many sources of variations, for instance, different temperatures in fabrication steps, different dopant concentrations, different thicknesses of growing oxide, etc.

1.5 Temperature Variations

The temperature coefficient (TC) of a reference voltage with respect to temperature is given by Eq. (1.6) [5, 9–11],

$$TC(V_{REF}) = \frac{1}{V_{REF}} \times \frac{\partial V_{REF}}{\partial T} \tag{1.6}$$

TC of reference is given by the deviation of the output voltage from its mean value in the tested temperature range. A BGR circuit in which the first-order temperature coefficient of V_{BE} has been compensated has an ideal TC of 15–20 ppm/°C [2].

1.6 Supply Variations

The sensitivity of supply variation V_{DD} with respect to voltage reference V_{REF} is given by Eq. (1.7) [10],

$$S_{V_{\mathrm{DD}}}^{V_{\mathrm{REF}}} = \frac{V_{\mathrm{DD}}}{V_{\mathrm{REF}}} \times \frac{\partial V_{\mathrm{REF}}}{\partial V_{\mathrm{DD}}}, \tag{1.7}$$

In some applications, supply variation will be high, like automotive devices where it will be in the range of 4.5–18 V or higher [11]. In designing a BGR for this range of supply voltages to get a constant V_{REF}, the supply variations have to accurately compensate. From Eq. (2.5), we can observe that the sensitivity of V_{REF} with respect to variations in V_{DD} can be minimized by isolating V_{REF} from V_{DD}. This can be obtained by using bootstrap bias technique [12], also referred to as self-biasing in current mirror-based BGR circuit and also by using cascade structure in CM_BGR circuit [2].

1.7 Process Variations

The process variation is accounted for both in design and layout of BGR circuit. Several device parameters such as junction capacitance and threshold voltage Vth [2] are used to estimate the impact of process variations. These models are called process corners. Generally, slow and fast corners are defined for PMOS and NMOS devices. Junction capacitance and threshold voltage Vth in slow corner case are larger than expected in NN corner (NN corner is a case without any process variations); digital circuits are slower in slow corner. On other hand, fast corner reduces Vth and junction capacitances and that is why digital devices are faster. The five corners are classified as SS, SF, FS, FF, and TT as presented.

1. SS (NMOS slow, PMOS slow),
2. SF (NMOS slow, PMOS fast),
3. FS (NMOS fast, PMOS slow),
4. FF (NMOS fast, PMOS fast), and
5. TT (NMOS typical, PMOS typical).

Technology variations have significant effect on analog precision and influence device performance. The geometrical size of manufactured MOS transistor differs from geometrical size of designed transistor as fabrication of transistor consists of a series of chemical and mechanical processes. Process variations will lead to mismatches in transistors because of which we will get nonzero TC [3–5, 8]. Due to impact of process variations on output reference voltage, post-processing trim techniques are used for their compensation.

Summary: Every design should be tested in different technology (process) corners, temperatures, and supply voltages [1]. For BGR voltage reference circuit, realization of CTAT and PTAT generation circuits is important as they are the main building blocks. Circuits used to realize CTAT and PTAT have been discussed [13]. The effect of PVT parameters on BGR circuit is discussed, and the relationship is mathematically formalized.

2   Methodology: Design and Implementation of Bandgap Reference in CMOS Technology

Design analysis of bandgap reference (BGR), implementation of bandgap reference, negative-TC voltage, positive-TC voltage, self-biasing circuit, implementation of BGR layout, and layout challenges are explained in brief in this unit.

2.1   Design Analysis of Bandgap Reference (BGR)

BGR circuit combines two voltages with opposite temperature coefficient to achieve a constant output voltage. The most common voltages used in this addition are the base–emitter voltage (V_{BE}) and thermal voltage (V_T) produced by BJTs. The current mirror (CM) bandgap reference (BGR) circuit designed to generate constant voltage is shown in Fig. 6 [10]. The CM-BGR was designed in 180 nm technology.

The CM_BGR circuit in Fig. 6 uses a "bootstrapped" current mirror to reduce the power supply dependency. The transistors $M1$ and $M2$ force nodes S_1 and S_2 to be of equal voltage. As the transistor $Q2$ is "N" times bigger than transistor $Q1$, a PTAT voltage appears across R_1 and a CTAT voltage at $Q3$. Therefore, the sum of PTAT and CTAT voltage gives constant output bandgap reference as shown in Fig. 6.

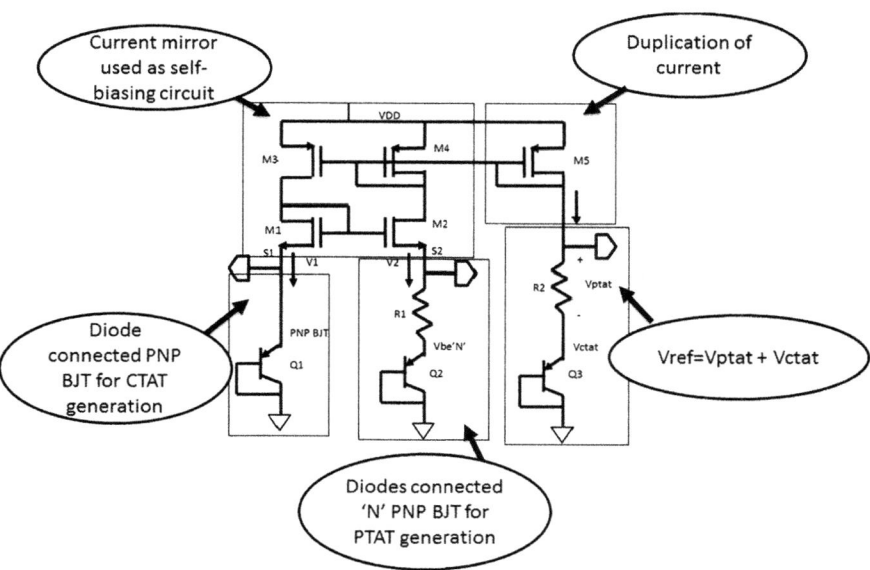

Fig. 6 First-order current mirror-based bandgap reference (CM_BGR)

2.2 *Design and Implementation of Bandgap Reference*

In this unit, we will look at the design of CTAT and PTAT voltage generation circuits and self-biasing circuit which are the building blocks of BGR as shown in Fig. 6.

(a) **Negative Temperature Coefficient (NTC) voltage**

   The base–emitter voltage of bipolar transistors or, more generally, the forward voltage of a pn junction diode exhibits a negative TC.

   Figure 7 [10] shows the CTAT voltage realization circuit. The circuit consists of a BJT and an ideal current source. In this configuration, the parasitic vertical PNP bipolar transistor with base–emitter junction is exploited to obtain a diode. Relationship between I-V of BJT is approximated by Eq. (2.1) [12] called Shockley's equation.

$$I_{\text{REF}} = I_S \left\{ \exp^{\left[\frac{V_{\text{be}}}{\eta V_T} \right]} - 1 \right\} \qquad (2.1)$$

where $I_{\text{REF}} \rightarrow$ Current flowing through the diode, $I_s \rightarrow$ Reverse saturation current and its value is in range of $10^{-12} - 10^{-18}$, $V_{\text{be}} \rightarrow$ Base–emitter voltage across diode-connected PNP BJT, $V_T \rightarrow$ Thermal voltage proportional to temperature, and $\eta \rightarrow$ Ideality factor which is typically around 1. We first have to analyze the expression for constant current I_E of vertical PNP BJT device. The emitter current in PNP BJT is given in Eq. (2.2) [6]

$$I_E = I_S \exp(V_{\text{BE}}/V_T) \qquad (2.2)$$

Fig. 7 Single diode-connected VPNP BJT

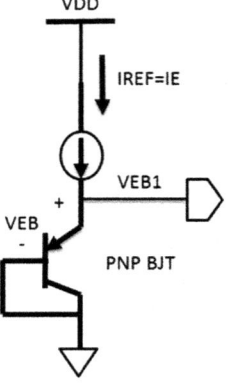

Here, we know that the thermal voltage is given by Eq. (2.3):

$$V_T = \frac{kT}{q} \tag{2.3}$$

where $k \rightarrow$ Boltzmann constant (1.38×10^{-23}) J/K, $T \rightarrow$ Absolute temperature in Kelvin $(273 + {}^\circ C)$, and $q \rightarrow$ Magnitude of the electronic charge (1.602×10^{-19}) Coulomb.

The saturation current I_S is given in Eq. (2.4) [3],

$$I_S \propto \mu k T n_i^2 \tag{2.4}$$

The mobility and intrinsic carrier concentration n_i^2 are given in Eqs. (2.5) and (2.6):

$$\mu = \mu_0 T^m \tag{2.5}$$

$$n_i^2 \propto T^3 \exp\left[\frac{-E_g}{kT}\right] \tag{2.6}$$

Substituting Eqs. (2.5) and (2.6) in Eq. (2.4), which gives Eq. (2.7)

$$I_S = \mu_0 T^m (T) k \left[T^3 \exp\left[\frac{-E_g}{kT}\right] \right] \tag{2.7}$$

where $I_S \rightarrow$ Reverse saturation current of diode, $\mu \rightarrow$ Mobility, $k \rightarrow$ Boltzmann constant, $T \rightarrow$ Absolute temperature, $n_i^2 \rightarrow$ Intrinsic concentration (minority carrier), $m \approx -1.5$, and $E_g \approx 1.12$ eV is the bandgap energy of silicon.

$$I_S = b T^{4+m} \exp\left[\frac{-E_g}{kT}\right] \tag{2.7}$$

$$\frac{\partial I_S}{\partial T} = I_S \left[\frac{(4+m)}{T} + \frac{E_g}{kT^2} \right] \tag{2.8}$$

In Eq. (2.7), b is a proportionality factor. Eq. (2.8) is the derivative of I_S with temperature. As shown in Fig. 7 [4], when a constant current I_{REF} is pumped through a single vertical PNP BJT, a CTAT voltage (V_{CTAT}) is obtained by measuring base–emitter voltage V_{BE} across VPNP BJT, which gives negative temperature coefficient (NTC), and the base–emitter voltage across diode-connected BJT equation is given by Eq. (2.9) [4],

$$V_{BE} = V_T \ln\left(\frac{I_0}{I_S}\right) \tag{2.9}$$

where V_{BE} is the base–emitter voltage across diode-connected BJT, V_T is the thermal voltage which represents the PTAT nature of curve, and $\ln\left(\frac{I_0}{I_S}\right)$ (I_0 is a constant current and I_S is a reverse saturation current) is the dominant parameter in Eq. (2.9) which represent the CTAT nature of the BJT. Differentiating Eq. (2.9) with temperature T gives Eq. (2.14)

$$\frac{\partial V_{BE}}{\partial T} = \frac{\partial V_T}{\partial T} \ln \frac{I_C}{I_S} \tag{2.10}$$

$$\frac{\partial V_{BE}}{\partial T} = \frac{\partial V_T}{\partial T} \ln \frac{I_C}{I_S} + V_T \frac{\partial \left[\ln \frac{I_C}{I_S} \right]}{\partial T} \tag{2.11}$$

$$\frac{\partial V_{BE}}{\partial T} = \frac{\partial V_T}{\partial T} \ln \frac{I_C}{I_S} - \frac{V_T}{I_S} \frac{\partial I_S}{\partial T} \tag{2.12}$$

$$\frac{\partial V_{BE}}{\partial T} = \frac{V_T}{T} \ln \frac{I_C}{I_S} - (4+m) \frac{V_T}{T} - \frac{E_g}{kT^2} V_T \tag{2.13}$$

$$\frac{\partial V_{BE}}{\partial T} = \frac{V_{BE} - V_T(4+m) - E_g/q}{T} \tag{2.14}$$

Therefore, Eq. (2.14) gives the derivative of base–emitter voltage with temperature T. Using this equation, the slope of base–emitter voltage across a diode-connected VPNP BJT which varies with temperature T can be calculated. And by using Eq. (2.9), the base–emitter voltage of vertical PNP bipolar transistor can be calculated as shown below.

$$V_{BE} = 26 \times 10^{-3} \ln \left(\frac{10 \ \mu A}{2.4 \times 10^{-18}} \right) = 755.511 \times 10^{-3} \ V$$

$$\frac{\partial V_{BE}}{\partial T} = \frac{755.511 \times 10^{-3} \ V - (4 - 1.5)26 \ mV - 1.2}{300} = -1.7 \ mV/K$$

The V_{BE} of VPNP BJT is calculated to be 755.511×10^{-3} V with I_0 constant current pumping $10 \ \mu A$ and reverse saturation current is given as 2.4×10^{-18} (Noted from 180 nm technology model parameter). When V_{BE} is equal to 755.511×10^{-3} V and E_g/q is equal to 1.5 with thermal voltage V_T of 26 mV at room temperature 300 K, then the calculated value of slope of a diode-connected base–emitter voltage of VPNP BJT with temperature T is estimated to be -1.7 mv/K. Generally, V_{BE} is equal to 0.7 V and E_g/q is equal to 1.2 with thermal voltage V_T of 26 mV at room temperature 300 K; then, the calculated value of slope of a diode-connected base–emitter voltage of BJT with temperature T is estimated to be -1.6 mv/K. This is further explained in the PTC of PTAT. The size of vertical PNP BJT is calculated using built-in voltage of emitter–base junction (EBJ) which is forward biased and its expression is shown in Eq. (2.15):

$$V_{bi} = V_{be} = \frac{K_B T}{e} \ln \frac{N_{A,e} N_{D,e}}{n_i^2} \tag{2.15}$$

where $V_{bi} \rightarrow$ Built-in voltage, $N_{A,e} \rightarrow 10^{17}$ cm^{-3} (Acceptor density of holes in emitter), $N_{D,b} \rightarrow 10^{16}$ cm^{-3} (Acceptor density of electrons in base).

$$V_{be} = 26 \text{ mV} \ln \left(\frac{10^{17} \text{ cm}^{-3} \times 10^{16} \text{ cm}^{-3}}{2.25 \times 10^{20}} \right) = 757.189 \text{ mV}$$

$$V_{be} = 757.189 \text{ mV}$$

Formula for depletion width for diode-connected base–emitter voltage is given by

$$W_{be} = \sqrt{\frac{2 \in \times (N_D + N_A)(Vbi - V)}{q \times N_A \times N_D}} \tag{2.16}$$

$$W_{be} = \sqrt{\frac{2 \times 1.04 \times 10^{-12} \times (10^{16} + 10^{17})(757.189 \text{ m} - 1.8)}{1.6 \times 10^{-19} \times 10^{16} \times 10^{17}}}$$

$$W = 1.22 \times 10^{-4} \text{ cm} \approx 1.3 \text{ μm}$$

Therefore, the width of diode-connected base–emitter voltage is calculated to be 1.22×10^{-4} which is approximately equal to be 1.3 μm which is the emitter size 1.3 μm × 1.3 μm used in designing CM_BGR.

(b) Positive temperature coefficient (PTC) voltage

In Fig. 8, when equal constant currents are pumped through transistor $Q1$ which is a single diode-connected BJT and through transistor $Q2$ which is N times bigger than $Q1$ and also which can be realized by connecting N $Q1$ transistors in parallel, the $Q2$ cancels the CTAT-dominated nature of curve resulting in PTAT nature of curve with positive temperature coefficient (PTC). The PTAT voltage (V_{PTAT}) is obtained by the potential difference ($V_{BE1} - V_{BE2}$), which is measured across $Q1$ and $Q2$. And also PTAT voltage can be measured across the resistor $R1$ which is connected in series with $Q2$ transistor which realizes the nature of PTAT. The change in base–emitter voltage across $Q1$ and $Q2$ is ΔV_{BE}, and its relation is given by Eq. (2.17),

$$\Delta V_{BE} = V_{BE1} - V_{BE}(N) = V_T \ln N \tag{2.17}$$

$$V_{BE1} = V_T \ln \left(\frac{I_0}{I_S} \right), \quad V_{BE}(N) = V_T \ln \left(\frac{I_0}{N I_S} \right)$$

$$V_{BE} - V_{BE}(N) = V_T \left[\ln \left(\frac{I_0}{I_S} \right) - \ln \left(\frac{I_0}{N I_S} \right) \right] \tag{2.18}$$

Fig. 8 PTAT nature
realization circuit

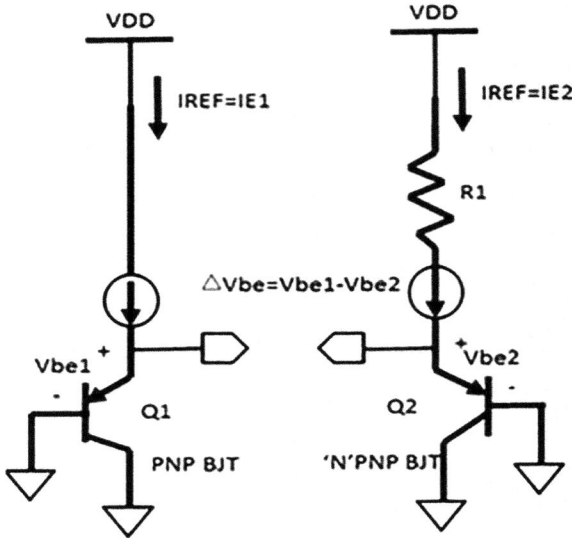

By substituting the logarithm property of division, i.e., $\log(A) - \log(B) = \log(A/B)$ and even I_S term which is temperature-dependent that is canceled by keeping $\partial I_0 / \partial T = 0$, then Eq. (2.18) is simplified to Eq. (2.19]

$$V_{BE} - V_{BE}(N) = V_T \left[\ln\left(\frac{I_0}{I_S}\right) \div \ln\left(\frac{I_0}{NI_S}\right) \right]$$
$$\Delta V_{BE} = V_{BE} - V_{BE}(N) = V_T \ln N$$
(2.19)

The derivative of Eq. (2.19) with temperature T is given by Eq. (2.20),

$$\frac{\partial V_{BE}}{\partial T} = \frac{\partial V_T}{\partial T} \ln N$$
(2.20)

where $\ln N$ is the PTAT term set to be constant and N is the number of diodes connected in parallel ($N = 2, 3, 8 \dots n$). And the variation of thermal voltage $V_T = KT/q$ with respect to temperature is derived in Eq. (2.21),

$$\frac{\partial V_T}{\partial T} = \frac{K}{q}$$
(2.21)

where K is the Boltzmann constant, i.e., 1.38×10^{-23} V and q is the charge of an electron, i.e., 1.6×10^{-19} C. Therefore, the derivative of thermal voltage with temperature ($\partial V_T / \partial T$) is the slope of PTAT which is equal to 86.25×10^{-6} V/K approximated to be 87×10^{-6} V/K.

Unit 3 gives the complete calculation and simulated result of proposed PTAT generation circuit. In our actual circuit simulation, we observe $\frac{\partial I_o}{\partial T} \neq 0$. As the variation is not zero, we are getting PTAT current in our BGR circuit. The variation in CTAT circuit due to variation in PTAT current w.r.t. temperature is theoretically derived by using Eq. (2.20). In Eq. (2.20), the V_T and (I_0/I_S) terms which are simplified to be N are all considered to be temperature-dependent. The modified Eq. (2.20) is given by Eq. (2.22),

$$\frac{\partial V_{BE}}{\partial T} = \frac{\partial V_T}{\partial T} \ln\left(\frac{I_0}{I_S}\right) \tag{2.22}$$

Differentiating Eq. (2.22), we get Eq. (2.23),

$$\frac{\partial V_{BE}}{\partial T} = V_T \left[\frac{1}{I_o}\cdot\frac{\partial I_o}{\partial T} - \frac{1}{I_s}\frac{\partial I_s}{\partial T}\right] + [\ln(I_o) - \ln(I_s)]\frac{\partial V_T}{\partial T} \tag{2.23}$$

In the above Eq. (2.23), familiar terms which are already been derived with temperature are thermal voltage V_T given in Eq. (2.21), and reverse saturation current I_S given in Eq. (2.8). Now, we need to find the $(\partial I_0/\partial T)$. In Fig. 8, the voltage across resistor V_{R1} is equal to $R_1 I_{REF}$ and resistor R_1 is the PTAT-dependent component which is given as $(V_T \ln(N)/I_o)$.

By this, the current $I_{REF} = I_o$ through resistor R_1 can be expressed as given in Eq. (2.24),

$$I_o = \frac{\ln(N)}{R_1} V_T \tag{2.24}$$

The derivative of Eq. (2.24) with temperature T is given in Eq. (2.25),

$$\frac{\partial I_o}{\partial T} = \frac{\ln(N)}{R_1}\frac{\partial V_T}{\partial T} \tag{2.25}$$

In Eq. (2.25), R_1 also has got small dependency on temperature but now it has been neglected for time being. Substituting Eq. (2.21) in Eq. (2.25), we get Eq. (2.26),

$$\frac{\partial I_o}{\partial T} = \frac{V_T \ln(N)}{R_1} \times \frac{1}{T} \tag{2.26}$$

Further simplifying Eq. (2.26), we get Eq. (2.27),

$$\frac{\partial I_o}{\partial T} = \frac{I_o}{T} \tag{2.27}$$

In Eq. (2.27), I_o generates the PTAT reference current. Then by substituting all three Eqs. (2.8), (2.21) and (2.27) in Eq. (2.23), i.e., given by

$$\frac{\partial V_{BE}}{\partial T} = V_T \left[\frac{1}{I_o} \cdot \frac{I_o}{T} - \frac{1}{I_s} \left\{ I_s \left[\frac{(4+m)}{T} + \frac{E_g}{KT^2} \right] \right\} \right] + [\ln(I_o) - \ln(I_s)] \frac{V_T}{T} \quad (2.28)$$

Therefore, Eq. (2.28) is simplified as shown in Eq. (2.29), i.e.

$$\frac{\partial V_{BE}}{\partial T} = \frac{V_{BE} - (3+m)V_T - E_g/q}{T} \quad (2.29)$$

From Eq. (2.29), we can extract the slope of CTAT considering I_o as I_{PTAT}; earlier, we had extracted slope of CTAT by keeping I_o as constant. Table 1 gives the comparison between the two stages of I_0.

Generally, by substituting (V_{BE} is equal to 0.7 V and E_g/q is equal to 1.2 with thermal voltage V_T of 26 mV at room temperature 300 K) in Eq. (2.14), the calculated value of slope of a diode-connected base–emitter voltage of BJT with temperature T is estimated to be -1.6 mv/K during which I_o is set to constant. Similarly, by substituting (V_{BE} is equal to 0.7 V and E_g/q is equal to 1.2 with thermal voltage V_T of 26 mV at room temperature 300 K) in Eq. (2.29), then the calculated value of slope of a diode-connected base–emitter voltage of BJT with temperature T is estimated to be -1.79 mv/K during which I_o is set to I_{PTAT}. The negative sign "$-$" indicates that when the voltage across diode increases, the temperature decreases. Therefore, the current through diode is PTAT in nature but the voltage across diode is CTAT in nature. The only difference between Eqs. (2.14) and (2.29) is the middle terms in slope of CTAT, $(4+m)V_T$ and $(3+m)V_T$ which is of lesser value actually varying from V_T (26–65 mV) and V_T (26–36 mV), respectively. This shows that the slope itself is dependent on temperature and still it can be used as a CTAT voltage. Unit 3 gives the simulation and result of all these estimated values obtained as stated.

(c) **Self-biasing circuit**

In Fig. 9 [8], self-biasing technique is designed to give the predetermined identical voltages with gate–source potential difference of V_{GS1} and V_{GS2} across $M1$ and $M2$ transistors, where $M1$ and $M2$ are two identical transistors of same size and are laid out to match well. $M3$ and $M4$ are the PMOS devices which are used as current mirror. The size of these transistors is large enough to prevent lambda effects from significantly degrading the accuracy of the PTAT voltage. Figure 9

Table 1 Comparison of slope of base–emitter voltage with temperature

Sl. No.	$(\partial V_{BE}/\partial T)$	Reference current $I_{REF}(I_0)$	Slope of CTAT (mV/K)
1	$\frac{V_{BE} - (4+m)V_T - E_g/q}{T}$	$I_o = $ constant	-1.6
2	$\frac{V_{BE} - (3+m)V_T - E_g/q}{T}$	$I_o = I_{REF}$	-1.79

Fig. 9 Current mirror used as self-biasing circuit

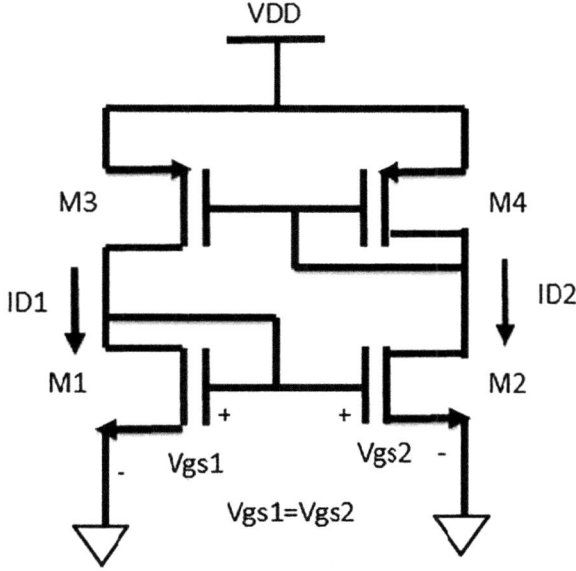

shows a part of BGR which is designed to convert current of 10 μA to a gate–source voltage of $V_{gs1} = V_{gs2}$ and the current is pumped by this self-biased circuit through the nodes V_{be1} and V_2 to the branches of CTAT and PTAT generation circuit spontaneously. Further, the size of each of the MOS devices is calculated as follows

Calculation of W/L ratio of self-biased circuit in Fig. 9.

(a) The oxide capacitance is computed by using Eq. (2.30)

$$C_{ox} = \frac{\varepsilon_{ox}}{t_{ox}} \tag{2.30}$$

where $t_{ox} \rightarrow$ Thickness of oxide, $\varepsilon_{ox}(= 3.79\varepsilon_o) \rightarrow$ Dielectric constant of silicon dioxide. $(\varepsilon_o = 8.854 \times 10^{-14}$ F/cm)

$$C_{ox} = \frac{3.9 \times 8.854 \times 10^{-12}}{4 \times 10^{-9}} = 8.63 \text{ mF} \tag{2.31}$$

$$\mu_n C_{ox} = 345 \text{ μA/V}^2, \quad \mu_p C_{ox} = 55 \text{ μA/V}^2 \tag{2.32}$$

(b) I_D, drain current in saturation region of operation, is given by Eq. (2.33),

$$I_D = \frac{\mu C_{OX} W}{2L} (V_{gs} - V_{th})^2 \qquad (2.33)$$

where $I_D \rightarrow$ Drain current, $V_{gs} \rightarrow$ Gate–source voltage, $V_{th} \rightarrow$ Threshold voltage, and $\left(\frac{W}{L}\right) \rightarrow$ Aspect ratio (width and length of MOS devices).

(c) Aspect ratio calculation of NMOS devices is given by Eq. (2.34)

$$\left(\frac{W}{L}\right) = \frac{2 * I_D}{\mu_n C_{ox} (V_{gs} - V_{thn})^2} \qquad (2.34)$$

$$\left(\frac{W}{L}\right)_N = \frac{2 \times 10\ \mu A}{345 \times 10^{-6} \times (600\ mV - 0.48)^2} \qquad (2.35)$$

$$\begin{aligned} w_N &= 20\ \mu m \\ L_N &= 5\ \mu m \end{aligned} \qquad (2.36)$$

(d) Aspect ratio calculation of PMOS devices is given by Eq. (2.38)

$$\left(\frac{W}{L}\right)_P = \frac{2 * I_D}{\mu_p C_{ox} (V_{sg} - V_{thp})^2} \qquad (2.37)$$

$$\left(\frac{W}{L}\right)_P = \frac{2 \times 10\ \mu A}{55 \times 10^{-6} \times (430.30 + 0.43)^2} \qquad (2.38)$$

$$w_P = 20\ \mu m,\ L_P = 5\ \mu m \qquad (2.39)$$

Equation (2.30) gives the gate oxide capacitance of the MOS device which is estimated to be 8.63 mF in 180 nm CMOS technology. The process-dependent constant of NMOS device $\left(k_n'\right)$ is given by $\mu_n C_{ox}$ which is equal to 345 $\mu A/V^2$, and the process-dependent constant of PMOS device $\left(k_p'\right)$ is given by $\mu_p C_{ox}$ which is equal to 55 $\mu A/V^2$. From Eq. (2.33), the drain current of the MOS devices is designed to fetch 10 μA of current by neglecting channel length modulation and body bias effect. The threshold voltage of NMOS and PMOS device is 0.48 and -0.43. And the gate—source voltage V_{gs} of NMOS device is equated to be 600 mV and the source–gate voltage V_{sg} of PMOS device is equated to be 430.30 mV. With these estimated values, the width of w_n and w_p is equated to be 20 μm and the length is estimated to be 5 μm. Unit 3 gives the simulated result of self-biased circuit.

(d) CM_BGR without start-up circuit

Figure 10 [3] shows the basic implementation of current mirror-based BGR circuit without start-up circuit, which consists of M_1-M_4 self-biased circuit transistors used to generate equal voltages ($V_{gs1} = V_{gs2}$) with supply of constant current ($I_{d1} = I_{d2}$) across both the branches of diode-connected BJT transistors. All the three bipolar transistors are characteristically executed by the diode-connected parasitic vertical PNP bipolar junction transistor in CMOS process with the emitter current I_E proportional to exponential of (V_{EB}/V_T), where V_{EB} is the base–emitter voltage and V_T is the thermal voltage. During constant I_{REF}, V_{EB} is toughly dependent on thermal voltage and temperature. The current mirror is shaped by PMOS devices in Fig. 10; $M1$ and $M2$ are designed to bias $Q1$ and $Q2$ with identical currents $I_{d1\ and}\ I_{d2}$. Since the nodes V_{EB1} and V_2 pointed are biased to generate equal voltages as shown in the above circuit representing the potential independent of supply variations, Eq. (2.40) gives the expression for $R1$ resistor voltage drop,

$$V_{R1} = [V_{EB1} - (N)V_{EB}] = V_T \ln(N) = V_T \ln\left(\frac{A_2}{A_1}\right) \qquad (2.40)$$

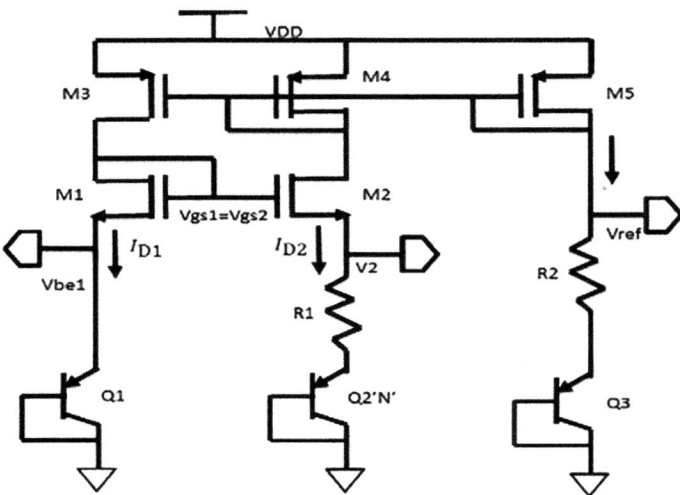

Fig. 10 Current mirror-based bandgap reference (BGR) without start-up circuit

where N is the number of diodes connected in parallel in $Q2$ and it also represents the emitter areas of $Q1$ and $Q2$ denoted by A_1 and A_2 in Eq. (2.40). It is noted that V_{EB1} exhibits a negative temperature coefficient and V_{R1} exhibits a positive temperature coefficient when A_2 is greater than A_1, i.e., in proposed design of BGR circuit, the ratio of $Q1$ and $Q2$ transistor is estimated to be in the range of [1:8]. Therefore, a constant current I_0 of 10 μA is pumped through parallel diodes of ratio 1:8 with thermal voltage V_T of 26 mV at room temperature 300 K, in which estimated values are used to calculate the $R1$ resistor as shown in Eq. (2.41)

$$R_1 = \frac{V_T \ln N}{I_0} = \frac{26 \text{ mV } \ln(8)}{10 \text{ μA}} = 5.4 \text{ K}\Omega \qquad (2.41)$$

The current mirror formed by $M5$ PMOS devices in Fig. 10 is intended to bias $Q3$ with base–emitter voltage of V_{EB3}, $N = 1$, and current I_{d5} represents the PTAT current generation I_{PTAT} that enters the $R2$ resistor with voltage drop of V_{R2} equating V_{PTAT}. Equation (2.42) gives the output voltage of proposed BGR circuit.

$$V_{REF} = \left(\frac{R2}{R1} \ln N\right) V_T + V_{BE3} \qquad (2.42)$$

In Eq. (2.42), the first term in the equation is proportional to absolute temperature (PTAT), which is used to compensate the second term with complementary to absolute temperature (CTAT) of V_{EB3}. In general, the PTAT voltage comes from the thermal voltage V_T with a temperature coefficient of about +0.087 mv/°C in CMOS technology, which is quite smaller than that of V_{EB}. From Eq. (2.3), the coefficient constant α_2 of V_{BE3} is set to one such that the coefficient constant of α_1 is equated to be $(R2/R1) \ln N$. By this analysis, the expression of V_{REF} is given by Eq. (2.43), i.e.,

$$V_{REF} = \alpha_1 V_{PTAT} + \alpha_2 V_{CTAT} \qquad (2.43)$$

$$\alpha_1 = (R2/R1) \ln N \qquad (2.44)$$

Using derivative of Eq. (2.43), the value of α_1 can be calculated by substituting α_2 to 1, $(\partial V_{EB}/\partial T)$ to -1.5 mv/K and the $(\partial V_T/\partial T)$ to 195.3 μV/K. Therefore, α_1 will be equal to 17.59 which is the coefficient value of first term shown in Eq. (2.43). Further, $R2$ resistor can be calculated using estimated value of $(\alpha_1 = 17.59, R1 = 5.4 \text{ K}\Omega$ and $N = 8)$ in Eq. (2.44), then $R2$ is calculated to be 48.87 KΩ. Similarly, using equation (2.44) of α_1, N can be calculated. After multiplying the thermal voltage in Eq. (2.42) with an appropriate factor $[(R_2/R_1) \ln N]$, we obtain PTAT voltage to be around 54.06 mV. Summing PTAT voltage with V_{EB3} in Eq. (2.42), we obtain V_{REF} equal to 1.12 V, which is the output voltage V_{ref} fixed/constant reference value of BGR circuit generated by independent of PVT at room temperature, 27 °C, and temperature coefficient of CM_BGR output voltage is 0.2 ppm/°C with −40 °C to +125 °C range of temperature under supply voltage V_{DD} of 1.8 V.

Table 2 Resistor and transistor dimensions used in the proposed BGR circuit without start-up block

Required components	Estimated values	Simulated values
$Q1$, $Q3$	Normalized area (N) = 1	Normalized area (N) = 1
$Q2$	Normalized area (N) = 8	Normalized area (N) = 8
$M1$, $M2$, $M3$, $M4$, $M5$	$W = 20 \, \mu m$, $L = 5 \, \mu m$	$W = 20 \, \mu m$, $L = 5 \, \mu m$
$R1$	5.4 K (Ω)	5.4 K (Ω)
$R2$	45.89 K (Ω)	40.9 K (Ω)

The component values required for calculating and simulating design analysis of current mirror BGR circuit without start-up block is tabulated in Table 2. The simulated results are updated in unit 3.

(e) **Implementation of BGR Layout**

As shown in Fig. 11, BGR layout consists of bipolar transistors (PNP BJT), poly-resistor and MOS devices (PMOS and NMOS) [14], where current mirror is designed using interdigitation matching technique which is preferable technique used to match the two current nets and they are surrounded with dummy devices

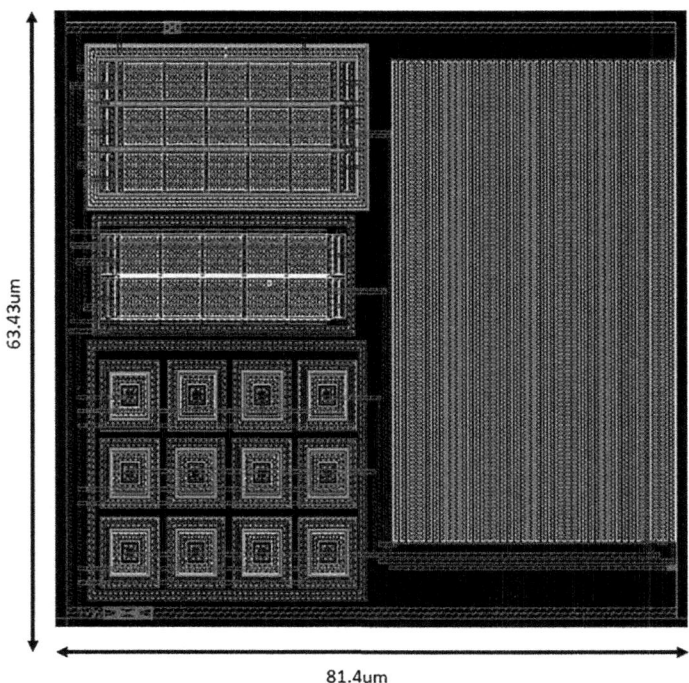

Fig. 11 Layout of current mirror bandgap reference (BGR)

which are placed in left and right edges to reduce stress during etching process. Equal current densities ($I1$ and $I2$) are maintained by placing both PMOS and NMOS devices nearby. In the block of bipolar transistors (PNP_BJT), $Q2$ is composed of $N = 8$ blocks while $Q1$ and $Q3$ are composed of $N = 1$. To achieve decent matching in current mirror block, common-centroid matching practice is applied. Figure 11 shows the layout of the whole bandgap reference circuit. The area of layout is measured to be 81.4 μm × 63.43 μm in 0.18 μm CMOS process technology.

(f) Drawback of the Designed BGR circuit

The drawbacks of designed BGR circuit can be analyzed as shown in Fig. 10 and the circuit exhibits an operating point with no current flowing in the bandgap core during $VX = VY = 0$ or VDD. Startup of the circuit may fail because of the mismatch in current mirror which pushes the bias point in the wrong direction. When CM_BGR without start-up circuit is simulated using transient analysis, it enters into ZCR which is the main drawback seen in the circuit. Therefore, in Fig. 12, BGR is designed using a start-up circuit which ensures a nonzero current state by sourcing or sinking a small amount of current to or from the current generator self-biasing circuit as shown in Fig. 12. This start-up circuit is fed into a low-impedance node, such as node "W" or "Z" in Fig. 12. But start-up current is normally low and therefore does not need to be precisely controlled. Moreover, it can flow continuously or only when the circuit approaches its zero current state.

The overall objective of designing a precision BGR reference is to achieve high accuracy over all working conditions. Most important are temperature variation, process variation, and supply variation [1]. Concepts of reducing the impact of

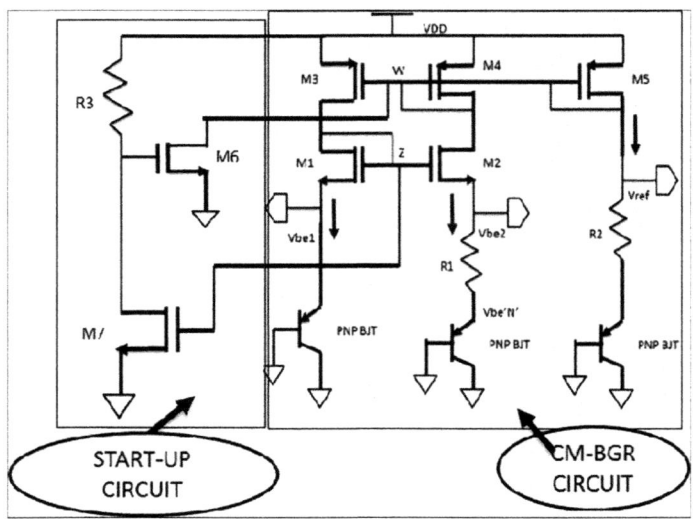

Fig. 12 Current mirror-based bandgap reference (BGR) with start-up circuit

those variations on reference voltage deviation, which were used in the proposed circuit, will be explained in the following. The base–emitter voltage of bipolar transistor or, more generally, the forward bias voltage of a pn junction diode exhibits negative temperature coefficient (TC). On the other hand, two bipolar transistors that operate at unequal current densities exhibit difference between their base–emitter voltages that is directly proportional to the absolute temperature [15]. Using these negative- and positive-TC voltages, a reference having a nominally zero temperature coefficient can be developed.

3 Design of Bandgap Reference with Start-up Circuit

Designs of the start-up circuit, its region of operations like normal operating region (NOR), and zero current region (ZCR) are demonstrated. BGR without start-up circuit, BGR with start-up circuit, and transient response of BGR with start-up circuit are explained in this unit.

3.1 Design of the Start-up Circuit

The CM_BGR circuit designed in Fig. 10 gets deactivated when it enters into an inadequate normal operating bias point. To overcome this problem incurred by designed CM_BGR circuit at the operating point of stability, an auxiliary circuit called the start-up circuit is added to the designed CM_BGR circuit. The start-up circuit should automatically get deactivated when BGR circuit enters normal operating point. BGR circuit will have the two basic stable operating regions: (a) Normal operating region (NOR) and (b) zero current region (ZCR). Normal operating region is the operating point that generates the stabilized output (voltage/current) of designed circuit, which is extracted according to design requirements [3]. Zero current region is another operating point that generates destabilized output of designed circuit. By the effect of zero operating point, the entire circuit automatically gets disturbed resulting in deactivation of the circuit at critical interval of time period. In order to avoid ZCR, start-up circuit has been designed in all the bandgap reference circuit. The start-up circuit acts as an instant energy booster for the main block of analog circuits.

3.2 Overview of CM_BGR with Start-up Circuit

Figure 12 [16] shows the complete circuit of proposed CM_BGR with start-up circuit, where the start-up circuit consists of two NMOS devices—M6, M7, and a resistor R3. The CM_BGR circuit in Fig. 12 is designed in such a way that the auxiliary start-up block disturbs the gate nodes at W and Z transistors and forces them to enter into conduction point for stabilizing the circuit. The voltages at node W and Z are used as the reference points for analyzing the functionality of start-up devices. The start-up circuit operates in two steady states: one is the normal working state (NOR) and other is the zero current state (ZCR). During these two conditions of circuit, the gate–source voltages (V_{sg3} and V_{sg4}) of PMOS current source devices $M3$ and $M4$ are represented by node "W" which generates supply-independent current that operates in two stable states ($W = V_{dd}$ and $W = $ GND) corresponding to the ZCR and NOR. Similarly, the gate–source voltages (V_{gs1} and V_{gs2}) of NMOS current sink devices $M1$ and $M2$ are represented by node "Z" which generates supply-independent current that operates in two stable states ($Z = $ GND, $Z = $ VDD) corresponding to ZCR and NOR. The proposed start-up circuit is designed in such a way that it monitors the bandgap reference circuit to produce a desired output voltage without disturbing the normal operating region of BGR. By this, when CM_BGR enters into ZCR (i.e., $W = $ VDD and $Z = $ GND) then automatically the start-up circuit will be turned ON and when CM_BGR reaches normal operating region (i.e., $W = $ GND and $Z = $ VDD) then automatically start-up circuit will be turned OFF. When the BGR circuit enters ZCR region ($W = $ VDD and $Z = $ GND), NMOS transistors $M6$ and $M7$ work like an active load. As the voltage at node W is high, the voltage at node Z is low. M6 transistor turns ON by pulling down the voltage at node W and simultaneously $M7$ transistor turns ON by pulling up the voltage at node Z. This action will lead to current flow into the main CM_BGR block with the voltage at nodes (W and Z) till $M7$ and $M6$ transistors gate voltage turn to low/high, respectively. This ensures that the CM_BGR circuit always start-up at the correct operation point. Thus, the start-up circuit will not affect the NOR of the proposed CM_BGR circuit. Therefore, any of the designed start-up circuit must encounter the stable performance, reliable start, small current consumption, and start by enabling signal which are the requirements to be satisfied as an secondary circuit. Further, construction and significance of start-up circuit introduced in CM_BGR circuit are explained in detail.

3.3 Design of CM_BGR with Start-up Circuit

Figure 12 shows the design of a CM_BGR with start-up block. In this design, a transistor $M6$ is introduced. The size of transistor is proportional to the on time of start-up circuit. To keep the on time of start-up circuit to a minimum, the width of $M6$ is designed to be small. The drain of $M6$ device is connected to the potential of node "W" and pulls down the gate–source voltages "W" of PMOS device to ground. The gate of $M6$ is controlled by drain of another device $M7$ and the gate of $M7$ device is connected to potential node "Z" which is the gate–source voltages of NMOS devices and along with that a resistor $R3$ of 10 K ohms is introduced to balance the gate–source voltage of $M6$ device. Initially, in Fig. 10, the circuit got stuck at zero current operating point (ZCR) in the case of CM_BGR without start-up circuit. But now in Fig. 12, the start-up circuit disturbs the ZCR of CM_BGR and encourages CM_BGR to perform NOR without start-up issues. With this illustration of CM_BGR circuit with start-up block, the calculation of each component used in designing is as follows,

1. Calculation of device dimension for $M7$ NMOS transistor is calculated by using drain current equation in linear region of operation, i.e., given in Eq. (3.1).

$$I_{D(\text{linear})} = \mu_n C_{\text{ox}} \left(\frac{W}{L}\right) \left[\left(V_{\text{gs7}} - V_{\text{th}}\right) V_{\text{DS}} - \frac{V_{\text{DS}}^2}{2} \right] \tag{3.1}$$

$$\left(\frac{W}{L}\right)_7 = \frac{I_{D7(\text{linear})}}{\mu_n C_{\text{ox}} \left[\left(V_{\text{gs7}} - V_{\text{th}}\right) V_{\text{DS}} - \frac{V_{\text{DS}}^2}{2} \right]}$$

$$W = 15 \times 10^{-6}, L = 0.5 \times 10^{-6}$$

By simplifying Eq. (3.1), the width (W_7) of $M7$ device is calculated to be 15×10^{-6} and its length (L_7) is designed to be 0.5×10^{-6}.

2. Calculation of device dimension for $M6$ NMOS transistor is calculated by using drain current equation in sub-threshold or weak inversion region of operation, i.e., given in Eq. (3.2).

$$I_{\text{ds6}} = I_{t6} \left(\frac{W}{L}\right)_6 e^{\frac{q\left(V_{\text{gs6}} - V_{\text{th}}\right)}{nKT}}. \tag{3.2}$$

$$I_{t6} = 2\mu_n C_{\text{ox}} \left(\frac{W}{L}\right) \left(\frac{KT}{q}\right)^2 \tag{3.3}$$

Using Eq. (3.3), I_{t6} is calculated to be 692.94×10^{-9} A which is the characteristic current that defines the current that leaks through the transistors $M6$. The aspect ratio of $M6$ is calculated using Eq. (3.4).

$$\left(\frac{W}{L}\right)_6 = \frac{I_{\text{ds6}}}{I_{t6} e^{\frac{q\left(V_{\text{gs6}} - V_{\text{th}}\right)}{nKT}}} \tag{3.4}$$

where W and L are the $M6$ transistor width and length, n is the slope factor (around 1–1.5) and defines the effect that the gate voltage has on the drain current, V_{th} is the threshold voltage, q is the magnitude of electron charge, and k is the Boltzmann's constant with absolute temperature T. Using Eq. (3.4), the width and length of $M6$ transistor are estimated to be 2×10^{-6} m and 0.5×10^{-6} m. Table 3 tabulates the

Table 3 Resistor and transistor dimensions used in the proposed CM_BGR with start-up

Components	Estimated values	Simulated values
$Q1$, $Q3$	Normalized area (N) = 1	Normalized area (N) = 1
$Q2$	Normalized area (N) = 8	Normalized area (N) = 8
$M1$, $M2$, $M3$, $M4$, $M5$	$W = 20$ µm, $L = 5$ µm	$W = 20$ µm, $L = 5$ µm
$M6$	$W = 2$ µm, $L = 0.5$ µm	$W = 2$ µm, $L = 0.5$ µm
$M7$	$W = 15$ µm, $L = 0.5$ µm	$W = 15$ µm, $L = 0.5$ µm
$R1$	5.4 K (Ω)	5.4 K (Ω)
$R2$	45.89 K (Ω)	40.9 K (Ω)
$R3$	10 K	10 K

designed resistor values and transistor dimensions of the proposed design of CM_BGR with start-up circuit.

3.4   Layout of the CM-BGR with Start-up Circuit

Figure 13 shows the layout of bandgap reference circuit with start-up block of area 81.4 µm × 72.56 µm in 0.18 µm CMOS process technology. Layout challenges are taken care using DRM (design rule manual) in the library of gpdk180 which

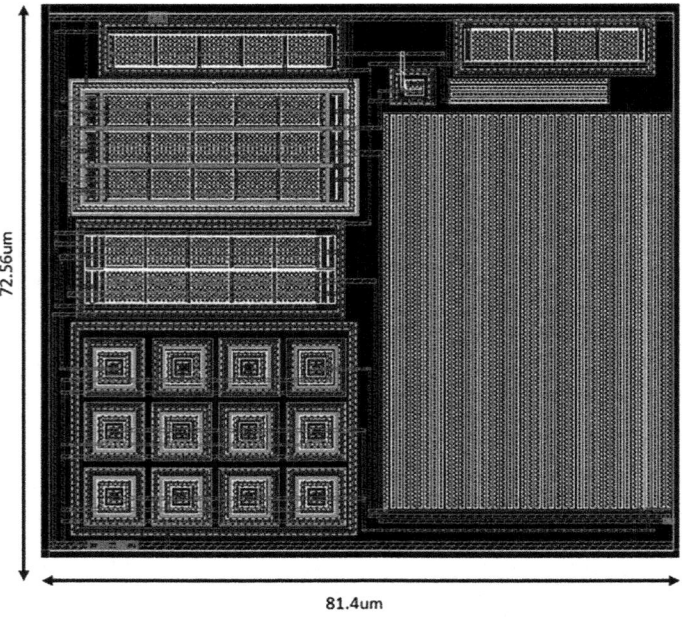

Fig. 13 Layout of current mirror bandgap reference (BGR) with start-up block

will be able to analyze the following tribulations to be taken care during layout designing process.

1. Current density rules, width of power supply net, and number of vias are considered as per DRM manual.
2. To avoid electron migration (EM) effect, current nets have to be taken care by increasing width of metal and more number of contacts are been used.
3. Latch-up effect and shielding of the devices are avoided by using guard ring.
4. Well proximity effect is taken care by well continuing process.
5. In the layout of 180 nm technology resistor (Poly-resistor), size is very large due to less sheet resistance and comparatively in 45 nm technology resistor (N + poly silicide resistor), size is lesser due to more sheet resistance within the resistor.
6. Metal orientation is followed (VHV) by placing metal1 vertically, metal2 horizontally, and metal3 vertically.

Summary: The summary of unit 3 illustrates the design and layout of improved CM_BGR circuit with start-up [17] block, produces an output voltage of 1.127643 V at 27 °C room temperature, gives a temperature coefficient of 0.3 ppm/°C, and is comparatively more than the temperature coefficient of CM_BGR without start-up block. But the significance of start-up circuit ensures that the correct operating point in any of the bandgap reference circuits starts up using start-up block [7] which are simulated thoroughly using transient analysis in unit 4 using Spectre ADE.

4 Simulation Results

The design of proposed bandgap reference circuit is simulated using Spectre ADE, its results, and tabular column of all CTAT, PTAT, and self-biasing circuit generations of the proposed bandgap reference circuit which is independent of the parameters such as temperature, supply voltage, and process technology (TVP) which are illustrated in this unit.

4.1 Simulated Results of CTAT Generation

The plot of voltage versus temperature is shown in Fig. 14a. Figure 14a illustrates the reduction in voltage across diode-connected vertical PNP BJT with an increase in temperature. From Fig. 14a, we observe that the voltage is CTAT in nature with reference voltage of about 729.8 mV at 27 °C. The derivative of the curve gives the slope of the graph as −1.509 mV with voltage variation of 249.88 mV resulting in negative temperature coefficient (NTC) of −2071.13 ppm/°C when temperature varies from −40 to 125 °C in design of CM_BGR circuit without start-up block in Fig. 10. Similarly, considering the design of CM_BGR circuit with start-up block in Fig. 12, the plot of voltage across diode-connected vertical PNP BJT versus temperature is shown in Fig. 14b with reference voltage of about 729.6 mV at 27 °C. The derivative of the curve gives slope of the graph as −1.505 mV with voltage variation of −249.53 mV resulting in negative temperature coefficient (NTC) of −2072.78 ppm/°C when temperature varies from −40 to 125 °C.

Table 4 gives the measured performance of voltage and temperature parameters of the CTAT generation circuit in the proposed design of CM_BGR without and with start-up block. In Fig. 10, simulated results of output voltage reference without start-up circuit are tabulated as 578.6 and 828.48 mV, which is the maximum Vctat(max) and minimum Vctat(min) voltage reference observed at the T(max) of 125 °C and T(min) temperature of −40 °C with the temperature variation ΔTemp is about 165 °C. Similarly, when Fig. 12 is simulated, the results of output voltage reference are tabulated to be 578.7 and 828.23 mV, which is the maximum and minimum voltage references observed under the temperature range of 125 and −40 °C with the temperature variation of about 165 °C. Comparatively, the output reference voltage variation and the temperature coefficient variation are observed during simulation between the circuits in Figs. 10 and 12 which are very small, about ±0.2% and ±1.65% at different temperatures, respectively. Therefore, when variations are minimum, we can conclude that introducing the start-up circuit will not have any effect on the CTAT generation circuit.

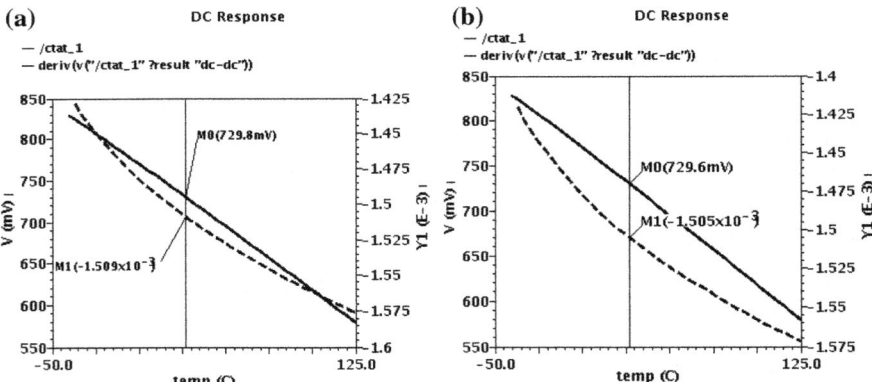

Fig. 14 **a** Plot of voltage v/s temperature CM_BGR without start-up circuit. **b** Plot of voltage v/s temperature CM_BGR with start-up circuit

Table 4 Measured performance of voltage and temperature parameters of the CTAT generation circuit without and with start-up

Performance parameters of CTAI generation	Simulated results of CM_BGR without start-up	Simulated results of CM_BGR with start-up
Voltage v/s temperature	Vctat (mV)	Vctat (mV)
Vctat @ 27 °C	729.8	729.6
Vctat (max) @ 125 °C	578.6	573.7
Vctat (max) @ 40 °C	823.48	828.23
ΔVctat	−249.88	−249.53
ΔVctat/Vctat	−342.39	−342.009
ΔVctat ΔTemp	−1.509	−1.505
Temperature coefficient (TC) ppm/°C	−2071.13	−2072.78

4.2 Simulated Results of PTAT Generation

In Fig. 15a, $V2$ is the source–substrate voltage V_{SB} (node potential) observed at $M2$ transistor to be around 729.24 mV and Vbe'N' is the parallel base–emitter voltage to be around 676.15 mV; both are simulated at 27 °C room temperature. And both node voltages ($V2$, Vbe'N') vary in complementary to absolute temperature (voltage decrease with increase in temperature). When these two node voltages are subtracted to each other, i.e., $V2$–Vbe'N', then a voltage of 53.09 mV is generated which is PTAT in nature. By this, PTAT voltage generation is achieved in the proposed design of CM_BGR in the absence of start-up block. In Fig. 15b, $V2$ is the node potential observed to be around 728.18 mV and Vbe'N' is the parallel base–emitter voltages to be around 675.26 mV; both are simulated at 27 °C room temperature. And both node voltages ($V2$, Vbe'N') vary accordingly with CTAT nature of curve (voltage decreases with increase in temperature). When these two node voltages are subtracted to each other, i.e., $V2$-Vbe'N', then a voltage of 52.92 mV is generated which is PTAT in nature. Similarly, PTAT voltage generation is achieved in the proposed design of CM_BGR in the presence of start-up block.

Table 5 illustrates the performance of voltage and temperature parameters of the PTAT voltage generation without and with start-up block. The maximum $V_{PTAT}(\text{max})$ and minimum $V_{PTAT}(\text{min})$ PTAT voltage variation measured with −40 to 125 °C temperature range is observed to be around 190.6 µV which is generated in the absence of start-up circuit and 199.27 µV is generated in the presence of start-up circuit.

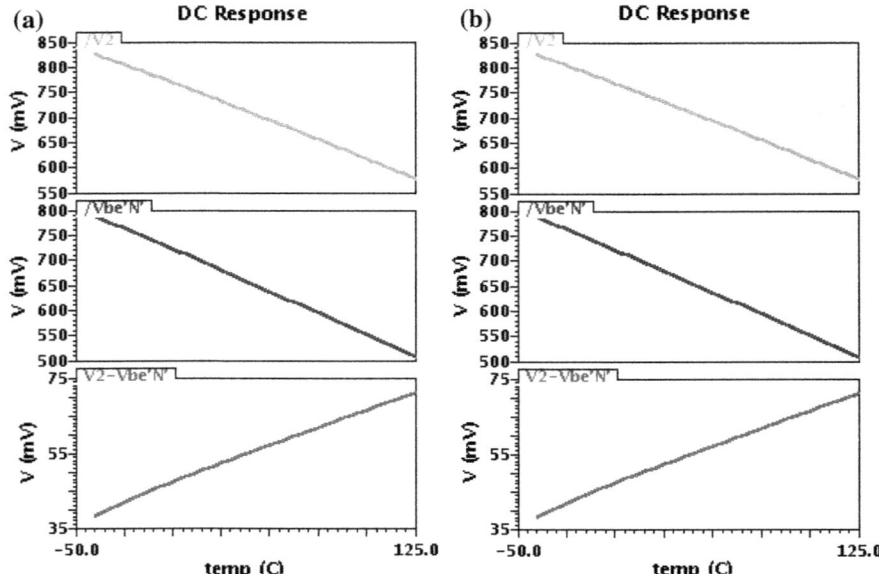

Fig. 15 a DC response plot of PTAT voltage generation without start-up circuit. **b** DC response plot of PTAT voltage generation with start-up circuit

Table 5 Measured performance of voltage and temperature parameters of the PTAT voltage generation without and with start-up

Parameters of PTAT	Simulated results of CM_BGR without start-up			Simulated results of CM_BGR with start-up		
Voltage v/s temperature	V2 (mV)	Vbe 'N' (mV)	[V2-Vbe'N'] (mV)	V2 (mV)	Vbe'X' (mV)	[V2-Vbe'N'] (mV)
VPTAT (max) @ 125 °C	577.78	507.13	70.65	578.42	507.35	71.07
VPTAT (min) @ −40 °C	826.2	787	39.2	824.65	786.46	38.19
VPTAT @ 27 °C	729.24	676.15	53.09	728.18	675.25	52.92
VPTAT/ΔT	−1.50	−1.69	190.6u	−1.49	−1.69	199.27u

4.3 Simulated Results of Self-biased Circuit Generation

Figure 16a shows the simulated result of proposed design of self-biased circuit in the absence of start-up block and Fig. 16b shows the simulated result of proposed design of self-biased circuit in the presence of start-up block. Both simulated results appear to be the same but with small variation in temperature and voltage parameters. In Fig. 10, $V1$ is the source–substrate voltage V_{SB} (node potential) pointed between $M1$ and $Q1$ transistor. Where its source–substrate voltage (V_{sb1}) is observed to be about 730.9 and 728.9 mV, it is the measured value of source–substrate voltage (V_{sb2}) pointed at $V2$ node potential between $M2$ transistor and $R1$ resistor. Basically, the self-biasing circuit is designed to generate equal node potential across nodes $V1$ and $V2$. But, in our proposed design of self-biasing circuit, there is a small variation of about 2 mV between nodes $V1$ and $V2$, due to supply voltage dependency. This dependency can be further reduced by using cascaded current mirror circuit connection in Fig. 10. Similarly, in Fig. 12, the voltage drop across $V1$ is found to be 731 mV and the voltage drop across $V2$ is found to be 730 mV with small voltage variation between $V1$ and $V2$ that is about 1 mV, which is comparatively lesser than 2 mV in Fig. 10. Therefore, the significance of start-up circuit has been illustrated with small variation in self-biased circuit and its simulated values are charted in Table 6.

In Table 6, the maximum and minimum voltage drop across node $V1$ is (578.10, 829.16 mV) and V2 is (577.78, 826.17) at temperature of about 125 to −40 °C which is illustrated in case of the absence of start-up circuit. Similarly, the maximum and minimum voltage drop across node $V1$ is (578.64, 827.37 mV) and V2 is (578.42, 824.61) at temperature of about 125 to −40 °C which is illustrated in the case of presence of start-up circuit.

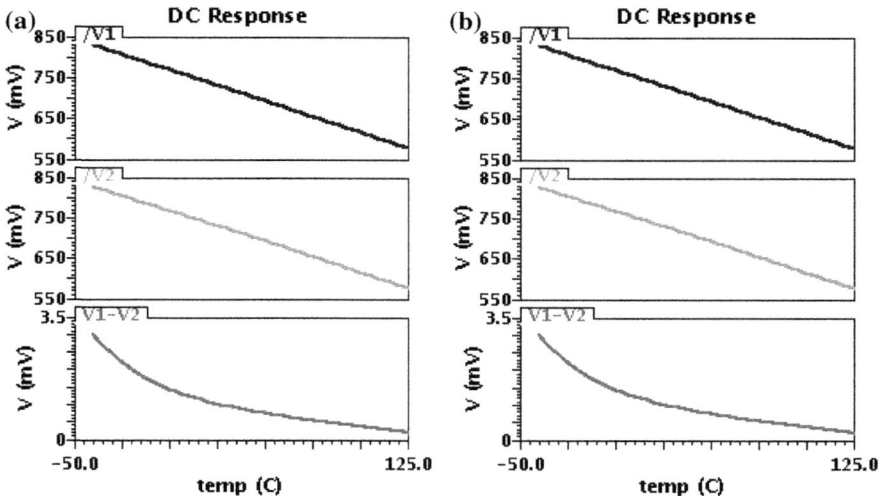

Fig. 16 **a, b** Plot of $V1$, $V2$ and ($V1$–$V2$) v/s temperature

Table 6 Measured performance of voltage and temperature parameters of the self-biasing circuit generation without and with start-up

Performance parameters of self-biasing circuit	Simulated results of CM_BGR without start-up			Simulated results of CM_BGR with start-up		
Voltage v/s Temperature	VI (mV)	V2 (mV)	ΔI = VI − V2	VI (mV)	V2 (mV)	ΔV = VI − V2
V (max) @ 125 °C	578.10	577.78	230 μV	578.64	578.42	245.97 mV
V (min) @ −40 °C	829.16	826.17	2.99 mV	327.37	324.61	248.73 mV
V (nominal) @ 27 °C	730.9	723.9	2 mV	731	730	1 mV

4.4 Temperature Variation Simulated Results of Output Voltage Reference of Without and with Start-up Circuit of CM_BGR

Figure 17 a, b gives the simulated output voltage reference of CM_BGR without and with start-up circuit. As seen from Fig. 17a, b, there is no significant difference seen during the DC analysis. Table 7 tabulates the measured performance of proposed design of BGR, which generates constant output voltage reference which is independent of temperature and voltage variations. Table 7 illustrates the output voltage reference V_{ref} to be 1.12764 V, when bandgap reference circuit is designed without start-up circuit and V_{ref} to be 1.12763 V when bandgap reference circuit is designed with start-up circuit. These two circuits of output voltage reference V_{ref} values are approximately equal to 1.12 V at 27 °C (300 K) room temperature within a temperature range of −40 °C to +125 °C (165 °C). When V_{ref} equals to 1.12764 V at 27 °C (300 K) room temperature, the temperature coefficient (TC) is found to be 0.2 ppm/°C in case of the absence of start-up circuit and similarly V_{ref} equal to 1.127643 V at room temperature with temperature coefficient (TC) of

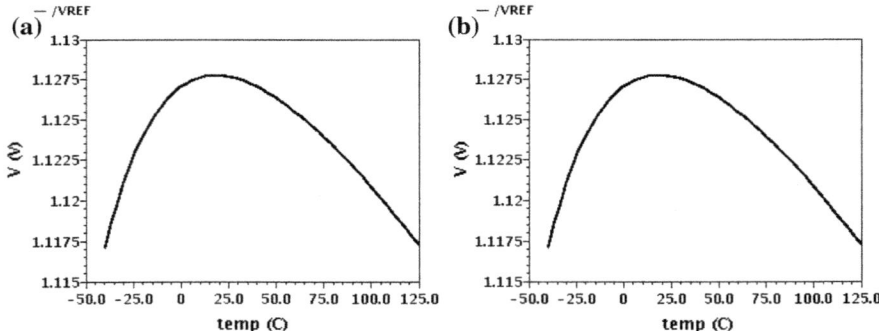

Fig. 17 a DC analysis of designed CM_BGR without start-up circuit. **b** DC analysis of designed CM_BGR with start-up circuit

Table 7 Measured performance of output voltage reference with temperature parameter of the proposed BGR circuit without and with start-up block

Performance parameters of	Simulated values Without start-up	Simulated values With start-up
Temperature range(/°C)	−40 to +125 °C	−40 to +125 °C
ΔT (°C) = (Tmax − Tmin)	165	165
V_{ref} (max) @ 125 °C	1.117226	1.117263
V (min) @ −40 °C	1.117186	1.117205
V_{ref} @ 27 °C	1.12764	1.127643
ΔV_{ref} = (V_{ref}_Temp (max) − V_{ref}_Temp (min))	0.24 μV	0.35 μV
Temperature coefficient (TC)	0.2 ppm/°C	0.3 ppm/°C

0.3 ppm/°C in the presence of start-up circuit. The maximum voltage reference V_{ref}(max) at 125 °C is observed to be 1.117226 V and the minimum voltage reference V_{ref}(min) at −40 °C is observed to be 1.117186 V in case of CM_BGR without start-up circuit. Similarly, the maximum voltage reference V_{ref}(max) at 125 °C is observed to be 1.117263 V and the minimum voltage reference V_{ref}(min) at −40 °C is observed to be 1.117205 V in case of CM_BGR with start-up circuit.

The change in voltage reference ΔV_{ref} of BGR circuit without start-up is around 0.24 μV and with start-up is 0.35 μV. Therefore, comparatively no big variations can be identified between without and with start-up circuit of CM_BGR in terms of V_{ref} with respect to temperature variation parameter.

4.5   Supply Variations Simulated Result Without and with Start-up Circuit of CM_BGR

Figure 18a, b shows the plot of V_{ref} versus V_{dd} with maximum and minimum supply voltage range of about 1.8 and 3.3 V where, at 3.3 V, maximum output voltage reference V_{ref}(max) is observed to be 1.16 and at 1.8 V the minimum voltage reference V_{ref}(min) is observed to be 1.12 V in case of the absence of start-up circuit in CM_BGR. Similarly, at 3.3 V, the maximum voltage reference V_{ref}(max) is observed to be 1.157 and at 1.8 V the minimum voltage reference V_{ref}(min) is observed to be 1.102 V in case of the presence of start-up circuit in CM_BGR. The change in voltage reference ΔV_{ref} of without start-up is around 0.033 mV and with start-up is 0.055 mV is measured and tabulated in Table 2. In this case, V_{ref} varies bit in case of the presence of start-up circuit due to its supply voltage variations which can be further resolved by using cascaded current mirror circuit which reduces the variations in output reference voltage with supply voltage parameter. In Table 8, V_{ref}(max) is equal to 1.16 at 3.3 V V_{DD}(max) and V_{ref}(min) is equal to 1.127 V at 1.8 V V_{DD}(min), and the change in supply voltage ΔV_{DD} is 1.5 V which gives the line regulation of 22 mV/V in case of the absence of start-up

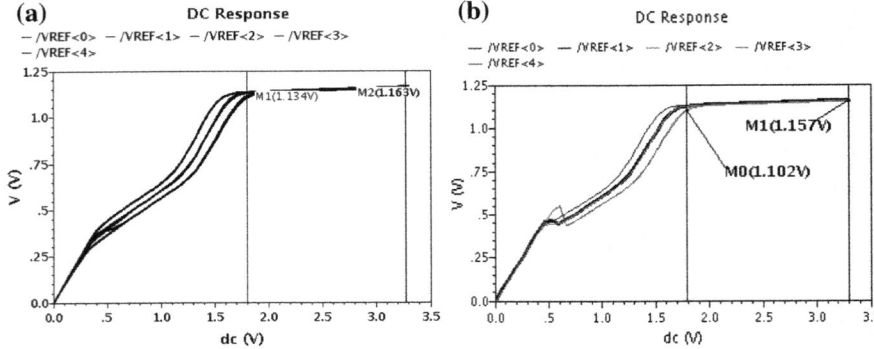

Fig. 18 **a** Plot of V_{ref} v/s V_{dd}. **b** Plot of V_{ref} v/s V_{dd}

Table 8 Measured performance supply voltage parameters of the proposed BGR circuit without and with start-up

Parameters	Simulated values Without start-up	Simulated values With start-up
Supply voltage range	1.8–3.3 V	1.8–3.3 V
$\Delta V_{DD} = V_{DD}$ (max) − V_{DD} (min)	1.5 V	1.5 V
V_{ref} (max) @3.3 V	1.16 V	1.157 V
V_{ref} (min)@ 1.8 V	1.127 V	1.102 V
$\Delta V_{ref} = V_{ref}$ (max) − V_{ref} (min)	0.033 mV	0.055 mV
Line regulation (LRE) $\Delta V_{ref\_}V_{DD} = \Delta V_{ref}/\Delta V_{DD}$	22 mV/V	36.66 mV/V

circuit. Similarly, V_{ref}(max) is equal to 1.157 V at 3.3 V V_{DD}(max) and V_{ref}(min) is equal to 1.102 V at 1.8 V V_{DD}(min) which gives the line regulation 36.66 mV/V in case of the presence of start-up circuit.

4.6 Simulated Result of Proposed CM_BGR Without and with Start-up Using Process Corner Parameter

Figure 19a, b shows the plot of Vref versus temperature which illustrates the simulated output voltage reference (Vref) of five different process corners (i.e., NN, SF, FS, SS, and FF) of the proposed bandgap reference(CM_BGR) without and with start-up circuit as a function of process and temperature variations over the range −40 to 125 °C.

Table 9 shows the output reference voltage characteristics over temperature at five process corners and also it can be seen that the absolute value of the voltage reference is different from those at the nominal NN process corner. Comparatively,

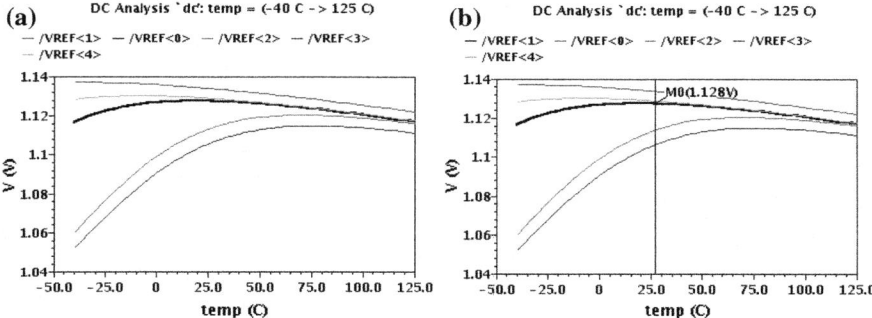

Fig. 19 **a** Plot of V_{ref} v/s temperature. **b** Plot of V_{ref} v/s temperature

during normal (NN) process corner, the voltage reference at room temperature in Fig. 19a is observed to be 1.127626 V with ZTC and zero voltage variations, but in case of Fig. 19b it is observed to be 1.127625 V with positive temperature coefficient (PTC) of about 0.44 ppm/°C and 82 µV of voltage variations. During slow SS process corner, the voltage reference at room temperature in Fig. 19a is observed to be 1.129201 V with negative temperature coefficient (NTC) about −60.91 ppm/°C and −0.01135 V of voltage variations, but in case of Fig. 19b it is observed to be 1.1068 V with PTC of about 320.33 ppm/°C and −0.0585 of voltage variations. During fast (FF) process corner, the voltage reference is observed to be 1.134061 V with NTC about −82.58 ppm/°C and −0.015453 V of voltage variations, but in case of analysis of process corners with start-up block, the voltage reference is observed to be 1.134104 V with NTC of about −78.51 ppm/°C and −0.0147 of voltage variations. During slow–fast (SF) process corner, the voltage reference is observed to be 1.1069 V with PTC about 321.940 ppm/°C and 0.0588 V of voltage variations, but in case of analysis of process corners with start-up block, the voltage reference is observed to be 1.1144 V with PTC of about 300.7 ppm/°C and 0.0553 V of voltage variations. During fast–slow (FS) process corner, the voltage reference is observed to be 1.1143 V with PTC about 301.86 ppm/°C and 0.0555 V of voltage variations, but in case of analysis of process corners with start-up block the voltage reference is observed to be 1.129216 V with NTC of about −57.15 ppm/°C and −0.01061 V of voltage variations. The above analysis and observation of the curvatures of the V_{ref} voltages which are tabulated are sensitive to process variations, which are the weakness of the proposed design approach. In general, process variation effect can be reduced by constructing cascade structure in self-biasing circuit. This process variation in voltage reference is due to the threshold voltage variation in the MOS transistors which is an important aspect that needs to be considered in the analysis of sensitivity to process variations.

Table 9 Measured performance of process corner parameters of the proposed BGR circuit without and with start-up block

Performance parameters	Process corners without start-up circuit					Process corners with start-up circuit				
PROCESS Corner	NN	SF	FS	SS	FF	NN	SF	FS	SS	FF
V_{ref} (min) T (min) @ −40 cG	1.117205	1.0528	1.0609	1.128586	1.137136	1.117205	1.0611	1.128586	1.0531	1.137136
V_{ref} @ 27 °C	1.127626	1.1069	1.1143	1.129201	1.134061	1.127625	1.1144	1.129216	1.1068	1.134104
V_{ref} (max) T (max) @ + 125 °C	1.117205	1.1116	1.1164	1.117236	1.121683	1.117287	1.1164	1.117973	1.1116	1.122444
TC (ppm/°C)	0	321.940	301.86	−60.91	−82.5S	0.44	300.7	−57.15	320.33	−78.51
ΔVref = (Vref_Temp (max)-Vref_Temp (min))	0	0.0588	0.0555	−0.01135	−0.015453	82uV	0.0553	−0.01061	−0.0585	−0.0147

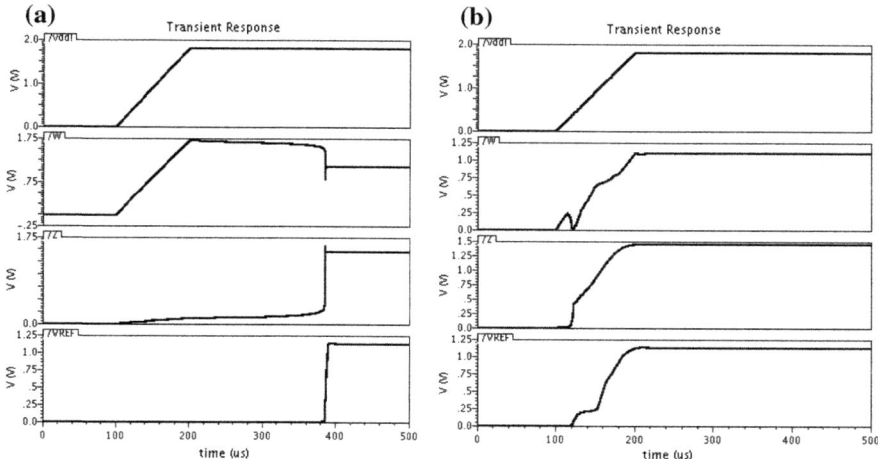

Fig. 20 **a** Transient analysis of designed CM_BGR circuit without start-up. **b** Transient analysis of designed CM_BGR circuit with start-up

4.7 Transient Response of CM_BGR Without and with Start-up

To see the significance of start-up circuit, we go for transient analysis of designed CM_BGR circuit. Figure 20a, b gives transient response analysis of CM_BGR without and with start-up circuit.

4.7.1 Observation of CM_BGR Without and with Start-up Circuit

With the construction of start-up circuit as shown in Fig. 12, now to run transient analysis of Fig. 12, save all currents of the start-up circuit and simulate it with transient analysis. The analysis of the current and voltage performance of start-up circuit and the plot of Id_6 and V_{gs6} versus time (µs) of $M6$ device is shown in Fig. 20b, which illustrates that when CM_BGR circuit enters ZCR then M6 device will turn ON automatically and start pumping minimum amount of current to the branches through potential nodes "W" and "Z" until CM_BGR enters into NOR. When the CM_BGR block enters into NOR, the gate voltage of $M6$ device is pulled down by the $M7$ device below its threshold voltage (20.85 m) resulting in turns OFF of start-up circuit spontaneously. Therefore, the start-up circuit is isolated completely from main CM_BGR block. The plot of "V_{dd}" and "V_{ref}" versus time in Fig. 20b gives output voltage reference of $V_{ref} = 1.128$ V without start-up issues. Hence, the significance of the start-up circuit plays major role in the CM_BGR circuit. In Fig. 19, it is hard to detect the necessity of start-up circuit with DC simulations because most of the simulator will avoid the ZCR. In Fig. 19, when we run DC simulation and sweep the temperature from −40 to 125 °C of range and simulate the circuit, we can observe in Fig. 20a a plot of output reference voltage (V_{ref} = 1.12 V) which is generated without start-up circuit.

But in case of DC analysis simulator, it will not show the significance of start-up circuit because simulator will automatically avoid ZCR. Therefore, by disabling the V_{dc} of DC analysis, the significance of start-up circuit can be analyzed by setting pulse signal (V_{pulse}, $V1 = 0$, $V2 = 1.8$, delay = 100 µs, rise time = 100 µs, and pulse width of 0.5 s) of transient analysis. Initially, when supply is zero, integrated circuit (IC) will be turned OFF and when supply is gradually increased, then IC will be turned ON with maximum supply voltage of 1.8 V. This transient analysis is set to run for 500 µs and its plot of output voltage reference (V_{ref}) response is simulated as shown in Fig. 20a. The observation of this plot shows the start-up issue in the CM_BGR circuit with small delay during ZCR. This delay is due to the negligible leakage current (sub-threshold current) which flow in the MOS (PMOS and NMOS) devices resulting in slower performance of the CM_BGR circuit as shown in Fig. 20a. At ZCR, the gate–source voltage (V_{gs1} and V_{gs2}) of NMOS devices will be less than their threshold voltage and gate–source voltage (V_{sg3} and V_{sg4}) of PMOS devices will be having no difference in their threshold voltage. Hence, all devices will be in sub-threshold region of operation. The plot of "V_{dd}" and "W" versus time (µs) in Fig. 20a illustrates the difference between V_{dd} and W with trifling amount of voltage (104 mV) and a trifling amount current (51.58 pA) which are tabulated in Table 10. Similarly, the plot of "V_{dd}" and "Z" versus time (µs) in Fig. 20a) illustrates the difference between "V_{dd}" and "Z" with trifling amount of voltage (118.3 mV), which will be generated in the CM_BGR resulting in deprived performance of the circuit. The plot of "V_{dd}" and "V_{ref}" versus time in Fig. 20a gives

Table 10 Transient analysis of performance parameters of the proposed BGR circuit without and with start-up

Performance parameters	Without start-up circuit		With start-up circuit	
Reference points in the circuit	Zero current region (ZCK)	Normal operators region (NOR)	Zero current region (ZCR)	Normal operating region (NOR)
PMO S (G) = W	−364.4a	1.696	1.19f	1.09v
NMOS (G) = Z	1.658f	118.6 m	623.1av	1.445v
V_{REF}	77.59z	1.398 m	2.29av	1.128v
PMOS(Is)	329.4 × 10^{-27} A	63.25pA	3.75 3 × 10^{27}	8.502pA
$V_{gs6}(M6)$	–	–	6.549z	20.83 m
$V_{gs7}(I)$	–	–	−67.95 × 10^{-27}	8.502pA
$V_{gs7}(M7) = z$	–	–	712.9a	1 446v
$V_{gs7}(I)$	–	–	−550.6 × 10^{-27}	177.9 µA
$R3(I)$	–	–	−575.4 × 10^{-27}	177.9 µA

output voltage reference with start-up issues. By selecting all the source terminals (Is) of PMOS devices, we can plot "V_{dd}" and "Is" versus time (μs) as shown in Fig. 20a which results in the pico-ampere range of current flow in all the three branches. With all the above transient analysis performance of CM_BGR circuit, it is showed that the CM_BGR circuit will not function (will drop into ZCR) well without start-up circuit. The significance of start-up circuit is observed and tabulated using transient analysis in Fig. 20a, b. The observation in Fig. 20 illustrates that from transient analysis, it is observed that the design of CM_BGR with start-up circuit is better than CM_BGR without start-up circuit, where the design may fail to respond during ZCR but with start-up circuit; comparatively, we will get good response during ZCR which is independent of temperature, supply, and process variations.

5 Conclusions

In bandgap reference, mismatches of resistors and bipolar transistors may seriously affect its performances [18]. But it can be eliminated by an appropriate layout design to a great extent [12]. In order to acquire accurate ratio of two resistors, a group of parallel sub-resistors with identical geometries are used, and the two resistors are arranged in the form of fork. In the layout of bipolar transistors, $Q2$ is composed of eight blocks, while $Q1$ and $Q3$ are one. Meanwhile, common-centroid layout is applied to achieve good matching [19]. Figure 13 shows the layout of the whole bandgap reference circuit with start-up circuit. The area of layout is 81.4 μm × 72.56 μm and major area is occupied by BJT that is 37.295 μm 28.275 μm and poly-resistor (37.64 μm × 54.11 μm) with a 0.18 μm CMOS process, DRC (design rule check), and LVS (layout versus schematic) are checked. The finalized result of proposed bandgap reference circuit is able to function properly under different temperature ranges (−40 to 125 °C), supply voltages (i.e., 1.8–3.3 V), and process corner conditions (i.e., NN, SF, FS, SS, and FF), which are required by the design specification.

References

1. Paul R. Gray, Paul J. Hurst, Stephen H. Lewis, Robert G. Meyer, "Analysis and Design of Analog Integrated Circuits", Fourth Edition.
2. Guijie WANG "CMOS Bandgap references and temperature Sensors and their applications" Master of Science in Electronics, Nankai University, China, Geboren Te Henan, China, 2005.
3. Ting-Chou Lu, Ming-Douker, Hsiao-Wen Zan, Chung-Hung Kuo, Chun-Huai Li, Yao-Jen Hsieh and Chun-Ting Liu, "Design of Bandgap voltage reference circuit with all TFT devices on glass substrate in a 3-um LTPS process" Proc. 2008 IEEE CICC, pp. 721–724, Oct. 2008.

4. P. K. T. Mok and N. L. Ka, "Design considerations of recent advanced low-voltage low-temperature-coefficient CMOS bandgap voltage reference," in Custom Integrated Circuits Conference, 2004. Proceedings of the IEEE 2004, ser. pp. 635–642, Custom 2004.

5. Rajarshi Paul and Amit Patra, "A temperature compensated bandgap voltage reference circuit for high precision applications," Proceedings of the IEEE INDICON, pp. 553–556, Dec. 2004.

6. Utthej Nukala, "Design of a temperature independent MOSFET only current reference", Master of Science thesis, Akron University, Dec 2011.

7. Y.-H. Lam and W.-H. Ki, "CMOS bandgap references with self-biased symmetrically matched current-voltage mirror and extension of sub-1-V design," IEEE Trans. Very Large Scale Integr. (VLSI) Syst., vol. 18, no. 6, pp. 857–865, 2010.

8. D. Colombo, G. Wirth, S. Bampi, and C. Fayomi, "Impact of noise on trim circuits for bandgap voltage references," in 14th IEEE International Conference on Electronics, Circuits and Systems, ICECS, pp. 775–778, 2007.

9. Boni, A. "Op-amps and startup circuits for CMOS bandgap references with near I-V supply", IEEE JSSC, Volume: 37, Issue: 10, Oct. 2002.

10. Jan Skoda, "Low voltage low power bandgap reference", Diploma thesis, kvetna 2015.

11. Dalton Martini Colombo. "Bandgap voltage references in sub-micrometer CMOS technology", M.S. Theses, Universidad Federal Do Rio Grande Do Sul, Porto Alegre, 2008.

12. Vishal Gupta, "An Accurate, Trimless, High PSRR, Low-Voltage, CMOS Bandgap Reference IC", Degree Doctor of Philosophy in the School of School of Electrical and Computer Engineering, Georgia Institute of Technology August 2007.

13. Gabriel A. Rincon Mora, "Voltage Reference from Diodes to Precision High Order Bandgap Circuits", Texa Instruments, Inc. Dallas, Texas.

14. Meijer G. C. M., Guijie Wang, Fruett F. "Temperature sensors and voltage references implemented in CMOS technology", IEEE sensors journal, Volume: 1, Issue: 3, Oct 2001.

15. B. Razavi, "Design of Analog CMOS Integrated Circuits", Tata McGraw-Hill Higher Education, international Edition 2001.

16. Siew Kuok Hoon, Jun Chen, Maloberti F., "An improved bandgap reference with high power supply rejection," Circuits and Systems, 2002. ISCAS 2002. IEEE International Symposium on, vol. 5, pp. V-833, V-836, 2002.

17. Amr I. Kamel, Ahmed Saad and Lee Seng Siong, "A high band PSSR and Fast start-up current mode bandgap reference in 130 nm CMOS technology ", IEEE 2016 Custom Integrated Circuits Conference, pp. 506–509, 2016.

18. Gabriel A. Rincon Mora, "Voltage Reference from Diodes to Precision High Order Bandgap Circuits", Texa Instruments, Inc. Dallas, Texas.

19. Juan Pablo Martinez Brito, Hamilton Klimach, Sergio Bampi, "A design methodology for matching improvement in bandgap references," unpublished.

Analytical Modelling of Room-Temperature GaAs/InAs$_{0.3}$Sb$_{0.7}$ Detector for H$_2$S Gas Detection

Trilok Kumar Parashar and Rajesh Kumar Lal

1 Introduction

The infrared region is a very favourite region for non-telecommunication applications, mainly in sensors. One of the potential applications of the photovoltaic detectors is optical gas sensors. Such gas sensor operates at signature wavelength of the corresponding pollutant and toxic gases as well as liquids. Many toxic gases and liquids having their characteristic absorption band lie in the NWIR and MWIR [1–10]. Signature characteristic wavelength of these toxic gases such as NH$_3$, H$_2$S, CH$_4$, CO$_2$, CO, etc. is nearly spaced and lies in LWIR region and lower part of MWIR. Extensive investigations have been carried out worldwide to develop photodetectors based on narrow bandgap materials. But so far, the existing gas sensors have had strong limitations such as poor selectivity and stability, low speed, high cost, large size and weight [11]. In this chapter an analytical model is proposed for H$_2$S gas similar to often used earlier by researchers of this field [5, 6, 9, 12] for other gases of MWIR and LWIR. Also, the effects of trap on the parameters of the gas detector are also presented in this paper. We observed that the R$_0$A peak is affected and slightly shifts with trap. In our study, we have selected very recently proposed GaAs/InAs$_{0.3}$Sb$_{0.7}$ detector structure for the first time for our modelling. The proposed photodetection can be utilised for H$_2$S gas detection. H$_2$S gas can harm human and animal life in different ways. Recently, Lihua Zeng et al. published a paper in which he reported that the H$_2$S gas concentration can affect the growth and even reproductive capacity of the chickens, which also threatens their health [13].

T. K. Parashar
University Polytechnic, Birla Institute of Technology, Mesra, Ranchi, India
e-mail: parashar303017@rediffmail.com

R. K. Lal (✉)
Department of ECE, Birla Institute of Technology, Mesra, Ranchi, India
e-mail: rklal@bitmesra.ac.in

© Springer Nature Singapore Pte Ltd. 2019
V. Nath and J. K. Mandal (eds.), *Proceeding of the Second International Conference on Microelectronics, Computing & Communication Systems (MCCS 2017)*, Lecture Notes in Electrical Engineering 476, https://doi.org/10.1007/978-981-10-8234-4_36

The III–V $InAs_{1-x}Sb_x$ material has now established as most important material utilised for optoelectronic devices specifically optical gas sensor applications. It is mainly because of its broad bandgap range, better stability when compared to competitor materials, and least alteration of energy gap bandwidth with composition variation. Furthermore, added advantage is the high electron mobility of the undoped InAsSb materials when compared to GaAs [14, 15]. Therefore, the InAsSb-based materials are the obvious choice, having potential in applications for high-speed device. The major practical difficulty is to grow very high-quality $InAs_{1-x}Sb_x$ epilayer, because no proper suitable substrate is available. Although some most commonly used substrates such as for III–V materials InSb, GaSb and InAs provide relative smaller lattice mismatches, they are expensive and provide poor quality compared to GaAs substrates [15]. The heteroepitaxial growth of $InAs_{1-x}Sb_x$ on GaAs substrate is best suited for such kind of device development as it provides the GaAs substrates with high quality and availability of advanced processing technology for it. However, due to very large lattice mismatch between InAsSb and GaAs, the growth of a high-quality material is still very difficult. During last few years, the successful growth of InAsSb on GaAs substrate using appropriate buffer layers has been reported by scientists from different laboratories [14–19]. Also, the direct growth of InAsSb on GaAs has been reported by Liu et al. [15] and Rao et al. [20].

In our paper, we describe a theoretical model for characterization of an InAsSb-based single hetrojunction P^+-n^0 $GaAs/InAs_{0.3}Sb_{0.7}$ detector grown on GaAs (100) substrate. The present work aims to explore the effect of trap energy on device parameters of the proposed gas detector. Proposed GaAs/InAsSb-based photovoltaic infrared detector provides good stability, selectivity and high speed at room temperature.

2 Modelling of Gas Detector

In our paper, we present a MATLAB-based theoretical model to study the end parameter of p^+-n^0 $GaAs/InAs_{0.3}Sb_{0.7}$. The structure under consideration is a single heterojunction p^+-n^0 $GaAs/InAs_{0.3}Sb_{0.7}$ photodiode grown on GaAs substrate directly by MBE as suggested by Liu et al. [15] and is shown in Fig. 1a. The importance of this material lies in the fact that the device based on this material can be used at room temperature in MWIR regions (~ 2 to 5 μm). The detector based on this material has a long wavelength cut-off just above 3.8 μm that makes it suitable for use in optical gas sensors for the detection of H_2S [15]. The energy band diagram for the active junction is shown in Fig. 1b.

Light incident on the top p^+ GaAs layer. The lightly doped n^0 InAsSb layer works as the active layer. The absorption of incident light takes place in neutral p^+, n^0 regions and also in the depletion region formed at the p^+-n^0 junction.

In our model of the dark current of the InAsSb photodetector, we have considered (i) the diffusion component of the thermally generated carriers of neutral

Fig. 1 **a** Schematic diagram of the proposed detector and **b** energy band diagram

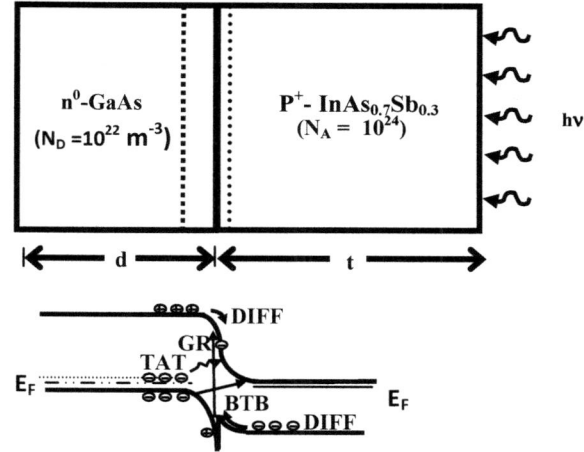

regions, I_{diff}; (ii) generation-recombination component of carriers in the depletion region, I_{gr}; (iii) the tunnelling of carriers through the barrier, I_{tun}.

In single heterojunction photodetector, the total tunnelling current, I_{tat}, arises due to (a) band-to-band tunnelling (I_{btb}) and (b) trap-assisted tunnelling (I_{tat}). We have not considered, I_{btb}, band-to-band tunnelling component in our modelling. Thus, $I_{\text{tun}} = I_{\text{tat}}$

Therefore, the net dark current can be written as

$$I_{\text{net}} = I_{\text{diff}} + I_{\text{gr}} + I_{\text{tat}} \tag{1}$$

2.1 Diffusion Component

For the structure under consideration, the diffusion current constitutes contributions from neutral n^0 and p^+ regions. The minority carrier diffusion current density under the application of a low bias voltage, V, can be analysed by solving 1-D diffusion equation under appropriate boundary conditions. Assuming the $p^+ - n^0$ metallurgical junctions as the reference, the diffusion current component due to holes and electron, respectively, can be obtained as

$$(I_{\text{diff}})_p = \frac{qAn_i^2}{N_D} \left[\frac{kT}{q} \frac{\mu_p}{\tau_{\text{eff}}} \right]^{\frac{1}{2}} \tanh \left(\frac{d - x_n}{L_p} \right) \left(\exp \left(\frac{qV}{kT} \right) - 1 \right) \tag{2a}$$

$$(I_{\text{diff}})_n = \frac{qAn_i^2}{N_A} \left[\frac{kT}{q} \frac{\mu_n}{\tau_{\text{eff}}} \right]^{\frac{1}{2}} \tanh \left(\frac{d - x_p}{L_n} \right) \left(\exp \left(\frac{qV}{kT} \right) - 1 \right) \tag{2b}$$

Here, in above equations, n_i denotes intrinsic carrier concentration of InAsSb. The diffusion current density can be written as

$$I_{\text{diff}} = (I_{\text{sn}} + I_{\text{sp}}) \left(\exp\left(\frac{qV}{kT}\right) - 1 \right) \tag{3}$$

2.2 Generation-Recombination Component

The main cause of generation-recombination (I_{gr}) component of the dark current is the indirect or SRH recombination of the thermally generated carriers through the recombination levels arises due to the defects within the depletion region. The GR component of current density can be approximated as [21]

$$I_{\text{gr}} = \frac{q n_i W \sigma N_f v_{\text{th}} A}{2} \exp\left(\frac{qV}{2kT}\right) \tag{4}$$

where V is the applied voltage, N_f is trap density at the P–n^0 heterointerface, n_i is the intrinsic carrier concentration and τ_{SRH} is the Shockley–Read–Hall generation-recombination lifetime.

2.3 Trap-Assisted Tunnelling Component

Trap-assisted tunnelling (TAT) occurs when carriers start tunnelling from trap states, populated by the minority carriers on the quasi-neutral side of the junction to the empty band states on the opposite side of the junction or tunnelling can also be through some trap sites that exist even in the depletion region of the junction. This tunnelling process is shown in Fig. 1b. These trap centres are the energy levels created between conduction band and valence band by the presence of donor impurities which may be intentionally doped and can be accidentally and unintentionally introduced during fabrication in the material. The TAT tunnelling component of current density can be obtained from simple 1-D model and can be expressed as [22]

$$I_{\text{tat}} = \frac{A q^2 m_p M^2 V N_f}{8 \pi \hbar^3} \exp\left(-\frac{4\sqrt{2 m_p (E_{g1} - E_t)^3}}{3 q \hbar E} \right) \tag{5}$$

where E_g is the energy bandgap of the semiconductor, E is the maximum electric field (E_{max}) across the depletion region, E_t is the trap centres energy (in eV), N_f denotes the trap density occupied by the electrons and M is the matrix element [22].

The resistance area product (RA) for the different individual components (diff, GR and TAT) can be obtained as

$$(RA)_x = \left(\frac{dJ_x}{dV}\right)^{-1} \tag{6}$$

where the suffix x corresponds to different components. Each component of R_0A product further comprises contributions from both neutral regions.

The R_0A product in each case is calculated using the following relation:

$$\frac{1}{(RA)} = \frac{1}{(RA)_p} + \frac{1}{(RA)_n} \tag{7}$$

The overall value of the resistance area product RA_{net} is given by

$$\frac{1}{RA_{net}} = \frac{1}{RA_{diff}} + \frac{1}{RA_{gr}} + \frac{1}{RA_{tat}}. \tag{8}$$

3 Lifetime Modelling

In our modelling, we have considered all three dominant recombination processes: radiative (which is band-to-band recombination), Shockley–Read–Hall and Auger recombination (which are non-radiative recombination mechanisms).

The lifetime of carriers for the radiative recombination can be expressed as [1]

$$\tau_{RAD} = \frac{n_i^2}{G_r\left(n_0 + \frac{n_i^2}{n_0}\right)} \tag{9}$$

where G_r is the recombination rate for bonds with spherical symmetry and can be expressed as [1]

$$G_t = \frac{8\pi}{h^3 c^3} \int_0^\infty \frac{\varepsilon(E)\alpha(E)E^2 dE}{\exp\left(\frac{E}{kT} - 1\right)} \tag{10}$$

The dependence of the absorption coefficient α on the incident photons energy of the can be expressed following Ref. [22] as

$$\alpha(E) = \frac{2^{2/3} m_0 q^2}{3\varepsilon_\infty^{1/2} h^2} \left(\frac{m_e^* m_h^*}{m(m_e^* m_e^*)}\right)^{3/2}$$
$$\times \left(1 + \frac{m_0}{m_e^*} + \frac{m}{m_e^*}\right)\left(\frac{E - E_g}{mc^2}\right)^{1/2} \tag{11}$$

The non-radiative Auger recombination process, on the other hand, is quite complex and a non-radiative mechanism. A narrowband semiconductor consists of a conduction band and light-hole and heavy-hole valence bands, and there can be a minimum of ten different kinds of Auger transitions. Among these transitions, the two major transitions which may occur at the minimum threshold energy ($E_T = E_g$) are the (i) CHCC (A-1) and (ii) CHLH (A-7). The A-1 and A-7 are dominant in n-type and p-type material, respectively [22–24]. Here, in the present analysis, the structure under consideration is taken as InAsSb, for which the spin split-off energy is compared to the bandgap of the material. So, one more type of Auger transition known as CHSH (A-S) process also becomes significant. Therefore, we also considered Auger-S in our computation of the effective lifetime of the carriers. The net Auger recombination lifetime (τ_{AU}) of the carriers can thus be written as [25]

$$\frac{1}{\tau_{AU}} = \frac{1}{\tau_{A-1}} + \frac{1}{\tau_{A-7}} + \frac{1}{\tau_{A-S}} \tag{12}$$

The lifetimes for these Auger mechanisms can be written, respectively, as

$$\tau_{A-1} = \frac{2\,\tau_{A-1}^i}{1 + n_0/p_0}$$

$$\tau_{A-7} = \frac{2\,\tau_{A-7}^i}{1 + p_0/n_0} \tag{13}$$

$$\tau_{A-S} = \frac{2\,\tau_{A-S}^i}{1 + p_0/n_0}$$

Here, τ_{A-1}^i, τ_{A-7}^i and τ_{A-S}^i indicate intrinsic recombination times for corresponding Auger transitions, respectively, can be expressed as [24]

$$\tau_{A-1}^i = \frac{3.8 \times 10^{-18}\varepsilon_s^2(1+\mu)^{1/2}(1+2\mu)\exp\left[\left(\frac{1+2\mu}{1+\mu}\right)\frac{qE_g}{kT}\right]}{\frac{m_e^*}{m_0}|F_1 F_2|^2 \left(\frac{kT}{qE_g}\right)^{3/2}} \tag{14}$$

$$\tau_{A-7}^i = \frac{m_e^*(E_{th})\left(1 - \frac{5qE_{th}}{4kT}\right)}{m_{e0}^*\left(1 - \frac{3qE_{th}}{2kT}\right)}\,\tau_{A-1}^i \tag{15}$$

$$\tau_{A-S}^i = \frac{5}{54}\,\frac{\varepsilon_s^2 m_{hh}^{*3} m_e^{*3/2} kT\Delta^2(E_g+\Delta)}{\pi^4(h/2\pi)^3 q^4\, n_i^2 m_s^{5/2}(\Delta - E_g)\,\exp\left[\frac{q(\Delta-E_g)}{kT}\right]} \tag{16}$$

where F_1 and F_2 are the overlap integrals of the periodic part of Bloch's functions. Here, $m_e*(E_{th})$ is the electron effective mass related to threshold energy for the Auger-7 transition. For this transition, $E_{th} \approx E_g$.

The lifetime of carriers arises from Shockley–Read–Hall recombination that can be modelled in terms of N_f and σ as

$$\tau_{\text{SRH}} = \frac{1}{\sigma N_f v_{\text{th}}} \tag{17}$$

where N_f and σ are the SRH trap density and capture cross section, respectively. v_{th} is the thermal velocity of the minority carriers in the active region, given by

$$v_{\text{th}} = \sqrt{\frac{3kT}{m^*}} \tag{18}$$

m^* being the effective mass of the charge carriers. The SRH recombination lifetime could be altered or improved by selecting pure- and good-quality material. The advanced processing techniques also improve SRH recombination lifetime.

4 Quantum Efficiency and Specific Detectivity

The quantum efficiency (η) of the photodetector has mainly three major contributions. These contributions are from the three regions, e.g. two from neutral n and p region (η_n and η_p, respectively) and the depletion region (η_{dep}). The net quantum efficiency can be obtained by

$$\eta = \eta_n + \eta_p + \eta_{\text{dep}} \tag{19}$$

These quantum efficiency components can be obtained as [6]

$$\eta_p = \frac{(1 - r_p)\alpha_p L_n}{\alpha_p^2 L_n^2 - 1} \left[\begin{array}{c} \frac{\alpha_p L_n + \gamma_n - \exp\left\{-\alpha_p(t - x_p)\right\}\left[\gamma_n \cosh\left(\frac{t - x_p}{L_n}\right) + \sinh\left(\frac{t - x_p}{L_n}\right)\right]}{\gamma_n \sinh\left(\frac{t - x_p}{L_n}\right) + \cosh\left(\frac{t - x_p}{L_n}\right)} \\ -\alpha_p L_n \exp\left\{-\alpha_p(t - x_p)\right\} \end{array} \right] \tag{20}$$

$$\eta_n = \frac{(1 - r_p)\,(1 - r_n)\,\alpha_n L_p}{\alpha_n^2 L_p^2 - 1} \exp\left\{-(\alpha_p t + \alpha_n w_n)\right\}$$

$$\left[\begin{array}{c} \frac{(\gamma_p - \alpha_n L_p)\,\exp\left\{-\alpha_n(d - x_n)\right\} - \left[\gamma_p \cosh\left(\frac{d - x_n}{L_p}\right) + \sinh\left(\frac{d - x_n}{L_p}\right)\right]}{\gamma_p \sinh\left(\frac{d - x_n}{L_p}\right) + \cosh\left(\frac{d - x_n}{L_p}\right)} \\ +\alpha_n L_p \end{array} \right] \tag{21}$$

where $\gamma_n = \frac{S_n L_n}{D_n}$ and $\gamma_p = \frac{S_p L_p}{D_p}$.

The quantum efficiency in the depletion region can be obtained as

$$\eta_{\text{dep}} = (1 - r_n)\,(1 - r_p) \left[\begin{array}{c} \exp\big(-\alpha_p\,(t - x_p)\big) \\ -\exp\big(-(\alpha_p t + \alpha_n x_n)\big) \end{array} \right] \tag{22}$$

The performance parameter of any detector used for non-telecommunication application is mainly described by its specific detectivity D^*. The detectivity of the detector can be estimated from the net effect of the R_0A product (due to all involved in recombination mechanisms). The specific detectivity of the photodetector which is a function of the applied voltage can be written as

$$D^* = \frac{q\eta\lambda}{hc} \sqrt{\frac{RA_{\text{net}}}{4kT}} \tag{23}$$

RA_{net} is the net or effective value of the component zero-bias resistance area products in different mechanisms.

5 Results and Discussion

Physics-based analytical model has been carried out on single hetrojunction p^+–InAs$_{0.3}$Sb$_{0.7}$/n^0–GaAs photodetector at room temperature (300 K). The light has been assumed to be incident from the top p^+–InAs$_{0.3}$Sb$_{0.7}$ side. The incident photons with energy higher than the band gap of InAs$_{0.3}$Sb$_{0.7}$ are absorbed in both p^+ and n^0 regions.

In the analytical modelling of the present photodetector, the assumed value of various parameters used in modelling is mainly taken from references [25, 26] and is listed in Table 1.

The various dark current components and the R_0A products are obtained using proposed theoretical model. In present work, the dependence of the performance parameters such as RA product, detectivity and quantum efficiency at (i) various applied voltages and (ii) trap densities has been estimated quantitatively.

Figure 2 shows the effect of applied bias voltage on net dark current together with its component from all three major recombination mechanisms (such as diffusion, generation-recombination and tunnelling) for the trap level $e_t = 1.33$ eV.

At low e_t, the bottom of the conduction band of InAsSb is always above the trap level. Thus, the possibility of tunnelling of the electrons from the valance band of GaAs to the conduction band of InAsSb is very low. Hence, low value of TAT current is obtained in forward as well as reverse bias; this clearly indicates that the TAT component has no significant effect on the detectivity of the device at this trap level. Also, the drift current component of the dark current is negligible in reverse bias and increase rapidly in forward bias as in normal diode. Therefore, the net detector current I_{net} is solely due to the generation-recombination (GR) component even up to forward bias <0.4 V. Rest of the two current components have negligible

Table 1 Device parameter

Parameter	Values
N_D	10^{22} m^{-3}
N_f	10^{20} m^{-3}
E_{g_GaAs}	1.424 eV
E_g (InAs$_{0.3}$Sb$_{0.7}$)	$0.411 - \dfrac{3.4 \times 10^{-4}T^2}{T+210}$ $-0.876x + 0.7x^2$ $+3.4 \times 10^{-4}x(1+x)T$
t	3 µm
d	1 µm
ε_r (InAs$_{0.3}$Sb$_{0.7}$)	$(15.15 + 1.65x)\,\varepsilon_0$
m_n^*(GaAs), m_p^*(GaAs)	$0.063m_0$, $0.51m_0$
m_n^*, m_p^* (InAs$_{1-x}$Sb$_x$)	$0.023 - 0.039x + 0.03x^2\ m_0$, $0.41 + 0.02 \times m_0$
Δ (InAs$_{1-x}$Sb$_x$)	0.516 eV

Fig. 2 Variation of the net dark current along with its component with the voltage at constant trap level (e_t = 1.33 eV)

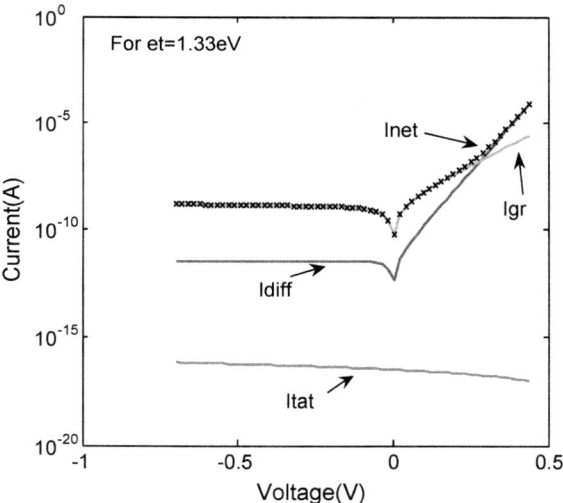

contributions on the net current. As a result, we get fully overlapped curve of I_{net} with I_{GR}. At forward bias $V > 0.4$ V, large total current is observed due to diffusion component. This is also reflected in figure for applied voltage above than 0.4 V.

In Fig. 3, the variation of R_0A_{net} and its component for the same e_t = 1.33 eV is shown with applied voltage. As obvious, R_0A_{net} value depends mainly on the GR component due to the small value of the tunnelling current.

However, when diffusion current comes in to picture in forward bias, R_0A_{net} decreases more rapidly due to R_0A_{diff}.

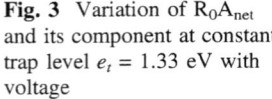

Fig. 3 Variation of R_0A_{net}
and its component at constant
trap level $e_t = 1.33$ eV with
voltage

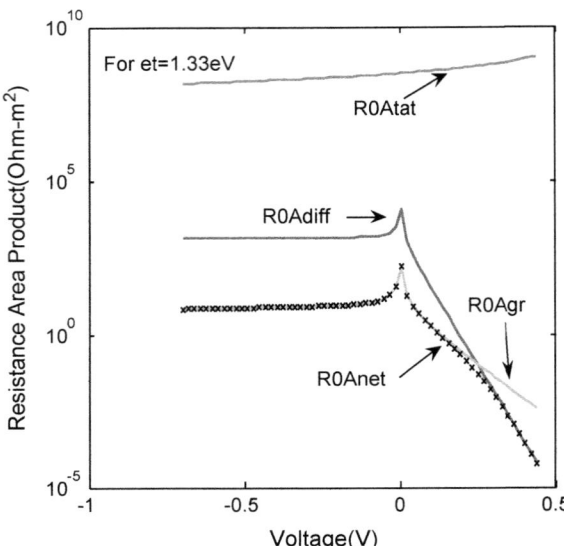

In the proposed GaAs/InAsSb photodetector, the nature of the I-V characteristic is largely dependent on the trap level e_t. These trap centres can arise while device fabrication. The contribution of the trap-assisted tunnelling component, however, not only depends on the number of trap centres but also depends on the location of the trap level in the forbidden band.

Effect of the trap level is beautifully illustrated in Figs. 4 and 5 in a sequential manner, where we have plots of the net R_0A along with its component.

(a) In the first group shown in Fig. 4, we have a plot of net R_0A and its components for the two different values of trap level: $e_t = 1.33$ and 1.36 eV. In these plots, we found that as the trap level (e_t) increases there will be more chance that empty states of n_0-InAsSb conduction band are in front of trap levels. Therefore, the R_0A_{net} increases but for both e_t values (1.33 and 1.36 eV), tunnelling component of R_0A has almost no effect on net R_0A. Hence, R_0A_{net} follows R_0A_{gr} and R_0A_{diff} components below $e_t = 1.37$ eV.

(b) At $e_t = 1.37$ eV, in Fig. 5, the tunnel current is almost equal to generation-recombination (GR) component, but still tunnelling component of R_0A does not affect the net resistance area product. However, at 1.374 eV, after crossover, tunnel components will be more than generation-recombination component. Figure 5 clearly demonstrates it. From this point ($e_t = 1.37$ eV), R_0A_{tat} also starts to affect R_0A_{net}.

(c) In the last group (above $e_t = 1.37$ eV) shown in Fig. 6, the plots of net R_0A and components for $e_t = 1.38$ and 1.41 eV are shown. Here, for both e_t values, the tunnelling current leads to other dark current components. Therefore, R_0A_{net} is dominated by R_0A_{tat} for all trap levels above $e_t = 1.37$ eV. However, R_0A_{diff}

Fig. 4 Variation of the net R_0A and its component with voltage for two different values of $e_t = 1.33$ and 1.36 eV

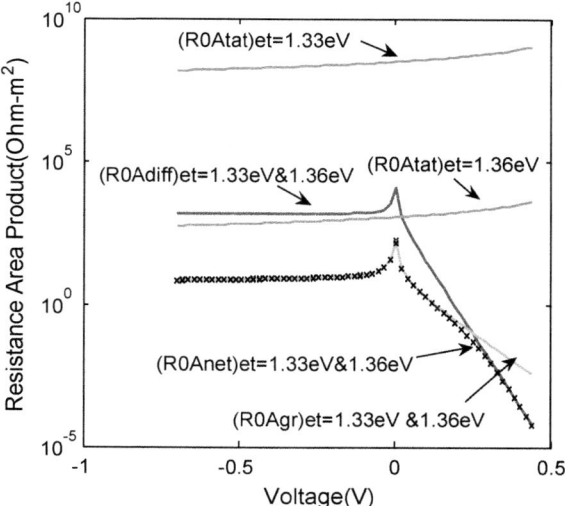

Fig. 5 Variation of the net R_0A and its component with voltage for two different values of $e_t = 1.374$ and 1.37 eV

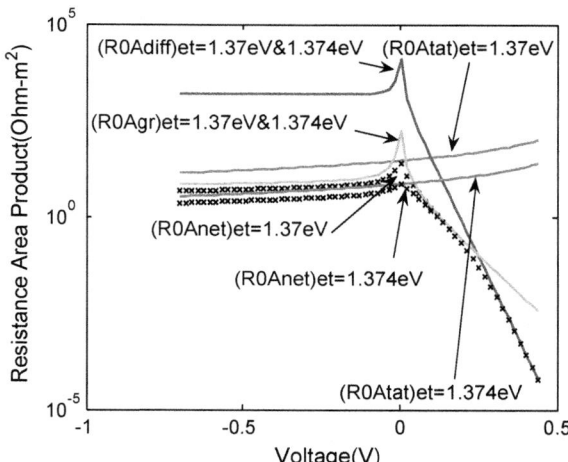

component remains to be dominating at forward bias. One more thing we can observe is that as tunnel current increases, the conductivity of the detector increases drastically. Thus, the value of net R_0A decreases almost 10^2 times for the change in e_t from 1.38 to 1.41 eV.

It is not only the trap level which affects the detector parameters but the carrier density (N_f) of trap level near P^+-n^0 heterointerface also largely alters the basic characteristic parameters of gas detector.

Figure 7 shows the variations of net resistance area product of the gas detector with applied voltage for different trap densities at the $P-n^0$ heterointerface. Here,

Fig. 6 Variation of the net
R_0A and its component with
applied voltage for two
different values of $e_t = 1.38$
and 1.41 eV

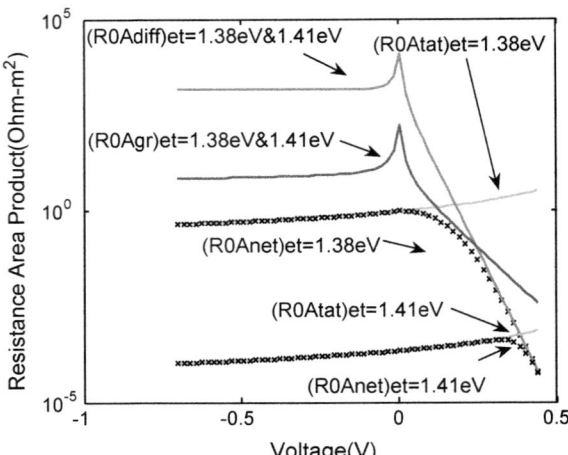

we found that for the lower effective trap density, N_f, the net R_0A peak and its position is shifted toward positive applied voltage. Also, one can see a slight increase in magnitude of R_0A_{net} with decrease of trap level carrier density. The increase of N_f resulted in decrease of net R_0A and same will result in detectivity of the detector. Thus, to improve the overall device performance, the trap level density must be as low as possible.

Figure 8 shows the efficiency with respect to the operating wavelength. Here, efficiency has its maximum value ~ 0.75 at 2.7 µm and decreases sharply on both sides. So the device operates well between the wavelength ranges 2.5–3.0 µm. As we know that 2.7 µm is the characteristic wavelength of H_2S gas, it is best suited for H_2S gas detection.

In Fig. 9, the Net R_0A with its component is plotted with respect to donor concentration at $e_t = 1.33$ eV. For lower donor concentration of n region smaller than 10^{24} m^{-3}, net R_0A is solely due to diffusion current but at higher concentration it becomes constant with TAT component. Detectivity increases with increase in doping below 10^{24}. Beyond this detectivity, it is almost not affected by the concentration, as the resistance of the detector is directly proportional to $(R_0A)^{\frac{1}{2}}$.

In the last two plots of Figs. 10 and 11, the effect of trap density (N_f) and position of the trap level (e_t) on detectivity of the device is shown, respectively.

In Fig. 10, large detectivity about 8×10^9 mHz$^{1/2}$/W is observed near wavelength 2.7 µm. The detectivity decreases with increase in trap density.

Similarly, in Fig. 11, the variation of the detectivity of the GaAs/InAsSb detector with operating wavelength for the different values of e_t is shown. Figure shows that the detectivity sharply decreases toward both sides. The value of the trap level affects the detectivity of the detector. With low e_t value, detectivity

Fig. 7 Variation of the net R_0A with applied voltage for different values of N_f (trap density at the P^+-n^0 heterointerface)

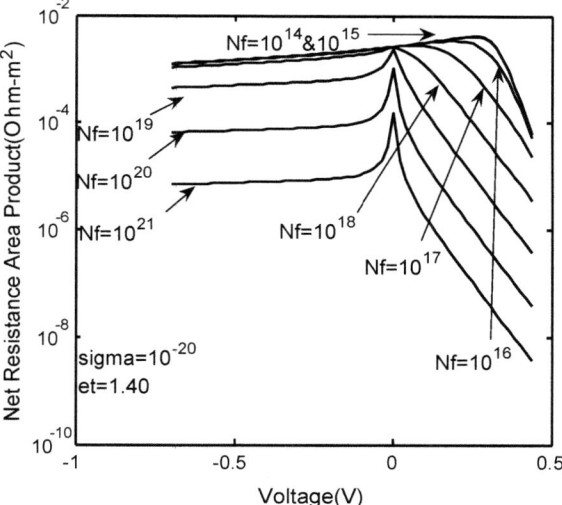

Fig. 8 Efficiency of the detector with respect to operating wavelength

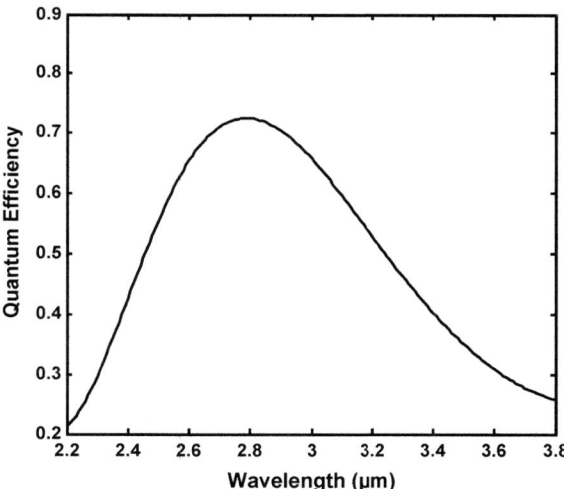

plot becomes slightly sharper and there is an increase in detectivity. In Figs. 10 and 11, the detectivity peak is obtained for wavelength 2.7 μm as said above, which is the signature wavelength of the H_2S gas. So, we can say that this detector is best suited for H_2S gas detection.

In this work, an analytical model of a mid-infrared gas detector has been developed. We applied this model on very recently proposed $GaAs/InAs_{0.3}Sb_{0.7}$ P^+-n^0 gas detector structure first time in our study. GaAs substrate provide the good lattice matching and good performance with InAsSb [27].

Fig. 9 Net R_0A with its component with respect to donor concentration at $e_t = 1.33$ eV

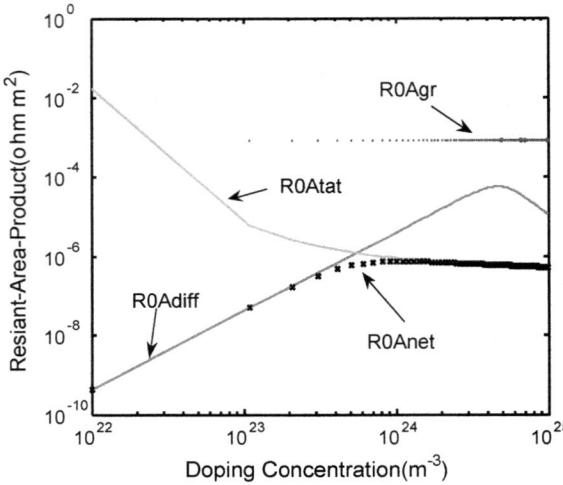

Fig. 10 Trap density effect on detectivity of the detector

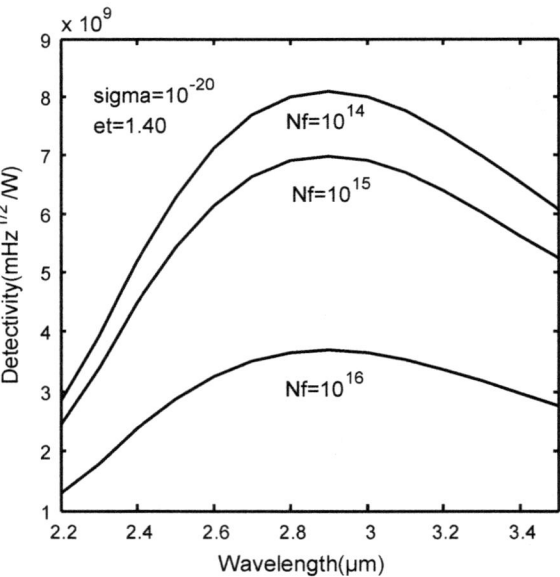

Fig. 11 Effect of trap level, e_t, on the detectivity of the GaAs/InAsSb detector

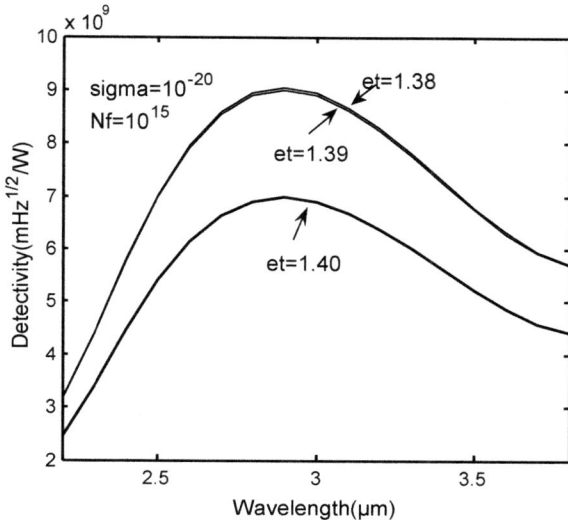

6 Conclusion

This structure provides high-speed and stable photodetector with improved performance. The study reveals that the basic parameters (such as Dark current, R_0A) as well as end parameters (detectivity) of the detector are largely affected by the presence of trap levels, its location and its carrier density (N_f). These traps may be present in the detector during fabrication. Results obtained on the basis of model clearly indicate that the device parameters and the performance of the detector can be slightly optimising by reducing trap level while fabrication. The trap level must be low for better performance of the device. Also, the result shows that R_0A varies with donor concentration in n^0 region. Detectivity increases with low-level doping. A very high detectivity $D > 8 \times 10^9$ is $mHz^{1/2}/W$ obtained at the wavelength 2.7 µm on the basis of the model. This shows that the proposed detector is well suited for H_2S gas detection. Design engineers can use the model presented here as a tool for detecting the effect of intentionally incorporated trap centre and can utilise it for better performance of the photodetector.

References

1. A. Rogalski, K. Adamiec and J. Rutkowski, "Narrow-Gap Semiconductor Photodiode", SPIE Press, Bellingham, USA, 2000.
2. M. A. Afrailov, "Photoelectrical characteristics of the InAsSbP based uncooled photodiodes for the spectral range 1.6–3.5 µm", Infrared Physics & Technology, 53, pp. 29–32, 2010.

3. Chichih Liao, Bing-Ruey Wu, K. C. Hsieh and K. Y. Cheng, "High electron mobility InAs$_{0.8}$Sb$_{0.2}$ grown on InP substrates by gas source molecular beam epitaxy", J. Vac. Sci. Technol. B, Vol. 26, No. 3, 2008.

4. H. Shao, W. Li, A. Torfi, D. Moscicka, and W. I. Wang, "Room-Temperature InAsSb Photovoltaic Detectors for Mid-Infrared Applications", IEEE Photonics Technology Letters, 18, NO. 16, 2006.

5. P. Chakrabarti, P. K. Saxena and R. K. Lal, "Analytical Simulation of an InAsSb photovoltaic detector for mid-infrared applications," International J. Infrared and Millimeter waves, **27**, pp 1119–1132, 2006.

6. R. K. Lal, M. Jain, S. Gupta and P. Chakrabarti, "An analytical model of a double-heterostructure mid-infrared photodetector", *Infrared Physics and Technology*, vol. 44, pp. 125–132, Feb. 2003.

7. Cengiz Besikci, Selcuk Ozer, Chris Van Hoof, Lars Zimmermann, Joachim John and Patrick Merken, "Characteristics of InAs$_{0.8}$Sb$_{0.2}$ photodetectors on GaAs substrates", Semicond. Sci. Technol. **16**, 992–996, 2001.

8. Y. Tian, B. Zhang, T. Zhan, H. Jiang, Y. Jin, "Theoretical analysis of the detectivity in N-p and P-n GaSb/GaInAsSb infrared photo-Detectors", IEEE Trans. Electron Dev., **ED-47**, pp. 544–551, 2000.

9. H. H. Gao, A. Krier and V. V. Sherstnev, "Room temperature InAs$_{0.89}$Sb$_{0.11}$ photodetectors for CO detection at 4.6 μm", Applied Phys. Lett., **77**, pp. 872–874, 2000.

10. A. Rogalski, "Third generation infrared detectors. Proceedings of the Symposium on Photonics Technologies for 7th Framework Program, Wroclaw 12–14 October, 2006.

11. Mello, M., Poti1, B., Risi, A., de Passaseo, A., Lomascolo M. and Vittorio, M. De: "GaN optical system for CO and NO gas detection in the exhaust manifold of combustion engines", J. Opt. A: Pure Appl. Opt. 8, S545–S549, 2006.

12. Gong, X. Yi., Yamaguchi, T., Kan, H., Makino, T., Iida, T., Kato, T., Aoyama, M., Hayakawa Y. and Kumagawa, M., "Room-Temperature Mid-Infrared Light-Emitting Diodes from Liquid-Phase Epitaxial InAs/InAs$_{0.89}$Sb$_{0.11}$/InAs$_{0.80}$P$_{0.12}$Sb$_{0.08}$ Heterostructures", Jap. J. Appl. Phys. 36 2614, 1997.

13. Lihua Zeng 1,2,3, Mei He 1,3, Huihui Yu 1,3 and Daoliang Li "An H2S Sensor Based on Electrochemistry for Chicken Coops", Sensor, 16, 1398, pp. 1–10, 2016.

14. Zotova, N. V., Kizhaev, S. S., Molchanov, S. S., Popova, T. B. and Yakovlev, Yu. P.: Long-wavelength Light-Emitting Diodes (λ = 3.4–3.9 mm) Based on InAsSb/InAs Heterostructures Grown by Vapor-Phase Epitaxy. Semiconductors, 34, 12, 1402–1405, 2000.

15. Xiaoming Liu, Hongtao Li, Fengyun Guo, Meicheng Li, Liancheng Zhao, " Effect of annealing on electrical properties of InAsSb films grown on GaAs substrates by molecular beam epitaxy", Physica E, 41, 1635–1639, 2009.

16. B. M. Nguyen, D. Hoffman, E. K. Huang, S. Bogdanov, P. Y. Delaunay, M. Razeghi and M. Z. Tidrow, "Demonstration of mid-infrared type-II InAs/GaSb superlattice photodiodes grown on GaAs substrate", Applied Physics Letters, Vol. 94, No. 22, 2009.

17. S. Nakamura, P. Jayavel, T. Kyama, Y. Hayakawa, J. Cryst. Growth 300, 497, 2007.

18. Hanchao Gao, Wenxin Wang, Zhongwei Jiang, Linsheng Liu, Junming Zhou, Hong Chen, "The growth parameter influence on the crystal quality of InAsSb grown on GaAs by molecular beam epitaxy", Journal of Crystal Growth, 308, 406–411, 2007.

19. Gao, H., Wang, W., Jiang, Z., Liu L., Zhou J., Chen, H., The growth parameter influence on the crystal quality of InAsSb grown on GaAs by molecular beam epitaxy. J. Crystal Growth, 308, 406–411, 2007.

20. B. V. Rao, D. Gruznev, T. Tambo, C. Tatsuyama, "Long wavelength infrared detectors. Gordon and Breach science publishers", J. Cryst. Growth 224, pp. 316, 2001.

21. R. Schoolar, S. Price, and J. B. Rosbeck, "Investigation of the generation-recombination currents in HgCdTe midwavelength infrared photodiodes", J. Vac. Sci. Technol., B 10, pp. 1507–1514, 1992.

22. R. K. Lal and P. Chakrabarti, "A comparison of dominant recombination mechanisms in n-type InAsSb materials" Progress in Crystal Growth and Characterization of Materials, Volume 52, pp. 33–39, 2006.

23. J. Bardeen, F. J. Blatt, L. H. Hall, "Indirect transitions from the valance to the conduction bands", Photoconductivity Conference, Atlantic City, pp. 146–154, 1954.

24. R. K. Lal, and P. Chakrabarti, "An analytical model of P+-InAsSbP/n0-InAs/n+-InAs single heterojunction photodetector for 2.4–3.5 µm region", Optical and Quantum Electronics, 36, 935–947, 2004.

25. V. Gopal, S. K. Singh and R. M. Mehra, 'Analysis of dark current contributions in mercury cadmium telluride junction diodes', Infrared Physics and Technology, **43**, pp. 317–326, 2002.

26. M. Levinshtein, S. Rumyantsev, and M. Shur (Eds.): "Hand book series on Semiconductor Parameters", vol. 1 and 2, World Scientific, London, 1999.

27. A. Rakovska, V. Berger, X. Marcadet, B. Vinter, K. Bouzehouane and D. Kaplan, "Optical charecterization and room temperature life time measurements of high quality MBE-grown InAsS on Gasb", Semiconductor Science and Technology, vol. 15, pp. 34–39, 2000.

Fuzzy Prediction Model for Water Temperature in Scheffler Solar Reflector

Om Prakash

1 Introduction

Concentrated solar collector system is a very useful system from now and for the future scenario. It has very high solar energy which is specifically used to generate electricity with the help of conventional turbine and generator [1].

Parabolic dish reflector installation to get solar energy with minimum cost can be optimized at rural areas in India. In today's scenario, single storage tank of Scheffler dish is being used. The focused light has maximum temperature heat, which is taken for heating, making food and producing electricity. The parabolic dish can be optimized for accommodating boiled water for household purposes. The device has one water storage tank, which is used to give dual functions, i.e. taking sun radiation and the second one is safe for the warmth of water. There are frequent methods known to keep water heat as it is. In Scheffler dish, there is a cylindrical water storage tank, which is placed at the centre point of the reflector [2]. The Scheffler dish is used for maintaining the temperature, because it gives effective results compared to others. The cylindrical absorber is thermally insulated to minimize the losses. The aim of the system is to fulfil the need of water about 100–200 L throughout the day, and the required area is around 2.0–2.5 m^2. The diameter of tank is up to 0.2–0.4 m [3]. In today's scenario, single storage tank of Scheffler dish is used and absorber is situated at the side of reflector. The design should be like that it will provide 5 L hot water per day [4].

O. Prakash (✉)
Department of Mechanical Engineering, Birla Institute of Technology,
Mesra, Ranchi, India
e-mail: omprakash@bitmesra.ac.in

© Springer Nature Singapore Pte Ltd. 2019
V. Nath and J. K. Mandal (eds.), *Proceeding of the Second International Conference on Microelectronics, Computing & Communication Systems (MCCS 2017)*, Lecture Notes in Electrical Engineering 476, https://doi.org/10.1007/978-981-10-8234-4_37

2   Design and Experimentation

The parabolic dish reflector water heater is different from the flat-plate solar collector, from both construction and experimentation. The SRWH has single and dual working; it takes solar energy and saves the heat, while the flat-plate heating only moves the fluid. The design is shown in Fig. 1 [5]. The depository tank is coloured black to concentrate maximum amount of radiation. At the inlet line, a non-back pipe is fitted, and pressure ease pipe is fitted at the outlet line. This type of heater is easier to fabricate, not costly and simple in working and maintenance. The efficiency is higher than the flat-plate collector [6].

From Table 1, it is seen that there is a temperature increase at the daytime and it get decreases when moved to the nighttime. The Scheffler reflector's area is about 2.7 m^2 [7]. The solar light coming on the reflector reflects the light to the focus point.

3   Fuzzy Logic

To perform and complete the function of encoding the fuzzy systems, fuzzy law and action should be taken. For encoding, apply a fuzzy inference tool that is MATLAB fuzzy logic toolbox [7] as shown in Fig. 2.

4   Result and Discussion

In this paper, a study is being carried out over the development of the fuzzy logic prediction model for the Scheffler solar collector. The prediction model was developed in the MATLAB software version 7. The simulated results were validated with experimental results. It is observed that model was able to predict with high level of accuracy. In this model, there are two input variables, which are global radiation (W/m^2) and wind speed (m/s). The output variable is only one which is temperature of the outlet water from the collector. Figure 3 shows the output indication of the model. This gives the output result of the fuzzy model.

Figure 4 shows the rules applied for the fuzzy model. There are three rules developed for this model.

Figure 5 shows the membership function for the input and output variables. The triangular membership function is being selected. There are three membership functions being selected for each input and output variables. The beam radiation is 784 W/m^2. The water temperature predicted is 61.5 °C and the actual one is 62 °C, so the output is with negligible error of 0.8%.

Fig. 1 **a** Photograph of
Scheffler reflector water
heater [2]. **b** Schematic
diagram of Scheffler reflector
water heater [2]

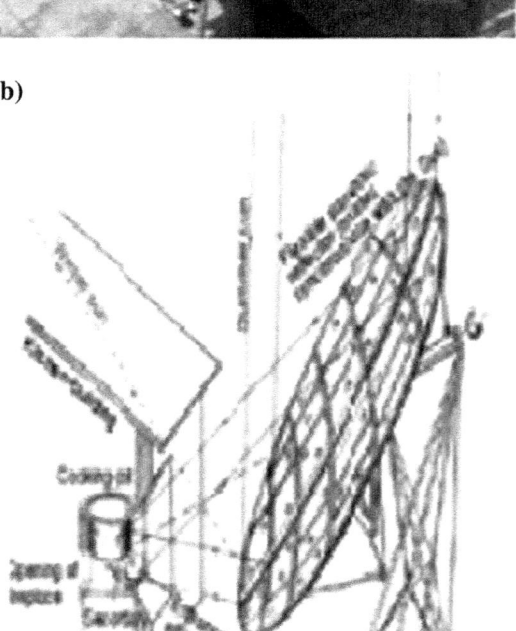

Table 1 Readings of water temperature, wind speed and ambient temperature [8]

S. No.	Time (h)	Generated water temperature (°C)	Wind speed (km h^{-1})	Ambient temperature (°C)
1	9	25	16	28
2	9.05	28	16	28
3	9.1	34	16	28
4	9.15	40	15	28.5
5	9.2	45	16	28.5
6	9.25	53	15	29
7	9.3	58	15	29
8	9.35	66	17	29
9	9.4	73	17	30
10	9.45	80	17	30
11	9.5	86	17	30
12	9.55	87	15	30
13	10	90	15	31
14	10.05	91	15	31
15	10.1	93	14	31

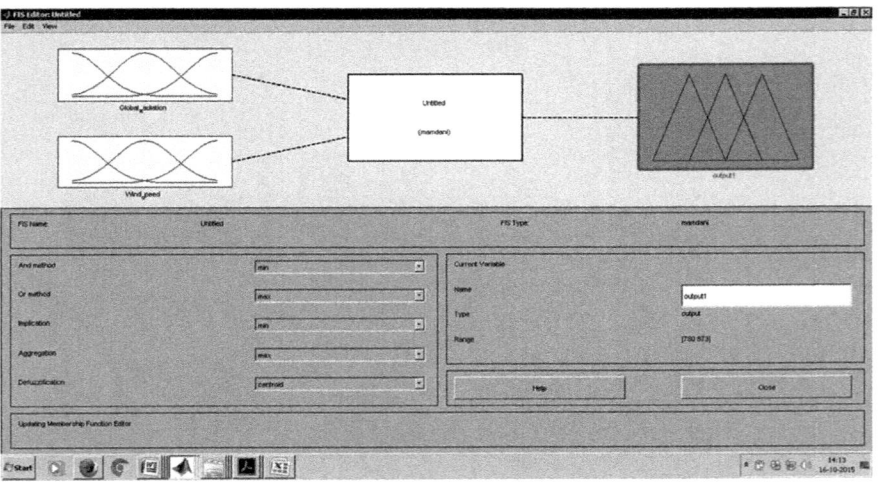

Fig. 2 MATLAB fuzzy toolbox

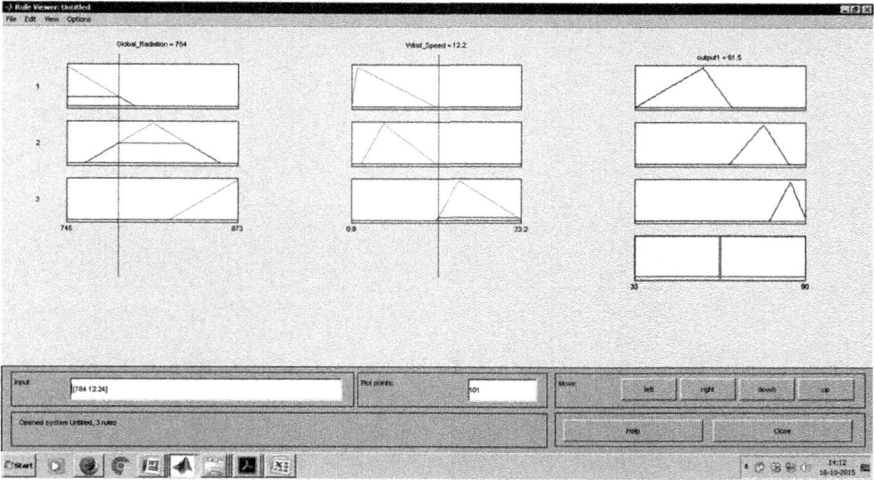

Fig. 3 Output indication

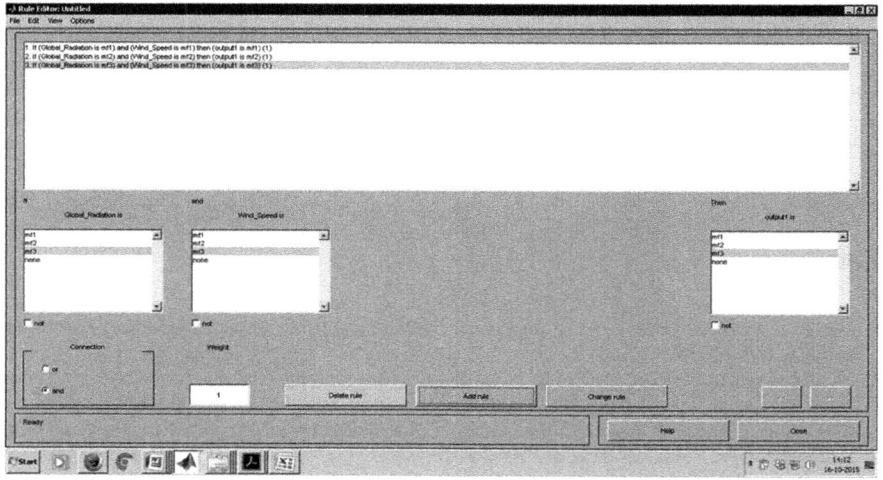

Fig. 4 Rules applied

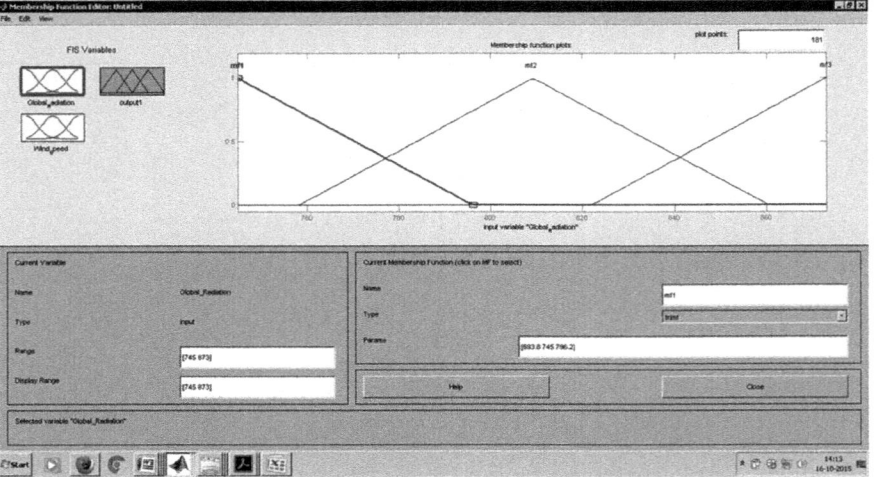

Fig. 5 Membership function and range for input

5 Conclusion

The present model gives the accurate temperature if the water is under ambient conditions, i.e. fuzzy is a very important tool to estimate the performance of Scheffler reflector water heating system.

References

1. Barlev D, Vidu R, Stroeve P. Innovation in concentrated solar power. Solar Energy Materials & Solar Cells 2011;95:2703–2725.
2. H Mousazadeh, et al. A review of principle and sun-tracking methods for maximizing solar systems output. Renewable and Sustainable Energy reviews 2009;13:1800–1818.
3. S Wagner, E Rubin. Economic implications of thermal energy storage for concentrated solar thermal power. Renewable Energy 2012:1–15.
4. J Sachin et al. Performance Analysis of Scheffler Reflector Using Approximate Generalised Model. International Journal of Emerging Trends in Engineering and Development 2014;6:2249–6149.
5. Shuang-ying W, et al. A parabolic dish/AMTEC solar thermal power system and its performance evaluation. Applied energy 2010;87:452–462.
6. Lovegrove K, et al. Developing ammonia based thermochemical energy storage for dish power plants. Solar energy 2004;76:331–337.
7. Prakash O, et al. Prediction of the Rate of moisture Evaporation from Jaggery in Greenhouse Drying using the Fuzzy logic. Heat Transfer Research 2015;46:923–935.
8. Patil R, et al. Experimental analysis of Scheffler reflector water heater. Thermal science 2011;15:599–604.

A Reversible Data Hiding Using Difference-Histogram Modification on Multi-directional in Two-Dimensional Histogram

Anil Pinapati and R. Padmavathy

1 Introduction

Reversible Data Hiding (RDH) also called as lossless compression is a technique which is used to recover Cover Image (CI) and the embedded data from Watermarked Image (WI) without any loss of data. Cover image data get altered due to embedded data. Retrieving cover image and embedded data from watermarked image, without any loss, is a challenging task. To solve that problem using RDH, RDH introduced a new trend in fields such as telemedicine, military, and law enforcement, where the original pixel should be recovered fully. Especially in medical images, getting error in recovered data is intolerable which may lead to miss diagnosis [1]. Earlier, different methods were proposed in RDH, such as Jun Tian introduced a new scheme in RDH using Difference Expansion (DE). Difference expansion embeds the data with the help of 1D Haar wavelet transform [2]. Different approaches have been proposed for RDH using histogram modification [3–5], prediction error expansion [6–8], and integer transform [9–11]. Among all these approaches, histogram modification got more popular in the field of RDH. The first proposal started with Zhicheng et al. [5]. Zhicheng et al. embedded data by using peak (P) and min (Z) pole from a histogram. This method fails to handle images that contain equal height poles in a histogram. A separate channel is required to send "P", "Z" values for successful restoration of embedded data and cover image. Tai et al. [3] introduced RDH using pixel difference. This method avoids the drawbacks in Zhicheng et al. Author used Binary Tree Structure (BTS) to introduce more redundancy in multiple levels. The height of the tree (L) is

A. Pinapati (✉) · R. Padmavathy
Department of C.S.E., National Institute of Technology, Warangal,
Warangal, Telangana, India
e-mail: ap@nitw.ac.in

R. Padmavathy
e-mail: rpadma@nitw.ac.in

© Springer Nature Singapore Pte Ltd. 2019
V. Nath and J. K. Mandal (eds.), *Proceeding of the Second International Conference on Microelectronics, Computing & Communication Systems (MCCS 2017)*, Lecture Notes in Electrical Engineering 476, https://doi.org/10.1007/978-981-10-8234-4_38

used for successful restoration of CI and embedded data. Hong [12] introduced RDH using Double Binary Tree (DBT) to increase more Embedding Capacity (EC) and PSNR compared with Tai et al.

Lee et al. [4] introduced RDH with Pixel Pair Mapping (PPM) by using D, where D is a matrix that contains difference values between two subsequent pixels. The points in D are categorized into RED, BLACK, and BLUE. The BLACK points are used to carry the data, BLUE points are used for shifting, and intact if it is RED. After embedding data, at most modification to a pixel is up or down (i.e., (x, y) is a pixel pair then up, down points are $(x, y + 1)$, $(x, y - 1)$), respectively. Li et al. [13] introduced RDH using Difference-Pixel Pair Mapping (DPPM) that modifies the original pixel along with x-direction (i.e., (x, y) is a pixel pair modified in four directions, those are $(x - 1, y)$, $(x + 1, y)$, $(x, y - 1)$, and $(x, y + 1)$) for more PSNR value compared with earlier RDH methods in histogram modification. RDH using histogram modification method chooses few pixel pairs to carry data, few for shifting, and remaining for intact.

The remaining paper is scattered into five sections. Introduction is covered in Sect. 1. Section 2 explains related work in RDH. Proposed method is covered in Sect. 3. Results and comparative study are explained in Sect. 4. Finally, paper concludes with Sect. 5.

2 Related Work in RDH

Embedding data with histogram modification was first introduced by Zhicheng et al. [5]. The method is as follows: identified peak (P) and min (Z) pole in a histogram (make sure $H(P) < H(Z)$). Shift bins right side by one unit from $P + 1$ to $Z - 1$ (including $P + 1$, $Z - 1$) and make $P + 1$ pole empty. Scan the image (I), check for P, if the pixel is P, and to be embedded bit is 1, then make it $P + 1$. Otherwise, P is intact, if the bit is 0.

Retrieving data from watermarked image (WI) is exactly the reverse process of embedding. The number of embedded bits is exactly equal to the size of pole P. To increase the embedding capacity more than "P" bits, author has taken more number of P and Z poles. This method requires a separate channel to send P, Z values for successful restoration. This method fails to handle images containing equal poles in a histogram. Zhicheng et al. method was enhanced by Tai et al. [3] using pixel difference. Author introduced Binary Tree Structure (BTS) to retrieve data successfully. This method carries data using high-frequency poles of a histogram generated using subsequent pixel difference. The method is as follows: H is the host image, N is the number of pixels in an image, and z_i is the pixel value in a grayscale image, $0 \leq z_i \leq N$ $1, z_i \in [0, 255]$.

1. Calculate deference values between z_{i-1} and z_i by scanning the image H in inverse s-order.

$$D_i = \begin{cases} z_i, & \text{if } i = 0 \\ z_{\{i-1\}} - z_i, & \text{Otherwise} \end{cases} \tag{1}$$

2. Draw the histogram using D_i and determine the maximum pole as "$M1$".
3. If $D_i > M1$, then shift z_i one unit by scanning the image H in inverse s-order like in Step 1.

$$w_i = \begin{cases} z_i & \text{if} & i = 0 \text{ or } D_i < M1 \\ z_i + 1 & \text{if} & D_i > M1 \text{ and } z_i \geq z_{i-1} \\ z_i - 1 & \text{if} & D_i > M1 \text{ and } z_i < z_{i-1} \end{cases} \tag{2}$$

where w_i is the watermarked pixel of ith location.

4. If $D_i = M1$, modify the pixel according to embedded bit b.

$$w_i = \begin{cases} z_i + b & \text{if} & D_i = M1 \text{ and } z_i \geq z_{i-1} \\ z_i - b & \text{if} & D_i = M1 \text{ and } z_i < z_{i-1} \end{cases} \tag{3}$$

The retrieving process is exactly reverse of embedding data. Those steps are as follows:

1. Retrieving cover image from watermark image,

$$w_i = \begin{cases} w_i + 1 & \text{if} & |w_i - z_{i-1}| > M1 \text{ and } w_i < z_{i-1} \\ w_i - 1 & \text{if} & |w_i - z_{i-1}| > M1 \text{ and } w_i > z_{i-1} \\ w_i & & \text{Otherwise} \end{cases} \tag{4}$$

2. Retrieving data from WI,

$$b = \begin{cases} 0 & \text{if} & |w_i - z_{i-1}| = M1 \\ 1 & \text{if} & |w_i - z_{i-1}| = M1 + 1 \end{cases} \tag{5}$$

Since the occurrence of $M1$ is more so, it was chosen to carry the data. To ensure reversibility, shift values above $M1$ and intact values below $M1$. This modification achieves more data for embedding with low distortion.

The pixel pairs which are involved in data embedding are called expansion. The pixel pairs which are used for increasing/decreasing are called shifting. The proposed work eventually will increase the embedding capacity by choosing more pixel pairs for embedding and fewer for shifting.

3 Proposed RDH Method

Adding another direction to DPM for more embedding is called as DHMMD. This section is categorized as follows: Lee's [4] and [13] procedures are introduced in Sect. 3.1. Parameters required for choosing prior pixel pairs with smooth region are explained in Sect. 3.2 (i.e., GAP and noise level). Evaluation of proposed method is explained in Sect. 3.3. The embedding and extraction procedures are explained in Sect. 3.4.

3.1 PPM, DPM for DHMMD

Consider the implementation of Lee's procedure from Fig. 1a; RED points are not used for embedding and shifting. BLACK points are mapped to its associate BLUE points for expansion. BLUE points are mapped to its associate BLUE points for shifting. Lee's procedure is as follows:

1. if $D = 0$ (i.e., (x, y) is RED point) where $D = y - x$, then the marked pixel is taken as itself.
2. if $D = \pm 1$ it is taken as BLACK point for carrying data based on bit to be embedded either 0 or 1

   - if to be embedded bit is 0, then no change in marked pixel (embedded pixel is itself).
   - if to be embedded bit is 1, then the marked pixel is mapped to its associate BLUE point.

3. if $D > \pm 1$, then it is taken as BLUE point those are mapped to its associate BLUE point.

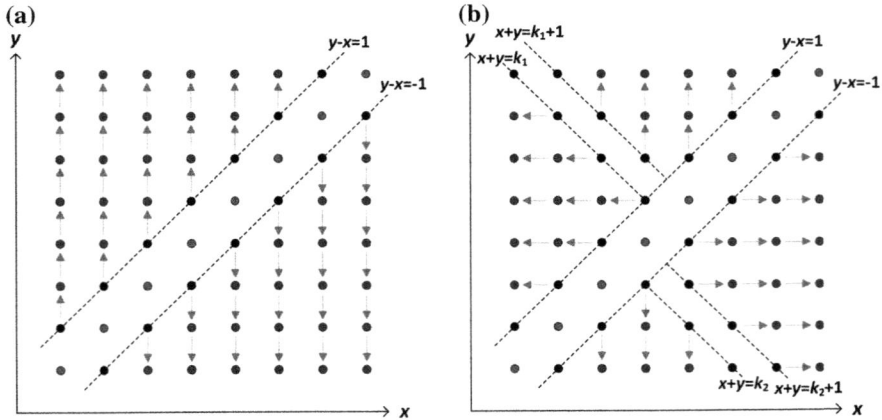

Fig. 1 Existing systems **a** Lee's method, **b** Li's method

This method uses BLACK points (i.e., (x, y) is mapped to either upper or lower pixel pair $(x, y + 1)$, $(x, y - 1)$, respectively) to carry the data. Lee's method was enhanced by Li's method, by adding one more direction (i.e., (x, y) is the pixel pair, possible modifications after embedding are $(x, y + 1)$, $(x, y - 1)$, $(x + 1, y)$, and $(x - 1, y)$) to carry the data. Proposed method added one more direction to DPM (i.e., if (x, y) is pixel pair, possible modifications after embedding are $(x, y + 1)$, $(x, y - 1)$, $(x + 1, y)$, $(x - 1, y)$, $(x + 1, y + 1)$, and $(x - 1, y - 1)$) to carry the data. This modification leads to more points that will be chosen for embedding and less points will be chosen for shifting. Especially in case of images like Baboon and Peppers, there is a drastic improvement in embedding capacity.

3.2 Choosing Prior Pixel Pairs in DHMMD

The following paragraph will explain the procedure for choosing prior pixel for embedding. For each pixel pair (x, y), compute $D_1 = x - y$ and $D_2 = y - z$ two difference values to form a two-dimensional difference histogram with (D_1, D_2) where z is the prediction of y calculated using GAP (Gradient Adjusted Prediction). For each (x, y), compute the perception of y based on the context of (x, y) for an accurate evaluation. First, the cover image can be divided into independent pixel pairs. For each individual pixel pair (x, y), compute the estimation of y to get z using GAP, based on Eq. (6).

$$z = \begin{cases} e_1 & \text{if } v_d - h_d > 80, \\ \frac{(e_1 + \text{adj})}{2} & \text{if } v_d - h_d \in (32, 80], \\ \frac{(e_1 + 3\text{adj})}{4} & \text{if } v_d - h_d \in (8, 32], \\ \text{adj} & \text{if } v_d - h_d \in [-8, 8], \\ \frac{(e_4 + 3\text{adj})}{4} & \text{if } v_d - h_d \in [-32, -8), \\ \frac{(e_4 + \text{adj})}{2} & \text{if } v_d - h_d \in [-80, -32), \\ e_4 & \text{if } v_d - h_d < -80. \end{cases} \tag{6}$$

where $\{e_1, e_2, ..., e_8\}$ are neighbor pixels of (x, y) (see Fig. 2), and $v_d = |e_1 - e_5| + |e_3 - e_7| + |e_4 - e_8|$ and $h_d = |e_1 - e_2| + |e_3 - e_4| + |e_4 - e_5|$ represent the vertical and horizontal measurements for gradients and adj $= \frac{e_1 + e_4}{2} + \frac{e_3 + e_5}{4}$. The value of z should be rounded with nearest integer. Compute the noise level of (x, y) using ten adjacent pixels $\{e_1, e_2, ..., e_{10}\}$ (see Fig. 2), for identifying suitable place for data embedding (noted as noise level (x, y)) calculated using Eq. (7).

$$\text{Noise Level} = \int_{(k',l') \in e} |\nabla I(k', l')| \tag{7}$$

Fig. 2 Adjacent pixels of (x, y)
to assess GAP predictor z from y

	j	j+1	j+2	j+3
i	x	y	e_1	e_2
i+1	e_3	e_4	e_5	e_6
i+2	e_7	e_8	e_9	e_{10}

where e is adjacent pixels of (x, y), each pixel pair has ten adjacent pixels except the last two rows and last two columns.

Here, ∇ denotes the gradient operator of the ten adjacent pixels. Noise level is calculated for a digital image by summing the horizontal and vertical differences of adjacent pixels. The maximum noise level is 13*255, the possible number of horizontal and vertical pairs are 13, and maximum difference value between the pixel pair is 255. Find the noise level for each pixel pair of (x, y) using Eq. (7). Threshold "T" is greater than the noise level, and compute the difference value (D_1, D_2); otherwise, omit the pixel pair for embedding. The less noise level will give smooth region for embedding. Embedding data on pixel pair (x, y) is done with the help of Table 1.

Table 1 Embedding data bit $b = \{0, 1\}$ with cover image

Condition on (D_1, D_2)	Embedding operation	DPM direction	PPM direction	Embedded value
$D_1 < 1$ and $D_2 < 0$ and $D_{1-1} = D_2$	Expansion embedding	Right	Right	$(x + b, y)$
$D_1 > 1$ and $D_2 < 0$ and $D_1 + 1 = D_2$	Expansion embedding	Left	Left	$(x - b, y)$
$D_1 = 0$ and $D_2 \leq 0$ $D_1 < 0$ and $D_2 = 0$	Expansion embedding	Upper left	Up	$(x, y + b)$
$D_1 = 0$ and $D_2 < 0$ $D_1 > 0$ and $D_2 = 0$ $D_1 10$ and $D_2 = -1$	Expansion embedding	Lower right	Down	$(x, y - b)$
$D_1 > 0$ and $D_2 > 0$ and $D_1 = D_2$	Expansion embedding	Up	Upper right	$(x + b, y + b)$
$D_1 < 0$ and $D_2 < 0$ and $D_1 = D_2$	Expansion embedding	Down	Lower left	$(x - b, y - b)$
$D_1 < -2$ and $D_2 < 0$ and $D_1 \geq D_2 + 2$	Shift	Right	Right	$(x + 1, y)$
$D_1 > 2$ and $D_2 < 0$ and $D_1 \leq D_{2-2}$	Shift	Left	Left	$(x - 1, y)$
$D_1 < 0$ and $D_2 > 0$	Shift	Upper left	Up	$(x, y + 1)$
$D_1 > 0$ and $D_2 < 0$ $D_1 = 1$ and $D_2 < -1$	Shift	Lower right	Down	$(x, y - 1)$
$D_1 > 0$ and $D_2 > 1$ and $D_1 \leq D_2 - 1$	Shift	Up	Upper right	$(x + 1, y + 1)$
$D_1 < 0$ and $D_2 < -1$ and $D_1 \leq D_2 + 1$	Shift	Down	Lower left	$(x - 1, y - 1)$

The BLACK points in DHMMD are mapped with associate BLUE points to carry data bits. The BLUE points are mapped with associate BLUE points for shifting. The detailed diagrammatic representation of proposed method is shown in Fig. 4. From Fig. 4, the points with high frequency are used to carry data. Choose some points as BLACK in difference histogram to carry data bits. Shift or omit remaining points according to the six allowable modification directions in Table 1.

The mapping of pixel pairs and difference values is shown in Fig. 3. From Fig. 3, the difference value (D_1, D_2) can be modified into six allowable modification directions based on difference in pixel pair. For instance, modification of y to $y + 1$, the modification direction to (x, y) is upper and the corresponding modification direction to (D_1, D_2) is "upper left", since D_1 changes to D_{1-1} and D_2 changes to $D_2 + 1$. The possible change in (x, y) to $(x + 1, y + 1)$ made modification in difference pixel D_1 no change and D_2 changed to $D_2 + 1$. Same modification is applied to pixel pairs $(x - 1, y - 1)$. The method is explained with an example, for a pixel pair $(x, y) = (136, 135)$ and $z = 134$, then the difference values $(D_1, D_2) = (1, 1)$, that is, a BLACK point is shown in Fig. 4. There are two possibilities depending on the data bit embedded.

1. if the data bit embedded is 0, then there is no change in $(x, y) = (136, 135)$.
2. if the data bit embedded is 1, then the point is associated with $(x + 1, y + 1) = (137, 136)$ modification to difference pair (D_1, D_2) is $(1, 2)$ means, and it points to the upper BLUE point in DHMMD method.

Another example to illustrate shifting is to assume that a pixel pair (x, y) is $(136, 135)$ and $z = 133$, then the difference values are $(D_1, D_2) = (1, 2)$ that is a BLUE point, map it with associate upper BLUE point (see Fig. 4). After modification, pixel pairs are $(x + 1, y + 1) = (137, 136)$; it leads to modify $(D_1, D_2) = (1, 3)$ that is associated with upper BLUE point, while retrieving embed data from WI is exactly inverse of data embedding procedure. To retrieve embed data, by calculating $D_1^w = x^w - y^w$ and $D_2^w = y^w - z$ for a pixel pair (x^w, y^w), where (x^w, y^w) are the pixel pairs of WI. The pixel pair (x^w, y^w) will produce an exact embedding data and cover image without any error.

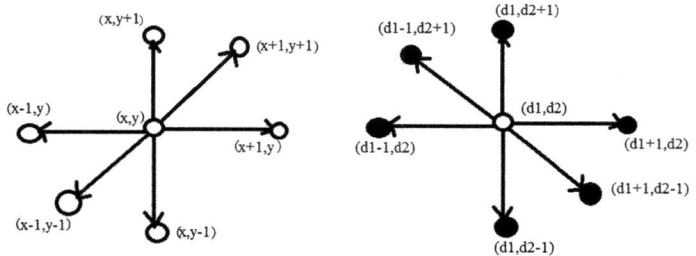

Fig. 3 Pixel pair (x, y) modification along with modified data in difference pair (D_1, D_2)

Fig. 4 Embedding data by
mapping difference-histogram
points

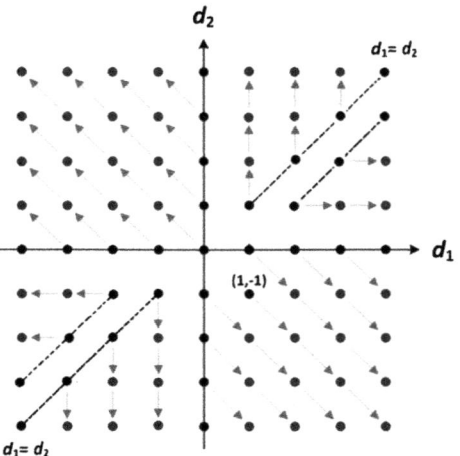

3.3 Evaluation of DHMMD

Performance of proposed system is evaluated with ratio of expanded pixels (noted as R_{exp}) between the N_{exp} and N_{shift}, where N_{exp} is the number of expanded pixel pairs and N_{shift} is the number of shifted pixel pairs, respectively.

$$R_{\mathrm{exp}} = \frac{N_{\mathrm{exp}}}{N_{\mathrm{exp}} + N_{\mathrm{shift}}} \tag{8}$$

Equation (8) shows the formula for calculating the embedding performance of a watermarked image in proposed method. The larger the ratio is less modification to cover image and leads to better performance. Evaluation of proposed DHMMD-based scheme is in terms of ratio of expanded pixels.

$$1 + \max_{(x,y)} \mathrm{NoiseLevel}\,(x, y) \tag{9}$$

Equation (9) shows the maximum threshold, when T is taken as its maximum noise level of all pixel pairs is disabled and the algorithm will work as usual means all pixel pairs will work for data embedding. The importance of the multi-dimensional difference histogram is the ratio becomes larger with smaller T.

3.4 Data Embedding and Extraction Procedures

The following paragraph will explain the data embedding and retrieving steps in proposed method. Divide the cover image (I) into independent pixel pairs. Embed the secret message into a part of cover image (noted as I'). Collect the LSBs of some pixels of I' and (noted as I'') will be recorded to get the binary sequence and embed this sequences into the part of I (i.e., $I - I'$). Using LSB replacement, embed the secondary information and the compressed location map into I''(use arithmetic coding for lossless compression of location map). The detailed data embedded procedure as follows:

Step 1: Leave last two rows from top to bottom and last two columns from left to right divide image into independent pixel pairs. Prepare location map (LM) of size is equal to half of I', to solve the problem of overflow/underflow. Pixel may alter either one bit less or more after embedding so. Location map will identify the pixel pair contain either [0, 255] by loading LM(i) = 1 otherwise LM(i) = 0. Along with overflow/underflow identify the smooth region suitable for embedding according to Eq. (7). While retrieving, data collected if LM(i) = 0 and noise level is less than the threshold T value.

Step 2: The secret message will be embedded according to location map LM, and pixel pair selection using T, for each $i \in \{1, \ldots, k\}$, considers the ith pixel pair in k number of pixel pairs:

1. if LM(i) = 1, do nothing because overflow/underflow might occur.
2. if LM(i) = 0, and its noise value is not less than T, do nothing (not suitable for embedding).
3. otherwise, the pixel pair will be used for expansion or shifting according to Table 1. Repeat these steps for all data bits, and the last data bit carrying pixel pair is denoted as k_{end}.

Step 3: The compressed location map and the secondary information will be embedded in this step for blind decoding. Those steps are as follows:

1. First, record LSBs of the first ($12 + 3 \lceil \log_2 k \rceil + 1$) image pixels to obtain a binary sequence S, where $\lceil . \rceil$ is the ceiling function.
2. Embed S into the remaining pixel pairs (i.e., pixel pair with index $k_{\text{end}} + 1$ to k using the same method in Step 2. After embedding data in $\{k_{\text{end}} + 1, \ldots, k^*\}$, k^* as the index of the last data carrying pixel pair of S.
3. Finally, replace LSB's of the first ($12 + 3 \lceil \log_2 k \rceil + 1$) pixels by the binary sequence S defined in Step 1 to recover the marked image.

   - pixel pair selection threshold: T (12 bits),
   - index of the last embedded pixel pair in Step 2: k_{end} ($\lceil \log_2 k \rceil$) bits),
   - compressed location map length: l ($\lceil \log_2 k \rceil$) bits),
   - index of the auxiliary data-embedded pixel pair:k^*($\lceil \log_2 k \rceil$) bits),

Remark of proposed system is that the "T" is minimum that was identified iteratively, because it will identify smooth pixel pairs which is suitable for embedding and more PSNR. The image restoration and data extraction process are summarized as follows.

Step 1: First read the LSB's of the first $(12 + 3 \lceil \log_2 k \rceil)$ pixels of watermarked image to determine the values of T, K_{end}, l, and k^*. Then, read the next l LSB to determine the compressed location map.

Step 2: Extract the LSB sequence S defined in Step 3 of embedding procedure. Same as embedding, except the last two columns and last two rows, divide the watermarked image into k-independent pixel pairs. Extract the binary sequence S from LSB's of k^* to $k_{end} + 1$ pixel pairs and restore the cover image.

1. Compute the prediction and noise level according to Eqs. (6) and (7). Notice that the pixel pair processing is inverse to that of embedding.
2. Then, if $L(i) = 0$ and the noise level is less than T, extract the cover image and embedded data using Table 2. Otherwise, there is no data embedded, and the pixel pair is just recovered as itself. It should be mentioned that, for the jth pixel pair with $j > k_{end}$, it is unchanged in embedding procedure, and thus the pixel pair can be recovered as itself.

Step 3: Replace LSB of the first $(12 + 3 \lceil \log_2 k \rceil + 1)$ image pixel by the binary sequence S is extracted in Step 2. Extract the embedded message from $\{k_{end}, ..., 1\}$ pixel pairs using Step 2.

Table 2 Retrieving data bit $b = \{0, 1\}$ from watermarked image

Condition on (D_1^w, D_2^w)	Extracted data bit b	Recovered value
$D_1 > 1$ and $D_2 \geq 0$ and $(D_{1-1} = D_2$ or $D_{1-2} = D_2)$	$\lfloor \frac{(D_1 - D_2)}{2} \rfloor$	$(x^w - b, y)^w$
$D_1 < -1$ and $D_2 < 0$ and $(D_1 + 1 = D_2$ or $D_1 + 2 = D_2)$	$\lfloor \frac{-(D_1 - D_2)}{2} \rfloor$	$(x^w + b, y^w)$
$(D_1 = 0$ and $D_2 \geq 0)$ or $(D_1 = -1$ and $D_2 \geq 1)$	$-(D_1)$	$(x^w, y^w - b)$
$(D_1 < 0$ and $D_2 = 0)$ or $(D_1 = 1$ and $D_2 = 1)$	D_2	$(x^w, y^w - b)$
$(D_1 = 0$ and $D_2 < 0)$ or $(D_1 = 1$ and $D_2 < -1)$	D_1	$(x^w, y^w + b)$
$(D_1 > 0$ and $D_2 = 0)$ or $(D_1 > 1$ and $D_2 = -1)$	$-(D_2)$	$(x^w, y^w + b)$
$(D_1 = 1$ and $D_2 = -1)$ or $(D_1 = 2$ and $D_2 = -2)$	D_{1-1}	$(x^w, y^w + b)$
$D_1 > 0$ and $D_2 > 0$ and $(D_1 = D_2$ or $D_1 + 1 = D_2)$	$D_2 - D_1$	$(x^w - b, y^w - b)$
$D_1 < 0$ and $D_2 < 0$ and $(D_1 = D_2$ or $D_{1-1} = D_2)$	$D_1 - D_2$	$(x^w + b, y^w + b)$
$D_1 > 0$ and $D_2 > 1$ and $D_1 + 1 < D_2$	No embedding data	$(x^w - 1, y^w - 1)$
$D_1 > 3$ and $D_2 > 0$ and $D_1 \geq D_2 + 3$	No embedding data	$(x^w - 1, y^w)$
$D_1 < -3$ and $D_2 < 0$ and $D_1 + 3 \leq D_2$	No embedding data	$(x^w + 1, y^w)$
$D_1 < 0$ and $D_2 < -2$ and $D_1 + 1 \geq D_2 + 2$	No embedding data	$(x^w + 1, y^w + 1)$
$D_1 < -1$ and $D_2 < 1$	No embedding data	$(x^w, y^w - 1)$
$(D_1 < 2$ and $D_2 < -1)$ or $(D_1 = 2$ and $D_2 < -2)$	No embedding data	$(x^w, y^w + 1)$

4 Results and Discussion

Tested grayscale images of size 512×512 including Lena, Baboon, Barbara, Airplane (F-16), Peppers, and Fishing Boat (see Fig. 5). Tables 3 and 4 show comparative results between methods of Sachnev et al. [14], Richard et al. [15], Hong [16, 17], and Li et al. [13] with proposed method. Proposed method is implemented experimentally with MATLAB R2013a, system configuration: Intel (R) core(TM) i5-2450 M CPU 2.50 GHz, 4 GB DDR3. Experimentally demonstrate that the DHMMD-based scheme can provide better embedding capacity than earlier RDH methods in PPM and DPM. Fallahpour et al., Hong et al., and Li et al. methods are based on the prediction error histogram with different predictors. Although these methods may exploit the spatial redundancy for a larger pixel context and perform better than Li et al. in few cases, the proposed scheme improves them by maintaining PSNR more than 50 dB with an embedding rate 0.03bpp.

Advantage of proposed method lies in the utilization of difference-histogram modification on multi-directional and priority pixel pair selection strategy. Li et al. method is based on pixel pair selection and limited modification directions for data embedding in difference-pixel pair mapping. The prediction error histogram strategy used in Sachnev et al. method makes embedding position selection scheme. Consider the case of few images when embedding capacity approaches its maximum (e.g., in case of Baboon and Airplane when embedding capacity reaches to 12,000, 45,000 bits); there are no pixel pairs for expansion in Li et al. The DHMMD method modifies more pixels into embedding less for shifting that leads to more embedding capacity compared with Li et al. [13].

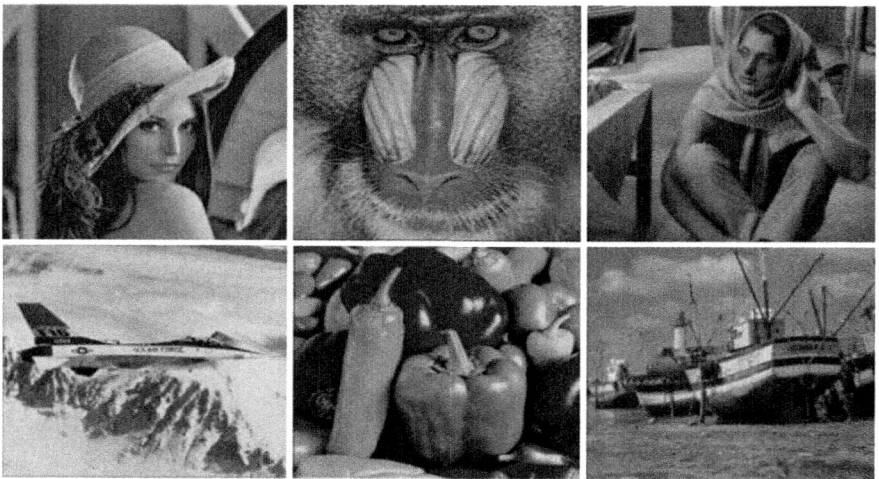

Fig. 5 Test images: Lena, Baboon, Barbara, Airplane, Peppers, and Fishing Boat

Table 3 Comparison of proposed and existing methods for an EC of 10,000 bits

Image	[10]	[11]	[17]	[12]	[1]	Proposed
Lena	58.18	58.20	58.78	58.41	59.78	58.59
Baboon	54.15	54.03	53.26	54.42	53.96	52.96
Barbara	58.15	58.61	58.36	58.21	59.67	58.15
Peppers	55.55	56.12	56.07	55.22	57.19	57.5
Fishing Boat	56.15	55.52	56.64	56.13	57.42	56.11
Average	56.44	56.5	56.62	56.48	57.6	56.65

Table 4 Comparison of proposed and existing methods for an EC of 20,000 bits

Image	[10]	[11]	[17]	[12]	[1]	Proposed
Lena	55.03	54.82	54.92	54.97	56.15	55.15
Barbara	55.04	55.29	54.89	55.18	56.24	54.36
Peppers	52.30	52.55	52.16	52.20	53.39	53.8
Fishing Boat	52.65	52.43	52.26	52.37	53.12	52.16
Average	53.76	53.77	53.55	53.68	54.72	53.87

Table 5 will give a better idea about EC of proposed method and Li et al. method. A graph is plotted for Baboon and Peppers in Fig. 6 with EC from 5,000 bits to its maximum with an increment of 1,000 bits. 5,000 bits of EC means an Embedding Rate (ER) of 0.019 bits per pixel (bpp).

It will give a better PSNR value as number of embedding bits increases; initially, it was low in both images but gradually increasing PSNR while increasing EC. Technically, prior pixel pairs were scattered more on diagonal in case of high-textured images like Baboon and Peppers. Pixel pairs were improved drastically only in DHMMD, because positive xy and negative xy plains are divided into two regions, so pixel pairs contain $(x + 1, y + 1)$ and $(x - 1, y - 1)$ which will help to improve more pixel pairs for embedding.

Table 5 Comparison of EC with Li [13] and proposed method

Image	[1] Bits	[1] BPP	Bits	PSNR	BPP
Baboon	12,830	0.049	13,608	50	0.052
Peppers	25,309	0.096	29,549	51.09	0.11
Average	19,069	0.0725	21,578	50.5	0.081

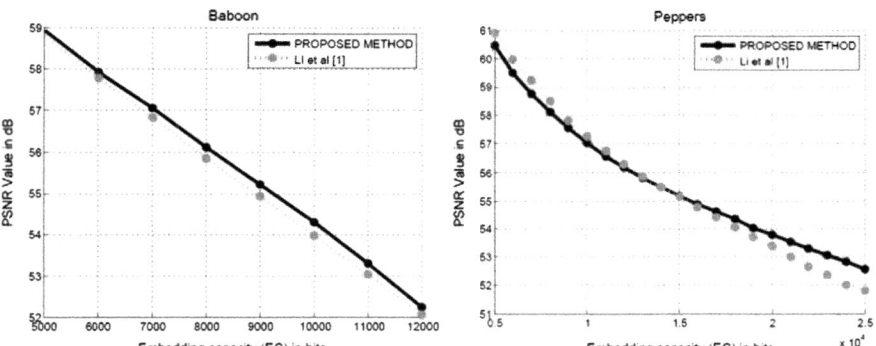

Fig. 6 Graphical representation of Li et al. and proposed method on Baboon and Peppers

5   Conclusion and Future Work

The proposed method achieved better embedding capacity compared with Li et al. [13]. The computational cost of proposed method is negligible, due to avoidance of complex operations like DCT (Discrete Cosine Transform), DFT (Discrete Fourier Transform), and DWT (Discrete Wavelet Transform). Good PSNR is achieved by choosing small "T" iteratively. There is a scope to enhance the embedding capacity and PSNR to another level by introducing error energy estimator. This method is blind watermarking, which is suitable for real-time purpose. There is no need of cover image to retrieve watermark.

References

1. G. Coatrieux, C. L. Guillou, J. M. Cauvin, and C. Roux, "Reversible watermarking for knowledge digest embedding and reliability control in medical images," IEEE Trans. Inf. Technol. Biomed, 2009, Vol. 13, pp. 158–165.
2. J. Tian, "Reversible data embedding using a difference expansion," IEEE Trans. Circuits Syst. Video Tech., 2003, Vol. 13, pp. 890–896.
3. W. L. Tai, C. M. Yeh, and C. C. Chang, "Reversible data hiding based on histogram modification of pixel differences," IEEE Trans. Circuits Syst. Video Technol., 2009, Vol. 19, pp. 906–910.
4. S. K. Lee, Y. H. Suh, Y. S. Ho, "Reversible image authentication based on watermarking," in Proc. IEEE ICME, 2006, pp. 1321–1324.
5. Z. Ni, Y.Q. Shi, N. Ansari, W. Su, "Reversible data hiding," IEEE Trans. Circuits Syst. Video Technol., 2006, Vol. 16, pp. 354–362.
6. H.T. Wu, J. Huang, "Reversible image watermarking on prediction errors by efficient histogram modification," Signal Process., 2012, Vol. 92, pp. 3000–3009.
7. D. M. Thodi, J. J. Rodriguez, "Expansion embedding techniques for reversible watermarking," IEEE Trans. Image Process., 2007, Vol. 16, pp. 721–730.
8. D. Coltuc, "Improved embedding for prediction-based reversible watermarking," IEEE Trans. Inf. Forensics Security., 2012, Vol. 6, pp. 873–882.

9. F. Peng, X. Li, B. Yang, "Adaptive reversible data hiding scheme based on integer transform," Signal Process., 2012, Vol. 92, pp. 54–62.
10. H.T. Wu, J. Huang, "Reversible image watermarking on prediction errors by efficient histogram modification," Signal Process, 2012, Vol. 92, pp. 3000–3009.
11. D. Coltuc, J. M. Chassery, "Very fast watermarking by reversible contrast mapping," IEEE Signal Process. Lett., 2007, Vol. 14, pp. 255–258.
12. W. Hong, T. S. Chen, C. W. Shiu, "Reversible data hiding for high quality images using modification of prediction errors," J. Syst. Software, 2009, Vol. 82, pp. 1833–1842.
13. Li. Xiaolong, Weiming Zhang, Xinlu Gui, and Bin Yang, "A Novel Reversible Data Hiding Scheme Based on Two Dimensional Difference Histogram Modification," IEEE Trans. Inf. Forensics and Security, 2013, Vol. 8, pp. 1091–1100.
14. V. Sachnev, H. J. Kim, J. Nam, S. Suresh, Y. Q. Shi, "Reversible watermarking algorithm using sorting and prediction," IEEE Trans. Circuits Syst. Video Technol., 2009, Vol. 19, pp. 989–999.
15. Richard Y. M. Li, Oscar C. Au, Carman K. M. Yuk, Shu-Kei Yip, Tai-Wai Chan, "Enhanced image trans-coding using reversible data hiding," in Proc. IEEE ISCAS, 2007, pp. 1273–1276.
16. W. Hong, "An efficient prediction and shifting embedding technique for high quality reversible data hiding," EURASIP J. Adv. Signal Process., 2010.
17. W. Hong, "Adaptive reversible data hiding method based on error energy control and histogram shifting", Opt. Commun., 2012, Vol. 285, pp. 101–108.

A Novel Single-Phase Multilevel Inverter Topology with Reduced Component Count

Chandan Kumar, Tanmoy Maity and K. C. Jana

1 Introduction

In 1970s, silicon-controlled rectifier came into existence and it gave a new path to the conventional VSI, whereas multilevel inverter (MLI) added some extra features to the conventional VSI in terms of efficiency, reliability, and modularity [1]. MLI has also extended its application to renewable energy sources, industrial application like oil and chemical industry, textile industry [2], aerospace application, marine propulsion, and hybrid vehicles [3]. The key point for choosing MLI as a better solution includes better output voltage waveform with less distortion in harmonics and lesser electromagnetic interference, low voltage stress (dv/dt) on switches [4]. Starting from medium-power to high-power range application, MLI stands superior comparatively choosing than any other conventional converters due to its efficient power quality, less maintenance cost, and less space requirement for installation [5].

The concept of MLI lies with large number of stepped output waveforms and to bring steps close to the sinusoidal waveform to reduce the THD as minimum as possible [6]. Thus, enhancing the voltage steps will reduce the voltage stress on each power semiconductor switch along with the betterment in output quality waveform, when load is operated at higher voltage. Classical topologies of MLI include neutral point clamped or diode clamped, flying capacitor, and cascade H-bridge abbreviated as NPC or DC, FC, and CHB, respectively [7]. There are various advantages and disadvantages of each configuration over the other. In NPC-MLI, with increase in output voltage levels, the clamping diode increases.

C. Kumar (✉) · T. Maity
Mining Machinery Engineering, IIT (ISM), Dhanbad, Dhanbad, Jharkhand, India
e-mail: chandanbit07@gmail.com

K. C. Jana
Electrical Engineering, IIT (ISM), Dhanbad, Dhanbad, Jharkhand, India

© Springer Nature Singapore Pte Ltd. 2019
V. Nath and J. K. Mandal (eds.), *Proceeding of the Second International Conference on Microelectronics, Computing & Communication Systems (MCCS 2017)*, Lecture Notes in Electrical Engineering 476, https://doi.org/10.1007/978-981-10-8234-4_39

489

Voltage balancing among capacitors is the major problem faced in this structure. In FC-MLI, it is easier to extend it to desired level compared to NPC-MLI. The CHB-MLI give a greater number of redundancies than the NPC and FC-MLI topologies, since every H-bridge has one excess switch state and the series association presents more redundancies, thus enabling the fault-tolerant capability [8].

Control strategies for MLI include various methods for low switching, medium switching, and high switching frequency: Sinusoidal pulse width modulation (SPWM) [9], selective harmonic elimination PWM (SHE-PWM) [10], space vector PWM (SVPWM) [11], and nearest level control (NLC) [12]. Application of NLC using single DC supply for open-end winding motor has been presented in [13]. Recently, more focus has been done regarding the lower switch for same output voltage level, thereby to reduce the cost, installation space, and efficient working capability. In [14], comparative study of different MLIs has been incorporated with various detailed descriptions. In [15–17], reduced switches topology for multilevel inverter has been studied.

The real-time digital simulator (RTDS) came to be a very useful tool which is applicable in power electronics, power system, industrial automation, robotics, and engineering institutes. The tools for offline simulation do not suit for industrial drives in modern time because these tools are not adequate and advanced enough for fixed-time step operation. The offline simulation cannot interface with the actual hardware and controller. Running time of real-time simulator is within the calculated time step, i.e., time step will be same as clock time [18].

Here, a new structure of 19-level MLI with reduced switches has been described. This structure has also been compared with the existing classical topology of cascade H-bridge multilevel inverter in terms of switches, diodes, power supplies, as only CHB-MLI follows the property of asymmetricity. % THD comparison for the proposed configuration has also been incorporated using phase disposition sinusoidal PWM (PD-PWM). The comparison has been done regarding number of DC sources, switches, and diodes, and simulation is done in MATLAB/SIMULINK software and after that, this simulation work is validated through the real-time digital simulator (OP-5600). This paper divided into four sections, i.e., Sect. 1 includes introduction, Sect. 2 contains the proposed topology and control strategy, Sect. 3 contains results and discussion that also contains the comparison tables, and concluded with Sect. 4.

2 Proposed Topology

The suggested topology of 19-level MLI is depicted in Fig. 1. It consists of five DC voltage supplies, in which four of them are of equal value of Vdc and remaining one is of value of 5Vdc and thirteen power electronic switches. In proposed topology, the power electronics switches (S_2, S_3), (S_6, S_7), (S_8, S_{11}), (S_9, S_{12}), and (S_{10}, S_{13}) should not be started together to prevent the short-circuit D.C. power supplies. The different switching states of power electronics switches for the

Fig. 1 Proposed topology of multilevel inverter

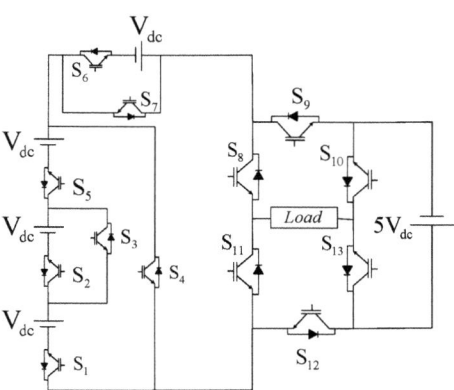

proposed topology have been shown in Table 1. From Table 1, it is clear that the introduced topology is able to generate nineteen different voltage levels, i.e., 9 V, 8 V, 7 V, 6 V, 5 V, 4 V, 3 V, 2 V, V, 0, −V, −2 V, −3 V, −4 V, −5 V, −6 V, −7 V, −8 V, and −9 V at the output.

This work uses the convergence of a sine waveform with a triangular waveform to create firing pulses. It is the average switching frequency on which the modulation algorithms are classified. The value of switching frequency above 1 kHz is considered here. Multi-triangular carrier PWM approach has been described as in [14, 19, 20]:

(a) Alternate Phase Opposition Disposition PWM (APOD-PWM): The comparison between the reference waveform (i.e., in this case, sine wave) and the triangular carrier waveforms, which are being displaced by 180°, has been done in this scheme.

(b) Phase Opposition Disposition PWM (POD-PWM): Here, the triangular carrier waveforms over zero level are in the phase, although beneath zero level the triangular carrier waveforms are opposite in phase to the carrier waves over zero level.

(c) Phase Disposition PWM (PD-PWM): The triangular carrier waveforms over and beneath of zero level are in the phase.

Here, only PD-PWM is chosen for generating the gate pulses (Fig. 2).

There are nineteen modes of operation in asymmetrical structure. The different output voltage steps generated with respective switching states are shown in Table 1.

Operating modes of asymmetrical structure:

Mode 1 (+9 V): In this mode, the switches numbered S_1, S_2, S_5, S_6, S_8, S_{10}, and S_{12} are ON, while rest other switches are OFF.

Mode 2 (+8 V): In this mode, the switches numbered S_1, S_3, S_5, S_6, S_8, S_{10}, and S_{12} are ON, while rest other switches are OFF.

Mode 3 (+7 V): In this mode, the switches numbered S_1, S_3, S_5, S_7, S_8, S_{10}, and S_{12} are ON, while rest other switches are OFF.

Table 1 Switching table for proposed 19-level inverter

Output voltage levels	Switching states												
	S_1	S_2	S_3	S_4	S_5	S_6	S_7	S_8	S_9	S_{10}	S_{11}	S_{12}	S_{13}
+9 V	1	1	0	0	1	1	0	1	0	1	0	1	0
+8 V	1	0	1	0	1	1	0	1	0	1	0	1	0
7 V	1	0	1	0	1	0	1	1	0	1	0	1	0
+6 V	0	0	0	1	0	1	0	1	0	1	0	1	0
+5 V	0	0	0	0	0	0	0	0	0	1	1	1	0
+4 V	1	1	0	0	1	1	0	1	0	0	0	1	1
+3 V	1	1	0	0	1	0	1	1	0	0	0	1	1
+2 V	1	0	1	0	1	0	1	1	0	0	0	1	1
+V	0	0	0	1	0	1	0	1	0	0	0	1	1
0	0	0	0	0	0	0	0	1	1	1	0	0	0
−V	0	0	0	1	0	1	0	0	1	1	1	0	0
−2 V	1	0	1	0	1	0	1	0	1	1	1	0	0
−3 V	1	1	0	0	1	0	1	0	1	1	1	0	0
−4 V	1	1	0	0	1	1	0	0	1	1	1	0	0
−5 V	0	0	0	0	0	0	0	1	1	0	0	0	1
−6 V	0	0	0	1	0	1	0	0	1	0	1	0	1
−7 V	1	0	1	0	1	0	1	0	1	0	1	0	1
−8 V	1	0	1	0	1	1	0	0	1	0	1	0	1
−9 V	1	1	0	0	1	1	0	0	1	0	1	0	1

Mode 4 (+6 V): In this mode, the switches numbered
S_4, S_6, S_8, S_{10}, and S_{12} are ON, while rest other switches are OFF.
Mode 5 (+5 V): In this mode, the switches numbered S_{11}, S_{10}, and S_{12} are ON,
while rest other switches are OFF.
Mode 6 (+4 V): In this mode, the switches numbered S_1, S_2, S_5, S_6 S_8, S_{13}, and S_{12}
are ON, while rest other switches are OFF.
Mode 7 (+3 V): In this mode, the switches numbered S_1, S_2, S_5, S_7, S_8, S_{13}, and S_{12}
are ON, while rest other switches are OFF.
Mode 8 (+2 V): In this mode, the switches numbered S_1, S_3, S_5, S_7, S_8, S_{13}, and S_{12}
are ON, while rest other switches are OFF.
Mode 9 (+V): In this mode, the switches numbered S_4, S_6, S_8, S_{13}, and S_{12} are ON,
while rest other switches are OFF.
Mode 10 (0): In this mode, the switches numbered S_8, S_9, and S_{10} are ON, while
rest other switches are OFF.
Mode 11 (−V): In this mode, the switches numbered S_4, S_6, S_9, S_{10}, and S_{11} are
ON, while rest other switches are OFF.
Mode 12 (−2 V): In this mode, the switches numbered S_1, S_3, S_5, S_7 S_9, S_{10}, and
S_{11} are ON, while rest other switches are OFF.
Mode 13 (−3 V): In this mode, the switches numbered S_1, S_2, S_5, S_7, S_9, S_{10}, and
S_{11} are ON, while rest other switches are OFF.

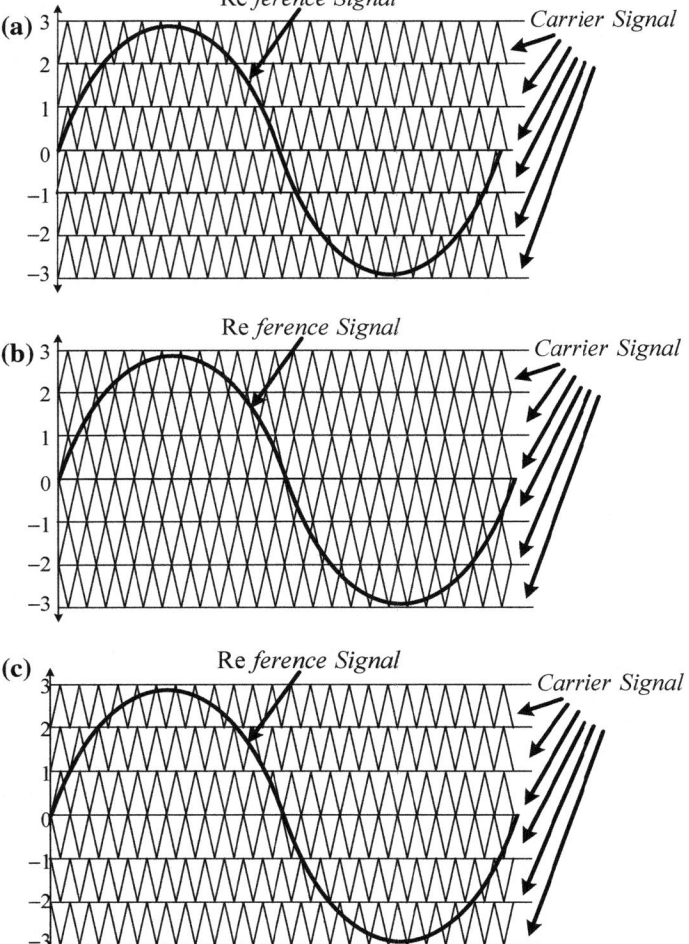

Fig. 2 **a** Arrangement of carrier waves for seven-level PD-PWM. **b** Arrangement of carrier waves for seven-level APOD-PWM. **c** Arrangement of carrier waves for seven-level POD-PWM

Mode 14 (−4 V): In this mode, the switches numbered S_1, S_2, S_5, S_6, S_9, S_{10}, and S_{11} are ON, while rest other switches are OFF.

Mode 15 (−5 V): In this mode, the switches numbered S_8, S_9, and S_{13} are ON, while rest other switches are OFF.

Mode 16 (−6 V): In this mode, the switches numbered S_4, S_6, S_9, S_{11}, and S_{13} are ON, while rest other switches are OFF.

Mode 17 (−7 V): In this mode, the switches numbered S_1, S_3, S_5, S_7, S_9, S_{11}, and S_{13} are ON, while rest other switches are OFF.

Mode 18 (−8 V): In this mode, the switches numbered S_1, S_3, S_5, S_6, S_9, S_{11}, and S_{13} are ON, while rest other switches are OFF.

Mode 19 (−9 V): In this mode, the switches numbered S_1, S_2, S_5, S_6, S_9, S_{11}, and S_{13} are ON, while rest other switches are OFF.

The load current path for the positive voltage levels is depicted in Fig. 3. In (*a*), ninth-level current path is shown; in (*b*), eighth-level current path is shown; and so on up to the first level as shown in (*i*).

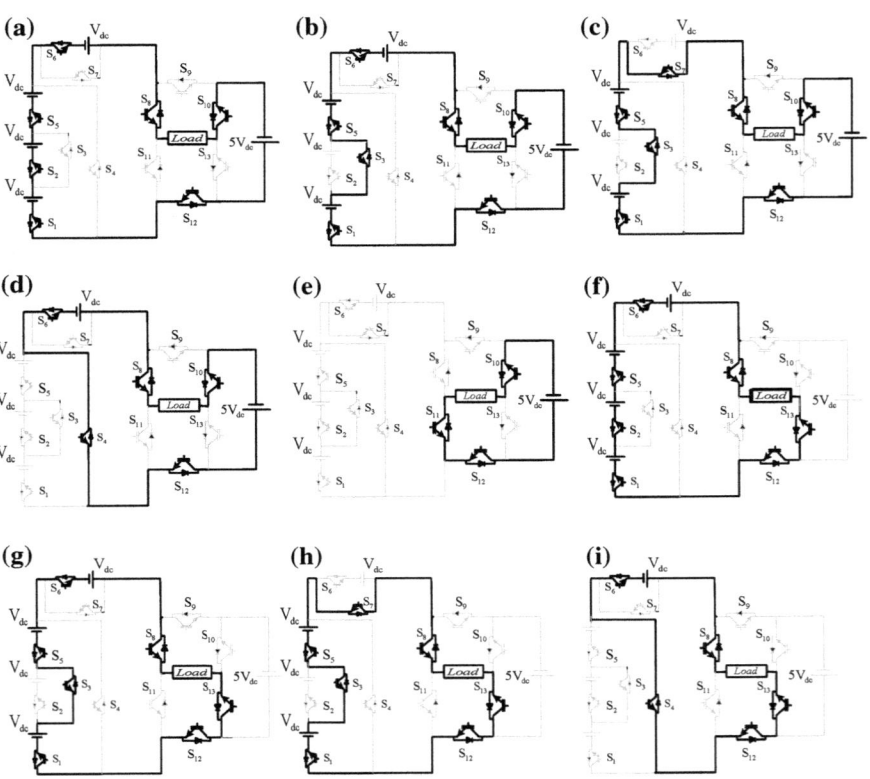

Fig. 3 Different modes of operation for positive voltage levels of proposed topology for 19-level inverter

Table 2 Comparison of component used and % THD for single-phase 19-level inverter

Parameters	Types of multilevel inverter	
	CHB	Proposed topology
Switches	20	13
Diodes	20	13
Isolated DC power supplies	5	5
Total components	45	31
% THD	6.83	6.83

3   Results and Discussion

The proposed multilevel inverter of 19-level MLI has been compared with the traditional topology of cascade H-bridge MLI. In Table 2, the comparison has been done regarding the total IGBT switches used, total diodes used, total isolated DC power supply used, and %THD for sinusoidal pulse width modulation (PD-PWM) for classical CHB-MLI with proposed topology of MLI. Cascade H-bridge inverter (CHB-MLI) has total harmonic distortion (in %) of 6.83 and the proposed topology also has % THD of 6.83 for PD-PWM. It clearly is shown from Table 2 that number of power electronics components needed in the proposed topology is reduced with respect to CHB-MLI as it shows 20 for switches and diode each for the classical CHB-MLI and 13 for switches and diodes for the proposed topology. Isolated DC supply is same for both topologies as four of them are same and equal to V_{dc} and remaining one is equal to $5V_{dc}$. So, it can be observed from Table 2 that the suggested topology needs significantly less number of total components when compared to traditional CHB-MLI.

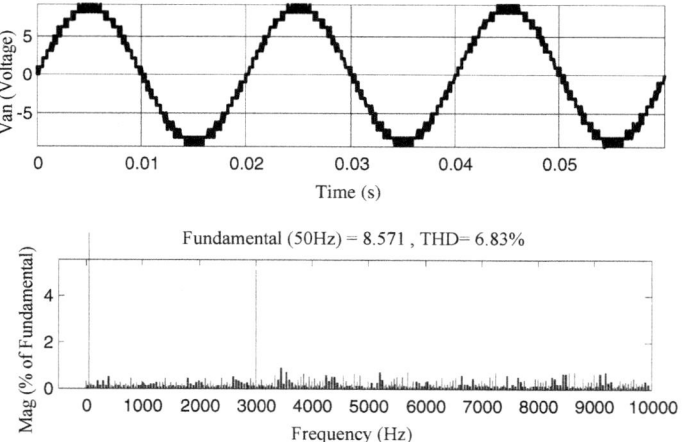

Fig. 4 Simulation and FFT result of single-phase output voltage for 19-level of proposed topology

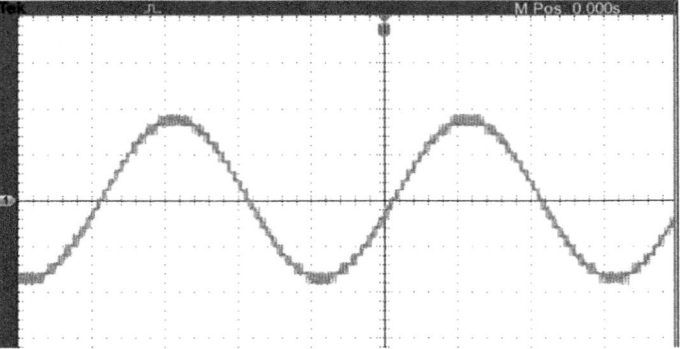

Fig. 5 RTDS result of phase output voltage for 19-level inverter of suggested topology (100 V/ division)

In Fig. 4, the 19-level output phase voltage (V_{an}) has been shown along with the %THD at fundamental frequency. It is noticed that in spite of having same % THD content, the other aspects of the proposed topology dominate the asymmetrical CHB with classical configuration (having same voltage ratio as proposed topology). Three cycles of output phase voltage have been incorporated in simulation result.

In Fig. 5, the RTDS output result by digital signal oscilloscope (DSO) of 19-level output phase voltage has been shown at fundamental frequency.

4   Conclusion

In this article, a structure of single-phase 19-level MLI is suggested. The working principles, control strategy, and switching states have been studied in details. Several types of comparisons have been done between the suggested topology of MLI and traditional topology of CHB-MLI regarding number of diodes, IGBT switches, isolated D.C. power supplies, and total components that have been shown in Table 2. It has been noticed that this suggested topology needs lesser number of overall components. The proposed topology of 19-level inverter has been simulated in MATLAB and validated by using the real-time simulator (OP-5600) as depicted in Fig. 5.

References

1. Rodríguez, J., Bernet, S., Wu, B., Pontt, J.O., Kouro, S.: 'Multilevel voltage-source-converter topologies for industrial medium-voltage drives' *IEEE Trans. Ind. Electron.*, 2007, 54, (6), pp. 2930–2945.
2. Abu-rub, H., Member, S., Holtz, J., Rodriguez, J., Baoming, G.: 'Medium-Voltage Multilevel Converters—State of the Industrial Applications' 2010, 57, (8), pp. 2581–2596.

3. Tolbert, L.M., Member, S., Peng, F.Z., Cunnyngham, T., Chiasson, J.N.: 'Charge Balance Control Schemes for Cascade Multilevel Converter in Hybrid Electric Vehicles' IEEE Trans. Ind. Electron., 2002, 49, (5), pp. 1058–1064.
4. Rodríguez, J., Lai, J.S., Peng, F.Z.: 'Multilevel inverters: A survey of topologies, controls, and applications' IEEE Trans. Ind. Electron., 2002, 49, (4), pp. 724–738.
5. Kouro, S., Malinowski, M.: 'Recent advances and industrial applications of multilevel converters' IEEE Trans. Ind. Electron., 2010, 57, (8), pp. 2553–2580.
6. Franquelo, L.G., Rodriguez, J., Leon, J.I., Kouro, S., Portillo, R., Prats, M. a M.: 'The age of multilevel converters arrives' IEEE Ind. Electron. Mag., 2008, 2, (2), pp. 28–39.
7. Wen, J., Ma Smedley, K.: 'Synthesis of multilevel converters based on single- and/or three-phase converter building blocks' IEEE Trans. Power Electron., 2008, 23, (3), pp. 1247–1256.
8. Malinowski, M., Gopakumar, K., Rodriguez, J., Pérez, M.A.: 'A survey on cascaded multilevel inverter' IEEE Trans. Ind. Electron., 2010, 57, (7), pp. 2197–2206.
9. McGrath, B.P., Holmes, D.G.: 'Multicarrier PWM strategies for multilevel inverters' IEEE Trans. Ind. Electron., 2002, 49, (4), pp. 858–867.
10. Li, L., Czarkowski, D., Liu, Y., Pillay, P.: 'Multilevel selective harmonic elimination PWM technique in series-connected voltage inverters' IEEE Trans. Ind. Appl., 2000, 36, (1), pp. 160–170.
11. Jana, K.C., Biswas, S.K.: 'Generalised switching scheme for a space vector pulse-width modulation-based N-level inverter with reduced switching frequency and harmonics' IET Power Electron., 2015, 8, (12), pp. 2377–2385.
12. Mahato, B., Raushan, R., Jana, K.C., 'Comparative Study of Asymmetrical Configuration of Multilevel Inverter for Different Levels' 3rd IEEE Int'. Conf. on Recent Advances in Information Technology (RAIT), pp. 300–303, 3–5 March, 2016.
13. Mahato, B., Raushan, R., Jana, K.C., 'Modulation and Control of Multilevel Inverter for an Open-end Winding Induction Motor with Constant Voltage Levels and Harmonics' IET Power Electron., D.O.I. https://doi.org/10.1049/iet-pel.2016.0105.
14. Kumar, C., Mahato, B., et. al., 'Comprehensive study of various configurations of three-phase Multilevel Inverter for different levels' 3rd IEEE Int'. Conf. on Recent Advances in Information Technology (RAIT), pp. 310–315, 3–5 March, 2016.
15. Ceglia, G., Guzmán V., et. al., 'A New Simplified Multilevel Inverter Topology for DC-AC Conversion' IEEE Trans. Power Electron., 2006, 21, (5), pp. 1311–1319.
16. Gautam S.P., Sahu L.K., Gupta S., 'Reduction in number of devices for symmetrical and asymmetrical multilevel inverters' IET Power Electron., 2016, 9, (4), pp. 698–709.
17. Kaliamoorthy, M., Rajasekaran V., et. al., 'Generalised hybrid switching topology for a single-phase modular multilevel inverter' IET Power Electron., 2014, 7, (10), pp. 2472–2485.
18. Prasad, H., Maity, T., Real-time simulation for performance evaluation of bidirectional quasi z-source inverter based medium voltage drives", COMPEL: The International Journal for Computation and Mathematics in Electrical and Electronic Engineering, Vol. 35 Iss: 3, pp. 1123–1135.
19. Manjrekar, M.D., Steimer, P.K., and Lipo, T.A.: 'Hybrid Multilevel Power Conversion System: A Competitive Solution for High Power Applications' IEEE Trans. Ind. Applicat., vol. 36, no. 3, pp. 834–841, May/June 2000.
20. Carrara, G., Gardella, S., Marchesani, M., Salutari, R., and Sciutto, G.: 'A New Multilevel PWM Method: A TheoreticalAnalysis' IEEE Trans. Power Electron., vol. 7, no.3, pp. 497–505,1992.

Energy-Efficient 64-Bit Asynchrobatic Adder

K. Srilakshmi, A. V. N. Tilak, K. Srinivasa Rao and Y. Syamala

1 Introduction

The requirements of low-power digital integrated circuits had impelled designers to explore new methodologies in the design of VLSI circuits. Adiabatic logic [1] or energy recovery is a new design outlook for attaining low power consumption. To recycle the energy of the circuits, adiabatic logic makes use of AC voltage as an alternative to the DC voltage as a power supply. If the time period of the AC source is greater than the time constant of charging path, using slowly varying voltage source requires less energy to charge the capacitor. The supply source will recover the energy when the supply voltage V_{DD} decreases due to which the load capacitor C_L starts discharging. The energy E required to perform an adiabatic switching depends on the operating clock frequency f_{clk} by the following relation:

$$E = \frac{1}{2} C_L \cdot V_{DD}^2 \cdot f_{clk} \qquad (1)$$

Increasing the clock period (T_{clk}) results in a decrease in energy necessary to perform any computation. Theoretically, any adiabatic logic circuit is realized with zero power consumption by ignoring the leakage power. But the leakage current increases significantly due to rapid device miniaturization and threshold voltage

K. Srilakshmi (✉) · A. V. N. Tilak · Y. Syamala
Gudlavalleru engineering college, Gudlavalleru, Andhra Pradesh, India
e-mail: slkaza06@gmail.com

A. V. N. Tilak
e-mail: avntilak@yahoo.com

Y. Syamala
e-mail: coolsyamu@gmail.com

K. Srinivasa Rao
TRR College of Engineering, Inole, Telangana, India
e-mail: principaltrr@gmail.com

© Springer Nature Singapore Pte Ltd. 2019
V. Nath and J. K. Mandal (eds.), *Proceeding of the Second International Conference on Microelectronics, Computing & Communication Systems (MCCS 2017)*, Lecture Notes in Electrical Engineering 476, https://doi.org/10.1007/978-981-10-8234-4_40

scaling becomes a considerable part of power consumption [2]. Hence, it is essential to reduce leakage power by exploring new devices as an alternative to bulk MOS devices. Among the various devices, FinFET, a double-gate device with quasi-planar structure, has been proposed as an alternative for addressing the challenges due to continued device miniaturization [3].

The datapath is the basic module of any microprocessor, data processing Application-Specific Integrated Circuit (ASIC), and Digital Signal Processor (DSP). The basic datapath element is the adder; hence, the overall performance of the system depends on the adder implementation. The simplest adder structure is carry ripple adder with $O(n)$ delay and area complexity, but the delay time is a linear function of the size of operands. This will limit the overall performance of the system. The look-ahead carry adder [4, 5] is having area and delay complexity of $O(n \log(n))$ and $O(\log(n))$, respectively. It has low power consumption as it has less number of logic levels compared to other structures and low fan-out nets. When all the three parameters, viz., delay, power, and area, are equally critical, CLA with carry-generate and carry-propagate blocks can be a good choice. To implement, generate as well as propagate blocks, the basic gates required are AND, XOR, and OR with as many inputs as $n + 1$, which is impractical to realize in hardware. To overcome this drawback, a group of four bits is used as a block with carry look-ahead generator.

In this work, a 64-bit carry look-ahead adder with radix-four structure is implemented using 45 nm technology node with CMOS and FinFET-based static and asynchrobatic logic. The results of the comparison between asynchrobatic circuits and static CMOS counterparts are presented by considering the Power Dissipation (PD), Power-Delay Product (PDP), and Energy Dissipation (ED) for various frequencies and supply voltages. This paper is structured in the following manner. The outlines of asynchrobatic logic and FinFET technologies are given in Sects. 2 and 3. Section 4 described the adder designs. The results of the work are given in Sect. 5. Lastly, the conclusion of the paper is presented in Sect. 6.

2 Asynchrobatic Logic

The word asynchrobatic is originated from asynchronous and adiabatic terms. This is one of the low-power design techniques that will combine the low energy advantage of both the asynchronous and adiabatic logic [6]. "Adiabatic" is a term with a Greek origin and is related to a thermodynamic process. It describes a module in which output change happens without energy being lost or gained. The "asynchronous" term is also originated from Greek; it means "not with time." However, in the VLSI design contest, its meaning is without a global time reference. Therefore, this logic is a design technique that will not use a reference time signal to synchronize various operations of the system. In general, most of the digital systems currently in existence are being designed using synchronous design approach. The asynchronous logic uses handshaking signals to provide

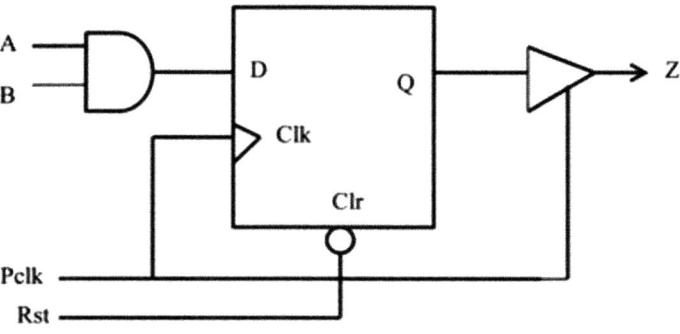

Fig. 1 Model of two-input AND gate using asynchrobatic logic [5]

communication between different components instead of a clock signal. Figure 1 shows the model of two-input AND gate with asynchrobatic logic. The output Z is evaluated during positive edge of the clock and is logic "0" initially.

This structure is different from the conventional logic in that the output is obtained whenever there is a change in the input, and in the remaining period the output is uninitialized or in high impedance state.

Different adiabatic logic families are reported [7] to synthesize various digital circuits. In this work, Positive Feedback Adiabatic Logic (PFAL) was chosen as it offers lowest energy consumption, a good robustness against technology parameter variations and also capable of operating in a fully reversible manner [8] compared to other logic families.

3 FinFET Technology

The FinFET device is having a three-dimensional structure, which comprises a very lean silicon body fabricated perpendicular to the wafer plane. Figure 2 illustrates the cross-sectional diagram and three-dimensional view of a FinFET. The gate electrodes enclose the conducting channel in all the three ways. So, the current direction is parallel to the plane of the wafer [9]. The Short Channel Effects (SCEs) are suppressed, and the device can provide strong control over the channel. Due to this, FinFET offers lower leakage, faster-switching speed, and higher on-state current.

The FinFETs offer incredibly high-performance and low-power characteristics. The three primary modes of operation of the FinFET are based on the configuration of back and front gates, viz., Short Gate (SG), Independent Gate (IG), and Low-Power (LP) modes [10].

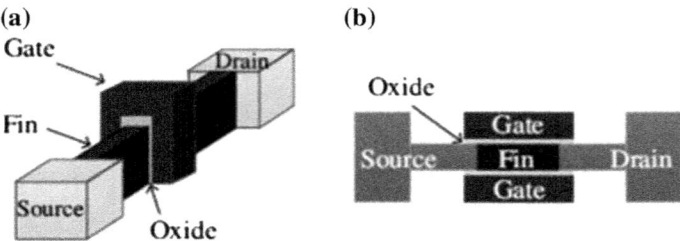

Fig. 2 **a** Three-dimensional view, **b** cross-sectional diagram of FinFET

4 Adder Design

Different adder structures like Skylansky and Kogge-stone are given in [4]. For adder modules, more than size 16-bit, fan-out is a problem with Skylansky structure. The wiring may be complicated if Kogge-Stone structure is used because of dual-rail nature of adiabatic logic. Hence, for larger sized adders, higher radix extension of Knowles adder is used. So, radix-four 64-bit CLA is considered in this work. Figure 3 shows the block diagram of a four-bit CLA, which is composed of four single-bit adders with four-bit carry look-ahead logic. Internally, each single-bit adder is implemented using basic adiabatic datapath cells consisting of buffer, AND, AND–OR, and XOR gates. These cells are asynchronously controlled by power clock signal.

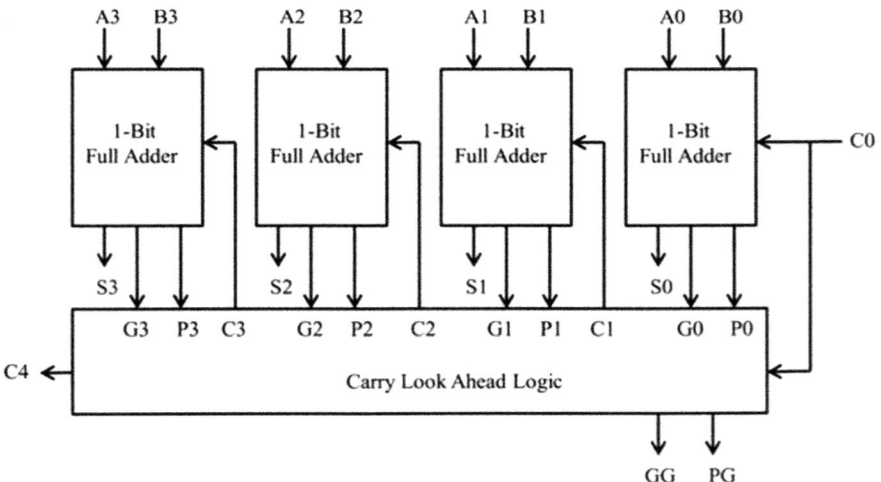

Fig. 3 Four-bit CLA structure

The generate and propagate signals can be calculated in constant time using

$$G_i = A_i \cdot B_i \tag{2}$$

$$P_i = A_i \oplus B_i \tag{3}$$

$$C_i + 1 = G_i + P_i \cdot C_i \tag{4}$$

The propagate and generate functions to the next stage radix-four CLA are

$$GG_0 = G_3 + P_3 \cdot G_2 + P_3 \cdot P_2 \cdot G_1 + P_3 \cdot P_2 \cdot P_1 \cdot G_0 \tag{5}$$

$$PG_0 = P_3 \cdot P_2 \cdot P_1 \cdot P_0 \tag{6}$$

The NMOS trees used to realize generate and propagate functions with bulk CMOS and FinFET are shown in Figs. 4 and 5, respectively. The carry out of the next stage is implemented by using an AND–OR structure. The NMOS tree for carry out of the first stage using bulk CMOS and FinFET is shown in Fig. 6.

The basic structure of radix-four 64-bit CLA is shown in Fig. 7. This uses four 16-bit CLAs, each developed using basic 4-bit cells. The intermediate carry signals C_{16}, C_{32}, and C_{48} are generated in constant time.

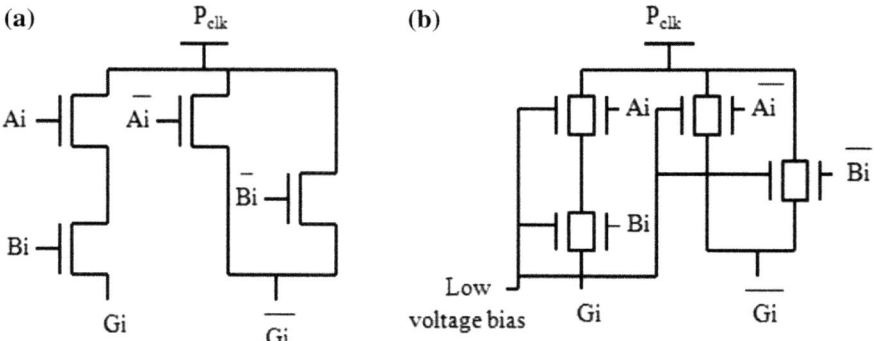

Fig. 4 NMOS tree for generate function using (**a**) bulk CMOS and (**b**) FinFET

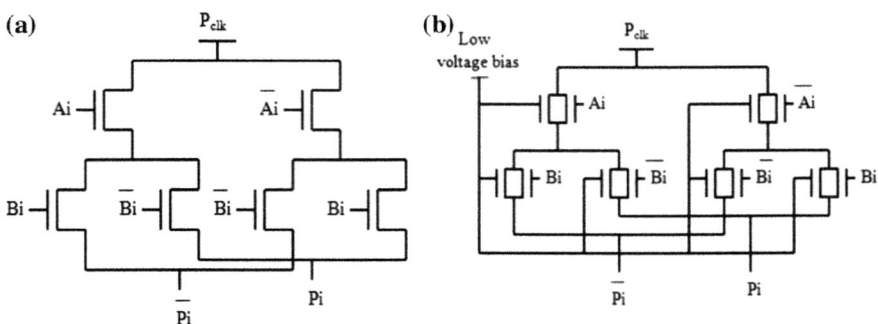

Fig. 5 NMOS tree for propagate function using (**a**) bulk CMOS and (**b**) FinFET

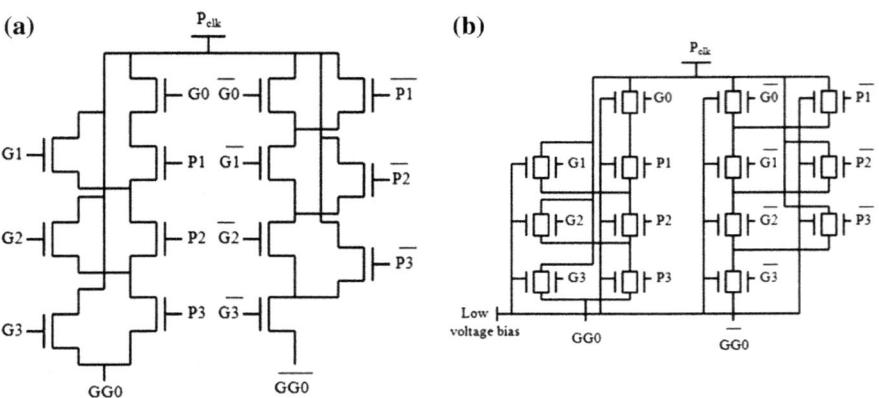

Fig. 6 NMOS tree for carry out of the first stage using (**a**) bulk CMOS and (**b**) FinFET

5 Results and Discussion

The designed circuits are subjected to extensive simulation using Cadence Spectre simulator and Power Dissipation (PD), Power-Delay Product (PDP), and Energy Dissipation (ED) at different frequencies and supply voltages are computed. The FinFET width parameter is calculated from the following relation:

$$W_{fin} = 2 \cdot H_{fin} + T_{fin} \tag{7}$$

The device parameters selected are, for the FinFET, $L = 45$ nm, $T_{fin} = 25$ nm, $H_{fin} = 30$ nm, and $T_{ox} = 1$ nm, and for the bulk CMOS are $L = 45$ nm and $W = 90$ nm.

Where T_{fin}, W_{fin}, and H_{fin} are thickness, width, and height of the FinFET, respectively.

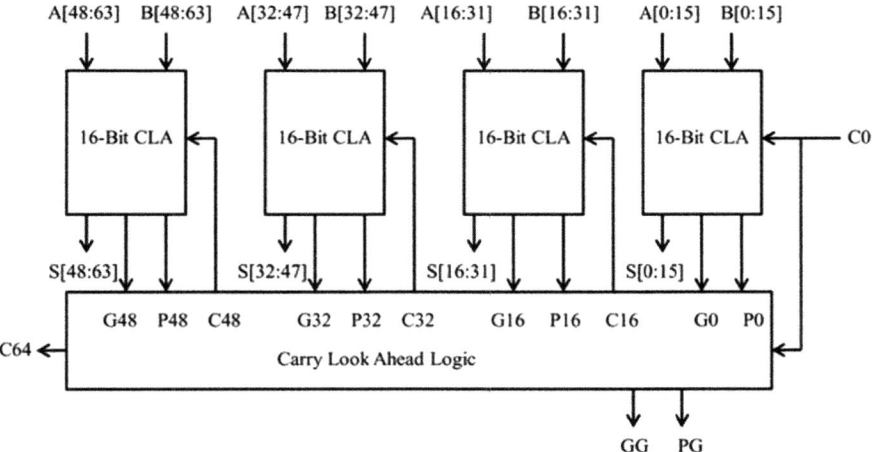

Fig. 7 Radix-four 64-bit CLA structure

Fig. 8 Effect of supply voltage scaling on power-delay product for radix-four 64-bit asynchrobatic and static CMOS adders

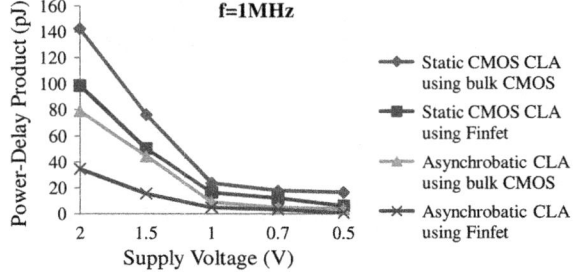

The effect of supply voltage scaling on the power-delay product for 64-bit static CMOS and asynchrobatic CLA using bulk CMOS and FinFET technologies are shown in Fig. 8.

An improvement of 61.47 and 67.87% in PDP was observed at a supply voltage of 1 V for asynchrobatic bulk CMOS and FinFET circuits, respectively, over static CMOS circuits. When the supply voltage is further scaled down to 0.5 V, the improvement was found to be 76 and 89%, respectively, for these circuits.

To simulate asynchrobatic logic circuits, a power clock signal with an amplitude of 0.7 V is used. The variation of average Power Dissipation (PD) over a frequency range of 50 KHz to 1000 MHz is shown in Fig. 9. For the asynchrobatic adder, maximum power savings are achieved at low frequencies due to dependancy of power dissipation on sub-threshold current, which decreases with frequency. In medium-frequency range, from 5 to 100 MHz, the power savings are moderate. An increase in power dissipation is found at higher frequencies as there is a voltage difference in the charging path. Combined use of asynchrobatic logic with

Fig. 9 Power dissipation versus frequency of asynchrobatic and static 64-bit CLA

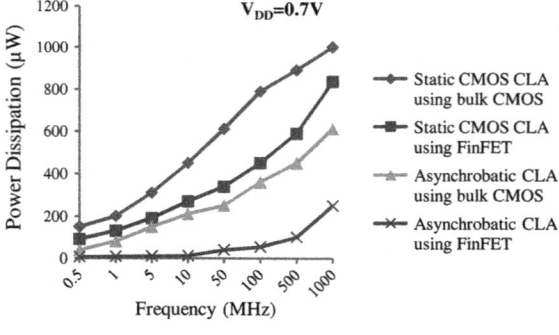

Fig. 10 Energy dissipation of asynchrobatic CLA using bulk CMOS and FinFET technologies. The static CMOS adder under the same conditions dissipates 37.26 and 34.5 pJ, respectively

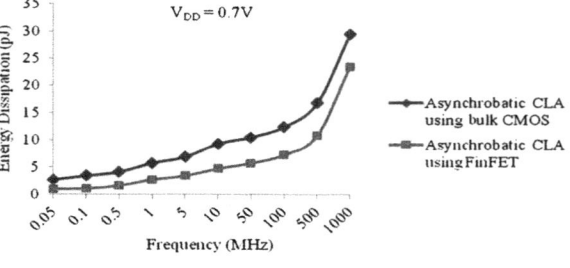

ultralow-power FinFET exhibits maximum power savings of 97.2% compared to bulk CMOS adder.

Figure 10 gives the energy dissipation over a cycle of asynchrobatic adder using bulk CMOS and FinFET technologies. The average energy dissipation of static CLA is 37.26pJ and 34.5pJ with bulk CMOS and FinFET, respectively, whereas for asynchrobatic adder using FinFET, this is found to be less by a factor of 20, 9.6, and 2, respectively in low (up to 500 KHz), medium (from 1 to 50 MHz), and high (above 100 MHz) frequency ranges. The energy loss increases with increase in frequency according to the relation given by Eq. (1).

Table 1 compares the power dissipation of asynchrobatic adder of the present work with other 64-bit adder circuits. 64-bit hybrid adder was designed based on radix-four structure [11, 12] at 90 nm technology node. In [13], high-performance

Table 1 Performance comparison of 64-bit adders

	MWSCAS [11]	ISC3 [12]	ICSCN [13]	This work
Year	2012	2014	2015	2016
Technology (nm)	90 (CMOS)	90 (CMOS)	45 (CMOS)	45 (FinFET)
Type of adder	Hybrid	Hybrid	Weinberger	Asynchrobatic
Bit size	64	64	64	64
Supply voltage (V)	1	1	1.1	0.7
Power dissipation (mW)	2.82	9.58	0.33	0.2

adders for different data widths based on Weinberger algorithm at 45 nm technology node were implemented. The asynchrobatic FinFET adder dissipates less power as compared to other adder circuits.

6   Conclusion

A 64-bit Carry Look-ahead Adder (CLA) with radix-four structure is implemented at 45 nm node with asynchrobatic logic using bulk CMOS and FinFET technologies. The asynchrobatic logic effectively minimizes the energy dissipation and improves the performance over the static CMOS logic. Simulation results show significant energy savings of the order of 20, 9.6, and 2, respectively, at low, medium, and high frequencies for asynchrobatic CLA using FinFET. Since the demand for low-power biomedical implants is increasing day by day, the implemented design can impact the energy savings of such implants.

References

1. J. Denker, "A review of adiabatic computing," *Proc. of the IEEE Int. Symp. on Low Power Electronics*, 1994. pp. 94–97.
2. S. Mukhopadhyay, A. Raychowdhury, and K. Roy, "Accurate estimation of total leakage in nanometer-scale bulk CMOS circuit based on device geometry and doping profile," *IEEE Trans. Comput.-Aided Des. Integr. Circuits Syst.*, Vol. 24, No. 3, pp. 363–381, 2005.
3. D. Hisamoto, Wen-Chin Lee, J. Kedzierski, H. Takeuchi, K. Asano, C. Kuo, E. Anderson, Tsu-Jae King, J. Bokor, and Chenming Hu, "FinFET: a self-aligned double gate MOSFET scalable to 20 nm", *IEEE Trans. Electron Devices*, Vol. 47, No. 12, pp. 2320–2325, 2000.
4. S. Knowles, " A family of Address," *Proc. of the 15th IEEE Symp. on Computer Arithmetic*, 2001, pp. 30–34.
5. David J. Willingham and Izzet Kale, "A ternary adiabatic logic (TAL) implementation of a four-Trit full-adder," *Proc. of IEEE Int. Conf. (NORCHIP), 2011*, November 14–15, Lund, Sweden.
6. D. J. Willingham and I. Kale, "Asynchronous, quasi-adiabatic (Asynchrobatic) logic for low-power, very wide data width applications," *Proc. of Int. Symp. on Circuits and Systems*, 2004, pp. 257–260.
7. Blotti, S. Pascoli, and R. Saletti, "Sample model for positive feedback adiabatic logic power consumption estimation," *Electronics Letters*, Vol. 36, No. 2, pp. 116–118, 2000.
8. Y. Ye and K. Roy, "Energy recovery circuits using reversible and partially reversible logic," *IEEE Trans. on Circuits Syst. I*, Vol. 43, No. 9, pp. 769–778, 1996.
9. Liao Nan, Cui XiaoXin, Liao Kai, Ma KaiSheng, Wu Di, Wei Wei, Li Rui, and Yu DunShan, "Low power adiabatic logic based on FinFETs," *J. of Sci. China Inform. Sciences*, Vol. 57, Issue 2, pp. 1–13, 2014.
10. D. E. Muller and W. S. Bartky, "A theory of asynchronous circuits," *Proc. of Int. Symp. on the theory of Switching*, 1959, pp. 204–243.
11. Shao-Kai Kai Chang and Chin-Long Wey, "A fast hybrid adder design in 90 nm CMOS process", *IEEE 55th International Midwest Symposium on Circuits and Systems (MWSCAS)*, 2012, pp. 414–417.

12. Shao-Hui Shieh, Der-Chen Huang and Ying-Yi Chu, "Low voltage and Low power 64-bit Hybrid adder design based on radix-4 prefix tree Structure", *IEEE International Symposium on Computer, Consumer and Control (IS3C)*, 2014, pp. 446–449.
13. R. Suganya and D. Meganathan, "High performance VLSI adders", *Proc. of IEEE Int. Conf. on signal processing, Communication and Networking (ICSCN)*, 2015, pp. 450–456.

PV Array's Resistance and Temperature Sensitivity Analysis with Shading Effects

Gourab Das, M. De and K. K. Mandal

1 Introduction

There has been a significant increase in the use of renewable energy sources. It is expanding rapid requirement in numerous innovation and application of modern society taking place in response to combustion of fossil fuels and redundant excessive increase in average temperature of the earth's atmosphere issue. Under such circumstances, solar cells are most feasible, eco-friendly, and durable energy sources for power generation. Solar cells usher considerable interests in new trends as well as distributed energy generation to provide clean and green energy. Non-conventional energy sources tend to be the most usable substitute for renewable energy amenity for raising energy requirement in response to diminution of combustible fuels and the increase in atmospheric temperature triggered by unrestrained ignition of combustible fuels. Owing to their emanation of low gamut and load-dependence fluctuating dc voltage, which are incorporated with barrier like minimal efficiency, slack reverberation to abrupt turn into current, lower utilization factor of cell are imperative to reinforced in arrangement to uplift the input dc low voltage transform into desired ac voltage. A *PSPICE* modeling of PV module subsisting several cells/modules in series was developed in order to carry out the study [1], where the work focused on the shadowed brightened cells under different illumination stratuma. Another detailed literature study of a unique MPPT that presented the synthesis of centralized as well as distributed MPPTs [2]. The scheme proposed in [3] appeases the distribution and the impact of shading over complete array and hence, reducing mismatches (losses) as a result of partial shading conditions. Such system performance was then scrutinized for various

G. Das (✉) · M. De · K. K. Mandal
Department of Power Engineering, Jadavpur University, Kolkata 700098, India
e-mail: gourabdas.ju@gmail.com

© Springer Nature Singapore Pte Ltd. 2019
V. Nath and J. K. Mandal (eds.), *Proceeding of the Second International Conference on Microelectronics, Computing & Communication Systems (MCCS 2017)*, Lecture Notes in Electrical Engineering 476, https://doi.org/10.1007/978-981-10-8234-4_41

shadowed conditions, and MATLAB/SIMULINK outcomes were represented to depict that power which is extricated out of the PV arrays under partial shading conditions is enriched significantly. A significant contribution was made in literature [4] where the developed software helps in simulation of a PV array panel for varieties of light and temperature circumstances, which includes diversified temperature and illumination in same time period. An innovative method to enhance forecasting in prevailing PV array and power generation in multifarious earthly situation was illustrated [5]. An approach based on genetic algorithm optimization for total cross-tied modules in a solar array was presented [6]. Among the PV applications discussed, building integrated photovoltaic, concentrated photovoltaic and photovoltaic thermal systems were seemed to be the most technically vibrant and exhibit solution for future energy challenges [7]. Evaluation of three schemes for betterment of relation between PV energy production and electricity demand was investigated in this work, which included optimally arranging PV modules, combination of geographically disseminated arrays, and utilizing the energy storage system [8]. PV array simulation strategy based on programming was proposed [9] to simulate the impact of partial shading condition and developed new MPPT algorithm for energy conversion systems. The main aim of this paper was focused on the investigations on the modeling and quality analysis of different types of solar PV model.

This research work is focused on the investigations on the modeling, simulation, and analysis of different types of solar PV model.

2 Basics of Solar PV System

The PV diode consists of photon-sensitive substance intersection that instantly transforms incoming radiation to electric current. Generally, PV with p-n junctions, anterior and posterior power connections, also with anti-glare layer, is utilized to retain increased incidents sunlight in PV as shown in Fig. 1. Incoming radiations on solar PV having energy higher from band gap of PV substance produces e-h conjugates. If these carriers prior recombination arrive adjacent diode junction, potential at intersection segregates holes to p-side and electrons to n-side. Charge separation produces electric potential of reverse nature w.r.t. electric field operating beforehand, suppressing total field of junction. Reducing electric field produces enhancement of diffusion amperes; here, a voltage forms over p-n junction. When PV diode is put to an external load, then extra carriers produced due to absorption in incoming radiation flows across load. When PV connection is opened, forward bias in PV diode reaches a level where amperes of PV equalize sunlight-produced amperes, to achieve net ampere null. In this instant, potential difference across cell is termed open-circuit voltage and presents highest voltage which is achieved from PV. If PV is short-circuited, potential difference of diode becomes null, and total light produced ampere passes across shorted lead. Sunlight-produced ampere equals in ampere of shorted lead. This ampere can be

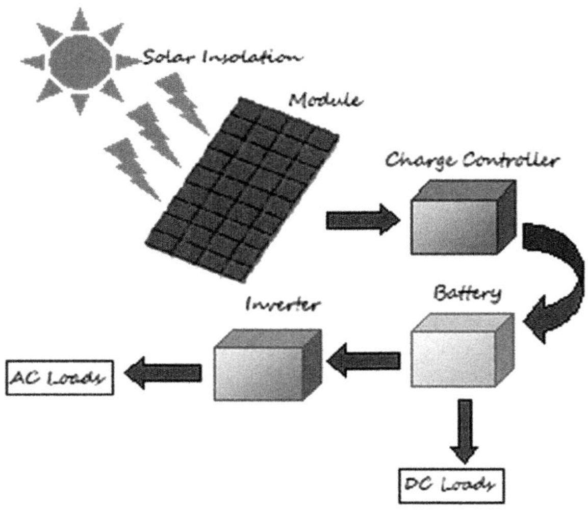

Fig. 1 Typical structure of
solar PV system

termed short-circuit ampere which presents highest current achieved from PV diode. Following research shows that stationary and dynamic modeling of equivalent power circuit of unit diode PV is done in order to reflect their electrical features, and also related power loss factors. The representations as shown Fig. 1 are utilized to analyze their sensitivity to circuit elements and also degree of heat. The equivalent circuits of both PVs are put together in series and parallel architectures to produce PV modules to achieve requisite quantity of produced power suitable for input to power conditioners. PV-relied power conditioners comprise converter circuits along with rectifier and produce steady potential. Fuel cell-relied power conditioners incorporate extended band produced and comprise dual boosting steps to establish a steady potential. PWM inverter with single and poly-phase circuits is examined which achieves alternating current resultant.

3 Photovoltaic Model and Its Governing Equation

The known characteristics of an Si p-n junction are shown in Fig. 2.
Mathematically, this can be expressed as

$$I = I_0\{\exp(V/V_T) - 1\} \tag{1}$$

Here, I_0 is the reverse amperes, V represents small module potential, I represents cell current, V_T is known as potential equivalent of temperature and at room temperature of 20 °C its approximate value is 26 mV and equalizes KT/Q when K represents the Boltzmann's constant, and T represents temperature in K and also Q presents electrons. When p-n junction is brightened, the characteristics got

Fig. 2 Schematic circuit of solar PV cell

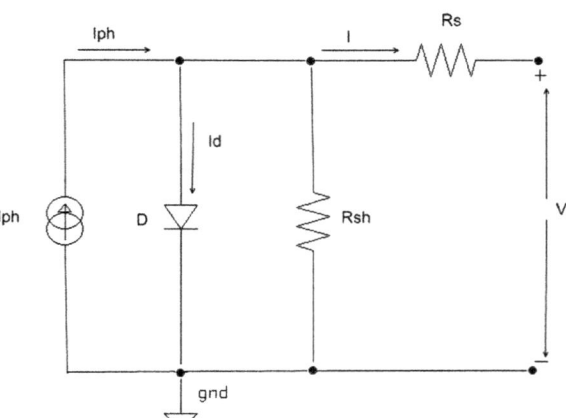

improved in shape and moving down as photon-originated element is appended with reverse leakage current.

The diode Eq. (1) is modified as follows:

$$I = -I_{SC} + I_0\{\exp(V/V_T) - 1\} \qquad (2)$$

If the terminal is short-circuited and the voltage V is zero, current $I = -I_{SC}$ which is flowing through the external path, arising from p-side, I_{SC} is represented as short-circuit current and its magnitude depends on the solar insulation.

Now, a voltage source is replaced in external path with polarity of p-side. Then, calculate the open-circuit voltage, V_{oc} in terms of, V_T, I_{SC}, and I_0 may be modified as

$$V_{oc} = V_T \ln\{(I_{sc}/I_0) + 1\} \qquad (3)$$

Typically, $I_{SC} = 2$ A, $I_0 = 1$ mA, and $V_{OC} = 0.55$ A (at room temperature). Mathematically, the I-V characteristic of solar cell may be written (as per standard sign convention) as

$$I = I_{SC} - I_0\{\exp(V/V_T) - 1\} \qquad (4)$$

Sun power presents electromagnetic radiation which gets transformed into electric power. Basically, solar module consists of solar cell [8] and solar cell is a p-n junction diode of semiconductor. The pv module is composed of silicon cells. In open-circuit condition, the silicon cells have operating voltage of 0.7 V. The current ratings depend on the area of the particular cell. The current ratings of the module are proportional to the two-dimensional surface in the PV and obtain the higher output. I-V characteristics given in (4) are determined for ideal condition, considering that internal series resistance of cell is zero, and the shunt resistance is infinite. In actual practice, both have finite values, which would alter the characteristics. The ideal and practical equivalent circuits of the solar cell are shown in Fig. 5. In practice, I_{SC} is no longer equal to light generated current but is less by shunt current through shunt resistance (R_{sh}). Mathematically, it can be presented as (Fig. 3),

$$I = I_L - I_0\{\exp(V + IR_S)/V_T) - 1\} - (V + IR_S)/R_{sh} \qquad (5)$$

V is PV voltage and I_L is the photoelectric current.
For typical high quality, one square inch silicon cell is used.
Series resistance (R_s) = 0.05–0.10 Ω,
Shunt resistance (R_{sh}) = 200-300 Ω.
T = Temperature of the module in Kelvin
I = PV current

$$I_L = I_{SC}$$

PV array connections are in series and shunt. In standard condition, modules are ∼22 V and ampere in module terminal is ∼5.5 A. Larger modules may be more cost effective. Figure 4 shows the equivalent model of solar cell and Fig. 5 shows the series-connected solar cell, while characteristics for the same are shown in Figs. 6, 7, 8, 9, and 10.

Fig. 3 Equivalent model of solar cell connected in series

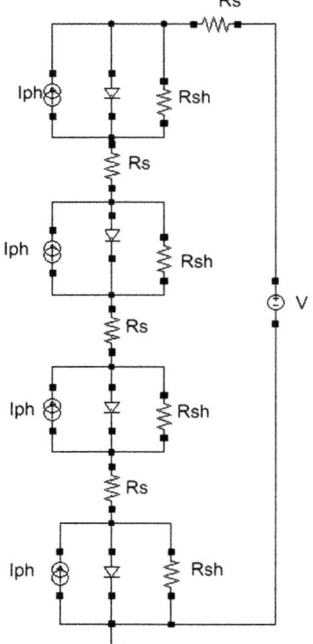

Fig. 4 Equivalent model of
solar cell

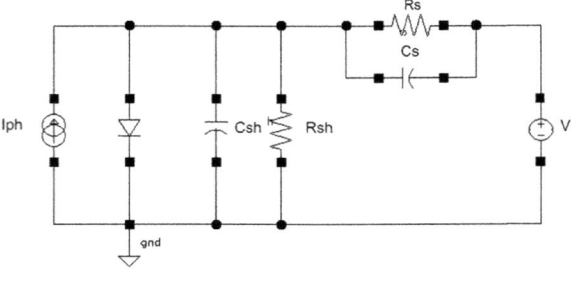

Fig. 5 Equivalent model of
series-connected solar cell

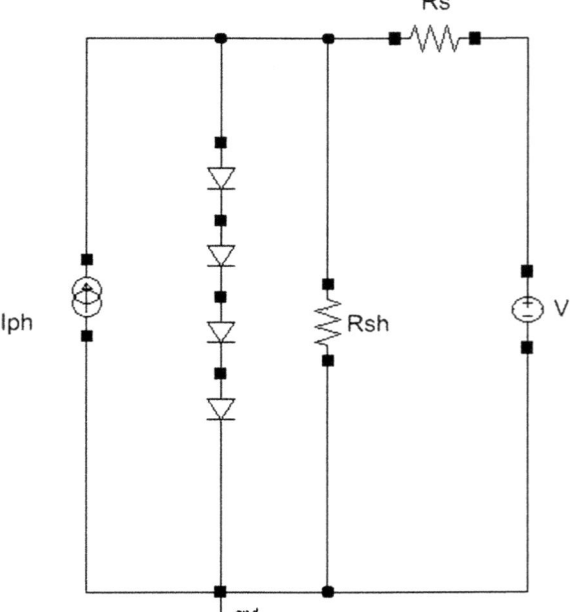

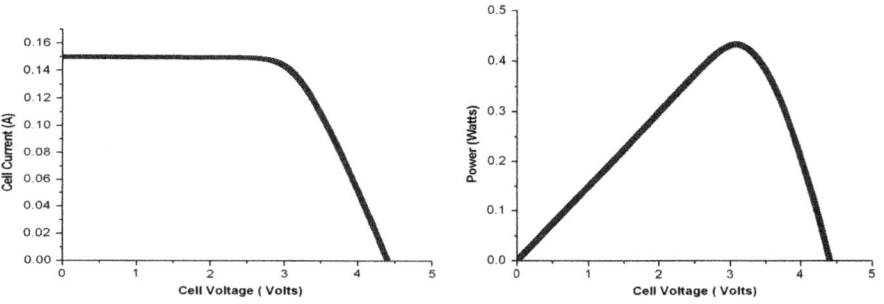

Fig. 6 Characteristics curve of series-connected solar cell

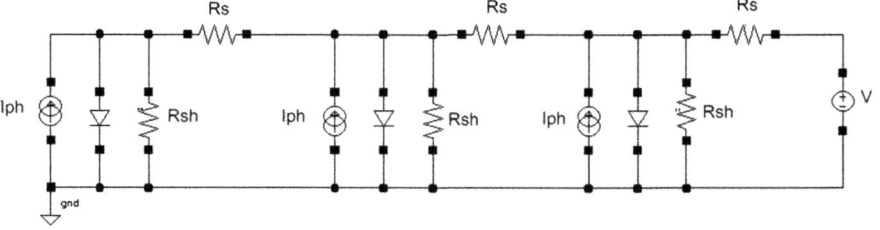

Fig. 7 Equivalent model of parallel-connected solar cell

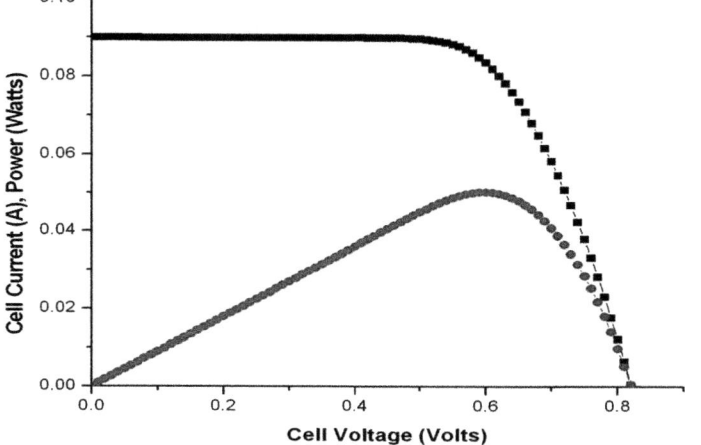

Fig. 8 Characteristics curve of parallel-connected solar cell

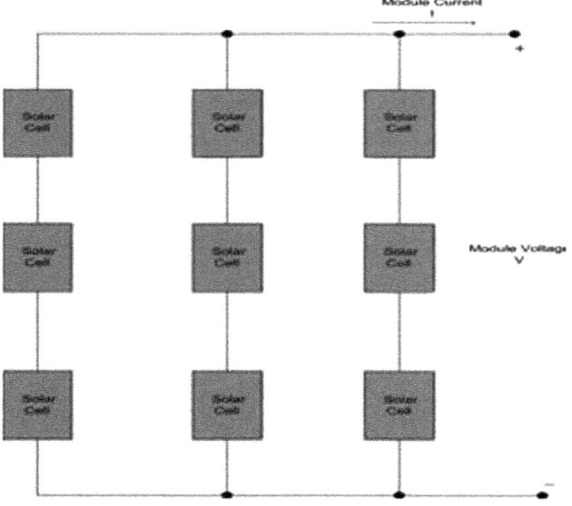

Fig. 9 Equivalent model of parallel–series-connected solar cell

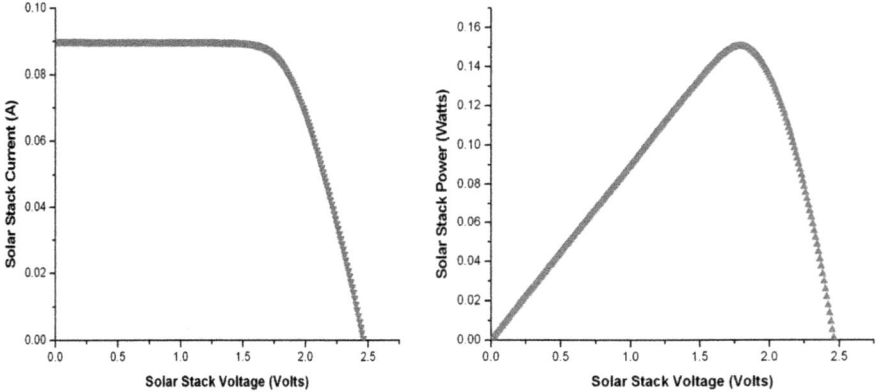

Fig. 10 I-V and PV characteristics of series–parallel-connected solar cell

4 Classification of Solar Cell

See Fig. 11.

5 Solar Cell Simulation

The simulated pv model with pv modules is connected in series and its output of PV and V-I characteristics shown in Fig. 12.

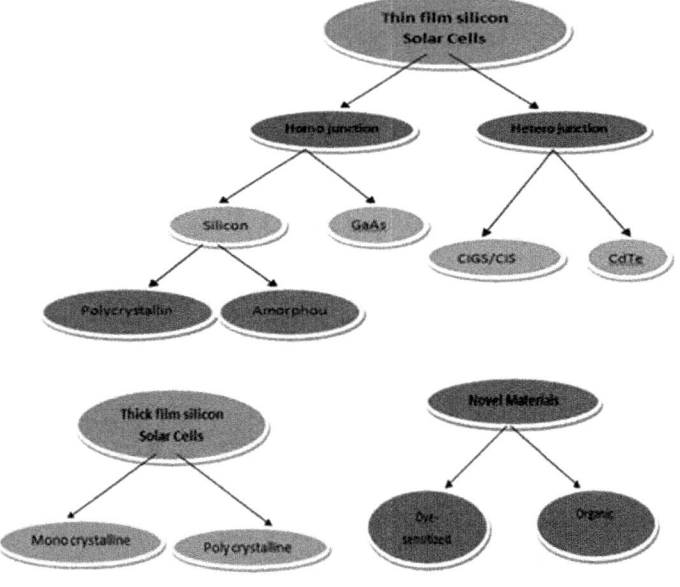

Fig. 11 Different types of solar cell

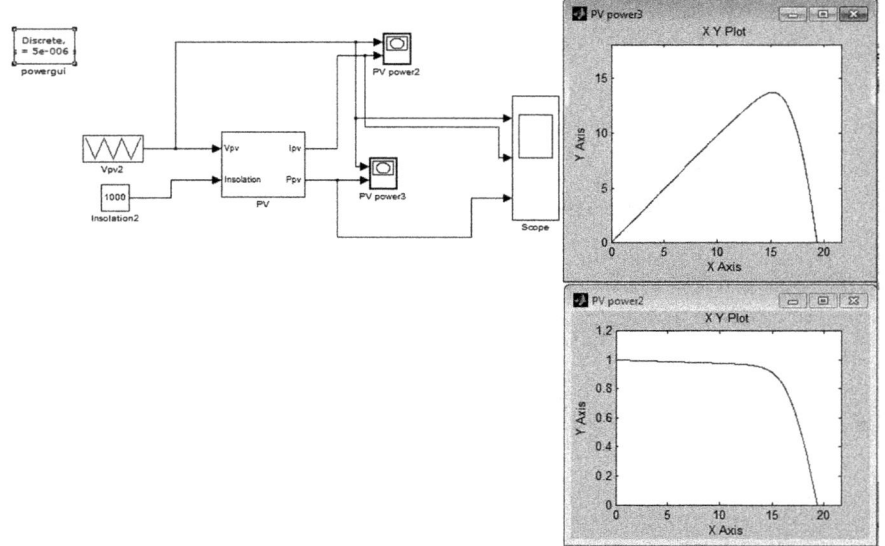

Fig. 12 Simulation model of PV module

Programing-based code simplifies the simulation block and increases its efficiency and readability. The PV simulation model has been implemented through
different stages (Figs. 13 and 14).

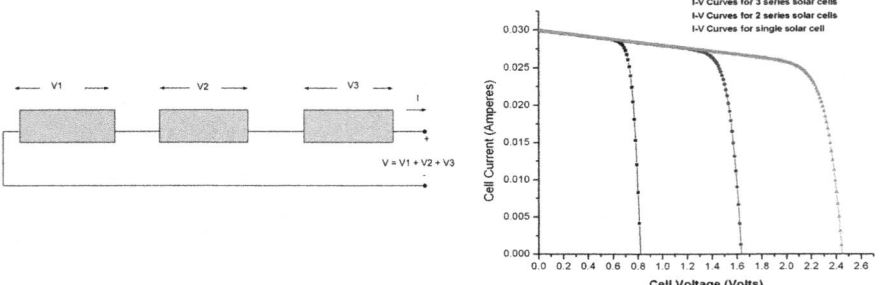

Fig. 13 Series connection of solar cell and their I-V curves

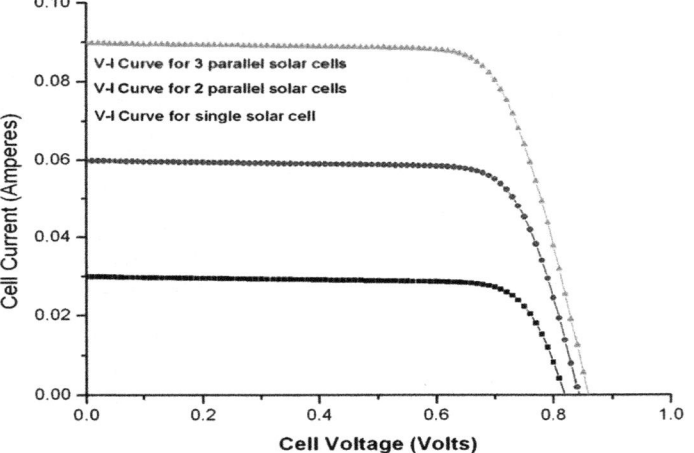

Fig. 14 Parallel connection of solar cell and their I-V curves

6 Flowchart of Simulation Algorithm

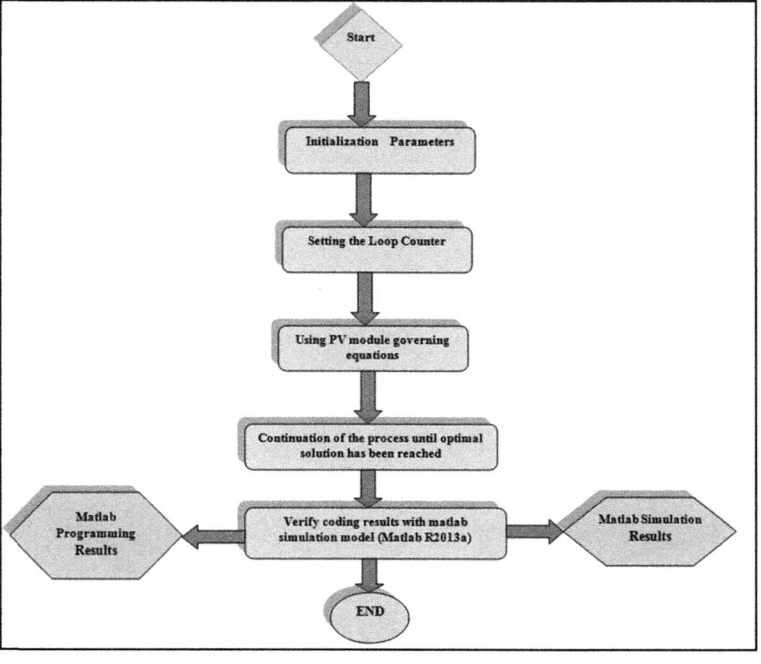

7 Resistance Sensitivity

Series resistance of solar cell provided to improve cell quality. The linear gradually improved in the resistance degrades the FF of current-to-voltage curve which measures the maximum power available from the cell. The effects of series resistances (1, 3.25, 5.5, 7.75, and 10 Ω) are shown in Figs. 15 and 16.

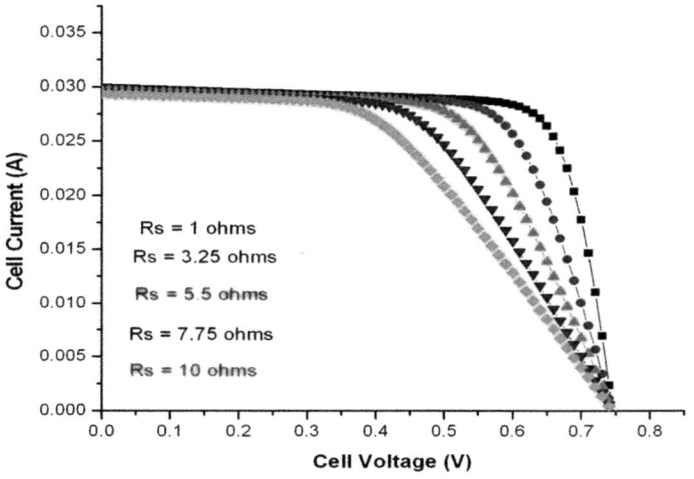

Fig. 15 Effects of series resistance of solar cell I-V characteristics

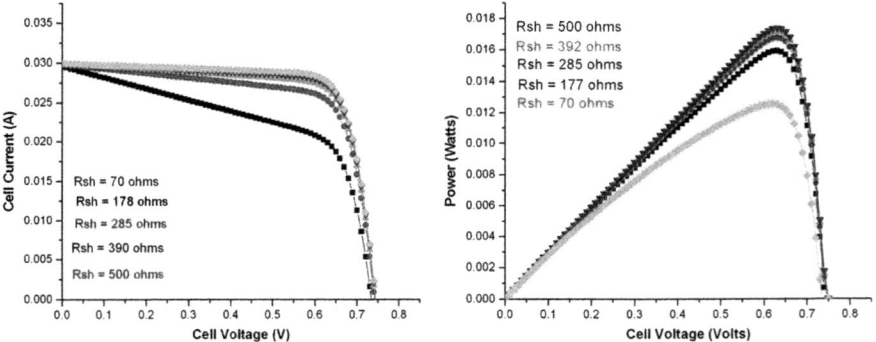

Fig. 16 Effects of shunt resistance of solar cell I-V and PV characteristics

8 Temperature Sensitivity

The operating temperature of solar cell is very sensitive. In semiconductor devices with the increase of temperature, band gap energy is decreased which affects a behavior of module characteristics (Figs. 17 and 18; Table 1).

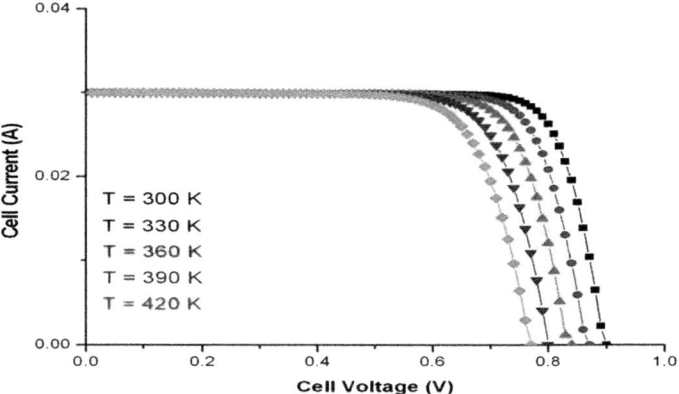

Fig. 17 Temperature effect of solar cell I-V Characteristics

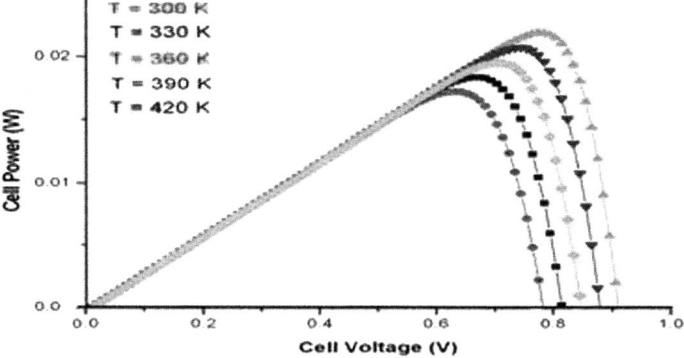

Fig. 18 Temperature effects of solar cell in PV characteristics

9   Shading Effects

The shading effect in solar cell has enhanced impact on efficiency. Series connection of PV panels is done in commercial sector. Protection of the cells from destructive reverse voltages in case of shadowing or other abnormalities is done by connecting a number of bypass diodes (Figs. 19, 20, and 21).

Table 1 Different parameters used in simulation

Simulation parameter values	Values
Varying solar radiation intensities (S)	200, 400, 500, 800, 1000 W/m^2
Temperature of cell (T)	25 + 273
Reference temperature (Tref)	40 + 273
Short-circuit temperature coefficient (k_i)	0.0023 mA/°C
Reverse saturation current (I_{rr})	21 × 10^{-10} A
Boltzmann constant (k)	1.38065 × 10^{-23}
Cell saturation current (I_{scr})	0.75 mA
Area of the module (A)	0.40 m^2
Temperature coefficient (Current) (α)	0.473 mA/°C
Temperature coefficient (V) (β)	637 V/°C
Number of parallel cells (Np)	4
Number of series cells (Ns)	100

Fig. 19 Bypass diode effects in solar cell

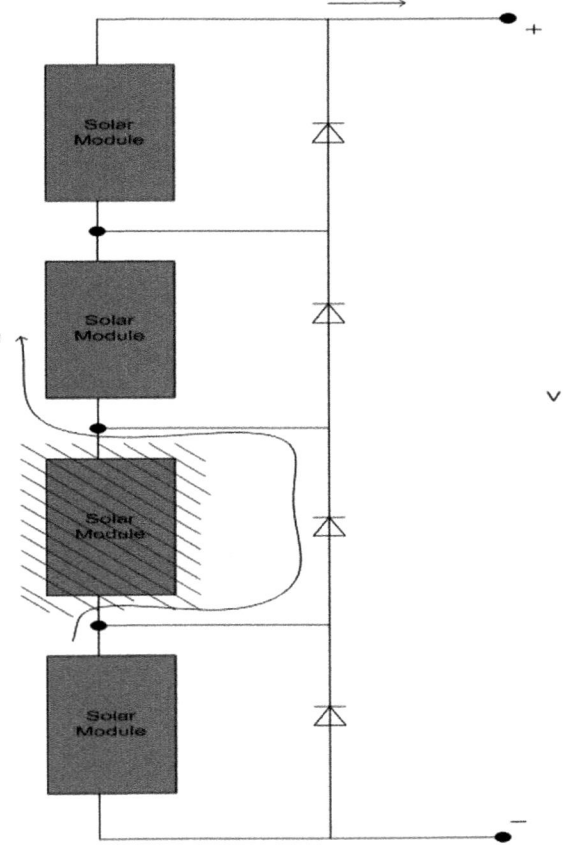

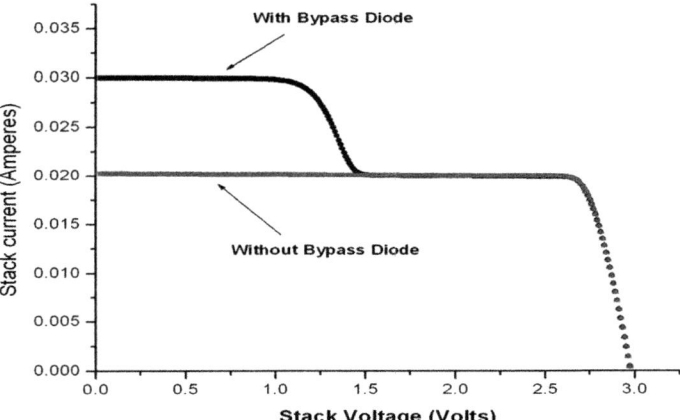

Fig. 20 Effects on bypass diode of I-V characteristics

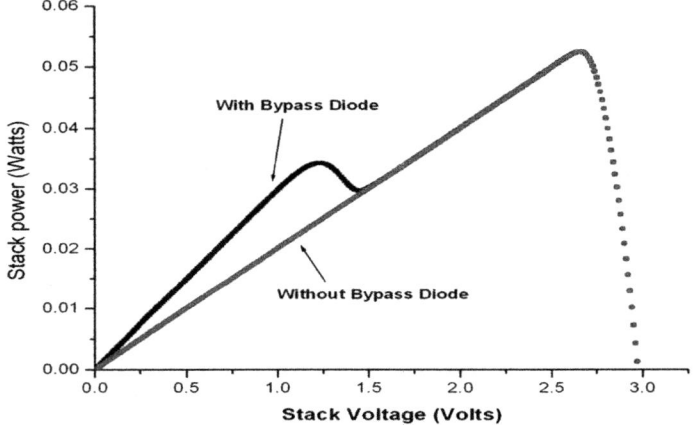

Fig. 21 Effects on bypass diode of PV characteristics

10   Conclusion

This paper dealt with different types of PV module characteristics under sensitivity parameter analysis. Matlab Simulink software is utilized here to implement the analytical model presented with different solar irradiations and varying temperatures. The results are obtained by vigorous analysis of the analytical method under different irradiances, temperatures, and different types of solar module parameters. The comparative study of sensitivity analysis of different solar modules and their shading effects is also presented in the present work.

References

1. Roberto Bruno, Piero Bevilacqua, Luigi Longo, Natale Arcuri, "Small size single-axis PV trackers: control strategies and system layout for energy optimization", Energy Procedia, Vol 82, 2015, pp. 737–743.
2. Nibedita Das, Nitai Pal, Sadhu K. Pradip, "Economic cost analysis of LED over HPS flood lights for an efficient exterior lighting design using solar PV" Building and Environment Vol 89, 2015, pp. 380–392.
3. P. Srinivasa Rao, G. Saravana Ilango, Chilakapati Nagamani "Maximum power from PV arrays using a fixed configuration under different shading conditions", IEEE Journal of photovoltaics, Vol. 4, No. 2, March 2014, pp. 679–686.
4. J. Mroczka, M. Ostrowski, "Photovoltaic array simulation technique for non-uniform insolation conditions" International Conference on Renewable Energies and Power Quality (ICREPQ' 14), 8th to 10th April, 2014.
5. D. Picault, B. Raison, S. Bacha, J. de la Casa, J. Aguilera, "Forecasting photovoltaic array power production subject to mismatch losses", Solar Energy Vol. 84, 2010, pp. 1301–1309.
6. Francesca De Rossi, Tadeo Pontecorvo, Thomas M. Brown, "Characterization of photovoltaic devices for indoor light harvesting and customization of flexible dye solar cells to deliver superior efficiency under artificial lighting", Applied Energy Vol. 156, pp. 413–422, 2015.
7. A.K. Pandey, V.V. Tyagi, Jeyraj A/L Selvaraj, N.A. Rahim, "Recent advances in solar photovoltaic systems for emerging trends and advanced applications", Renewable and Sustainable Energy Reviews Vol. 53, 2016, pp. 859–884.
8. David B. Richardson, L.D.D. Harvey, "Strategies for correlating solar PV array production with electricity demand", Renewable Energy Vol. 76, 2015, pp. 432–440.
9. Young-Hyok Ji, Jun-Gu Kim, Sang-Hoon Park, Jae-Hyung Kim, and Chung-Yuen Won, "C-language Based PV Array Simulation Technique Considering Effects of Partial Shading", IEEE International Conference on Industrial Technology, ICIT 2009, pp. 1–6.

Effect of Location of Piezoelectric Sensor Over a Smart Structure

Sukesha

1 Introduction

For 'Active Vibration Control' (AVC) of smart structure (instrumented with piezoelectric patches) understanding of fundamentals of piezoelectricity [1], finite element modelling and control techniques are required. Simulations and/or experiments can be performed to evaluate the performance of designed smart structure. To create a finite element model of the smart structure, various softwares are available. In MATLAB software, a computer code can be written to numerically integrate stiffness, mass, and damping matrices of a finite element model [2–5]. Thereafter, using assembly procedure global mass, damping and stiffness matrices can be obtained. Newmark-β method/Wilson-θ method/state space method can be employed to numerically solve these second-order coupled ordinary differential equations [6, 7]. Modal frequencies and mode shapes of the structure can be obtained using 'Eigenvector calculation' method [6, 8, 9]. Modal displacements and modal velocities of the structure can be computed and plotted in MATLAB [10–13]. AVC scheme can be constructed using numerous control laws, such as negative modal velocity feedback method [1, 14–17], LQR [18], fuzzy [19], neural network [20], etc. It is very difficult to simultaneously control all the modes of the continuous structure. In a typical vibration signal of a continuous structure, only first few modes are usually dominant. So, by controlling first few modes, vibrations of a structure can be controlled effectively.

For experimental verification of a smart structure, a test rig consisting of host structure, sensors, actuators, signal conditioner, processor, and amplifier is erected. Cantilevered smart plate instrumented with piezoelectric patches is widely used as test structure. To control the vibrations of a structure effectively, placement of

Sukesha (✉)
UIET, Panjab University, Chandigarh, India
e-mail: Sukesha@pu.ac.in

© Springer Nature Singapore Pte Ltd. 2019
V. Nath and J. K. Mandal (eds.), *Proceeding of the Second International Conference on Microelectronics, Computing & Communication Systems (MCCS 2017)*, Lecture Notes in Electrical Engineering 476, https://doi.org/10.1007/978-981-10-8234-4_42

sensors and actuators over the structure is very important [21–25]. In this paper, finite element modelling of smart plate is done. Response of the sensor voltage in open loop is observed and compared for different element locations. Organization of the paper is as follows: in Sect. 2, finite element model of the section is given, results are discussed in Sect. 3, and in Sect. 4 conclusions are drawn.

2 Finite Element of Smart Plate

In finite element technique, whole structure is discretized into small finite elements and equations are written for individual elements using Hamilton's principle or Galerkin's approach. Thereafter, assembly procedure is employed to find out global stiffness matrix, global mass matrix and global force vector. According to Hamilton's Principle [21]:

$$\delta \int_{t1}^{t2} L \mathrm{d}t = 0 \tag{1}$$

where 'δ' is the variation operator, '$L = T_e - V_e$' is called Lagrange, 'T_e' is the kinetic energy of system at a particular instant of time and 'V_e' is the potential energy of system at a particular instant of time. Expression of kinetic energy and potential energy of system can be substituted in Hamilton's equation so as to obtain equation of motion of one finite element. For finite element modelling of smart piezo-structure, electromechanics of piezoelectric sensor/actuator is required to be included in Eq. (1). Electromechanics of piezoelectric materials is expressed in the form of constitutive equations of piezoelectricity [17, 23]. Constitutive equations relate to stress, strain, electric field, electric displacement, temperature, humidity, etc. of a piezoelectric material.

Plate is divided into 64 elements having 81 nodes as shown in Fig. 1a–d. Shaded portion shows placement of piezoelectric sensor (Fig. 1a–d). Equation of motion of single element is given as

$$\left(\left[m_s^e\right] + \left[m_p^e\right]\right)\{\ddot{u}_e\} + \left(\left[k_s^e\right] + \left[k_p^e\right]\right)\{u_e\} = \{F^e\} \tag{2}$$

where '$\left[m_s^e\right]$' is the elemental mass matrix of substrate, '$[m_p^e]$' is the elemental mass matrix of piezoelectric, '$[k_s^e]$' is the elemental stiffness matrix of substrate, '$\left[k_p^e\right]$' is the elemental stiffness matrix of piezoelectric, '$\{F^e\}$' is the force vector, '$\{u_e\}$' is the elemental displacement vector and '$\{\ddot{u}_e\}$' is the elemental acceleration vector. Following the assembly procedure, the global equation of motion is given as [21–24]:

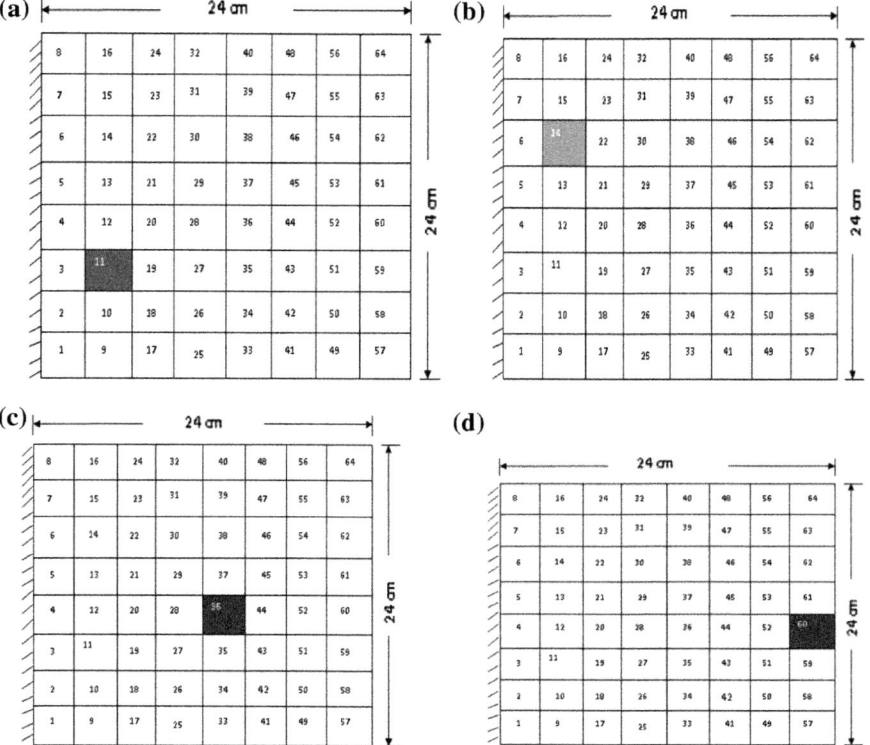

Fig. 1 Smart cantilevered plate divided into 64 elements and 81 nodes where shaded portions show placement of piezoelectric sensor. **a** 11th location, **b** 14th location, **c** 36th location, **d** 60th location

$$\left[M \right]_{216 \times 216} \{\ddot{x}\}_{216 \times 1} + \left[C \right]_{216 \times 216} \{\dot{x}\}_{216 \times 1} + \left[K \right]_{216 \times 216} \{x\}_{216 \times 1} = \{F\}_{216 \times 1} \quad (3)$$

where '$[M]$' is the mass matrix, '$[C]$' is the damping matrix, '$[K]$' is the stiffness matrix, '$\{F\}$' is the force vector, '$\{x\}$' is the displacement vector, '$\{\dot{x}\}$' is the velocity vector and '$\{\ddot{x}\}$' is the acceleration vector. Matrix Eq. (2) has coupled second-order ordinary differential equations. These equations can be uncoupled using modal truncation. Thereafter, response of the structure is observed for different element locations (Fig. 1a–d). Sensor voltage at one element location is given as [24]:

$$\text{Sensor voltage} = \frac{\text{electromechanical interaction matrix}}{\text{capacitance of piezoelectric patch}} \{u_e\} \quad (4)$$

3 Results and Discussions

Square plate of size 24 cm × 24 cm × 0.6 mm instrumented with piezoelectric patches of size 3 cm × 3 cm × 1 mm is considered in this work. Simulations are performed in MATLAB software. Newmark-β method is used to solve second-order differential equation (Eq. 3). Response of the sensor voltage (open loop) is observed separately for piezoelectric patches pasted at 11th, 14th, 36th and 60th element locations. As shown in Fig. 2a, maximum peak value of sensor voltage is 2.4 V when piezoelectric patch is pasted at 11th element location. Sensor voltage keeps on decreasing, as sensor location is moved away from cantilevered end (Fig. 2b–d). This difference is also clearly visible from Fig. 3, where

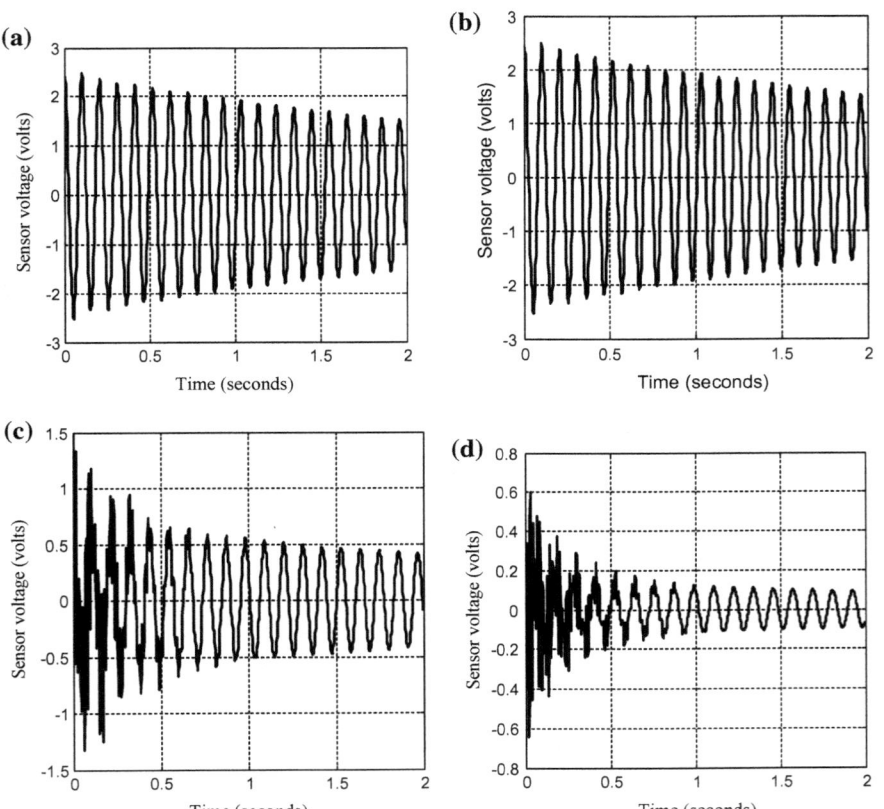

Fig. 2 Time response of sensor voltage when sensor is placed at **a** 11th location, **b** 14th location, **c** 36th element location, **d** 60th element location

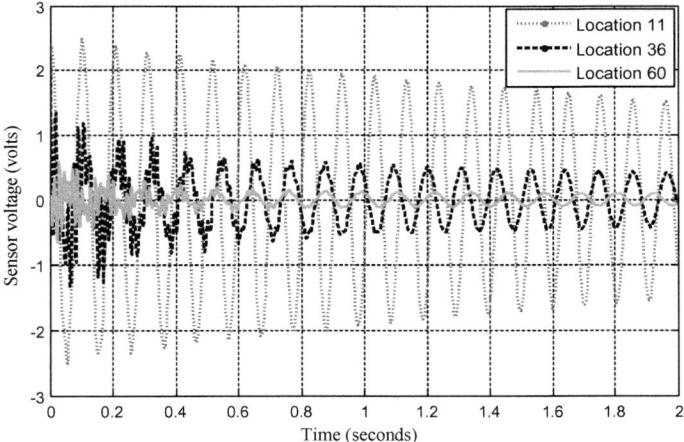

Fig. 3 Comparison of time response of sensor voltage when sensor is placed at 11th element location, 36th element location, and 60th element location

Table 1 Comparison of sensor voltage

Piezoelectric sensor location	Peak sensor voltage (V)
11th element	2.4
14th element	2.4
36th element	1.2
60th element	0.6

comparison of three different element locations is done. It can be observed from Fig. 3 that more peak value of sensor voltage is achieved when sensor is pasted at 11th element location as compared to the case when sensor is pasted at 36th element location. Least value of sensor voltage is obtained when sensor is pasted at 60th element location as compared to 11th and 36th element locations. Reason for this difference is that piezoelectric sensor is a strain sensor. Maximum strain is available at the element location near to cantilevered end. Therefore, maximum sensor output is obtained when sensor is placed at 11th and 14th element locations, which are more near to cantilevered end. Higher modes are also visible in the time response of sensor voltage for the element location far from cantilevered end (Fig. 2c–d; Table 1).

Peak sensor voltage at different element locations is also shown in Table 1. As can be seen in this table, maximum peak sensor voltage can be obtained if sensor is near to cantilevered end.

4   Conclusions

Response of cantilevered square smart plate is observed in this paper. Finite element modelling is done using Hamilton's principle. Time response of sensor voltage is observed for different element locations. It is observed that highest value of sensor voltage is obtained at the element location near to cantilevered end as compared to the element locations, which are far from cantilevered end. Therefore, element locations near to the cantilevered end are best element locations for placement of piezoelectric sensors to obtain maximum sensor response.

References

1. S. Sharma, R. Vig, and N. Kumar, "Effect of electric field and temperature on dielectric constant and piezoelectric coeffcient of piezoelectric material: A review," Integrated Ferroelectrics, Vol. 167, pp. 154–175, 2015.
2. V. Gupta, M. Sharma, N. Thakur, "Optimization criteria for optimal placement of piezoelectric sensors and actuators on a smart structure: A Technical Review," Journal of Intelligent Materials Systems and Structures, Vol. 21, pp. 1227–1242, 2010.
3. M. Sharma, S. P. Singh and B. L. Sachdeva, "Modal control of a plate using fuzzy logic controller," Smart Materials and Structures, Vol. 16, pp. 1331–1341, 2007.
4. S. Sharma, R. Vig, and N. Kumar, "Active vibration control: considering effect of electric field on piezoelectric patches," Smart Structures and Systems., Vol. 16, No. 6, pp. 1091–1105, 2015.
5. S. Sharma, R. Vig, and N. Kumar, "Finite element modelling of smart piezo structure: considering dependence of piezoelectric coefficients on electric field," Mechanics Based Design of Structures and Machines, 2015. https://doi.org/10.1080/15397734.2015.1076728.
6. J. E. Mottershed and Y. M. Ram, "Inverse Eigen value problems in vibration absorption: passive modification and control," Mechanical Systems and Signals Processing, Vol. 20, pp. 5–44, 2006.
7. S. Sharma, R. Vig, and N. Kumar, "Temperature compensation in smart structure by application of DC bias on piezoelectric patches," Journal of intelligent materials systems and structures, Vol. 27, pp. 2024–2035, 2016.
8. B. M. Dutta and V. Sokolov, "A solution of the affine quadratic inverse Eigen value problem," Linear Algebra and its Applications, Vol. 434, pp. 1745–1760, 2011.
9. V. Gupta, M. Sharma, N. Thakur, "Active vibration control of a smart plate using a piezoelectric sensor-actuator pair at elevated temperatures," Smart Materials Structures, Vol. 20, pp. 105023-1–105023-13, 2011.
10. M. A. Ahmad, M. S. Ramli, A. N. K. Nasir, R. Ismail and M. S. Saealal "Performance assessment of active vibration control using nominal characteristics trajectory following NCTF controller," 4th Asia International Conf. Mathematical/Analytical Modelling and Computer Simulation (AMS), pp. 414–419, 2010.
11. U. H. Diala and G. N. Ezeh, "Nonlinear damping for vibration isolation and control using semi active methods," Academic Research International, Vol. 3, No. 3, pp. 141–152, 2012.
12. S. M. Hasheminejad and M. Vahedi, "Active vibration control of a thick piezolaminated beam with imperfectly integrated sensor and actuator layers," International Journal of Automation and Control, Vol 8, No. 1, pp. 58–87, 2014.

13. K. Knipe, "Structural analysis and active vibration control of tetraform space frame for use in microscale machining," Doctoral dissertation, University of Central Florida Orlando, Florida, 2007.

14. H. S. Bouomy "Active vibration control of a dynamical system via negative linear velocity feedback," Nonlinear Dynamics, Vol. 77, pp. 413–423, 2014.

15. Y. K. Kang, H. C. Park, J. Kim, and S. B. Choi, "Interaction of active and passive vibration control of laminated composite beams with piezoelectric sensors/actuators," Materials and Design, Vol. 23, pp. 277–286, 2002.

16. F. M. Li, K. Kishimoto, Y. S. Wang, Z. B. Chen and W. H. Huang, "Vibration control of beams with active constrained layer damping," Smart Materials and Structures, Vol. 17, No. 6, pp. 065036, 2008.

17. W. G. Cady, "Piezoelectricity," New York: Dover, 1964, Vol. 1.

18. K. R. Kumar and S. Narayanan, "Active vibration control of beams with optimal placement of piezoelectric sensor/actuator pairs," Smart Materials and Structures, Vol. 17, No. 5, pp. 055008, 2008.

19. Zheng, K. (2008) 'Active vibration control of adaptive truss structure using fuzzy neural network,' Chinese Control and Decision Conference, pp. 4872–4875.

20. Zheng, K. (2010) 'Intelligent vibration control of piezo-electric truss structure using GA-based fuzzy neural network,' Proc. 8th World Congr. Intell. Control Auto., pp. 5136–5139, Jinan, China.

21. M. Petyt, "Introduction to finite element vibration analysis," Cambridge University Press, New York, 2nd Ed., 2010.

22. Z. C. Qiu, X. M. Zhang, H. X. Wu and H. H. Zhang, "Optimal placement and active vibration control for piezoelectric smart cantilevered plate," Journal of Sound and Vibrations, Vol. 301, pp. 521–543, 2007.

23. Ikeda T, "Fundamentals of piezoelectricity," Oxford University Press, UK, 1996.

24. V. Gupta, M. Sharma, N. Thakur, "Active structural vibration control: robust to temperature variations," Mechanical Systems and Signals Processing, Vol. 33, pp. 167–180, 2012.

25. J. H. Han and I. Lee, "Optimal placement of piezoelectric sensors and actuators for vibration control of a composite plate using genetic algorithm," Smart Materials and Structures, Vol. 8, No. 2, pp. 257–267, 1999.

Design and Implementation of a Reaction Timer Using CMOS Logic

Varun Bohra, Neha Nidhi, Sumit Singh, Deepak Prasad,
Anand Kr. Thakur, Ajay Kumar and Vijay Nath

1 Introduction

In 1786, Maskelyne of Royal Astronomer of Greenwich Observatory and Kinnebrooke observed the times of star movements. They predict the delay difference in order of second. Such error calculation was too tough at that time. Such an error was un-correctable because the calibration of clocks to determine the standard time depended on the correct observation of the place and time when the stars were seen. The personal difference was calculated by German astronomer, Bessel in 1819 who found an average difference of 1.041 s between him and another observer. Kinnebrooke's error prediction was of the order of 0.8 s. This difference is known as personal equation and was found to vary with different

V. Bohra · N. Nidhi · S. Singh · D. Prasad (✉) · V. Nath
VLSI Design Group, Department of Electronics and Communication Engineering,
Birla Institute of Technology, Mesra, Ranchi 835215, Jharkhand, India
e-mail: prasaddeepak007@gmail.com

V. Bohra
e-mail: varunbohra777@gmail.com

N. Nidhi
e-mail: nehanidhi199@gmail.com

S. Singh
e-mail: mr.sumitsingh007@gmail.com

V. Nath
e-mail: vijaynath@bitmesra.ac.in

A. Kr. Thakur
SSMC Ranchi University, Ranchi, India
e-mail: fmruanand@gmail.com

A. Kumar
ARTTC BSNL, Ranchi 835215, Jharkhand, India
e-mail: ajaykumararttc@gmail.com

© Springer Nature Singapore Pte Ltd. 2019
V. Nath and J. K. Mandal (eds.), *Proceeding of the Second International Conference on Microelectronics, Computing & Communication Systems (MCCS 2017)*, Lecture Notes in Electrical Engineering 476, https://doi.org/10.1007/978-981-10-8234-4_43

individuals between 0.770 and 1.021 s. Chronograph and chronoscope were invented to minimize the errors. Hipp chronoscope was used in 1892. This timing device measures time intervals in thousandths of a second or milliseconds. After that, it was realized that at the human level, there is a measurable time lag between the appearance of a signal and the response to it. This research gives new dimensions what today known as the name of reaction time or response time [1]. In different psychological institutes, a specific system is used to detect reaction time of the persons. In some places, it is used as IQ tester.

Today, human lives in a world where everything around them is just a click away from them and all this has been possible because of some great inventions that have proved to be a boon for their lives. They are so in a habit of getting things done fast that any delay causes them annoyance. Even response has become quick response these days. But, despite these advancements, there are places where technology cannot do all the work. For example, people talk about the human body where the detection and cure of diseases are readily available but the human body takes its own time to adapt itself to an environment, whether external or internal, or to get healed from an injury naturally. Thus, the reaction and stimuli speed play a vital role in their day-to-day lives. The measurement of time taken to respond to a stimulus by an organism can be countered as reaction time. The average reaction time for a visual stimulus is 200–250 ms; for hearing, it is 150–200 ms; and for touch, it is 130–170 ms [2]. The stone for the measurement of the reaction time was laid down by F. C. Donder who performed the first experiment of calculating reaction time by giving a shock to the foot of a patient and found the response time of 1/15 of a second [3]. In this paper, reaction timer is designed using CMOS for VLSI technology. CMOS has its own list of advantages but the primary reason of using CMOS in VLSI technology [4–7] is that it eases fabrication and thereby helps in making device very much compact and consumes low power. Reaction timer is very important circuit that can be utilized to measure muscles response to a stimulus. It is highly applicable in smartphone, smart grid, and Internet of Things (IoTs) for getting frequent response, through which it can measure human cognitive abilities. Human can also emit a tone from the built-in speaker or externally with headphone. It is also possible to experiment with multiple stimuli at once for more complex experimentation. For easy analysis of EMG recording, put a tick mark in line [8].

2 Design of Master–Slave J-K Flip-Flop

Master–slave J-K flip-flop uses seven 2-input NAND gates, two 3-input NAND gates, and three 2-input AND gates. Each gate is designed using CMOS logic using optimum conditions. Figure 1 is constructed with two PMOS as load and two NMOS as driver. The aspect ratios of these transistors in Fig. 1 are given as PM0

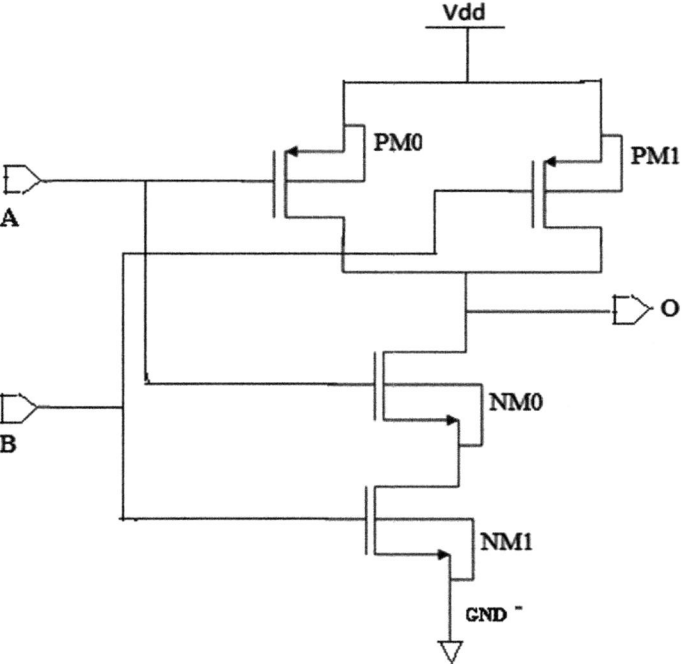

Fig. 1 Two-input NAND gate using CMOS technology

125 nm/45 nm, PM1 125 nm/45 nm, NM0 200 nm/45 nm, and NM1 200 nm/45 nm. 2-input AND gate is created using 2-input NAND gate (Fig. 1). Figure 2 shows PM2 250 nm/45 nm and NM2 100 nm/45 nm. Figures 1 and 2 show proper balanced circuit with 0 delayed. Figure 3 is designed using three PMOS as load and three NMOS as driver. The aspect ratios of these transistors are PM0, PM1, PM2 250 nm/3 * 45 nm, and NM0, NM1, NM2 300 nm/45 nm [9].

Triggering can be provided in two ways [4]:

Level triggering: In this triggering circuit, if input is changing in given clock interval, output also changes accordingly and if input is constant then output also remains constant.

Edge triggering: In edge triggering, output does not change with input. When first input rises, output also rises and after that output continuously rises, no matter of input.

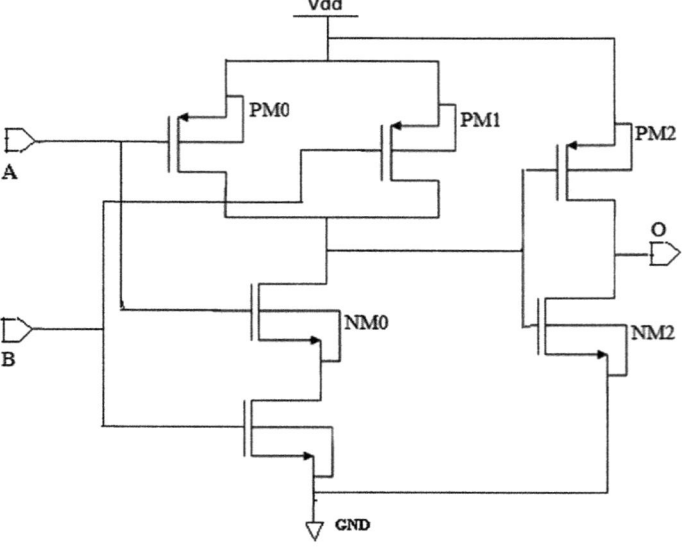

Fig. 2 Two-input AND gate using CMOS technology

2.1 Race-Around Condition

It occurs only in level triggering type trigger, J-K flip-flop when $J = K = 1$ and when

$$t_{\text{pd}} = t_{\text{pw}}$$

To avoid the race-around condition, master–slave flip-flop has been used and

$$t_{\text{pd}} = t_{\text{pw}} = t_{\text{CL}\kappa}$$

where t_{pw} = Pulse width, t_{pd} = Pulse delay, and t_{clk} = clock time pulse [5, 6].
Setup time (t_{su}): It is the minimum time required to keep data at proper level before applying clock.
Hold time (t_h): It is the minimum time required to keep same data after applying clock to store data properly.

$$t_{\text{su}} > t_h$$

Master–slave J-K flip-flop has been used to avoid race-around condition. Here, at one time, either master (M) changes or a slave (S) changes, which means that race-around condition will never come [7]. Slave never goes to 1, 1 while it may go to 1, 0 or 0, 1. Therefore, race-around condition vanishes. Condition 1, 1 might come in master which put no effect.

In master–slave flip-flop, output will change when slave output is changing. In this, input clock is applied to master and inverted clock is applied to slave due to which master and slave will not change at a time [10]. Hence, there is no race-around condition (Figs. 4 and 5).

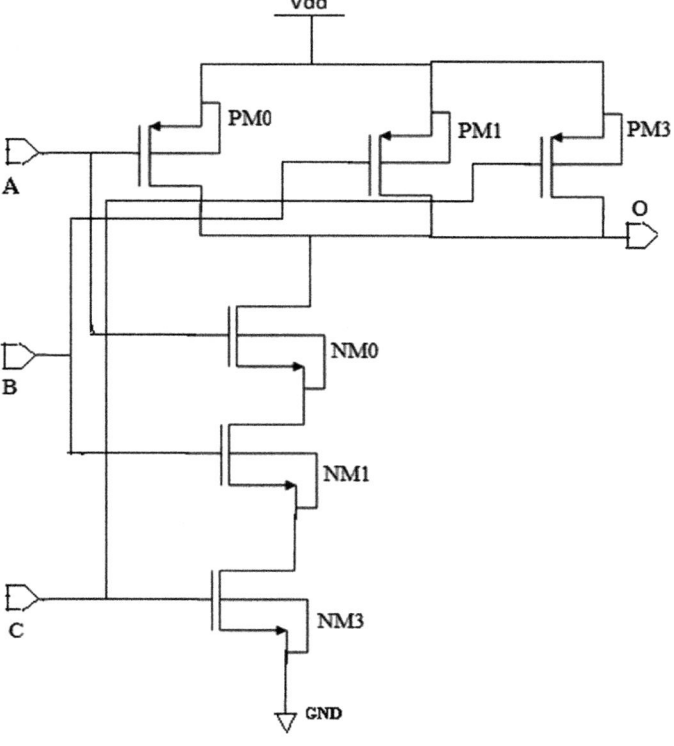

Fig. 3 Three-input NAND gate using CMOS technology

Fig. 4 Master–slave J-K flip-flop

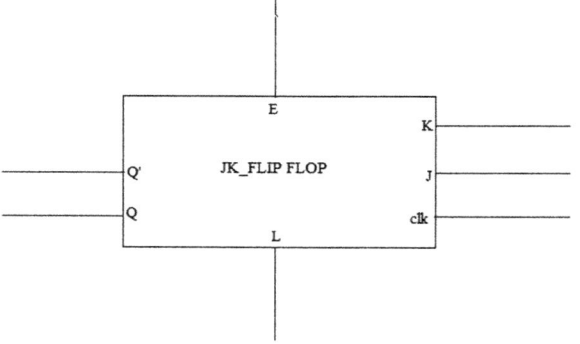

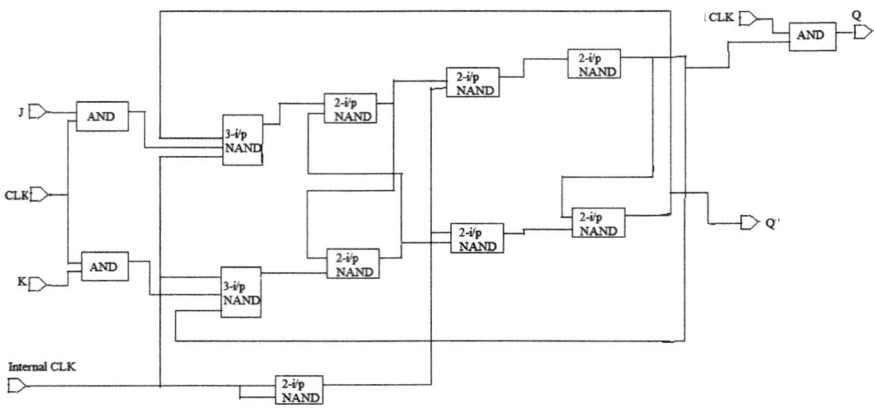

Fig. 5 Master–slave J-K flip-flop

The above-connected circuit is stored as a block and four of these blocks are used to form a sequential circuit that forms a decade counter.

Here, decade counter design and implementation using master–slave J-K flip-flop is shown in Fig. 6. A logical decade counter counts straight upward on each trailing edge of the input clock transition signal beginning from 0000 till it reaches the output value 1001 (decimal 9). Both outputs QD and QA are now equal in value and equal to logic "1". On the verge of the next clock pulse, the output transition and execution of the NAND gate changes state from logic "1" to a logic "0" level. As the signal from the NAND gate is interfaced to the CLEAR (CLR) inputs of all the J-K flip-flops, this signal enforces and results in all of the Q outputs to be reset back to binary 0000 on the very count of 10.

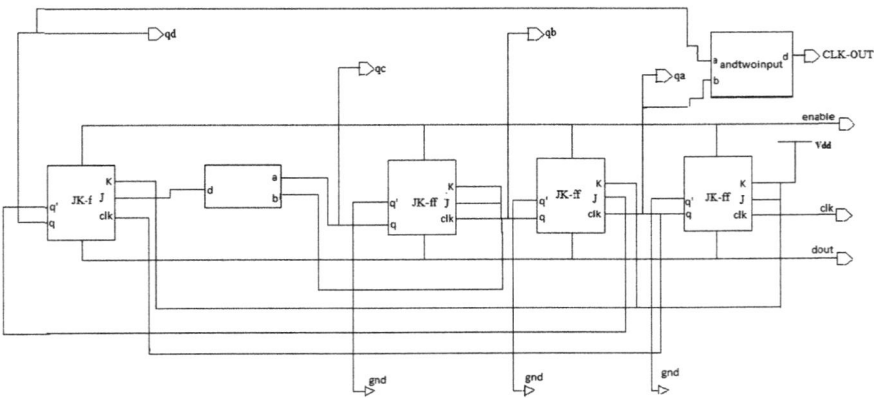

Fig. 6 Decade counter using J-K flip-flop

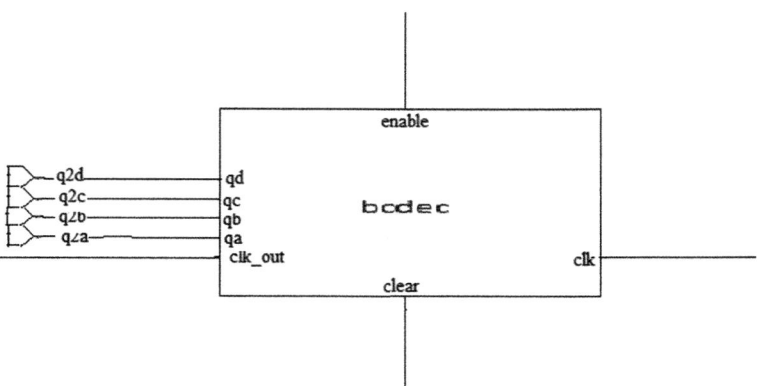

Fig. 7 BCD decade counter

As outputs QD and QA are now both equal and have the value logic "0", the flip-flops had undergone reset, the output from the NAND gate returns back to a logic level "1", and the counter again begin from 0000. Thus, it achieved Modulo-10 up counter (Fig. 7).

3   Design of Reaction Timer Using Decade Counter

The reaction timer, for accuracy up to two decimal places, is designed using three decade counters, connected to satisfy a sequential logic block. The above decade counter is saved as a block to be used further. There are three inputs: Enable, Reset, and Clear. And four outputs $q1$, $q2$, $q3$, and $q4$ which collectively give the BCD representation of a digit [11], e.g., if the reaction time is 0.02 s, the outputs of the decade counters would be 0000, 0000, and 0010.

The final form of the reaction timer is represented in Fig. 8. It is an asynchronous, since a common clock is not provided to all the components. To the clock

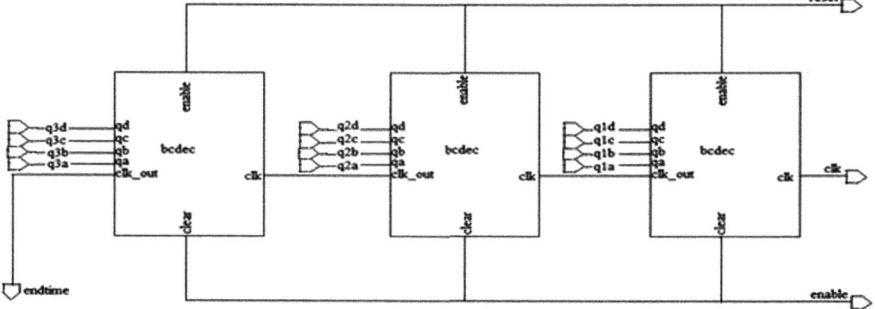

Fig. 8   Reaction timer

input, in the figure, a delay and a clock input are to be provided, so that the output is genuine. The controls of the external clock can be called START and STOP for logic "1" and logic "0", respectively. The Clear pins are given a common input Reset and the Enable pins are given an Enable signal that is equivalent to a POWER ON/OFF switch. A pin EXTTIME is also provided so that the number of decimal places can be increased. This helps improve the accuracy of the timer.

4 Result and Conclusion

In the present paper, a reliable CMOS reaction timer is demonstrated. Figure 9 shows the change in output pulse w. r. t. given input to J-K flip-flop, while Fig. 10 depicts the output waveform of BCD decade counters. On another hand, Figs. 11, 12, and 13 show the output waveform of $q1$ (a, b, c, d), $q2$ (a, b, c, d), and $q3$ (a, b, c, d). The proposed reaction timer is supplied with a power supply of 1 V.

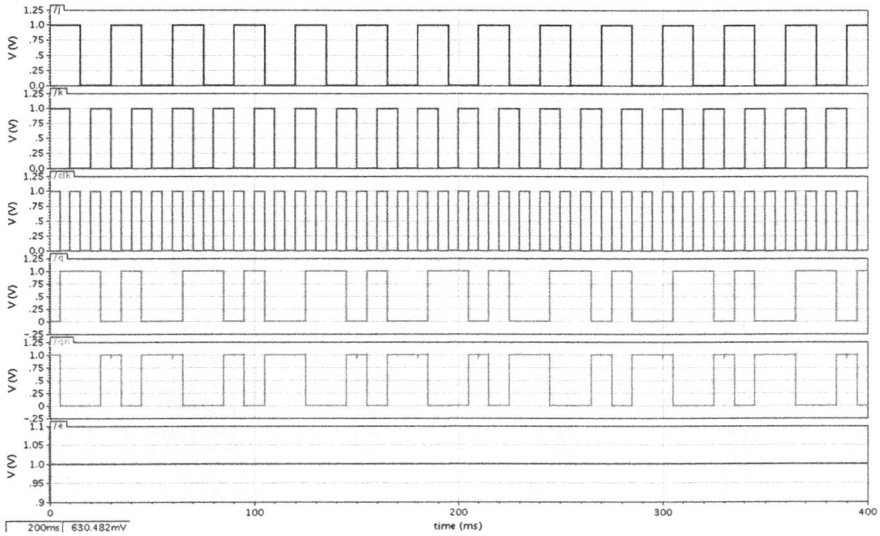

Fig. 9 Output waveform of J-K flip-flop

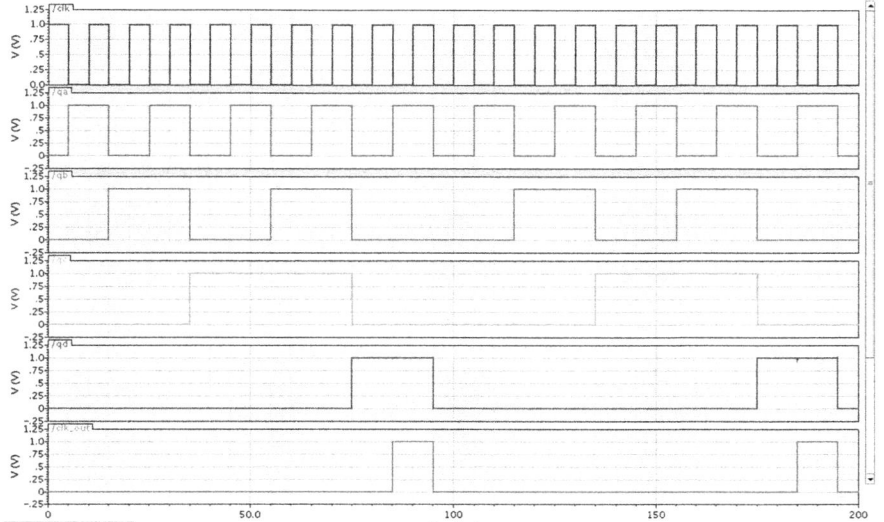

Fig. 10 Output waveform of BCD counter

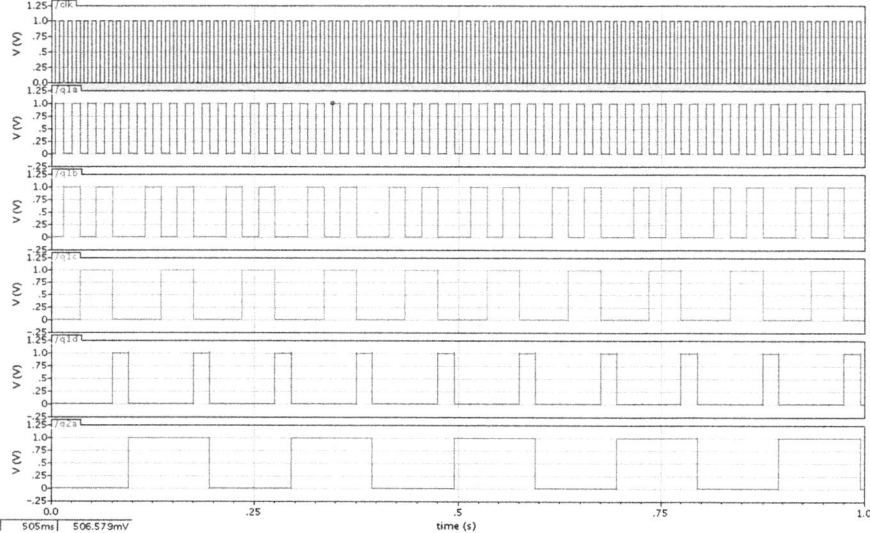

Fig. 11 Output waveform of reaction timer

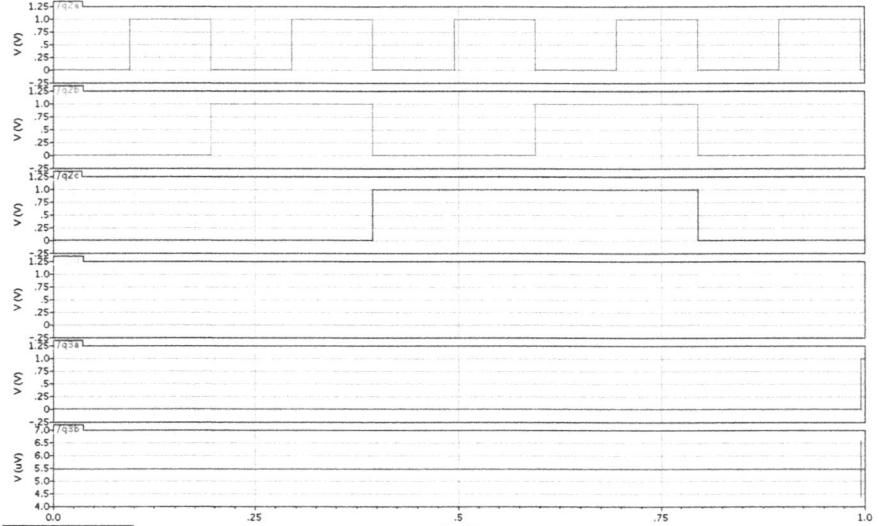

Fig. 12 Output waveform of reaction timer

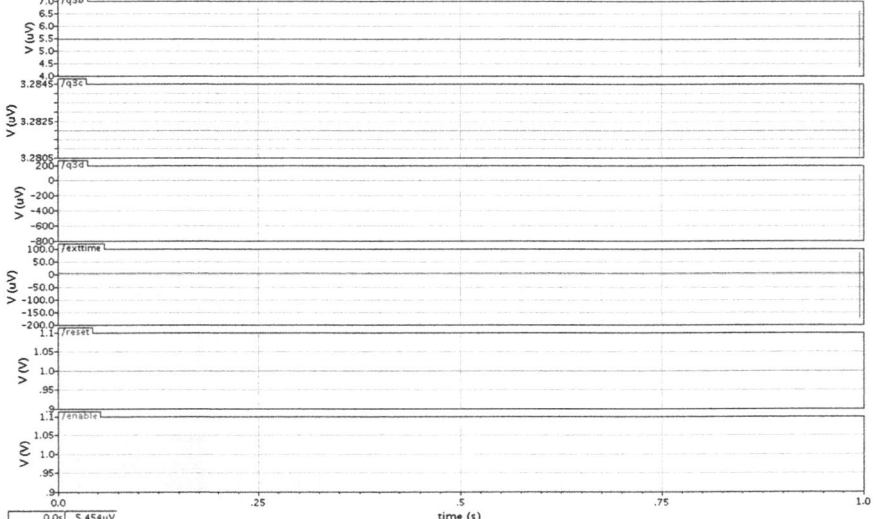

Fig. 13 Output waveform of reaction timer

Acknowledgements Authors are thankful to Prof. V. R. Gupta HOD ECE, Prof. T. Ghosh Dean Sponsored Research, Prof. R. Sukesh Kumar Dean Faculty Affairs, and Prof. M. K. Mishra Vice-Chancellor, BIT Mesra Ranchi for providing infrastructure facility to carry out this research work. Authors are also thankful to RESPOND ISRO for funding this project.

References

1. http://medisystemsindia.in/audio-visual-reaction-timer.ph.
2. G. D. D. Gajski, Principle of Digital Design, (Prentice-Hall: Upper Saddle River, N.J., 1997).
3. http://www.carolina.com/teacher-resources/Interactive/reactions-and-reflexes%3F/tr23009.tr, accessed on 17/07/2016.
4. Kang. S. M., Leblebici. Y, "CMOS Digital Integrated Circuits: Analysis and Design", third edition, Tata McGraw-Hill edition 2003.
5. Phillip E. Allen, Douglas R. Holberg, "CMOS Analog Circuit Design", Oxford University Press, Inc. 2002.
6. Kumar. M. A, Tumamala. V, "Design of a Low Power Variable-Resolution Flash ADC", 22nd IEEE, International Conference on VLSI Design, 2009.
7. Razavi B. (1998) Design of CMOS Analog Integrated Circuits, McGraw Hill, California.
8. https://backyardbrains.com/products/ReactionTimer.
9. Vijay Nath. http://www.ide.iitkgp.ernet.in/Pedagogy_view/example.jsp?USER_ID=210.
10. Allen P. E., Holberg D. R. (2009) CMOS circuit design, oxford Indian Edition.
11. Geiger Randall L., Allen Phillip E. and Strader Noel R. (2010) VLSI design techniques for analog and digital circuits. p. 486.

A Novel Hybrid Resource Allocation Scheme for Maximum Fairness Among Multiple Services

Shubham Goswami, Sagnik Mukherjee, Iti Saha Misra,
Deepan Mukherjee and Biswanath Chakraborty

1 Introduction

A resource sharing problem is one where a fixed resource needs to be allocated among a variable number of services in the fairest manner possible. Here, fairness implies that none of the services should suffer from starvation and the delay and throughput constraints should not be violated. One of the aims of the work was to determine the capacity of the system, which is the maximum number of services that the system can schedule.

The initial step was to decide upon an appropriate scheduling algorithm. There are several standard scheduling algorithms like round robin (RR), earliest deadline first (EDF), priority queuing (PQ), weighted fair queuing (WFQ) [1], etc. which are used in different cases depending on the requirement (e.g., RR/WFQ for intra-class scheduling and EDF/PQ for interclass scheduling). It was observed that none of the algorithms individually could satisfy the requirements of fairness and efficiency to the desired degree. So there was a necessity to explore a combination of two or more algorithms giving rise to what are called hybrid algorithms [2]. The problem demanded absolute priority for one of the services ensuring fair allocation among all the other services. The above premise propelled us toward a hybrid algorithm combining PQ and WFQ.

The Jain's fairness index [3] or JFI (discussed later) serves as an accurate performance parameter indicating the fairness of the resource allocation scheme. The objective of the designer is to maximize JFI while ensuring that all other system constraints are adhered to. A schematic of such a server is shown in Fig. 1. It is shown that the requests from the different services which have been admitted come in the form of packets with corresponding arrival rates (λ_i) and are served at a

S. Goswami (✉) · S. Mukherjee · I. S. Misra · D. Mukherjee · B. Chakraborty
Electronics and Telecommunication Engineering Department, Jadavpur University,
Kolkata 700032, India
e-mail: goswamishubham6@gmail.com

© Springer Nature Singapore Pte Ltd. 2019
V. Nath and J. K. Mandal (eds.), *Proceeding of the Second International Conference on Microelectronics, Computing & Communication Systems (MCCS 2017)*, Lecture Notes in Electrical Engineering 476, https://doi.org/10.1007/978-981-10-8234-4_44

Fig. 1 Schematic of a server
scheduling M services

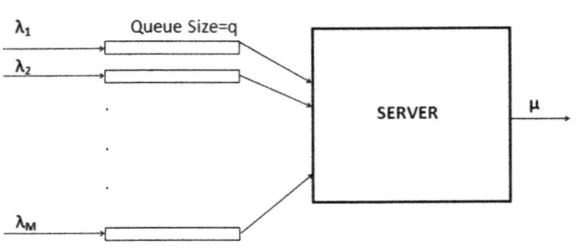

constant service rate (μ). To provide a mathematical structure to the problem, stochastic modeling was performed using Markov chain [4] (as explained in Sect. 2). The mathematical analysis provides us with the different steady-state probabilities (SSP) associated with the different states of the Markov chain. The SSPs provide us with the different QoS performance parameters (see Sect. 2). It can be shown that these SSPs are functions of the weights assigned to the different services by the WFQ algorithm. Thus, the problem of maximizing JFI translates to a problem of finding the optimum weight distribution. Finally, two different implementations of the server are considered. The feasibility and time complexity of these approaches are studied and compared.

2 Hybrid Packet Scheduling Algorithm

Before we dive into the analysis, it is important to understand the hybrid algorithm employed. The packets are sorted by service class into separate queues. The highest priority queue is served first (Priority Queuing) and once this queue is empty, the remaining queues are served for different durations of time in the ratio of the weights assigned to them (Weighted Fair Queuing). The algorithm steps are given below.

{

Step Action

1. Is the priority queue empty?
2. If yes, go to step 5
   Else go to step 3
3. Serve the next packet in the priority queue.
4. Go to step 1.
5. Serve the remaining queues for a total time T where the nth queue is served for a time $w_n T$.
6. Go to step 1.
   [w_n is the weight assigned to the nth service]}

The mathematical model is discussed next.

3   Mathematical Model

3.1   Markov Chain Analysis

A mathematical model representing the dynamics of the system has been developed. The call admission is performed beforehand. Thus, our analysis is restricted to packet scheduling only. The packet arrivals may be assumed to follow the Pareto or the Poisson distribution [4]. For simplicity, it is assumed that the admitted calls try to access the server in a steady, non-bursty manner maintaining independence. Hence, the Poisson distribution is considered.

Our stochastic model comprises a server characterized by a constant service rate, μ. There are M-independent services with their corresponding Poisson arrival rates $(\lambda_1, \lambda_2, \lambda_3 \ldots \lambda_M)$. Each service has its own queue of the same finite queue length (see Fig. 1). The service time distribution is exponential [4] (with mean service time $= 1/\mu$).

Each state of the system is described by a vector of length N $(n_1, n_2, \ldots n_M)$ where n_i represents the number of packets waiting in the ith service queue and $0 \le n_i \le q$. The arrival of packets is completely random and their service depends only on the present state of the queue.

Hence, the state transition is memoryless and can be modeled by a birth–death process involving multivariable states [4]. This calls for *Markov Chain Analysis*.

The total number of states in the system is $T = (q + 1)^M$. Each state has a corresponding steady-state probability (π_s where $s \in S$; S = state vector of size T).

The system is governed by the equation:

$$\frac{\mathrm{d}(\pi s)}{\mathrm{d}t} = 0 \ \forall \ s \in S \tag{1}$$

In other words, the rate of arrival equals the rate of departure. The following constraints hold true:

$$\text{i) } 0 < \pi_s < 1 \quad \text{ii) } \sum_{s \in} S(\pi_s) = 1$$

Each state of the Markov chain has a steady-state equation that may be derived following the principle stated in Eq. (1). The state transition diagram corresponding to a particular state s_i is given below (Fig. 2).

π_{si} represents the steady-state probability of state s_i. The states which lead to s_i on the arrival and service of a single request are represented by the sets s_j and s_k, respectively. Applying the principle in Eq. (1), the steady-state equation for state s_i is given by

Fig. 2 Markov chain for *M* services

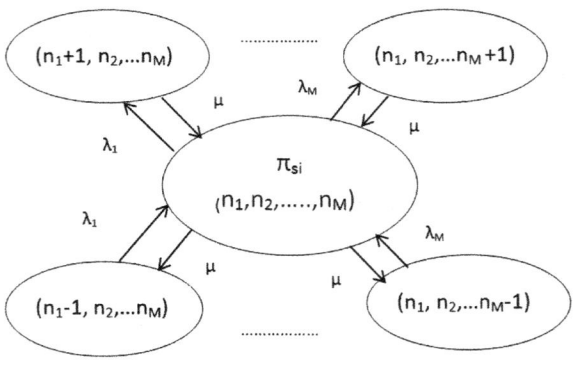

$$\sum_{j=1}^{M} (\pi_{sj}) * \lambda j + \sum_{k=1}^{M} (\pi_{sk}) * \mu = \sum_{j=1}^{N} (\pi_{si}) * \lambda j + M\mu * \pi_{si}$$

$$\sum_{j=1}^{M} (\pi_{sj}) * \lambda j + \sum_{k=1}^{M} (\pi_{sk}) * \mu - \sum_{j=1}^{M} (\pi_{si}) * \lambda j + M\mu * \pi_{si} = 0 \qquad (2)$$

The total number of steady-state equations is equal to the number of states, i.e., $T = (q + 1)^M$. To solve this system of equations, any one of the T equations is replaced by the linear combination of the remaining $T - 1$ equations. The solution to the problem is given by

$A \cdot X = B$ where A = coefficient matrix of the system or

$$X = A^{-1}B \qquad (3)$$

where $X = (\pi_{s1} \ \pi_{s2} \ _{...} \ \pi_{sT})^T$ and $B = (0 \ 0 \ ... \ 1)^T$.

3.2 QoS (Quality of Service) Parameters

The QoS performance parameters are of great importance as far as design of any system is concerned. They may be treated as measures of performance of the algorithm used in the design. The designer's aim can be to maximize a parameter like throughput or fairness index or to minimize a parameter like queuing delay or packet loss probability. To define the parameters, the concept of marginal probability needs to be introduced. $\pi_{(n1, \ n2, \ ..., \ nM)}$ is the steady-state probability corresponding to $s_{(n1, \ n2, \ ..., \ nM)}$. The marginal probability, π_{ni}, the probability that n number of packets are present in the *i*th service queue (where $i \in 1(1) \ N$ and $n \in 0(1) \ q$) is calculated for each service as follows:

$$\pi_{ni} = \sum \sum \sum \cdots \sum \pi_{(n1, n2,\ldots,ni,\ldots,nM)} \quad \forall j = 1(1)M \text{ and } j \neq i \qquad (4)$$

Based on the marginal probabilities, the QoS performance parameters are defined as follows:

Mean Marginal Number of Packets (*mmnp*) waiting in the queue is the average (statistical expectation) number of packets waiting in the particular service queue.

$$\text{mmnp}_i = \sum_{n_i=0}^{q} n_i * \pi_{n_i} \; \forall i = 1(1)M \qquad (5)$$

Mean Marginal Throughput (*mmt*) is defined as the average number of packets being served from a particular service queue in steady state.

$$\text{mmt}_i = \lambda_i * \sum_{n_i=0}^{q-1} \pi_{n_i} \; \forall i = 1(1)M \qquad (6)$$

Mean Marginal Queuing Delay (*mmqd*): Using *Little's Law*, mmqd is defined as the ratio of mmnp and mmt.

$$\text{mmqd}_i = \text{mmnp}_i / \text{mmt}_i \quad \forall i = 1(1)M \qquad (7)$$

Marginal Packet Loss Probability (*mplp*) is the probability that the corresponding service queue is full.

$$\text{mplp}_i = \pi_{qi} \; \forall i = 1(1)M \qquad (8)$$

Jain's Fairness Index (*JFI*) provides the measure of fairness of the resource allocation strategy. It gives an estimate of whether the resources have been adequately allocated to each of the services trying to access the server.

$$\text{JFI} = \frac{\left(\sum_{i=1}^{M} \text{mmti}\right)^2}{M * \sum_{i=1}^{M} \text{mmti}^2} \qquad (9)$$

This parameter is central to our analysis.

4 Markov Chain Analysis and the Determination of Optimum Weights

The vector B was already known. The coefficient matrix A was determined following which the steady-state probabilities were evaluated with the help of Eq. (3). It was assumed that one service (say service p) has absolute priority. To incorporate this feature into our model, Eq. (2) was modified. The service p is served at the rate of μ, while the other services have their service rates modified in accordance with their assigned priorities. If the weights are given by the matrix $W = [w_1, w_2, \ldots, w_{M-1}]$, such that $\sum w_j = 1$ $\forall j \neq p$, then the individual service rates become $\mu_j = w_j * \mu$.

The problem now reduces to an optimization problem intended to find the weight distribution that maximizes JFI. At this juncture, it is important to mention that the process that was used to arrive at the optimum weight distribution was two-level process. At first, the performance of the server was observed for different weight distributions. The best weight distribution was chosen and swarm optimization techniques were used to fine-tune the weight distributions and improve the JFI.

4.1 Optimization

The optimization technique used was the particle swarm optimization (PSO) [5]. The objective function is the JFI, which can be expressed as a function of the weights. A constraint used in the optimization is the condition $\sum w_i = 1$, i.e., the sum of the weights is equal to 1. Since the updating of the weights in PSO has an element of randomness associated with it, the sum of the weights is not exactly equal to 1. Thus, the constraint used was $0.95 \leq \sum w_i \leq 1.05$.

Because of this approximation, some errors creep into the optimization. However, this is taken care of later on using a weight correction.

In particle swarm optimization, a random population or a swarm is defined. Each particle in the swarm is associated with a position and a velocity which are updated in a systematic manner. The swarm has a leader and the position of every particle is governed by the position of the leader. The updated position of a particle is determined by two variables called the global best (gbest) and present best (pbest) along with a degree of randomness. This process is continued until the optimum weights are reached.

5 Simulation Results

At first, the performance of the algorithm was observed for various weight distributions.

We introduced a parameter k, which is equal to the arrival rate (λ_M) of the least priority service. Since k denotes rate of arrival of a service, its unit is packets per second (pps). For simplicity, our analysis has been described completely in terms of this parameter.

To test our algorithm, it was necessary to assume reasonable packet arrival rates for the services. For $M = 3$ (say), the packet arrival rates were assumed to be in the ratio 3:2:1. In other words, the arrival rate of the highest priority service was $3k$ pps while that of the second highest one was $2k$ pps. For $M = 4$, the rates were assumed to be in the ratio 4:3:2:1, and so on. Moving from a lower value of M to the immediate higher value may be looked upon as the addition of a single higher priority service to the system.

The variation of JFI was observed by (i) varying the value of k from 1 to 10 pps for a constant value of M and (ii) varying M from 3 to 6 for a constant value of k. These observations are presented in this section.

Figure 3 shows the variation of the fairness index with traffic rate for varied number of services M. It can be seen that JFI is good (85%) and almost fixed as long as number of services remains below 4. For around 50 pps of traffic arrival rate

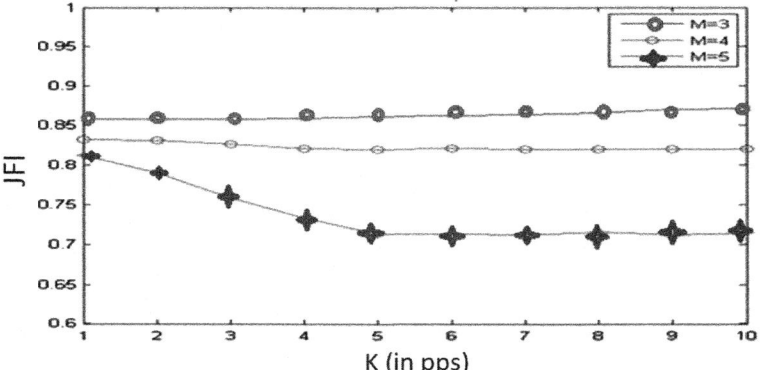

Fig. 3 JFI versus k with M as parameter

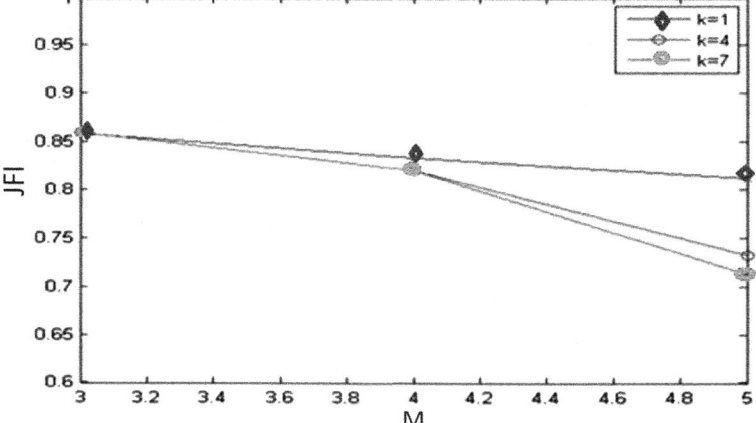

Fig. 4 JFI versus M with k (in pps) as parameter

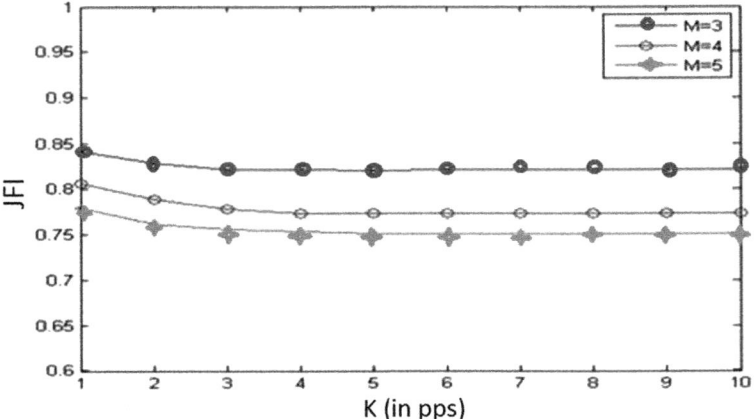

Fig. 5 JFI versus M for a queue length of 2

with $M \geq 5$, JFI falls around 70%, which is an indication of starvation of lower priority services. So the analysis for observing blocking and dropping probability of individual service type is a necessity. Also, as the arrival rates increase, JFI decreases owing to high packet loss probability and high queuing delay.

Figure 4 shows the variation of JFI against M in a precise scale. Now if we decrease queue size, the packet loss is bound to increase and JFI falls as a result. The following result is shown for $q = 2$. As observed from Fig. 5, the JFI values have decreased. Readings for higher queue sizes were not recorded due to the memory constraints of the computational system. By inductive logic, it can be inferred that JFI increases with increasing queue length.

6 Practical Issues Related to the Implementation of the Server

Although swarm intelligence was used to find the optimum weight distribution, such evolutionary algorithms are not suitable for live systems because of their time complexities. This calls for an alternative implementation strategy. Optimum weights for various different scenarios are determined beforehand. The results of the analysis are then either stored in the form of a lookup table for the server to consult or used as regression dataset to predict optimum weights.

The pros and cons of each implementation have been discussed. In a real system, the proper choice needs to be made by keeping in mind the various system requirements.

6.1 Lookup Table Method

After the optimum weight distributions for different values of λ_M and μ that have been found out, they may be stored in a database called a lookup table. It should be noted that the service rate μ is generally constant for a server. However, servers with adjustable service rates may be designed.

A typical server is equipped with the ability to sense the rate of flow of traffic of the least priority service (k) at any given time. Once k is determined, the server uses it as a key in a search process. The other key is the value of μ which may be constant or variable. Now these two keys are used to search for the required field which is a weight vector of size $M - 1$ containing the optimum weights.

In our analysis, integral values of k were considered. However, in real systems, the incoming traffic will, more often than not, be a floating point number.

The server approximates $k = I$ whenever $k \in (I - \Delta, I + \Delta]$ where I is an integer and $\Delta = 0.5$.

So if $k = 2.7$ the server uses the weights for $k = 3$. Similarly, if $k = 4.3$ the server uses the weights for $k = 4$. Because of the approximation, the weights used may not be very close to the optimum. Thus, the system suffers from some inaccuracies. However, these inaccuracies may be remedied by incorporating more data into the table and varying k in smaller steps.

6.2 Regression Method

Recall that by performing PSO, the optimum weight distribution was determined for different values of k and μ. Using this as a dataset, regression may be used to express each weight as a function of k for a constant value of μ.

Thus, for constant μ, $w_i = f(k)$, where $i = 1(1)\ M - 1$. During the analysis, it should be ensured that $\sum_{i=1}^{M-1} wi = 1$ is satisfied to a high degree of accuracy.

Given a set of values (x, y), we may model $y(x) = a_n x^n + a_{n-1} x^{n-1} + a_{n-2} x^{n-2} + \cdots + a_0$. The coefficients are determined by performing least square analysis. Given a set of readings (k, w_i), the weights may be modeled as $W_i = a_n k^n + a_{n-1} k^{n-1} + a_{n-2} k^{n-2} + \cdots + a_0$.

Table 1 shows a small portion of the dataset that has been used in the analysis. The linear regression equations derived for $M = 5$ are tabulated in Table 2. To choose a suitable degree for the polynomial, regression analyses were performed by

Table 1 Optimum values of W_1 for different k values for a system with five services

k	W_1	W_2	W_3	W_4
1	0.4702	0.2582	0.1543	0.1512
2	0.4543	0.2749	0.1633	0.1536
⋮	⋮	⋮	⋮	⋮
10	0.4518	0.267	0.1634	0.1634

Table 2 Regression equations for W_i using a fifth-degree polynomial

Dependent variable	Regression equation (using polynomial of degree 5)
W_1	$W_1 = 0.0006\ k^4 + (-0.0060)\ k^3 + 0.0263\ k^2\ (-0.0474)\ k + 0.4848$
W_2	$W_2 = 0.0009\ k^4 + (-0.0093)\ k^3 + 0.0439\ k^2 + (-0.0918)\ k + 0.3181$
W_3	$W_3 = 0.0001\ k^4 + (-0.0012)\ k^3 + 0.0086\ k^2 + (-0.0220)\ k + 0.1782$
W_4	$W_4 = 0.0001\ k^5 + (-0.0020)\ k^4 + 0.0198\ k^3 + (-0.0895)$ $k^2 + 0.1740\ k + 0.0531$

Table 3 Variation of the mean absolute percentage error (MAPE) with the degree of the polynomial used

Degree of polynomial	10	5	3	2	1
MAPE (%)	>100	34.41	5.66	0.89	1.2

taking polynomials of different degrees and evaluating the mean absolute percentage errors (MAPEs) in each case. Table 3 shows the MAPEs for polynomials of different degrees. It is evident from the table that a robust model may be built by considering either linear or quadratic regression equations. Table 4a, b presents, respectively, the linear and quadratic regression equations that were derived for $M = 5$. Figure 6a, b shows graphical representations of the performances of these regression models.

6.3 Weight Correction

The optimum weights are determined beforehand using swarm intelligence. So, the sum of the weights is very close to 1. However, it is not exactly equal to 1, thus creating a negligible error in the system. To eliminate this error, the following correction may be performed on the weights:

$$W_i| = W_i / \left(\sum_{(i=1)}^{(M-1)} wi \right)$$

It can be easily verified that $\sum W_i| = 1$.

6.4 Pros and Cons of the Methods Proposed

The lookup table method can achieve higher accuracy since the optimum weights for the nearest k value are fetched directly from the table. However, this method

Table 4 **a** Linear regression equations obtained for $M = 5$. **b** Quadratic regression equations for $M = 5$

μ	W_1	W_2	W_3	W_4
(a)				
10	$0.0014\,k + 0.4647$	$-0.0001\,k + 0.2614$	$0.0018\,k + 0.1550$	$0.0012\,k + 0.1541$
30	$-0.0010\,k + 0.4628$	$0.0017\,k + 0.2509$	$0.0013\,k + 0.1601$	$-0.0009\,k + 0.1649$
50	$-0.0001\,k + 0.4571$	$-0.0009\,k + 0.2659$	$0.0019\,k + 0.1554$	$-0.0011\,k + 0.1671$
(b)				
20	$-0.0001\,k^2 + 0.0002\,k + 0.4604$	$0.0001\,k^2 - 0.0008\,k + 0.2602$	$-0.0002\,k^2 + 0025\,k + 0.1603$	$0.0002\,k^2 - 0.0023\,k + 0.1639$
40	$0.0003\,k^2 - 0.0049\,k + 0.4799$	$0.0008\,k^2 + 0.0079\,k + 0.2501$	$0.0006\,k^2 - 0.0045\,k + 0.1673$	$0.0005\,k + 0.1514$

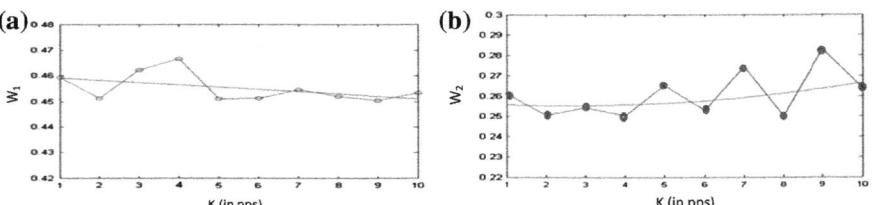

Fig. 6 a Fitting of a linear curve. **b** Fitting of a quadratic curve

requires large storage space for storing the lookup tables. Where storage space is a concern, the regression method should be used, since it is much more memory-efficient. But in the absence of memory constraints, lookup tables should be employed since they are likely to be more accurate.

7 An Overview of the Operation of the Server

The server first senses the traffic in the system by using some sophisticated technology. From the traffic rate, the variable k is determined. The variable k can then be used either as a key in a lookup table or as the independent variable for a regression process. The lookup table or the regression equations (as the case may be) are determined beforehand. By using either process, the optimum weight distribution is obtained. The corresponding weights are allocated to the different services by the server and the scheduling is carried out. This continues for a fixed time Δt, after which the server senses the traffic once again and a new cycle begins. The stepwise flow of the operations is given below.

{
Step action

1   Start.
2   Sense the packet arrival rate k of the least priority service.
3   Determine optimal weights by regression or lookup table.
4   Perform weight corrections.
5   Serve the system for time Δt.
6   Go to Step 1}.

7.1 Selection of Δt

Another important design issue is the value of Δt, which is the time for which the system functions under a certain weight distribution. Intuitively, the parameter Δt must be chosen by keeping the following in mind:

(a) Abrupt changes in packet arrival rates should be detected. (b) The system should be power efficient. (c) The server should get enough time to determine the new set of weights, i.e., $1/\Delta t$ should be less than the processor speed. It was decided that the parameter Δt should be of the order of the mean marginal queuing delays (i.e., the average time spent by a packet in the queue). The following heuristic may be proposed: $\Delta t = $ max (processor latency, min (mmqd for UGS, RTPS, ERTPS, NRTPS, BE)

8 Bandwidth Allocation in BWA Networks

The resource allocation problem is a stringent one for most of communication engineering cases whether wired or wireless. For wireless communication, it is more challenging for multimedia services as radio resource for wireless communication is scarce. We take WiMAX as an example case and try to allocate bandwidth among its services in the fairest manner possible for five types of services. Most of the literature with WiMAX handled three QoS service classes for ugs, ertps, and rtps [6–8]. Our intention is to see five different QoS classes considering Nrtps and BE also as non-real-time services. They are of great significance since non-real-time services like internet surfing are the most common activities on the internet. In fact, most of the internet traffic can be classified as BE.

Due to the utmost importance of UGS (unsolicited grant scheme), it is assigned absolute priority, which means that if there is any UGS packet in the queue, it would be served first. For the remaining four services, weighted fair queuing has been used in the manner described in previous sections.

$$\mu_{ugs} = \mu; \mu_{ertps} = w_{ertps} * \mu; \mu_{rtps} = w_{rtps} * \mu; \mu_{nrtps} = w_{nrtps} * \mu; \mu_{be} = w_{be} * \mu$$

$$\lambda_{ugs} = 5 * k; \lambda_{ertps} = 4 * k; \lambda_{rtps} = 3 * k; \lambda_{nrtps} = 2 * k;$$

$$\lambda_{be} = k \text{ where } k \text{ is an integer which is varied from 1 } to \text{ 10}$$

$$\sum w_i = 1 \& w_{ertps} \geq w_{rtps} \geq w_{nrtps} \geq w_{be}$$

At first, the performance of the system was observed for different weight distributions. It was found that the JFI was significantly high when the weight distribution was [0.45 0.25 0.15 0.15]. The remaining analysis is the same as that of the general resource sharing problem with $M = 5$. Figures 7, 8, 9, and 10 show the various QoS parameters plotted against the variable k.

Figure 7 shows that the mean marginal throughput of a service has a direct variation with the weight assigned to it. So, UGS has the highest throughput since it gets the greatest attention from the server. BE's throughput is the least since the priority assigned to it is the lowest.

Figure 8 shows the mean marginal queuing delays (in seconds) of the various services. The observations are consistent with intuition. The highest priority service

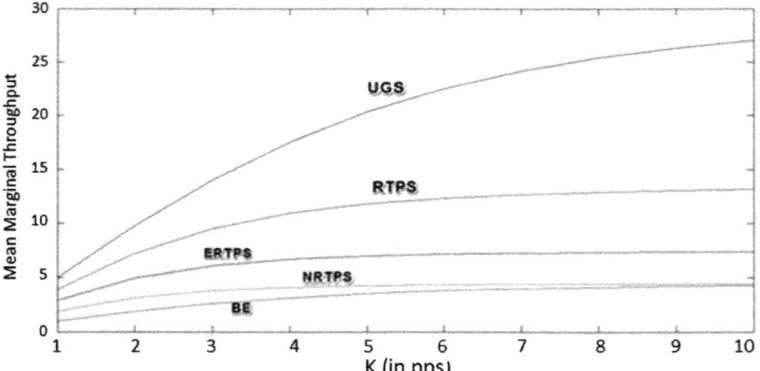

Fig. 7 Variation of mean marginal throughput with k for the different services

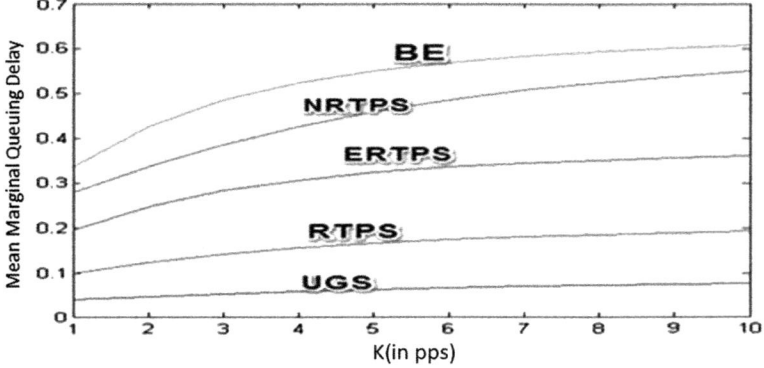

Fig. 8 Mean marginal queuing delay (in seconds) plotted against k

or UGS suffers the least amount of delay, whereas the least significant BE service suffers from considerable delay.

Figure 9 shows the variation of the mean marginal packet loss probabilities for the various services. Here, it is observed that the traffic loss has a direct variation with the value of traffic load or λ/μ. The service having highest traffic load (viz., NRTPS) suffers from packet loss to the greatest extent.

Figure 10 shows the JFI variation for the WiMAX server with traffic arrival rate. With increasing traffic arrival rate, the performance falls as expected. Similar to our previous analysis, PSO was performed to arrive at the optimum weight distributions. After updating the weight distribution by using optimization, the JFI variation was reduced significantly.

Thus, we have seen an application of the hybrid resource sharing algorithm in a wireless telecommunication problem. This also proves that the five QoS service

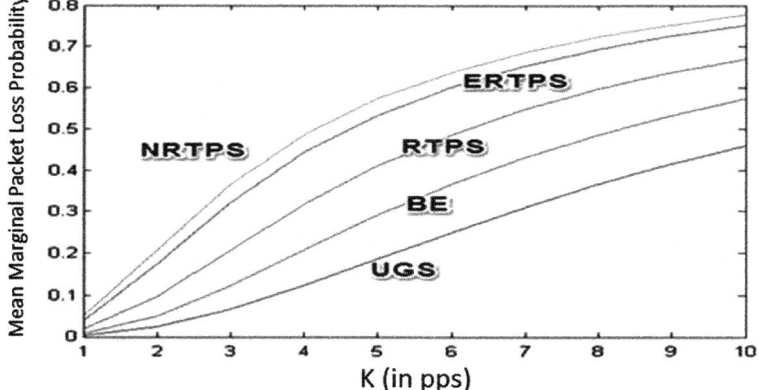

Fig. 9 Mean marginal packet loss probabilities plotted against k

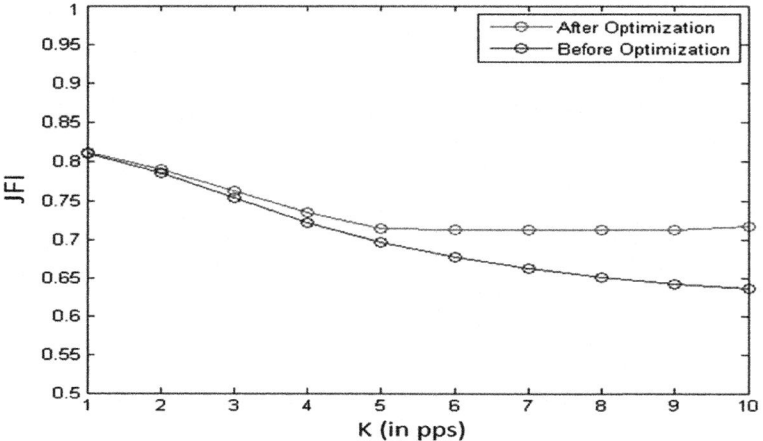

Fig. 10 JFI variation with traffic arrival rate and its improvement after optimization

classes of WiMAX can be successfully scheduled by using a hybrid scheduling algorithm. Previously, this WiMAX scheduling problem had been solved by for three QoS classes. One may refer to [6] for the treatment done with three classes. Using our algorithm, it has been possible to solve the same problem for five QoS classes. Thus, a WiMAX server can schedule five QoS classes using this algorithm while maintaining fairness, avoiding starvation, and adhering to delay and throughput constraints.

9 Conclusion

A hybrid packet scheduling algorithm employing PQ and WFQ can efficiently schedule multiple services while ensuring a high degree of fairness and keeping the QoS constraints unviolated. The hybrid algorithm is first tested for varied number of services and it is found that the JFI falls after five services. This finding is directly applied to WiMAX BWA networks taking five different QoS classes as WiMAX generally provides. Previous literature on scheduling algorithms for WiMAX incorporated at most three QoS classes. However, the present algorithm has been shown to successfully schedule five QoS service classes of WiMAX in terms of mean queuing delay, packet loss, and throughput. The performance of the algorithm was seen to improve with increase in queue length. However, the results for higher queue lengths could not be reported because of processor limitations. For a similar reason, the results for a higher number of services could not be reported. However, this limitation of the algorithm will be overcome eventually since processor speeds will increase with rapid advancement in computing technology. The optimized weight distribution fairly increases JFI (an improvement of 12.6% for $k = 10$ pps) for increased traffic arrival to the system indicating that fair resource distribution is there even for the lowest priority classes. Two practical implementation strategies of weight tuning are discussed for real-time operation with their pros and cons. Since there is very limited literature on scheduling of multiple services using a similar approach, a comparison with other publications could not be provided.

This resource scheduling scheme is a generic one and it can be applied to domains beyond telecommunication and computer networks as well. The most suitable applications would be in the fields of traffic control (management of vehicular traffic at busy road intersections) and operations research (allocation of man, material, or financial resources across different verticals within organizational structures).

References

1. Chak chai So-In, Raj Jain and Abdel-Karimi Tamimi, "Scheduling in IEEE 802.16e Mobile WiMAX Networks: Key Issues and a Survey", IEEE Journal on Selected on Selected Areas in Communications, Vol. 27, No. 2, February 2009.
2. Maxim S Martynom, "Design and Implementation of Hybrid Packet Scheduling Algorithms for High Speed Networks", Purdue University, December 2007.
3. Raj Jain, Dah-Ming W. Chiw, William R Hawe, "A Quantitative Measure of Fairness and Discrimination for Resource Allocation in Shared Computer System".
4. Thiagarajan Viswanathan, "Telecommunication Switching Systems and Networks".
5. James Kennedy and Russel Eberhart, "Particle Swarm Optimization", Washington DC, 20212.

6. Prasun Chowdhury, Iti Saha Misra and Salil K Sanyal, "An Integrated Call Admission Control and Uplink Packet Scheduling Mechanism for QoS Evaluation of IEEE 802.16 BWA Networks", Canadian Journal on Multimedia and Wireless Networks Vol. I, No. 3, April 2010.
7. Kuowei Hwang and V Rao, "Broadband Wireless Access (BWA) Networks: A Tutorial", University of California, Davis.
8. Tzu-Chieh Tsai, Vhi-Hong Jiang and Chuang-Yin Wang, "CAC and Packet Scheduling Using Token Bucket for IEEE 802.16 Networks", Journal of Communications, vol. 1, No. 2, pp. 30–37, 2006.

A Novice Approach to Implementation of System on Chip Based Smart CMOS Sensor for Quantum Computing Based Applications

Rajinder Tiwari

1 Introduction

The advent of the sensor-dependent image processing technology has come into existence late back to 1970s with the use of MOS devices. The functioning of this system primarily depends on the utilization of array of image sensors which in turn rely on the proper functioning of the array of CCD technology so as to provide better and enhanced quality imaging arrays. With further advancement in this technology, i.e., with the availability of CMOS-based smart sensors, the analysis of the image becomes quite simple, cost-effective, and enhanced accuracy of the results. In this system, CMOS readouts with millions pixel capacity are used in addition to the detector arrays with the help of flip chip fabrication technology, i.e., the most important feature of a SoC design based system.

Figure 1 shows the schematic arrangement, i.e., cross-sectional view of the CMOS sensing system. The system on chip design based systems normally follows the Moore's law which utilized CMOS technology so as to produce higher performance sensors in various ranges of visibility. Due to this reason, CMOS technology without any doubt has now emerged the most common approach for the design of both infrared and visible range sensors that in turn finds enormous applications in the domain of MSP. Based on this feature, it has been found that most of the mixed signal processing based industrial applications are still dependent on CMOS image sensors (CIS) technology as compared to CCD approach [2–8].

R. Tiwari (✉)
Department of Electrical & Electronics Engineering, ASET,
Amity University, Lucknow, India
e-mail: trajan@rediffmail.com

R. Tiwari
Department of Electronics & Communication Engineering,
Model Institute of Engineering & Technology (MIET), Jammu, India

© Springer Nature Singapore Pte Ltd. 2019
V. Nath and J. K. Mandal (eds.), *Proceeding of the Second International Conference on Microelectronics, Computing & Communication Systems (MCCS 2017)*, Lecture Notes in Electrical Engineering 476, https://doi.org/10.1007/978-981-10-8234-4_45

Fig. 1 Crossectional view of
CMOS image sensing device
[1]

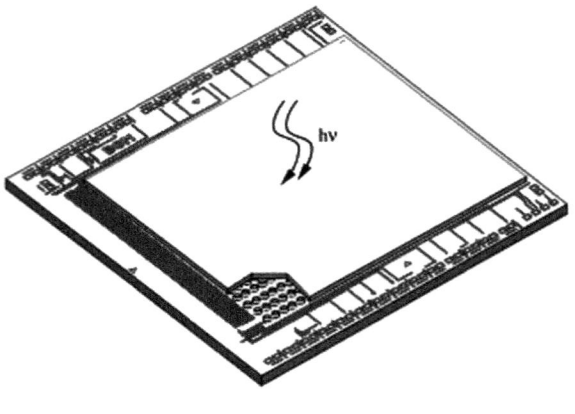

This smart sensor requires to be designed and implemented using system on chip (SoC) design approach because most of the MSP-based applications require the portability features so that it can be utilized in remote areas. The basic requirement of this SoC design approach is embedded processors, memories, and various hardware modules. In addition to this, this system requires the design flow of the functional executable model that meets the required specifications of the system from hardware and software perspective. The basic and dominant software module of this system has been implemented and analyzed with the help of certain tools with best performance, i.e., using the tools of LabVIEW software. The hardware module of this system has been tested and analyzed with the help of mathematical expressions, i.e., mathematical modeling using MATLAB software. In this type of design flow based approach, the system requires a programming tool that spans all the levels of the model, i.e., a system specification language (UML) with other supporting software tools. With this approach, one can study and discuss the desired level of the performance of the system in terms of the dominant parameters, especially, input, output, and the processed resultant signal. In addition to this, the SoC design approach requires System C or Embedded C as the system implementation language [9–16].

2   Basics of SoC Design System

Most of the application of the present technology is based on system on chip (SoC) platforms, since it provides the desired performance of the system with better growth in terms of time, speed, size, power dissipation, etc. These blocks are available from vendors of intellectual property (IP) as hard cores or soft cores. A hard core, or hard IP block, is one where the circuit is available at a lower level of abstraction such as the layout level; it is impossible to customize a hard IP to suit the requirements of the embedded system. As a result, there are limited opportunities in optimizing the cost functions by modifying the hard IP. For example, if some functionality included in the IP is not required in the present application, we cannot remove the function to save area. Soft IP refers to circuits which are available at a higher level of abstraction, such as register-transfer level. It is possible to customize the soft IP for the specific application. The designer of an embedded SoC integrates the IP cores for processors, memories, and application-specific hardware to create the SoC.

Figure 2 illustrates the architecture of an embedded system on chip (SoC) which consists of basic four modules, i.e.,

- An analog front end which includes the analog/digital and digital/analog converters.
- Programmable components which include microprocessors, microcontrollers, and DSPs. The number of embedded processors is increasing every year. The microcontroller/microprocessor is useful in handling interrupts, housekeeping, and performing timing-related functions. The DSP is used for processing the audio and video information, e.g., compression and decompression of audio and video information. The application software is normally preloaded in the memory and is not user programmable, unlike general-purpose processor-based systems.
- Application-specific components—these include hardware accelerators for compute-intensive functions. Examples of hardware accelerators include digital image processors which are useful in cameras

The memory subsystem of SoC consists of the following type:

- On-chip memory organization,
- Cache-based memory organization, and
- Scratchpad memory based organization.

The design of an embedded system begins with a behavioral description as shown in Fig. 3. In the present scenario, there are many languages available to capture the system behavior, e.g., System Verilog, System C, and so on.

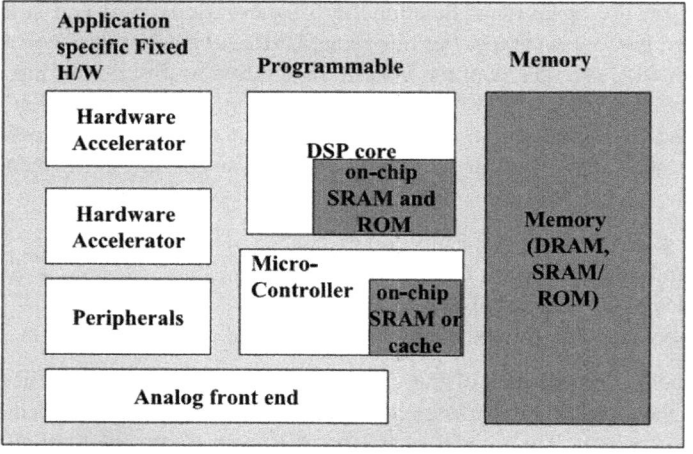

Fig. 2 Block diagram of an embedded SoC system [17]

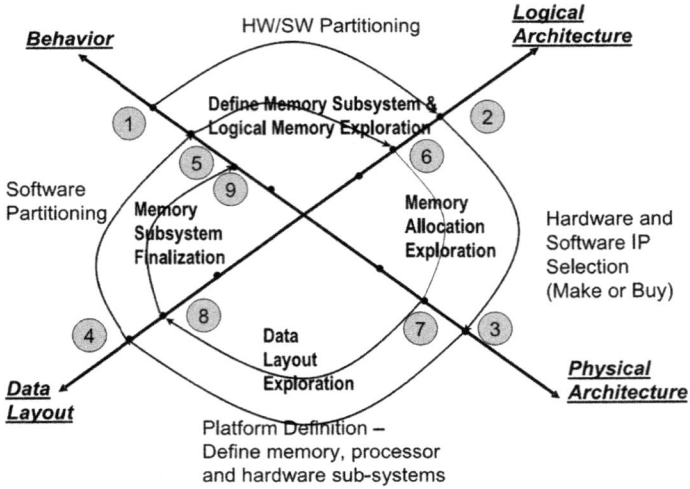

Fig. 3 Application-specific SoC design flow [18]

The hardware–software partitioning is performed so as to identify which functionalities of the description are best performed in hardware and which are best implemented in software. Hardware implementation is cost-intensive but improves the performance [19–25].

3 Design and Implementation of the SoC-Based Smart CMOS Pixel Sensor Array

The design and implementation approach of the proposed system is entirely dependent on the operational functionality of the various hard and soft modules with desired level of accuracy. For this same, UML 2.0 has been utilized along with the Embedded C to implement the system. In addition to this, author has also gone through the feasibility of the use of other programming and simulation tools that can be used to implement this system with enhanced level of the performance. Following steps are required to follow the analysis of the performance of the system:

- Define a UML profile for Embedded C;
- Structural and behavioral features of the system;
- Extended state diagram of the system;
- Behavioral model with extended state machine diagram.

These behavioral models of the system are conceived for code generation and modeling the operational performance with an isomorphic Embedded C implementation approach. This model of the smart system is obtained simply with the proper design of the interconnections between the sub-modules of the system using the help of block diagrams [26–30].

Figure 4 shows the design environment of the system. This approach of the design and implementation of the system is still under discussion with the development of the executable models. It can be categorized into two modules, i.e., a development module and a runtime environment module (RE).

The logical design and implementation of the mathematical model of the smart system can be categorized in various sub-modules which are further implemented as interfacing programs for RTOS environment. The UML profile of the system has been implemented with the use of Embedded C that elaborates the behavioral functionality of the smart system in terms of state diagram.

Figure 5 represents the arrangement of the distribution of CMOS-based array sensors. With this approach, we are interested in implementing the device with technical aspects of image processing of the signal within a system on chip. Although, with SoC approach, the system has to compromise within various characteristics, i.e., versatility, parallelism, processing speed, and resolutions. The author has implemented the proposed system by keeping these limitations in mind and provides the mechanism with an overall increase in the system performance.

Figure 6 shows the pixel diagram, i.e., pixel area of the CMOS sensor array. This diagram depicts the storage capability of the system within the desired range, i.e., $40 \times 40 \ \mu m^2$ and a fill factor of 9%.

Figure 7 shows the schematic arrangement of the designed model for the proposed smart system that has got tremendous utilization in mixed signal processing based applications. This model of the system provides an enhanced level of

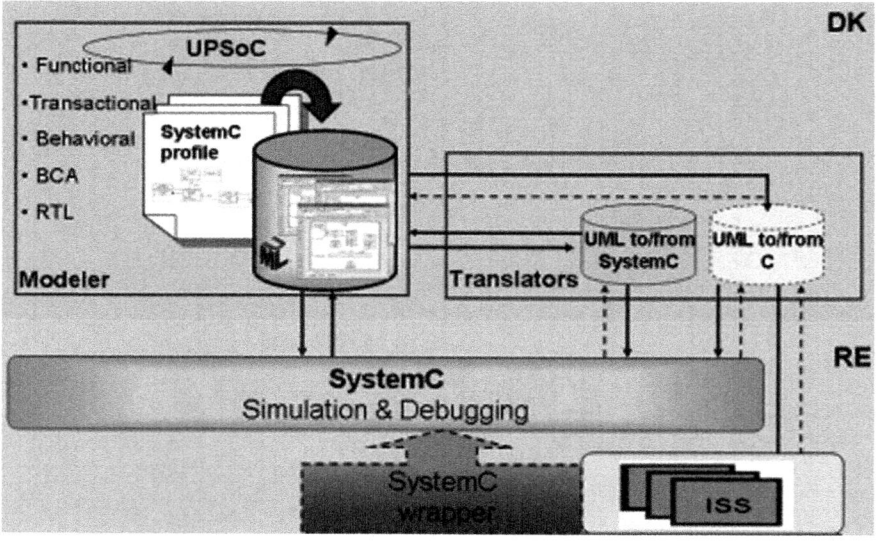

Fig. 4 Functional and behavioral model of the system [31]

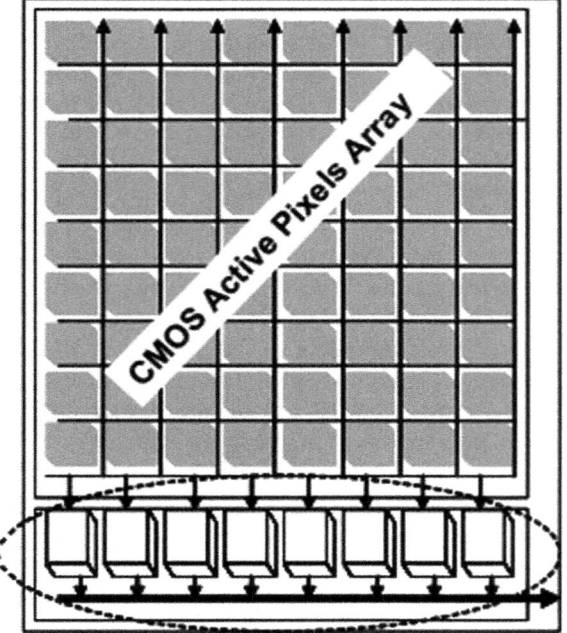

Fig. 5 Schematic diagram of CMOS pixels array [32]

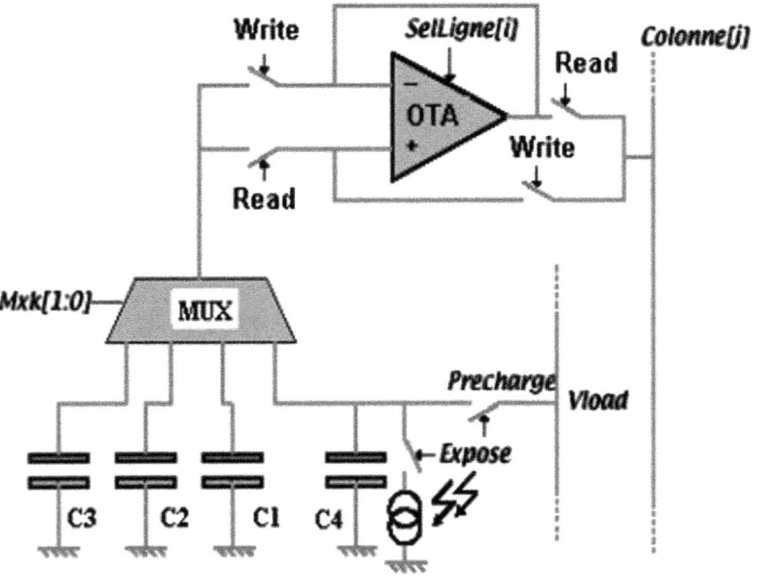

Fig. 6 Pixel diagram of CMOS sensor [33]

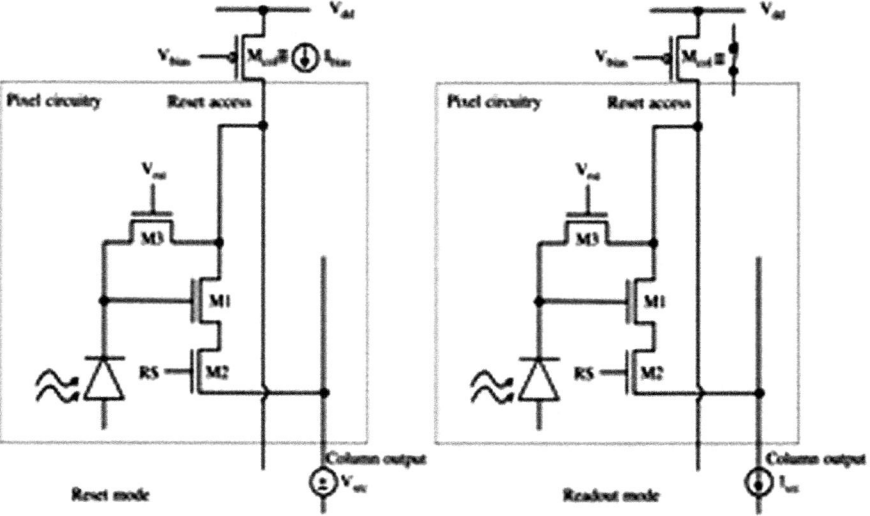

Fig. 7 CMOS-based pixel array amplifier [34]

accuracy at the output with less amount of unwanted noise signal. The most affected parameter of this system is the channel resistance which in turn deteriorates the overall performance of the system [35–46].

4 Simulation Results and Discussion

The simulation work of the CMOS-based smart pixel sensor circuit has been carried out with MATLAB software. The mathematical model of the system is first of all designed and then the various plots are obtained so as to undergo the analysis of the system. This simulation work can also be done with the help of either LabVIEW or PSpice software but the author has made the use of MATLAB 2014 version since it is quite easy to model and analyze the performance of the system.

Figure 8 presents the behavioral model of the unwanted signal that deteriorates the performance of the proposed system, i.e., the behavior of the noise has been plotted with respect to the capacitance of the CMOS transistor. Thus, from this model of the system, we can deduce that the unwanted signal interfering the performance of the overall system can be minimized with the use of cascaded transistors operating in the active region (Fig. 9).

The basic motive in analyzing the model of this system is to optimize the presence of the noise at the output so as to enhance the performance of the proposed system. Due to this, the noise parameter has been plotted with different dominant parameters of the proposed system that determines the performance of the system.

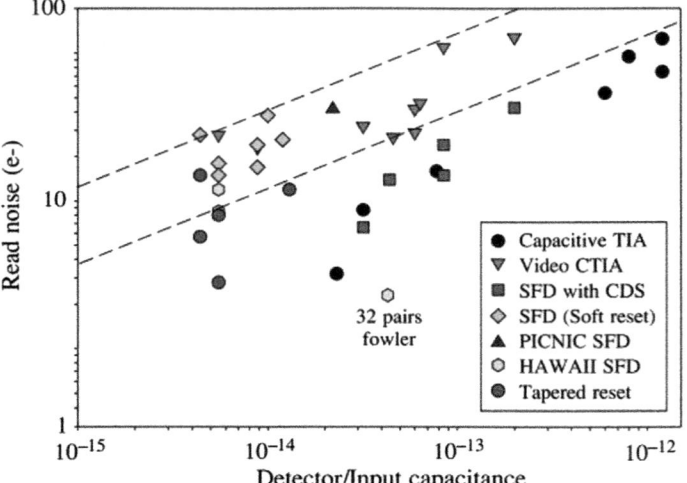

Fig. 8 Noise model behavior of SoC-based CMOS smart image sensors

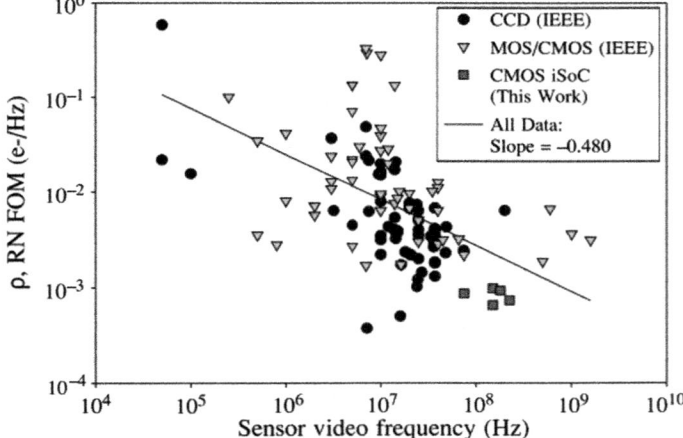

Fig. 9 Normalized FoM of SoC-based CMOS smart image sensors

It means that the most important and effective types of noise signal have been introduced in the system so as to understand the effect of them on the resultant output of the system. After this process, it becomes quite easy to eliminate or minimize the presence of the noise signal with the proper use of the suitable filters (Fig. 10).

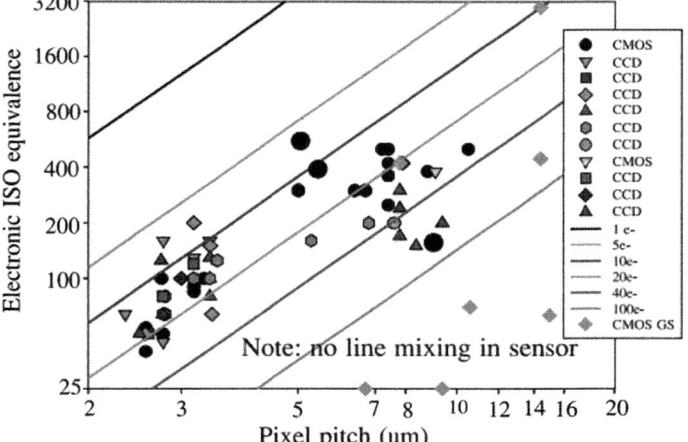

Fig. 10 Speed versus pixel pitch of SoC-based CMOS smart image sensors

Table 1 Comparative study data

S. No.	SoC reference data	SoC theoretical data	SoC obtained data
1	100	105	098.50
2	150	145	148.00
3	200	195	199.25
4	250	245	249.00

5 Conclusion

Table 1 presents the resultant data which has been obtained from the analyses of the system on chip (SoC) methodology based proposed smart system. Based on the comparative analysis of the data obtained from three sources, i.e., reference data, theoretical data, and the data obtained from the proposed system, the author has concluded that the performance of the proposed system is quite appreciable since it minimizes the effect of the distorting signals thereby enhancing the performance. Another effective aspect of the proposed system is that it improves the loss of power with the use of the CMOS-based array of transistors for the execution of the signals.

Acknowledgements The authors are thankful to Hon'able C—VI, Mr. Aseem Chauhan (Additional President, RBEF and Chancellor AUR, Jaipur), Maj. General K. K. Ohri (AVSM, Retd.) Pro-VC Amity University, Uttar Pradesh Lucknow, Prof (Dr) Ankur Gupta, Director, MIET Jammu, Wg. Cdr. Dr. Anil Kumar, Retd. (Director, ASET), Prof. S. T. H. Abidi (Professor Emeritus), Brig. U. K. Chopra, Retd. (Director AIIT), Prof. H. K. Dwivedi (Director, ASAP), Prof. O. P. Singh (HOD, Electrical & Electronics Engg.), and Prof. N. Ram (Dy. Director ASET) for their motivation, kind cooperation, and suggestions.

References

1. E. Riccobene, P. Scandurra, A. Rosti and S. Bocchio. A UML 2.0 Profile for System C. ST Microelectronics Technical Report, 2004.
2. E. Riccobene, P. Scandurra, A. Rosti, S. Bocchio. A SoC Design Methodology Based on a UML 2.0 Profiles for System C. In DATE 05.
3. E. Riccobene, P. Scandurra. Modelling SystemC Process Behavior by the UML Method State Machines. In RISE 04, Springer-Verlag 2004.
4. P.-A. Muller, P. Studer, F. Fondement, and J. Bézivin. Platform independent Web Application Modeling and Development with Netsilon. Journal SoSym, 2005.
5. P. Dudek, J. Hicks, "A CMOS General-Purpose Sampled-Data Analogue Microprocessor". Proc. of the 2000 IEEE International Symposium on Circuits and Systems. Geneva, Suisse.
6. Y. Ni, J.H. Guan, "A 256x256-pixel Smart CMOS Image Sensor for Line based Stereo Vision Applications", IEEE, J. of Solid State Circuits, Vol. 35 No. 7, Juillet 2000, pp. 1055–1061.
7. A. Elouardi, S. Bouaziz, R. Reynaud, "Evaluation of an artificial CMOS retina sensor for tracking systems". Proceeding of IEEE Intelligent Vehicles Symposium 2002, Versailles, France.
8. A. Dupret, J. O. Klein, A. Nshare "A DSP-like Analog Processing Unit for Smart Image Sensors", International Journal of Circuit Theory and Applications 2002. 30: pp. 595–609.
9. A. Dupret, E. Belhaire, J.C Rodier "A high current large bandwidth photo sensor on standard CMOS Process" presented at EuroOpto'96, AFPAEC, Berlin, 1996.
10. D. Litwiller "CCD vs. CMOS: Facts and Fiction". The January 2001 issue of PHOTONICS SPECTRA, Laurin Publishing Co. Inc.
11. R. Deriche, "Fast algorithms for low level-vision". IEEE Transaction of Pattern Analysis and Machine Intelligence, vol 12-1, 1990.
12. F. Garcia Lorca, L. Kessal, D. Demigny "Efficient ASIC and FPGA implementation of IIR filters for real time edge detections", Proc. International Conference on Image Processing, IEEE ICIP 1997.
13. A. Dupret, J.O Klein, A. Nshare, "A programmable vision chip for CNN based algorithms". CNNA 2000, Catania, Italy: IEEE 00TH8509.
14. A. Elouardi, S. Bouaziz, A. Dupret, J.O Klein, R. Reynaud, "On Chip Vision System Architecture Using a CMOS Retina". Proceeding of IEEE Intelligent Vehicle Symposium, IV'04. Pages 206–211. ISBN 0-7803-8311-7. June 14–17, 2004. Parma, Italy.
15. Drumea, Al. Vasile, P. Svasta, M. Blejan, System on Chip Signal Conditioner for LVDT Sensors, 1st Electronics System integration Technology Conference ESTC06, Dresden, Germany, September 2006, pp. 629–633.
16. Drumea, Al. Vasile, P. Svasta, I. Ilie System on Chip Signal Conditioner for LVDT Sensors, 2nd Electronics System integration Technology Conference ESTC08, September 2008.
17. W. Guimei,et al., "Mine Pump Comprehensive Performance Testing System Based on Labview," In Measuring Technology and Mechatronics Automation 2009. ICMTMA'09, International Conference on 2009, pp. 300–303.
18. G. Beitao, et al., "'Application of LabVIEW for Hydraulic Automatic Test System," in Industrial and Information Systems, 2009. IIS '09. International Conference on, 2009, pp. 348–351.
19. Embedded control handbook—Volume 1, Microchip Inc., 1997.
20. K. Astrom, B. Wittenmark, Computer—Controlled Systems. Theory and Design, 3rd Edition, Prentice Hall ,1997, pp. 324–369.
21. L. Bierl, Das grosse MSP430 Praxis Buch, Franzis, 2004, pp. 110–190.
22. J. Travis and J. Kring, "LabVIEW for Everyone: Graphical Programming Made Easy and Fun," 2006.
23. National Instrument's PID Control Toolset User Manual.
24. Active pixel sensors: Are CCD dinosaurs? ER Fossum IS&T/SPIE's Symposium on Electronic Imaging: Science and Technology, 2-14A.

25. http://www.vision-systems.com/articles/2015/03/sony-rumored-to-discontinue-production-of-ccd-sensors.html.
26. CCD vs. CMOS, Dave Litwiller, Photonics Spectra, 2001.
27. Determination of the optimal electrical bandwidth in CCD- and CMOS-based image detector applications, Robert H. Philbrick, SPIE 5499, Optical and Infrared Detectors for Astronomy, 2004.
28. CMOS vs. CCD: Changing Technology to Suit HDTV Broadcast, Lester J. Kozlowski, 2003.
29. Fundamental performance differences between CMOS and CCD imagers: Part 1, James Janesick et al., SPIE 6276, High Energy, Optical, and Infrared Detectors for Astronomy II, 62760M, 2006.
30. A 0.7 e-rms Temporal Readout Noise CMOS Image Sensor for Low Light Level Imaging, Y. Chen et al., IEEE International Solid-State Circuits Conference (ISSCC), 2012.
31. http://www.caeleste.com/caeleste_publications/2011CNES/20111207_CNES_caeleste_presentation.pdf.
32. L²CMOS Image Sensor for Low Light Vision, Pierre Fereyre et al., International Image Sensor Workshop, 2011.
33. Night Vision CMOS Image Sensors Pixel for SubmilliLux Light Conditions Amos Fenigstein, International Image Sensor Workshop, 2015.
34. A Review of the Pinned Photodiode for CCD and CMOS Image Sensors Eric R. Fossum, et al., IEEE Journal of The Electron Devices Society, Vol. 2, no. 3, may 2014.
35. No image lag photodiode structure in the interline CCD image sensor, N Teranishi et al., Electron Devices Meeting, 1982 International (Volume:28), 1982.
36. http://www.sony.net/Products/SC-ZP/new_pro/may_2014/icx825_e.html.
37. A 3D stacked CMOS image sensor with 16Mpixel global-shutter mode and 2Mpixel 10000fps mode using 4 million interconnections, Symposium on VLSI Circuits (VLSI Circuits), T. Kondo et al., pages C90–C91, 2015.
38. M. S. Robbins and B. J. Hadwen, "The noise performance of electron multiplying charge-coupled devices," IEEE Transactions on Electron Devices, vol. 50, no. 5, pp. 1227–1232, May 2003.
39. Electron Multiplying Device Made on a 180 nm Standard CMOS Imaging Technology, Pierre Fereyre et al., International Image Sensor Workshop, June 2015.
40. First Measurements of True Charge Transfer TDI (Time Delay Integration) Using a Standard CMOS Technology, F. Mayer et al., International Conference on Space Optics, 2012.
41. CMOS long linear array for space application G. Lepage, Proc. SPIE 6068, Sensors, Cameras, and Systems for Scientific/Industrial Applications VII, 606807, 2006).
42. Time-Delay-Integration Architectures in CMOS Image Sensors G. Lepage et al., IEEE Transactions on Electron Devices, vol. 56, no. 11, November 2009.
43. R. Shimizu and Al., "A Charge-Multiplication CMOS Image Sensor Suitable for Low-Light-Level Imaging" IEEE Journal of Solid-State Circuits, vol. 44, no. 12, pp. 3603–3608 December 2009.
44. The lock-in CCD-two-dimensional synchronous detection of light, T. Spirig, P. Seitz et al., IEEE Journal of Quantum Electronics, Vol. 31, Iss. 9, p. 1705–1708, Sep 1995.
45. Demodulation pixels in CCD and CMOS technologies for time-of-flight ranging Robert Lange et al., Proc. SPIE 3965, Sensors and Camera Systems for Scientific, Industrial, and Digital Photography Applications, 177 (May 15, 2000).
46. 320x240 Oversampled Digital Single Photon Counting Image Sensor N. AW. Dutton, VLSI Circuits Digest of Technical Papers, 2014.

Novel Approach for Sleep Transistor Sizing to Suppress Power and Ground Bouncing Noise in MTCMOS Clustering Technique

Neha Gupta, Priyanka Parihar, Vaibhav Neema and Praveen Singh

1 Introduction

In recent year, high-performance portable applications, such as mobile phones and laptops, are designed. For designing of such portable system, battery life is the most challenging task. High power dissipation for the portable battery will reduce the long life of battery.

To design a desirable circuit leads to reduced feature size of transistors and suppressing the sub-threshold leakage current to extend the battery lifetime. The most common strategy for the leakage power and delay reduction is multi-threshold CMOS [1, 2]. In MTCMOS circuit, high threshold voltage sleep transistors are used between low threshold voltage logic circuits. When more than one low threshold voltage logic circuits are connected in series, the number of sleep transistors is also increased which increases the overall area and delay of the circuit [3, 4]. To reduce area, delay, and leakage power dissipation of transistor, only one high $|v_{th}|$-pmos and high v_{th}-nmos sleep transistor are connected between low threshold voltage logic circuits. PMOS sleep transistor is attached between real power line and virtual power line, and NMOS sleep transistor is attached between real ground line and virtual ground line. More than one low threshold voltage logic circuits are con-

N. Gupta · P. Parihar · V. Neema (✉) · P. Singh
Electronics and Telecommunication Engineering Department,
IET—Devi Ahilya University, Indore 452017, India
e-mail: vaibhav.neema@gmail.com

N. Gupta
e-mail: nehagupta121192@gmail.com

P. Parihar
e-mail: priyankapariharpriyanka@gmail.com

P. Singh
e-mail: psinghdavv@yahoo.com

© Springer Nature Singapore Pte Ltd. 2019
V. Nath and J. K. Mandal (eds.), *Proceeding of the Second International Conference on Microelectronics, Computing & Communication Systems (MCCS 2017)*, Lecture Notes in Electrical Engineering 476, https://doi.org/10.1007/978-981-10-8234-4_46

nected between a virtual power line and virtual ground line; this technique is called clustering technique [5].

When the sleep transits from sleep to active mode, high instantaneous current flows through the circuit and large voltage fluctuation occur on virtual power and virtual ground line which are called power bouncing noise and ground bouncing noise, respectively. When sleep transits from sleep to active and active to sleep, at that time, wakeup and sleep time are also calculated.

This paper proposed the technique for different W/L ratios of more than one inverter circuits connected in series to form a clustering design. Calculated parameters are delay, dynamic power dissipation, leakage power dissipation, wakeup time, sleep time, virtual power and ground line voltages, virtual power, and ground line discharging current.

The paper is organized as follows: Characterization of the circuit has been proposed in Sect. 2. Simulation parameters have been done in Sect. 3. Experimental results are presented in Sect. 4, and the paper is concluded in Sect. 5.

2 Characterization of Circuit

MTCMOS circuit is used to degrade the total delay of the circuit and increased overall circuit speed [6]. Sleep transistors are used in high V_{th} and all other transistors are used for low threshold voltage [7]. When the number of inverters in the circuit is increasing, the number of sleep transistors is also increased with inverters. In this paper, clustering techniques are used to reduce overall area.

Two inverters are connected in cascade so that the output of one inverter is an input of another inverter and final output "Out" is shown in Fig. 1b. Three and four inverters are connected in cascade so that the output of one inverter is an input of another inverter as shown in Fig. 1c, d. When pulse is applied to the input and sleep transistors are active to sleep mode, large fluctuation occurs in virtual power and ground line. This fluctuation is also called power bouncing noise and ground bouncing noise [8, 9]. When transistor sizing is increased, these fluctuations are reduced or approach to zero. Ground bouncing noise can be reduced by flowing large leakage current through sleep transistors during the transition from sleep to active mode [10].

3 Simulation Parameters

A. *Delay*: Delay is calculated between trigger (input) and target (output) value. It is a time difference between input transition (50%) and 50% output transition. The t_{PLH} defines the response time of the gate for a low to high output transition, while t_{PHL} refers to a high to low transition. The propagation delay [11] is the average of the two

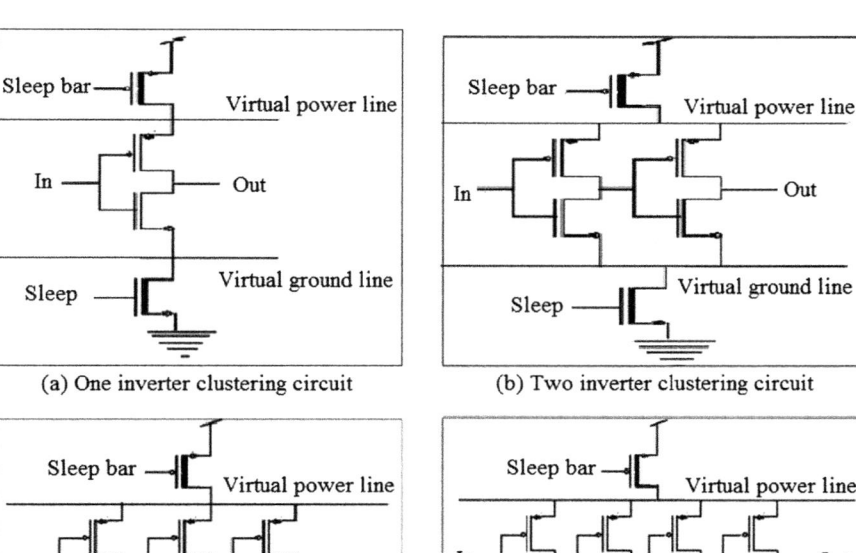

(a) One inverter clustering circuit   (b) Two inverter clustering circuit

(c) Three inverter clustering circuit   (d) Four inverter clustering circuit

Fig. 1 High V_{th} MTCMOS inverters

$$T_{delay} = \frac{t_{PHL} + t_{PLH}}{2}$$

B. ***Dynamic power dissipation***: Due to logic transitions causing logic gates to charge/discharge load capacitance, this is the power dissipation that occurs when the sleep is in active mode. In this mode, PMOS sleep signal is logic low and NMOS sleep signal is logic high so that both sleep transistors are in on state and circuit behaves like a basic CMOS circuit [12].

$$P_{dynamic} = V_{dd}^2 * f * C_L * \alpha$$

where

$P_{dynamic}$ Dynamic power
C_L Load capacitance
α Driving factor
f Frequency
V_{dd} Supply voltage

Fig. 2 Wakeup and sleep
time setup

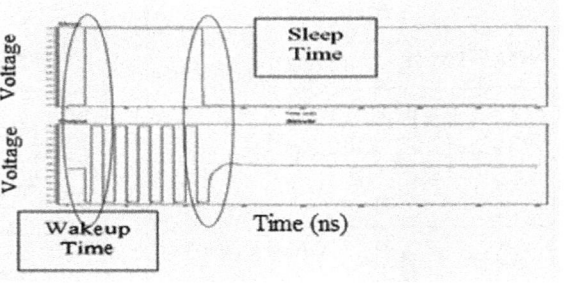

C. **Leakage power dissipation**: This is the power dissipation that occurs when the sleep is in standby mode or idle mode. In this mode, PMOS sleep signal is logic high and NMOS sleep signal is logic low so that both sleep transistors are in off state and very small leakage current flow through the transistor. It will reduce leakage power dissipation [11] of the circuit.

$$P_{\text{leakage}} = I_{\text{leakage}} \cdot V_{\text{dd}}$$

where

P_{leakage} Leakage power dissipation
I_{leakage} Leakage current
V_{dd} Supply voltage

D. **Wakeup time**: It is a time difference of sleep signal to go high and output signal to be stable, when sleep transistors transit from sleep to active mode [13].
E. **Sleep time**: It is the time difference of sleep signal to go low and output signal to be stable, when sleep transistors transition from active to sleep mode. The waveform is shown in Fig. 2.
F. **Virtual power and ground line voltage**: When sleep transistors are in active mode, sudden fluctuations occur because of an internal node in virtual power and virtual ground line [8, 9]. These fluctuations are calculated which is the highest peak voltage on both virtual lines.
G. **Virtual power and ground line discharging current**: This is drain current drawn from PMOS sleep and NMOS sleep transistors. Calculate fluctuation is the highest peak current on virtual lines [8, 9].

4 Simulation Result

The TSPICE simulator is used for the comparison of various parameters. All simulation parameters were obtained at 70 nm technology node. Channel width and channel length are same for NMOS transistor which is same as technology node.

Table 1 Best results for all parameters in different inverter circuits

Logic Circuits ↓	Parameters ↓						
	Delay [5] (ps)	Leakage power [9] (pW)	Dynamic power [9] (nW)	Wakeup time [4] (μs)	Sleep time (ns)	Virtual power line voltage [10] (mV)	Virtual power line current [10] (μV)
Inverter 1	31	1.1	72.8	20	4.98	25.7	18.4
Inverter 2	54.1	2.11	232	200	2.49	34.9	20.8
Inverter 3	62.1	2.14	338	260	7.42	52.2	22.6
Inverter 4	12	2.1	416	240	9.04	52	19.3

Channel width of PMOS transistors is double of technology node or channel length. But sleep transistors use variable width-to-length ratio. The width-to-length ratio of NMOS sleep transistor is $\frac{W}{L} = 1, 3, 5, 10$ and PMOS sleep transistor is $\frac{W}{L} = 2, 6, 10, 20$, respectively.

The threshold voltage of low V_{th} NMOS is 0.2 V, low V_{th} PMOS is -0.22 V, high V_{th} NMOS is 0.4 V, and high V_{th} PMOS is -0.44 V. 1.2 V supply voltage has been taken for the simulation. The complete table of all parameters is shown in Table 2.

Delay is studied from the reference paper [5], leakage power, and dynamic power dissipation are studied from conventional paper [9], wakeup time is studied from the reference paper [4], and virtual power line voltage and virtual power line current are studied from the existing paper [10]. The best result from these inverters is shown in Table 1 and the complete graph of all parameters is shown in Fig. 12 and Table 2.

A. **Delay** is reduced with the increasing width-to-length ratio. Delay is reduced due to high threshold voltage sleep transistors and switches are connected in series as compared to a low threshold voltage CMOS circuit. When delay is reduced, the circuit speed of the transistor is increased and graph of delay with different sizes is shown in Fig. 3.

B. **Leakage current** is increased with the increase of W/L ratio. Because sub-threshold leakage current is the drain-to-source leakage current, sleep transistor is in off state. The following equation relates sub-threshold leakage current I_{sub} [11].

$$I_0 \propto \mu C_{ox} \left(\frac{W}{L} \right)$$

where W and L are width and length of the transistor and μ is the carrier mobility. It is observed that with the increase of W/L ratio, leakage current is increased so that leakage power is also increased, because of direct relation to leakage power and leakage current. When the number of inverters in series (clustering) increases, in that case, leakage power dissipation is also increased in the circuit. The leakage power dissipation with different sleep transistors is shown in Fig. 4.

Table 2 Comparison of all the parameters with different W/L ratios

Parameters ↓	Logic designs ↓															
	Inverter 1				Inverter 2				Inverter 3				Inverter 4			
	W/ L = 1	W/ L = 3	W/ L = 5	W/ L = 10	W/ L = 1	W/ L = 3	W/ L = 5	W/ L = 10	W/ L = 1	W/ L = 3	W/ L = 5	W/ L = 10	W/ L = 1	W/ L = 3	W/ L = 5	W/ L = 10
Delay (pS)	45.4	35.03	32.76	31.03	54.12	57.69	65.97	90.8	84.78	67.92	64.39	62.11	12.02	96.73	89.96	84.44
Dynamic power (µW)	0.07	0.08	0.08	0.08	0.23	0.27	0.28	0.28	0.34	0.38	0.39	0.41	0.42	0.47	0.50	0.52
Leakage power (pW)	1.09	1.356	1.524	1.704	2.112	3.408	4.565	7.58	2.136	3.792	5.616	10.43	2.102	3.7	5.496	10.32
Wakeup time (nS)	0.1	0.05	0.04	0.02	0.32	0.18	0.24	0.2	0.42	0.33	0.29	0.26	0.46	0.28	0.57	0.24
Sleep time (nS)	5.04	5.02	5.01	4.98	33.13	25.17	25.06	24.9	19.92	10.13	11.49	7.42	9.04	17.19	18.84	16.9
Virtual power line voltage (mV)	257	81.74	47.4	25.69	201.9	99.05	64.11	34.9	199.2	138.7	95.11	52.2	248.8	133.1	95.56	51.97
Virtual power line current (µA)	18.4	20.5	22.82	24.45	20.79	31.23	37.04	38.9	22.64	34.49	45.05	43.99	19.34	46.72	55.36	74.22

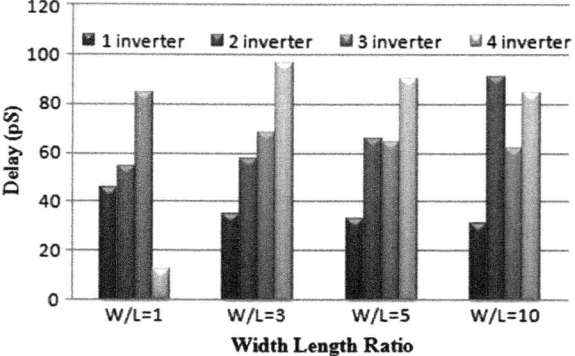

Fig. 3 Delay analysis with different sleep transistors sizing

C. **Dynamic power** is increased with the increasing the size of sleep transistors and an increasing number of inverters series. It is observed that the width is proportional to the current and current is proportional to power, so that power is increased with increasing size of sleep transistors. Dynamic power analysis graph is shown in Fig. 5.

D. *Wakeup time* is reduced with the increase of *W/L* ratio in all the clustering circuits. Because when increasing *W/L* ratio, the output is stable either zero or one logic sharply, so that time difference of sleep signal when transitioning from sleep to active mode and output signal is decreased when increasing the width-to-length ratio. The wakeup time graph with different inverter clustering circuits is shown in Fig. 6.

E. *Sleep time* is reduced with the increase of width-to-length ratio and increasing a number of inverter in the clustering circuit. When sleep signal transits from active to sleep, the output signal is stable, taking less time, so that time difference of sleep signal when transition from active to sleep mode and the output signal is reduced and the graph is shown in Fig. 7.

F. *Virtual power and ground line voltage* are reduced with the increase of the size of sleep transistors. When sleep transistor size is small that is W/L = 1, W/L = 3 large fluctuation occurs in virtual power and ground line after sleep

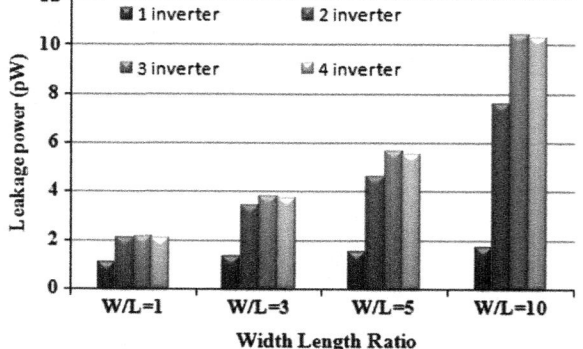

Fig. 4 Leakage power analysis with different sleep transistors sizing

Fig. 5 Dynamic power
analysis with different sleep
transistors sizing

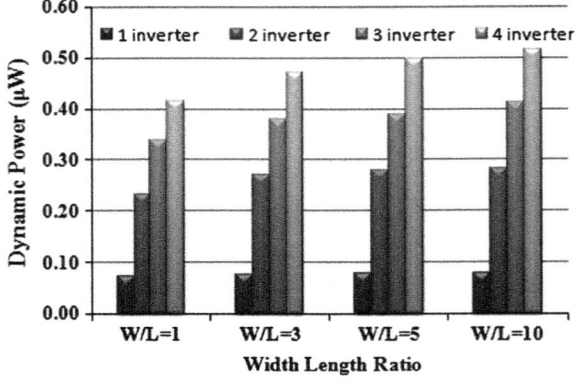

Fig. 6 Wakeup time analysis
with different sleep transistors
sizing

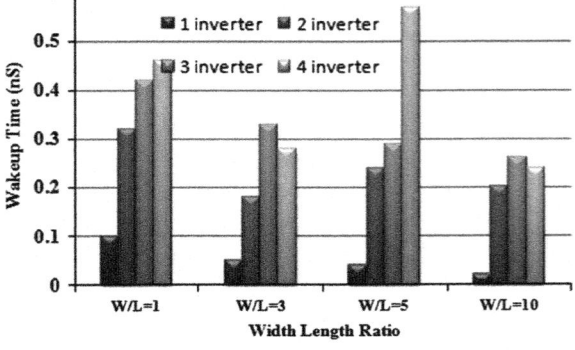

Fig. 7 Sleep time analysis
with different sleep transistors
sizing

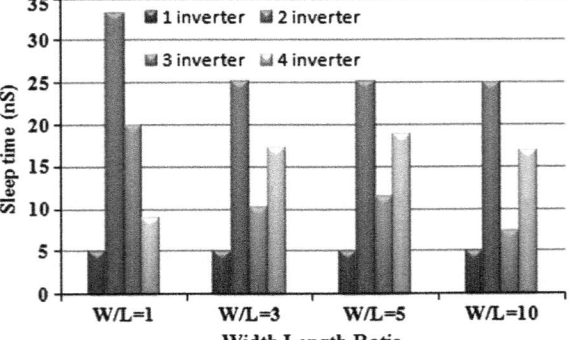

transistor transition from sleep to active mode. When sleep transistor size is increased that is W/L = 5, W/L = 10 fluctuation is reduced and signal is smooth as shown in Fig. 8 and the graph is shown in Fig. 9.

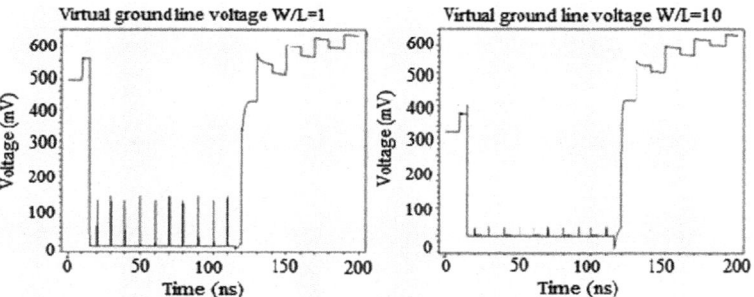

Fig. 8 Comparison of virtual ground line voltage with *W*/*L*

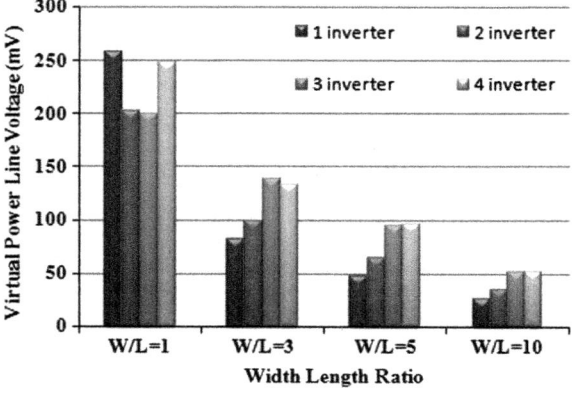

Fig. 9 Graph of virtual power line voltage

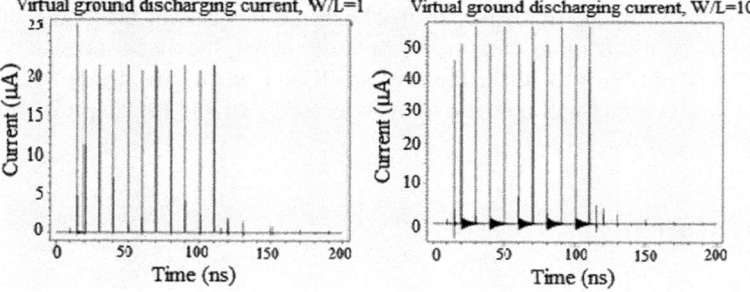

Fig. 10 Virtual ground line discharging current

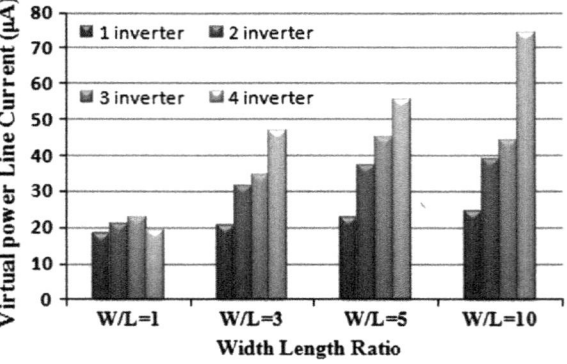

Fig. 11 Graph of virtual power line discharging current

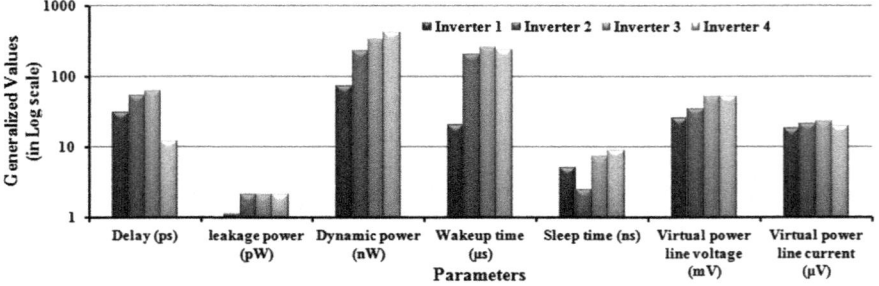

Fig. 12 Complete graph of best results for all parameters in different inverter circuits

G. *Virtual power and ground line discharging current* are increased with increasing the size of sleep transistors in the ratio of $W/L = 1$, $W/L = 3$, $W/L = 5$, and $W/L = 10$. It discusses before that current is directly proportional to the width of the transistor, so that with the increase of the sizing of sleep transistors, discharging current is increased. When the sleep transistor size is $W/L = 1$ or $W/L = 3$, the current is small and when increasing $W/L = 5$ or $W/L = 10$ current will increase as shown in Fig. 10 and the graph is shown in Figs. 11 and 12.

5   Conclusion

In this paper, different sizes of sleep transistors with an increasing number of inverters using clustering technique are presented. Simulation result shows that delay is reduced up to 0.68 times when the width-to-length ratio is increased by 10 times in the circuit. Dynamic power is increased up to 1.24 times when W/L ratio is increased by 10 times in circuit; leakage power is increased up to 1.54 times when size is increased by 10 times in circuit; wakeup time is reduced up to 0.52 times when size is increased by 10 times in circuit; and sleep time is reduced up to 0.75 times when size is increased by 10 times in the clustering circuit.

Virtual power line voltage is reduced up to 0.20 times when size is increased by 10 times in circuit. Virtual ground line voltage is reduced up to 0.15 times when size is increased by 10 times in one inverter circuit.

Virtual power line discharging current is reduced up to 3.84 times when size is increased by 10 times in the clustering circuit. Virtual ground line discharging current reduced up to 1.86 times when size is increased by 10 times in the clustering circuit.

Acknowledgements We are thankful to M. P. Council of Science and Technology, Bhopal, India, for finical support under R&D project scheme.

No: 1950/CST/R&D/Phy & Engg Sc/2015: August 27, 2015.

References

1. V. Kursun and E. G. Friedman, "Multi-Voltage CMOS Circuit Design," John Wiley & Sons Ltd., ISBN 0-470-01023-1, 2006.
2. S. M. Kang and Y. Leblebic, "CMOS Digital integrated circuits Analysis and Design," Tata McGraw Hill, New Delhi, India, 2013.
3. J. C. Park and V. J. Mooney, "Sleepy Stack Leakage Reduction," IEEE Transaction on Very Large Scale Integration (VLSI) Systems, vol. 14, no. 11, pp. 1250–1263, November 2006.
4. Vaibhav Neema, "A Circuit Technique for Leakage Power Reduction in Sleep Mode of Operation", Journal of circuits, systems and computers word scientific publication, 2007.
5. Hamada M. et al. "A top-down low power design technique using clustered voltage scaling with variable supply-voltage scheme," Proceedings of the IEEE Custom Integrated Circuits Conference, pp. 495–498, 1998.
6. Shashikant Sharma, Anjan Kumar, ManishaPattanaik, and Balwinder Raj, "Forward Body Biased Multimode Multi-Threshold CMOS Technique for Ground Bounce Noise Reduction in Static CMOS Adders," International Journal of Information and Electronics Engineering, Vol. 3, No. 6, November 2013.
7. Kao JT, Chandrakasan "Dual-threshold voltage techniques for low-power digital circuits," IEEE Journal of Solid-State Circuits, vol. 35 (7), pp. 1009–1018, July 2000.
8. Hailong Jia and Volkan Kursun, "Ground-Bouncing-Noise-Aware Combinational MTCMOS Circuits," IEEE transactions on circuits and systems–I regular papers, vol. 57, no. 8 August 2010.
9. H. Jiao and V. Kursun, "Ground Bouncing Noise Suppression Techniques for MTCMOS Circuits," in Proc. IEEE Asia Symposium on Quality Electron Design, pp. 64–70, July 2009.
10. V. Neema, S.S. Chouhan, S. Tokekar, "Novel Circuit Technique for Reduction of Leakage Current in Series/Parallel PMOS/NMOS Transistors Stack", IETE Journal of Research, vol. 56, no. 6, pp. 350–354, 2010.
11. B.S. Deepaksubramanyam and Adrian Nunez, "Analysis of Subthreshold Leakage Reduction in CMOS Digital Circuits", Proceedings of the 13th NASA VLSI symposium.post falls, Idaho, USA, June 5–6, 2007.
12. Chandrakasan AP, Brodersen RW "Minimizing power consumption in digital CMOS circuits," Proceedings of the IEEE, vol .83, no. 4, pp. 498–523, April 1995.
13. Z. Liu and V. Kursun, "Characterization of wake-up delay versus sleep mode power consumption and sleep/active mode transition energy overhead tradeoffs in MTCMOS circuits," Proceedings of the IEEE International Midwest Symposium on Circuits and Systems, pp. 362–365, August 2008.

Optimal Placement of Distributed Generation Using Genetic Algorithm Approach

Mohan Kashyap, Ankit Mittal and Satish Kansal

1 Introduction

The power demand of electrical distribution system is uniformly rising day by day which in turn cause the increase in load burden and degradation of voltage profile [1]. The node voltages of distribution system are reduced at a far distance from the substation than at nearby nodes. Thus there is always a demand for reactive power compensation for improvement in voltage profile [1, 2]. While comparing with transmission systems, the distribution systems have low X/R ratio which result large losses of power and drops in voltage. There are a lot of possibilities for power loss reduction and voltage profile improvement such as allocation of DG, capacitor, etc. [1, 3]. The allocation of DG in distribution system provides active power and hence decreases the load burden. It will result the peak demand loss reduction, energy loss reduction, voltage profile improvement, power factor and stability improvement in distribution system [4–6].

The various DG technologies used in distribution system may be broadly categorized into two types called non-conventional energy sources and conventional or fuel based energy sources [7–9]. There are certain factors such as technical, economic and environmental factors which should be considered for DG placement in distribution systems [10–14]. If DG is not placed optimally, it may give rise to fault currents, power losses, voltage disturbances, operating and capital costs [13].

M. Kashyap (✉)
I.K.G. Punjab Technical University, Jalandhar, Punjab, India
e-mail: mohan_kashyap80@rediff.com

A. Mittal
GIMT, Kurukshetra, Haryana, India
e-mail: ankitm675@gmail.com

S. Kansal
B.H.S.B.I.E.T., Lehragaga, Sangrur, Punjab, India
e-mail: kansal.bhsb@gmail.com

© Springer Nature Singapore Pte Ltd. 2019
V. Nath and J. K. Mandal (eds.), *Proceeding of the Second International Conference on Microelectronics, Computing & Communication Systems (MCCS 2017)*, Lecture Notes in Electrical Engineering 476, https://doi.org/10.1007/978-981-10-8234-4_47

587

Thus the problem of DG placement is a challenging issue and demands the optimization of a number of objectives such as minimization of total power losses, voltage changes, line loading, installation costs and maximization of system stability etc. [15]. The purpose is to find optimal DG location and size in distribution system considering the constraints of nodes voltage limit, branches thermal/current limit and DG maximum size [14, 15]. The PSO based approach is employed to obtain the optimal allocation of different types of DGs [16]. The voltage index (VI) based approach is used to determine the optimal DG allocation [17]. The Loss Sensitivity Index (LSI) is employed to obtain the DG optimal location [18]. A plenty of optimization techniques has been employed in literature for solving the DG placement problem for instance analytical approach [19], genetic algorithm [20–25], Particle Swarm Optimization (PSO) techniques [26–29], and evolutionary programming [30].

This paper is arranged as follows. Section 2 introduces the basic Genetic Algorithm approach. Section 3 presents the methodology for solving the optimal DG placement problem. It describes the load flow technique, formulation of optimization problem and proposed Genetic Algorithm approach. In Sects. 4 and 5, the simulation results are discussed and conclusions are drawn respectively.

2   Genetic Algorithm

Genetic Algorithm is an optimization technique based on Darwinian principle of selection and survival of fittest. It is a search algorithm that uses natural genetic operations for instance crossover and mutation. It is an artificial intelligence (AI) method used for solving the optimal DG placement problem for distribution systems. A simple genetic algorithm in most of practical problems consists of three operators: reproduction, crossover and mutation. Figure 1 provides the basic Genetic Algorithm approach [20–25].

Following are certain points/facts which differentiate the genetic algorithm approach from other optimization approaches:

- GA uses the coding of a set of parameters instead of coding of parameters themselves.
- GA develops the use of objective function values instead of using another extra knowledge.
- GA always performs the population search instead of search for an individual point.
- GA always applies the probabilistic rules instead of deterministic rules.

Fig. 1 Basic GA approach

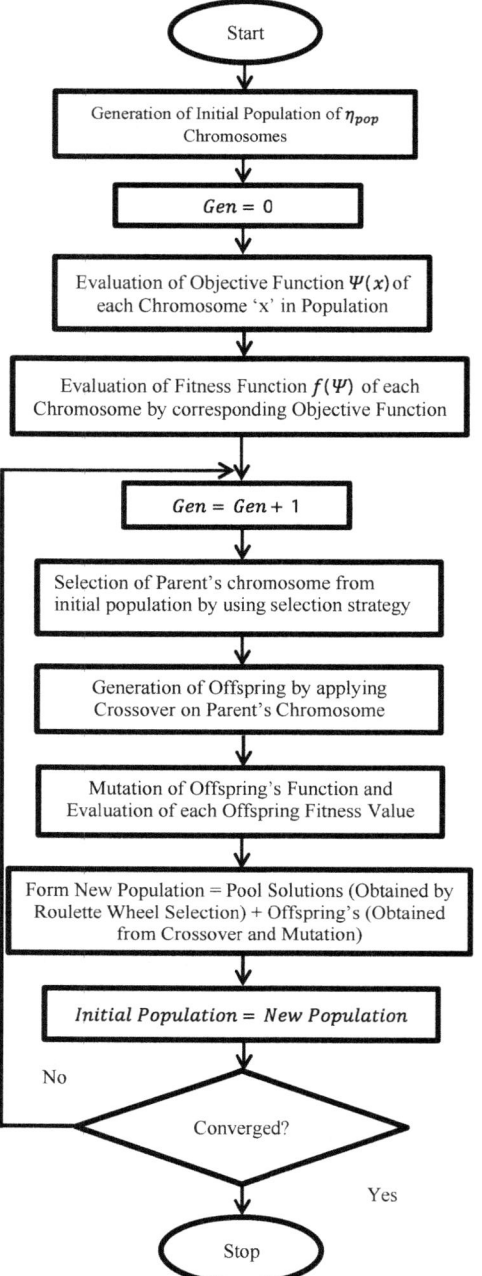

3 Proposed Methodology

3.1 Load Flow Technique

The forward backward sweep method of load flow is employed in present work. This technique of load flow consists of two steps called backward sweep and forward sweep. It is based on equivalent current injections (ECI) [31]:

Backward Sweep: This is first step used for calculating the load current of N number of nodes in distribution system as follows:

$$\overrightarrow{I_L}(m) = \frac{P_L(m) - jQ_L(m)}{\overrightarrow{V^*}(m)} \tag{1}$$

In above equation, $m = 1, 2, 3 \ldots N$. $P_L(m)$ and $Q_L(m)$ denote active and reactive power demands at m node. $\overrightarrow{I_L}$ and $\overrightarrow{V^*}$ represent phasor values of load current and voltage. Each branch current is determined by:

$$\vec{I}(mn) = \overrightarrow{I_L}(n) + \sum_m \overrightarrow{I_L}(m) \tag{2}$$

Forward Sweep: This is next step used for calculating each node voltage in distribution system as given below:

$$\vec{V}(n) = \vec{V}(m) - \vec{I}(mn)Z(mn) \tag{3}$$

In above equation, m and n denote the sending and receiving end nodes for mn branch. $Z(mn)$ denotes the impedance of mn branch.

Equivalent Current Injections: This technique of load flow is stated by equivalent current injection of the node in distribution system. Complex load S_i of a node in distribution system is described as:

$$S_i = P_i + jQ_i \tag{4}$$

where $i = 1, 2 \ldots N_B$.

Therefore ECI may be modeled as:

$$I_i = I_i^r(V_i) + jI_i^i(V_i) = \left(\frac{P_i + jQ_i}{V_i}\right)^* \tag{5}$$

For solving the load flow, ECI of ith node for kth iteration is determined by:

$$I_i^k = I_i^r(V_i^k) + jI_k^i(V_i^k) = \left(\frac{P_i + jQ_i}{V_i^k}\right)^* \tag{6}$$

3.2   Problem Formulation

The main objective is to minimize the total active power losses (P_{loss}) without violating the voltage limits and other constraints in distribution system. The proposed Genetic Algorithm technique will be utilized to compute the minimum distribution loss and determine the optimal sizing and siting of DG. The objective function mathematically formulated as follows:

$$\text{Min}\Psi\left(\Upsilon_p\right) = \left(\sum_i^{N_L} P_{\text{loss}}^{\text{with}_{\text{DG}}}\right) \tag{7}$$

$$\text{Min}\Psi\left(\Upsilon_p\right) = \left(\frac{\sum_i^{N_L} P_{\text{loss}}^{\text{with}_{\text{DG}}}}{\sum_i^{N_L} P_{\text{loss}}^{\text{without}_{\text{DG}}}}\right) \tag{8}$$

The optimization problem is subjected to following constraints:
The voltage limits at all nodes are as follow

$$v^{\min} \leq v_i \leq v^{\max} \tag{9}$$

The thermal limits are given by

$$I_{L\_i} \leq I_{L\_\max\_i} \tag{10}$$

DG power limits are as follow

$$P_{\text{DG}}^{\min} \leq P_{\text{DG}_i} \leq P_{\text{DG}}^{\max} \tag{11}$$

The fitness function $f(\Psi)$ is evaluated as follows:

$$f(\Psi)_i = \frac{\Psi\left(\Upsilon_p\right)_i}{\sum \Psi\left(\Upsilon_p\right)_i} \tag{12}$$

where $i = 1, 2 \ldots$popsize.
The proposed optimization approach using Genetic Algorithm to find optimal location and size of DG in radial distribution system is shown in Fig. 2.

Fig. 2 Proposed genetic
algorithm approach for
optimal placement of DG

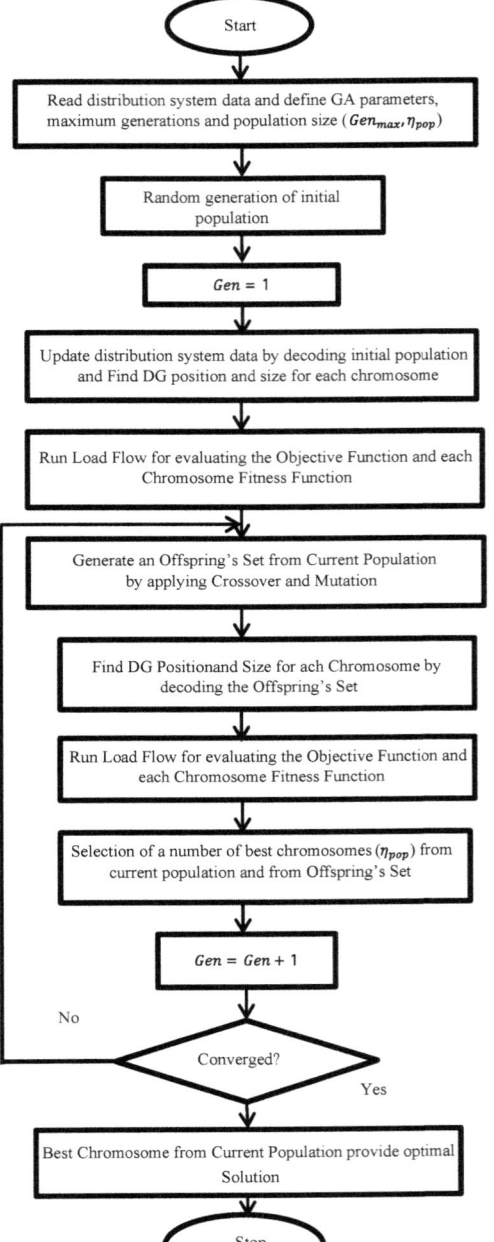

4 Simulation Results

As already discussed that for solving the DG placement problem a useful algorithm is required to calculate the active power losses and node voltages and find optimal location and sizing of DG. So the algorithm as shown in Fig. 2 is proposed and tested on IEEE 33 and 69 bus test systems. The Forward-Backward Sweep load flow is employed in this work. A computer program/code is written in MATLAB R2014a to obtain the simulation results for IEEE 33-bus and 69-bus test systems.

For 33 bus test system, total active generation and total reactive generation is 3926 kW and 2443 KVAr respectively. Total active power loss is 211 kW and reactive power loss 143 KVAr. The minimum voltage is 0.9038 without DG. For 69-bus test system, total active generation and total reactive generation is 4027 kW and 2797 KVAr respectively. Total active power loss is 225 kW and total reactive power loss 102 KVAr respectively. The minimum voltage is 0.9092 without DG.

After application of DG using proposed GA approach, the results obtained are presented in Tables 1 and 2 for 33-bus and 69-bus distribution systems respectively. Figures 3 and 4 describe the voltage profiles for 33-bus and 69-bus distribution systems respectively.

Table 1 Optimal DG placement for 33 bus test system

DG parameters	Without DG	With DG
DG optimum location	–	6
DG optimal size (kW)	–	2600
Total active power loss (kW)	211	111.03
Reduction in active power loss (%)	–	47.39
Minimum bus voltage (pu)	0.9038	0.9425

Table 2 Optimal DG placement for 69 bus test system

DG parameters	Without DG	With DG
DG optimum location	–	60
DG optimal size (kW)	–	1850
Total active power loss (kW)	225	63.12
Reduction in active power loss (%)	–	72
Minimum bus voltage (pu)	0.9092	0.9682

Fig. 3 Voltage profile for 33
bus distribution system

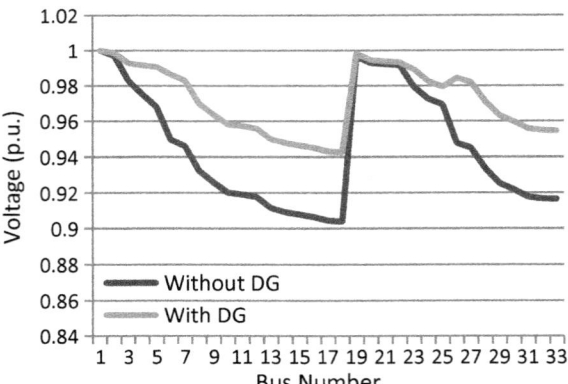

Fig. 4 Voltage profile for 69
bus distribution system

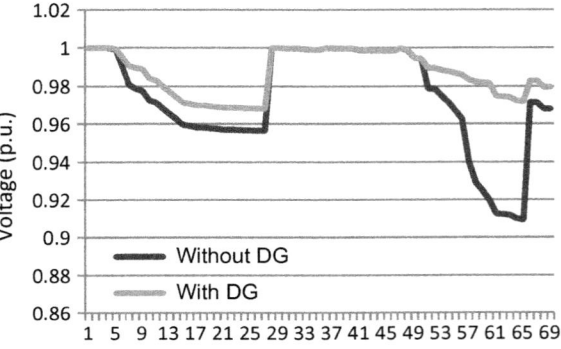

Although the proposed GA approach has been tested on 33-bus and 69-bus test systems, the results of proposed GA approach are also compared with the results obtained from PSO algorithm, improved analytical (IA) method, Hybrid approach, repeated load flow method of 33-bus test system for comparative study. Table 3 summarizes the optimal solutions obtained by these approaches. It is observed that proposed GA approach provides the maximum percentage of active power loss reduction as compared to all other approaches.

Table 3 Comparison of results for 33 bus test system

DG parameters	With DG (proposed GA)	With DG (PSO) [16]	With DG (IA) [32]	With DG (hybrid approach) [32]	With DG (repeated load flow) [5]
DG optimum location	6	6	6	6	6
DG optimal size (kW)	2600	3150	2600	2490	3150
Total active power loss (kW)	111.03	115.29	111.10	111.20	115.20
% Reduction in active power loss	47.38	45.36	47.34	47.29	45.40

5 Conclusion

DG installation in distribution systems is playing an important role in power generation. It permits the distribution system to withstand with heavy loading, reduces the losses and improves the voltage profile. This paper proposes an innovative approach using Genetic Algorithm to find optimal location and size of DG in the radial distribution networks where total active power loss of system is employed as the objective to be minimized. After analyzing the results, it is concluded that optimal DG placement is beneficial as it minimizes the power loss and improves the voltage profile of distribution system to a large extent. Also as resulted from comparative study, it is concluded that proposed GA approach provides the maximum percentage of active power loss reduction as compared to all other approaches.

Acknowledgments We gratefully acknowledge our deep indebtedness to IKG Punjab Technical University, Jalandhar, Punjab, India, which helped us during the whole period of the work.

References

1. R. E. Brown, Electric Power Distribution Reliability, CRC Press, 2008.
2. S. H. Horowitz, A. G. Phadke, Power System Relaying, 2nd ed. Baldock, Research Studies Press Ltd, 2003.
3. T. Ackermann, G. Andersson, and L. Sder, "Distributed generation: a definition," Elect. Power Syst. Res., vol. 57, pp. 195–204, 2001.
4. T. Ackerman and V. Knyazkin, "Interaction between distributed generation and the distribution network: operation aspects," IEEE PES Transm. and Distrib. Conference and Exhibition, vol. 2, pp. 1357–1362, 2002.
5. S. Kansal, B. B. R. Sai, B.Tyagi, V. Kumar, "Optimal placement of distributed generation in distribution networks," Int J Engineering, Science and Technology, vol. 3(3), pp. 47–55, 2011.

6. Mohd Zamri Che Wanik, Istvan Erlich, and Azah Mohamed, "Intelligent management of distributed generators reactive power for loss minimization and voltage control," MELECON 2010-15th IEEE Mediterranean Electro-technical Conference, pp. 685–690, 2010.
7. P. S. Georgilakis and N. D. Hatziargyriou, "Optimal distributed generation placement in power distribution networks: models, methods, and future research," IEEE Trans. Power Syst., 28 (3), pp. 3420–3428, 2013.
8. Y. A. Katsigiannis and P. S. Georgilakis, "Effect of customer worth of interrupted supply on the optimal design of small isolated power systems with increased renewable energy penetration," IET Gener. Transm. Distrib., 7 (3), pp. 265–275, 2013.
9. J. A. Peças Lopes, N. Hatziargyriou, J. Mutale, P. Djapic, and N. Jenkins, "Integrating distributed generation into electric power systems: a review of drivers, challenges and opportunities," Elect. Power Syst. Res., 77 (9), pp. 1189–1203, 2007.
10. R. Jabr, and B. Pal, "Ordinal optimization approach for locating and sizing distributed generation," IET Gener. Transm. Distrib., 3 (8), pp. 713–723, 2009.
11. A. Gopi, and P. A. Raj, "Distributed generation for line loss reduction in radial distribution system," International Conference on Emerging Trends in Electrical Engineering and Energy Management (ICETEEEM-2012), pp. 29–32, 2012.
12. N. C. Sahoo, S. Ganguly and D. Das, "Recent advances on power distribution system planning: a state-of-the-art survey," Energy Systems, 4, pp. 165–193, 2013.
13. A. Pecas Lopes, N. Hatziargyriou, J. Mutale, P. Djapic and N. Jenkins, "Integrating distributed generation into electric power systems: a review of drivers, challenges and opportunities," Elect. Power Syst. Res., vol. 77, pp. 1189–1203, 2007.
14. A. Keane, "State-of-the-Art techniques and challenges ahead for distributed generation planning and optimization," IEEE Trans. Power Syst., vol. 28 (2), pp. 1493–1502, 2013.
15. A. Alarcon-Rodriguez, G. Ault and S. Galloway, "Multi-objective planning of distributed energy resources: A review of the state-of-the-art," Renewable and Sustainable Energy Reviews, vol. 14, pp. 1353–1366, 2010.
16. S. Kansal, V. Kumar, B. Tyagi, "Optimal placement of different type of DG sources in distribution networks," Int J Electr Power and Energy Syst, vol. 53, pp. 752–760, 2013.
17. K. Vinoth kumar and M.P. Selvan, "Planning and operation of distributed generations in distribution systems for improved voltage profile," IEEE PES Power Systems Conference and Exposition (PSCE 2009), March 2009, Washington, USA.
18. J.B.V. Subrahmanyam and C. Radhakrishna "Distributed generator placement and sizing in unbalanced radial distribution system," International Journal of Electrical and Electronics Engineering, vol. 3, pp. 746–753. 2009.
19. D. Q. Hung, N. Mithulananthan and R.C. Bansal, "Analytical expressions for DG allocation in primary distribution networks," IEEE Trans. Energy Conversion, vol. 25 (3), pp. 814–820, 2010.
20. C.L.T. Borges and D.M. Falcao, "Optimal distributed generation allocation for reliability, losses and voltage improvement," International Journal of Power and Energy Systems, vol. 28 (6), pp. 413–420, July 2006.
21. G. Celli, E. Ghiani, S. Mocci and F. Pilo, "A multi-objective evolutionary algorithm for the sizing and siting of distributed generation," IEEE Trans. Power Syst., vol. 20 (2), pp. 750–757, 2005.
22. G. Carpinelli, G. Celli, S. Mocci, F. Pilo and A. Russo, "Optimization of embedded generation sizing and siting by using a double trade-off method," IEE Proc. Gener. Transm. Distrib., vol. 152 (4), pp. 503–513, 2005.
23. D. Singh, and K. S. Verma, "Multi-objective optimization for DG planning with load models," IEEE Trans. Power Syst., vol. 24 (1), pp. 427–436, 2009.
24. D. Das, "Reactive power compensation for radial distribution networks using genetic algorithm," International Journal of Electrical Power & Energy Systems, vol. 24, issue 7, pp. 573–581, October 2002.

25. Nien-Che Yang and Tsai-Hsiang Chen, "Dual genetic algorithm based approach to fast screening process for distributed generation interconnections," IEEE Trans. Power Delivery, vol. 26(2), pp. 850–858, April 2011.
26. A. M. El-Zonkoly, "Optimal placement of multi-distributed generation units including different load models using particle swarm optimisation," IET Gener. Trans. Distrib., vol. 5 (7), pp. 760–771, 2011.
27. A. H. Mantway and M. M. Al-Muhaini, "Multi-objective BPSO algorithm for distribution system expansion planning including distributed generation," IEEE/PES Transmission and Distribution Conference and Exposition, pp. 1–8, 2008.
28. N. C. Sahoo, S. Ganguly and D. Das, "Simple heuristics-based selection of guides for multi objective PSO with an application to electrical distribution system planning", Eng. Appl. Artif. Intell., 24, pp. 567–585, 2011.
29. S. Ganguly, N.C. Sahoo and D. Das, "A novel multi-objective PSO for electrical distribution system planning incorporating distributed generation," J. Energy Syst., 1, pp. 291–337, 2010.
30. B.A. De-Souza and J.M.C. De-Albuquerque, "Optimal placement of distributed generators networks using evolutionary programming," Proc. of IEEE/PES Transmission & Distribution Conference and Exposition: Latin America: 6, 2006.
31. Jen-Hao Teng, "A direct approach for distribution system load flow solutions," IEEE Trans. Power Delivery, vol. 18(3), pp. 882–887, July 2003.
32. S. Kansal, V. Kumar, B. Tyagi, "Hybrid approach for optimal placement of multiple DGs of multiple types in distribution networks" Int J Electr Power and Energy Syst, vol. 75, pp. 225–236, 2015.

Efficient Design and Simulation of Novel Exclusive-OR Gate Based on Nanoelectronics Using Quantum-Dot Cellular Automata

Mukesh Patidar and Namit Gupta

1 Introduction

QCA is an emerging nanotechnology to fabricate electronic circuits for computing systems and applicable for forthcoming generation [1]. The main advantages of QCA are very high devices operating speed, reduction the size, and high packing density [2]. It is a way to create digital electronic devices without using transistors that has potential to be faster and extremely low heat energy consumption with more energy efficient than CMOS (complementary metal oxide semiconductor) technology. The sizes of CMOS transistor keep shrinking to enhance the devise density on chips by law of Gordon Moore [3].

The CMOS affects the device's performance due to constraints as dissipation of energy and consumption of power [4]. In addition, defect in QCA manufacturing may well manifest themselves differently at logic level than CMOS. In the QCA devices, the information flow between inputs to output through the quantum cell by interactions between electrons in neighbor cells by (Coulomb's law) [5, 6].

In the 1993, Lent et al. introduced new QCA technology. The primary element in this technology is a "quantum cell" which is composed of four quantum dots. The QCA design fabrication at NTP becomes practicable in this technology [7].

This work is connected as follows: Sect. 2 explains QCA architecture, Sect. 3 explains exclusive-OR gate implementation with different logical layouts designed, Sect. 4 explains the simulation results, Sect. 5 is the conclusion of proposed work, and last section is the references.

M. Patidar (✉) · N. Gupta
Department of Electronics Engineering, Shri Vaishnav Vidyapeeth Vishwavidyalaya,
Indore, Madhya Pradesh, India
e-mail: mukesh.omppatidar@gmail.com

N. Gupta
e-mail: namitg@hotmail.com

© Springer Nature Singapore Pte Ltd. 2019
V. Nath and J. K. Mandal (eds.), *Proceeding of the Second International Conference on Microelectronics, Computing & Communication Systems (MCCS 2017)*, Lecture Notes in Electrical Engineering 476, https://doi.org/10.1007/978-981-10-8234-4_48

2 QCA Architecture

The QCA cell initially developed at the "ATIPS laboratory" [8] by University of Calgary (U of C), Canada. The QCA design is built as an array of quantum cells [9] or bistable cells, and the ideal size of square cells (18 nm × 18 nm) was composed of four dots at the corners [10]. Each quantum cell implants two mobile electrons which are permitted to tunnel between adjacent sites of cell [11].

Quantum dot is made using Si/SiO_2 and GaAs/AlGaAs material system [12]. In nanoelectronics, the information representation with binary logic is an innovatory idea by means of polarization state of QD [13]. The binary information is applied by polarization $P = -1$ and $P = +1$ [13], its quantum-dot cell polarizations is represented in Fig. 1, and polarization of a quantum cell based on its charge densities is given by Eq. 1 [14].

$$P = \frac{(\rho_1 + \rho_3) - (\rho_2 + \rho_4)}{\rho_1 + \rho_2 + \rho_3 + \rho_4} \tag{1}$$

where ρ_i is the charge density at site "i" within the cell, e.g., on inner sites corner "1 and 3", the cell is "+1" polarization. If the excess electrons are completely on inner sites corner "2 and 4", the cell is "−1" polarization [11, 14].

In irreversible calculation, it is proved that the one-bit information is lost, generates heat energy, and dissipates $K_bT \times \log_e 2$ joules, where K_b is Boltzmann's constant approximately $1.3806505 \times 10^{-23} JK^{-1}$ and T is the operating temperature [14–16].

2.1 Basic Elements of QCA Circuit Design

The key elements of QCA logic circuit design through wire, inverter, and three-input majority voter [14, 17]. The wire is built-up series of quantum cells demonstrated in Fig. 2a, the QCA inverter structure demonstrated in Fig. 2b, and it is a simple device that inverts an input signal. The majority voter (MV) consists of four cells around a center cell: three input QCA cells and one output cell [15]. It

Fig. 1 Quantum cells with polarizations ($P = +1$ and $P = -1$)

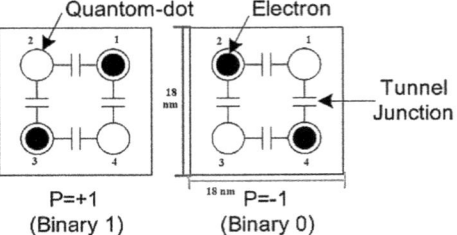

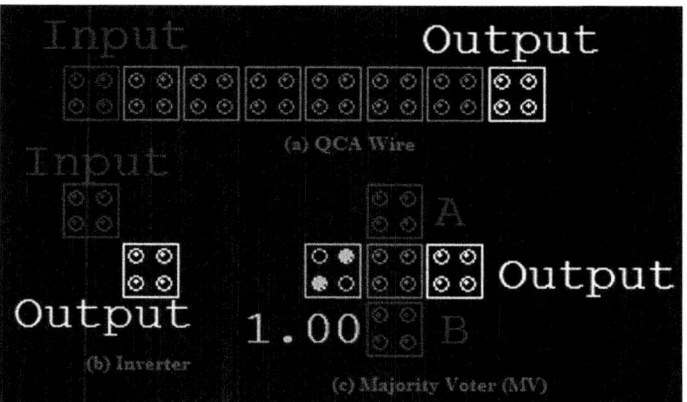

Fig. 2 Basic elements for QCA logic design

implements the logic function in Eq. 2 and demonstrated in Fig. 2c. Logical gates are AND for fixed any one input −1 and OR for fixed any one input +1 in majority voter [18].

$$M(A, B, C) \text{ or } F = AB + BC + AC. \tag{2}$$

2.2 QCA Clocking Concepts

In QCA technology, signal synchronization is required at the wire layout level due to the nature of signal travel in quantum cells [19].

The signal travel in QCA is determined at the logic layout level by the clocking structure since a group of cells in a particular clocking zone (each zone consists of four clocks) acts as a clock cycle or delay element [19] with four phases [20]. It is shown in Fig. 3, where clock 0: Switch/Latch, clock 1: Hold, clock 2: Release, and clock 3: Relax.

3 Exclusive-OR Gate Implementation

The hybrid logic gates are called as the exclusive-OR (or equivalence) gate and its complement the exclusive-NOR (Ex-NOR) gate; the binary EXOR gates are very useful in system requiring, parity checking, and error correction–detection

Fig. 3 QCA clocking phases
within clock zones [21]

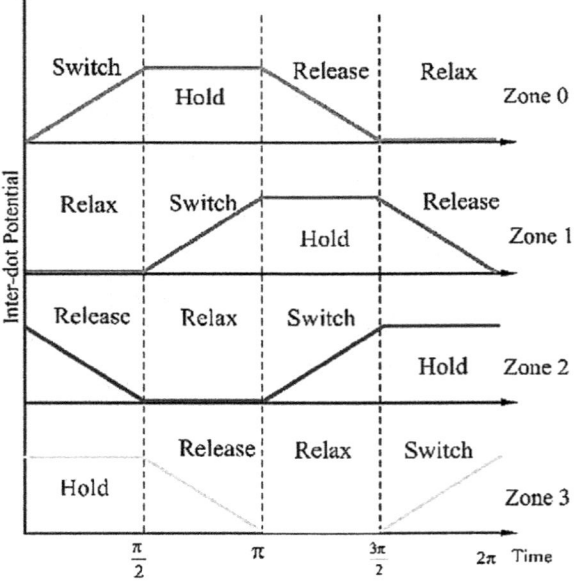

Fig. 4 Exclusive-OR gate

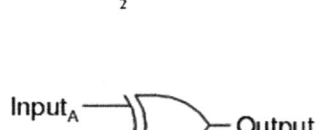

Table 1 Truth table for
EXOR

A	B	Output
0	0	0
0	1	1
1	0	1
1	1	0

techniques [1]. The QCA implementation for EXOR logic symbol in IEEE standards is shown in Fig. 4 [17], Eq. 3, and result in verification through the truth table in Table 1:

$$A \oplus B = A'B + AB' \tag{3}$$

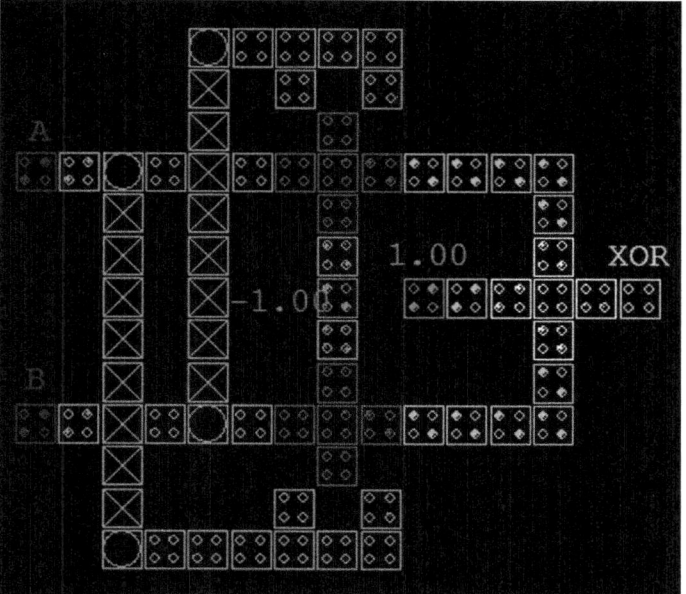

Fig. 5 Conventional QCA XOR gate [22–24]

3.1 Conventional Design Approach for EXOR

The conventional schematic layout representation of exclusive-OR gate is demonstrated in Fig. 5 [22–24]. This layout is implemented by using multilayer (via and crossover connection).

3.2 Novel Exclusive-OR Design Approach

This novel exclusive-OR layout design approach uses one three-input MV, one inverter, and QCA wire. These proposed layouts designed are more efficient in terms of number of cells, design area, latency, power dissipation, cell cost, and

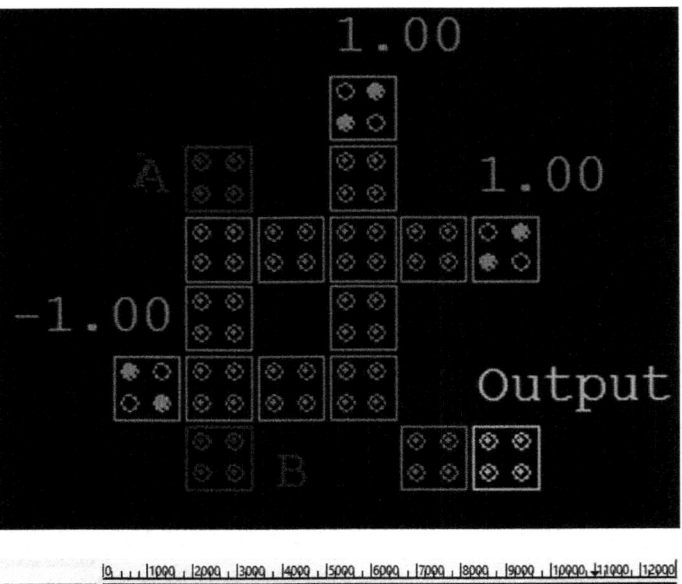

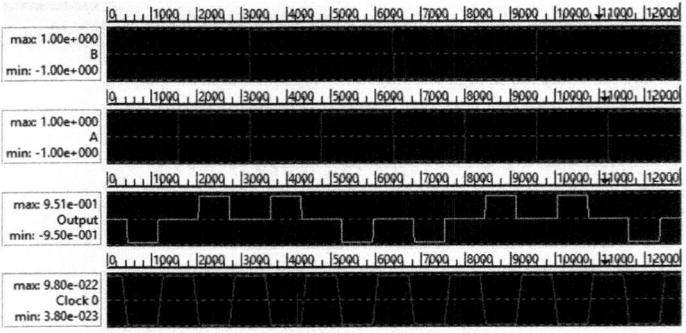

Fig. 6 First QCA EXOR layout and result

complexity. The simulation results are shown in Figs. 6, 7, 8, 9, 10, 11, and 12 and used parameters in Table 2.

Each one of the circuits is examined through bistable engine approximation simulator and similar outcomes have been achieved. A comparison between previous and our proposed novel EXOR gate results is shown in Tables 3 and 4, respectively, with different aspects.

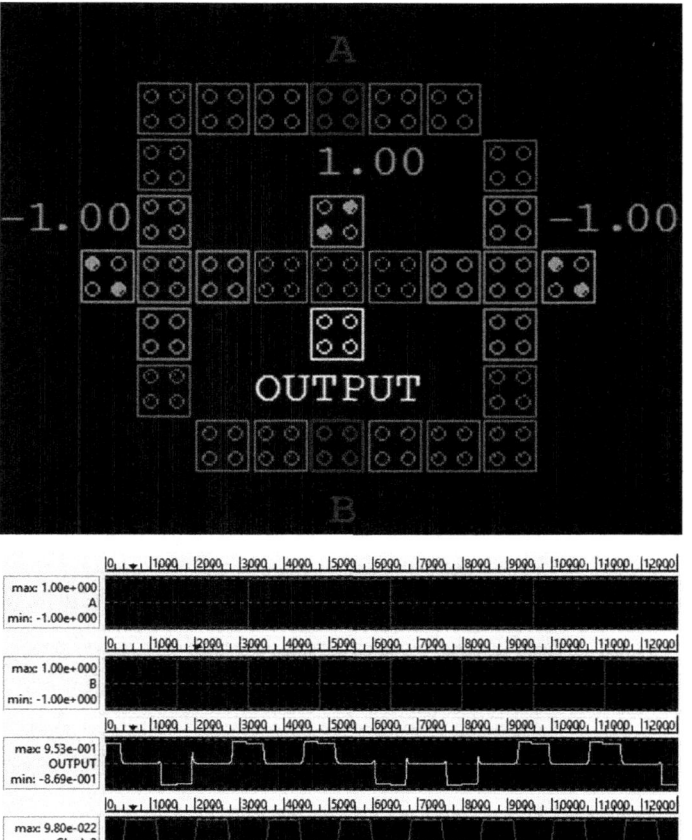

Fig. 7 Second QCA EXOR layout and result

4 Simulation Results

To verify our proposed novel, EXOR gate layout designs on using specific resolution tool QCA designer version (V2.0.3) with the bistable simulation engine. This proposed design results in significant improvements in QCA cell count, overall area (μm^2), used cell area, delay (latency) in clock cycle, and power consumption and its comparison of previous and proposed and are graphically represented, respectively, in Figs. 13 and 14.

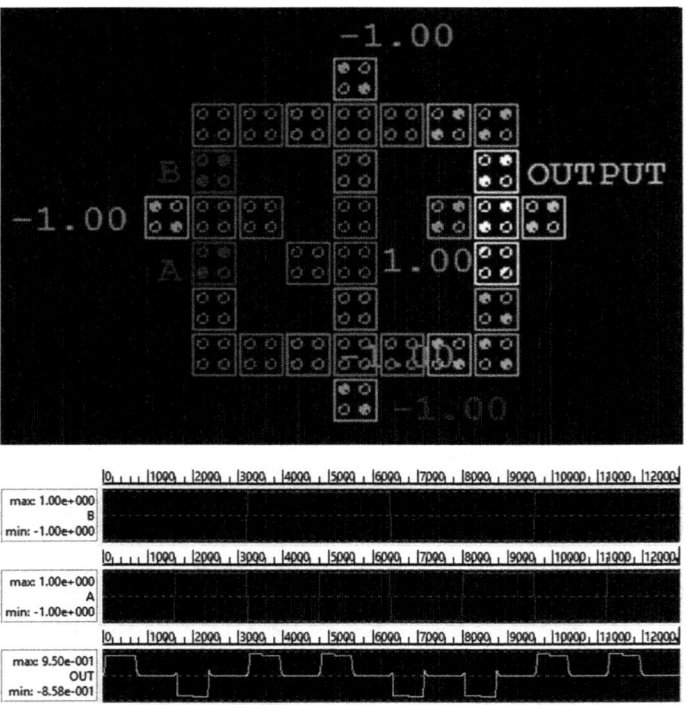

Fig. 8 Third QCA EXOR layout and result

The proposed exclusive-OR layout design consists of 17 cells, overall area of 0.03 μm², used cell area of 0.006 μm², latency of 0.25 clock delays, and power consumption of 0.049 meV. The other six proposed layout design results in comparison graph are demonstrated in Fig. 14; it has consists of 31 cells, 33 cells, 33 cells, 40 cells, 42 cells, and 44 cells, respectively; cell area in μm² are 0.03, 0.03, 0.04, 0.05, 0.05, and 0.05 μm² used.

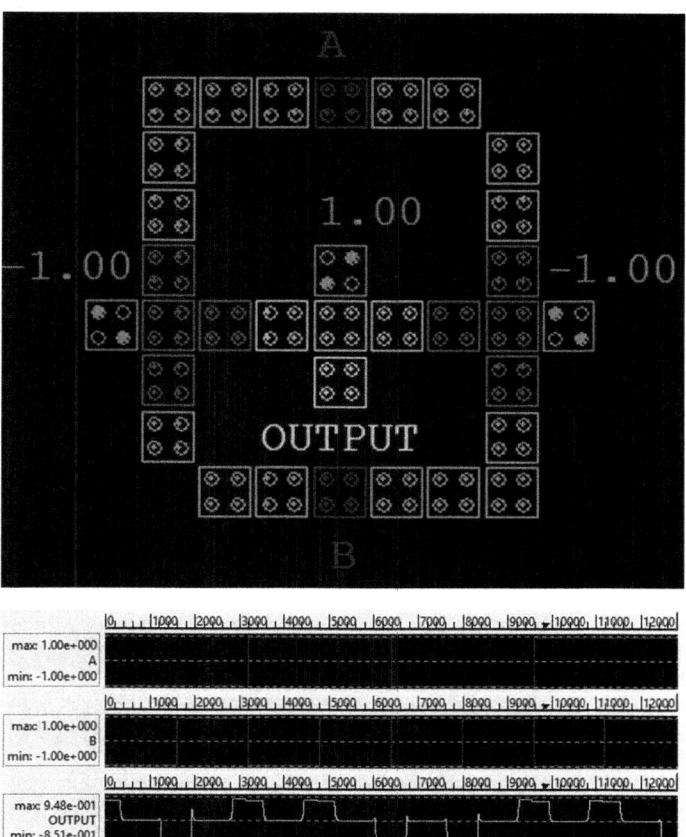

Fig. 9 Fourth QCA EXOR layout and result

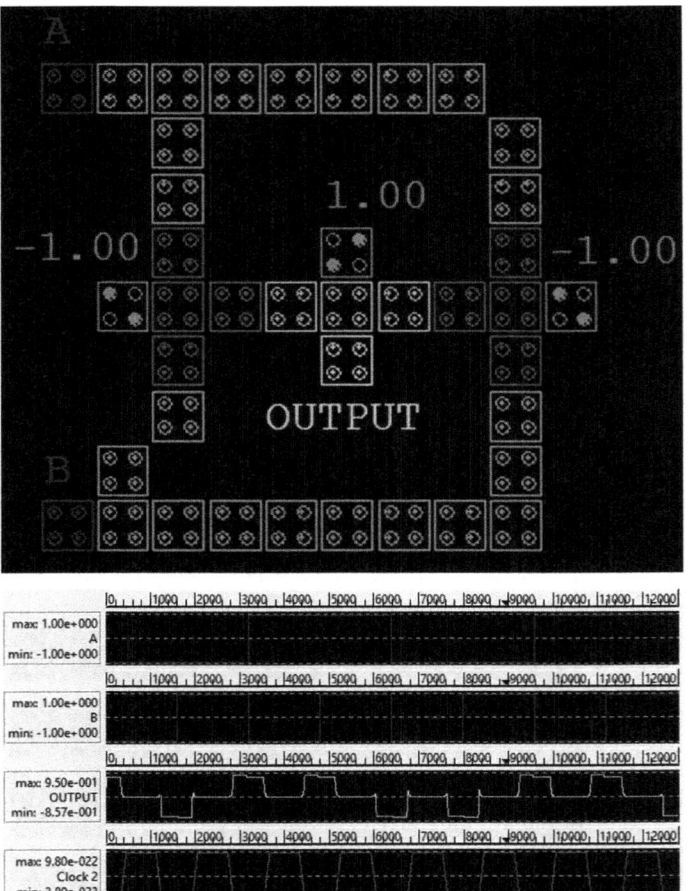

Fig. 10 Fifth QCA EXOR layout and result

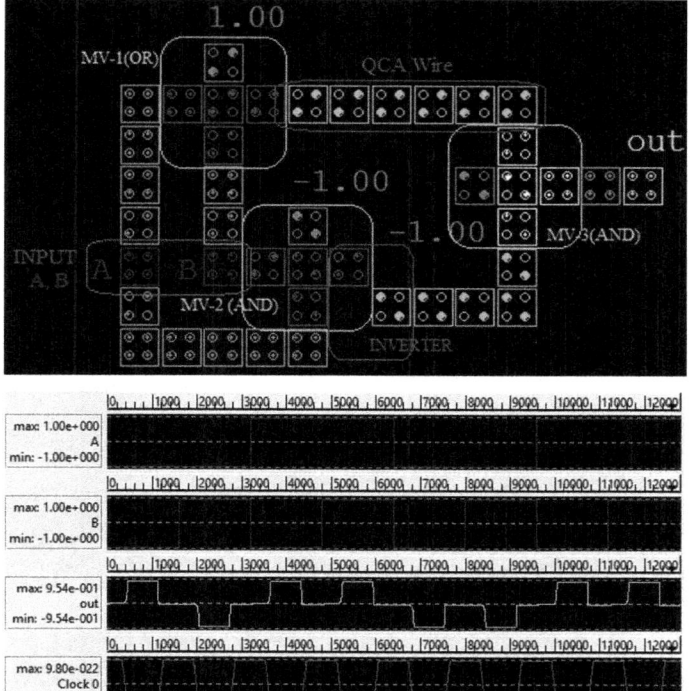

Fig. 11 Sixth QCA EXOR layout and result

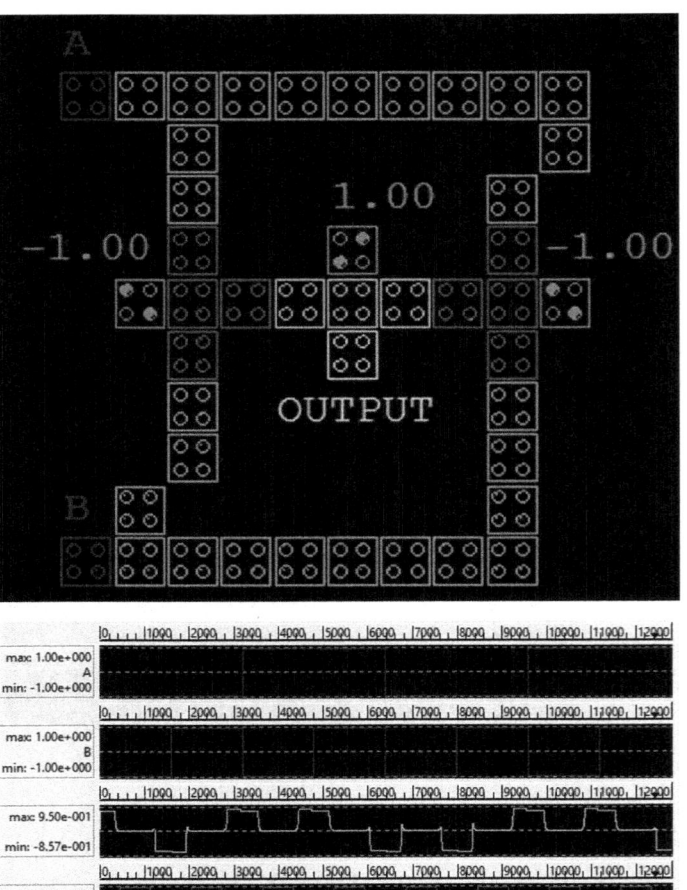

Fig. 12 Seventh QCA EXOR layout and result

Table 2 Simulation parameters [16, 25]

Simulation parameter	Value	
Cell width	18 nm	Layer properties
Cell height	18 nm	
Dots diameter	5 nm	
Default clock	Clock 0	
Time step	1.0e-16 s	Bistable engine setup approximation
Relaxation time	1.0e-15 s	
Simulation time	7.0e-11 s	
Low-clock	3.8e-023 J	
High-clock	9.8e-022 J	
Amplitude of clock	02.00	
Cell radius	65 nm	
Permittivity	12.90	
Separation of layer	11.50 nm	

Table 3 Comparative results of the previous designs EXOR gate

Reference number: Exclusive-OR layout design	QCA cells	Area (μm^2)	Delay
Santanu Santra (2014) et al. [1]	30	0.02	4
Firdous Ahmad (2014) et al. [29]	30	0.03	0.5
Gurmohan Singh (2016) et al. [31]	32	0.04	1
Firdous Ahmad (2014) et al. [28]	34	0.06	1
Manisha G. Waje (2014) et al. [12]	36	0.03	0.75
Mrinal Goswami (2014) et al. [27]	40	0.04	3
Young-Won You (2016) et al. [23]	44	0.05	4
Kianpour (2014) et al. [26]	45	0.05	1
Firdous Ahmad (2014) et al. [29]	46	0.06	1
Peer Zahoor Ahmad (2014) et al. [30]	49	0.07	2
Santanu Santra (2014) et al. [1]	51	0.05	5
Mustafa (2013) et al. [24]	55	0.09	2
Young-Won You (2016) et al. [23]	83	0.08	4
Manisha G. Waje (2014) et al. [12]	87	0.08	1

Table 4 Comparative results analysis of proposed exclusive-OR design

Proposed: Exclusive-OR layout	Cells count	Overall area (μm^2)	Delay clock	Power consumption (meV)
Figure 6	17	0.03	0.25	0.049
Figure 7	31	0.03	0.75	0.047
Figure 8	33	0.03	1.00	0.05
Figure 9	33	0.04	0.75	0.052
Figure 10	40	0.05	0.75	0.05
Figure 11	42	0.05	0.75	0.05
Figure 12	44	0.05	0.75	0.05

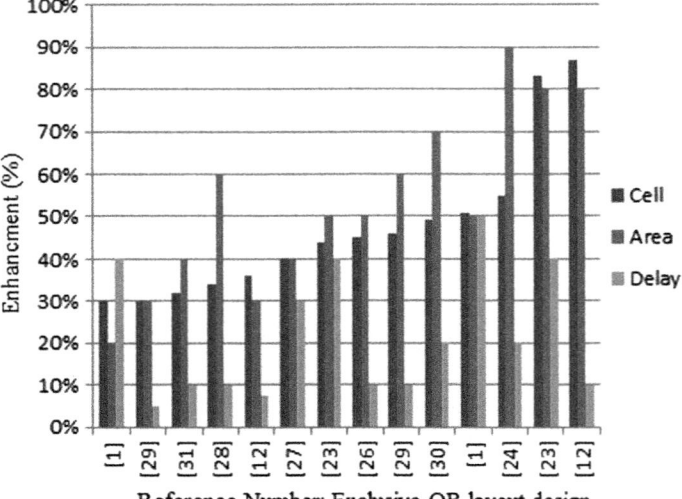

Fig. 13 Previous result analysis of QCA exclusive-OR layout

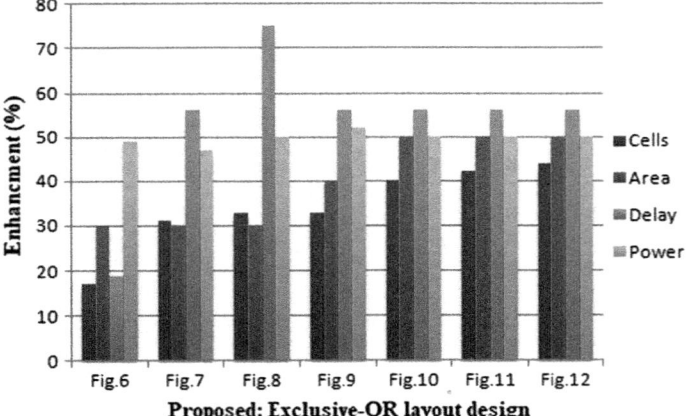

Fig. 14 Result analysis of QCA exclusive-OR layout design

5 Conclusion

QCA simulation layout design is challenged with unique features at digital system. In this paper, several layout functions were implemented and the results were verified by simulation QCA design tool—2.0.3. The novel exclusive-OR in QCA design with proposed comparison has 43% improvement in cell count, least area in μm^2, and required low power as compared to past ideas. This design is very useful for future ultralow power consumption in digital circuits and quantum computation.

References

1. Santanu Santra and Utpal Roy "Design and Implementation of Quantum Cellular Automata Based Novel Adder Circuits", *International Scholarly and Scientific Research & Innovation*, Vol. 8, No. 1, 2014.
2. Tamoghna Purkayastha, Debashis and Tanay Chattopadhyay "Universal shift register implementation using quantum dot cellular automata", *Faculty of Engineering, A in Shams Engineering Journal, Production and hosting by Elsevier*, 2016.
3. David Alan Grier, "The innovation curve [Moore's law in semiconductor industry]" *Reach Higher Computer*, vol. 39, pp. 8–10, Feb. 2006.
4. Mahalakshmi K S, Shiva Hajeri, Jayashree H V and Vinod Kumar Agrawal "Performance Estimation of Conventional and Reversible Logic Circuits using QCA Implementation Platform", *IEEE-ICCPCT*, 2016.
5. Dharmendra Kumar, Debasis Mitra "Design of a practical fault-tolerant adder in QCA" *ELSEVIER, Microelectronics Journal* Vol. 53, pp. 90–104, 2016,.
6. C.S. Lent, "Quantum cellular automata", *Nanotechnology*, Vol. 4, pp. 49–57, 1993.
7. Craig S. Lent, Beth Isaksen, and Marya Lieberman "Molecular Quantum-Dot Cellular Automata", *J Am Chem Soc Articles*, Vol. 125, No. 4, pp. 1056–1063, 2003.
8. Rumi Zhang, Konrad Walus, Wei Wang, Member, and Graham A. Jullien "A Method of Majority Logic Reduction for Quantum Cellular Automata" *IEEE Transactions on Nanotechnology*, Vol. 3, No. 4, December 2004.
9. Hema Sandhya Jagarlamudi, Mousumi Saha, and Pavan Kumar Jagarlamudi "Quantum Dot Cellular Automata Based Effective Design of Combinational and Sequential Logical Structures" *International Scholarly and Scientific Research & Innovation*, Vol. 5, No. 12, 2011.
10. Heumpil Cho and Earl E. Swartzlander "Adder Designs and Analyses for Quantum-Dot Cellular Automata" *IEEE Transactions on Nanotechnology*, Vol. 6, No. 3, May 2007.
11. Nilesh Patidar, Namit Gupta, Amita Khabia, Sumant Katiyal and K.K. Choudhary "A Novel 4-Bit Arithmetic Logic Unit Implementation In Quantum-Dot Cellular Automata" *International Journal of Nanotechnology and Application (IJNA)*, Vol. 3, No. 2, pp. 1–8, Jun 2013.
12. Manisha G. Waje and Dr. P.K. Dakhole "Design and Simulation of New XOR Gate and Code converters using Quantum Dot Cellular Automata with reduced number of wire crossings" *IEEE, International Conference on Circuit, Power and Computing Technologies [ICCPCT]*, 2014.
13. Kunal Das, Debashis De and Mallika De "Modified Ternary Karnaugh Map and Logic Synthesis in Ternary Quantum Dot Cellular Automata" *Taylor & Francis group, IETE Journal of Research*, 2016.

14. Ali Newaz Bahar, Sajjad Waheed and Nazir Hossain "A new approach of presenting reversible logic gate in nanoscale" *Springer Plus*, Vol. 4, No. 153, pp. 1–7, 2015.
15. Namit Gupta, Nilesh Patidar, Sumant Katiyal and K.. K. Choudhary "Design of Hybrid Adder-Subtractor (HAS) using Reversible Logic Gates in QCA" *International Journal of Computer Applications*, Vol. 53, No. 15, September 2012.
16. Jadav Chandra Das and Debashis De "Reversible Comparator Design Using Quantum Dot-Cellular Automata" *Taylor & Francis group, IETE Journal of Research*, 2015.
17. Angona Sarker, Ali Newaz Bahar, Provash Kumar Biswas, Monir Morshed "A Novel Presentation of Peres Gate (PG) in Quantum-Dot Cellular Automata (QCA)" *European Scientific Journal*, Vol. 10, No. 21, pp 101–106, July 2014.
18. Rashmi Pandey, Namit gupta, Nilesh Patidar "Design and Implementation of 16-bit Arithmetic Logic Unit using Quantum dot Cellular Automata (QCA) Technique" *Int. Journal of Engineering Research and Applications*, Vol. 4, No. 9, pp. 10–16, September 2014.
19. Rahul Singhal "Logic Realization Using Regular Structures in Quantum-Dot Cellular Automata (QCA)" Dissertations and theses, Portland State University, 2011.
20. J. Timler and C. S. Lent "Power gain and dissipation in quantum dot cellular automata", *Journal of Appl. Phys.* Vol. 91, pp. 823–830, 2016.
21. Shadi Sheikhfaa, Shaahin Angiz, Soheil Sarmadi and Samira Sayedsalehi "Designing efficient QCA logical circuits with power dissipation analysis" Elsevier, *Research Gate Microelectronics Journal*, pp. 1–19, 2016.
22. Mohammad Rafiq Beigh, Mohammad Mustafa and Firdous Ahmad "Performance Evaluation of Efficient XOR Structures in Quantum-Dot Cellular Automata (QCA)" *Scientific Research, Circuits and Systems*, Vol. 4, 2013.
23. Young-Won You and Jun-Cheol Jeon "Coplanar Wire Crossing Based QCA XOR Gate Using Nand-Nor-Inverter Gate" *Asia-pacific Proceedings of Applied Science and Engineering for Better Human Life*, Vol.6, pp. 41–44, 2016.
24. M Mustafa and M R Beigh "Design and implementation of quantum cellular automata based novel parity generator and checker circuits with minimum complexity and cell count" *Indian Journal of Pure & Applied Physics*, vol. 51, pp. 60–66, 2013.
25. QCA Designer Tool Version 2.0.3, Available at hyperlink: *http://* www.mina.ubc.ca/qcadesigner_downloads.
26. M. Kianpour1, R. Sabbaghi-Nadooshan2 "Novel Design of n-bit Controllable Inverter by Quantum-dot Cellular Automata" *Int. J. Nonsocial. Nanotechnology*, Vol. 10, No. 2, pp. 117–126, June 2014.
27. Mrinal Goswami, Brajendra Kumar, Harsh Tibrewal and Subhra Mazumdar "Efficient Realization of Digital Logic Circuit using QCA Multiplexer" *2nd International Conference on Business and Information Management (ICBIM)*, 2014.
28. Firdous Ahmad, M. Mustafa, Nisar Ahmad Wani and Feroz A Mir "A novel idea of pseudo-code generator in quantum-dot cellular automata (QCA)" *Int. J. Simul. Multisci. Des, Optim.* Vol. 5, 2014.
29. Firdous Ahmad1, Ghulam Mohiuddin Bhat and Peer Zahoor Ahmad "Novel Adder Circuits Based on Quantum-Dot Cellular Automata (QCA)" *Scientific Research, Circuits and Systems*, Vol. 5, pp. 142–152, 2014.
30. Peer Zahoor Ahmad, Firdous Ahmad and Hilal Ahmad Khan "A new F-shaped XOR gate and its implementations as novel adder circuits based Quantum-dot cellular Automata (QCA)" *IOSR Journal of Computer Engineering (IOSR-JCE)*, Vol. 16, No. 3, pp. 110–117. May-Jun 2014.
31. Gurmohan Singh, R. K. Sarin and Balwinder Raj "A novel robust Exclusive-OR function implementation in QCA nanotechnology with energy dissipation analysis" *Springer, J Comput Electron*, 2016.

Design Strategy for Smart Toll Gate Billing System

Varun Bohra, Deepak Prasad, Neha Nidhi, Anup Tiwari
and Vijay Nath

1 Introduction

The continuous advancements in technology have always proved to be a boon for human life. Almost every individual in today's world is well acquainted with the word called "Automation". Automation has not only reduced human efforts but has also provided for humans to lead their life in such a way which was never ever imagined before. Due to its numerous features, automation has been welcomed in almost every field. The paper presents the design of a smart toll and billing system which helps in reducing the delay at tollbooths which can be observed at almost all the major national and state highways which are the lifeline of villages, towns, and cities. The present manual tradition at tollbooths is not only a time taking process but also leads to an increase in congestion, thereby causing inconvenience to the riders and harming the environment due to the fuel consumption during the wait time. To add to this, human error is also making this manual tradition of collecting toll taxes an ineffective decision. Hence, considering all these flaws of manual operation at toll plazas, an approach to design smart toll and billing system has been made.

V. Bohra · D. Prasad · N. Nidhi · V. Nath
Department of Electronics and Communication Engineering, Birla Institute of Technology,
Mesra, Ranchi 835215, Jharkhand, India
e-mail: prasaddeepak007@gmail.com

A. Tiwari (✉)
Department of Electronics and Communication Engineering, Jharkhand Rai University,
Ranchi, India
e-mail: anuptiwari_ece@yahoo.com

© Springer Nature Singapore Pte Ltd. 2019
V. Nath and J. K. Mandal (eds.), *Proceeding of the Second International Conference
on Microelectronics, Computing & Communication Systems (MCCS 2017)*, Lecture Notes
in Electrical Engineering 476, https://doi.org/10.1007/978-981-10-8234-4_49

2 Aim of Project

Here, we are going to present a few points in order to demonstrate the motive behind choosing this topic and the application and need of such a mechanism in our life.

(a) *Avoiding the fuel loss,*
(b) *Relieving congestion,*
(c) *Saving time in collecting toll, and*
(d) *Avoiding financial loss.*

According to a survey conducted by the Maharashtra Government in September 2010, the estimated amount of annual toll tax collection is 1500 crore rupees a year. But, owing to the present scenario, only 1200 crore rupees is added to the state revenue through toll tax collection which means the state government is incurring a loss of 300 crore rupees every year due to faults in manual operation. With the present system operating at toll plazas on the highways, it takes few minutes to complete the toll collection process for one vehicle but with the proposed automatic procedure, it will take 40–42 s to complete the whole process thereby saving many seconds in each cycle.

As projected by the proposed design, a reduction in time for completion of the process can be observed so it can be figured out that comparatively little to no traffic congestion can be assumed. Also, this can be seen from a different perspective of fuel wastage and hence, depressurizing the environment.

3 Design Strategy

When the vehicle enters the toll zone, the camera pre-fixed on the barrier gates comes into action and captures the image of the license plate of the vehicle and other related information. This camera then sends the image to the processing system or a smart system which scans the image of the queued vehicle and generates the appropriate bill as well as displays the same at the display segment. The scanned image can also be stored for future use such as tracing the lost vehicle during journey. The vehicle riders can either pay the calculated billed amount by using a smart card or the amount may be automatically deducted from the user account. The whole procedure can be monitored by the local police stations in case of any safety issue. Once the transaction is successful, the barrier gate lifts up automatically providing an exit to the vehicles. The complete design strategy is shown in Fig. 1.

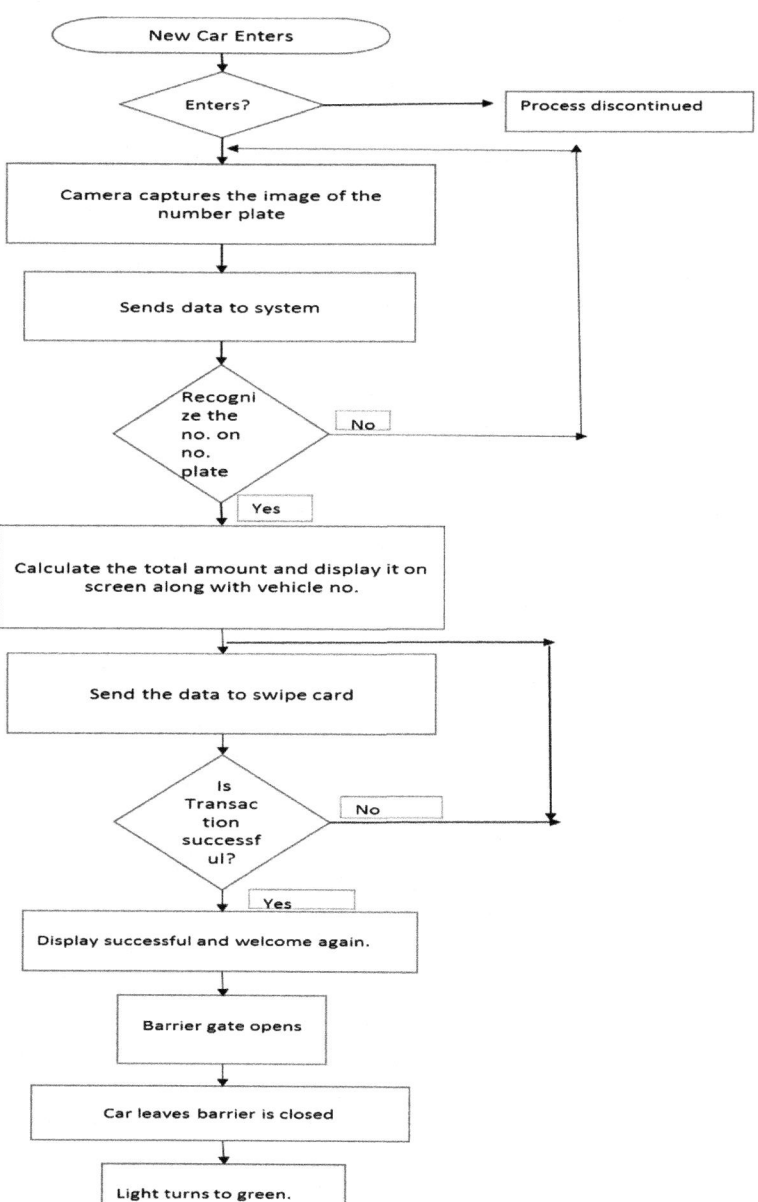

Fig. 1 Flowchart of design strategy

4 Proposed System

4.1 Input Image

The whole process starts with the image capturing process where the image of the number plate is captured by electronics device like digital camera or webcam which is placed at a distance of 5–6 m. Since the image is captured in JPEG color format, it will be converted into gray image scale for further process (Fig. 2).

4.2 Preprocessing

The reason behind conversion of color state is to enhance the quality of the image captured. Grayscale image consists of only two-color channel. The process of sharpening the edge of image, manipulation in contrast, brightness, reducing noise, and segmentation of image comes under image enhancement technique. Then, after the area of number plate will be cropped, it discards the unwanted area of the image captured. The cropping process helps in increasing the speed of image processing technique (Fig. 3).

4.3 Character Segmentation

Our proposed method consists of very simple mathematical operation along with localization of number plate and character recognition technology so that the exact number shall be extracted. It is basically geometrical approach for simple analysis of image. Structuring element plays an important role in mathematical morphology where it measures the shape and size of image which can be further manipulated for process (Fig. 4).

Fig. 2 Model vehicle

Fig. 3 Grayscale image

Fig. 4 Character segmentation

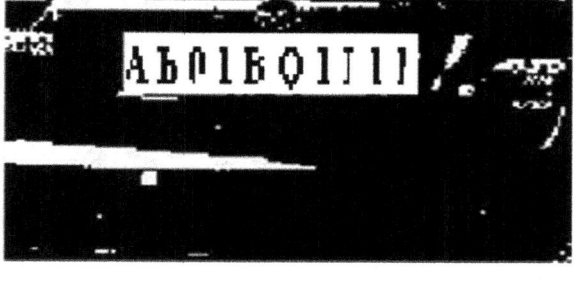

4.4 Character Recognition

Here, the image of the different texts is being allocated a specific character. All the different steps like character segmentation, extractions, etc. are included in Optical Character Recognition (OCR). The process of normalization takes place before the recognition algorithm which helps reducing data size of character. To make the match perfect, the input image should be equal sized with the database character. The next step is to compare the character with the database character and best similarity is extracted out. Correlation method is adopted to check the similarity where it measures the correlation coefficient between a number of a known image with the unknown image of same size (Fig. 5).

Fig. 5 Character reorganization

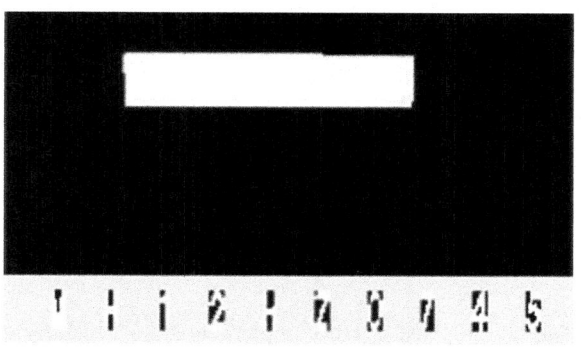

Fig. 6 Extracted result

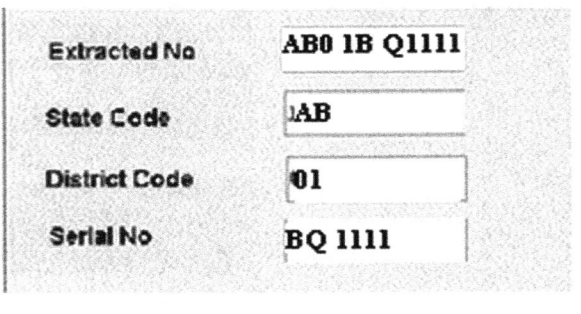

Extracted No	AB0 1B Q1111
State Code	1AB
District Code	01
Serial No	BQ 1111

5 Result

Finally, the vehicle number is extracted and it has been subdivided into state code, district code, and serial code. It shall be recorded for further documentation (Fig. 6).

6 Conclusion

The above system when implemented successfully can significantly contribute to improvement in traveling by addressing the delay factor caused by both recurring and non-recurring congestion at the toll plazas. The proposed system would collect toll tax from the vehicles driving on toll roads almost without making the vehicle stop at the tollbooths. This can be accomplished by installing a wireless mechanism in both the vehicles and the tollbooths to mutually exchange toll-related information using different data transfer techniques such as via cable, infrared, radio frequency, Bluetooth, etc. The system includes benefits for both toll authorities and users and facilitates them in terms of time saving and cost saving, improved security, increased capacity, and greater convenience for the riders. The system thus provides a broad overview of automated toll tax collection mechanism and its added advantages for both the toll authorities and the motorists.

Acknowledgements We are thankful to Department of Electronics and Communication, BIT, Mesra, India for their kind cooperation. We are also thankful to the Vice Chancellor Dr. M. K. Mishra and the head of department Dr. V. Gupta for their encouragement and support. We are also thankful to RESPOND ISRO Ahmadabad for funding the project.

References

1. T. Horie, T. Saida (2000), Hitachi Makes a significant contribution to the construction of secure and reliable ETC systems in Japan: 49, no. 3, April 2000.

2. S. Shankar, S. M. Mahmud., An intelligent architecture for intelligent area parking and toll collection, IEEE 2005
3. Mitretek Systems, Intelligent transportation system Benefit: June 2001
4. S. R. Reddy., Wireless Bluetooth Technology: IEEE communication Magazine, June 2002.
5. Assimakopoulos Nikitas A., Anastasis N. Riggas & Giorgos K. Kotsimpos, "A Systemic Approach for an Open Internet Billing System", 2003, http://www.afscet.asso.fr/resSystemica/ Crete02/Assimakopoulos,%20Riggas,%20Kotsibos.pdf
9. Ghosh Anup K., "E-commerce Security": Weak Links, Best Defenses, Wiley Computer Publishing, 1998.
10. NN Murthy, BM Mehtre, KPR Rao, GSR Ramam, PKB Harigopal, and KS Babu, "Technologies For E- Commerce: An Overview", CMC Center-R&D, CMC Limited Old Mumbai Highway, Gachibowli Hyderabad 500 019, Andhra Pradesh, 2000.
11. Sumanjeet Singh, "Emergence of Payment Systems In The Age Of Electronic Commerce: The State Of Art", Global Journal of International Business Research Vol, 2, No, 2, 2009.
12. Chan Henry, Raymond Lee, Tharam Dillon and Elizabeth Chang "E-commerce Fundamental and Applications", Baffins Lane, Chichester, West Sussex, PO19 lUD, England, 2001.
13. Crookes J, "Multiservice Billing System-a platform or the future", BT Technol J Vol 14 No 3 July 1996.
14. Mostafa hatem, "Billing System: Introduction", codeproject, 2005, http://www.codeproject. com/KB/architecture/billing.aspx

Novel Cell Search Method in Long-Term Evolution System

Smita A. Lonkar, K. T. V. Reddy and Sanket P. Singhania

1 Introduction

LTE is a trending communication scheme designed for high speed, minimum latency with optimization of radio access technology resource [1]. The goal of LTE is to support traffic based on packet switching with seamless mobility maintaining the quality of service. LTE System based on MIMO can support a maximum data rate of 300 Mb/s within a bandwidth of 20 MHz. To support a flexible bandwidth Single-Carrier Frequency Division Multiple Access (SC-FDMA) technology for uplink and Orthogonal Frequency-Division Multiple Access (OFDM) for downlink is used in LTE [2].

UE starts the procedure of cell search in order to access an LTE cell. The process for cell search at the physical layer involves synchronization phases to decide frequency and time factors necessary for downlink demodulation and transmission of uplink signals with the precise timings [3].

S. A. Lonkar
U.M.I.T., Mumbai, India
e-mail: smitalonkar@gmail.com

S. A. Lonkar · K. T. V. Reddy
IETE, New Delhi, India
e-mail: drktvreddy@gmail.com

S. A. Lonkar · S. P. Singhania (✉)
SSJCOE, Mumbai, India
e-mail: singhania12@gmail.com

K. T. V. Reddy
MGMCET, Navi Mumbai, India

© Springer Nature Singapore Pte Ltd. 2019
V. Nath and J. K. Mandal (eds.), *Proceeding of the Second International Conference on Microelectronics, Computing & Communication Systems (MCCS 2017)*, Lecture Notes in Electrical Engineering 476, https://doi.org/10.1007/978-981-10-8234-4_50

Table 1 Comparative study of various methods used by leading authors published in reputed journals

Paper and publication	Method
Reference [4]	A technique with the cyclic-based auto-correlation is employed to perform time domain-based time and coarse frequency offset synchronization. PSS and SSS signals are used to identify cell, sector, and integer frequency offset
Reference [5]	To identify CP type, lag-based auto-correlation is first implemented estimating symbol timing and frequency offset. To decide symbol and frame timing auto-correlation based on primary synchronization signal is employed
Reference [6]	The author has proposed two new detectors Almost-Half Complexity (AHC) and Central-Self-Correlation (CSC) detectors to lower the complexity with reliable PSS detection. The detectors are based on central-symmetric property
Reference [7]	To reduce computational load, the author proposes a correlation technique between the received sequence and the sum of three groups of locally created PSS sequences
Reference [8]	The authors have simulated the schemes for PSS and SSS recognition of cell identity and sub-frame index using linear correlation and circular correlation. The probability of PSS and SSS detection for each method is compared using Monte Carlo simulation method

PSS and SSS are transmitted in every cell to facilitate cell search procedure. When the UE detects the synchronization signals, it synchronizes time and frequency with the cell and obtains physical layer identity of the cell and mode of deployment (FDD or TDD).

The transmission of PSS and SSS is done in the central six resource blocks for allowing bandwidth independent mapping. Through this, UE synchronizes to the system without any previous information of assigned bandwidth.

The PSS uses Zadoff–Chu (ZC) sequences which are complex exponential sequences satisfying a Constant Amplitude Zero Autocorrelation (CAZAC) property.

The SSS is built on M-sequences which consist of maximum length sequences. These are generated by a shift register, which results in $2^n - 1$ length of the sequence.

Table 1 describes the comparative study of various methods used by the leading authors published in reputed journals.

For this work, we focus on the method for PSS and SSS generation and recognition of cell parameters using Cross-Correlation using MATLAB as a simulation environment.

2   LTE Downlink Frame Structure

The downlink procedure is designed on sub-frame level. Each sub-frame is of 1 ms length and contains two slots which are of 0.5 ms length. A radio frame is formed by 10 sub-frames hence of length 10 ms. From the 72 subcarriers, the middle 6 resource blocks are employed to broadcast PSS and SSS.

To ease the task of network planning, physical layer cell identities (NID) are used, this facilitates the neighboring cells to use different cell IDs. In LTE, 504 unique NID's are used. An NID is calculated from $N_{\text{ID}}^{(1)}$, which can of the value 0–167 and $N_{\text{ID}}^{(2)}$ in the range 0–2 [2].

The NID is calculated by the equation as follows:

$$N_{\text{ID}}^{\text{cell}} = 3N_{\text{ID}}^{(1)} + N_{\text{ID}}^{(2)} \tag{1}$$

3   Primary Synchronization Signal

- PSS generation

The Primary Synchronization Signals are generated using any of the three possible Zadoff–Chu sequences. PSS is generated from the following ZC sequence equation:

$$d_n(n) = \begin{cases} e^{-j\frac{\pi u n(n+1)}{63}}, & n=0,1,\ldots,30 \\ e^{-j\frac{\pi u(n+1)(n+2)}{63}}, & n=31,32,\ldots,61. \end{cases} \tag{2}$$

PSS in downlink frame appears twice, first in slot 0 and the then in slot 10. Both the slots carry the same sequence. PSS provides information about physical layer identity, which can have 3 different values (0–2) depending upon the root index of the Zadoff–Chu sequence (see Table 2).

The generated PSS with $u = 25$ is shown in Fig. 1.

Figure 1 is a plot representing real part on X-axis and imaginary part on Y-axis of the generated PSS with $u = 25$ in the frequency domain. It can be clearly seen in the plot that the sequence is a constant amplitude signal as it lies on a unity circle.

Table 2 Zadoff–Chu root index (u) corresponding to physical layer identity

N_{ID}^2	Root index (u)
0	25
1	29
2	34

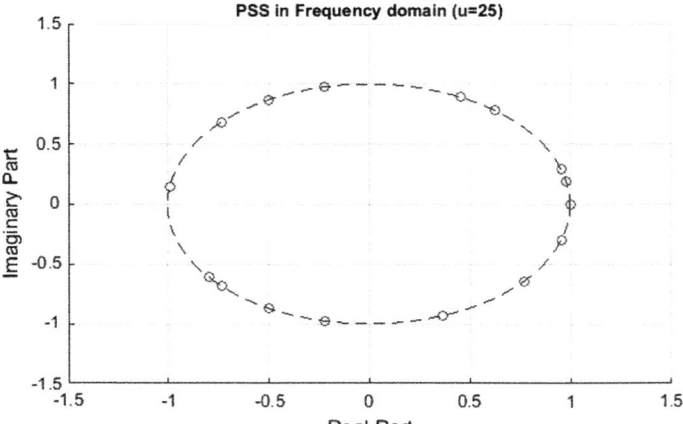

Fig. 1 PSS with root $u = 25$

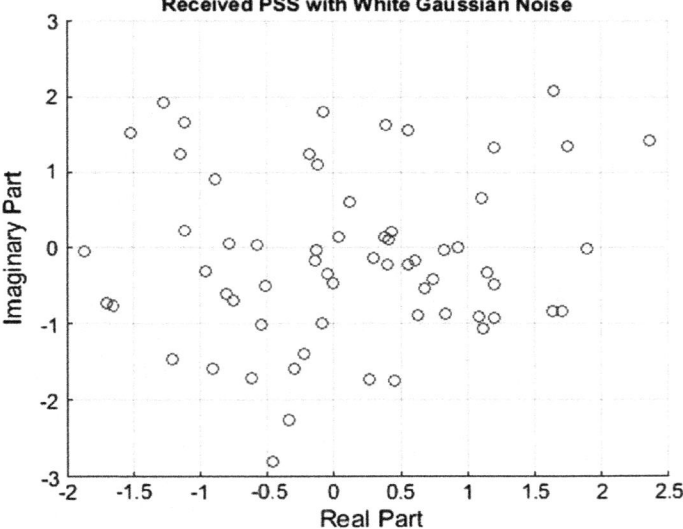

Fig. 2 Received PSS

- PSS detection

The generated PSS is corrupted by Additive White Gaussian Noise (AWGN) to stimulate the effect of channel distortions. Figure 2 shows the received PSS signal, distorted by the channel AWGN.

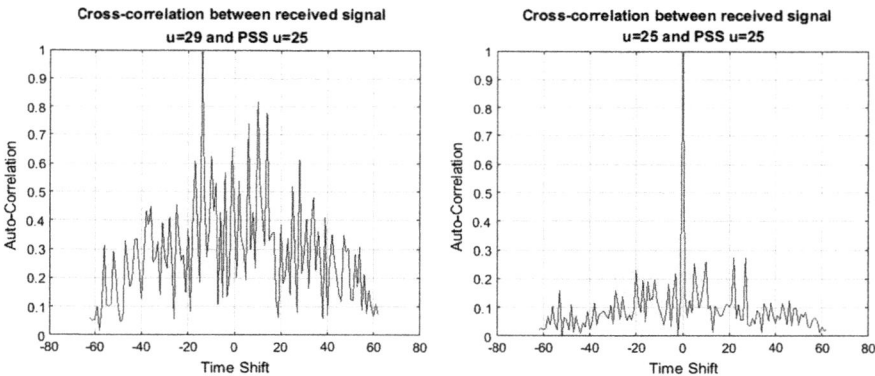

Fig. 3 Cross-correlation of the received signal ($u = 29$) with PSS ($u = 25$) and the received signal ($u = 25$) with PSS ($u = 25$)

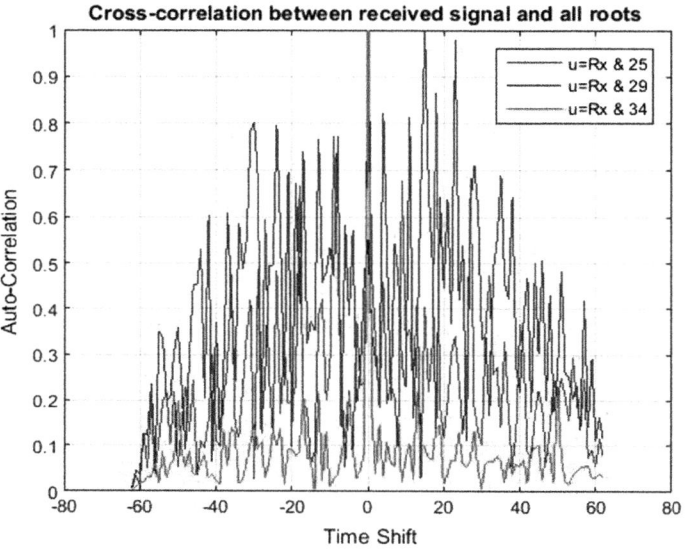

Fig. 4 Correlation between the received signal and all generated roots

The user equipment has a local copy of each of all the three roots of the PSS sequences. To detect which of the three possible N_{ID}^2 is received, it performs cross-correlation of all likely PSS sequences with the received signal.

As can be seen in Figs. 3 and 4, the cross-correlation of the received signal when root index 25 was transmitted shows an auto-correlation peak at zero lag, whereas no such peak at zero lag can be seen when correlated with other root index signal ($u = 29$).

By the above method, the UE can detect which of the PSS was transmitted and hence can know N_{ID}^2 as shown in Table 1.

4 Secondary Synchronization Signal

- SSS generation

The interwoven combination of two binary sequences of length 31 is used as SSS sequences. The PSS is used to scramble this SSS sequence using scrambling equations.

The even and odd part of the SSS sequence is generated from the below equation:

$$
d(2n) = \begin{cases} s_0^{(m_0)}(n)c_0(n) & \text{in subframe 0} \\ s_1^{(m_1)}(n)c_0(n) & \text{in subframe 5} \end{cases} \tag{3}
$$

$$
d(2n+1) = \begin{cases} s_1^{(m_1)}(n)c_1(n)z_1^{(m_0)}(n) & \text{in subframe 0} \\ s_0^{(m_0)}(n)c_1(n)z_1^{(m_1)}(n) & \text{in subframe 5} \end{cases} \tag{4}
$$

where $0 \le n \le 30$.

The simulated SSS with $N_{\mathrm{ID}}^2 = 0$ and $N_{\mathrm{ID}}^1 = 83$ for slot 0 and slot 10 is depicted in Fig. 5.

- SSS detection

Figure 6 shows the suggested SSS decoding scheme [9]. The received signal is de-interleaved into the odd part and the even part. Since N_{ID}^2 is known from the PSS, the scrambling code $c(n)$ is already identified to the user equipment and is

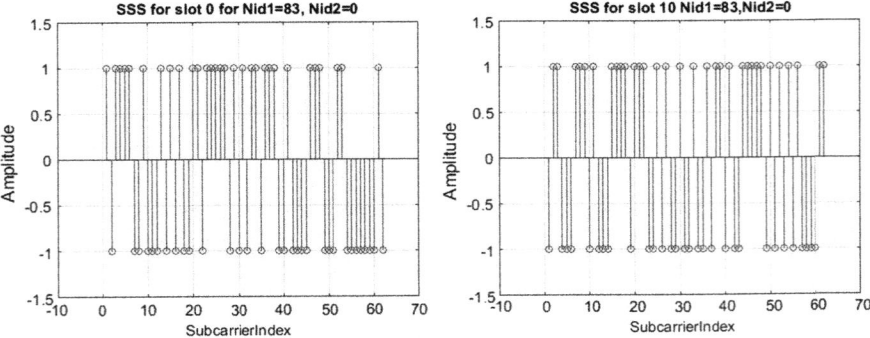

Fig. 5 SSS with $N_{\mathrm{ID}}^2 = 0$ and $N_{\mathrm{ID}}^1 = 83$

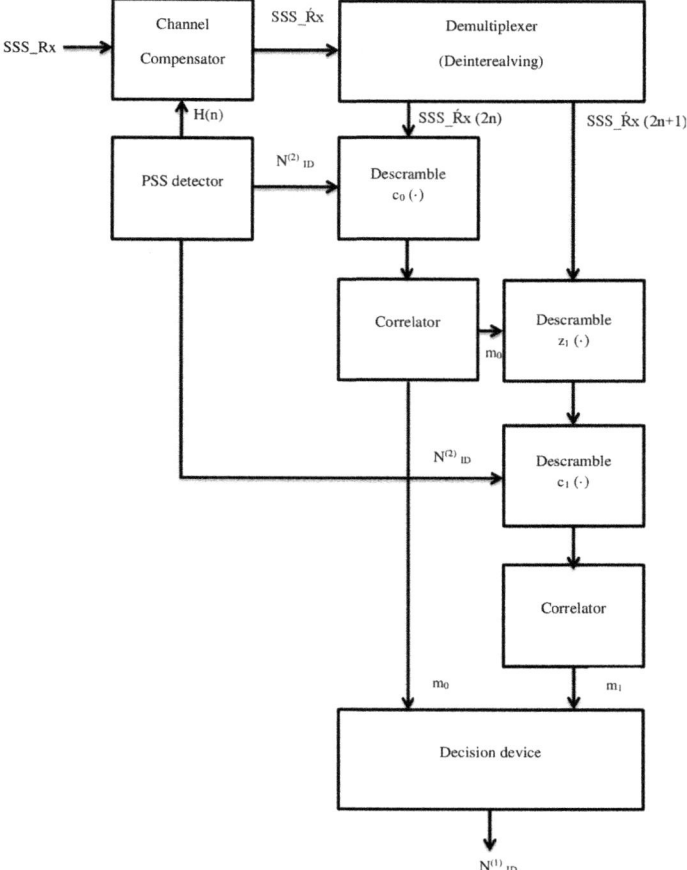

Fig. 6 SSS detection block diagram

extracted from the signal that is received. Now that the UE knows one value of m_0 or m_1, it can use it to descramble the sequence z.

Now, c_1 also can be descrambled using the known value of N_{ID}^2. s_1 can be correlated with the 30 sequences at the receiver to obtain the right match and this gives m_1. Now as both the values of m_0 and m_1 are known the N_{ID}^1 can be determined using the relation,

$$m_0 = m' \bmod 31 \qquad (5)$$

$$m_1 = \left(m_0 + \frac{m'}{31} + 1 \right) \bmod 31 \qquad (6)$$

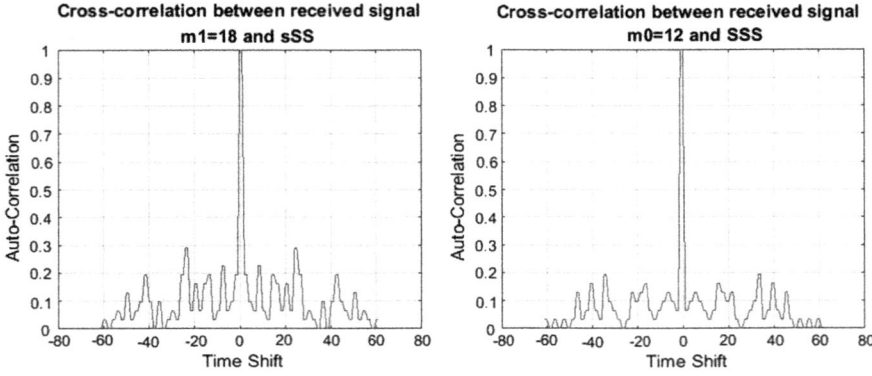

Fig. 7 m_0 and m_1 extraction using correlation

$$m' = N_{\text{ID}}^1 + \frac{q(q+1)}{2} \tag{7}$$

Figure 7 shows the correlation of the extracted m_0 and m_1 from the received signal using cross-correlation.

Cross-correlation is performed on the received signal with all likely values of m_0 and m_1. These values are used to derive the N_{ID}^1 using a decision device in Fig. 6, which lists a mapping matrix of m_0 and m_1 values for all possible values of N_{ID}^1 (0–167).

The cell search procedure is completed by applying the two parameters N_{ID}^1 and N_{ID}^2 to Eq. (1).

5 Conclusion

A method has been identified for generation and detection of PSS and SSS for LTE in this paper. The suggested method uses cross-correlation to identify the received signal with images of all likely expectations. The possibility of false detection is unlikely due to the simple nature of the technique as the matched sequences give peak value at zero lag. Using the proposed method, the UE can seamlessly synchronize with the eNodeB.

References

1. 3GPP TSG RAN TR 25.913 v7.3.0, Requirements for Evolved Universal Terrestrial Radio Access (UTRA) and Universal Terrestrial Radio Access Network (UTRAN).
2. F. Khan, LTE for 4G Mobile Broadband, Air Interface Technologies and Performance, Cambridge University Press, 2009.
3. S. Sesia, I. Toufik, M. Baker, LTE—The UMTS Long Term Evolution From Theory to Practice, 2nd Edition, John Wiley & Sons Ltd, 2011.
4. K. Manolakis, D. M. G. Estevez, V. Jungnickel, W. Xu, and C. Drewes, "A closed concept for synchronization and cell search in 3GPP LTE systems," in Proc. IEEE Wireless Commun. Netw. Conf., Budapest, Hungary, Apr. 2009, pp. 1–6.
5. W. Xu and K. Manolakis, "Robust synchronization for 3GPP LTE systems," in Proc. IEEE GLOBECOM, Miami, FL, 2010, pp. 1–5.
6. Zhongshan Zhang, Jian Liu and Keping Long, "Low-Complexity Cell Search With Fast PSS Identification in LTE", in IEEE Transactions On Vehicular Technology, Vol. 61, No. 4, May 2012, pp 1719-1728.
7. "A Novel Time Synchronization for 3GPP LTE Cell Search" Yongzhi Yu Qidan Zhu 2013 8th International Conference on Communications and Networking in China (CHINACOM) IEEE 978-1-4799-1406-7 [7].
8. Smita A. Lonkar, Amey C. Uchagaonkar, K T V Reddy "Comparative Analysis of Cell Search Schemes in Long Term Evolution Systems" International Conference on Communication, Information & Computing Technology (ICCICT), Jan. 16–17, Mumbai, India 978-1-4799-5522-0/15/ ©2015 IEEE.
9. Wassal Amr G., Elsherif Ahmed R., Efficient Implementation of Secondary Synchronization Symbol Detection in 3GPP LTE, IEEE International Symposium on Circuits and Systems (ISCAS) 2011, pp. 1680–1683, May 2011.

Various Feature Extraction and Classification Techniques

Dalvir Kaur and Sukesha Sharma

1 Introduction

Feature extraction plays an important role in image processing [1]. In the feature extraction procedure, different techniques can be used to extract the feature from the raw data. Feature extraction and selection are important for classifying the system [2]. Feature extraction is important for the system performance [3]. The goal of feature extraction is to select the most relevant features from the raw data [4]. Extracted features are known as feature vector [5]. Feature extraction is co-related with reduction of dimension. This can reduce the high-dimension data into low dimension [6]. High dimensionality of data increases the complexity. To remove this complexity, feature extraction is required [7]. PCA- and ICA-based feature selection and for classification process SVM is introduced in this paper [8]. Organization of paper is as follows. In Sect. 2 feature extraction technique is explained, and in Sect. 3 feature selection techniques are discussed. In Sect. 4 classification techniques are discussed. Finally, in Sect. 5 conclusions are drawn.

2 Feature Extraction

Feature extraction is a general term for strategies for determining values expected to be useful from measured information. The arrangement of separated components is known as a feature vector. Feature extraction is identified with dimensionality

D. Kaur · S. Sharma (✉)
UIET, Panjab University, Chandigarh, India
e-mail: sukeshauiet@gmail.com

D. Kaur
e-mail: nancysidhu007@gmail.com

© Springer Nature Singapore Pte Ltd. 2019
V. Nath and J. K. Mandal (eds.), *Proceeding of the Second International Conference on Microelectronics, Computing & Communication Systems (MCCS 2017)*, Lecture Notes in Electrical Engineering 476, https://doi.org/10.1007/978-981-10-8234-4_51

lessening [6]. Feature extraction is achieved over uniform signals. Different feature extraction techniques are used for performing the feature extraction process such as Discrete Cosine Transform (DCT), Discrete Fourier Transform (DFT), and Wavelet Transform (WT) which are explained in this paper [3].

2.1 Discrete Cosine Transform (DCT)

DCT is based on linear transformation related to Fourier transform. DCT is the same as discrete Fourier transform (DFT). Real numbers are used by DCT [9]. Combination of linear cosine function expresses a finite-length discrete time sequence $x[n]$. DCT is mainly used in image processing for lossy data compression, it has solid information compaction property [3].

The mathematical equation of DCT is in the following equation:

$$X_k = \sum_{n=0}^{N-1} x_n \cos\left[\frac{\pi}{N}\left(n+\frac{1}{2}\right)k\right] \tag{1}$$

where $k = 0, 1, \ldots N - 1$

This function is used because this can collect the useful information in lower frequency components.

2.2 Discrete Fourier Transform (DFT)

DFT is a technique to acquire information in frequency domain of discrete time signals. This is generally utilized in problems of ultrasound testing. The equation for DFT in finite-length discrete time signal $x[n]$ can be expressed as [3]

$$X\left(e^{\frac{j2\pi k}{N}}\right) = \sum_{n=0}^{N-1} X[n]e^{\frac{-j2\pi kn}{N}} \quad 0 \le k \le N - 1 \tag{2}$$

2.3 Wavelet Transform (WT)

Wavelet transform can represent the signal in a time domain [3]. This can divide the input signal into a set of values that can explain the signal-frequency component at given period of time. Wavelet transform is also known as 'mother wavelet'. Wavelet transform is classified in two categories:—Continuous Wavelet Transform (CWT) and Discrete Wavelet Transform (DWT) [10–12].

(1) *CWT (Continuous Wavelet Transform)*

CWT of a time domain signal $f(t)$ is calculated using the following equation

$$F(x, y) = \int\limits_{-\infty}^{+\infty} f(t)\varphi_{x,y}(t)\mathrm{d}t \qquad (3)$$

where x and y are scale and shift parameters, respectively $(\varphi(t))$ is a mother wavelet function. Mother wavelet function must have to satisfy the following equation

$$\int\limits_{-\infty}^{\infty} \varphi(t)\mathrm{d}t = 0 \qquad (4)$$

The parameter 'y' moves the wavelet so that neighborhood data around time $t = y$ is held in the transformed function. The window size is controlled by parameter 'x'. The statistical analysis of signal by using CWT gives an abundance of data [13, 14].

(2) *DWT (Discrete Wavelet Transform)*

Discrete Wavelet Transform is any wavelet transform for which the wavelets are discretely sampled. In DWT signals are preceded by low- and high-pass filters in various stages. In the first stage, the signal is divided into estimation coefficients (using low-pass filter) and into constituent coefficients (using high-pass filters). In the next level, the decomposition may be carried out by low-pass estimation coefficient got during the past level. This procedure is replicated until final stage may be attained [13, 14].

3 Feature Selection

The procedure of choosing features for particular recognition issue is essential to final framework execution. It may be referred to that the size of the features vector needs an immediate effect on the execution of the performance of classification system. Though the utilized dimension size may be too small, relevant data might be neglected reducing the system performance. Our objective is to select the most relevant features from the complete set of features. In this section, different techniques used for feature selection is discussed in order to attain ideal information compaction. PCA and ICA are discussed to select relevant features [10–12, 15].

3.1 Principal Component Analysis (PCA)

PCA is a well-known technique used for feature selection. PCA is a scientific methodology used for orthogonal transformation to change a set of perceptions of potentially associated variables under a set about uncorrelated variables known as principal components. Principal components are less than the original components. The principal component contains greater information by preserving the maximum deviation of features and by minimizing error. Each principal component is named as eigenvector. By computing eigenvalues of the input covariance matrix, PCA continuously transforms large structural input values into small structural values, whose values are uncorrelated. Principal components are arranged by their corresponding eigenvalues. The first principal component contains the greatest significant value and the second principal component contains the consequent significant value. Eigenvectors are used as features in this work. These are determined by applying following steps [5, 6] (Fig. 1).

Fig. 1 Practical module in a typical signal processing system

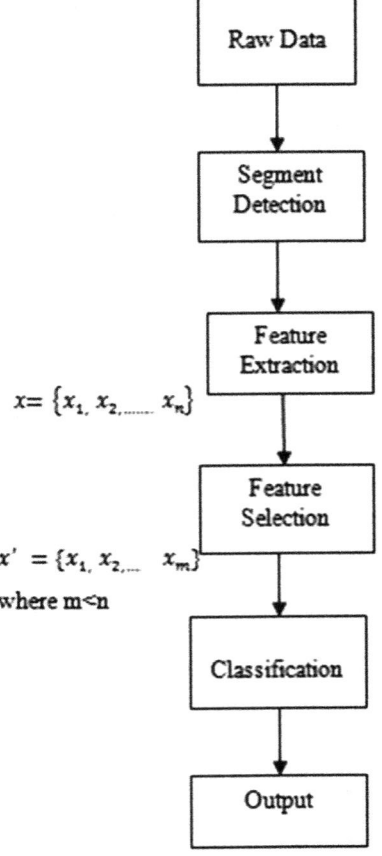

(a) First, the mean of the information will be subtracted from each of the information extends to process an information set with zero mean. After this, covariance matrix may be calculated.

For M observations and N variables, the average is calculated as

$$\bar{X} = \frac{1}{M} \sum_{n=1}^{M} X_n \tag{5}$$

X_n is the dimensional column vector of nth observations.

(b) Covariance matrix is calculated by

$$C = \frac{1}{M} \sum_{n=1}^{M} \left[(X_n - \bar{X})(X_n - \bar{X})^T \right] \tag{6}$$

Since the information is 'n' dimensional, the covariance matrix will have a chance to be $N * N$ [6].

(c) Extraction of eigenvectors from covariance matrix.

If V_i is the eigenvector of $A^T \times A$ the Eigenvalues are

$$A^T A V_i = U_i V_i \tag{7}$$

Then the eigenvector for C can be calculated by

$$U_i = A v_i \quad \text{where } C = A \times A^T \tag{8}$$

(d) Once eigenvectors are discovered from covariance matrix, the consequent step is to arrange them by eigenvalue from high to low. Along with this, it may disregard the components which are lesser significant and the last information situated will have less dimension than the original dataset [3].

3.2   Independent Component Analysis (ICA)

ICA is originally formed to visually impaired blind source whose objective is with recuperate commonly unnamed source signal in distinction to their linear mixtures. It is a computational technique to divide a multivariate signal into additive subcomponents. Subcomponents calculate the free parts (ICs) by the amplification from claiming non-Gaussianity alternately or minimization of the common majority of the data between signals. It could provide finer signal representational toward looking for measurable independence.

Over ICA, consider that arbitrary x variable is measured with the help of sensor system as follows:

$$x_t = Ay_t \tag{9}$$

where y_t are irregular variables and commonly statistically separate with mean value of zero. The y_t are estimated by

$$y_t = Wx_t \tag{10}$$

ICA transforms the entered signals into independent components subspace. ICA components are more partially local than PCA. ICA local features can offer better information representational. In addition, contrasting for PCA calculation, this technique does not determine the order of Independent Components (ICs) [16, 17].

4 Classification

In this section, a classification technique, Support Vector Machine (SVM) is discussed [4]. The important step before classification is feature extraction and feature selection. PCA and ICA performances are compared in SVM. Classification is proposed for detecting defects. SVM is a widely known technique to identify and classify defects [8]. SVM is progressively constantly utilized because of its better performance and good experimental execution [18]. The principal of this algorithm is it is dependent upon the Structural Risk Minimization (SRM) [19]. SVM has been effectively used in various practical problems correspondent to regression and classification problems [20].

The classification method based on SVM is indicated in Fig. 2. Before preceding experiments, it is not realized that which PCs and ICs are best for classification. In this task, first two PCs and all of the ICs have utilized likewise information of the SVM. First, the PCs alternately ICs need to be scaled in the range of $[-1, +1]$ to get the training data. The primary advantage of scaling is it should keep away from attributes with a greater range. With the help of distinct training label, defect types

Fig. 2 Classification using SVM

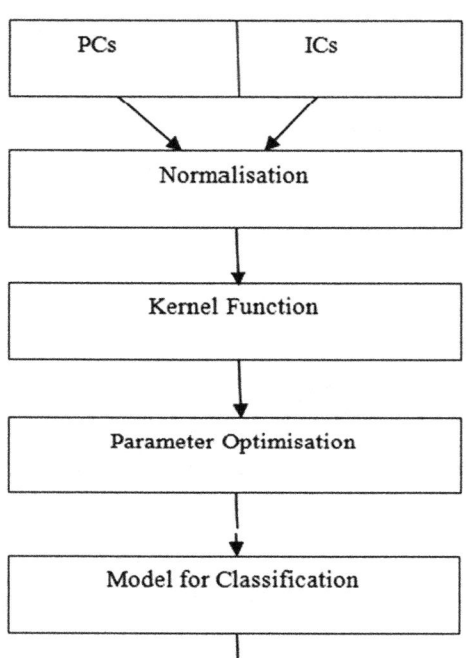

are set up. Different kernel functions are used for different classification problem to achieve the best classification result [16, 17].

S. No.	Author	Publication	Remarks
1	F. C. Cruz et al.	Elsevier, 2017	• Feature extraction and selection are done • For feature extraction Fourier, wavelet, cosine transforms are used • PCA technique is used for feature selection [3]
2	Gianni D' Angelo et al.	Elsevier, 2016	• In this paper, PCA and LDA techniques are used for to reduce the dimensionality of the data [5]
3	Bin Gaoa et al.	Elsevier, 2016	• A quantitative analysis strategy is proposed for validating detection performance of various thermal feature extraction techniques based on Eddy Current-Stimulated Thermography [25]
4	Chi Qin Lai et al.	Springer, 2016	• Feature extraction is done using histogram bins • PCA technique is used for feature extraction and SVM is used as a classifier in this [26]

(continued)

(continued)

S. No.	Author	Publication	Remarks
5	Rajdeep Chatterjee et al.	IEEE, 2016	This paper proposed the technique of feature extraction on EEG signals • SVM and MLP are used for classification [27]
6	Wattana Punlumjeak	IEEE, 2015	• Proposed the comparison between SVM, Genetic algorithm, and information gain. • Decision tree, neural network, naïve Bayes, and k-nearest neighbor are used as classifier [29]
7	M. P. Paulraj	Elsevier, 2013	• Discusses about the time domain-based feature extraction algorithms developed to extract features from the vibration signals [7]

After classification is done, this can be found that PCA-based method is better than the ICA-based method. Because, ICA-based method is more time consuming than the PCA. PCA-SVM gives better and accurate result for signals in the field of feature selection and classification [16, 17].

Name	Applications
DCT	DCT has the capacity to pack the majority of the data in a small no. of coefficients [21]
DFT	DFT is generally used for ultrasound testing problems in feature extraction [3]
DWT	DWT has the ability to differentiate the fine points in a signal. This can decompose an input signal into component wavelets [22]
PC A	PCA is an absolute method for dimensionality minimization and multivariate analysis [23]
ICA	ICA is a signal processing method used for dividing statistically autonomous signals. It is generally used in signal processing, wireless communication, machinery fault diagnosis. Vibration analysis [24]

5 Conclusion

In this paper, review of signal transformation and features is done. These transformations are studied in both time and frequency domain because transformations show a different type of representation in both the manner. In the same context, feature extraction methods like DWT and FFT show different behaviors of signals. These different analyses are helpful for classification of signal with minimum error. After classification, it can be found that PCA-based method is better than the ICA because, ICA method is more time consuming than PCA. PCA-SVM gives better and accurate result for signals in the field of feature selection and classification.

References

1. Asutosh Kar and Leena Das, "A Technical Review on Statistical Feature Extraction of ECG signal," *IJCA Special Issue on 2nd National Conference—Computing, Communication and Sensor Network (CCSN)* (2):35– 40, 2011.
2. Übeyli, Elif Derya, "Statistics over features: EEG signals analysis." *Computers in Biology and Medicine* 39.8: 733–741, 2009.
3. Cruz, F. C, et al. " Efficient feature selection for neural network based detection of flaws in steel welded joints using ultrasound testing" Ultrasonics 73: 1–8, 2017.
4. Khodayari-Rostamabad, Ahmad, et al. "Machine learning techniques for the analysis of magnetic flux leakage images in pipeline inspection. "*IEEE Transactions on magnetics*" 45.8: 3073–3084, 2009.
5. D' Angelo, Gianni, and Salvatore Rampone. "Feature extraction and soft computing methods for aerospace structure defect classification." *Measurement* 85: 192–209, 2016.
6. Eremenko, V. S., A. V. Pereidenko, and E. F. Suslov. "Neural Network Based System for Nondestructive Testing of Composite Materials Using Low-Frequency Acoustic Methods." *Universal Journal of Engineering Science* 1.3: 95–109,2013.
7. Paulraj, Murugesa Pandiyan, et al. "Structural steel plate damage detection using non destructive testing, frame energy based statistical features and artificial neural networks." *Procedia Engineering* 53: 376–386, 2013.
8. He, Yunze, et al. "Support vector machine and optimised feature extraction in integrated eddy current instrument." *Measurement* 46.1: 764–774, 2013.
9. Ehteram, Saeedreza, et al. "A New Pattern Recognition Technique in Non Destructive Testing by the Use of Linear Discriminate Analysis." *Modern Applied Science* 3.5: 118, 2009.
10. Simas Filho, Eduardo F., et al. "Decision support system for ultrasound inspection of fiber metal laminates using statistical signal processing and neural networks." *Ultrasonics* 53.6: 1104–1111, 2013.
11. Cerrillo, C., et al. "New contributions to granite characterization by ultrasonic testing." *Ultrasonics* 54.1: 156–167, 2014.
12. Diniz, Paulo SR, Eduardo AB Da Silva, and Sergio L. Netto. *Digital signal processing: system analysis and design.* Cambridge University Press, 2010.
13. Rioul, Olivier, and Martin Vetterli. "Wavelets and signal processing." *IEEE signal processing magazine* 8.LCAV-ARTICLE-1991–005: 14-38, 1991.
14. Liang, Wei, and Pei-wen Que. "Optimal scale wavelet transform for the identification of weak ultrasonic signals." *Measurement* 42.1: 164–169, 2009.
15. Simas Filho, Eduardo F., and José M. Seixas. "Unsupervised statistical learning applied to experimental high-energy physics and related areas." *International Journal of Modern Physics C* 27.05: 1630002, 2016.
16. G. Yang, G.Y. Tian, P.W. Que, T.L. Chen, Independent component analysis-based feature extraction technique for defect classification applied for pulsed eddy current NDE, Research in Nondestructive Evaluation 20 230–245, 2009.
17. M. Cacciola, G. Ripepi, G. Yang, G.Y. Tian, F.C. Morabito, ICA based Algorithms for Flaw Classification in Pulsed Eddy Current data: A Study 226: 162–171, 2011.
18. Hassan, Muhsin, et al. "Pipeline defect classification by using non-destructive testing and improved support vector machine classification." *Int J Eng Innovative Technol (IJEIT)* 2.7: 85–93, 2013.
19. Khelil, Mohamed, et al. "Classification of Defects by the SVM Method and the Principal Component Analysis (PCA)." *World Academy of Science, Engineering and Technology, International Journal of Electrical, Computer, Energetic, Electronic and Communication Engineering* 1.9: 1446–1451, 2007.
20. Liu, Baoling, et al. "An improved PSO-SVM model for online recognition defects in eddy current testing." *Nondestructive Testing and Evaluation* 28.4: 367–385, 2013.

21. Kaur, Amanjot. Et al. "Comparison of DCT and DWT of image compression techniques." *ISSN: 2278-067X, Volume 1, Issue 4 (June), PP. 49–52, 2012.*
22. Sifuzzaman, M., M. R. Islam, and M. Z. Ali. "Application of wavelet transform and its advantages compared to Fourier transform." (2009).
23. Labib, Khaled, and V. Rao Vemuri. "An application of principal component analysis to the detection and visualization of computer network attacks." *Annales des télécommunications.* Vol. 61. No. 1–2. Springer-Verlag, 2006.
24. Uddin, Zahoor, et al. "Applications of Independent Component Analysis in Wireless Communication Systems." *Wireless Personal Communications* 83.4: 2711–2737, 2015.
25. Bin Gao, et al. "Quantitative validation of eddy current stimulated thermal features on surface crack" NDT& E International vol. 85 P No. 1–12, 2017.
26. Chi Qin Lai, "Efficiency improvement in the extraction of histogram oriented gradient feature for human detection using selective histogram bins and PCA" Vol. 398 no. 267–275. Springer 2016.
27. Chatterjee, Rajdeep, Tathagata Bandyopadhyay, and Debarshi Kumar Sanyal. "Effects of Wavelets on Quality of Features in Motor-Imagery EEG Signal Classification." 2016.
28. Sophian, Ali, et al. "A feature extraction technique based on principal component analysis for pulsed Eddy current NDT." *NDT & E International* 36.1: 37–41, 2003.

Path Tracking Method of ALV Model Based on ADRC Strategy and Differential Flatness Theory

Abhinav Kumar and Sushma Kamlu

1 Introduction

In the past, path tracking of the autonomous land vehicle means to keep the steering of a vehicle in the centre of its lane using state feedback control [1]. Later many path tracking methodologies are introduced like fuzzy controller with a parameter self-tuning modules was proposed to adjust the quantization factor and scale factor according to current velocity and heading declination [2, 3]. The latest work is done only using ADRC on different ALV models. The main aim of the path tracking problem is to control the unpredictable yaw motion of the vehicle since the system is robust and underactuated which may cause the vehicle spinning and deviating from the main path. Lateral tire force sensors to estimate vehicle sideslip angle to improve the yaw motion stability has been implemented in [4].

Differential flatness theory is introduced in this paper in which the flatness of the model is proved and a flat output is found. Fliess et al. first proposed the concept of differential flatness in [5]. With the help of this property, the complex integration process is no longer necessary since it calculates the state, input and output in terms of flat output and its derivatives

An ADRC is a controller first introduced by Han [6] which maintains the basic work of PID controller and it is very superior in estimating and rejecting the internal and external disturbances. The important aspect of this controller is that it requires very less knowledge of the system, in fact, the order and output of the system. In ADRC the concepts of the extended state observer (ESO) [7], tracking differentiator and nonlinear feedback controller are introduced that are utilized as to find out the tracked path, desired state and the optimized nonlinear controlled input after rejecting the disturbances. With the properties of less dependency on the system and strong robustness, the ideas of ADRC have been used in different areas.

A. Kumar (✉) · S. Kamlu
Department of EEE, Birla Institute of Technology, Mesra, Ranchi, India
e-mail: abhinav.ee14@gmail.com

© Springer Nature Singapore Pte Ltd. 2019
V. Nath and J. K. Mandal (eds.), *Proceeding of the Second International Conference on Microelectronics, Computing & Communication Systems (MCCS 2017)*, Lecture Notes in Electrical Engineering 476, https://doi.org/10.1007/978-981-10-8234-4_52

The paper is structured as follows. In Sect. 2, a linearized "bicycle" model of the ALV is introduced and also writes it in the state-space form. In Sect. 3, the flat output and its derivatives in the form of states and inputs of the linearized ALV model are found. In Sect. 4, ADRC strategy is applied on the flat output and the parameters tuning is applied for proper tracking. In Sect. 5, the results have been shown and comparison with standard trackers has been presented and finally, the concluding remarks have been summarized in Sect. 6.

2 Bicycle Model of ALV

ALV is an autonomous land vehicle which is highly nonlinear, robust and under-actuated system. The path followed by ALV is very tedious to control.

Make the following assumptions.

(1) Ignore the longitudinal force generated by the tires.
(2) Longitudinal velocity is high enough so that slip angles cannot be ignored.
(3) The slip angles are so small that the lateral tire forces are proportional to the slip angles.
(4) The vehicle uses only front wheels to steer.

The model can be written as follows:

$$\begin{aligned}
\dot{y} &= v_x \sin\theta + v_y \cos\theta \\
\dot{\theta} &= r \\
\dot{v}_y &= \frac{1}{m}(F_{yf} + F_{yr}) - v_x r \\
\dot{r} &= \frac{1}{I_z}(F_{yf}l_f - F_{yr}l_r)
\end{aligned} \tag{1}$$

where

v_x	longitudinal velocity in XOY;
v_y	lateral velocity in XOY;
Y	lateral displacement in XOY;
$\dot{y}$	lateral velocity in XOY;
θ	yaw angle;
$\dot{\theta} = r$	vehicle yaw rate;
F_{yf}	lateral tire force of front wheel;
F_{yr}	lateral tire force of rear wheel;
S_f	front tire cornering stiffness;
S_r	rear tire cornering stiffness;
l_f	distance between the center of gravity and the front axle;
l_r	distance between the center of gravity and the rear axle;
m	vehicle mass;
I_z	yaw moment of inertia.

With above-mentioned assumptions, the following relationship has been obtained between tire force and tire slip angle [4]:

$$\left.\begin{array}{l} F_{yf} = -S_f a_f \\ F_{yr} = -S_r a_r \end{array}\right\} \tag{2}$$

where a_f and a_r have been considered as the slip angle at the front and rear tires respectively. The physical explanation of slip angle is given as follows:

$$\left.\begin{array}{l} a_f = \frac{v_y + r l_f}{v_x} - \delta_f \\ a_r = \frac{v_y - r l_r}{v_x} \end{array}\right\} \tag{3}$$

where δ_f is front wheel steering angle.

Substituting (2) and (3) into (1), the vehicle lateral dynamic model is obtained as

$$\left.\begin{array}{l} \dot{y} = vx \sin(\theta) + vy \cos(\theta) \\ \dot{\theta} = r \\ \dot{vy} = -\frac{Sr + Sf}{mvx} vy + \left(\frac{Srlr - Sflf}{mvx} - vx\right) r + \frac{Sf}{m} \delta f \\ \dot{r} = \frac{-lfSf + lrSr}{Izvx} vy - \frac{lf^2 Sf + lr^2 Sr}{Izvx} r + \frac{lfSf}{Iz} \delta f \end{array}\right\} \tag{4}$$

where

$$\alpha_1 = -\frac{S_r + S_f}{m}, \alpha_2 = \frac{S_r l_r - S_f l_f}{m}$$

$$\alpha_3 = \frac{S_r l_r - S_f l_f}{I_z}, \alpha_4 = -\frac{S_f l_f^2 + S_r l_r^2}{I_z}$$

$$\beta_1 = \frac{S_f}{m}, \beta_2 = \frac{S_f l_f}{I_z}, v = v_x, u = \delta_f$$

With small angle approximation $\sin\theta \approx \theta, \cos\theta \approx 1$, and assuming $x_1 = y, x_2 = v, x_3 = \theta, x_4 = r$, the state-space representation of the system above can be written as

$$\left.\begin{array}{l} \dot{x} = Ax + Bu \\ y = Cx \\ x \in R^4, u \in R \end{array}\right\} \tag{5}$$

$$A = \begin{bmatrix} 0 & 1 & 20 & 0 \\ 0 & -3.8851 & 0 & -19.7787 \\ 0 & 0 & 0 & 1 \\ 0 & 0.1394 & 0 & -4.2685 \end{bmatrix}$$

$$B = \begin{bmatrix} 0 \\ 91.2162 \\ 0 \\ 60.3191 \end{bmatrix}$$

Parameters	Nominal values
m	1480 kg
I_z	2350 kg m^2
l_f	1.05 m
l_r	1.63 m
S_f	67,500 N/rad
S_r	47,500 N/rad

Table 1 Vehicle parameters

$$C = \begin{bmatrix} 1 & 0 & 0 & 0 \\ 0 & 0 & 1 & 0 \end{bmatrix}$$

The uncertain parameters S_f, S_r, m, I_z, which are difficult to model accurately will limit the performance of control methods. The values parameters make the system disturbances high which require the robust controller ADRC that estimate and reject the disturbances using ESO (Table 1).

3   Differential Flat System and Flat Output

3.1   Differential Flatness Theory

The important property of flat systems is that a flatness of the system is proved when the set of outputs can be found and in terms of these outputs and their time derivatives, a complete parameterization of all system variables including state, inputs and outputs can be realized. The number of outputs is equal to the number of control inputs.

Consider a system

$$\dot{x} = g(x, u), x \in Rn, u \in Rm$$

$$y = g(x), y \in Rm \tag{6}$$

where x is state variable and u is input variable.

If the output $F \in Rm$ is found of the form then system (6) can be regarded as a differential flat system and F is called flat output. In all cases using differential flatness theory, the most important and difficult issue is to find a proper flat output, since no systematic method exists.

3.2 Differential Flat Output of ALV Model

The flat output of the linear controllable is proportional to the last row of the inverse of the Kalman controllability matrix, multiplied by the state x [6]. More precisely, for a linear system a flat output can always be obtained using

$$F = \begin{bmatrix} 0 & 0 & \ldots & 1 \end{bmatrix} \begin{bmatrix} b & Ab & \ldots & A^{n-1}b \end{bmatrix}^{-1} x. \tag{7}$$

For ALV system as (6), the controllability matrix is shown in (7). With some algebraic manipulations, it is easy to conclude that the rank of U_c is four. Hence, the linear system (6) is controllable. Note that the flat output depends only on the state variables of the system. In the lateral path tracking system (6), two outputs y and ψ are considered simultaneously. Since the differential flat system is independent of the output matrix C, differentially flat theory can be used here to analyze the system. In order to simplify the calculation, the flat output is assumed in the following form since $\alpha \alpha_3 \beta_1 - \beta_2 \alpha_1$ is constant

$$F = 0.001(\alpha_3 \beta_1 - \beta_2 \alpha_1) \begin{bmatrix} 0 & 0 & 0 & 1 \end{bmatrix} \begin{bmatrix} b & Ab & A^2b & A^3b \end{bmatrix}^{-1} x \} \tag{8}$$

By expanding

$$
\begin{aligned}
F &= a_1 x_1 + a_2 x_2 + a_3 x_3 + a_4 x_4 \\
\dot{F} &= b_1 x_1 + b_2 x_2 + b_3 x_3 + b_4 x_4 \\
\ddot{F} &= c_1 x_1 + c_2 x_2 + c_3 x_3 + c_4 x_4 \\
\dddot{F} &= d_1 x_1 + d_2 x_2 + d_3 x_3 + d_4 x_4
\end{aligned}
\tag{9}
$$

Note that $b_0 = c_0 = d_0 = 0$, which means x can be obtained in terms of F and its time derivatives only. Then, the expressions of state variables can be readily obtained as follows:

$$
\begin{aligned}
x_1 &= p_1 F + p_2 \dot{F} + p_3 \ddot{F} + p_4 \dddot{F} \\
x_2 &= q_1 F + q_2 \dot{F} + q_3 \ddot{F} + q_4 \dddot{F} \\
x_3 &= r_1 F + r_2 \dot{F} + r_3 \ddot{F} + r_4 \dddot{F} \\
x_4 &= s_1 F + s_2 \dot{F} + s_3 \ddot{F} + s_4 \dddot{F}
\end{aligned}
\tag{10}
$$

3.3 Controller Design Based on Differential Flatness and ADRC Strategy

Assume,

$$F = g_1, \dot{F} = g_2, \ddot{F} = g_3, \dddot{F} = g_4 \tag{11}$$

the system which is equivalent to (6) using differential flatness theory is as follows:

$$\left. \begin{aligned} \dot{g}_1 &= g_2 \\ \dot{g}_2 &= g_3 \\ \dot{g}_3 &= g_4 \\ \dot{g}_4 &= g(g_1, g_2, g_3, g_4) + bu \\ y &= g_1 \end{aligned} \right\} \tag{12}$$

where

$$g = -\left(\alpha_3 + \frac{\alpha_1 \alpha_4 - \alpha_2 \alpha_3}{v^2}\right) g_3 + \frac{\alpha_1 + \alpha_4}{v} g_4$$
$$b = 0.001\left(\alpha_1 \beta_2 - \alpha_3 \beta_1\right)$$

In this way, the underactuated problem in lateral path tracking control of an ALV can be solved by controlling the differential flat output function g to track the reference one.

In an ADRC, the ESO has been introduced to deal with nonlinearity since it can estimate the total unknown term besides the state variables. In (12), g has been treated as an extended state g_5, i.e., $g(g_1, g_2, g_3, g_4) = g_5$. In this way, (12) can be rewritten as

$$\left. \begin{aligned} \dot{g}_1 &= g_2 \\ \dot{g}_2 &= g_3 \\ \dot{g}_3 &= g_4 \\ \dot{g}_4 &= g_5 + bu \\ \dot{g}_5 &= k(t) \\ y &= g_1 \end{aligned} \right\} \tag{13}$$

where $k(t)$ is the derivative of g_5. Note that $k(t)$ has been treated as bounded in practice.

Assuming the sampling time is h, using forward Euler rule, (14) can be discretized in the form of

$$\left.\begin{array}{l} g_1(k+1) = g_1(k) + hg_2(k) \\ g_2(k+1) = g_2(k) + hg_3(k) \\ g_3(k+1) = g_3(k) + hg_4(k) \\ g_4(k+1) = g_4(k) + h(g_5(k) + bu(k)) \\ g_5(k+1) = g_5(k) + hk(k) \\ y(k+1) = g_1(k+1) \end{array}\right\} \quad (14)$$

where k means the kth sampling period. Based on (14), an ADRC controller has been designed. The order of the system is four, which is comparatively high. Therefore, linear ADRC has been selected which consist of four steps [6].

First, design a TD to arrange the transition process

$$\left.\begin{array}{l} v_1(k+1) = v_1(k) + hv_2(k) \\ v_2(k+1) = v_2(k) + hv_3(k) \\ v_3(k+1) = v_3(k) + hv_4(k) \\ v_4(k+1) = v_4(k) + hf(k) \\ f(k+1) = -r(r(r(r(v_1(k) - v_0(k+1)) + 4v_2(k)) + 6v_3(k)) + 4v_4(k)) \end{array}\right\} \quad (15)$$

where k means the kth sampling period and v_0 is the set value of flat output g_1. Second, use an ESO to track the states and estimate the system's disturbances

$$\left.\begin{array}{l} e(k+1) = z_1(k) - y(k+1) \\ z_1(k+1) = z_1(k) + h(z_2(k) - \beta_{01}e(k+1)) \\ z_2(k+1) = z_2(k) + h(z_3(k) - \beta_{02}e(k+1)) \\ z_3(k+1) = z_3(k) + h(z_4(k) - \beta_{03}e(k+1)) \\ z_4(k+1) = z_4(k) + h(z_5(k) - \beta_{04}e(k+1) + \hat{b}u(k)) \\ z_5(k+1) = z_5(k) + h(-\beta_{05}e(k+1)) \end{array}\right\} \quad (16)$$

where k means the kth sampling period, $\hat{b}$ is the estimator of b, and β $\beta_{01}, \beta_{02}, \beta_{03}, \beta_{04}, \beta_{05}$ are the observer gains.

Third, the following linear feedback (LF) is chosen as follows:

$$\left.\begin{array}{l} e_1(k+1) = v_1(k+1) - z_1(k+1) \\ e_2(k+1) = v_2(k+1) - z_2(k+1) \\ e_3(k+1) = v_3(k+1) - z_3(k+1) \\ e_4(k+1) = v_4(k+1) - z_4(k+1) \\ u_0(k+1) = \tau_1 e_1(k+1) + \tau_2 e_2(k+1) + + \tau_3 e_3(k+1) + \tau_4 e_4(k+1) \end{array}\right\} \quad (17)$$

where k means the kth sampling period and $\tau_1, \tau_2, \tau_3, \tau_4$ are the controller gains.

Finally, compensating the disturbances (Fig. 1),

$$u = \frac{u_0 - z_5}{\hat{b}} \quad (18)$$

Fig. 1 Block diagram of
ALV model based on ADRC
and differential flatness theory

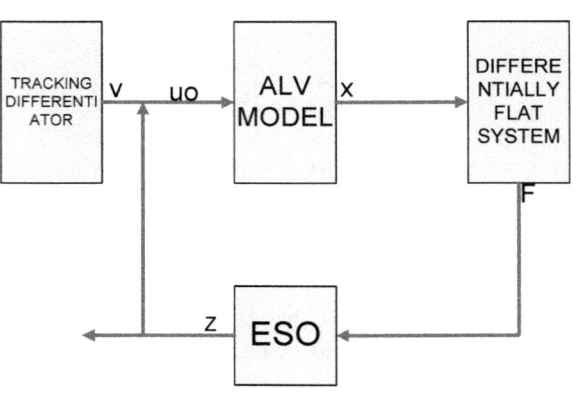

4 Parameters Tuning for the ADRC Controller

In [9], a fundamental method of tuning and optimizing the gains of the controller is given. The value of 'r' for tracking differentiator is chosen which gives best-optimized value. Value of 'h' for Euler's theorem and ESO is arbitrarily taken as 0.001. Tune the observer gains and controller gains using the method of bandwidth parameterization to estimate and compensate for the disturbances.

$$\beta_{01} = 5\omega_0, \beta_{02} = 10\omega_0^2, \beta_{03} = 10\omega_0^3, \beta_{04} = 5\omega_0^4, \beta_{05} = \omega_0^5 \qquad (19)$$

Similarly, place all four poles of the closed-loop characteristic polynomial of the controller at $-\omega_c$

$$s^4 + \tau_4 s^3 + \tau_3 s^2 + \tau_2 s + \tau_1 = (s + \omega_c)^4 \} \qquad (20)$$

The controller gains are obtained as follows:

$$\tau_1 = 4\omega_c, \tau_2 = 6\omega_c^2, \tau_3 = 4\omega_c^3, \tau_4 = \omega_c^4 \qquad (21)$$

By tuning ω_0 and ω_c, proper parameters of ESO and controller have been selected.

5 Simulation and Results

5.1 Reference Value

To testify the validity of the lateral control method, simulations based on double lane-changing trajectory are conducted with fixed longitudinal velocity. Assume that the double lane change completes in 120 m in longitude displacement. The reference trajectory consisting of reference lateral displacement y_r and reference yaw angle θ_r is function of time 't' as (Figs. 2 and 3)

Fig. 2 Reference yaw angle

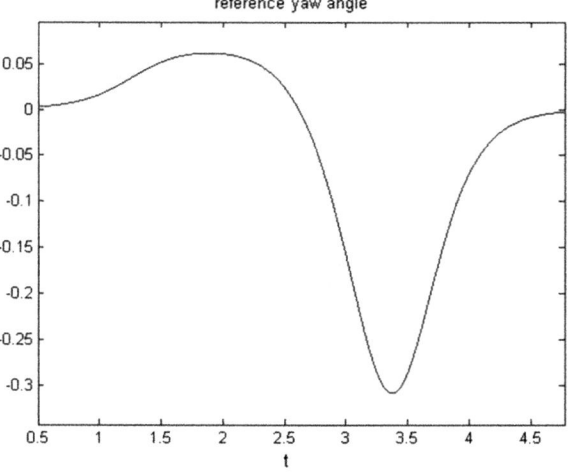

Fig. 3 Reference lateral displacement

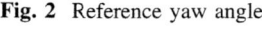

$$\left.\begin{array}{l} yr = \frac{4.05}{2}\left(1+\tanh(p)\right) - \frac{5.7}{2}\left(1+\tanh(q)\right) \\ \theta r = acr\tan(4.05(\frac{1}{\cosh(p)})^2(\frac{1.2}{25}) - 5.7(\frac{1}{\cosh(q)})^2(\frac{1.2}{21.95}) \\ vy = \frac{\mathrm{d}(y)}{\mathrm{d}t} - (v\theta) \\ r = \frac{\mathrm{d}(\theta)}{\mathrm{d}t} \\ p = 2.4\frac{(vt-27.19)}{25} - 1.2 \\ q = 2.4\frac{(vt-56.46)}{21.95} - 1.2 \end{array}\right\} \qquad (22)$$

5.2 PID and LQT

PID is the proportional integral derivative in which the proportional is used to maintain the overshoot of the system, integral part is used to reduce the steady-state error and derivative part is used to increase the speed of the response [10]. Proper tuning of gains of PID controller leads to tracking the path of the ALV model. Since, the system is underactuated it is hard to track the exact path. The linear quadratic tracker tracks the reference path [11] with the proper selection of gains and compensator used. The desired value has been subtracted from the reference signal to get the tracking error, which along with compensator and regulator gain in forward and feedback path give the input of plant. The controller gains have been chosen to minimize performance index. It is solved using the Simulink model in MATLAB (Figs. 4, 5, 6, 7, 8, 9, 10 and 11).

Fig. 4 Block diagram of LQT

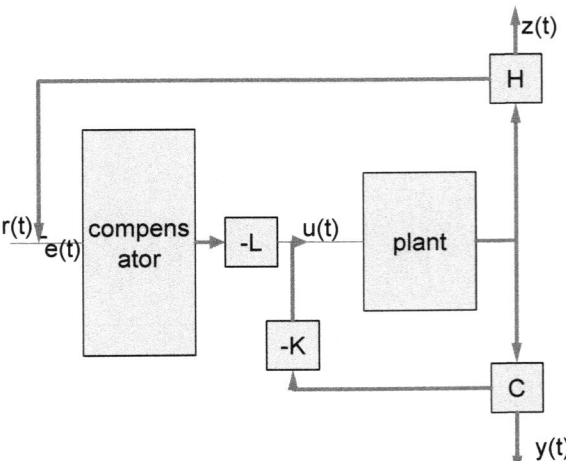

Fig. 5 Yaw angle tracking using ADRC and LQT

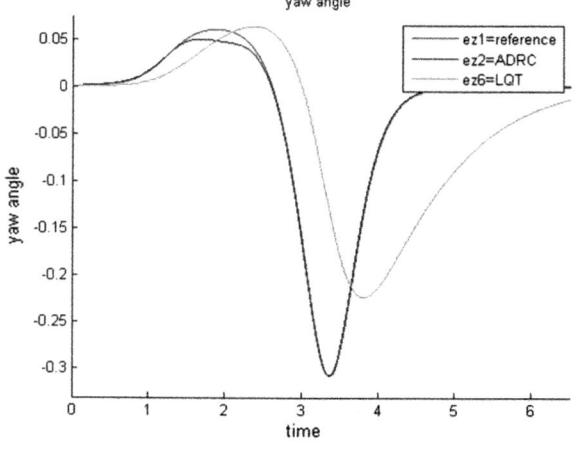

Fig. 6 Yaw angle tracking using PID controller

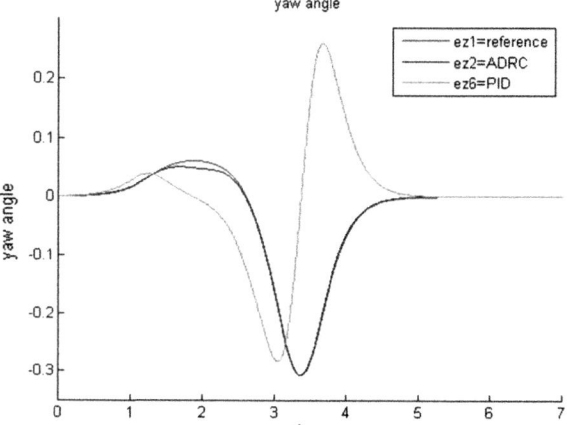

Fig. 7 Yaw angle error of
LQT and ADRC

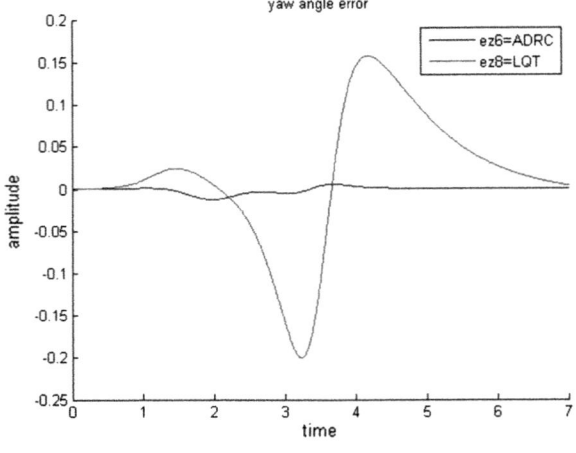

Fig. 8 Lateral displacement
using PID

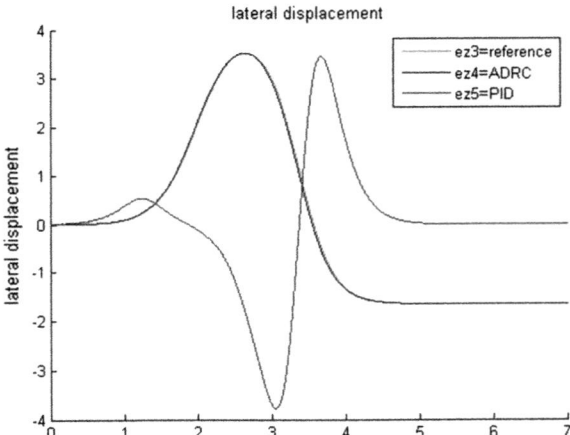

Fig. 9 Lateral displacement tracking using LQT

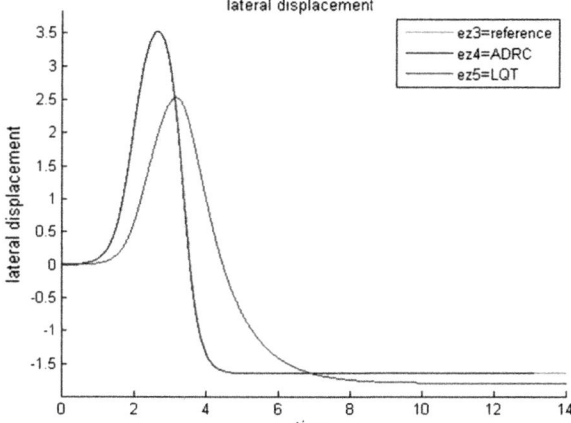

Fig. 10 Lateral displacement error of ADRC and LQT

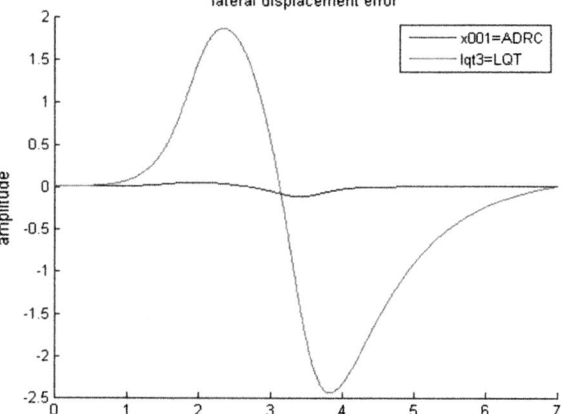

Fig. 11 Controlled input
using ADRC

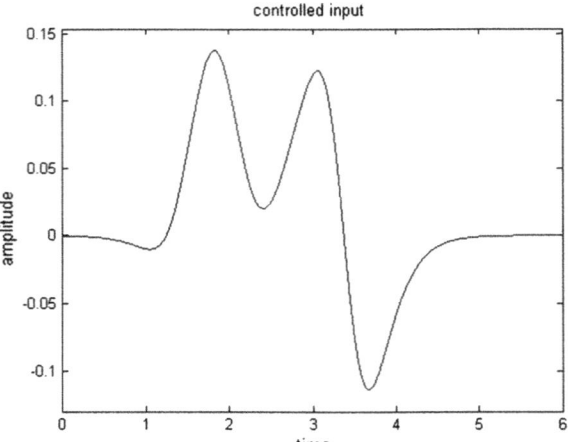

6 Brief Discussion on Comparing the Result with PID and LQT

Table 2 shows the quantitative evaluation to find the maximum and minimum value of yaw angle and lateral displacement with the ADRC, LQT and PID controllers. PID shows the deviation and having large overshoot. The same case is happening with LQT but ADRC is very good in tracking in the case of underactuated problem. Talking about the error part we get the negligible error in ADRC and in the same above Figs. 7 and 10, we get the high peak error in LQT

Table 2 Approx. minimum value and maximum value of tracking path

	y_{max}	y_{min}	ψ_{max}	ψ_{min}
Reference	3.5	−1.5	0.05	−0.3
ADRC	3.5	−1.5	0.048	−0.3
PID	3.5	−3.9	0.25	−0.29
LQT	2.8	−1.6	0.06	−0.22

7　Conclusion

We have found out the tracked path of the given output which is lateral displacement and yaw angle of ALV model using ADRC strategy and differential flatness theory. ADRC is better in tracking purpose as compared to LQT and PID since ADRC reduces the overshoot portion of the tracking path but LQT and PID are unable to hide the spikes as given by above plots. The results shown by ADRC are very accurate, however there are still some problems such as path planning, and systems with delay issues need to be observed in the near future.

References

1. R. Rajamani, "Vehicle Dynamics and control. New York, NY, USA: Springer, 2011, ISBN 978-1-4614-1433-9.
2. S. Krishna, S. Narayanan, and S. Ashok, "Fuzzy logic based yaw stability control for active front steering of a vehicle," J. Mech. Sci. Technol., vol. 28, no. 12, pp. 5169–5174, 2014.
3. Yuan Gao, Yuanqing Xia "Lateral path tracking control of Autonomous Land Vehicle based on Active Disturbance Rejection control" Proceedings of the 32nd Chinese Control Conference pp. 5467–5472, 2013.
4. S. You, J. Hahn, and H. Lee, "New adaptive approaches to real-time estimation of vehicle side slip angle," Control Eng. Pract., vol. 17, no. 12, pp. 1367–1379, 2009.
5. M. Fliess, J. Lévine, and P. Martín, "Flatness and defect of nonlinear systems: Introductory theory and examples," Int. J. Control, vol. 61, no. 9, pp. 1327–1361, 1995.
6. J. Han, "From PID to active disturbance rejection control," IEEE Trans. Ind. Electron., vol. 56, no. 3, pp. 900–906, Mar. 2009.
7. S. Li, J. Yang, W. Chen, and X. Chen, "Generalized extended state observer based control for systems with mismatched uncertainties," IEEE Trans. Ind. Electron., vol. 59, no. 12, pp. 4792–4802, Dec. 2012.
8. J. Han, "Auto-disturbance rejection control and its applications," Control Decis., vol. 13, no. 1, pp. 19–23, 1998.
9. Z. Gao, "Scaling and bandwidth-parameterization based controller tuning," in Proc. Amer. Control Conf., 2003, vol. 1, no. 6, pp. 4989–4996.
10. Auday Al-Mayyahi, William Wang and Phil Birch "Path Tracking of Autonomous Ground Vehicle Based on Fractional Order PID Controller Optimized by PSO" IEEE 13th International Symposium on Applied Machine Intelligence and Informatics January 22–24, 2015.
11. Bohner, M. and Wintz, N. (2011) "The Linear Quadratic Tracker on time scales', Int. J. Dynamical Systems and Differential Equations, Vol. 3, No. 4, pp. 423–447.

Performance Analysis of Conventional SRAM with Higher Order SRAM Topologies

Ravneet Kaur, Garima Joshi, Maninder Kaur Saggu and Vishal Sharma

1 Introduction

In modern CMOS circuit, the Static Random Access Memory (SRAM) is widely used as on-chip cache. In nanometer technologies, embedded SRAM occupies a main portion of System on Chips (SOCs). Stability, high speed and low power are major concerns for nanoscaled SRAM. Various popular methods like voltage scaling, sizing of device and architectural techniques are used to improve the overall performance of cell. However, voltage scaling results in a reduction of power dissipation in a cell, but on the other hand, low-voltage operation degrades the cell stability. So there is a trade-off between these performance parameters.

A conventional 6T SRAM cell is an arrangement of two CMOS inverters connected back to back. Inverters are formed with (M2, M4 pMOS and M1, M3 nMOS transistors) and two nMOS access transistors (M5 and M6). Each bit in a SRAM cell is stored on four transistors and this forms two cross-coupled inverters. There are two additional access transistors which help to store data during read and write operation.

R. Kaur (✉) · G. Joshi · M. K. Saggu · V. Sharma
Department of Electronics, U.I.E.T, Panjab University, Chandigarh, India
e-mail: ravneetkaurdua@yahoo.com

G. Joshi
e-mail: joshi_garima5@yahoo.com

M. K. Saggu
e-mail: manindersaggu92@gmail.com

V. Sharma
e-mail: vishaluiet@yahoo.co.in

© Springer Nature Singapore Pte Ltd. 2019
V. Nath and J. K. Mandal (eds.), *Proceeding of the Second International Conference on Microelectronics, Computing & Communication Systems (MCCS 2017)*, Lecture Notes in Electrical Engineering 476, https://doi.org/10.1007/978-981-10-8234-4_53

2 SRAM Performance Parameters

2.1 Power Dissipation

SRAM cell consumes two types of power, i.e. static and dynamic. The amount of power that is consumed by the unit to retain the data is static power. The static power is leakage power in digital circuits and it is due to the leakage current passing through the circuit when there is no activity:

$$P_{\text{Total}} = P_{\text{Dynamic}} + P_{\text{Static}} \tag{1}$$

Dynamic power consumption depends upon the switching activity of the cell. The reduction in the swing voltage reduces the dynamic power dissipation during switching activity.

$$P = \alpha C_{\text{Interconnect}} f \, V_{DD}^2 \tag{2}$$

where f is the frequency of operation, $C_{\text{Interconnect}}$ is the interconnect capacitance, α is switching activity and V_{DD} is the supply voltage.

Among different mechanisms for leakage current, the sub-threshold leakage of the transistor is the prominent source of the leakage current in an SRAM cell. The Gate-Induced Drain Leakage (GIDL) which is usually more than one order of magnitude smaller than the sub-threshold leakage.

$$P_{\text{static}} = P_{\text{Leakage}} = I_{\text{Leakage}} \cdot V_{\text{dd}} \tag{3}$$

2.2 Delays

Access times are the times that SRAM cell takes for its read and write operations to complete. Two access times that are important are the read access and the write access times. Read access time is the time required to read the data from the cell and is calculated as the time delay between 50% word line activation to when the sense amplifier has reached 90% of its full swing. For write operation, the write delay is defined as the time between the activation 50% of word line to when one of the output nodes is 90% of its full swing. For example, in write "0", delay is the difference between time when WL reaches 50% of its activation (i.e. T1) to when output reaches 90% of its full voltage swing (i.e. T2).

The timing diagram is shown in Fig. 1.

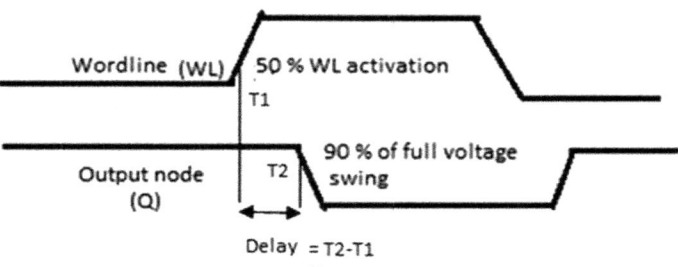

Fig. 1 Timing diagram of word line and output to find delay

2.3 Static Noise Margin

SNM is defined as the maximum DC noise voltage required for flipping cell data. First time, Seevinck et al. in 1987 explained the way of calculating the SNM of cross-coupled inverters [1]. To simulate the SNM of memory cell, two bit lines and the word line of the cell are kept at V_{DD}. This SNM is also called Read SNM and the setup for simulation is shown in Fig. 2 is for read operation. V_N are two DC sources placed between inverters. These voltages are swept between 0 and $V_{DD}/2$ or more until the cell's storage data flips. The butterfly curves obtained by plotting the DC Voltage Transfer Characteristics (VTC) curve of the inverters of the SRAM cell gives the valuable information about the stability of the SRAM cell during the read and hold modes. The eye of the butterfly curve during the read mode is always less than the case when the SRAM is held in the hold mode. The SNM is the length of a maximum possible square that can be embedded in butterfly curves as shown in Fig. 3.

Fig. 2 Simulation setup for read SNM [1]

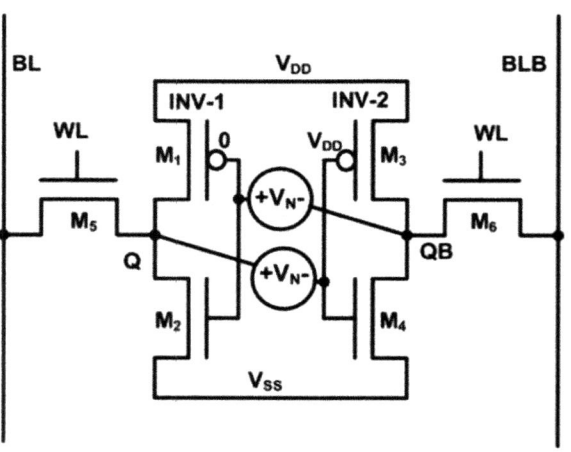

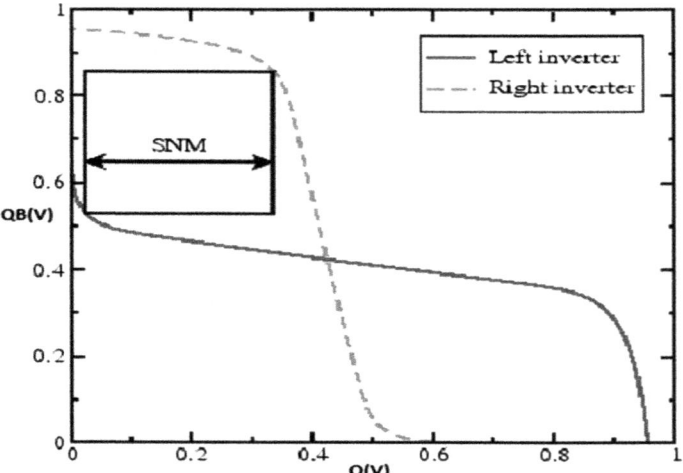

Fig. 3 Butterfly curve of two cross-coupled inverter

Different methods such as bit line sweep, word line sweep and *N* curve, etc., are explored in [2] used to evaluate write SNM.

3 Review of Existing SRAM Cell Topologies

The four designs of 7T SRAM cell are presented in the paper. Each design has different performance parameters and is useful in different applications.

3.1 7T_a SRAM Cell

The single-ended 7T SRAM cell [3] with transmission gate for improved read SNM is shown in Fig. 4. Before the read operation, bit line is pre-charged to supply voltage V_{DD}. During the read mode, M7 transistor is turned ON by activating word line (WL) signal. The transmission gate M5–M6 is also maintained ON. If logic 0 is stored at node Q, then bit line gets discharged through M5–M6, M7 and M2 to ground. In write operation, the transmission gate M5–M6 is deactivated thereby cutting the cell feedback loop. This leads to better write SNM.

Fig. 4 7T$_a$ SRAM cell [3]

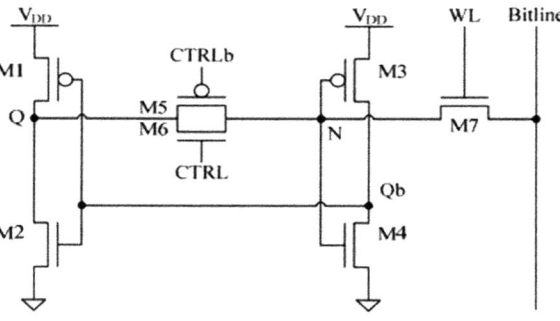

3.2   7T$_b$ SRAM Cell

The 7T SRAM cell [4] shown in Fig. 5 is single bit line with separate read and write path for a very high read SNM. During write operation, WL is held ON and the data is loaded from the bit line. Strong access transistor N3 allows bit line to overpower the cell, so that the data can be written into the cell. Before the read operation the bit line is pre-charged to V_{DD} and during read operation RL is activated and this separates the bit line from the output node Q. This leads to very high read SNM.

3.3   7T$_c$ SRAM Cell

The 7T SRAM cell [5] as shown in Fig. 6 has better read stability than conventional 6T SRAM cell, as it eliminates the chance of cell data being flipped. There are two ground rails in this design shown in Fig. 5. In read operation, the signals W_b and R are made high, thus the data at node Q is read by MR, read assist transistor which is connected to virtual ground 2. As there is separate read bit line to read data so it has better read SNM. During a write operation, MW transistor (connected to virtual

Fig. 5 7T$_b$ SRAM cell [4]

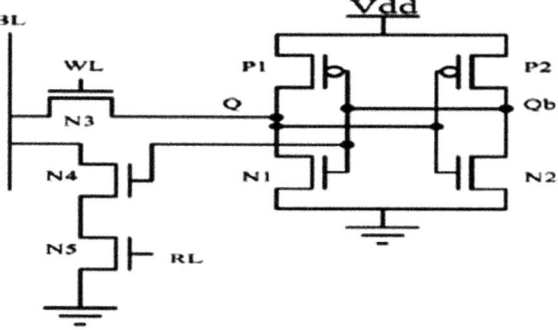

Fig. 6 7T$_c$ SRAM cell [5]

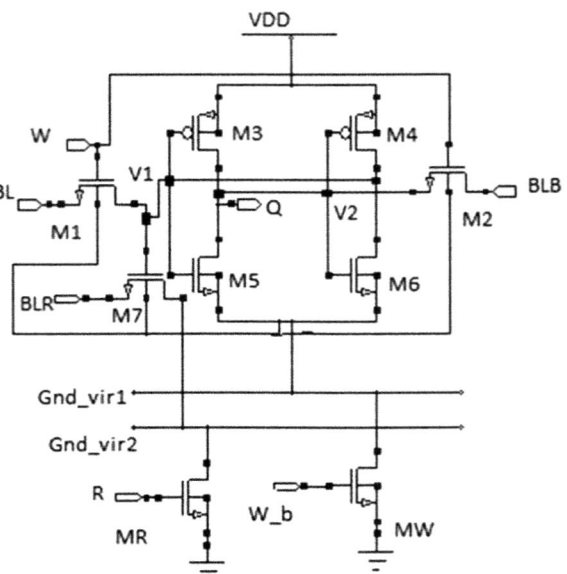

ground) 1 is made off so cell gets disconnected from ground rail 1, hence weakening the feedback and speeding up the write operation. During hold operation, there is strong feedback in the cell as MW transistor is held ON.

3.4   7T$_d$ SRAM Cell

This design [6] shown in Fig. 7 has single-sided write operation, unlike conventional 6T SRAM. The M5 transistor and Write bit line are used to load data into the cell. The dynamic power consumption is reduced due to the single-ended write operation. Read SNM gets increased due to separate read path from storage nodes. M6 and M7 transistors are used to read the cell data. The M7 connected to virtual ground minimizes standby leakage.

Three designs of 8T SRAM topologies are presented below. These designs have different characteristics which suit for different applications.

3.5   8T$_a$ SRAM Cell

The 8T SRAM cell [7] shown in Fig. 8 has different read and write paths, which help in eliminating read upset failure mode. This cell has same write operation as conventional 6T SRAM cell. In read operation, read bit line and read word line are

Fig. 7 7T$_d$ SRAM cell [6]

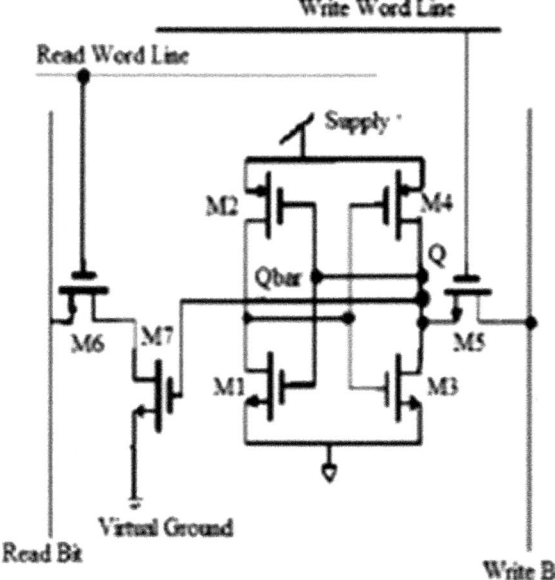

Fig. 8 8T$_a$ SRAM cell [7]

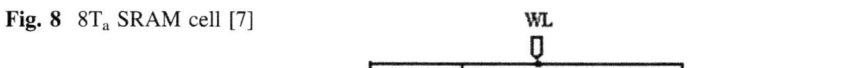

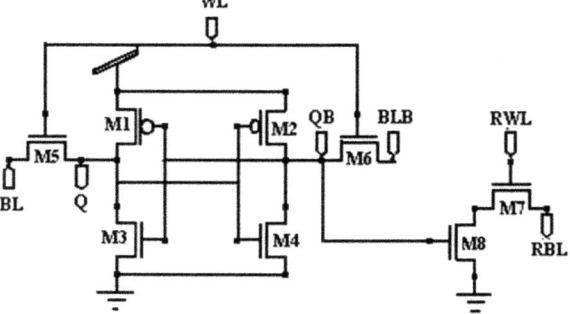

made ON. The value stored on the storage node is read through read bit line RBL when RWL is made high.

3.6 8T$_b$ SRAM Cell

The 8T SRAM cell [8] shown in Fig. 9 makes use of memristors which are directly connected to the nodes Q and QB and enables the storage of complementary backup data by maintaining the nonvolatile characteristics. During read operation, the switch line is held low to turn off the RSWs (RSWL and RSWR), so as maintain the

Fig. 9 $8T_b$ SRAM cell [8]

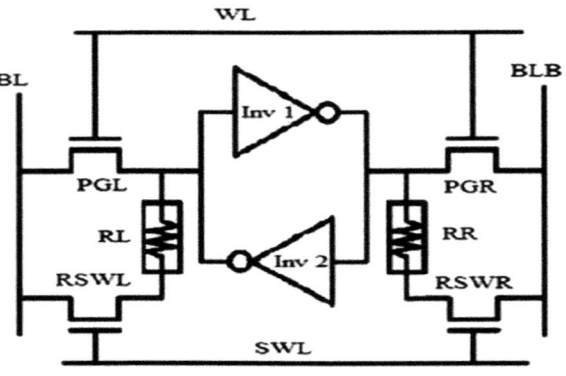

stability of the cell. In write operation, both word line and SWL are made ON. The data to be written on the BL/BLB is written into the SRAM cell using pass gates and RSWs.

3.7 8T_c SRAM Cell

The 8T SRAM cell [9] shown in Fig. 10 is double-ended read decoupled and single-ended write. As there is separate read port and also read current does not flow through storage nodes, thus read SNM gets enhanced. Before the read operation RBL and RBLB are pre-charged to V_{DD}. WL is set to ground and WLB is set to V_{DD}. RT, read assist transistor is turned ON and MP3 and MN3 are turned OFF. The voltage difference between two bit lines during read operation is sensed and amplified through sense amplifier circuit. During write operation, the write transistors MP3 and MN3 are turned ON by holding the word line WL high and WLB low.

Three designs of 9T SRAM cell are presented in the paper. These three designs have different performance parameters to meet the requirements of different applications.

3.8 9T_a SRAM Cell

The 9T SRAM cell [10] shown in Fig. 11 has write operation similar to that of conventional 6T SRAM. In read operation, the read signal RD is held high, WR is low and reading is done through N5, N6 and N7 transistors. During write operation N3 and N4 transistors are ON as WR is maintained ON and RD is held low. As N3 and N4 are isolated from storage nodes in read operation so this design has better read stability. This circuit design has the problem in it that is high bit line capacitance with more pass transistors on the bit line.

Fig. 10 8T$_c$ SRAM cell [9]

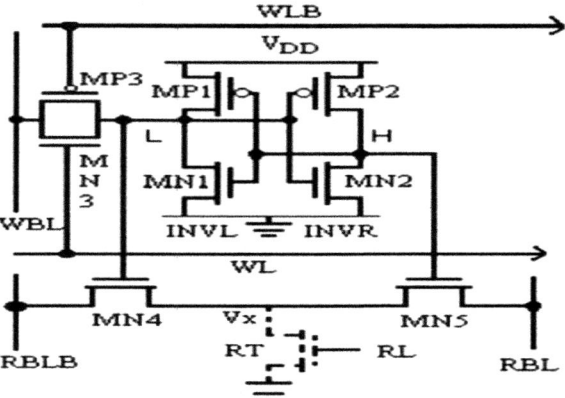

Fig. 11 9T$_a$ SRAM cell [10]

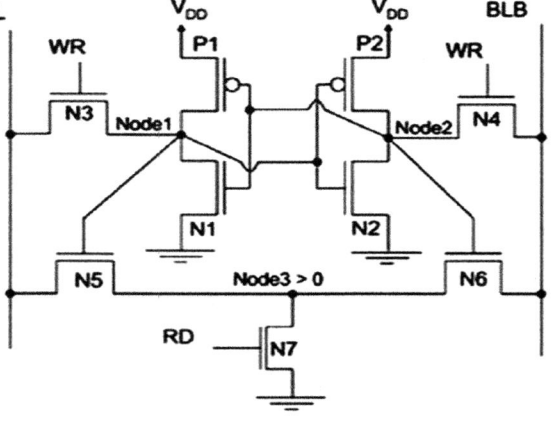

3.9   9T$_b$ SRAM Cell

The design [11] presented in Fig. 12 presents a single-ended disturbance free 9T cell in sub-threshold region. During write, the signal WR is made low thus turning off M9 transistor resulting in weak feedback and hence faster writes. In read operation, word line WL is kept low and read bit line (RBL) is kept high so as to maintain the stability of data at nodes. Word line WL is made low and read bit line (RBL) is pre-charged to VDD. For instance to read 0 at node Q, read word line (RWL) is made high. During the hold state, WL and RWL are made low with WR held high so as to maintain data.

3.10 9T$_c$ SRAM Cell

The 9T SRAM cell [12] shown in Fig. 13 includes a read/write port M5–M9. Bit line (BL), Write word line A (WWLA), Write word line B (WWLB), respectively, are column based. During hold state, WL, both the write word lines are disabled and VVSS is kept at V_{DD}. Data is kept by cross-coupled inverters M1–M4 is decoupled from bit line.

There is one design of 10T cell presented in the paper. This architecture has different characteristics suited for various applications.

3.11 10T$_a$ SRAM Cell

The design [13, 14] shown in Fig. 14 is based upon NMOS transistor with gated-ground technique. This gated-ground railing reduces the leakage and thus static power dissipation. This cell has a tail transistor in series connection. The output of the XOR gate controls this tail transistor. The inputs to the XOR gate are WWL and RWL.

3.12 11T SRAM Cell

The circuit shown in Fig. 15 contains 11T transistors instead of 6 but due to low power consumption and faster access time, this hardware burden can be tolerated. The bit line and bit line bar controls two tail transistors N7 and N9 respectively. During write, RWL is hold zero while asserting WWL. For write 0, BL = 0, N9 goes OFF, Node *B* turns to high making Node *A* = 0. For write 1, BL = 1, N7 is OFF, Node *A* pushes to high resulting low at node *B*. In write, BL or BL bar is neither charging nor discharging thus saving a lot of power. During read operation,

Fig. 12 9T$_b$ SRAM cell [11]

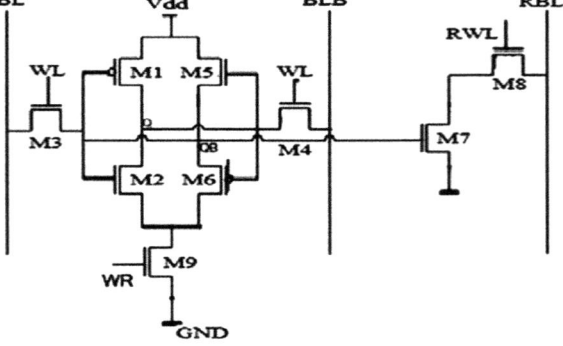

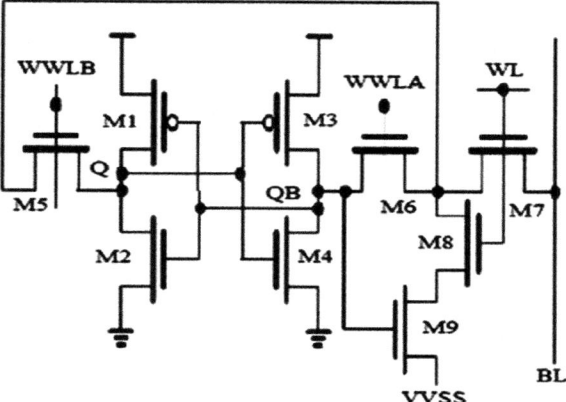

Fig. 13 9T$_c$ SRAM cell [12]

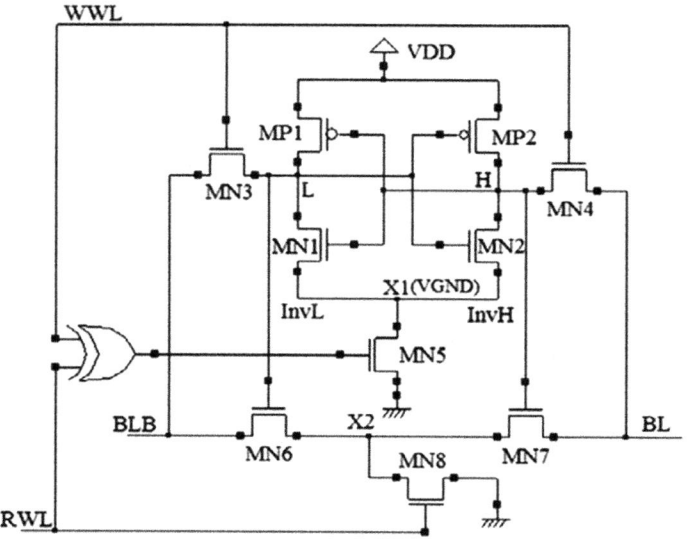

Fig. 14 10T$_a$SRAM cell [13]

RWL is enabled and WWL is disabled. There is less read power consumption due to the usage of one single bit line and two NMOS transistors. In this cell, 0–1 and 1–0 transitions are not allowed that means lot of power saving. [15]

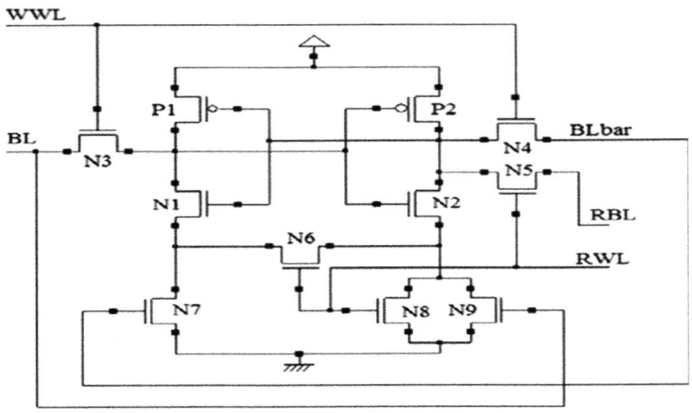

Fig. 15 11T SRAM cell [15]

3.13 12T SRAM Cell

The MTCMOS-based 12T SRAM [14] shown in Fig. 16 is proposed to obtain low dynamic power at high frequencies, low leakage current, low static power dissipation due to charge recycling and better stability. Two voltage sources V1 and V2 are connected to outputs of the bit line and it bit line bar respectively. Two NMOS transistors VT1 and VT2 connected with bit line and bit line bar, respectively, to switch on and off the voltage sources while write mode of operation. Two high V_t S1 and S2 (sleep transistors) are used in the cell. Sleep transistors disconnect the logic cells from the supply or ground to reduce leakage in sleep mode. The low threshold transmission gate is connected between the two virtual nodes M and N to provide charge sharing. To activate sleep transistors and transmission gate, control signals ST and CS (charge sharing control signal) are used. Table 1 shows the

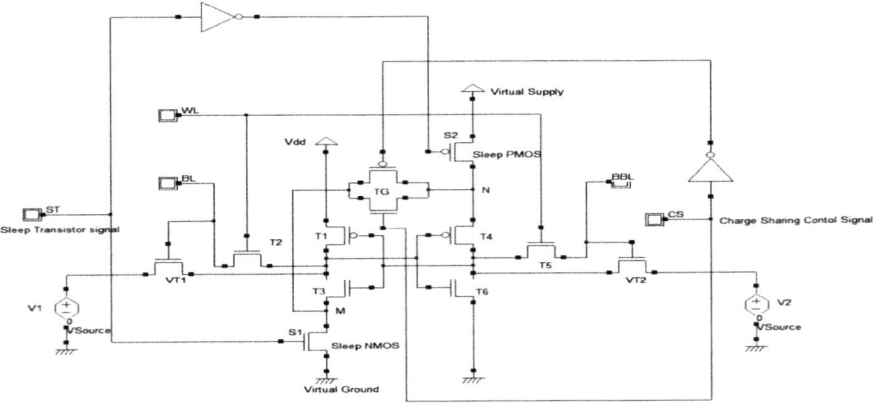

Fig. 16 12T SRAM cell [14]

Table 1 Qualitative comparison of power, delay and SNM of various SRAM cell topologies

SRAM cells	Power						Delay		SNM	
	Read			Write			Read	Write	Read	Write
	Static	Dynamic	Total	Static	Dynamic	Total				
6T	High	Medium	High	High	Medium	High	Low	Low	Low	High
7T$_a$	–	High	–	–	Medium	–	Low	Low	Medium	Medium
7T$_b$	–	–	**Low**	–	–	**High**	Medium	Low	**High**	Low
7T$_c$	Medium	Low	Low	–	–	High	Medium	Low	Low	Low
7T$_d$	Low	Low	Low	Low	Low	Medium	High	Medium	Low	Low
8T$_a$	Medium	Medium	Medium	High	Low	High	High	High	Low	Low
8T$_b$	**Low**	**Low**	**Low**	**Low**	–	**Medium**	**High**	**High**	**Low**	**Medium**
8T$_c$	–	–	–	–	–	–	Very low	Very low	Very high	Medium
9T$_a$	High	Medium	Medium	High	Medium	High	High	High	Low	Low
9T$_b$	**Low**	**Low**	**Low**	**High**	**Low**	**Medium**	**High**	**High**	**Low**	**Low**
9T$_c$	Low	Low	Low	Medium	Low	Medium	High	High	Low	Low
10T$_a$	Low	Low	Low	Medium	Low	Medium	Medium	Medium	Low	Medium
11T	Low	Low	Low	Low	Low	Low	High	High	Medium	High
12T	Low	Low	Low	Low	Low	Low	Medium	Medium	High	High

qualitative overview of power, delay and SNM of various SRAM cell topologies discussed above. The generalized conclusion that can be drawn from Table 1 are:

- 7T SRAM cell architectural designs are better than conventional 6T SRAM cell in terms of read power consumption. $7T_a$ SRAM cell has better write stability (WSNM) and $7T_b$ has better read stability (RSNM).
- $8T_b$ SRAM cell has low total power dissipation as compared to conventional 6T SRAM cell but has higher read and write delay. $8T_c$ SRAM cell is better in RSNM and delay than 7T and 6T SRAM cells.
- $9T_b$ has low total read power than $8T_a$.
- 9T and 10T SRAM cell architectural designs are better for low-power applications.
- 11T and 12T SRAM cells are suitable for low-power and high-SNM applications.
- Going onwards from conventional 6T to 12T SRAM cells, though parameters such as power, delay and SNM have been improved but transistors count has also been increased that may lead to more area consumption.

4 Simulation and Results

In this work, SRAM cell architectural designs are simulated on TANNER EDA tool and their performance is compared in terms of power consumption, delay, RSNM, WM, HSNM and Power-Delay-Area Product (PDAP). PDAP should be minimum. Table 2 shows the simulation results on different parameters. Figure 17 shows the HSNM and RSNM curve of 6T and $7T_b$ SRAM cells

$$PDAP = P_{average} \times Delay_{max} \times Area_{total} \qquad (4)$$

Table 3 shows the comparative analysis of $7T_e$ SRAM cell with differential 8T (D8T), conventional 6T and 8T with enhanced write stability (WRE8T) on basis of leakage power, delay and SNM. So $7T_e$ with less number of transistors is better than 6T and D8T in terms of write SNM, best in read SNM, better in read delay than D8T and has less leakage power than 6T and WRE8T.

Table 2 Comparative analysis of power, delay, SNM and power–delay product between different SRAM cells architectural designs at 1.2 V at 90 nm (W_{min} = 120 nm)

Cell	Power (µW)		Delay (ps)		SNM (mV)			PDAP
	W	R	W	R	R	H	W	($x10E - 12$)
6T ($\beta = 2$)	38.2	38	172	43	180	420	480	6.27
7Tb	14	13.3	231	85	420	430	476	**3.62**
8Tb	34	32.3	244	90	185	–	378	9.13
9Tb	26	30.5	184	64	194	–	328	5.64

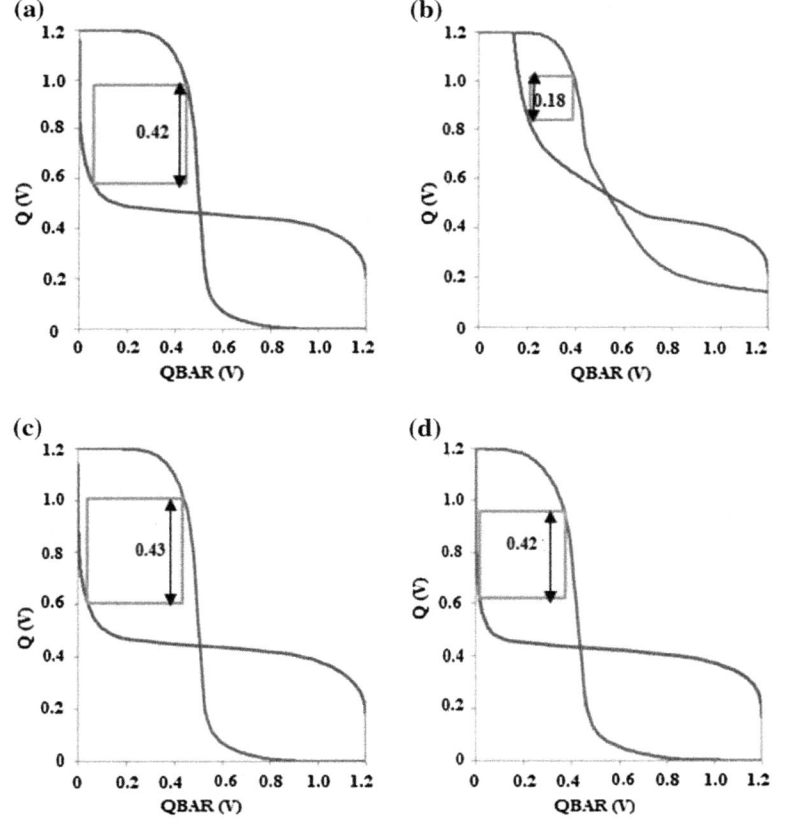

Fig. 17 Butterfly curves to calculate SNM. **a** HSNM 6T, **b** RSNM 6T, **c** HSNM 7T, **d** RSNM 7T

Table 3 Comparative analysis of leakage power, SNM and delay of different architectural designs at 0.5 V at 90 nm. Present publication [16]

Cell	Leakage power range (nW) (from graph)	Delay (ps)		SNM range (from graphs) (mV)	
		W	R	W	R
6T	10–15	27.5	47.2	100–150	0–50
7T$_e$	10	284.8	66.2	200–250	150–200
D8T	5–10	20.9	71.4	150–200	50–100
WRE8T	10–15	235.5	42.9	250–300	100–150

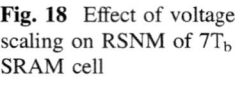

Fig. 18 Effect of voltage scaling on RSNM of 7T$_b$ SRAM cell

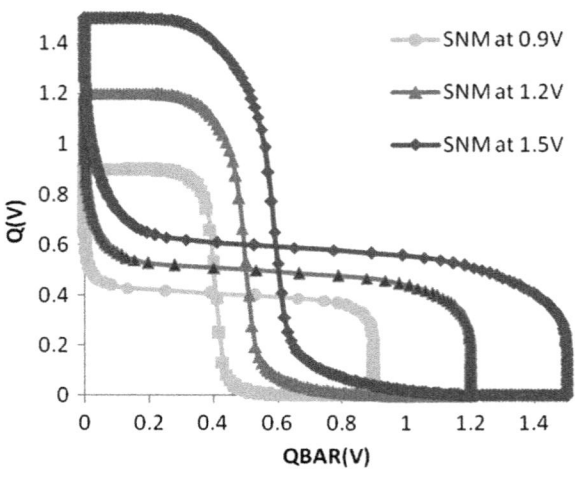

Fig. 19 Effect of temperature variations on SNM

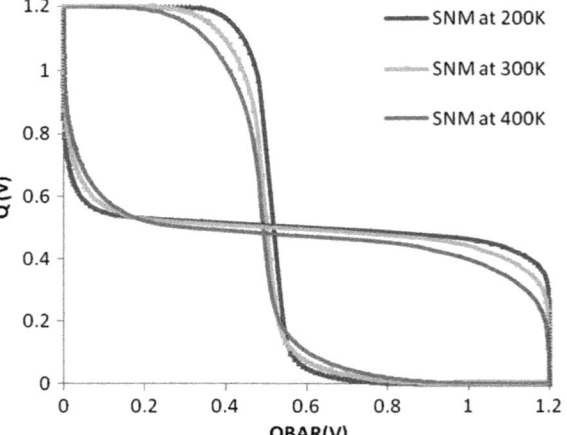

Figure 18 shows the effect of voltage scaling on SNM in case of 7T$_b$ SRAM cell. As voltage increases, SNM also increases and thus stability increases.

The effect of operating temp is quite visible on SNM. Figure 19 shows the results of SRAM cell operated at 200, 300 and 400 K. SNM degrades as we move from 200 to 400 K. As the temperature increases the static current noise margin decreases, thus the current required to flip the state of the cell decreases. Hence, the cell becomes less stable as the operating temperature increases.

Overall stating, the conventional 6T SRAM has the least delay it consumes the maximum power. 7T$_b$ SRAM cell consumes the least power and also in terms of SNM the 7T SRAM performs best. Also, considering the Power-Delay-Area product 7T is the best topology. It has the less number of transistors as compared to 8T and 9T configurations.

References

1. Evert Seevinck, Frans J. List and Jan Lohstroh, "Static-Noise Margin Analysis of MOS SRAM Cells", *IEEE Journal of Solid-State Circuits*, Vol. SC-22, No.-5,1987.
2. Wang J, Nalam S and Calhoun B H 2008 Analyzing Static and Dynamic Write Margin for Nanometer SRAMs. IEEE *International Symposium on Low Power Electronics and Design*, pp. 129–134.
3. Khawar Sarfraz and Volkan Kursun "Characterization of a Low Leakage Current and High-Speed 7T SRAM Circuit with Wide Voltage Margins*", IEEE Computer Society Annual Symposium on VLSI*, pp. 64–69,2013.
4. Madiwalar B and Kariyappa B S 2013 Single Bit-line 7T SRAM cell for Low Power and High SNM. *IEEE International Multi Conference on Automation, Computing, Communication, Control and Compressed Sensing*, pp. 223–228.
5. Azam T, Cheng B, Cumming DRS, "Variability resilient low power 7T-SRAM design for nano scaled technologies" *11th international symposium on quality electronic design (ISQED)*, pp 9–14, 2010.
6. S. A. Tawfik and V. Kursun, "Low power and robust 7T dual-Vt SRAM circuit," *Proceedings of the IEEE International Symposium on Circuits and Systems*, pp. 1452–1455, 2008.
7. Chen G, Sylvester D, Blaauw D, Mudge T, " Yield-driven near-threshold SRAM design", *IEEE Trans Very Large Scale Integration (VLSI) Systems*, 18:1590–8. No. 11, 2010.
8. Chiu P, Chang M, Wu C, Chuang C, Sheu S, Chen Y, *et. al.*, "Low store energy, low VDDmin, 8T2R nonvolatile latch and SRAM with vertical stacked resistive memory (memristor) devices for low power mobile applications", *IEEE J Solid-State Circuits*; 47(6): 1483–96, 2012.
9. Soumitra Pal and Shahnawaz Arif, " A Single Ended Write Double Ended Read Decoupled 8-T SRAM with Improved Read Stability and Writability", *International Conference on Computer Communication and Informatics*, pp. 1–4, 2015.
10. Liu Z, Kursun V, "Characterization of a novel nine-transistor SRAM cell", *IEEE Trans Very Large Scale Integration (VLSI)* Syst; 16:488–92. No. 4, 2008.
11. Ramani AR, Choi K., "A novel 9T SRAM design in subthreshold region" *IEEE international conference on electro/information technology (EIT)*, pp. 1–6, 2011.
12. Tu M, Lin J, Tsai M, Lu C, Lin Y, Wang M, *et. al.*, "A single ended disturb-free 9T subthreshold SRAM with cross-point data-aware write word-line structure, negative bit-line, and adaptive read operation timing tracing", *IEEE J Solid-State Circuits*; 47(6):1469–82, 2012.
13. S. Singh, N. Arora, N. Gupta, M. Suthar, "Leakage reduction in differential l0T SRAM cell using gated VDD control technique" *International conference on computing, electronics and electrical technologies*, pp. 610–4, 2012.
14. P. Upadhyay, R. Kar, D. Mandal and S.P. Ghoshal, "A design of low swing and multi threshold voltage based low power 12T SRAM cells", *Computers & Electrical Engineering*, Volume 45, pp 108–121, July 2015.
15. A.K.Singh, C.M.R. Prabhu, Soo WP, Hou TC. A proposed symmetric and balanced 11-T SRAM cell for lower power consumption, *IEEE TENCON region 10conference*, pp. 1–4, 2009.
16. M. Moghaddam, M.H. Moaiyeri and M.Eshghi, "Ultra Low Power 7T SRAM cell Design Based on CMOS", 23rd Iranian Conference on Electrical Engineering (ICEE), pp. 1357–1361, 2015.

Performance Analysis of TRDMA Under Multi-path Rician Fading Channel

Shashank Shekhar, Shanidul Hoque, Ashraf Hossain and Wasim Arif

1 Introduction

The evolution of wireless communication technologies escalates the use of connected personal digital devices in manifold. Almost every device are accessing the integrated network for recording, transferring, viewing or even monitoring data which in turn increases the requirement of spectrum bandwidth to support such appetite for data and content access.

In February 2016, CISCO Virtual Networking Index (VNI) anticipated the growth in overall mobile traffic to 30.6 exabytes per month by 2020, which is around five times than the present traffic in 2016 [1]. Presently, the mobile service providers are using sub-6 GHz spectrum with various complex access techniques in order to achieve better spectral efficiency [2, 3]. Since, a very small room is left for further improvement in spectral efficiency, it is inevitable to use millimeter wave (mmWave) spectrum to meet the extreme data traffic demands [4, 5]. The mmWave spectrum will offer a large bandwidth, which is the primary advantage of 5G over 4G. In [6], the authors demonstrated that the time-reversal (TR) system offers better achievable rate than the OFDM system in case of large bandwidth. Time-reversal promises to be an emerging solution for wideband communication systems where higher throughput with reduced complexity is the fundamental requirement [6].

S. Shekhar (✉) · S. Hoque · A. Hossain · W. Arif
Electronics and Communication Engineering, NIT Silchar, Silchar, India
e-mail: shashankshekhar1991@hotmail.com

S. Hoque
e-mail: shanidulhoque257@gmail.com

A. Hossain
e-mail: ashraf@ece.nits.ac.in

W. Arif
e-mail: arif.ece.nits@gmail.com

© Springer Nature Singapore Pte Ltd. 2019
V. Nath and J. K. Mandal (eds.), *Proceeding of the Second International Conference on Microelectronics, Computing & Communication Systems (MCCS 2017)*, Lecture Notes in Electrical Engineering 476, https://doi.org/10.1007/978-981-10-8234-4_54

Along with its low complexity [7], time-reversal (TR) signal is also able to harvest energy from the rich multi-path propagation which is not possible for the available communication models in force and that makes TR an absolute model for the low energy green wireless communications.

Time-reversal (TR) creates a spatio-temporal resonance effect, *focusing effect*, from unavoidable but rich multi-path radio propagation environment. In TR object *B* sends a pilot impulse which is received by object *A* after propagating through a scattering and multi-path environment. After receiving the signal, object A transmits the time-reversed signal to object *B* using the same transmission medium. In [8–10], the authors investigated spatial and temporal focusing properties for transmission of EM signal with time-reversal. In [11], the authors projected a multiple access technique by using TR considering *Rayleigh fading* channel in a multi-user broadband environment.

In this paper, the performance of TRDMA is investigated with multiple base station antennas and single antenna user system over a multi-path *Rician fading* channel. A comprehensive analysis of the model is carried out and simulation results are obtained for various performance measuring matrices under diverse system configuration. In Sect. 2, MISO TRDMA and *Rician fading* channel models are introduced. The performance measuring parameters to evaluate the system performances are presented in Sect. 3. The simulation results are discussed in Sect. 4. Finally, in Sect. 5, conclusions of the paper are drawn along with the future scope.

2 System Model

The system model is consisting of multiple-input-single-output (MISO) with Time-Reversed Division Multiple Access (TRDMA). The base stations (BS) are equipped with M_T number of antennas and there are N users with a single-antenna transceiver system.

2.1 Channel Model

In this paper, downlink network is considered and the channel model is approximated to be as *Rician fading* which incorporates the effect of line-of-sight component. In our presented model, we consider that all the users are well appointed with single antenna in their communication devices. The channel impulse response (CIR) of the path between *m*th BS antenna and the *i*th user can be written as

$$h_i^{(m)}[k] = \sum_{l=0}^{L-1} h_{i,l}^{(m)} \delta[k - l], \tag{1}$$

where $h_i^{(m)}[k]$ denotes the kth tap of the channel impulse response, L is the *length*, and $\delta[.]$ is the unit impulse signal. We assume that for every single communication link, $h_i^{(m)}[k]$ are independent complex Gaussian random variables (RVs), i.e.

$$h_i^{(m)}[k] = \Re\left(h_i^{(m)}[k]\right) + \Im\left(h_i^{(m)}[k]\right)$$

where

$$\Re\left(h_i^{(m)}[k]\right) \sim \mathcal{N}\left(\mu_k, \frac{1}{2} e^{-\frac{kT_s}{\sigma_T}}\right)$$

$$\Im\left(h_i^{(m)}[k]\right) \sim \mathcal{N}\left(\mu_k, \frac{1}{2} e^{-\frac{kT_s}{\sigma_T}}\right)$$

and

$$E\left[\left|h_i^{(m)}[k]\right|^2\right] = (K + 1) e^{-\frac{kT_s}{\sigma_T}}, \ 0 \le k \le L - 1 \tag{2}$$

where K is known as *Rice factor* [12] and modelled as

$$K = \frac{2\mu_k^2}{e^{-\frac{kT_s}{\sigma_T}}} \tag{3}$$

and σ_r denotes the RMS delay spread of the communication passage and T_S is the sampling interval of the system, where $1/T_S$ is equal to the bandwidth (B) of the system [13]. In this model, one channel is assumed to be ergodic, reciprocal as well as block-wise constant having tap values static during one duty cycle. This consists of two phases (i) transmission phase (γ), and (ii) recording phase $(1 - \gamma)$ with $\gamma \in (0, 1)$ and is proportional to change in the channel over time.

2.1.1 Phase 1: Recording Phase

A down link MISO TRDMA configuration is shown below which consist of N users receiving statistically independent messages X_i $i(1{:}N)$ *from* the BS. In Fig. 1, a device named as a *time-reversal mirror* (TRM) which records the received waveform and time-reverse it which is convolved in the next transmission phase. In the recording phase, BS receives N impulse signals from intended users. In the intervening period, the TRMs at the BS registers the CIR of all the links and

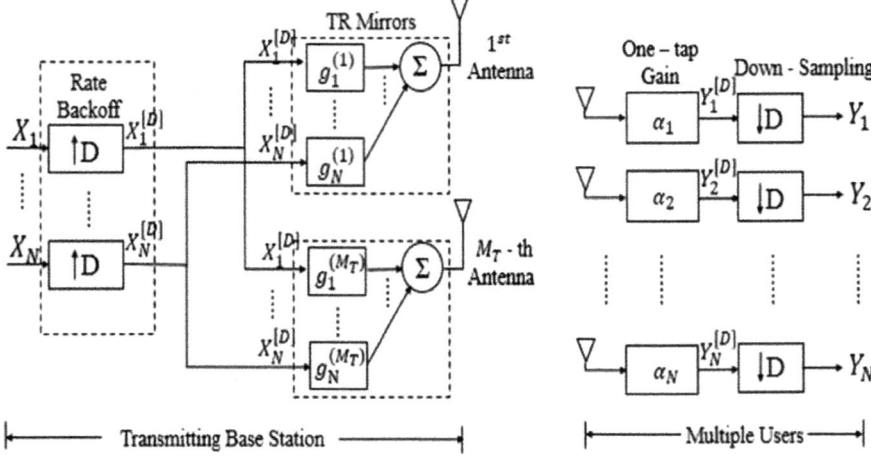

Fig. 1 MISO TRDMA multiuser downlink system

cumulate the time-reversed and conjugated description of each CIR to be used for transmission phase. Here, we neglect the thermal and quantization noise and hence may consider the recorded waveform as original CIR.

2.1.2 Phase 2: Transmission Phase

The transmission phase starts with the completion of the recording phase. Let X_i represent the string of information symbol where i varies from 1 to N and each X_i is modelled as an independent RV with zero mean and variance θ. Also, for each X_i, any two symbols of sequence are also independent, i.e.

$$E\{X_i[k]X_i[l]\} = \begin{cases} 0 & k \neq l \\ \theta & k = l \end{cases} \tag{4}$$

Each sequence X_i is upsampled by rate back-off factor D at the BS. Therefore, the ith upsampled sequence may be written as

$$X_i^{[D]}[k] = \begin{cases} X_i[k/D], & \text{if } k \bmod D = 0 \\ 0, & \text{if } k \bmod D \neq 0 \end{cases} \tag{5}$$

Thereafter, the $X_i^{[D]}[k]$ are sent to the TRMs. The ith upsampled sequence and the TR waveform $g_i^{(m)}[k]$ are convolved to generate an output of the ith TRM of mth antenna, where

$$g_i^{(m)}[k] = h_i^{(m)*}[L-1-k] \Big/ \sqrt{E\left[M_T \sum_{l=0}^{L-1} \left|h_i^{(m)}[l]\right|^2\right]} \qquad (6)$$

$g_i^{(m)}[k]$ is a normalised complex conjugate of time-reversed $h_i^{(m)}[k]$. The normalisation is obtained by the average total energy of MISO channels. Then, the outputs from the TRM banks are added at each transmitting antenna and the combined signal $S^{(m)}[k]$ from each mth antenna is transmitted into wireless channel, where

$$S^{(m)}[k] = \sum_{i=1}^{N}\left(X_i^{[D]} * g_i^{(m)}\right)[k] \qquad (7)$$

The convolution above offers an option to embed location-specific detail of each channel into the transmitted signal for the users intended for.

The ith user received the following signal:

$$Y_i^{[D]}[k] = \sum_{m=1}^{M_T}\sum_{j=1}^{N}\alpha_i\left(X_j^{[D]} * g_j^{(m)} * h_i^{(m)}\right)[k] + \alpha_i\tilde{n}_i[k] \qquad (8)$$

The ith receiver (user i) simply executes some gain adjustment of the incoming received signal before downsampling with D. Therefore, the signal $Y_i[k]$ is obtained as

$$Y_i[k] = \alpha_i\sum_{j=1}^{N}\sum_{m=1}^{M_T}\sum_{l=0}^{\frac{2L-2}{D}}\left(h_i^{(m)} * g_j^{(m)}\right)[Dl]X_j[k-l] + \alpha_i n_i[k] \qquad (9)$$

where $n_i[k] = \tilde{n}_i[k]$ is a white Gaussian additive noise.

3 Performance Parameter

Signal-to-interference ratio (SINR) and achievable sum rate are two important parameters for performance analysis of the wireless system. This section probes into the performance analysis of MISO TRDMA in terms of the above-mentioned parameters under *Rician fading* channel.

3.1 Effective SINR

Effective SINR is a very crucial performance measuring parameter for the systems in which the interferences are inherited from either the symbol (ISI) or other users (IUI) in a multi-user network. The maximum-power central peak can be obtained as

$$\left(h_i^{(m)} * g_j^{(m)}\right)[L-1] = \sum_{l=0}^{L-1} \left|h_i^{(m)}[l]\right|^2 \Bigg/ \sqrt{E\left[M_T \sum_{l=0}^{L-1} \left|h_i^{(m)}[l]\right|^2\right]} \qquad (10)$$

Since the receiver is having only one tap, the ith receiver is intended to predict $X_i\left[k - \frac{L-1}{D}\right]$ exclusively based on observation $Y_i[k]$. The remaining components of Y_i are then arranged as three components representing ISI, IUI and noise as

$$Y_i[k] = a_i \sum_{m=1}^{M_T} (h_i^{(m)} * g_i^{(m)})[L-1]X_i\left[k - \tfrac{L-1}{D}\right] \qquad \text{(signal)}$$

$$+ a_i \sum_{\substack{l=0 \\ l \neq (L-1)/D}}^{(2L-2)/D} \sum_{m=1}^{M_T} (h_i^{(m)} * g_i^{(m)})[Dl]X_i[k-l] \qquad \text{(ISI)}$$

$$\qquad\qquad\qquad\qquad\qquad\qquad\qquad\qquad\qquad (11)$$

$$+ a_i \sum_{\substack{j=i \\ j \neq i}}^{N} \sum_{l=0}^{(2L-2)/D} \sum_{m=1}^{M_T} (h_i^{(m)} * g_j^{(m)})[Dl]X_j[k-l] \qquad \text{(IUI)}$$

$$+ a_i n_i[k]. \qquad \text{(Noise)}$$

Note that the effective SINR does not affect by the one-tap gain α_i so we take it as $\alpha_i = 1$. Considering random CIRs, the signal power P_{sig} can be expressed as

$$P_{\text{sig}}(i) = \theta \left| \sum_{m=1}^{M_T} \left(h_i^{(m)} * g_i^{(m)}\right)[L-1] \right|^2 \qquad (12)$$

The powers related to ISI and IUI can also be expressed as

$$P_{\text{ISI}}(i) = \theta \left| \sum_{\substack{l=0 \\ l \neq (L-1)/D}}^{(2L-2)/D} \sum_{m=1}^{M_T} \left(h_i^{(m)} * g_i^{(m)}\right)[Dl] \right|^2 \qquad (13)$$

$$P_{\text{IUI}}(i) = \theta \sum_{\substack{j=i \\ j \neq i}}^{N} \sum_{l=0}^{(2L-2)/D} \left| \sum_{m=1}^{M_T} \left(h_i^{(m)} * g_j^{(m)}\right)[Dl] \right|^2 \qquad (14)$$

The average *effective SINR*, $\mathrm{SINR}_{\mathrm{avg}}(i)$ at user i *is expressed as*

$$\mathrm{SINR}_{\mathrm{avg}}(i) = \frac{E\left[P_{\mathrm{sig}}(i)\right]}{E[P_{\mathrm{ISI}}(i)] + E[P_{\mathrm{IUI}}(i)] + \sigma^2} \tag{15}$$

where each term is specified in (12), (13) and (14) and $E[.]$ represents the average value.

4   Achievable Sum Rate (R)

The achievable sum rate measures the total information which is an important parameter for assessment of efficiency in wireless downlink scheme under the constraint of total transmit power P, which can be given as

$$P = \frac{N \times \theta}{D} \tag{16}$$

The instantaneous effective SINR of user i with symbol variance θ for the random channels presented in the previous section, may be obtained as

$$\mathrm{SINR}(i, \theta) \triangleq \frac{P_{\mathrm{sig}}(i)}{P_{\mathrm{ISI}}(i) + P_{\mathrm{IUI}}(i) + \sigma^2} \tag{17}$$

All the terms in the above equation are explored in (12), (13) and (14).

Therefore, considering total power P, the instantaneous achievable rate $R(i)$ of user i can be formulated as

$$\begin{aligned} R(i) &= \frac{\gamma}{T_S \times B \times D} \log_2(1 + \mathrm{SINR}(i, PD/N)) \\ &= \frac{\gamma}{D} \log_2(1 + \mathrm{SINR}(i, PD/N)) \, (\mathrm{bps/Hz}) \end{aligned} \tag{18}$$

where γ denotes discount factor which indicates the fractional quantity of the transmission phase in the complete duty cycle. The information rate succeeded per unit bandwidth is referred as *spectral efficiency*.

Also, the instantaneous achievable sum rate can be formulated as

$$R = \sum_{i=1}^{N} R(i) \tag{19}$$

Averaging (19), we obtain the average value of R which is used as a comprehensive measure of the long-term performance. The average value of R can be written as

$$R_{\text{avg}} = E\left[\frac{\gamma}{D}\sum_{i=1}^{N}\log_2(1 + \text{SINR}(i, PD/N))\right] \qquad (20)$$

5 Results and Discussion

Here, we simulate the performance measuring metrics discussed in Sect. 3 for a MISO TRDMA multi-user downlink scheme. Simulations are performed for the system and channel models described in Sect. 2 with various system and channel configuration such as M_T antennas at transmitting BS and N users. Also, for each system configuration diverse channel are used with a different value of K. The length of CIR, L and RMS delay spread σ_T are taken as 257 and $128T_S$ respectively. We define a modified *SNR per symbol* to prevent the potential multi-path gain for a fair comparison and which is given as

$$\rho = \frac{P}{\sigma^2}E\left[\sum_{l=0}^{L-1}\left|h_i^{(m)}[l]\right|^2\right] = (K+1)\frac{P}{\sigma^2}\frac{1 - e^{-\frac{LT_s}{\sigma_T}}}{1 - e^{-\frac{T_s}{\sigma_T}}} \qquad (21)$$

First, we show the average Effective SINR (SINR_{avg}) varying w.r.t. ρ under various system and channel configurations in Figs. 2, 3 and 4. Then we show Average Achievable Sum Rate (R_{avg}) as a function of ρ under same system configuration in Figs. 5, 6 and 7. We consider $\gamma \approx 1$, and neglect the overhead caused by the recording phase under a slow fading channel. In all the results, solid line curves, dash–dash curves and dash-dot curves are obtained for $K = 0, 0.02$ and 0.04 respectively.

Fig. 2 Impact of M_T and K on average effective SINR when $D = 8$, $N = 5$

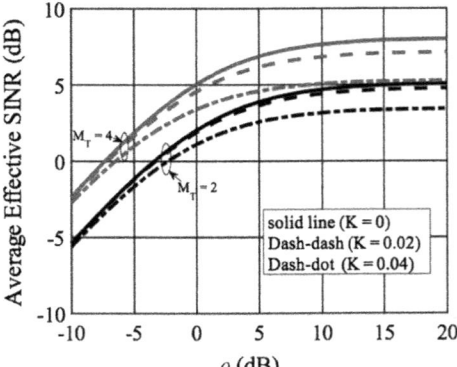

Fig. 3 Impact of D and K on average effective SINR when $M_T = 4$, $N = 5$

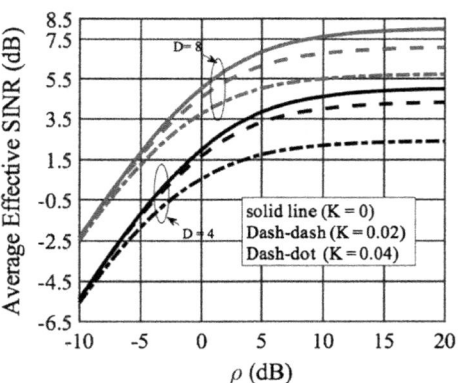

Fig. 4 Impact of N and K on average effective SINR when $D = 8$, $M_T = 4$

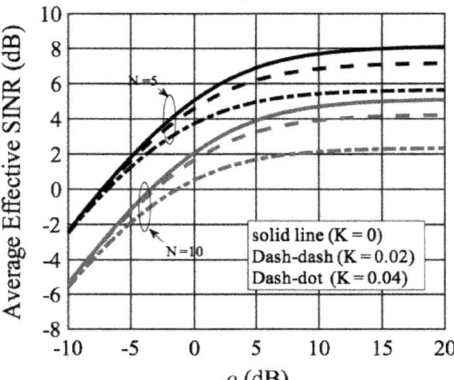

Fig. 5 Impact of M_T and K on average achievable sum rate when D = 8, N = 5

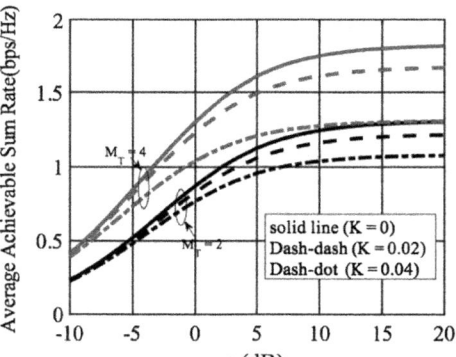

Fig. 6 Impact of D and K on average achievable sum rate when $M_T = 4$, $N = 5$

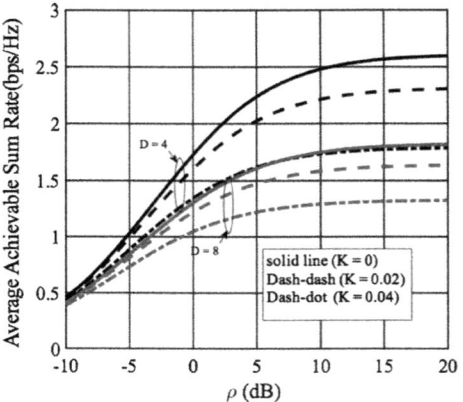

Fig. 7 Impact of N and K on average achievable sum rate when $D = 8$, $M_T = 4$

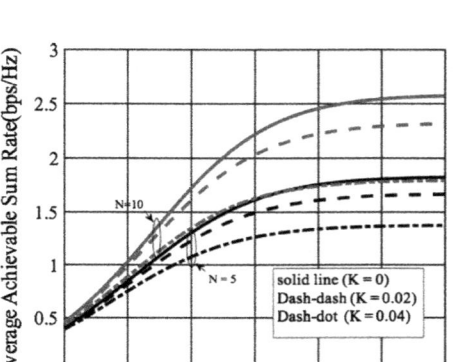

In Fig. 2 results are shown with $D = 8$ and $N = 5$ for $M_T = 2, 4$ and $K = 0, 0.02, 0.04$. Figure 2 represents that average effective SINR increases as the number of antenna at BS increases for same channel characteristics, i.e. same K value. Rice factor K represents the ratio of specular power and scatter power so as the part of specular power increases over scatter power the performance of system degrades as K value increases.

In Fig. 3, we present $SINR_{avg}$ with $M_T = 4$ and $N = 5$ for $D = 4, 8$ and $K = 0, 0.02, 0.04$. It is observed that for the same value of K, a higher D value results in better system performance, which reduces ISI and IUI. For the same value of D, as K value increases the $SINR_{avg}$ value decreases.

Figure 4 presents the results of $SINR_{avg}$ with $D = 8$ and $M_T = 4$ for different number of user and $K = 0, 0.02, 0.04$. From Fig. 4, one can confirm that due to presence of IUI, as number of user increases, the Average Effective SINR value decreases for same value of K, which creates a situation of trade-off between number of users (can be a metric for network capacity) and quality of signal reception for individual user.

In Fig. 5, results of R_{avg} is depicted with $D = 8$ and $N = 5$ for $M_T = 2, 4$ and $K = 0, 0.02, 0.04$. As we have seen, for increasing number of M_T, effective SINR value increases. Therefore, according to (20), for the same value of D and channel realization, the sum rate increases with increasing M_T. As mentioned earlier with increasing value of K for a particular system configuration, achievable sum rate performance also degrades with K.

In Fig. 6, we have plotted the results of R_{avg} with $N = 5$ for $D = 4, 8$ and $K = 0, 0.02, 0.04$ under the impact of D. From Fig. 3, it is clear that a larger D improves the effective SINR for a particular value of K but as D in denominator of (20) and SINR inside the logarithm function, so overall sum rate decreases with D.

In Fig. 7, the results of R_{avg} are plotted with $D = 8$ and $MT = 4$ or different number of users ($N = 5, 10$) and $K = 0, 0.02, 0.04$. We can see that a larger N for a particular value of K, results in higher achievable sum rate, because with increase in N number of multiplexed data stream increases and as a result in (20), achievable sum rate increases with N.

6 Conclusion

In this paper, MISO TRDM Access scheme in multi-user downlink system is investigated under multi-path *Rician* fading channels. We analysed the system performance parameters such as effective SINR, achievable sum rate with a various value of K and various system configuration. Finally, we observed that system performance improves monotonically with a number of antennas at BS. A larger value of rate back-off factor improves the system performance with respect to effective SINR but degrades in terms of achievable sum rate and vice versa is true in case of a number of users. System performance declines with an increase in Rice factor in all the system configurations. So, it is inferred that MISO TRDMA works better in rich scattering environment than the environment with the existence of specular power. A comprehensive study of TRDMA system with multiple antenna at each user (i.e. MIMO TRDMA) is included in our future work.

References

1. "Cisco visual networking index: Mobile data traffic forecast update, 2015–2020," CISCO, San Jose, CA, USA, Feb. 2016.
2. F. Khan, *LTE for 4G Mobile Broadband: Air Interface Technologies and Performance*, Cambridge Univ. Press, 2009.
3. D. Tse and P. Viswanath, *Fundamental of Wireless Communication*. Cambridge University Press, 2005.
4. Z. Pi and F. Khan, "An introduction to millimeter-wave mobile broadband systems," *IEEE Commun. Mag.*, vol. 49, no. 6, pp. 101–107, Jun. 2011.

5. T. S. Rappaport *et al.*, "Millimeter Wave Mobile Communications for 5G Cellular: It Will Work!," in *IEEE Access*, vol. 1, no., pp. 335–349, 2013.

6. Y. Chen, Y.-H. Yang, F. Han, and K. J. R. Liu, "Time-reversal wideband communications," *IEEE Signal Process. Lett.*, vol. 20, no. 12, pp. 1219–1222, Dec. 2013.

7. B. Wang, Y. Wu, F. Han, Y. H. Yang and K. J. R. Liu, "Green Wireless Communications: A Time-Reversal Paradigm", in *IEEE Journal on Selected Areas in Communications*, vol. 29, no. 8, pp. 1698–1710, September 2011.

8. G. Lerosey, J. de Rosny, A. Tourin, A. Derode, G. Montaldo, and M. Fink, "Time reversal of electromagnetic waves," Physical Review Letters, vol. 92, pp. 193904(3), May 2004.

9. B. E. Henty and D. D. Stancil, "Multipath enabled super-resolution for rf and microwave communication using phase-conjugate arrays," Phys. Rev. Lett., vol. 93, no. 24, p. 243904(4), 2004.

10. M. Emami, M. Vu, J. Hansen, A. J. Paulraj and G. Papanicolaou, "Matched filtering with rate back-off for low complexity communications in very large delay spread channels," *Conference Record of the Thirty-Eighth Asilomar Conference on Signals, Systems and Computers, 2004.*, 2004, pp. 218–222 Vol. 1.

11. F. Han, Y. H. Yang, B. Wang, Y. Wu and K. J. R. Liu, "Time-Reversal Division Multiple Access over Multi-Path Channels," in *IEEE Transactions on Communications*, vol. 60, no. 7, pp. 1953–1965, July 2012.

12. G. L. Stuber, *Principles of Mobile Communications*, 2nd edition. Kluwer, 2001.

13. A. J. Goldsmith, *Wireless Communication*. Cambridge University Press, 2005.

Analysis of Software Development Life Cycle Models

Amninder Singh and Puneet Jai Kaur

1 Introduction

Nowadays the computer technology has rapidly changed, everyone wants the best software system that is fast, cheap, bug free, maintainable, and time saving that can fulfill all demands. Such type of product needs a working scenario that covers all the aspect. Best way to create an application is to choose software development life cycle also known as application development life cycle. It creates software in systematic manner and assures the product meets the needs of customer. The products are economical and highly excellent. It provides better resource management and plan for workflow. There are five models based on SDLC: Water fall model, Incremental model, RAD model, Spiral model, V model, Agile model. There are two types of software development methodologies, heavy weight and light weight. There is no model best or worst among both type of methodologies in different circumstances. The selection of best methodology depends on its nature, skills of team members and management criteria [1]. In SDLC software usability is an important attribute to assure quality. Scarcity in usability cause loss in way of cost, trust, and reputation. For this purpose, a new technique called U-SDLC is introduced. It concentrates on key attribute of every phase that gives best quality product with less number of errors [2]. Software maintenance is always a big issue for SDLC models to overcome from problem there is a new model created that concern with maintainability tasks called M-SDLC. It gives best maintenance practices in each phase of SDLC. That helps to create products that need low maintenance [3]. Management of any project providing better supremacy and

A. Singh (✉) · P. J. Kaur
UIET, Panjab University, Chandigarh, India
e-mail: amninder.saini001@gmail.com

P. J. Kaur
e-mail: puneet@pu.ac.m

© Springer Nature Singapore Pte Ltd. 2019
V. Nath and J. K. Mandal (eds.), *Proceeding of the Second International Conference on Microelectronics, Computing & Communication Systems (MCCS 2017)*, Lecture Notes in Electrical Engineering 476, https://doi.org/10.1007/978-981-10-8234-4_55

minimize the workload by diving complex task into small ones. But none of the SDLC models provide management for this purpose a new hypothetical SDLC model introduce that facilitate change, incident and release management. It creates three-dimensional model containing user, owner, and developer of the project. It helps to find more bugs and to cure them. It also gives better accuracy [4]. In this paper Sect. 2 contains description about various SDLC models. Section 3 is about analysis or comparison between models. The last section concludes all the paper's contents and at last references.

2 Software Development Life Cycle Models

SDLC: Software development life cycle model is a technique to create a software product that assures quality and easily fulfill the needs of the customer. It works in a systematic manner that is economical and efficient for the project management technique. There are six phases used in it that are Planning, Analysis, Design, Implement, Test, and Maintain.

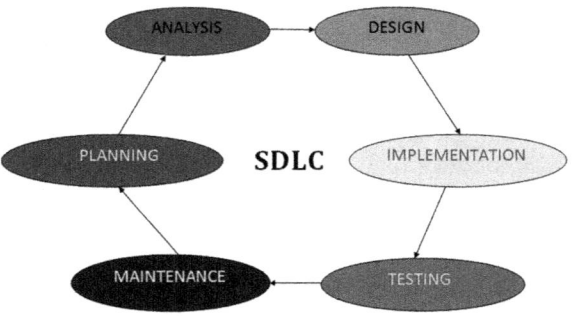

Waterfall Model: It is a linearly processed model, in which workflow stably goes downward. The model is processing in one direction only and very simple to use. Once a phase is processed it does not reuse again for further processing and each phase must be executed fully before next phase originates. Every other model considers it as a base model. After completion of every phase or stage, a review takes place to check if the project is on right way. If it does not show results as desired, taking a decision whether it to continue or eliminate the project is important. It is used for small project where all requirements should be clear. There are several phases introduce in waterfall model which are requirement analysis, system design, implementation, testing, deployment, and maintenance [5].

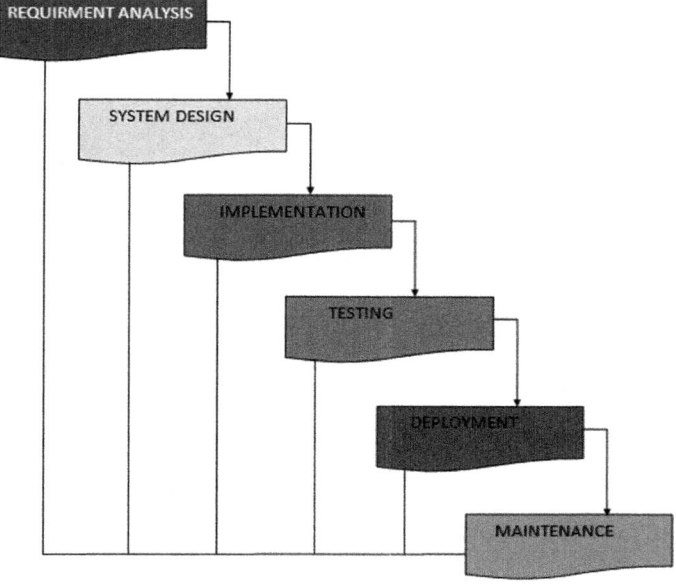

The initial phase of the waterfall is requirement analysis for gathering information about project; find the main purpose or objective of user or customer. Several brainstorming sessions, interviews are conducted during this phase. Next is design phase where all strategies, layouts, and operational details are created. It provides two types of design high-level design and low-level design. The third phase is implementation phase where we write code for design. Then testing phase begins for quality check to find errors or bugs in the product. And then checking whether product's result is according to objective for customer acceptance. Next is deployment phase to check whether test criteria met, whether the product is error free after test and the environment of the product is up to date. Maintenance phase is to fix bugs and problem arise during test phase, make sure the product will work properly, if any change is done write a code, update it, etc.

Incremental Model: It is an updated version of waterfall model. The number of modules created are called build. First of all the requirements are gathered by thorough analysis. Then a first build starts which does design development, testing, implementation. The results of first build are used as input for next build, so further improvement has done on the basis of previous results. All the processing are done incrementally until the product is finished. Incremental model provides flexibility to user. It promotes the maintainability in products. Because each build process small module, it is easy to use and simple to understand. Try to make working model first and analyse their results after that you can build the another models as per needs.

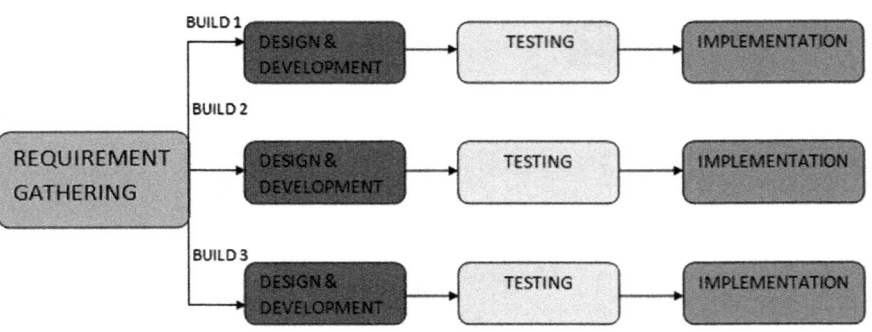

RAD Model: It stands for Rapid Application Development model. It is an advanced version of incremental model. It uses parallel processing which means all the builds are run at same time. Main purpose of the model is to deliver something to see and use. It also provides feedback on each build. There are five phases introduced by the model. The initial phase is business modeling phase which recognizes the data flow between different business methodologies.

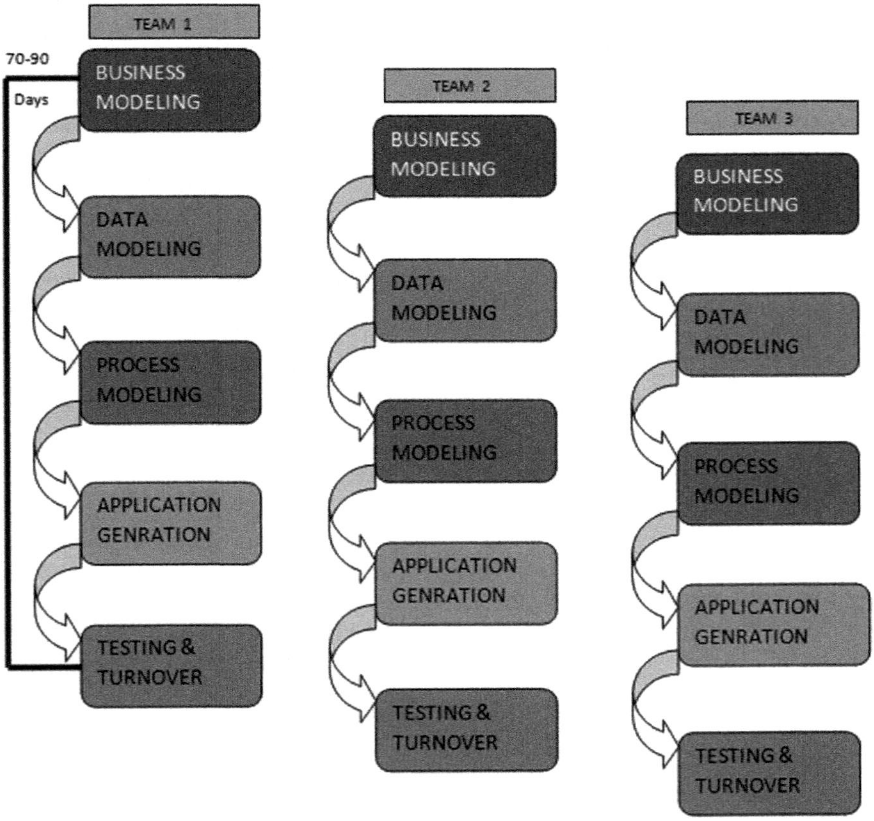

All the facts are gathered and documented during this phase. Next phase is data modeling phase and is used to describe resources according to information collected. Third phase is process modeling phase which uses resources that are described during the second phase and convert them to achieve the objective. Next phase is application generation phase which generates code on the basis of the process model. The last phase is testing phase to find the bugs and check whether the product is according to desire or not for acceptance. The model used for large project and provide more flexibility as compared to other.

Spiral Model: It is alike to incremental model which is more dominant risk analysis. There are four phases in spiral model: Analysis phase, Evaluation, Development, and Planning. The product passes through each phase repeatedly in iterations. The results of first iteration are used for evaluation so that product is free from anomalies and best changes should be done to get the best product [6].

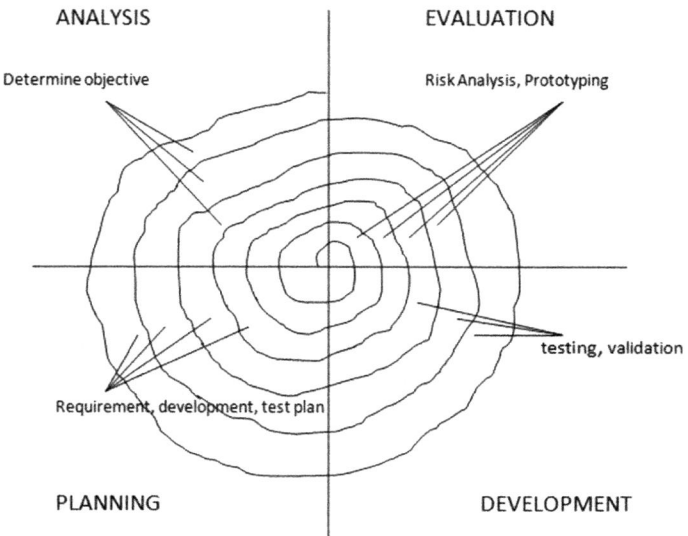

V Model: It is similar to waterfall model and in addition, every phase is simultaneously tested. Verification and Validation make the V shape of the model. Every stage has its own test case. Once a stage completes its process, its results are validated at the same time. Each phase must be completed before next phase starts [7]. There are five stages in v model and each consists of their own testing criteria.

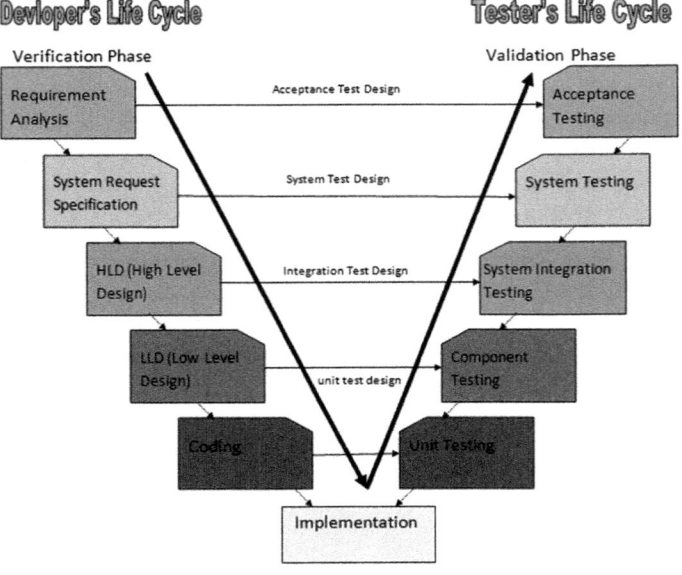

Agile Model: It is an advanced type of incremental model. Software system is created in progressive, rapid cycle. These results in small progressive deliveries with each delivery built on previous functionality. Agile methodology is more flexible as compared to other. There are some other models that come under agile methodology such as Scrum, Extreme programming (XP), and Rational unified process (RUP) [7]. It is economical, efficient, and provides better quality as compared to the traditional model.

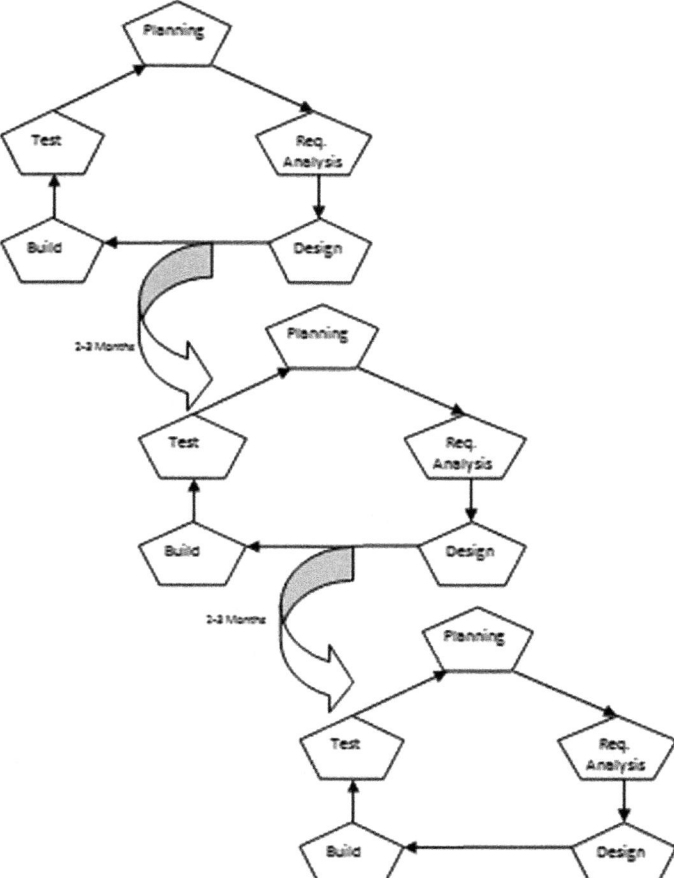

3   Comparison Between Methodologies

As we studied, there are a number of SDLC models. Each has its own specification and working scenario. During the survey, a comparison is done between 4 models on the basis of some characteristics [8], so other remaining models are added to it with more characteristics and a table is made. Most of the models do planning in early stages expect RAD model because it uses small iterations that help to provide results in small time for feedback then further improvements are done on basis of them. Next is requirement specifications that are considered as the first step in each and every model. Further, if we talk about framework type only two models operate in a linear manner, Waterfall and V model, and other models use iterations. The traditional model like waterfall, incremental, V, and spiral model need a big team size. But agile methodology and RAD model can be handled by the small team. Maintenance is a big issue in SDLC models except for RAD model, every other model can provide maintenance up to very little extent. The flexibility of models depends on there working scenarios the waterfall model and v model are working in a linear manner if one stage is complete then we cannot access it again which means they are irreversible. On the other hand, remaining models use iteration which provides too much flexibility to users. Next, the capabilities of the models, only spiral or V model is able to handle large project rest of the model are used for small or medium type projects. Next is detailed documentation, there are three models that is waterfall, spiral and V model which require full documentation and remaining need little of it. Time frame or duration of models depends on there working era. Only RAD and V model complete its processing in short time because they use parallel processing (Table 1).

4   Conclusion

In this paper, analysis of different system development life cycle models has been done by considering various parameters. During the review, we studied several models each has its own working scenario and benefits. Every model minimizes the anomalies came in the previous model. There are two types of model Traditional (waterfall, incremental, RAD, spiral, and V) and Agile (Scrum, RUP, XP). Nowadays, most of the companies use V model because it takes less time to create a software system and parallel testing facilitates better quality that gives a sound product. Large size projects are easily handled by V model and that's why most of the organizations prefer it. An agile terminology helps to handle the medium type of projects and provides more flexibility as compared to the traditional model. Although we have several models but there are some factors that need some improvement, for example, maintenance, improvement in management factors, and need to create a hybrid model by combining existing ones. That should give a better result.

Table 1 Comparison between different SDLC models

S. No.	Properties of model	Water-fall model	Incremental model	Spiral-model	Rad-model	V-model	Agile-model
1.	Planning in early stage	Yes	Yes	Yes	No	Yes	Yes (but little)
2	Returning to an earlier phase	No	Yes	Yes	No	Yes	Yes
3	Handle large project	Not Appropriate	Not appropriate	Appropriate	Not appropriate	Appropriate	Not appropriate
4	Detailed documentation	Necessary or full	Yes but not much	full	Low	Necessary	low
5	Delivery time	Late	Early	Early	Early	Late	Early
6	Cost	Medium	High	Expensive	Low	Low (as compared to waterfall model)	Expensive (as compared to waterfall model)
7	Requirement specifications	Beginning	Beginning	Beginning	Time-boxed release	Beginning	Beginning
8	Flexibility to change	Difficult	Easy	Easy	Easy	Difficult	Easy
9	User involvement	Only at beginning	Intermediate	High	Only at the beginning	Medium	Medium
10	Maintenance	Least	Promotes maintainability	Typical	Easily maintained	Least	Yes provide little bit
11	Duration	Long	Very long	Long	Short	Short	Very long
12	Risk involvement	High	Low	Low	Low	High	High
13	Framework type	Linear	Linear + iterative	Linear + iterative	Linear	V shape	Linear + iterative

(continued)

Table 1 (continued)

S. No.	Properties of model	Water-fall model	Incremental model	Spiral-model	Rad-model	V-model	Agile-model
14	Testing	After completion of coding phase	After every iteration	At the end of the engineering phase of coding	After completion of coding	At each and every phase	In each Iteration
15	Overlapping phases	No	Yes (as parallel development is there)	No	Yes	Yes (as parallel testing is there)	No
16	Re-usability	Least possible	To some extent	To some extent	Yes	Least possible	To some extent
17	Timeframe	Very Long	Long	Long	Short	Short (as compared to waterfall model)	Long
18	Working software availability	At the end of the life cycle	At the end of every iteration	At the end of every iteration	At the end of the life cycle	At the end of the life cycle	At the end of every iteration
19	Objective	High assurance	Rapid development	High assurance	Rapid development	High assurance	Rapid development
20	Team size	Large team	Large team	Large team	Small team	Large team	Small or 10 people
21	Customer control over administrator	Very low	Yes	Yes	Yes	Low	Yes
22	Prototype build	No	Yes	Yes	No	No	No

References

1. Ben-Zahia, Mohamed A., and Ibrahim Jaluta. "Criteria for selecting software development models." *Computer & Information Technology (GSCIT), 2014 Global Summit on*. IEEE, 2014.
2. Velmourougan, S., et al. "Software development Life cycle model to build software applications with usability." *Advances in Computing, Communications and Informatics (ICACCI), 2014 International Conference on*. IEEE, 2014.
3. Velmourougan, S., et al. "Software Development Life Cycle Model to Improve Maintainability of Software Applications." *Advances in Computing and Communications (ICACC), 2014 Fourth International Conference on*. IEEE, 2014.
4. Ragunath, P. K., Velmourougan, S., Davachelvan, P., Kayalvizhi, S., & Ravimohan, R. (2010). Evolving a new model (SDLC Model-2010) for software development life cycle (SDLC). International Journal of Computer Science and Network Security, 10(1), 112–119.
5. Davis A.M, Bersoff E. Hm Comer E.R, A strategy for comparing alternative oftware development life cycle models, IEEE transactions on Software Engineering, Volume: 14, Issue: 10 Publication Year: 1988, Page(s): 1453–1461.
6. S Velmourougan, P Davachelvan, and R Baskaran, Software Reliability Qualification Model, International Journal of Performability. 8(2012) 437–446.
7. Boehm B, "A Spiral of software development and Enhancement", ACM SIGSOFT Software Engineering Notes, ACM, 1986.
8. Thitisathienkul, Patra, and Nakornthip Prompoon. "Quality assessment method for software development process document based on software document characteristics metric." Digital Information Management (ICDIM), 2014 Ninth International Conference on. IEEE, 2014.

Design of Narrowband 2.69 GHz CMOS Low-Noise Amplifier for WiMAX Application

Sumit Singh, Namrata Yadav, Abhishek Pandey, Deepak Prasad and Vijay Nath

1 Introduction

High data rate and high mobility have proved to be the backbone of modern wireless communication. On the other hand, the introduction to WiMAX in the wireless communication takes it to some other level due to its low sensitivity and high dynamic range in the receiver system.

To provide larger area coverage, technology has moved from WPAN, WLAN to WMAN which offers mobility in a larger area. IEEE standard for WMAN is IEEE 802.16 [1], established in the year 1999. Due to higher coverage and high throughput performance, it can be used for cellular backhaul support. It can effectively work in a disaster situation [2].

IEEE 802.16a is an evolution of IEEE 802.16 which is known as WiMAX. It is an appropriate technology for last mile connection. It offers higher range, wider bandwidth, and high data rate. However, system design in this frequency becomes more complex due to interference. On the other hand, IEEE 802.16e gives freedom to work on the frequency range of 2–6 GHz within three bands, namely, (1) 2.5–2.9 GHz, (2) 3.4–3.6 GHz and (3) 5.2–5.9 GHz. WiMAX offers a data rate of 70 Mbps and covers up to 50 km of range [3].

The evolution and advancement in CMOS technology raise many hopes in recent technology. Due to CMOS technology, it was possible to integrate all the elements of the transceiver in a single chip [4]. A low-noise amplifier (LNA) is the first key module of the long chain receiver which is used to chop off the noise at the

S. Singh · N. Yadav · A. Pandey · D. Prasad (✉) · V. Nath
VLSI Design Group, Department of Electronics and Communication Engineering,
Birla Institute of Technology, Mesra, Ranchi 835215, Jharkhand, India
e-mail: prasaddeepak007@gmail.com

V. Nath
e-mail: vijaynath@bitmesra.ac.in

© Springer Nature Singapore Pte Ltd. 2019
V. Nath and J. K. Mandal (eds.), *Proceeding of the Second International Conference on Microelectronics, Computing & Communication Systems (MCCS 2017)*, Lecture Notes in Electrical Engineering 476, https://doi.org/10.1007/978-981-10-8234-4_56

receiver frontend [5]. The features of low-noise figure, high gain and high linearity made it usable for WiMAX applications.

The front end of a transceiver is mainly subdivided into two parts, i.e. receiving part and transmitting part. The RF signal received by the receiver is too weak to use. The surrounded noise and other interferes need to be removed from the received RF signal. In general, the received RF signals are around −100 dBm which makes signal-to-noise ratio almost negligible. As LNA comes first in the receiver chain, its gain and noise figure hold significant importance in the overall performance of the receiver. Hence, it is advised to keep low-noise figure of LNA to maintain the system's noise figure low. On the other hand, the gain of LNA should neither be too low which may increase the noise contribution nor be too high to degrade the linearity of the system [6].

In superheterodyne receiver, the RF filter and image rejection filter which is matched to be 50 Ω is placed before and after the unit, LNA. Impedance matching is highly required in LNA, where the discrete components are used, which are too expensive. Hence, an alternate less expensive architecture should be proposed. The introduction of VLSI reduces the power consumption up to milliwatt. Hence, a low power LNA has been proposed in this research paper. The features and specifications of an ideal LNA are listed in Table 1 [7].

The remaining of the paper is organized as follows. Section 2 demonstrates description of cascode LNA, Sect. 3 demonstrates the result and discussion. Finally, Sect. 4 binds the conclusion on the proposed narrowband low-noise amplifier (LNA).

2   Cascode LNA Circuit

Figure 1 shows the schematic of proposed LNA. CS and CG are two widely used transistor configuration in CMOS LNA circuit. CS LNA has high gain and good noise performance [8]. In case of CG which gives low power, robust against parasitic and stable circuit but has weak noise performance [9]. So, in this proposed LNA CS configuration has been used over CG. By placing an inductor in the course of a CS stage, which is well known as inductive source degenerated is obtained.

Features	Specification
Less noise	NF < 2.5 dB
Good linearity	IIP3 > −10 dBm
Moderate gain	Av > 15 dB
Less power consumption	P < 25 mW
Narrowband design	F = 2.69 GHz
Other characteristics	Power supply of 1.5 V 0.45 μm process 50-Ω input impedance

Table 1 Characteristic and specification of the LNA

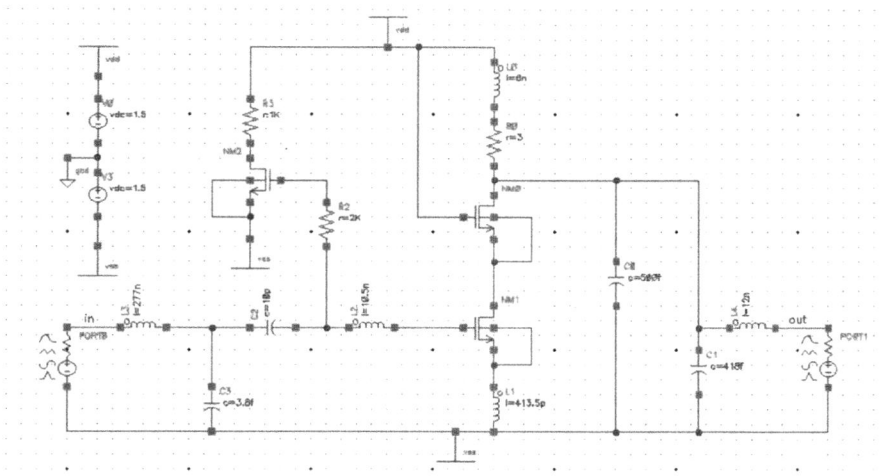

Fig. 1 Proposed low-noise amplifier

The reason behind adding an inductive source degenerated is to offer a perfect match without affixing any noise. It is used an inductor as a source degeneration device and an additional inductor connected to its gate.

Proposed LNA is designed by cascoding of two transistors. Cascode LNA guarantees high-power gain, low power consumption, good noise figure and high reverse isolation [10]. In lower band of microwave frequencies, the noise sources of upper cascode transistor are degenerated by the lower transistor output impedance [11]. Consequently, cascode stage has superior noise performance.

M1 is the CS transistor with the high width-to-length ratio (approximately 1000) and larger g_m (transconductance). M0 is a cascode transistor which isolates its output from the input, giving a good stability to circuit whereas the length of the entire transistors is fixed for this LNA. Table 2 shows the MOSFET parameters or transistor ratio (W/L) as used in proposed LNA circuit. In this paper, LNA is designed as described in Shaeffer et al. [13]. Without considering cascode transistor, W could be defined as follows

$$W = \frac{3}{2} \frac{C_{gs}}{C_{ox}L_{min}}$$

Transistor	Width (W) (μm)	Length (L) (nm)
M0	5	90
M1	3.5	90
M2	0.5	90

Table 2 Transistor ratio (W/L)

where

C_{ox} Oxide capacitance ($C_{ox} = \varepsilon_{ox}/T_{ox}$)
L_{min} Minimum length (which is 90 nm in our case)
W Channel width

From Noise theory, it is clear that the noise performance of any given cascade arrangement of multipoint network is highly dependent on first few stages of that network which is famously known as 'Friis formula' [11] is shown below

$$\text{NF front}_{end} = \text{NF}_{LNA} + \left(\frac{\text{NF}_{subsequent} - 1}{G_{LNA}} \right) \tag{1}$$

$\text{NF}_{subsequent}$ Referred input noise figure
NF_{LNA} Noise figure of LNA
G_{LNA} Gain of LNA

The LNA gain can reduce noise produced by subsequent stages in the front end. The noise of the LNA is directly fed to the receiver front end. That is why LNA should have high gain, low noise and low power consumption for wireless applications.

2.1 Stability Analysis

The main concern about any circuit is its stability. If a circuit is not stable, then that circuit would not function properly. Stability of LNA determined by K (Rollet's factor) and Δ_s. To make a circuit stable it must satisfy following criteria. Where

$$K > 1$$

$$\text{and } \Delta_s < 1$$

where K and Δ_s can be defined as follows:

$$K = \frac{1 - |S_{22}|^2 - |S_{11}|^2 + |\Delta_s|}{2|S_{12}S_{21}|} \tag{2}$$

$$\Delta_s = S_{11}S_{22} - S_{12}S_{21} \tag{3}$$

where

S_{11} input reflection coefficient with the output matched.
S_{21} forward transmission gain or loss.
S_{12} reverse transmission or isolation.
S_{22} output reflection coefficient with the input matched.

2.2 Small Signal Analysis

The small signal model for proposed LNA circuit is shown in Fig. 2.

Where C_{gs} represents gate to source capacitance of the transistor M1, which is identified as C. V_{gs} comprises as gate to source voltage of the transistor M1. g_m specified as transconductance of transistor M1. After applying the Kirchhoff's Voltage Law (KVL) at the input, it can be written as

$$V_{in} = I_{in}jw(L_g + L_s) + \frac{I_{in}}{jwC} + I_o jwL_s \tag{4}$$

At the output, it is found that

$$V_{gs} = \frac{I_{in}}{jwC} \tag{5}$$

$$I_o = g_m V_{gs} = \frac{g_m I_{in}}{jwC} \tag{6}$$

By substituting Eq. (6) in Eq. (4), we get

$$V_{in} = I_{in}jw(L_g + L_s) + \frac{I_{in}}{jwC} + \frac{g_m I_{in}}{jwC}jwL_s \tag{7}$$

$$V_{in} = I_{in}\left[jw(L_g + L_s) + \frac{1}{jwC} + \frac{g_m L_s}{C}\right] \tag{8}$$

So, we get input impedance as

$$Z_{in} = \frac{V_{in}}{I_{in}} = \left[jw(L_g + L_s) + \frac{1}{jwC} + \frac{g_m L_s}{C}\right] \tag{9}$$

For impedance matching, there are certain conditions which must be certified

$$Re[Z_{in}] = \frac{g_m L_s}{C} = R_s \tag{10}$$

Fig. 2 Small signal model

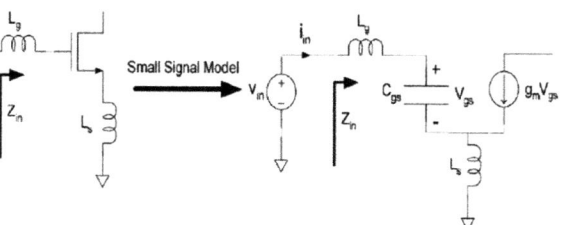

$$\text{Im}[Z_{\text{in}}] = jw(L_g + L_s) + \frac{1}{jwC} = 0 \tag{11}$$

In real part, it needs to be equal with 50 Ω. In case that input impedance matching operated at centre frequency, then imaginary term has to be nullified. So, we can write as

$$W_c = 1 \Big/ \sqrt{C(L_g + L_s)} \tag{12}$$

At lower significance values, if
$W < W_c$, it will be capacitive in nature.
$W > W_c$, it will be inductive in nature.

Calculation of NF is an important consideration for any LNA circuit. Calculation of noise performance of any circuit is called as Noise Factor (F). Generally, it exhibits in decibel and commonly cited as Noise Figure (NF).

$$NF = 10 \log_{10} F$$

where F is defined as

$$F = \frac{\text{SNR}_{\text{in}}}{\text{SNR}_{\text{out}}} \tag{13}$$

Or

$$F = \frac{\text{SNR}_{\text{in}}}{\text{SNR}_{\text{out}}} = \frac{P_{\text{sig}}/P_{\text{RS}}}{\text{SNR}_{\text{out}}} \tag{14}$$

where

SNR_{in} signal-to-noise ratios at i/p.
SNR_{out} signal-to-noise ratios at o/p.
P_{sig} input signal power.
P_{RS} source resistance noise power.

It follows that

$$P_{\text{sig}} = P_{\text{RS}} \cdot F \cdot \text{SNR}_{\text{out}} \tag{15}$$

and

$$P_{\text{RS}} = \frac{4kTR_s}{4} \frac{1}{R_{\text{in}}} = kT \tag{16}$$

3 Result and Discussion

In this paper, the proposed LNA circuit has been simulated in Cadence Virtuoso environment gpdk 45 nm CMOS technology. This proposed LNA is operated at 1.5 V power supply. Before moving to analysing the further results, there are two things to keep in mind for this proposed LNA:

(1) Any parasitic capacitance for this circuit will not be considered. It will be considered during the layout design of this LNA.
(2) This circuit will function in RF range and then uses the RF CMOS for RF domain, but here in this paper considered a normal CMOS. Because RF CMOS design is the substantial variability of device and temperature and circuit parameters with process [12]. Also, gives extremely small size to avoid parasitic elements [12]. So normal CMOS (proficiently work up to frequency 60 GHz) is used which is available in gpdk 45 nm CMOS technology.

The circuit became stable for K > 1 and $\Delta_s < 1$. The graph of stability factors is shown as follows in Figs. 3 and 4.

Characteristics of S-parameters have been obtained from the circuit. The result of proposed LNA gives voltage gain $S_{21} = 27.7$ dB. The input return loss $S_{11} = -7.957$ dB. The reverse transmission $S_{12} = -50.1$ dB which is demonstrated in bellow Figs. 5, 6 and 7.

The Noise Figure of LNA has been demonstrated as follows (Fig. 8).

Linearity of system can be determined by maximum allowable signal to its input. But in practical every system shows some degree of nonlinearity. The nonlinear behaviour of any system is directly proportional to distortion of signal. 1 dB compression point (P_{1dB}) and third-order interception point (IP3) are the most commonly used measure of nonlinearity [12, 13].

Figure 9 shows the gain compression. At the point where power gain drops below 1 dB from the idealized curve is referred as 1 dB compression point. To avoid nonlinearity in a system it must work several decibels below this level. In our proposed LNA circuit, it is −10.8802 dBm as shown in Fig. 9.

Fig. 3 *K*-factor plot of LNA

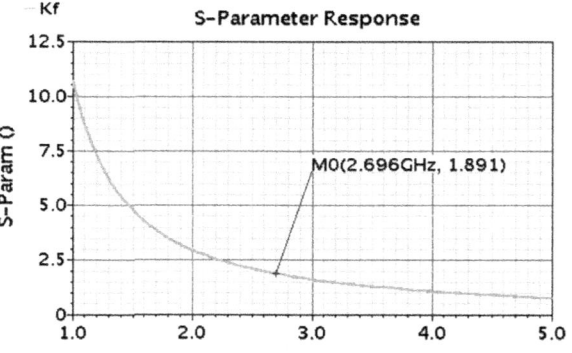

Fig. 4 Δ_s plot of LNA

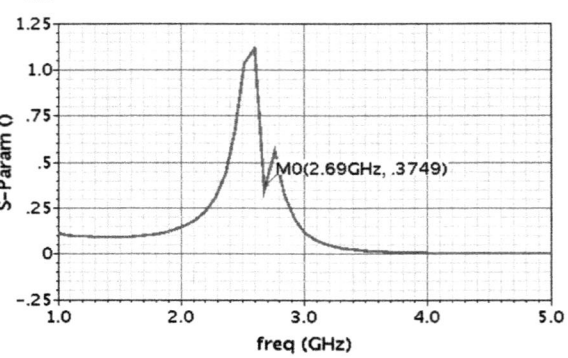

Fig. 5 S_{21} (Gain) plot of LNA

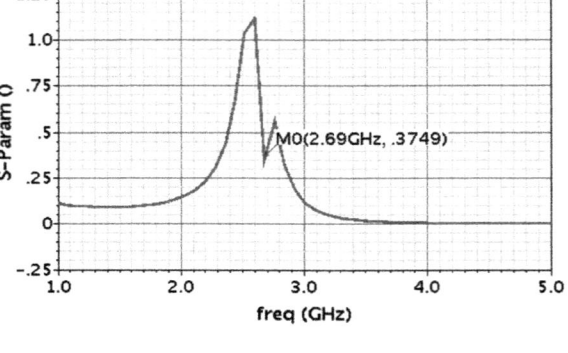

Fig. 6 Input returns loss S_{11} versus frequency

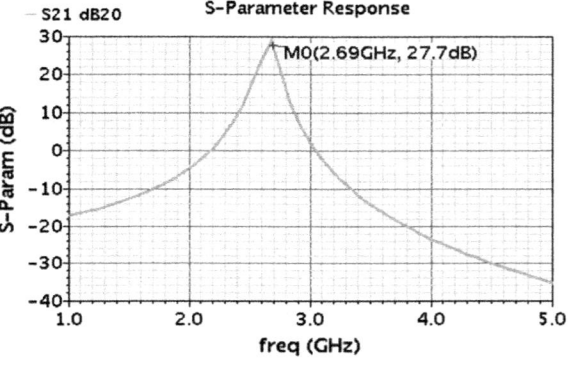

Fig. 7 Revers transmission S_{12} versus frequency

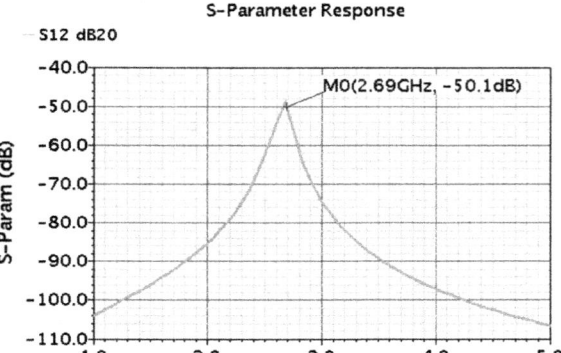

Fig. 8 NF of LNA with 1.5 V in power supply

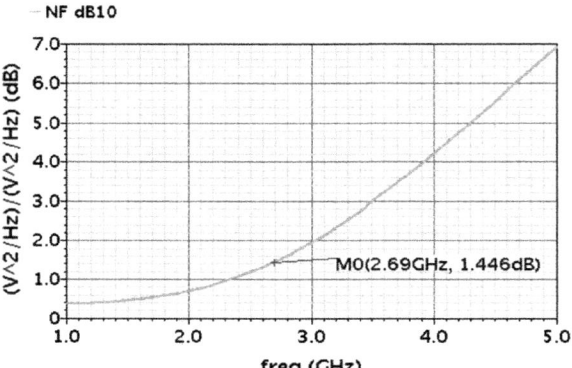

Fig. 9 1-dB compression point (P_{1dB})

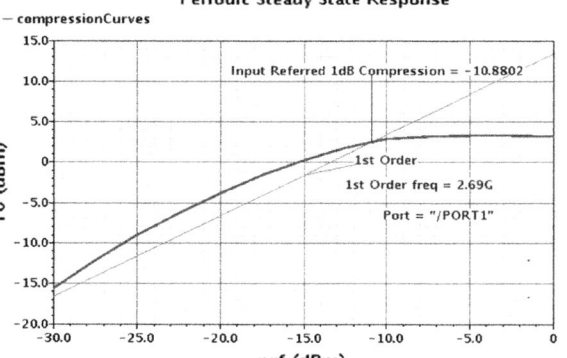

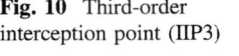

Fig. 10 Third-order
interception point (IIP3)

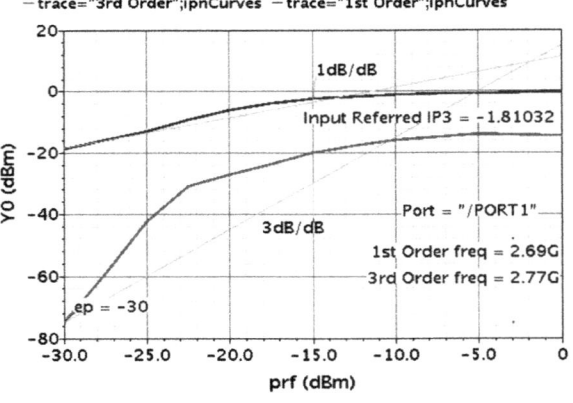

If a nonlinear system is fed with two signals with different frequencies, then output result shows that for the input frequencies some components are not harmonic. Nonlinearities of analog circuits often describe by harmonic distortion, certain cases in RF system require other measures of nonlinearity behaviour. Commonly used is the 'third-order interception point measured by a two-tone'. Third-order input intercept point is shown in Fig. 10. For proposed LNA circuit, it is −1.81 dBm.

The total power consumption of the proposed LNA circuits is 18.06 mW for 1.5 V power supply as shown in Fig. 11.

Process corners define when a circuit work in a real environment then what are the different kind of effects and variations in gain and NF of a circuit. By considering in different temperature, inaccuracies and other parameters variation. Usually, all process corners are located in the model library. For example, process corners include corners for: (TT, FF, SS, NN). These three corners (TT, FF, SS) are called even corners, because both types of devices are affected evenly, and

Fig. 11 Total power
consumption of the proposed
LNA

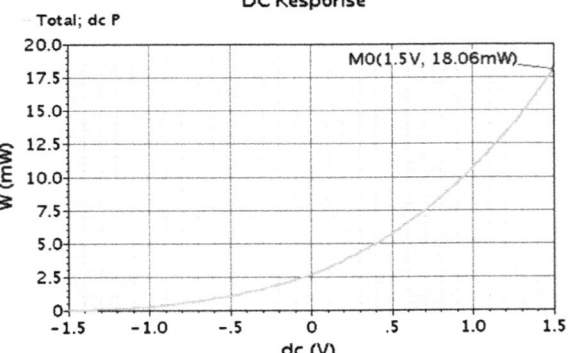

Table 3 Process corner summary of the proposed LNA

Process corners	S_{11} (dB)	S_{12} (dB)	S_{21} (dB)	S_{22} (dB)	NF (dB)
SS	−4.06	−51.62	26.79	−2.633	2.574
FF	−8.85	−49.68	28.15	−1.97	2.962
TT	−6.45	−50.65	27.47	−2.301	2.76
NN	−7.13	−49.88	27.63	−2.45	2.77

Table 4 Performance summary comparison with published works

Specification	Ref. [13]	Ref. [12]	Ref. [15]	This work
Technology (nm)	90	130	180	45
Operating frequency (GHz)	2.5	0.2–3.8	2.69	2.69
S_{11} (dB)	<−15	<−9	<−7	−7.95
S_{21} (dB) gain	16.5	19	21–25	27.7
Noise figure (dB)	1.8–2	2.8–3.4	2.6–3	1.44
Power supply (V)	1.2	1	1.8	1.5
IIP3 (dBm)	0.4	−4.2	–	−1.81
Power consumption (mW)	–	5.7	14.5	18.06

generally do not adversely affect the logical correctness of the circuit [14]. The result of process corners of proposed LNA is summed up in Table 3, which is represented for the S-parameters and NF.

Performance summary of LNA is summed up and represented in Table 4, which is compares with some previously published works.

4 Conclusion

In this paper, a narrowband CMOS LNA is designed, dedicated for the IEEE 802.16a (WiMAX) standard, which is studied and tested several times to realize an approach that can help to deliver a high gain, stable, low NF, low power consumption and fully integrated LNA design. It is designed and simulated through the Cadence Virtuoso analog design environment using gpdk 45 nm CMOS technology. Since this proposed LNA shows suitable power gain and minimal NF at 2.69 GHz, this is adding another feather to design a better LNA for a WiMAX application which can be useful in a practical scenario.

Acknowledgments The authors are thankful to Prof. V. R. Gupta, HOD ECE, BIT Mesra and our Vice-Chancellor Prof. M. K. Mishra for their constant support and research encouragement. They are also thankful to RESPOND ISRO Ahmadabad for funding this project.

References

1. M. A. G. Lorenzo and M. T. G. de. Leon, "Comparison of LNA Topologies for WiMAX Applications in a Standard 90-nm CMOS Process," *2010 12th International Conference on Computer Modelling and Simulation*, Cambridge, 2010, pp. 642–647.
2. R. Kundu, A. Pandey, S. Chakraborty and V. Nath, "A CMOS low noise amplifier based on common source technique for ISM band application," Microsystem Technologies, Vol 22 (11)2016.
3. N. Yadav, A. Pandey and V. Nath, "Design of CMOS low noise amplifier for 1.57 GHz," IEEE International Conference Microelectronics, Computing and Communications (MicroCom), 23–25 Jan 2016, Durgapur, India.
4. Kumar. V, Singh. K.K, Pandey. A, Nath. V, (2017) Design of Comparator in Sigma-Delta ADC Using 45 nm CMOS Technology. In: Nath V. (eds) Proceedings of the International Conference on Nano-electronics, Circuits and Communication Systems. Lecture Notes in Electrical Engineering, Vol 403, Springer.
5. S. Kumar, "Wireless Communication Fundamental and Advance Concepts," River publisher series in communication, vol. 1, pp. 117–121,2015.
6. Telecom Regulatory Authority of India, "consultation Paper on Allocation and pricing of spectrum for 3G services and Broadband Wireless Access," consolations Paper no. 9/2006.
7. T. H. Lee, "The Design of CMOS Radio-Frequency Integrated Circuit", ed. 1st, New York: Cambridge University Press, Chap. 11, p. 277–289,1998.
8. J. Le Ny, B. Thudi, J. McKenna, "A 1.9 GHz Low Noise Amplifier," EECS 522 Analog Integrated Circuits Project, Winter 2002.
9. S. Shekhar, J. S. Walling, S. Aniruddhan and D. J. Allstot, "CMOS VCO and LNA Using Tuned-Input Tuned-Output Circuits," in *IEEE Journal of Solid-State Circuits*, vol. 43, no. 5, pp. 1177–1186, May 2008.https://doi.org/10.1109/jssc.2008.920360.
10. X. Fan, H. Zhang and E. SÁnchez-Sinencio, "A Noise Reduction and Linearity Improvement Technique for a Differential Cascode LNA," in *IEEE Journal of Solid-State Circuits*, vol. 43, no. 3, pp. 588–599, March 2008. https://doi.org/10.1109/jssc.2007.916584.
11. A. A. Abidi, "On the operation of cascode gain stages," in *IEEE Journal of Solid-State Circuits*, vol. 23, no. 6, pp. 1434–1437, Dec 1988. https://doi.org/10.1109/4.90043.
12. B. Razavi, "CMOS technology characterization for analog and RF design," in *IEEE Journal of Solid-State Circuits*, vol. 34, no. 3, pp. 268–276, Mar 1999. https://doi.org/10.1109/4.748177.
13. D. K. Shaeffer and T. H. Lee, "A 1.5-V, 1.5-GHz CMOS low noise amplifier," in *IEEE Journal of Solid-State Circuits*, vol. 32, no. 5, pp. 745–759, May 1997. https://doi.org/10.1109/4.568846. B. Razavi, "RF Microelectronics": Prentice Hall PTR, Chap. 2, 2006.
14. F. Ghadimipoor, H. G. Garakani, "A Noise-Cancelling CMOS Low-Noise Amplifier for WiMAX," International Conference on Electronic Device, Systems and Application (ICEDSA), 2011.
15. http://www.edaboard.com/thread193273.html.

Sensitivity Analysis of Various Diamagnetic and Paramagnetic Materials Based on Faraday Rotation Principle

Sarita Kumari and Sarbani Chakraborty

1 Introduction

In 1945, Michael Faraday found that a glass block becomes optically active if kept under the strong magnetic field. The Faraday rotation takes place in the presence of magnetic field. The plane of polarization of any linearly polarized light, passing through any Faraday-active material in the presence of magnetic field, changes. The magneto-optic (MO) sensors are based on Faraday effect. It has become very popular due to noncontact-type sensing means and offers many advantages over traditional sensors such as immunity from electromagnetic interference, stray capacitance as well as environmental effect. The response time is quick, data loss in transmitting and receiving the signal is minimum and is highly reliable. The loss in data processing is minimum with minimum effect of noise compared to other means of measurement. MO effect is applicable for all mediums, i.e., solid, liquid as well as gaseous. For effective MO device, it must have high Verdet constant and high optical transparency. The magneto-optical sensor provides new perspective in sensor designing and testing.

The material is called Faraday-active material or magneto-optic material and the rotation of polarization plane is called Faraday rotation (θ) [1–3]. Faraday rotation is proportional to change in magnetic field (B) and optical path traveled (L). The proportionality constant is called as Verdet constant (V_{verdet}) which is function of wavelength. Usually V_{verdet} is very small for all the materials except for the materials that have paramagnetic ions which include Terbium such as TDG, TGG, etc. The rotation of any MO material varies with a change in wavelength.

The use of magnetic materials are gaining importance in all the fields such as industry, defense, and more research is being carried out in material science. Any

S. Kumari (✉) · S. Chakraborty
EEE Department, Birla Institute of Technology, Mesra, Ranchi 835215,
Jharkhand, India
e-mail: gs.sarita@gmail.com

© Springer Nature Singapore Pte Ltd. 2019
V. Nath and J. K. Mandal (eds.), *Proceeding of the Second International Conference on Microelectronics, Computing & Communication Systems (MCCS 2017)*, Lecture Notes in Electrical Engineering 476, https://doi.org/10.1007/978-981-10-8234-4_57

data can be processed and stored via magnetic medium and can be retrieved easily. The most commonly used application of MO effect is in smart card and magnetic strip scanning system. There are a variety of magnetic materials available such as diamagnetic, paramagnetic, ferromagnetic, and ferrimagnetic. Magnetic materials can be converted to magneto-optically sensitive material by adding rare earth ions to it. There are 17 rare earth elements are available but most commonly used are Yttrium (Y), Neodymium (Nd), Gadolinium (Gd), Terbium (Tb), and Dysprosium (Dy) [4–7].

Many researchers have reported various properties of MO materials such as Verdet constant, temperature dependency, and size scalability. Verdet constant of diamagnetic materials such as fused silica, borosilicate glass, and YAG crystal were calculated by Munin et al. for different wavelengths in the visible region [8]. Zhe Chen and team have fabricated Dy^{3+}-doped TGG and Tm^{3+} doped TGG by Czochralski (Cz) method [4, 5]. The Verdet constant and magneto-optical characteristics of the materials can be enhanced remarkably by adding rare earth ions. The pure TGG shows 20–30% less Verdet constant value than after it is doped with Ce^{3+}, Pr^{3+}, and Nd^{3+}. Verdet constant of MO materials are measured by different techniques such as a pulsed magnetic field for TGG, Faraday effect in the diamagnetic glass, ellipsometry technique in a different medium, i.e., flint glass and water [6, 9, 10]. Faraday rotation of rare earth garnet material is usually inversely proportional to the wavelength of light except for ytterbium iron garnet (YIG) where the Verdet constant increases above 1300 nm wavelength [7]. In 800–1700 nm wavelength, Zhao [11] has found the higher rotation and low temperature dependency of YbBi: YIG material. It exhibits high linearity, sensitivity, and accuracy than YIG. Huang and Zhang [12] have doped Y^{3+} and Yb^{3+} ions in Bi-substituted iron garnet to achieve large rotation and high temperature stability at 1550 nm. Koerdt et al. [13] found the Faraday effect in photonic crystals is filled with transparent liquid at 573 nm. Characterization of MO properties of various MO materials is explained by Donati [14] for a wide range of VIS-NIR wavelength. Chakraborty et al. designed current sensor using null detection technique and magnetic field sensor in visible region using TDG crystal as sensing element [15, 16]. Deeter et al. have designed YIG-based magnetic field sensor [17–20]. Based on the literature survey, it is found that MO sensors are used in many industrial applications such as measurement of displacement (linear and angular), magnetic field, current, temperature, vibration, and so on. Several types of research are being reported on MO material development and its properties [4, 5, 7, 8, 15, 16].

In this context, analysis of sensitivity can be regarded as one of the important performance criteria in designing sensors. It shows how the instrument or material behaves with the change in particular parameters. Since Terbium-doped materials exhibit very high rotation, these are very popular and widely used in sensor designing. In this paper, we are focusing on various MO materials at a different wavelength and analyzing its sensitivity under the different relative orientation of polarizer and analyzer.

2 Theory

Figure 1 shows the block diagram of schematic magneto-optic measurement setup. It consists of the laser (source of light), polarizer, MO sensor, analyzer, and photodetector to analyze the rotation. A linearly polarized light beam passes through the magneto-optic material in the presence of magnetic field, and after passing through the analyzer, the beam is received by the photodetector.

The polarization of light can be expressed in many forms such as Muller matrix, Jones matrix and Poincare sphere. In this paper, Mueller matrix approach has been considered for the analysis [21]. The output intensity (I_{out}) can be expressed in terms of Mueller matrix of Faraday rotator (M_{rot}), Mueller matrix of Faraday analyzer (M_{ana}) and S_{in}. S_{in} is the Stokes vector of the linearly polarized input beam (I_0) as written in Eq. (1).

$$I_{out} = I_0 M_{ana} M_{rot} S_{in} \tag{1}$$

$$M_{ana} = \frac{1}{2} \begin{bmatrix} 1 & \cos 2\alpha_a & \sin 2\alpha_a & 0 \\ \cos 2\alpha_a & \cos^2 2\alpha_a & \cos 2\alpha_a \sin 2\alpha_a & 0 \\ \sin 2\alpha_a & \cos 2\alpha_a \sin 2\alpha_a & \sin^2 2\alpha_a & 0 \\ 0 & 0 & 0 & 0 \end{bmatrix} \tag{2}$$

$$M_{rot} = \begin{bmatrix} 1 & 0 & 0 & 0 \\ 0 & \cos 2\theta & -\sin 2\theta & 0 \\ 0 & \sin 2\theta & \cos 2\theta & 0 \\ 0 & 0 & 0 & 1 \end{bmatrix} \tag{3}$$

$$S_{in} = \begin{bmatrix} 1 \\ \cos 2\alpha_p \\ \sin 2\alpha_p \\ 0 \end{bmatrix} \tag{4}$$

where α_a and α_p are the transmission angle of the analyzer and polarizer. From Eqs. (1)–(4), the expression of the detected intensity of the output beam can be written as in Eq. (5).

$$I_{out} = I_0 \left[\cos^2 \left(\theta + \alpha_p - \alpha_a \right) \right] \tag{5}$$

$$\theta = V_{verdet} BL \tag{6}$$

Fig. 1 Schematic diagram of magneto-optic measurement system

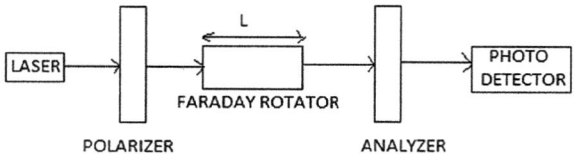

Equation (6) gives relationship of Faraday rotation with a change in magnetic field (B) and optical path traveled (L) [9, 10, 13, 15, 16, 22, 23]. Faraday rotation (θ) is proportional to magnetic flux density (B), the path length (L) and the Verdet constant of material which is the function of wavelength. The expression of Verdet constant (V_{verdet}) [1–3, 15, 16] is given in Eq. (7) in terms of optical constants, i.e., charge on electron (e), mass on electron (m), speed of light (c). λ denotes the wavelength and n is refractive index of the material [15].

$$V_{verdet} = -\frac{e\lambda}{2mc}\frac{dn}{d\lambda} \tag{7}$$

MO sensor is becoming popular because of many advantages over traditional sensors. Several works have been done on material analysis and properties. Sensitivity analysis of the material is an very important aspect of sensor designing. This study focuses on the response of detected intensity which is dependent on magnetic field, wavelength, and total optical path length. We have considered fixed optical path length and Verdet constant of the material is dependent on wavelength. Since the response is nonlinear, sensitivity does not remain constant throughout the entire range. In order to analyze, piecewise linearization needs to be considered for certain operation point. In our analysis, the linear region was considered between 4 and 6 mT in the observed response curve in Figs. 3, 4 and 5. For analysis, we have chosen magnetic field at 5 mT as the operating point about which the sensitivity has been calculated.

$$\text{Sensitivity} = \left|\frac{dI_{out}}{dB}\right| \tag{8}$$

Sensitivity is derived from Eq. (8) and can be written as shown in Eq. (9).

$$\text{Sensitivity} = I_0\left[-\sin 2\left(V_{verdet}BL + \alpha_p - \alpha_a\right) * VL\right] \tag{9}$$

Table 1 lists the Verdet constant of different MO materials at different wavelengths [4, 5, 8, 9, 15, 16, 24, 25]. Ferromagnetic materials show higher Faraday rotation which is nonlinear.

In this paper, we have compiled the Verdet constant of different MO materials and studied the detected intensity response with variation in magnetic flux density and several observations have been made for the analysis.

Table 1 Verdet constant of various MO materials at different wavelength

Magnetic materials	MO material	λ (nm)	V (rad/Tm)	Reference
Diamagnetic	Dense flint glass	505	33.6	Thamaphat (2006), [9]
		525	30.4	
	YAG	632.8	5.86	Munin (1992), [8]
	BK 7 Glass	632.8	4.30	
	Dynasil 1001	632.8	3.48	
Paramagnetic	TDG	543.5	101.8	Chakraborty (2008), [15]
		589.3	90.17	
		632.8	78.54	
		632	70	Chakraborty (2015), [16]
	TGG	532	190	Chen (2016), [4]
		632.8	134	Chen (2016), [5]
		780	82	Weller (2012), [25]
		1053	36.4	Yasuhara (2007), [24]
		1064	40	Chen (2016), [4]
	Dy^{3+}-doped TGG and Tm^{3+}-doped TGG	532	256.8	Chen (2016), [4, 5]
		633	178.6	
		830	102.3	
		1064	60.2	
		1330	28.3	
	Ce^{3+}-doped TGG	532	242.2	
		1064	53.2	
	Pr^{3+}-doped TGG	532	264.5	
		1064	62.7	
	Nd^{3+}-doped TGG	532	225	
		1064	41	
	Ce^{3+}-doped TAG ceramic	632.8	199.55	Chen (2016), [5]
Ferrimagnetic	YIG	780	380	Weller (2012), [25]
		1310	2200	Chen (2016), [4]
	Bi-doped YIG	1550	1700	

3 Result and Discussion

Figure 2 shows Verdet constant (rad/Tm) versus wavelength (nm) plot of various MO material such as TDG, pure TGG and doped TGG (Dy^{3+}, Ce^{3+}, Pr^{3+} and Nd^{3+} doping) [4, 5, 15, 16, 24, 25]. It is observed that the Faraday rotation increases by approximately 20–25% when TGG is doped with rare earth ions. Pr^{3+}-doped TGG gives highest Faraday rotation among any other rare earth doping. In this analysis, the range of magnetic field is considered from 0 to 10 mT and total optical path length is 30 mm. Equation (5) is used to plot detected intensity with variation in the magnetic field as shown in Figs. 3, 4 and 5.

The detected intensity responses with variation in magnetic field density have been plotted using MATLAB 8.0 software. Here we have considered different relative orientation between analyzer and polarizer angle. In this analysis, a linearly horizontally polarized ($\alpha_p = 0$) monochromatic beam is considered as input beam and analyzer angle is varied between 40° and 50° with a change of 1°. Various plots were achieved and analyzed for different materials for different wavelengths i.e. 532, 632.8 and 1064 nm. Sensitivity was analyzed at operating point 5 mT.

Figures 3, 4 and 5 show response of detected intensity with change in magnetic field density of different MO materials at 532, 632.8 and 1064 nm respectively. Figure 3 shows detected intensity versus magnetic field plot for TGG, Dy^{3+} doped TGG, Ce^{3+} doped TGG, Pr^{3+} doped TGG and Nd^{3+} doped TGG at 532 nm wavelength. Figure 4 shows the response of detected intensity with change in magnetic field density for YAG, BK-7 Glass, fused glass, TGG, TDG, and Ce^{3+}-doped TAG ceramic at 632.8 nm wavelength. Figure 5 shows the intensity versus magnetic field curve of different materials such as TGG, Dy^{3+}-doped TGG, Ce^{3+}-doped TGG, Pr^{3+}-doped TGG and Nd^{3+}-doped TGG at 1064 nm wavelength. Tables 2, 3, and 4 show the list of MO materials with their corresponding Verdet constant, the relative orientation of the analyzer and polarizer angle and obtained maximum sensitivity value at 532, 632.8, and 1064 nm respectively.

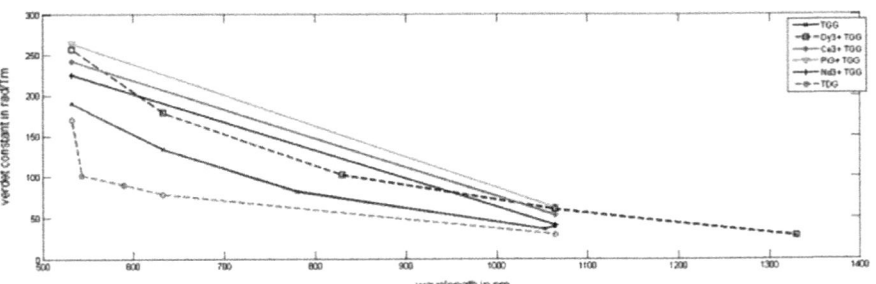

Fig. 2 Verdet constant (rad/Tm) versus wavelength (nm) plot for TDG, pure TGG, Dy^{3+}-doped TGG, Ce^{3+}-doped TGG, Pr^{3+}-doped TGG and Nd^{3+}-doped TGG

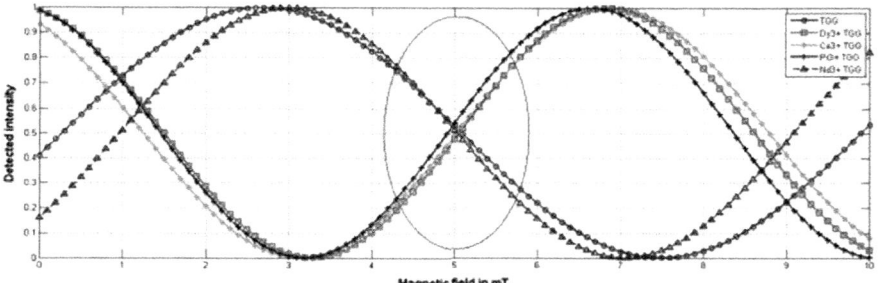

Fig. 3 Variation of detected intensity (I_{out}) with magnetic field (mT) at 532 nm wavelength of TGG, Dy^{3+}-doped TGG, Ce^{3+}-doped TGG, Pr^{3+}-doped TGG, and Nd^{3+}-doped TGG

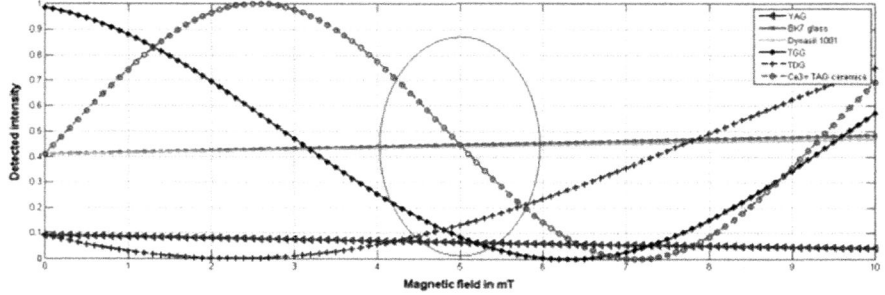

Fig. 4 Detected intensity versus magnetic field curve at 632.8 nm wavelength of YAG, BK-7 glass, fused glass, TGG, TDG, and Ce^{3+}-doped TAG ceramic

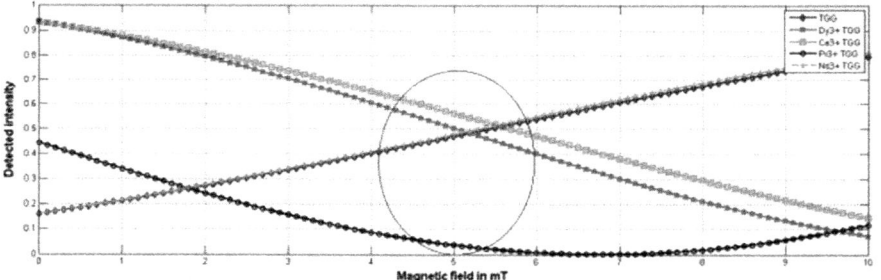

Fig. 5 Variation of detected intensity (I_{out}) with magnetic field (B) wavelength of TGG, Dy^{3+}-doped TGG, Ce^{3+}-doped TGG, Pr^{3+}-doped TGG and Nd^{3+}-doped TGG at 1064 nm

From Table 2, it is observed that Pr^{3+}-doped TGG materials give maximum sensitivity of 0.3932 at 532 nm wavelength which is 8.99% higher than pure TGG. Ce^{3+} doped TAG ceramic exhibits maximum sensitivity value of 0.3149 at 632 nm wavelength than any other material in same wavelength as shown on Table 3.

Table 2 Maximum sensitivity achieved at 532 nm wavelength

S. No.	Material	V_{verdet} (rad/Tm)	Relative angle between α_a and α_p (°)	Maximum sensitivity value
1	TGG	190	48	0.3033
2	Dy^{3+}-doped TGG	256.8	47	0.3857
3	Ce^{3+}-doped TGG	242.2	50	0.3697
4	Pr^{3+}-doped TGG	264.5	47	0.3932
5	Nd^{3+}-doped TGG	225	42	0.344

Table 3 Maximum sensitivity achieved at 632.8 nm wavelength

S. No.	Material	V_{verdet} (rad/Tm)	Relative angle between α_a and α_p (°)	Maximum sensitivity value
1	YAG	5.86	49	0.04915
2	BK-7 glass	4.30	48	0.00735
3	Dynasil 1001	3.48	48	0.00595
4	TGG	134	47	0.1333
5	TDG	78.54	49	0.22065
6	Ce^{3+}-doped TAG ceramic	199.55	48	0.3149

Table 4 Maximum sensitivity achieved at 1064 nm wavelength

S. No.	Material	V_{verdet} (rad/Tm)	Relative angle between α_a and α_p (°)	Maximum sensitivity value
1	TGG	40	42	0.0684
2	Dy^{3+}-doped TGG	60.2	50	0.10275
3	Ce^{3+}-doped TGG	53.2	50	0.0902
4	Pr^{3+}-doped TGG	62.7	40	0.4303
5	Nd^{3+}-doped TGG	41	42	0.0702

Table 4 shows that Pr^{3+}-doped TGG materials give maximum sensitivity value of 0.4303 at 1064 nm wavelength which is 36.19% higher than pure TGG. Table 5 shows the comparative study of different materials by various researchers.

In visible region, YAG material can be used for designing the optical isolator due to high optical quality [8]. Dy^3-doped TGG is a highly transparent MO material and can be used as isolators and for advanced optical communication applications in visible and near infrared region. Also YIG and Bi-doped YIG are applicable in VIS-NIR region because below 1100 nm it shows poor transparency [4]. Dy^{3+} doped and Tm^{3+} doped TGG shows paramagnetic properties below 10 K and between 10 and 300 K temperatures respectively [5]. Tm^{3+} doped TGG also find applications in VIS-NIR region. Ferrimagnetic materials such as iron, nickel, and

Table 5 Work done by various authors on magneto-optical properties of different materials

S. No.	Reference	Materials	Properties
1	Donati [14]	All fiber sensor (Silica Fiber)	Sensitivity, bandwidth, dynamic range, linearity
2	Deeter [17]	YIG	Sensitivity, speed, directionality
3	Deeter [19]	YIG	Sensitivity, frequency response
4	Chakraborty [21]	TDG	Sensitivity, range, and resolution
5	Proposed in this paper	Diamagnetic and Paramagnetic at 532, 632.8, and 1064 nm wavelengths	Sensitivity analysis, compilation of Verdet constant

gadolinium show nonlinear response though they exhibit high Faraday rotation [16]. Bismuth added iron garnet also proves large Faraday rotation than iron garnet in 0.8–1.7 μm. YIG are mainly used in microwave communication as well as designing of optical-based devices like rotator [22]. It shows high stability for temperature change.

Terbium is highly transparent but expensive. It shows high resistance to laser damage as well as high thermal conductivity [4, 5]. TDG sensor has been used for measurement of current [15], displacement [21], magnetic field [16], high pulse energy lasers [24], and so on. Size scalability is poor in TDG whereas size scalability is not a problem for TGG [21]. TGG can be used for high power applications due to excellent thermal conductivity (4.5–7.4 W/mK) [24]. Large size TGG can be developed by different techniques [4, 5].

4 Conclusion

TGG crystals have high Verdet constant and excellent thermal stability. It is useful for Faraday isolator application, because of size scalability. Pr^{3+} doped TGG shows highest Faraday rotation as well as highest sensitivity than other doped TGG materials. Paramagnetic materials can be used for visible and near infrared region and Ferrimagnetic materials are useful for infrared region as it shows poor transparency below 1100 nm. At 532 and 1064 nm wavelength, the maximum sensitivity was calculated with Pr^{3+} doped TGG and at 632.8 nm Ce^{3+} doped TAG shows maximum sensitivity. The magneto-optic materials are used as micro-optic devices such as switches, optical/Faraday isolator, modulator, deflector for fiber optic devices. Also, it is applicable for measurement of various parameters like the current, magnetic field, displacement, and so on.

References

1. Eugene Hecht, Optics, 2nd Edition, Low price edition.
2. Francis A. Jenkins, Harvey E. White, Fundamental of Optics, 4th Edition, McGraw-Hill International Edition.
3. William A. Shurcliff, Polarized Light Production And Use, Harvard University Press, Cambridge, Massachusetts, 1962.
4. Zhe Chen, Lei Yang, Yin Hang, Xiangyong Wang, "Faraday effect improvement by Dy^{3+} doping of terbium gallium garnet single crystal", Journal of Solid State Chemistry 233, 277–281, 2016.
5. Zhe Chen, Lei Yang, Yin Hang, Xiangyong Wang, "Improving characteristic of Faraday effect based on the Tm^{3+} doped terbium gallium garnet single crystal", Journal of Alloys and Compounds 661, 2016.
6. A Balbin Villaverde, DA Donatti, DG Bozinis, "Terbium gallium garnet Verdet constant measurements with pulsed magnetic field", J. Phys. C: Solid State Phys., Vol. 11, 1978.
7. RC Booth and EAD White, "Magneto-optic properties of rare earth iron garnet crystals in the wavelength range 1.1–1.7 pm and their use in device fabrication", J. Phys. D: Appl. Phys., 17, 579–587, 1984.
8. Egberto Munin, JA Roversi and A BalbinVillaverde, "Faraday effect and energy gap in optical materials", J. Phys. D: Appl. Phys. 25, 1635–1639, 1992.
9. Kheamrutai Thamaphat, Piyarat Bharmanee and Pichet Limsuwan, "Measurement of Verdet Constant in Diamagnetic Glass Using Faraday Effect", Kasetsart J. (Nat. Sci.) 40: 18–23, 2006.
10. S. Suchat, P. Viriyavathana, P. Jaideaw, N. Haisirikul, W. Kerdsang, and S. Petcharavut, "Measurement of the Verdet Constant in Different Mediums by Using Ellipsometry Technique", Progress In Electromagnetics Research Symposium Proceedings, Suzhou, China, Sept. 12–16, 2011.
11. Weizhong Zhao, "Magneto-optic properties and sensing performance of garnet YbBi:YIG", Sensors and Actuators, 250–254, 2001.
12. Min Huang, Shouye Zhang, "Growth and characterization of rare-earth iron garnet single crystals modified by bismuth and ytterbium substituted for yttrium" Materials Chemistry and Physics 73, 314–317, 2002.
13. C. Koerdt and G. L. J. A. Rikken and E. P. Petrov, "Faraday effect of photonic crystals", Applied Physics Letters Volume 82, Number 10, 10 March 2003.
14. S. Donati, V. Annovazzi-Lodi, T. Tanbosso, "Magneto-optical fiber sensors for electrical industry: analysis of performance", IEEE proceedings, Vol. 135, Pt. J., No. 5, Oct 1988.
15. Sarbani Chakraborty and Satish Chandra Bera, "Magneto-Optic Over-Current Detection with Null Optical Tuning", Sensors & Transducers Journal, Vol. 87, Issue 1, pp. 52–62, January 2008.
16. Sarbani Chakraborty and Sarita Kumari, "Design and Development of a Magneto-Optic Sensor for Magnetic Field Measurements", Sensors & Transducers, Vol. 184, Issue 1, pp. 153–158, January 2015.
17. M.N. Deeter, A.H. Rose, G.W. Day, "Fast, sensitive magnetic-field sensors based on the Faraday effect in YIG", J. Lightwave Technol., pp. 1838–1842, 8 (12) (1990).
18. R. Wolfe, E. M. Gyorgy, R. A. Lieberman, V. J. Fratello, S. J. Licht, M. N. Deeter, and G. W. Day, "High Frequency Magnetic Field Sensors Based On The Faraday Effect In Garnet Thick Films," International Conferences on Optical Fiber Sensors 1983–1997 (Optical Society of America, 1992).
19. M. N. Deeter, S. Miliin Bon, and G. W. Day, "Novel Bulk Iron Garnets for Magneto-optic Magnetic Field Sensing", IEEE Transactions on Magnetics, Vol. 30, No. 6, November 1994.
20. Merritt N. Deeter, "High sensitivity fiber-optic magnetic field sensors based on iron garnets" IEEE transactions on Instrumentation and Measurement, vol 44, no. 2, April 1995.

21. S.C. Bera, S. Chakraborty, "Study of Magneto-Optic Element as a Displacement Sensor", Measurement 44, 1747–1752, Elsevier Publication, 2011.
22. G.J. Chena, H.M. Lee, Y.S. Chang, Y.J. Lin, Y.L. Chai, "Preparation and properties of yttrium iron garnet microcrystal in P_2O_5–MgO glass", Journal of Alloys and Compounds 388, 297–302, 2005.
23. Matthieu Aerssens, Andrei Gusarov, Benoît Brichard, Vincent Massaut, Patrice Mégret and Marc Wuilpart, "Faraday Effect Based Optical Fiber Current Sensor for Tokamaks", 2nd International Conference on Advancements in Nuclear Instrumentation Measurement Methods and their Applications (ANIMMA), 2011.
24. Ryo Yasuhara, Shigeki Tokita, Junji Kawanaka, Toshiyuki Kawashima, Hirofumi Kan, Hideki Yagi, Hoshiteru Nozawa, Takagimi Yanagitani, Yasushi Fujimoto, Hidetsugu Yoshida and Masahiro Nakatsuka, "Cryogenic temperature characteristics of Verdet constant on terbium gallium garnet ceramics", OSA, Vol. 15, No. 18, OPTICS EXPRESS 11257, 3 September 2007.
25. L Weller, KS Kleinbach, MA Zentile, S Knappe, IG Hughes and CS Adams, "An optical isolator using an atomic vapor in the hyperfine Paschen-Back regime", Optics letters, Vol. 37, Issue 16, pp. 3405–3407, 2012.

A Secure Three-Way Handshake Authentication Process in IEEE 802.11i

Anil Kumar and Partha Paul

1 Introduction

In the preceding era, wireless networks expanded in an extensive momentum. One in every of the foremost advantageous options of Wi-Fi networks is they assist user mobility in an appropriate manner others are wide availability of hardware at reasonably less price. The hitch is that Wi-Fi networks are greater susceptible to assaults than their wired counterparts. This improved vulnerability mainly originates from the absence of bodily connections and also the transmission nature of radio communications. Hence, it is significant to furnish suitable security measures for Wi-Fi networks, which make sure the lustiness in their procedure even in case of malicious assaults [1].

IEEE has developed different security standards and IEEE 802.11i is one of the latest standards for wireless local area network. Several improvements were done in 802.11 for the perfection of range, functionality, bandwidth, and furthermost prominently its security.

Wired equivalent privacy is one of the simplest and earliest forms of security approved in the initial phases of wireless LAN. However, after few years major deficiencies have been recognized in authentication and encryption mechanisms in WEP [2–4].

To overcome these problems without altering the existing hardware, the Wi-Fi Alliance suggested a Temporal Key Integrity Protocol (TKIP) which furnishes robust security via a keyed cryptographic Message Integrity Code (MIC), an Extended Initialization Vector with a key partying function. Moreover, an

A. Kumar (✉) · P. Paul
Department of Computer Science and Engineering, Birla Institute of Technology,
Mesra, Ranchi 835215, India
e-mail: anilbitme@gmail.com

P. Paul
e-mail: p_india@rediffmail.com

© Springer Nature Singapore Pte Ltd. 2019
V. Nath and J. K. Mandal (eds.), *Proceeding of the Second International Conference on Microelectronics, Computing & Communication Systems (MCCS 2017)*, Lecture Notes in Electrical Engineering 476, https://doi.org/10.1007/978-981-10-8234-4_58

authentication mechanism primarily based on EAP/802.1X/RADIUS has been established to substitute the deprived Shared Key and Open System authentication.

As a lasting resolution to secure wireless LAN, the recent IEEE standard 802.11i was ratified on June 24, 2004.

This paper is organized as follows. Section 2 describes a brief introduction of IEEE 802.11i, modes of operations and different keys. Section 3 explains a proper analysis of the four-way handshake protocol and the problem in the four-way handshake protocol. Section 4 discusses related work done to elevate the problem present in the four-way handshake protocol. Section 5 result and discussion which summarizes our experimental results. Section 7 concludes this paper.

2   IEEE 802.11i

IEEE 802.11i also acknowledged as Wi-Fi Protected Access Version 2 (WPA2). It entirely implements IEEE 802.11i standard and is an improvement over Wi-Fi Protected Access (WPA). The noteworthy improvement was, introductions of Counter Mode with Cipher Block Chaining Message Authentication Code Protocol (CCMP) which makes use of Advanced Encryption Standard for data encryption, but TKIP (Temporal Key Integrity Protocol) is available for backward compatibility with current WAP hardware [5]. IEEE 802.11i authentication has two modes of operations:

(1) *Enterprise mode*: Authentication relies on IEEE 802.1X standards. It includes three major constituents which are Authentication Server, Authenticator, and Supplicant.

(2) *Personal mode*: Authentication does not require an authentication server; it is performed between the supplicant and the access point generating a 256-bit Pre-Shared Key (PSK) from passphrase (8–63 characters). PSK along with the Service Set Identifier (SSID) length and Service Set Identifier is used to calculate Pairwise Master Key (PMK) which is used in later stages for key generation.

2.1   IEEE 802.1X

WPA2 uses IEEE 802.1X to authenticate users and AES-based CCMP to encrypt messages. An 802.1X wireless arrangement involves three vital constituents:

1. Supplicant (stations or wireless client).
2. Authenticator (the Access Point).
3. Authentication server (typically a RADIUS server) (Fig. 1).

Fig. 1 IEEE 802.1X authentication and key management process

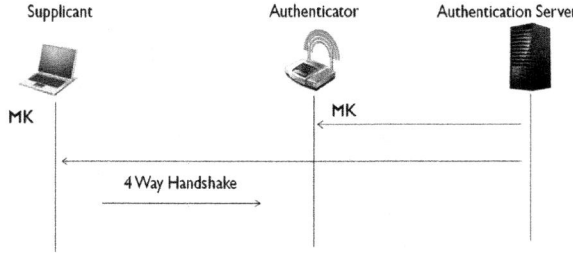

IEEE 802.11i authentication process involves handshakes between Supplicant and Authenticator as well as between Authenticator and Authentication Server [6]. After these handshakes, following keys are generated:

Pairwise Keys

- Master Key (MK)—It characterizes the optimistic access decision
- Pairwise Master Key (PMK)—It characterizes approval to access 802.11 mediums

Pairwise Transient Key—It is the pool of operational keys:

- Key Confirmation Key (KCK)—It binds the Pairwise Master Key to the Access Point, Supplicant; used to validate possession of the Pairwise Master Key.
- Key Encryption Key (KEK)—It allocates the Group Transient Key (GTK).
- Temporal Key (TK)—It provides secure data movement.

After the authentication among supplicant and authentication server, they generate a shared communal secret key known as Master Key (MK). Supplicant uses it to generate the PMK [7]. Authorization, Authentication, and Association (AAA) key on the server side is firmly conveyed to the authenticator to derive the same PMK [8, 9].

Otherwise, the supplicant and also the authenticator may configure to use a static Pre-Shared Key (PSK) for the Pairwise Master Key (PMK).

Auxiliary, in order to decrease the computational time throughout a re-association, a cached PMK are often used directly on the authentication server during recurrent authentication appeals from the identical user.

2.2 Four-Way Handshake

As soon as a shared Pairwise Master Key is established between the supplicant and the authenticator, then authenticator may initiate four-way handshake by itself or upon appeal from the supplicant [10]. The message altercation is shown in Fig. 2.

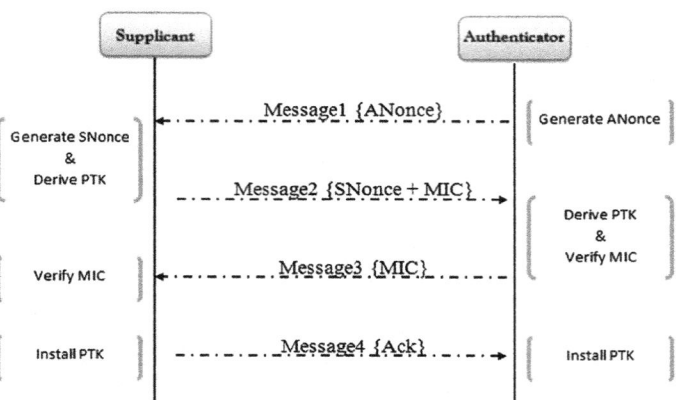

Fig. 2 IEEE 802.11i Four-Way Handshake Protocol

3 Problem in Four-Way Handshake

The trick is that this four-way handshake in Personal Mode can tactlessly be observed by a third party. If an attacker captures the handshake, then with the assistance of two MAC addresses, nonce values, and the cipher suite used can be obtained. The PTK will then be generated by the attacker with the help of captured information. A brute force or dictionary attack can then be executed against the PTK to find the corresponding original PSK it was derived from. Thus, selecting a weak password considerably reduces the effectiveness of WPA2 and greatly increases the chances that you're PSK will be discovered.

4 Related Works

Zha and Ma [11] presented an amended two-way handshake protocol, conferring two random numbers ANonce and BNonce will be produced by the access point, and then it will be encrypted along with supplicant MAC address by using PMK as key. Access Point then encapsulates this within Message1 and directs it to the supplicant. As Message1 is received, it'll be decrypted by Supplicant with PMK and computes PTK. Afterward, the supplicant will encrypts the BNonce and generated SNonce by the similar PMK and further encapsulates this within Message2 and directs it to access point. On receiving Message2, it's decrypted by the AP with PMK as key and confirms BNonce value and once confirmed calculates PTK with the assistance of same technique. It somehow can prevent key regaining but it increases computation power and it involves many modifications in the present hardware. It mostly focuses on prevention of DOS attack.

Jani et al. [12] proposed a mechanism called Multiple Packet System (MPS). MPS operates at the application layer. This system uses five different algorithms AES, RSA, IDEA, BLOWFISH, RC4 for encryption/decryption. This technique uses 256 packets in which one combination of five algorithms is inserted, there are 4-bit to differentiate between the combinations of algorithms in place of actual names. MPS stores key list for each algorithm. MPS header consists of key selector, packet selector, and MIC shuffle selector. MPS encrypts messages with different algorithms as well as different keys. In this mechanism shuffled tables are used to shuffle MIC bits with the original message. Access point (AP) will comprise all the configuration files and handle the encryption/decryption process. Different MPS files are maintained for each user. This proposed mechanism handles the MIC problem with its shuffle mechanism. The entire configuration is maintained at the Access Point. If server crashes the entire configuration will be lost since for all users it is stored at AP. More computation time will be required at AP.

Kumar et al. [13] proposed a mechanism consists of the use of DES cipher block chaining (DES-CBC) is used to encrypt ANonce value before directing it to the client. Before the authentication process, the PMK should be known to both the side, i.e., AP and the supplicant. PMK is used as a key in the encryption/decryption process. After obtaining ANonce value it is used for computing PTK value for proceeding steps. Brute force attack or dictionary attack could be done on the captured packet to and PMK can be obtained. The strength of the passphrase should be kept high and one more important point is how the supplicant verifies the ANonce value after it decrypts the message 1.

Eum et al. [14] proposed a cookie-based mechanism to derive the key. Two solutions for key derivation are a hash cookie and encryption key. In the proposed mechanism the client doesn't stores the PTK and ANonce values instead it creates these parameters into a cookie and directs them to AP. After receiving Message2 AP derives PTK and verifies it with MIC value. Cookie is again sent to the client and it decrypts it with its secret key also verifies the MIC. In second solution is somewhat same as first; when the client receives PTK it generates the hash value of PTK and the hash value of PTK is conveyed in the form of cookie to AP. The client does not retain SNonce and Pairwise Transient Key (PTK). In this approach no key is need. However, the first suggestion requests surplus computation power for encryption and decryption. While in the second the client doesn't stock SNonce and PTK, though it has to calculate PTK again. Besides, Message3 needs an extra SNonce, as when client receives the message, it is unaware of SNonce.

Liu et al. [15] investigated on the various freely available online tools like Aircrak, MDK3, Kismet, Netgear, Wireshark, River, etc., used for cracking WPA/WPA2 protocols and showed how easily through simple tools WPA2 can be cracked and deliberated the different vulnerabilities of WPA2 protocol. Analysis has also been done about the awareness, to the societies about the security issues related to WEP, WPA, WPA2 with their deployment in network security protocols.

5 Proposed Methodology

To remove this causality of easily accessible credentials, we encrypt ANonce of Message1 and SNonce of Message2 with the PMK generated on both the sides. Since the PMK is the secret value known to both the sides so we will use a strong encryption and decryption algorithm. This will make the first two messages of the handshake secure. And moreover, the number of handshakes will be decreased from four to three which results in an improved performance as Message4 is not much of importance.

For this we will go through two steps: (i) Selection for an appropriate algorithm for Encryption and Decryption process (ii) Implementation of the selected algorithm

Selection for an Appropriate Algorithm for Encryption and Decryption Process

Different algorithms are available to encrypt ANonce and SNonce. But we will go for the best which is to be measured in terms of performance metrics. For this, we will simulate the encryption/decryption process and with the help of simulation result and performance metric, one algorithm is selected from DES, 3DES, AES, RC6, and BLOWFISH.

5.1 Performance Metric

There are certain criteria for the assessment of the efficiency of any system these parameters are known as performance metrics. There are three varieties of performance metrics used to determine the performance of an algorithm.

1. Encryption Time: Time taken to generate a cipher text from a plain text with the help of an encryption algorithm [16].
2. Decryption Time: The inverse of encryption time.
3. Throughput: It is calculated by dividing the total text file size in kilobytes by the total encryption/decryption time in milliseconds.

Ten files of distinct size are used to conduct this simulation, where a comparison of 5 algorithms DES, 3DES, AES, RC6, and Blowfish are carried out [17]. Parameters metrics such as encryption, decryption time and throughput are used to measure the performance of selected algorithms.

Performance Evaluation Based on Encryption Time:

Table 1 shows text files of different size and corresponding encryption time taken by DES, 3DES, AES, RC6, and Blowfish algorithms in milliseconds.

On the basis of which we have, Fig. 3 which shows a Comparison of encryption time among DES, 3DES, AES, RC6, and BLOWFISH. By analyzing Fig. 3, Blowfish executes faster than DES, 3DES, AES, RC6.

Table 1 Comparative encryption time among DES, 3DES, AES, RC6, and Blowfish

File size	DES	3DES	AES	RC6	Blowfish
49	29	54	56	41	36
59	33	48	38	24	36
100	49	81	90	60	37
247	47	111	112	77	45
321	82	167	164	109	45
694	144	226	210	123	46
899	240	299	258	162	64
963	150	283	208	125	66
5345.28	1296	1466	1237	695	122
7310.336	1695	1786	1366	756	107
Average time	386.5	452.1	373.9	217.2	60.4
Throughput	4.136511	3.536301	4.275907	7.360781	26.46956

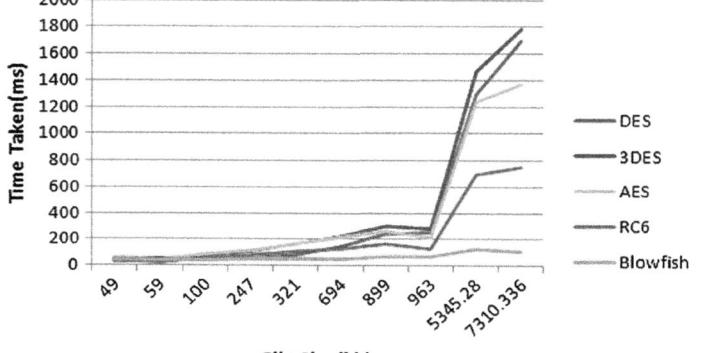

Fig. 3 Assessment of encryption time among DES, 3DES, AES, RC6, and BLOWFISH

Performance Evaluation Based on Decryption Time:

Table 2 shows text files of dissimilar size and corresponding decryption time taken by DES, 3DES, AES, RC6, and Blowfish algorithms in milliseconds.

Based on Table 2 and Fig. 4 represents a Comparison of decryption time among DES, 3DES, AES, RC6, and BLOWFISH. By analyzing Fig. 4, Blowfish executes faster than DES, 3DES, AES, RC6.

Performance Evaluation Based on Throughput:

Figure 5 shows a Comparison of the throughput for the encryption algorithm. While Fig. 6 is a Comparison of the throughput for decryption algorithm.

Table 2 Comparative decryption time among DES, 3DES, AES, RC6, and Blowfish

File size	DES	3DES	AES	RC6	Blowfish
49	50	53	63	35	38
59	42	51	58	28	26
100	57	57	60	58	52
247	72	77	76	66	66
321	74	87	149	100	92
694	120	147	142	119	89
899	152	171	171	150	102
963	157	177	164	116	80
5345.28	783	835	655	684	149
7310.336	953	1101	882	745	140
Average time	246	275.6	242	210.1	83.4
Throughput	6.499031	5.801022	6.606453	7.609527	19.1698

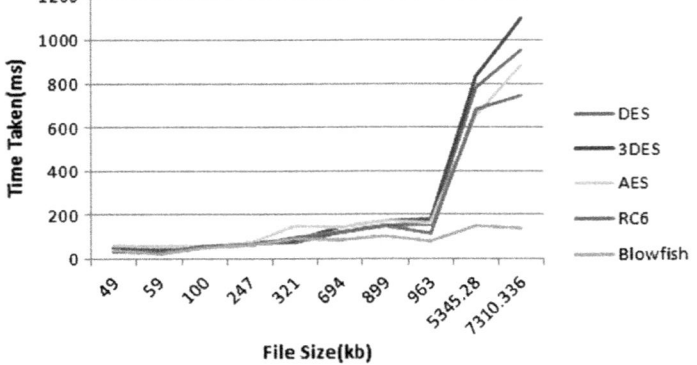

Fig. 4 Assessment of decryption time among DES, 3DES, AES, RC6, and BLOWFISH

6 Result and Discussion

The above results demonstrate the dominance of Blowfish algorithm with others in terms of the time and throughput. Increase in throughput value decreases the power consumption [18].

Blowfish is not only the fastest in terms of performance, but also offers strong protection through robust key size, that empowers it to be used in numerous applications like the internet-Based Security, Packet Encryption, Random Bit Generation, Bulk Encryption, and so many of the applications.

As it can be seen that Blowfish has not any well-known security weak points so far, this makes it an admirable candidate to be esteemed as a standard encryption algorithm for our proposed solution.

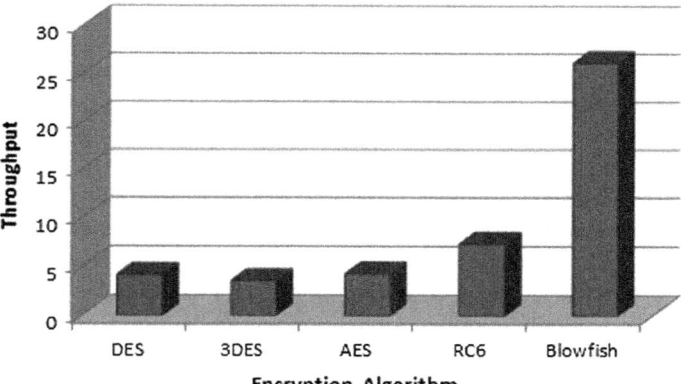

Fig. 5 Assessment of throughput among DES, 3DES, AES, RC6, and BLOWFISH

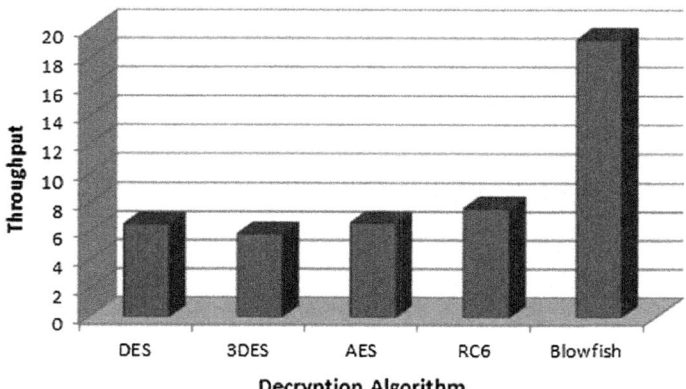

Fig. 6 Assessment of throughput among DES, 3DES, AES, RC6, and Blowfish

So our next step is to use the Blowfish algorithm for encrypting ANonce (Message1) and SNonce (Message 2) of the four-way handshake. and we will also reduce the handshake from 4- to 3-way handshake, because message 4 is not much of importance [19].

Before the initiation of the four-way handshake authentication process, PMK should be known to both supplicant and authenticator [20–22].

This Pairwise Master Key (PMK) can be calculated as [20, 23]

$$PMK = PBKDF2\,(passphrase, SSID, SSID\ length, 4096, 256)$$

This PMK is then used for the formation of another key so-called as Pairwise Transient Key (PTK). The PMK is employed as a key for encrypting the information which takes between two entities [23, 24]. Once the authenticator initially calculates the ANonce, this will use the PMK as the key to encrypt the ANonce (Fig. 7 frame first, AP's MAC address, and encrypted ANonce) before it'll be sent to the supplicant [25].

Once the supplicant receives Message1, it will be competent to decrypt ciphertext to achieve ANonce because of the knowledge of pre-shared PMK. Now the supplicant will generate its SNonce and with the help of decrypted ANonce, and SNonce it will derive Pairwise Transient Key (PTK).

PTK = PRF-X (PMK, Pairwise key expansion,

Min (Authenticator_MAC, Supplicant_MAC)
|| Max(Authenticator_MAC, Supplicant_ MAC)
|| Min(ANonce, SNonce)
|| Max(ANonce, SNonce)) [26, 27]

Supplicant will now encrypt SNonce (Fig. 7 frame second, Client's MAC address, and encrypted SNonce) of Message2 and send it to the Access Point. Now AP will perform the identical task and derive PTK with this it will generate

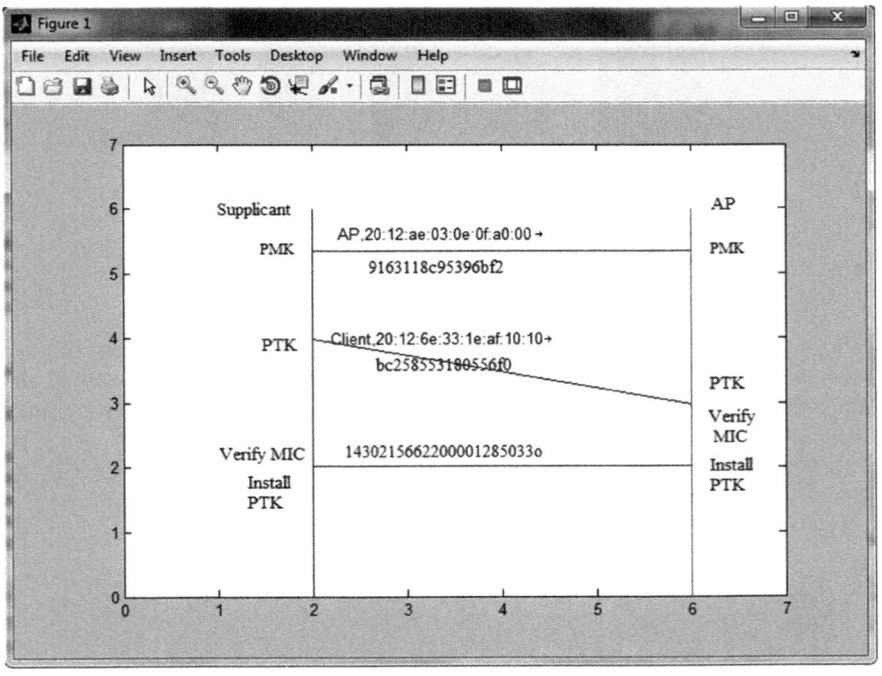

Fig. 7 Proposed Three-Way Handshake authentication process using Blowfish algorithm

MIC and verify its MIC with the Supplicant's MIC. Finally, after verification of MIC, it will install the PTK and send MIC (Fig. 7 frame 3rd) to Supplicant for his verification. If he finds both MIC similar he will also install the PTK thus resulting in a secure authentication process. Simulation of the above process is depicted in Fig. 7.

7 Conclusion

This paper shows a performance analysis of selected symmetric encryption algorithms. DES, 3DES, AES, RC6, Blowfish are the selected algorithms [28]. We overviewed the essential flow of these algorithms and analyzed the safety. We also recommend a three-way handshake protocol, a substitute for the contemporary four-way handshake protocol. Suggested three-way handshake protocol eliminates the last message from the standard protocol. Thus, the proposed protocol not only decreases the computational load but also the channel contention by dropping the number of messages in the handshake to establish the Pairwise Temporal Key between an access point and the supplicant. It too sustains the similar common authentication process as the four-way handshake protocol. However, the strength of this proposed solution is solely based on the Blowfish algorithm [29].

References

1. W. Liu, H. Duan, P. Ren, and J. Wu, "Weakness analysis and attack test for WLAN," International Conference on Green Circuits and Systems (ICGCS), pp. 387–391, June 2010.
2. Changhua He, John C. Mitchell, "Analysis of the 802.11i 4-way handshake," Proceedings of the 3rd ACM workshop on Wireless security, pp. 43–50, 2004.
3. Xinghua Li, Jianfeng Ma, Shen Yulong, An efficient WLAN initial access authentication protocol, Global Communications Conference (GLOBECOM), pp. 1035–1040, 2012.
4. Xinghua Li, Fenye Bao, Shuxin Li, Jianfeng Ma, "FLAP: An Efficient WLAN Initial Access Authentication Protocol," IEEE Transactions on Parallel and Distributed Systems, Vol. 25, Issue 2, pp. 488–497, 2013.
5. Jafri, A. Izani, and Y. Li, "ANonce encryption in 802.11i 4-way handshake protocol," 7th International Conference on Advances in Mobile Computing and Multimedia, pp. 475–480, Dec. 2009.
6. Ahmed, Idris and Anne James, "ESKIMO 2-Way Handshake," 2013 International Conference on Advances in Computing, Communications and Informatics (ICACCI), pp. 437–441, Aug. 2013.
7. J. Liu, X. Ye, J. Zhang, and J. Li, "Security verification of 802.11i 4-Way handshake protocol," 2008 IEEE International Conference on Communications, pp. 1642–1647, May 2008.
8. J. Liu, X. Ye, J. Zhang, J. Li, "Security Verification of 802.11i 4-Way Handshake Protocol," IEEE International Conference on Communications, pp. 1642–1647, 2008.

9. Snehasish Parhi, "Attacks Due to Flaw of Protocols Used in Network Access Control (NAC), Their Solutions and Issues: A Survey," International Journal of Computer Network and Information Security (IJCNIS), Vol. 4, No. 3, pp. 31–46, 2012.
10. Xiaodong Zha, Maode Ma, "Security improvements of IEEE 802.11i 4-way handshake scheme," IEEE International Conference on Communication Systems (ICCS), pp. 667–671, 2010.
11. Xiaodong Zha, Maode Ma, "Security improvements of IEEE 802.11i 4-way handshake scheme," IEEE International Conference on Communication Systems, pp. 667–671, Nov. 2010.
12. R. Jani and S. Jani, "Multiple Packet System: A Security Approach for Wireless Networks", International Journal of Advanced Research in Computer Science, Vol. 4, No. 3, pp. 200–203, 2013.
13. P. Kumar et al., "Encryption using DES of ANonce in 4-Way Handshake Protocol for Authentication in Wpa2," International Journal of Advanced Research in Computer Science and Software Engineering, Vol. 3, Issue 10, pp. 552–557, 2013.
14. S. H. Eum et al., "A Secure 4-Way Handshake in 802.11i Using Cookies," International Journal of Principles and Applications of Information Science and Technology, Vol. 2, No. 1, pp. 40–48, 2008.
15. L. Liu, T. Stimpson et al., "An Investigation of Security Trends in Personal Wireless Networks", Springer Science + Business Media, pp. 1669–1687, 2014.
16. Y. Alkady, M. I. Habib, and R. Y. Rizk, "A new security protocol using hybrid cryptography algorithms," 9th International Computer Engineering Conference (ICENCO), pp. 109–115, Dec. 2013.
17. M. Thomas and V. Panchami, "An encryption protocol for end-to-end secure transmission of SMS," 2015 International Conference on Circuits, Power and Computing Technologies [ICCPCT-2015], Mar. 2015.
18. Daemen, J and Rijmen, V, "AES Proposal: Rijndael", Banksys/Katholieke Universiteit Leuven, Belgium, submission, Jun 1998.
19. N. Manivannan, Dr. P. Neelameham, "Alternative Pair-wise Key Exchange Protocols (IEEE 802.11i) in Wireless LANs," Proceedings of International Conference on Wireless and Mobile Communications, pp. 52–58, 2006.
20. Aidil Izani Jafri, Yean Li Ho, "ANonce encryption in 802.11i 4-way handshake protocol," Proceedings of the 7th International Conference on Advances in Mobile Computing and Multimedia, pp. 475–480, 2009.
21. A. Boudguiga and M. Laurent, "Authentication in wireless mesh networks," in Security for Multihop Wireless Networks. Informa UK, pp. 419–446, 2014.
22. J. Liu, X. Ye, J. Zhang, and J. Li, "Security verification of 802.11i 4-Way handshake protocol," IEEE International Conference on Communications, 2008.
23. Z. Yang, A. C. Champion, B. Gu, X. Bai, and D. Xuan, "Link-layer protection in 802.11i WLANS with dummy authentication," Proceedings of the 2nd ACM Conference on Wireless Network Security, pp. 131–138, Mar. 2009.
24. A. Alabdulatif, Xiaoqi, and L. Nolle, "Analysing and attacking the 4-way handshake of IEEE 802.11i standard," in proc. 8th International Conference for Internet Technology and Secured Transactions (ICITST), pp. 382–387, Dec. 2013.
25. J. H. Abawajy, S. Mukherjea, S. M. Thampi, and A. Ruiz-Martínez, Eds., Security in computing and communications in Communications in Computer and Information Science. Springer International Publishing, 2015.
26. L. Ge, L. Wang, L. Xu, and S. Yang, "Password recovery for WPA/WPA2-PSK based on parallel random search with GPU," in Wireless Communications, Networking and Applications. Springer Science + Business Media, pp. 1149–1159, 2015.
27. Q.-A. Zeng, Ed., Wireless communications, networking and applications in Lecture Notes in Electrical Engineering. Springer India, 2016.

28. S. Kansal and M. Mittal, "Performance evaluation of various symmetric encryption algorithms," 2014 International Conference on Parallel, Distributed and Grid Computing (PDGC), pp. 105–109, Dec. 2013.
29. H. Altunbasak and H. Owen, "Alternative pair-wise key exchange protocols for robust security networks (IEEE 802.11i) in wireless LANs," Proceedings IEEE Southeast Con, pp. 77–83, March 2004.

BrowserGuard2: A Solution for Drive-by-Download Attacks

Gireesh Joshi, R. Padmavathy, Anil Pinapati and Mani Bhushan Kumar

1 Introduction

We propose a method named as BrowserGuard2, a behaviour based solution to drive-by-download attacks. Drive-by-download attacks happen when users open a malicious website, which downloads and executes malware in user computers. These attacks succeed because of various vulnerabilities present in web browsers or web plug-ins. These attacks can totally compromise the user's system by granting complete access to the attacker. Many solutions have been proposed for these attacks but most of these solutions have impractical false negatives, false positives or high performance overhead.

Since it is a behaviour based run-time solution, it does not analyse webpage source code or script code. It does not take website ranking into consideration and does not maintain exploit code sample. BrowserGuard2 analyses download process of the file and based on the conclusions it draws, it successfully blocks the malware execution. Malware is a file downloaded to the system without user's permission.

Hsu et al. [1] invented the basic model to prevent BrowserGuard attacks. Experiments conducted by Hsu et al. with respect to BrowserGuard, it has low negligible false negatives and false positives, low performance overhead (less than 2.5%).

G. Joshi · R. Padmavathy · A. Pinapati (✉) · M. B. Kumar
Department of C.S.E., National Institute of Technology Warangal, Warangal,
Telangana, India
e-mail: anilcse583@gmail.com

G. Joshi
e-mail: gireeshjoshi2@gmail.com

R. Padmavathy
e-mail: rpadma@nitw.ac.in

M. B. Kumar
e-mail: manipg13@gmail.com

© Springer Nature Singapore Pte Ltd. 2019
V. Nath and J. K. Mandal (eds.), *Proceeding of the Second International Conference on Microelectronics, Computing & Communication Systems (MCCS 2017)*, Lecture Notes in Electrical Engineering 476, https://doi.org/10.1007/978-981-10-8234-4_59

Here, we present some of the optimizations to BrowserGuard, which would give us a performance win. We also list the newly listed exploits in Microsoft Bulletin, which is also successfully blocked by BrowserGuard2 (as well as BrowserGuard).

The remainder of the paper is scattered into 7 sections, they are as follows. Section 2 discusses related works in BrowserGuard. Section 3 discusses sufficient background knowledge required to understand BrowserGuard2. Section 4 gives the review of the original work of Hsu et al. [1]. Section 5 explains design and implementation of BrowserGuard2 which is an enhanced method of BrowserGuard followed by comparison with BrowserGuard while providing sufficient reasoning to justify the optimizations we made. Section 6, evaluates BrowserGuard2 again while comparing it with BrowserGuard. Section 7 follows the conclusion.

2 Related Work

Many of the drive-by-download attacks use vulnerability present in Active-X controls. Microsoft's kill-bit [2] solves this problem. It blocks the use of certain known vulnerabilities in Active-X controls. Attacker can still do drive-by-download attack by using non-Active-X control vulnerability. Also, not all the vulnerabilities of Active-X control are listed.

Hsu et al. [3] proposed a method to use HSP to avoid heap spray. HSP keep track the location and number of int 80 instructions in a process and hides the information about legal int 80 instructions, hence it makes more complex to incorporate system calls by an attacker. However, it is a compiler based solution and does not provide solution for static linked libraries.

Many of the drive-by-download attacks happen by heap spray. Ratanaworabhan et al. [4] proposed nozzle detection heap spray attacks by analysing the shell code which usually has prepended long NOP sled. It yields less number of false negatives and false positive. However, an attacker can still attack using heap spray by writing NOP sleds and shell code into a heap string after Nozzle's examination. Egele et al. [5] analyse ×86 instructions to detect a shell code presenting a code in JavaScript. It leads more performance overhead.

Lu et al. [6] introduced Blade to prevent Drive-by-Download attacks. The blade is similar to BrowserGuard; it blocks execution of suspicious files. It sandboxes all downloaded files in a secure zone and prohibits processes from accessing unauthorized files. However, it may introduce considerable performance overheads to non-browser process because if a user manipulates a file, OS checks to see, whether the file is secure or not in the zone. Moreover, it produces zero error rates.

Gadaleta et al. [7] proposed a method named as Bubble, which solves heap spray attacks, by inserting random interrupt instructions in string variables of JavaScript before storing it into the heap. Bubble get backs the updated string to the exact original string before using a string. Hence, a shell code stored in JavaScript, string variable cannot be executed.

Sometimes, browser reputation or ranking systems are used to defend against drive-by-download attacks. Before opening a webpage, browser queries a remote database to verify the popularity of the website. Consider the websites with popularity above mansion threshold (user-defined value) are displayed. Many antivirus software's are adopted this approach. However, not all low popular websites are malicious and consistent query is an overhead.

JSAND [8] uses honey browsers of high interaction to analyse the behaviour of web pages in the OS for detecting malicious content. It can simulate any Active-X control or plug-in needed for a web page. However, this solution is not integrated into browsers and therefore does not yield real-time security to the browsers. Also, it cannot identify malicious content, once the honey browser is not vulnerable to the malicious page.

Song et al. [9] introduced the IMC, which matches inter-module communication events with predefined vulnerability-based signatures to detect drive-by-download attacks. However, it is hard to identify zero-day attacks because of its signature-based property.

Cao et al. [10] introduced a JShield, which is based on vulnerability-based approach. JShield uses a novel opcode vulnerability signature and a deterministic finite automaton (DFA) with a variable pool at opcode level, to match drive-by-download vulnerabilities. it primarily is for vulnerabilities of web browsers with JavaScript engine.

3   Background

This section discusses in detail about the drive-by-download attacks, the APIs used in IE7, API Hooking, and BHO.

3.1   Drive-by-Download Attacks

Whenever a user visits a site that injects the malware into the user's system leads to drive-by-download attack. Now, this webpage with malicious content inserted by an attacker or can be a website that allows users to put their content and that content is malicious. These attacks succeed because of various vulnerabilities present in the browser or its plug-ins [11].

The common type of drive-by-download attack is buffer overflow attack [12]. It occurs when a downloaded malicious program sends more data to a target application's (e.g., a browser) memory buffer than the buffer can handle—which can help an attacker to create backdoor to the system through which he can gain access to the system. The types of buffer overflow attacks include stack attack, heap sprays, etc.

3.2 APIs Responsible for File Download in IE

The download of any legal or benign file follows following steps. First, IE calls API InternetConnect, which establish communication and returns a handle. This handle is used by IE then calls the API HttpOpenRequest for assigning a name to the file. Next, to read the file from Internet IE calls the API InternetReadFile. Finally, API WriteFile is called to save the file to hard disk.

A new file is created when we click the right mouse button and clicking the hyperlink to download. In this case, IE calls the API DoFileDownload which opens a popup to ask for a location to save a file. After all the above steps, the file is saved to the specified directory.

3.3 APIs Responsible for File Execution in IE

To execute a file, IE calls API CreateProcessW which internally calls API CreateProcessInternalW for loading the file image. IE then forks a new child process.

3.4 Browser Helper Object

A BHO [13] is DLL module installed in IE as a plug-in to provide additional functionalities. It is loaded into IE's address space and sits there until the browser exits. Now since host browser and BHO share same address space, BHO is allowed to have access to the API and can also manipulate the execution flow. Therefore, the primary component in BrowserGuard2 is BrowserGuard2-BHO.

3.5 API Hooking

API hooking [14] is a technique to alter the execution flow by interrupting a function call in a program and diverting it to another program function. BrowserGuard2 uses MS Detours library [15] to implement API hooking. Detours place a jump to the user-provided detour function in place of the first few instructions in the target function. Detour function can update the target function or call it back again using the pointer to the target function.

4 Review

In this section, we discuss the design and implementation of the original work [1], BrowserGuard, of which this paper is an extension. Following the download steps of a file in IE, BrowserGuard hooks some of these steps to detect and block the malware.

4.1 Structure of BrowserGuard

Figure 1 shows the structure of BrowserGuard. In user space, it consists of BrowserGuard-BHO (which runs in address space of IE) and List Server process. List Server maintains two lists, black-list and white-list. White-list keeps hash of benign executable files and URLs of benign files. Black-list stores hash values of malicious files. There is only one list server per host. A most may have more than one IE window opened; hence white-list and black-list are separately monitored in list server which is common for all the windows.

BrowserGuard-BHO uses the function before_navigate (which is triggered by BeforeNavigate2 event) to store URL of benign files in white-list. It also hooks detour functions to APIs in IE's file download and file execution components. The hooked file download APIs are DoFileDownload, InternetReadFile, WriteFile. Hooked file execution API is CreateProcessInternalW. To communicate with list server process, BrowserGuard-BHO uses named pipe. BrowserGuard-kernel component does two tasks. First, it makes sure that file execution is issued by CreateProcessInternalW which is hooked by BrowserGuard-BHO. Second, it does not allow any modification to the white-list, unless it is issued from before_navigate or detour function of DoFileDownload. We would not be discussing the working of BrowserGuard-kernel in this paper.

Fig. 1 Key components and key data structures of BrowserGuard

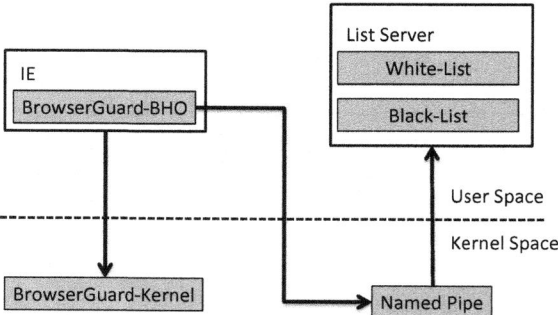

4.2   Working of BrowserGuard

Its working can be separated into two phases, *filtration phase* and *prohibition phase*. Filtration phase separates benign files from nonbenign ones. Prohibition phase then blocks execution of filtered nonbenign files.

1. *Filtration phase*: When benign files are downloaded, BeforeNavigate2 event gets fired which in turn invokes function before_navigate. When files are saved using right click, DoFileDownload API is called which is hooked in BrowserGuard-BHO. Both these functions receive URL of the file which is then stored in white-list (Fig. 2).
   File is downloaded from the Internet by InternetReadFile. The detour function of InternetReadFile in BrowserGuard-BHO checks if the URL of the file to be downloaded is present in white-list. If it is in white-list and file's first two bytes are '*MZ*' (.exe file) then the file's hash is added in white-list. Otherwise, if URL is not present in white-list and file's first two bytes are '*MZ*' then file's hash is added in black-list. Note that file data can only be read after the file is downloaded; hence after URL verification, file is downloaded and then checked if it is executable or not. WriteFile is used to write the downloaded file to disk. Detour function of WriteFile checks if first two of the file to be written is '*MZ*'. If yes

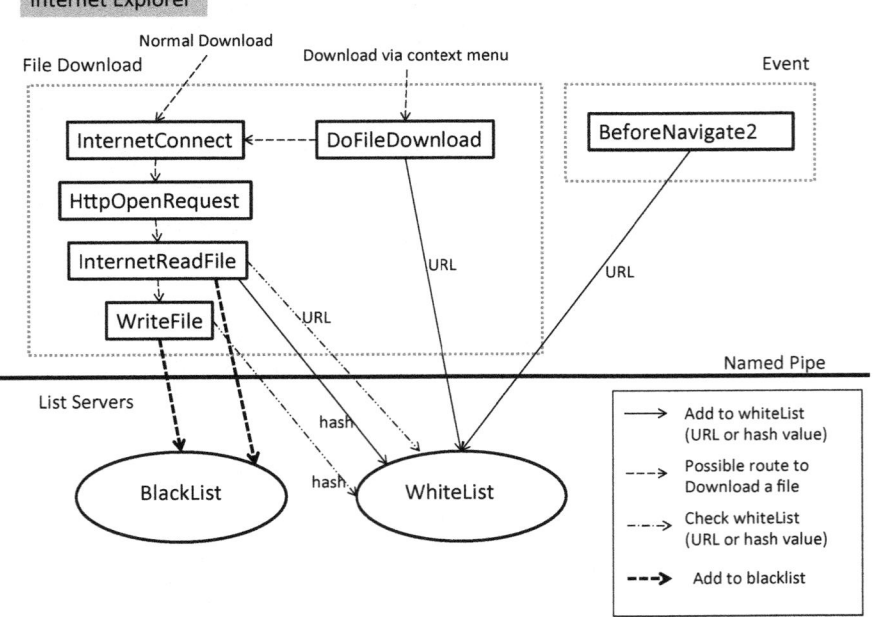

Fig. 2 Methods, operations, and data structures in filtration phase of BrowserGuard

then the file's hash is looked up in white-list. If it is not present in white-list, the hash value is added to the black-list. In both cases file is written to disk.

2. **Prohibition phase**: API CreateProcessInternalW executes an executable file. The detour function of CreateProcessInternalW first calculates the hash of the file to be executed. If the hash is present in white-list and not in black-list then file is allowed to execute; otherwise, execution is blocked.

5 Implementation

This section discusses the implementation of BrowserGuard2. Now since BrowserGuard2 is an extension of BrowserGuard, in this section we describe the changes we made and sufficient explanations to justify the changes.

5.1 Structure of BrowserGuard2

Figure 3 shows the structure of BrowserGuard2. In BrowserGuard, list server maintained two lists, white-list and black-list. Here in BrowserGuard2, we just maintain one list, white-list. Below we explain it with reasoning.

Explanation and reasoning: In BrowserGuard, detour function of InternetReadFile checks if URL of the file to be downloaded is in white-list. If yes and the file is an executable file, then its hash value is added to white-list. Otherwise, the hash value is added to black-list. Next in detour function of WriteFile, if the file to be written is executable then BrowserGuard checks if its hash is present in white-list. If not, it calculates hash and adds it to black-list. From above, it can be concluded that white-list and black-list are disjoint sets in the sense that the hash value of an executable file would either be present in white-list or black-list but not in both. As the final step in BrowserGuard, detour function of CreateProcessInternalW checks both white-list and black-list before executing a file. Now since the two lists are disjoint, checking one of the lists would be sufficient. Hence, we do not maintain

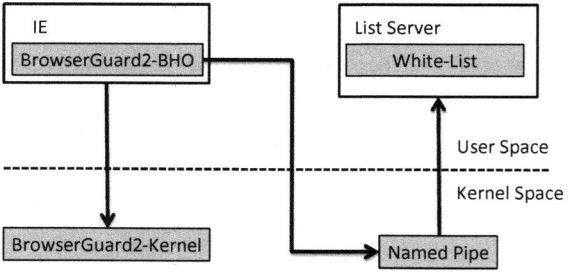

Fig. 3 Key components and key data structures of BrowserGuard2

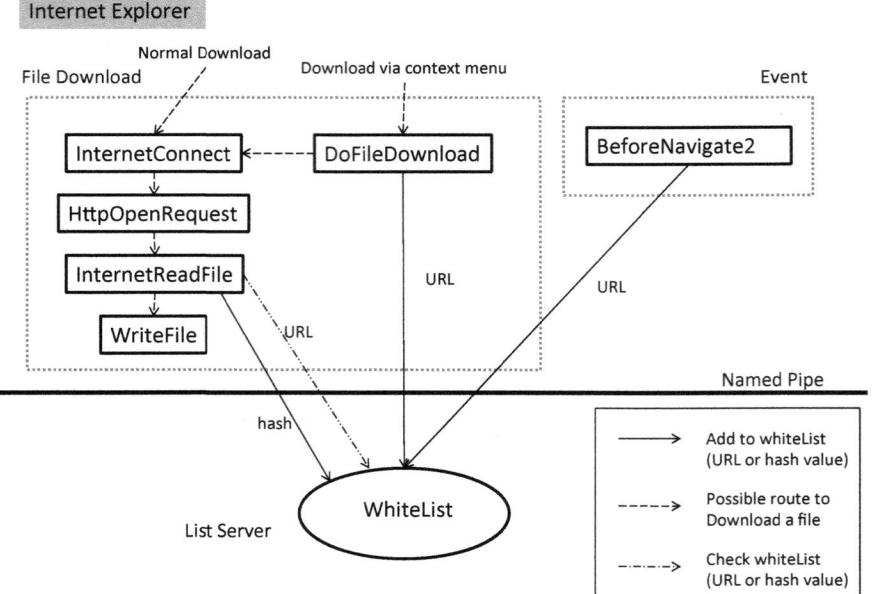

Fig. 4 Methods, operations, and data structures in filtration phase of BrowserGuard2

black-list in BrowserGuard2. Since we don't need black-list, we also need not hook the API WriteFile, which gives us another performance win (Fig. 4).

5.2 Working of BrowserGuard2

The working of BrowserGuard2 is similar to BrowserGuard except that there is no detour function of WriteFile and we use only one list, white-list. Here also working can be divided into two phases, filtration and prohibition phase.

1. *Filtration phase*: Here also URL received by before_navigate (which is invoked by event BeforeNavigate2) and detour function of DoFileDownload are saved to white-list. The detour function of InternetReadFile checks if URL of the file to be downloaded in present in white-list. If URL is in white-list and the file is executable, then the hash value of the file is added in white-list.
2. *Prohibition phase*: Detour function of CreateInternalProcessW first assess the hash value of file to be executed. If hash is present in white-list, it allows execution of the file; otherwise, execution is blocked.

6   Evaluation

This section clearly explains in detail about the comparison of BrowserGuard2 and BrowserGuard. Below we list the configuration of hosts, operating systems, and browsers which were used to verify BrowserGuard2. Host machine using Oracle VirtualBox and allows executing Guest machine also.

- Oracle VirtualBox 4.3.14 (**Memory: 1 GB**)
- Guest machine: (**OS: MS Windows Vista (32 bit), Browser: IE 7**)
- Host machine (**OS: MS Windows 8 (64 bit), CPU: Intel Core i5-3210M CPU P8600 2.5 GHz, RAM: 3 GB**)

6.1   *Effectiveness*

To determine false positives of BrowserGuard2, we visited top 150 ranked websites by Alexa [16]. These websites are not reported as malicious websites by Google. BrowserGuard2 also issued no alert for malicious content on these websites; therefore BrowserGuard2 has zero false positives for these websites. To determine false negatives we used Metasploit [17] to create malicious web pages based on the exploits for IE7. These vulnerabilities are listed on Microsoft bulletin and Metasploit can generate malicious web pages based on these vulnerabilities. If these web pages are accessed by the vulnerable target (in this case IE 7), the whole system (in this case guest machine) gets compromised. The vulnerabilities listed in the original paper [1] are listed in Table 1.

These exploits, except MS10-018, were no longer able to compromise the guest machine we installed. We also found some more recently listed exploits which were be blocked by BrowserGuard2 (as well as BrowserGuard). These are listed in Table 2.

Table 1 False negative test results of BrowserGuard [1] (MSB = Microsoft Security Bulletin [18])

No.	MSB	CVE-ID	Description	Result
1	MS06-014	2006-0003	RDS. Dataspace Active-X Control Vulnerability	Blocked
2	MS06-055	2006-4868	VML Fill Method Buffer Overflow	Blocked
3	MS06-067	2006-4777	Daxctle.ocx Keyframe Function Heap Overflow	Blocked
4	MS07-017	2007-0038	ANI LoadAniIcon Function Buffer Overflow	Blocked
5	MS07-017	2007-0038	The Malicious Executable is encoded in a jpg file	Blocked

Table 2 False negative test results of BrowserGuard2

No.	MSB	CVE-ID	Description	Result
1	MS10-018	2010-0806	DHTML Behaviours Use-after-free	Blocked
2	MS11-050	2011-1260	mshtml!CObjectElement Use-after-free	Blocked
3	MS12-043	2012-1889	MSXML Uninitialized Memory Corruption	Blocked
4	MS12-063	2012-4969	execCommand Use-after-free Vulnerability	Blocked

We deployed these malicious web pages on the host machine. These web pages contained both exploit code and shell code which is used to carry out drive-by-download attacks. When these pages were accessed by IE7 in the guest machine with BrowserGuard2 disabled, the guest machine got compromised. However with BrowserGuard2 enabled, the execution of the shell code got blocked, even though the malware was still downloaded to the guest machine. Hence BrowserGuard2 has zero false negatives for these malicious web pages.

Table 3 compares detection accuracy of BrowserGuard and similar works. Now since design and working principle of BrowserGuard2 are similar to BrowserGuard, we can safely claim that Table 3 holds for BrowserGuard2 too (where data is not provided by the corresponding work, N/A is written). IMC [9] is a signature-based solution; therefore if the database does not contain the vulnerability signatures, false negative rates rise to up to 48%.

6.2   Performance Overhead

In this subsection, we theoretically prove our claim that BrowserGuard2 is more efficient that BrowserGuard. From code analysis and as claimed in [1], following are the factors that impose performance overhead in BrowserGuard. Memory access is required to add URL in white-list and to check if URL is present in white-list. After file is downloaded or when file is to be written to disk, its first two bytes are to be checked and if the file is executable, its hash has to be calculated.

Memory accesses are required to modify or search through white-list and black-list. Extra overhead of executing code in detour functions and before_navigate is also present.

Table 3 Performance overhead introduced by BrowserGuard2

	False positives (%)	False negatives (%)
BrowserGuard2	0	0
Nozzle (50% threshold)	0	0
Bubble	N/A	0
JSAND	10.9	0.2
IMC	0	48 (0)

Now in BrowserGuard2 we just have one list which means lesser memory accesses and also one less detour function which means lesser code overhead. Therefore it is safe to claim that BrowserGuard2 is more efficient than BrowserGuard.

7 Conclusions

Drive-by-download attacks have major threats to web infrastructure these days. BrowserGuard2 presents a behaviour-based run-time solution against Drive-by-download attacks. BrowserGuard2 analyses download process of a file to be downloaded and based on the conclusions it draws, it successfully blocks malware execution. It is a run-time, behaviour based solution, therefore, it does not need to analyse the source code of a webpage or run-time states of a script code. It also does not take into account the reputation value of a website and does not need to maintain exploit code sample. Since no code analysis is done, it introduces very less overhead. Experimental results show that BrowseGuard2 has no false positives for top 150 ranked websites by Alexa. It also has no false negatives for specially crafted WebPages based on listed exploits on Microsoft Bulletin. BrowserGuard2 was added to and tested on IE 7.

References

1. Fu-Hau Hsu, Chang-Kuo Tso, Yi-Chun Yeh, Wei-Jen Wang, and Li-Han Chen, "BrowserGuard- a behaviour-based solution to Drive-by-Download attacks," in IEEE Journal on Selected Areas in Communications (vol: 29, issue: 7), 2011, pp. 1461–1468.
2. "Microsoft Security Research & Defense," [Online]. Available: https://blogs.technet.com/srd/archive/2008/02/06/The-Kill_2D00_BitFAQ_3A00_-Part-1-of-3.aspx.
3. F.-H. Hsu, C.-H. Huang, C.-H. Hsu, C.-W. Ou, L.-H. Chen, and P.-C. Chiu, "HSP: A solution against heap sprays," Journal of Systems and Software, vol. 83, 2010, pp. 2227–2236.
4. P. Ratanaworabhan, B. Livshits, and B. Zorn, "NOZZLE: a defense against heap-spraying code injection attacks," in Proc. 18th conference on USENIX security symposium. USENIX Association, 2009, pp. 169–186.
5. M. Egele, P. Wurzinger, C. Kruegel, and E. Kirda, "Defending browsers against drive-by downloads: Mitigating heap-spraying code injection attacks," in Proc. 6th International Conference on Detection of Intrusions and Malware, and Vulnerability Assessment, ser. DIMVA'09. Springer-Verlag, 2009, pp. 88–106.
6. L. Lu, V. Yegneswaran, P. Porras, and W. Lee, "BLADE: an attack-agnostic approach for preventing drive-by malware infections," in Proc. 17th ACM conference on Computer and communications security, ser. CCS'10. ACM, 2010, pp. 440–450.
7. F. Gadaleta, Y. Younan, and W. Joosen, "Bubble: A javascript engine level countermeasure against heap-spraying attacks," in Engineering Secure Software and Systems, ser. Lecture Notes in Computer Science, vol. 5965. Springer, 2010, pp. 1–17.

8. M. Cova, C. Kruegel, and G. Vigna, "Detection and analysis of driveby download attacks and malicious javascript code," in Proc. 19th international conference on World Wide Web, ser. WWW'10. ACM, 2010, pp. 281–290.

9. C. Song, J. Zhuge, X. Han, and Z. Ye, "Preventing drive-by download via inter-module communication monitoring," in Proc. 5th ACM Symposium on Information, Computer and Communications Security, ser. ASIACCS'10. ACM, 2010, pp. 124–134.

10. Y. Cao, X. Pan, Y. Chen, J. Zhuge, "Jshield: towards real-time and vulnerability-based detection of polluted drive-by download attacks", *Proceedings of the 30th Annual Computer Security Applications Conference*, pp. 466–475, 2014.

11. D. Harley and P.-M. Bureau, "Drive-by downloads from the trenches," in Malicious and Unwanted Software, 2008. MALWARE 2008. 3rd International Conference on, 2008, pp. 98–103.

12. An Zhiyuan and Liu Haiyan, "Realization of buffer overflow," in Information Technology and Applications (IFITA), 2010 International Forum on, 2010, pp. 347–349.

13. D. Esposito, "Browser Helper Objects: The Browser the WayYouWantIt." [Online]. Available: https://msdn.microsoft.com/enus/library/bb250436(v=vs.85).aspx.

14. Alex Abramov, "API Hooking with MS Detours" [Online]. Available: https://www.codeproject.com/Articles/30140/API-Hooking-with-MS-Detours.

15. "Detours Framework." [Online]. Available: https://www.microsoft.com/enus/research/project/detours/.

16. "Alexa Internet." [Online]. Available: https://www.alexa.com.

17. "Metasploit Framework." [Online]. Available: https://www.metasploit.com.

18. "Microsoft Security Bulletins" [Online]. Available: https://technet.microsoft.com/enus/library/security/dn631937.aspx.

GA-Based Energy Optimization in Traffic Grooming WDM Optical Mesh Network

Nishat Aafreen and Partha Paul

1 Introduction

Wavelength division multiplexing (WDM) is one of the most promising transport technologies for the optical network in today's information era to facilitate and bear the growth of Internet traffic. The Internet traffic increases with the increase in demand of multicast application. Multicasting technique is one of the techniques in which source node transmits a message to multiple destinations, e.g., video conferencing, HDTV, interactive learning, etc. In WDM technology, several wavelengths simultaneously send the data through a single fiber and it requires the bandwidth connection request which is very huge speed Terabit/second (e.g., OC192, OC-768 etc.). The traffic requested by the individual connection in practical networks is very low, i.e., in the range of megabits/second. Because of this difference in the wavelength bandwidth, a huge portion of transmission channel capacity was wasted. So, for the appropriate convention of wavelength, a traffic grooming technique is used. The traffic grooming is a mechanism which is used in the optical network where low rate traffic streams are multiplexed or groomed to the high-speed wavelength channels, which produce the low-cost network by reducing the number of higher layer electronic ports $(T_x \& R_x)$ and required wavelength.

In WDM network, the specific wavelength is used which is used to carry the traffic between the single source and destination which must establish the light-path [1, 2]. A light-path is responsible for a dedicated point-to-point network connection between two ends. Traffic grooming with routing and wavelength assignment problem (RWA) are more effective for optimization of the connection. But while

N. Aafreen · P. Paul (✉)
Department of Computer Science and Engineering, Birla Institute of Technology,
Mesra, Ranchi 835215, India
e-mail: p_india@rediffmail.com

N. Aafreen
e-mail: nishat.aafreen@gmail.com

© Springer Nature Singapore Pte Ltd. 2019
V. Nath and J. K. Mandal (eds.), *Proceeding of the Second International Conference
on Microelectronics, Computing & Communication Systems (MCCS 2017)*, Lecture Notes
in Electrical Engineering 476, https://doi.org/10.1007/978-981-10-8234-4_60

delaying and resolving the problem of the bandwidth in a network will not utilize the network resources more resourcefully. Due to this reason, the light-tree [3, 4] concept came into existence. It is a natural extension of the light-path where it is for the effective utilization of resources. A light-tree concept supports the point-to-multipoint (PtTM) connections in one single logical hop-tree. Hence, the average packet of the hop distances is reduced as well as the total number of T_x & R_x deployed in a network. The light-tree serves as a channel for upper layer traffic and low bandwidth for the connections which is routed by combining the multiple light-trees to reach the entire destination. Hence, the inconsistency between the bandwidth can be determined by combining the concept of light-tree with grooming. On the other hand, the light-tree is a natural topology for supporting multicast applications and considerable research has been studied on approaches for light-tree-based multicast traffic grooming.

In early days, the traffic grooming is implemented with the ring network [5, 6] and then after it motivated with the WDM mesh networks [7]. Several groups of researchers is planned that the traffic grooming problem in WDM mesh network for distinct traffic condition is static, incremental or dynamic. Also, the wavelength is taken as a significant feature for the cost for the network in the early days of WDM. Reducing the wavelengths in increasing the network traffic. But nowadays, the wavelength is not to be considered as the vital cost factor. Instead of it, higher electronic ports T_x & R_x become the central cost factor. T_x & R_x are used for transmitting and receiving the optical signals, respectively, so the number of T_x & R_x utilized in the networks must be equal to the number of IP router ports. Here in this paper, the primary focus on designing an energy-efficient optical WDM mesh network with the multicast static traffic grooming which optimizes the transmitter and receiver as well as the number of wavelengths used.

2 Related Work

WDM uses the energy-efficient traffic grooming networks using the most favorable number of network resources has been the topic of research in the recent times. Depending on the light-path request the traffic grooming is categorized into two: (i) Static traffic grooming and (ii) Dynamic traffic grooming. In static traffic grooming, the entire light-path is known precisely. Whereas, in dynamic traffic grooming, light-path is changing continuously.

Once the light-path has been established it remains throughout the network until the connection is terminated, and hence no alteration is done in the traffic matrix. In this case an optimal solution for traffic grooming with the objective of minimizing the resources used, e.g., wavelength, wave-links or add-drop ports or maximizing the network throughput can be obtained [1–3] using an Integer Linear Programming (ILP) method.

Lin et al. [4] presented a ILP formulation based on light-tree to optimize the network cost in terms of network resources such as higher layer electronic ports $(T_x \& R_x)$ and wavelength used. They also proposed heuristic algorithm; sub-light-tree saturated grooming (SLTSG) to achieve scalability. Also, a heuristic algorithm, i.e., saturated light-tree-based multicast traffic grooming (SLTMTG), has been proposed by Pradhan and De [5] for cost-effective design of WDM networks with multicast traffic grooming. They reduced the number of transceiver and receiver as well as minimize the cost of grooming and wavelength requirement ad tries to satisfy all connection requests.

Yang and Liao [8] have proposed an integer linear programming (ILP) formulation of multicast traffic grooming in which the number of add/drop multiplexers (ADMs) and the number of used wavelengths is minimized for the optimization of network cost. Modiano and Lin [7] and Zhu et al. [9] have proposed an improved efficiency and reduce network cost with an extension of the light-path concept to a light-tree and design a light-tree-based logical topology with delay bounds.

Zhang et al. [10] proposed a scheme to save energy in IP over WDM static networks according to the traffic variations during the hours of the day. They proposed a method based on mixed integer linear programming (MILP) to minimize the energy cost by shutting down idle line cards and chassis of routers when a load of traffic is low. Atta et al. [11] proposed a genetic-based algorithm method to minimize the EOE conversions needed as well as restricting the bandwidth wastage to satisfy all the traffic requests in a mesh network.

3 Mathematical Approaches

Given:

N	Number of vertices.
W	Wavelengths for optimization
λ	Wavelength index, range from [1 to W].
C	Wavelength of channel.
λ_m	Wavelength for multicast request 'm'
Z	Integer numeral
M	Multicast 'm' request.
m	Multicast request m lies in between $(1 \leq m \leq M)$.
f_m	Bandwidth request by a multicast request m
α	Cost of transmitter and receiver
β	Cost of wavelength in network
s_m	Source of request 'm'
D_m	Destination set in request m $\{d_1, d_2, d_3 \ldots\}$
$\{s_m; D_m; f_m\}$	A 3-tuple representing the multicast request 'm'.

Variables:

T_{xi} Transmitter positioned at 'i' node.
R_{xi} Receiver positioned at 'i' node.
ψ Wavelength channel for the network.
Y_λ Wavelength 'λ' in a binary variable lies in between 1 and 0.
L_{iD} LOHT (i, D) for ith node to the end node set $[D, D \in D_i]$.
L_{iD}^λ λ for the LOHT (i, D) for the ith node at set D.
$L_{iD}^\lambda > 1$ Node 'i' at set D using wavelength λ.
λ_{iD}^m Binary variable if '1' multicast request 'm' at LOHT (i, D) otherwise 0.

4   Objective Function

$$\text{Minimize: } \alpha * \sum_{i=1}^{M} (T_{xi} + R_{xi}) + \beta * \psi \tag{1}$$

This equation of objective function demonstrates the optimization of network cost in terms of $T_x \& R_x$ and wavelength used in the network. Here α is the cost of transmitter and receiver whereas β is the cost of the wavelength channel.

5   Constraints

Equation (2) states that number of transmitters must be less than or equal to number of receivers used in the network.

$$\sum_{i=1}^{M} T_{xi} \leq \sum_{i=1}^{M} R_{xi} \tag{2}$$

As the objective function is to optimize ψ, Eqs. (3) and (4) together ensure that the highest index for the wavelength 'λ' is used by the any light-tree.

$$\psi \geq \lambda \cdot Y_\lambda \quad \forall \lambda \tag{3}$$

$$Y_\lambda \geq \sum_i \sum_{D \in D_i} L_{iD}^\lambda / Z \quad \forall \lambda \tag{4}$$

Equation (5) ensures that the capacity of all multicast requests on LOHT (i, D) must be equal to or less than the total capacity of the network (i, D).

$$\sum_m f_m \cdot \lambda_{iD}^m \leq L_{iD} \cdot C \quad \forall i, D \in D_i \tag{5}$$

6 Proposed Method

The proposed genetic algorithm is to optimize the cost of energy for the high layer electronic ports, i.e., we use the transmitter and receiver for the network with different wavelengths channels.

Outline of the genetic algorithm-based multicast traffic grooming (GAMTG) procedure used is as follows:

- Create the AdjMat();
- Create the Load();
- Send the Scheme Order();
- Initialize the Chromosome:
  for GenLimit > 0
  for MSR > 0
- Selection of the session and wavelength
- Check the fitness value
- do Crossover()
- do Mutation()
- if fitness value all over again
- offspring further used for the population.
- Run the algorithm once more and generate the new population
- Return set of the successfully groomed trees
- Calculate the transceiver and receiver cost of the network

The cost of number of transceiver and receiver for a specified number of multicast session requests in the network is determined as follows:

Cost in the network:

$$\frac{(\text{Total multicast session request} - \text{Successful groomed trees})}{\text{Total multicast session request}} * 100$$

7   Heuristic Flowchart

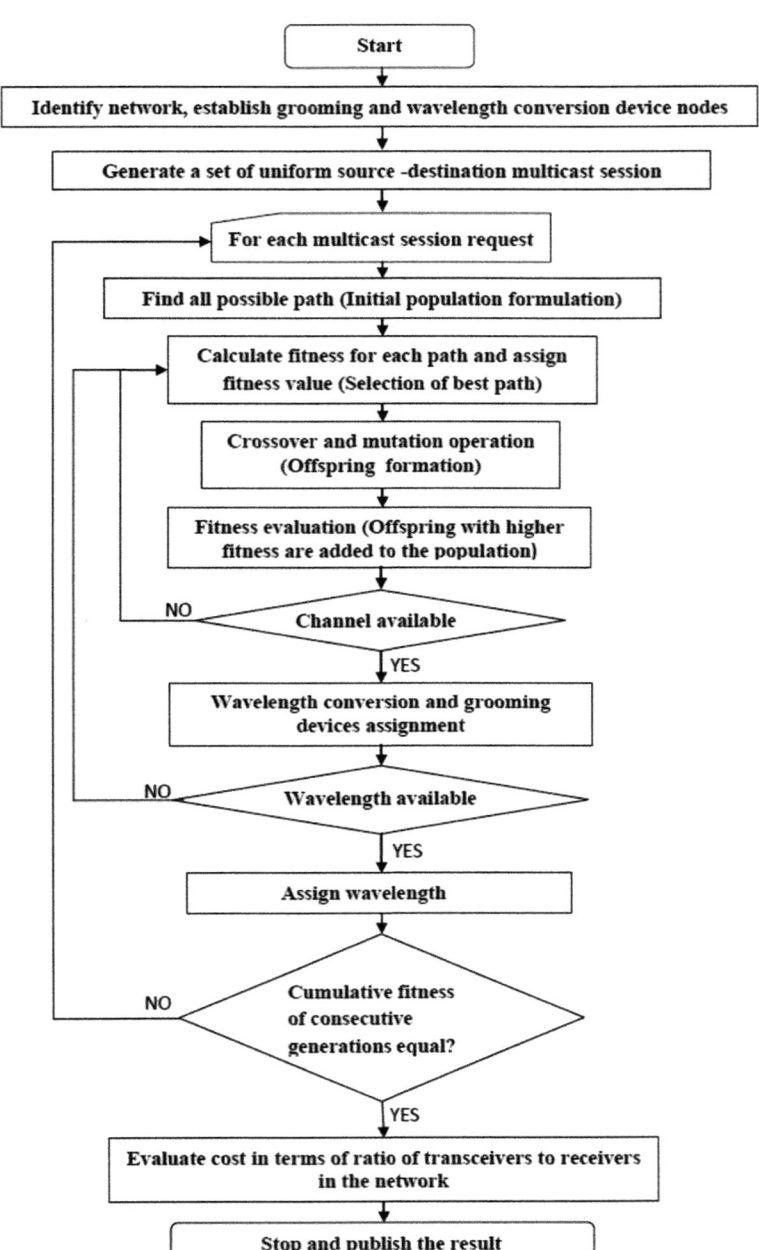

8   Result and Discussion

Here, we compare the simulation results of proposed genetic algorithm multicast traffic grooming (GAMTG) with multicast traffic grooming (MTG) and saturated light-tree multicast traffic grooming (SLTMTG) on 14-node NSF network. Here, multicast requests, i.e., chromosomes having a single source and multiple destinations are randomly generated. We have made an assumption that the channel capacity 'C' of wavelength is OC-48, and required channel bandwidth granularities are selected randomly from any one of OC-1, OC-3, OC-12, or OC-48 based on the highest fitness value of chromosomes which can be calculated from the objective function. We set relative cost parameters value to $\alpha = 3$ and $\beta = 1$. The mean value of 100 generations of simulations on generated multicast session request of proposed GAMTG algorithm is shown in this section (Fig. 1).

Figure 2 shows the result of wavelength requirement using genetic algorithm multicast traffic grooming (GAMTG). The number of session sizes varies here from [10, 100], taking maximum session size is 8. From our result, it is clear that GAMTG required the lowest wavelength than MTG (multicast traffic grooming) and SLTMTG (saturated light-tree multicast traffic grooming), hence having higher efficiency of grooming multicast request with respect to existing multicast traffic grooming (MTG) and saturated light-tree multicast traffic grooming (SLTMTG). Multicast requests are satisfied as per randomly generated requests in MTG algorithm. In MTG, multicast session requests are not arranged in any manner either by increasing or decreasing order of their session size or bandwidth requirement. But in GAMTG multicast requests having higher probability of matching and having higher fitness value will groom first which reduces the bandwidth wastage. This results in higher grooming and efficient utilization of bandwidth with fewer numbers of wavelengths used in the network.

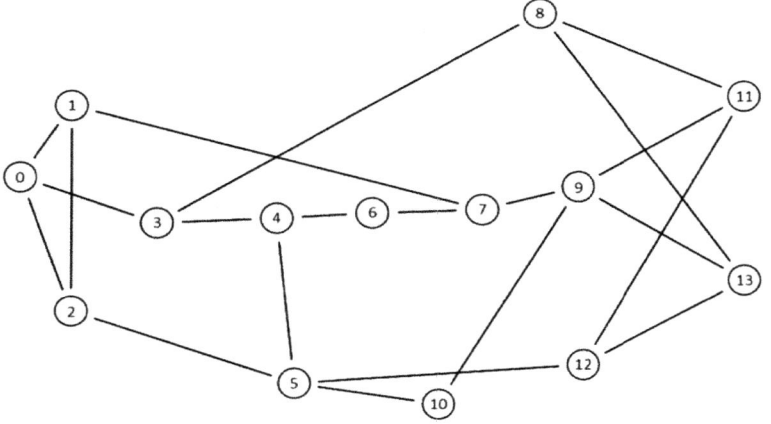

Fig. 1   14-node NSF network topology

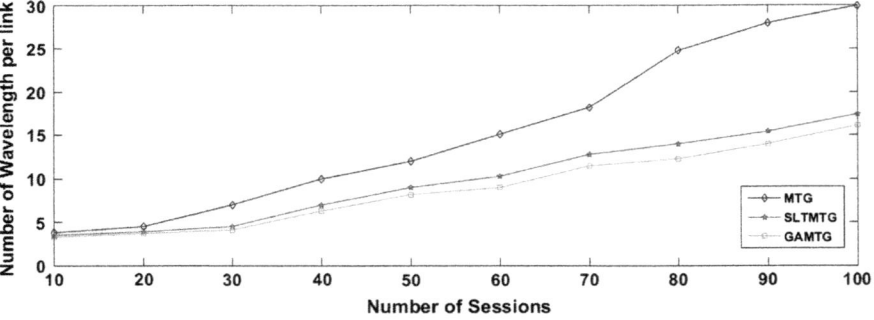

Fig. 2 Number of wavelengths per link versus number of sessions

Figures 3 and 4 compare the number of transmitter and receiver require per node respectively of the three algorithms, i.e., MTG, SLTMTG, and GAMTG. Here in NSF network destination size varies from [2, 12] for maximum session size 10. It shows that the number of transmitter and receiver required in GAMTG is less than in MTG and SLTMTG. MTG uses the largest number of transmitters. Also, SLTMTG shows slightly higher number of transmitters than GAMTG. This is because, in MTG, a light-path can reach only one destination and require more light paths to support requests. Hence, more number of transmitters is required as every light-path would use a transmitter. Whereas, GAMTG uses grooming for increasing the utilization of sub-light-trees having the highest fitness values so that more requests is benefitted and allows covering a large portion of the destinations. Hence, a lesser number of transmitters and receivers would be required to construct light-trees in order to cover the remaining destinations.

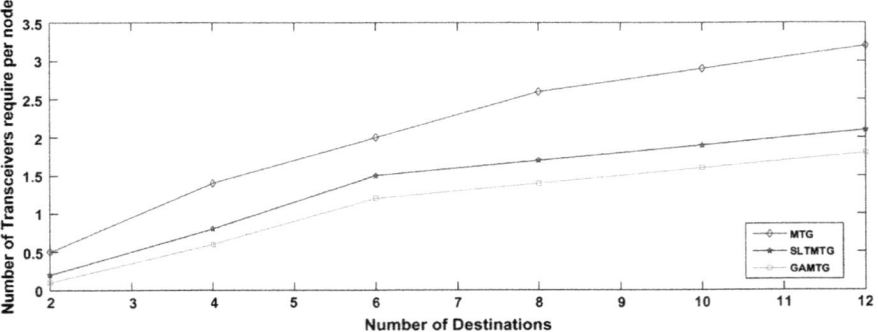

Fig. 3 Number of transceivers required per node versus number of destination

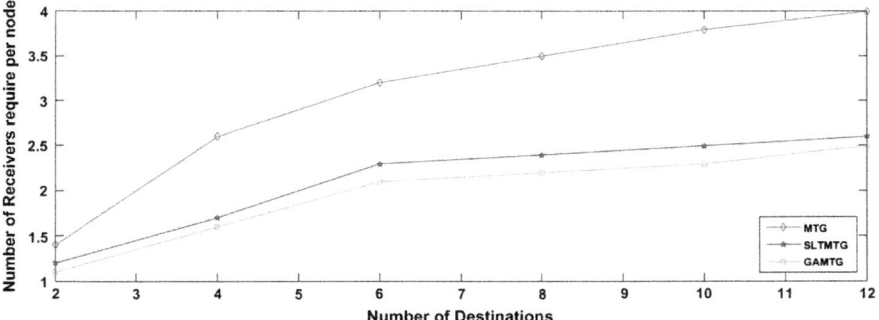

Fig. 4 Number of receivers required per node versus number of destination

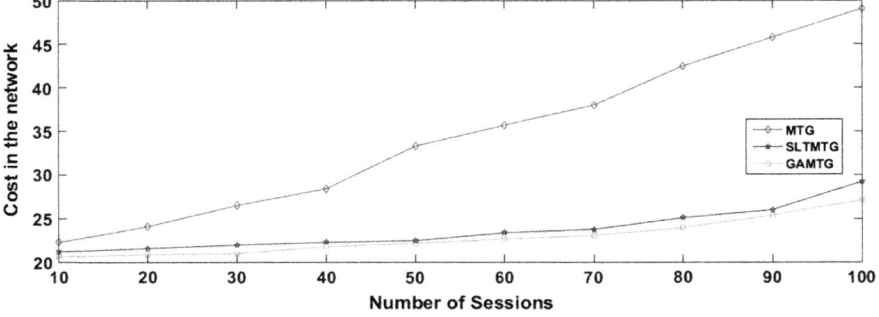

Fig. 5 Cost per node versus number of sessions

Figure 5 compares the result of cost per node of the three algorithms. The cost of GAMTG is lesser than the existing MTG and SLTMTG algorithm. This is because GAMTG constructs sub-multicast trees based on the fitness value of the light-path requests which can be fully or partially groomed.

However, in MTG multicast trees are groomed one by one and every single multicast tree is fragmented into smaller ones which cannot be shared by other multicast session requests. Hence, GAMTG requires a lesser number of electronic equipment, i.e., transmitter and receiver than MTG and SLTMTG. Thus, by using lesser number of electronic equipment, i.e. transmitter and receiver than MTG and wavelength in GAMTG, the cost of the network minimized and results in better performance than MTG as well as SLTMTG.

9 Conclusion

The main objective of traffic grooming is to provide the efficient resource utilization in an optical network by efficiently padding the low rate connections with the high rate light-paths. For achieving this, we have applied modified genetic algorithm for multicast static traffic grooming problem in WDM optical mesh network. The objective of this is to optimize the energy consumption cost with the higher layer electronic ports such as (tunable transmitters and receivers) having the number of wavelength by proposed GAMTG (genetic algorithm-based multicast traffic grooming) approach. As in result section, where genetic algorithm multicast traffic grooming (GAMTG) algorithm was compared with the normal grooming of light-trees and heuristic approach grooming based on the light-tree as explained in multicast traffic grooming (MTG) and saturated light-tree multicast traffic grooming (SLTMTG), respectively, as shown in Table 1. The performance of proposed GAMTG and existing MTG and SLTMTG has been observed in a standard 14 node NSF network. The heuristic GAMTG algorithm based on sub-multicast light-tree grooming outperforms the light-path based on MTG as well as SLTMTG. Because of the advantages of light-trees for multicast session requests, the GAMTG is recommended for using the resource efficiently than MTG and SLTMTG as GAMTG and create sub-multicast requests which is having the highest fitness value which can be utilized by several multicast session requests either partially or fully in order to optimize the network cost. Our result makes evident that our proposed GAMTG approach reduces the numbers in the optical network of tunable transceivers and receivers and also, reduces the number of wavelengths required which incurs the lowest cost than MTG and SLTMTG algorithm. This means that the traffic grooming technique having the better utilization of energy consumption cost in GAMTG algorithm which is based on grooming technique of multicast traffic to constrained light-trees.

Table 1 Comparative study table with MTG, SLTMTG, and GAMTG

MTG				SLTMTG				GAMTG			
Wavelength	Transceiver	Receiver	Cost (for session size 100)	Wavelength	Transceiver	Receiver	Cost (for session size 100)	Wavelength	Transceiver	Receiver	Cost (for session size 100)
3.8	0.5	1.4	22.3	3.5	0.2	1.2	21.2	3.3	0.1	1.1	20.6
4.5	1.4	2.6	24.1	3.9	0.8	1.7	21.6	3.7	0.6	1.6	20.9
7	2	3.2	26.5	4.5	1.5	2.3	22	4.1	1.2	2.1	21
10	2.6	3.5	28.4	7	1.7	2.4	22.3	6.3	1.4	2.2	21.8
12	2.9	3.8	33.3	9	1.9	2.5	22.5	8.2	1.6	2.3	22.2
15.1	3.2	4	35.7	10.3	2.1	2.6	23.4	9	1.8	2.5	22.7
18.2			38	12.8			23.8	11.5			23.1
24.8			42.5	14			25.1	12.3			24
28			45.8	15.4			26	14			25.4
30			49.1	17.4			29.2	16.1			27.1

References

1. K. Zhu, B. Mukherjee, (2002). Traffic grooming in an optical wdm mesh network, IEEE Journal on Selected Areas in Communications, 20 (1), pp. 122–123.
2. K. Zhu, B. Mukherjee, (2002). On-line approaches for provisioning connections of different bandwidth granularities in wdm mesh networks, Optical Fiber Communications Conference and Exhibit, OFC, pp. 549–551.
3. R. Ul-Mustafa, A. Kamal, (2006). Design and provisioning of WDM networks with multicast traffic grooming, IEEE Journal on Selected Areas in Communications 24 (4), pp. 37–53.
4. R. Lin, W.D. Zhong, S.K. Bose, M. Zukerman, (2011) "Design of wdm networks with multicast traffic grooming", Journal of Light wave Technology, 29(16), pp. 2337–2349.
5. A. K. Pradhan, Tanmay De, (2013) Multicast Traffic Grooming based Light-Tree in WDM Mesh Networks, Procedia Technology, 10, pp. 900–909.
6. Y. Xu, S.C. Xu, B.X. Wu, (2002). Traffic grooming in unidirectional wdm ring networks using genetic algorithm. Computer Communications, 25:1185–1194.
7. E. Modiano and P.J. Lin, (2001). Traffic grooming in wdm networks, IEEE Communication. Mag. Vol. 39, no. 7, pp. 124–129, 2001.
8. D.N. Yang, W. Liao, (2003) "Design of light-tree based logical topologies for multicast streams in wavelength routed optical networks, Twenty-Second Annual Joint Conference of the IEEE Computer and Communications. IEEE Societies, INFOCOM 2003. Vol. 1, pp. 32–41.
9. H. Zhu, H. Zang, K. Zhu and B. Mukherjee, (2003, Apr.) A novel generic graph model for traffic grooming in heterogeneous wdm mesh networks, IEEE/ACM Trans. Networking, vol. 11, no. 2, pp 285–299.
10. Y. Zhang, et al, (2011). Energy optimization in IP-over-WDM Networks, Optical Switching and Networking, 10, pp. 900–909.
11. Soumen Atta, Anirban Mukhopadhyay, (2012). Power-aware Traffic Grooming in WDM Optical Mesh Networks for Bandwidth Wastage Minimization: A Genetic Algorithm-based Approach. IEEE National Conference on Computing and Communication Systems (NCCCS).
12. Y.R. Yoon, T.J. Lee, M.Y. Chung, H. Choo, (2006). Traffic Groomed multicasting in sparse-splitting wdm backbone networks, in: Proceedings of the international conference on Computational Science and Its Applications- Volume Part 2. ICCSA, Springer-Verlag. pp. 534–544.

Piezoelectric Energy Harvesting: A Developing Scope for Low-Power Applications

Oshin Garg, Sukesha Sharma, Preeti and Pardeep Kaur

1 Introduction

With the progress in wireless technology and low-power electronics, wireless sensors can be used in almost every field. Some of the applications of wireless sensor networks are structure health monitoring, automation, robotics swarm, and military applications. However, these devices require their own energy source which in most cases is the conventional battery. The lifetime of these batteries is limited and they need to be replaced time to time. The task to replace the batteries is very challenging because most of the wireless sensors are placed in remote areas. This hinders the working of wireless sensors and restricts its use. Energy harvesting is one of the best possible solutions to extend the lifetime of wireless sensors and other portable low-power electronic devices. Energy harvesting will make it possible to use these devices in near future to much greater extent [1]. Energy harvesting is the technique to use the available energy from the environment and convert it into electrical energy to power the low-power electronic devices such as wireless sensors. The diverse sources of energy can be solar energy, electromagnetic waves, thermal energy, wind energy, kinetic energy, etc. Energy harvesting is an attractive alternative for the batteries and can be used for various energy applications ranging from low power to high power [2, 3]. In this paper, various energy harvesting techniques have been discussed. Organization of the paper is as follows: in Sect. 2 various energy harvesting techniques have been discussed. In Sect. 3, piezoelectric energy harvesting technique is introduced. In Sects. 4–6 transducer design, rectifier circuit, and storage units have been discussed respectively. Section 7 summarizes the study.

O. Garg (✉) · S. Sharma · Preeti · P. Kaur
UIET, Panjab University, Chandigarh, India
e-mail: osh.scorpion@gmail.com

© Springer Nature Singapore Pte Ltd. 2019
V. Nath and J. K. Mandal (eds.), *Proceeding of the Second International Conference on Microelectronics, Computing & Communication Systems (MCCS 2017)*, Lecture Notes in Electrical Engineering 476, https://doi.org/10.1007/978-981-10-8234-4_61

2 Different Approaches to Energy Harvesting

The energy harvesting techniques can be classified based on the sources of energy. The different energy sources for energy harvesting and their working principle have been shown in Fig. 1.

The energy can be harvested from solar energy by photovoltaic principle. The photovoltaic principle is the creation of voltage and current in the semiconductor material when it is exposed to light [4]. In solar energy harvesting, photovoltaic cells are used that converts the incoming light or photon into electricity. This type of energy harvesting is known as photovoltaic energy harvesting [5]. The energy can also be harvested from temperature difference by the principle of 'seeback effect', which states that when there is a difference in the temperature between two junctions made up of conducting materials a voltage is developed between them. This type of energy harvesting technique is known as thermal energy harvesting [6]. The energy can also be harvested from RF energy available in the environment due to various mobile communication services such as GSM by electromagnetic principle. This type of energy harvesting is known as RF energy harvesting [7]. Another approach to harvest energy is from motion or vibrations. The vibrational energy can be harvested by three principles, i.e., electrostatic, electromagnetic, and piezo-electric. The electrostatic principle is based on changing the capacitance of a capacitor using a source of vibration. Vibration separates the plates of a charged variable capacitor, and hence the mechanical energy is converted into electrical energy. The energy harvesting using this principle is known as electrostatic energy harvesting [8]. The electromagnetic principle is based on the fact that a motion in the magnet inside a copper coil induces a voltage in the coil due to change in the magnetic field. The energy harvesting using this principle is known as

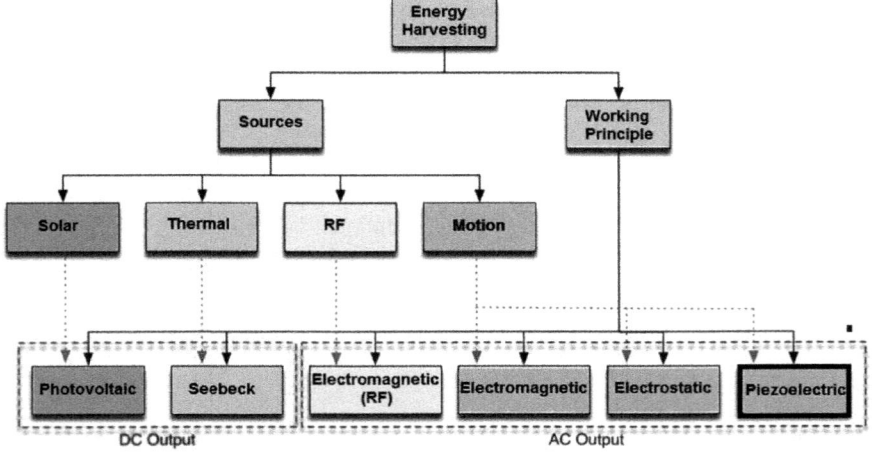

Fig. 1 Energy harvesting techniques and their working principle [38]

electromagnetic energy harvesting [9]. The piezoelectric principle is based on the fact that when a force is applied to piezoelectric materials they generate an electrical potential due to charge separation and vice versa. The energy harvesting using this principle is known as piezoelectric energy harvesting [10].

The above-discussed energy harvesting techniques can be classified as high-power energy sources and low-power energy sources based upon the power output. High-power energy sources (or macro sources) are the sources that give output power in kilowatts are, e.g., solar, wind, etc. [11]. Low-power energy sources (or micro) are the sources that produce an output power in the range of milliwatts, e.g., vibration, motion, heat, etc. [11]. Depending upon the power requirement of any application, one can select one of the approaches mentioned above. Piezoelectric energy harvesting is classified under low-power energy harvesting as it produces power in the range of milliwatts. Piezoelectric energy harvesting circuit is least complex and is easy to analyze as compared to other low-power vibrational energy harvesting approaches, hence it can be used to provide energy to low-power wireless devices. This paper is focused on piezoelectric energy harvesting.

3 Piezoelectric Energy Harvesting

Piezoelectric crystals are made up of materials, which are also called ferroelectric which have a property that when a force is applied on such materials they generate an electrical potential due to charge separation. This effect is known as the piezoelectric effect. Piezoelectric energy harvesting is based on the principle of the piezoelectric effect. It is a reversible process, i.e., when a voltage is applied across such crystals their shape is changed [12]. Some examples of piezoelectric materials are PZT (lead zirconium titanate), PVDF (polyvinylidene fluoride). Figure 2 shows the piezoelectric effect.

Piezoelectric energy harvesting circuit consists of three main components: transducer, rectifier, energy storage unit. The first step of piezoelectric energy harvesting circuit is to convert mechanical force or vibrations into electrical energy by means of transducer or energy harvester. It consists of piezoelectric ceramic,

Fig. 2 Piezoelectric effect [12]

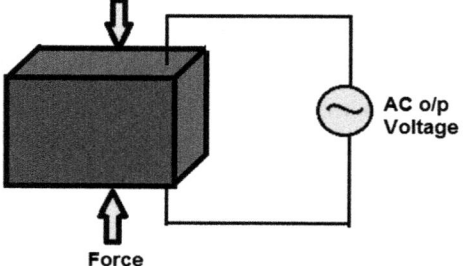

elastic body, and a tip mass. This simple structure of the harvester produces large strain when subjected to vibrations and can convert mechanical energy into electrical energy. This output is AC, i.e., alternating current and it needs to be converted to DC, i.e., direct current by using a rectifier. The last step of the harvesting circuit is to store this energy for future use to power electronic devices [13]. There are various factors such as the physical design of transducer, operating frequency, etc., that decides the performance of the harvesting circuit. These parameters have been discussed in detail. Piezoelectric transducer has been discussed in next section.

4 Piezoelectric Transducer

The transducer is a device that converts one form of energy into another. A piezoelectric transducer converts the mechanical vibrations or forces into electrical energy. It gives an AC output voltage whose magnitude is proportional to the applied force. There are some parameters that need to be considered while designing a piezoelectric transducer to maximize its output power. These parameters are discussed below.

4.1 Choice of Material

The material of a piezoelectric crystal is an important parameter that needs to be considered before designing a piezoelectric transducer. The piezoelectric material includes bulks, thin films, and nanostructure materials. Bulk piezoelectric materials include PZT (lead zirconate titanate) and related compositions. They have excellent piezoelectric properties but they are brittle in nature. Mostly single-crystal piezoelectric materials have better piezoelectric behavior than polycrystalline ceramics. Some of the single crystals are PZN-PT and PMN-PT, which have outstanding piezoelectric properties and can produce much higher power output than ceramics but they have been hardly used for energy harvesting because of high cost and difficulty in fabrication [14]. MEMS technology is another widely used technology because it could provide high flexibility and strain in the material. Also, materials fabricated by MEMS technology can be used in microelectronics and require very less power [15–18]. Flexibility tends to increase the usage of piezoelectric crystals and they can be used under large strain. To achieve flexibility nano- and microstructure materials and thin films have been used with polymers. In 2006, Zhong Lin Wang first introduced piezoelectricity in ZnO which were highly flexible [19]. PZT material has much larger piezoelectric coefficient hence generate larger output power [20, 21]. Hence, researchers worked on flexible piezoelectric structures using PZT but their synthesis was difficult. Some of the PZT-based flexible materials are nanotubes, nanorods, nanowires, thin films, etc. Different piezoelectric materials have different behaviors, for example, some are brittle but provide large output power, some are flexible but generate less power. Depending upon the application, working environment of transducer and power requirements material for the piezoelectric transducer can be selected. Various piezoelectric materials have been investigated and many research papers have been published. PZT material has been most widely used for energy harvesting because of its advantages over other practically available piezoelectric materials and it produces maximum power for random vibrations. Although PZT continues to be the best choice it contains lead which is not environment-friendly. Many researchers are researching for the lead-free piezoelectric material that could replace the PZT material. None of the presently available non-lead materials match the performance of PZT but several materials have been considered having advantages of low densities, high coupling factor, high mechanical strength. Some of the lead-free piezoelectric materials are barium titanate (BT), sodium bismuth titanate (NBT), potassium bismuth nitrate (KBT), bismuth ferrite (BFO), quartz (SiO_2), etc. [22].

4.2 Choice of Structural Design and Loading Mode

Another parameter that should be considered while designing a piezoelectric transducer is its structure and the loading mode. Based on the number of active layer, i.e., layers of piezoelectric material there are two possible structures namely, unimorph and bimorph. Unimorph structure of a piezoelectric transducer has one layer of piezoelectric material (active layer) and another of elastic material (passive layer). A bimorph structure of a piezoelectric transducer has two layers of piezo material (active layers) and a layer of elastic material (passive layer in between). Figure 3 shows unimorph and bimorph structures of a piezoelectric transducer [23].

Loading mode defines the direction of force applied to the piezoelectric crystal and the direction of which output is taken or measured. Although there are many possible loading modes there are two main loading modes which are used for energy harvesting purpose, i.e., 3-1 loading mode and 3-3 loading modes. In 3-1 loading mode, the direction of electric field is perpendicular to the direction of force applied. In 3-3 loading mode, the direction of electric field is parallel to the direction of force applied. Figure 4 shows 3-1 and 3-3 loading modes for a piezoelectric transducer [24].

With different loading modes and structures, maximum power is obtained is always different [25]. This comparison has been shown in Table 1. 3-3 loading mode always produces more power output [26].

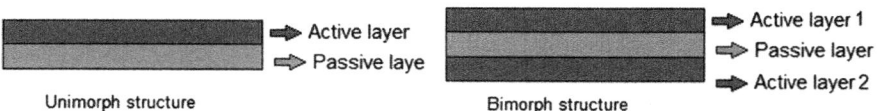

Unimorph structure   ⇒ Active layer   ⇒ Passive laye

Bimorph structure   ⇒ Active layer 1   ⇒ Passive layer   ⇒ Active layer 2

Fig. 3 Unimorph and bimorph structures [23]

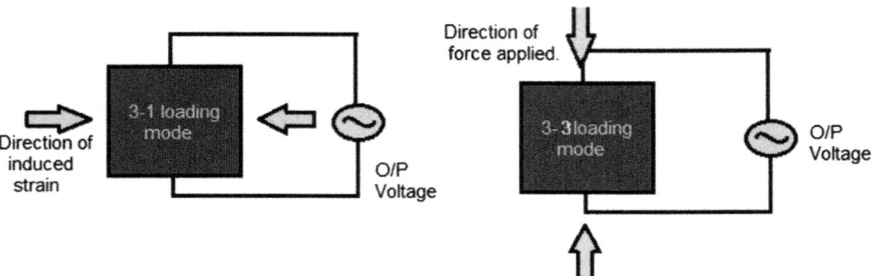

Fig. 4 3-1 and 3-3 loading modes [24]

Table 1 Power obtained at different loading modes and structures [25, 40]

Loading mode	Structural design	Average output power (nW)
3-1	Unimorph	3.9
3-1	Bimorph	9.6
3-3	Unimorph	21.42
3-3	Bimorph	22.47

4.3 Type of Fixation

Another parameter for the design of transducer is the type of fixation which defines the placement of piezoelectric crystal. There are mainly two types of fixations simple beam fixations, cantilever beam fixations. In simple beam, both ends are supported and the force is applied at the center of the beam. In cantilever beam, one end is fixed and another end is free and is set to vibrations/force is applied at the free end. The cantilever beam is widely used to obtain high power from piezoelectric transducer because it produces larger strain [27]. Figure 5 shows simple beam and cantilever beam fixations.

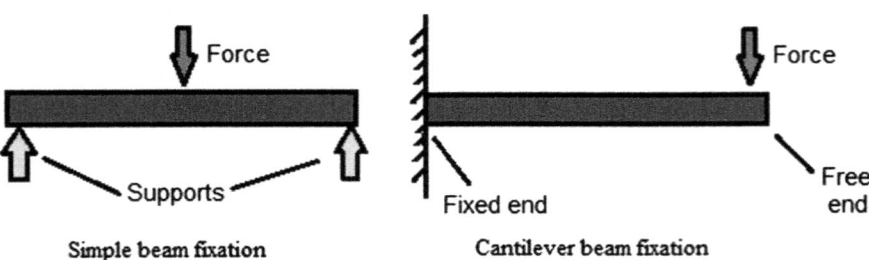

Fig. 5 Simple beam and cantilever beam fixations [27]

4.4 Impedance and Resonance Frequency Matching

The internal impedance of the piezo transducer is given by Eq. (1).

$$z = 1/(2\pi f_n C_{Piezo}) \tag{1}$$

where z is the input impedance, f_n is the natural frequency, C_{Piezo} is the capacitance of piezo transducer. According to the maximum power transfer theorem, this impedance should be matched with load impedance to transfer the maximum power to the load [25, 28].

To maximize the output power, the cantilever beam is operated on resonance frequency or the natural frequency of oscillations. Resonance frequency for a cantilever beam is given by Eq. (2).

$$f_r = v_n^2/2\pi L^2 \sqrt{(YI/mw)} \tag{2}$$

where f_r is the resonance frequency, $v_n = 1.875$ is the eigenvalue, L, m, w are the length, mass, width of the cantilever beam, respectively, I is the moment of inertia, Y is the Young's modulus [18, 24].

4.5 Addition of Mass at the Free End of Cantilever Beam

It has been shown experimentally that on increasing the mass at the tip of the cantilever beam the output power increases. This can also be seen from Eq. (2), that resonance frequency is inversely proportional to the mass at the tip of the cantilever beam. Hence, the resonance frequency can be decreased by adding a mass at the free end of the cantilever which is easier to achieve and hence the output power will increase. When the cantilever beam is excited at the resonance frequency the output

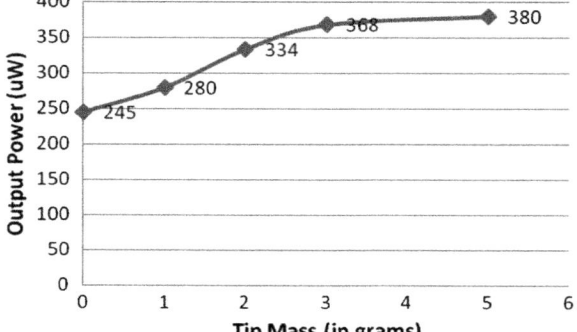

Fig. 6 Effect of addition of mass [28]

power will be maximum. Figure 6 shows the effect of adding mass at the tip of the
cantilever beam on the output power.

5 Rectifier Circuit

Rectifier circuits converts alternating current/voltage (AC) into direct current/
voltage (DC). Alternating current (AC) is the current which reverses its direction
periodically and direct current is the current that flows only in one direction (DC).
The output of the transducer is AC and it needs to be converted into DC for further
processing or storage thus rectifier circuit is required. The rectifier converts AC
signal into DC. This process is called rectification. For low-power level from the
piezoelectric microgenerator, rectification scheme must be chosen carefully so that
output from the piezoelectric transducer is efficiently transferred to the load.
Conventional rectifier circuits have been discussed below.

5.1 Bridge Rectifier

It is the most widely used rectifier circuit as it produces full-wave rectification and
output of the bridge rectifier gives smaller output voltage ripples (i.e., unwanted AC
variations in the DC signal). It consists of four diodes placed in the form of a bridge.
Figure 7 shows a conventional bridge rectifier. There are a pair of two diodes
(D_1, D_2, and D_3, D_4) connected in series and only one pair diodes allows current to
pass through them during one cycle of AC input [29]. BAT 754 series Schottky diode
is most suitable for bridge rectifier circuit as it has very low forward voltage and also
has low diode capacitance. The normal diode has a voltage drop of around 0.6–1.7 V
whereas Schottky diode has a voltage drop between 0.15 and 0.45 V [30].

Fig. 7 Full-wave bridge
rectifier [29]

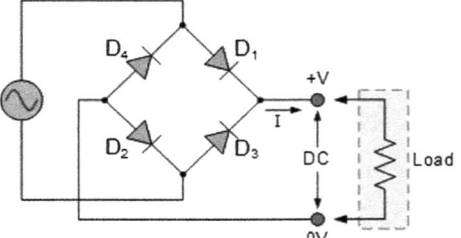

5.2　Voltage Doubler Rectifier

It is a rectifier circuit which takes the AC input and produces twice the input voltage at the output in the ideal case. The circuit of double voltage rectifier consists of two diodes (D_1, D_2), two capacitors (C_1, C_2), and an oscillating AC input voltage (V_{ac}). The diodes and capacitors work together efficiently to give almost double output than the peak to peak value of the input. Figure 8 shows a voltage doubler rectifier circuit [31].

5.3　Synchronous Rectifier

Although Schottky diode continues to be the best choice for the design of rectifiers, synchronous rectifiers are being studied because it provides very high efficiency. Different circuits using active devices such as MOSFET and FETs has been proposed. One of the proposed synchronous rectifiers is active cross-coupled rectifier with the synchronized switch. It has two active diodes instead of conventional passive diodes [32]. Figure 9 shows the efficiency comparison of synchronous rectifier using FETs and bridge rectifier using Schottky diode. Synchronous rectification is used for voltages above 5 V or currents below 10 A and it is normally used in the applications which work at the frequency less than 300 kHz. The Schottky diode can be used for high voltages and high output current [33]. It has been found that the power conversion efficiency of the conventional diode bridge rectifier is 38.08% and active rectifier with synchronous switches is 93.3% respectively [34].

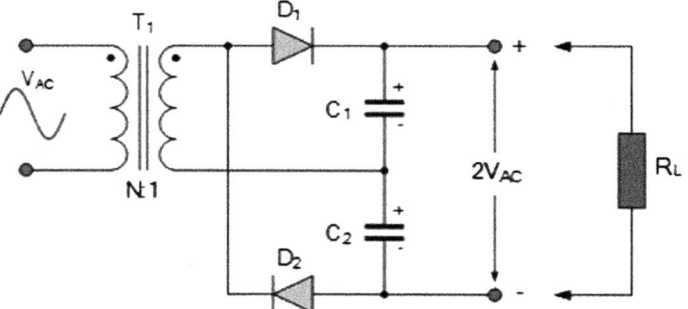

Fig. 8 Full-wave voltage doubler rectifier [31]

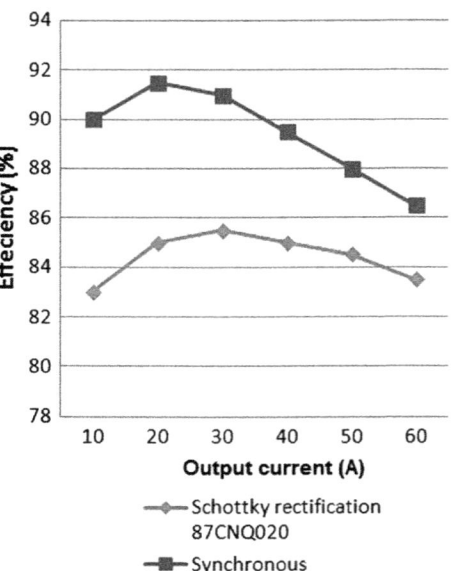

Fig. 9 Efficiency versus output current for FET and Schottky diodes [33]

6 Energy Storage Unit

There are mainly two types of energy storage units that are being used namely batteries and supercapacitors. Many batteries have been made till now that can be used to store energy to power mobile electronic devices. There are three main cells/batteries that are used in wireless sensor networks namely Nickel–Metal Hydride (NiMH), Lithium Ion (Li-Ion), and Lithium Polymer (Li-polymer). Different batteries have different characteristics such as charge density, voltage, charging time, etc. These parameters have been discussed in Table 2. Based upon these parameters, we can choose a battery for desired application.

Figure 10 shows the comparison between different batteries namely lead–acid (2 V), nickel cadmium (1.2 V), nickel–metal hydride (1.2 V), lithium-ion (3.6 V) and lithium polymer (3.7) based on the energy density and specific energy of different batteries.

Another type of storage unit used in energy harvesting circuit is supercapacitors. Capacitors are designed with different power levels and different materials example glass, ceramics, metal films etc. Supercapacitors are the capacitor with very high power density and extremely high energy density hence they are sometimes called a mechanical battery. They have very small size as compared to conventional batteries. They have a very less charging time. Supercapacitors can be classified as faradic and non-faradic supercapacitors [35]. Faradic capacitors charge faradically by transferring charge between the electrolyte and electrode. There two widely used conducting materials for making electrode namely conducting polymers such as polyacetylene

Table 2 Parameters of a battery [41]

S. No.	Parameter	Explanation
1	Battery voltage	It is the voltage of the cell
2	Capacity	It is the amount of electrical charge that can be stored in the battery
3	Specific energy	It is related to the volume and is measured in energy per weight
4	Energy density	It is the amount of energy that can be stored in the battery. It is measured in energy per volume
5	Internal Resistance	It specifies the ability of a battery to manage a specific load
6	Self-discharge	It points to the aging effect of the battery and leakage with time
7	Recharge cycles	It specifies the no. of times a battery can be charged till its performance degrades

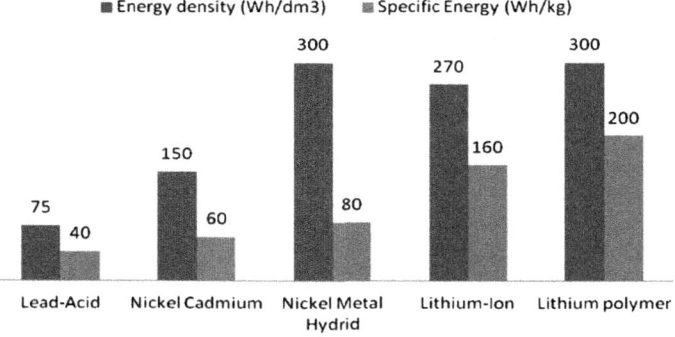

Fig. 10 Energy density and specific energy of various batteries [39]

and metal oxide such as ruthenium oxide [36]. Non-faradic capacitors charge non-faradically that is there is no transfer of charge between the electrolyte and the electrode. Non-faradic capacitors are made using carbon electrodes [37].

7 Conclusion

The following conclusions can be drawn from the literature survey:

1. Cantilever beam transducer design using bimorph structure, 3-3 loading mode produces more output power. The output power can be further increased by adding mass at the tip of the cantilever beam. To maximize the power output, load impedance should be matched with the input impedance and the cantilever should be excited at the resonant frequency.

2. There are three major types of rectifier circuits: voltage doubler rectifiers, bridge rectifiers, and synchronous rectifiers. Although conversion efficiency of

synchronous rectifiers is more than the conventional bridge rectifiers, diode bridge rectifies are most commonly used in energy harvesting circuits due to its simplicity. In diode bridge rectifier BAT 754 series Schottky diode is used because of its very low voltage drop.

3. In comparison between the rechargeable batteries and supercapacitors, the rechargeable batteries are preferred as they have higher energy stored per unit weight and slow discharging rate. Supercapacitors are used where fast charging is required.

References

1. S. Ulukus, K. Huang, R. Zhang, N. B. Mehta, and L. Tassiulas, "Special issue on energy harvesting in wireless networks," J. Commun. Networks, vol. 14, no. 2, pp. 115–120, 2012.
2. R. J. M. Vullers and R. Van Schaijk, "Energy Harvesting for Autonomous Wireless Sensor Networks," IEEE SOLID-STATE CIRCUITS Mag., pp. 29–38, 2010.
3. S. Basagni, M. Y. Naderi, and C. Petrioli, "Wireless Sensor Networks with Energy Harvesting," semantic scholar, pp. 7–11, 2013.
4. Y. Qiu, C. Van Liempd, P. G. Blanken, and C. Van Hoof, "5 uW-to-10 mW Input Power Range Inductive Boost Converter for Indoor Photovoltaic Energy Harvesting with Integrated Maximum Power Point Tracking," Solid-State Circuits, pp. 300–301, 2011.
5. P. B. P.T.V. Bhuvaneswari, R. Balakuma, V. Vaidehi, "Solar Energy Harvesting For Wireless Sensor Networks," First Int. Conf. Comput. Intell. Commun. Syst. Networks Sol., 2009.
6. G. Sebald, S. Pruvost, D. Guyomar, G. Sebald, D. Guyomar, and A. Agbossou, "On thermoelectric and pyroelectric energy," SMARTMATERIALS Struct. Sci., 2009.
7. D. Patel, R. Mehta, R. Patwa, S. Thapar, and S. Chopra, "RF Energy Harvesting," ijett journal, vol. 16, no. 8, pp. 382–385, 2014.
8. F. T. Fisher, "Energy harvesting vibration sources for Microsystems applications," Meas. Sci. Technol., vol. 17, no. 12, 2006.
9. A. Marin, J. Turner, D. Sam, S. R. Anton, and H. A. Sodano, "A micro electromagnetic generator for vibration energy harvesting," J. of Micromechanics and microengineering, vol. 17, pp. 1257–1267, 2007.
10. S. Chalasani and J. M. Conrad, "A Survey of Energy Harvesting Sources for Embedded Systems," IEEE Southeast Con, pp. 442–447, 2008.
11. T. Dikshit, D. Shrivastava, A. Gorey, A. Gupta, and P. Parandkar, "Energy Harvesting via Piezoelectricity," Int. J. Inf. Technol., vol. 2, no. 2, pp. 265–270, 2010.
12. S. M. Taware and S. P. Deshmukh, "A Review of Energy Harvesting From Piezoelectric Materials," IOSR J. Mech. Civ. Eng., pp. 43–50, 2013.
13. D. Kumar, P. Chaturvedi, and N. Jejurikar, "Piezoelectric Energy Harvester Design and Power Conditioning," IEEE Students' Conf. Electr. Electron. Comput. Sci. Piezoelectric, pp. 1–6, 2014.
14. J. Z. Zhengbao Yang, "Comparison of PZN-PT, PMN-PT single crystals and PZT ceramic for vibration energy harvesting," Elsevier, vol. 122, pp. 321–329, 2016.
15. H. Kim, S. Priya, H. Stephanou, and K. Uchino, "Consideration of Impedance Matching Techniques for Efficient Piezoelectric Energy Harvesting," IEEE Trans. Ultrason. Ferroelectr. Freq. Control, vol. 54, no. 9, pp. 1851–1859.
16. P. W. S. Roundy, "A piezoelectric vibration based generator for wireless electronics," Smart Mater. Struct., 2004.

17. M. Renaud, K. Karakaya, T. Sterken, P. Fiorini, C. Van Hoof, and R. Puers, "Fabrication, modeling and characterization of MEMS piezoelectric vibration harvesters," Sensors and Actuators- Elsevier, vol. 146, pp. 380–386, 2008.
18. Q. Wang, Z. P. Cao, and H. Kuwano, "Metal-based piezoelectric energy harvesters by direct deposition of PZT thick films on stainless steel," IET Micro Nano Lett., vol. 7, pp. 1158–1161, 2012.
19. X. Wang, J. Zhou, J. Song, J. Liu, and N. Xu, "Piezoelectric Field Effect Transistor and Nanoforce Sensor Based on a Single ZnO Nanowire," Nano Lett., 2006.
20. B. G. Xu, Z. Ren, P. Du, and W. Weng, "Polymer-Assisted Hydrothermal Synthesis of Single-Crystalline Nanowires," Adv. Mater., no. 7, pp. 907–910, 2005.
21. X. Y. Zhang, X. Zhao, C. W. Lai, J. Wang, X. G. Tang, and J. Y. Dai, "nanowire arrays," Appl. Phys. Lett., vol. 85, no. 18, pp. 4190–4192, 2004.
22. J. Rödel, K. G. Webber, R. Dittmer, W. Jo, and M. Kimura, "Feature Article Transferring lead-free piezoelectric ceramics into application," J. Eur. Ceram. Soc., vol. 35, no. 6, pp. 1659–1681, 2015.
23. G. N. Wahied G. Ali, "Design Considerations for Piezoelectric Energy Harvesting Systems," Eng. Technol., 2012.
24. A. Townley, "Vibrational Energy Harvesting Using MEMS Piezoelectric Generators," Citeseer, 2009.
25. S.-J. Y. Min-Gyu Kang, Woo-Suk Jung, Chong-Yun Kang, "Recent Progress on PZT Based Piezoelectric Energy," Actuators, 2016.
26. M. A. L. Ahmad and H. N. Alshareef, "Modeling the Power Output of Piezoelectric Energy Harvesters," J. Electron. Mater., vol. 40, no. 7, 2011.
27. H. A. Kim and S. Bowen, "Piezoelectric and ferroelectric materials and structures for energy harvesting applications," Energy Environ. Sci, vol. 320963, no. 320963, 2014.
28. S. W. Ibrahim and W. G. Ali, "Power Enhancement for Piezoelectric Energy Harvester," Proc. World Congr. Eng., vol. II, pp. 6–11, 2012.
29. M. Robert C. Genesi, Sterling, "Integrated full wave diode bridge rectifier," United States Pat., 1977.
30. A. Mustapha, N. M. Ali, and K. S. Leong, "Piezoelectric Microgenerator Rectifying Circuit Simulation using LTspice," Proc. Second Intl. Conf. Adv. Electron. Devices Circuits, pp. 978–981, 2013.
31. T. Kashiwao, I. Izadgoshasb, Y. Yan, and M. Deguchi, "Optimization of rectifier circuits for a vibration energy harvesting system using a macrofiber composite piezoelectric element," Microelectronics J., vol. 54, pp. 109–115, 2016.
32. S.-G. L. Xuan-Dien Do, Chang-Jin Jeong, Huy-Hieu Nguyen, Seok-Kyun Han, "A High Efficiency Piezoelectric Energy Harvesting System," IEEE, pp. 389–392, 2011.
33. Carl Blake, Alberto Guerra "Schottky diodes vs. FET synchronous," Electronics Engineer, 2000.
34. S. S. P. Baby, R. S. Edward and C. A. A. Allwyn, "Performance Analysis of an Efficient Active Rectifier for Powering LEDs using Piezoelectric Energy Harvesting Systems," 2013 International Conference on Circuits, Power and Computing Technologies (ICCPCT), Nagercoil, 2013, pp. 376–380.
35. C Chukwuka, KA Folly, "Batteries and supercapacitors." Power Engineering Society Conference and Exposition in Africa (PowerAfrica), 2012 IEEE. IEEE, 2012.
36. N. Khan, N. Mariun, M. Zaki and L. Dinesh, "Transient analysis of pulsed charging in supercapacitors," 2000 TENCON Proceedings. Intelligent Systems and Technologies for the New Millennium (Cat. No.00CH37119), Kuala Lumpur, 2000, pp. 193–199 vol.3.
37. Z. Li and F. Wu, "Diagnostic Identification of Self-Discharge Mechanisms for Carbon-Based Supercapacitors with High Energy Density," 2011 Asia-Pacific Power and Energy Engineering Conference, Wuhan, 2011, pp. 1–5.
38. M. M. R Caliò, UB Rongala, D Camboni, "Piezoelectric Energy Harvesting Solutions," sensors, pp. 4755–4790, 2014.

39. M. K. Stoj, M. R. Kosanovi, and L. R. Golubovi, "Power Management and Energy Harvesting Techniques for Wireless Sensor Nodes," Telecommun. Mod. Satell. Cable, Broadcast. Serv., 2009.
40. S. Kim, H. Park, S. Kim, H. C. Wikle, J. Park, and D. Kim, "Comparison of MEMS PZT Cantilevers Based on d 31 and d 33 Modes for Vibration Energy Harvesting," J. MICROELECTROMECHANICAL Syst., vol. 22, no. 1, pp. 26–33, 2013.
41. J. Eliasson, "Low-Power Design Methodologies for Embedded Internet Systems," Dep. Comput. Sci. Electr. Eng. Luleå Univ. Technol., 2008.

An Overview of Temperature Sensors

Deepak Prasad and Vijay Nath

1 Introduction

Temperature sensor technology has been longstanding. As early as the 1950s, the active electronics device was the vacuum tube developed by the U.S. government [1]. During that period, vacuum tube was serving in different forms for different temperature sensors. Approximately about the same time, silicon carbide was emerging as a semiconductor for high-temperature sensor [2]. Early motivation for the much-awaited development of high-temperature sensor was well logging continued to 1980, oil exploration and geothermal energy became the new motivation. Later, the development of energy and air force system was also one of the reasons to enhance the technologies. It was estimated that more than 40,000 wells are logged per year by oil industry where the temperature can range from 445 K and reach up to 585 K after several hours during logging while geothermal well logging occurs at a rate of 800 wells per year. It is analyzed that operational temperatures are generally worse in the field of oil well exploration. The temperature can touch 670 K when drilled to depth of 11 km. Success of both oil exploration and geothermal well logging highly depends on the accuracy of temperature sensor. In aircraft industry too, the temperature sensor plays an important role. By placing monitor and control electronics could forecast dangers and give benefit to, especially military aircraft. As the temperature increases, the situation becomes worse because with increase in temperature the thermal conductivity of good conductors usually decreases. The application of temperature sensor also ranged to automotive industry where it is needed to sense from 410 to 970 K in space technologies.

D. Prasad (✉) · V. Nath
VLSI Design Group, Department of Electronics and Communication Engineering,
Birla Institute of Technology, Mesra, Ranchi 835215, Jharkhand, India
e-mail: prasaddeepak007@gmail.com

V. Nath
e-mail: vijaynath@bitmesra.ac.in

© Springer Nature Singapore Pte Ltd. 2019
V. Nath and J. K. Mandal (eds.), *Proceeding of the Second International Conference on Microelectronics, Computing & Communication Systems (MCCS 2017)*, Lecture Notes in Electrical Engineering 476, https://doi.org/10.1007/978-981-10-8234-4_62

Material	Bandgap (eV)
Silicon	1.12
Gallium arsenide (GaAs)	1.43
Gallium phosphide (GaP)	2.24
Aluminum gallium arsenide (AlGaAs)	1.43–2.15
Aluminum gallium phosphide (AlGaP)	2.24–2.45
β-SiC	2.3
Germanium	0.7437
Diamond	5.5

Table 1 Material and respective bandgap

At present, semiconductor technology provides the only practical basis for high-temperature electronics systems. Consequently, the mainstay continues to be Si-based devices and technology because of existing broad technology base and wide availability of Si devices of all types. The prevailing opinion is that for signal processing applications, Si will satisfy needs up to 300 °C. For temperature above 300 °C, the electrical performance of Si devices declines rapidly and basing commercial devices on Si is probably impractical. Each semiconductor material has a fundamental maximum operating temperature that is usually determined by the magnitude of its bandgap energy. The bandgap energy, E_g, is a measure the measure of the amount of thermal energy needed to ionize the particular semiconductor material. Thus, the higher the bandgap, the higher the temperature at which devices based on this semiconductor material can continue to function. Semiconductor materials most commonly being preferred for high-temperature applications include silicon carbide (SiC), silicon, aluminum gallium arsenide (AlGaAs), gallium arsenide (GaAs), and diamond. Other than Si, semiconductor materials do not have technological maturity [3] (Table 1).

In most cases, these devices have failed because of their metallization or packaging and not because of the semiconductor material. The development of reliable, high-temperature electronics will not only open new markets, but will also offer significant advantages in device reliability.

The present paper shows and describes the different kinds of temperature sensor available.

2 Temperature Sensor

2.1 Thermocouple

Thermocouple is extensively used for measuring temperature. Basically, it is working on the principle of the Seeback effect that it must have two metal joined together to form two junctions, one is connected to body whose temperature is to measure, i.e., hot and other is connected to body of known temperature, i.e., cold.

Fig. 1 Working process of
thermocouple

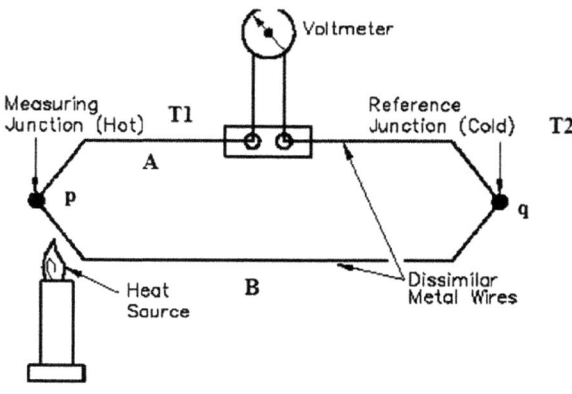

When two different or unlike metals are joined together at two junctions, an emf is generated at two junctions. The amount of emf generated is different for a different combination of metal. The working of thermocouple cannot be performed in the absence of two junctions [4]. In other words, two junctions are absolutely required for thermocouple. The same temperature generation across the two junctions result in a generation of equal and opposite emf and the net current becomes zero (null) across them. The amount of emf developed within the thermocouple is very small usually in mV. Figure 1 shows the working process of it.

2.2 Thermistor

Thermistor is also classified in the list of temperature sensor. It is a type of resistor whose resistance can vary according to temperature. In most of the metals, the resistance increases with temperature. The thermistor responds negatively to the temperature and their resistance decreases with the temperature. Thermistor is made up of ceramic-like semiconductor. Since the resistivity of the thermistor is very high, the resistance of circuit in which they are connected for measurement of temperature can be easily measured [5]. By passing a small known current, resistance can be easily measured and observing the voltage drop across it.

2.3 Resistance Temperature Detector (RTD)

The main principle behind the operation of an RTD is that the resistance of the corresponding material is proportional to the change in temperature, i.e., when temperature increases or decreases, the corresponding resistance will also increase or decrease. For making RTD, metal should be pure, else a deviation from conventional resistance–temperature graph will be noticed. Copper has a temperature

Fig. 2 Graph of RTD

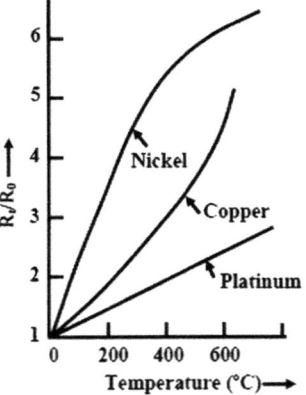

range 120 °C, nickel has 300 °C while platinum has 650 °C [6]. The graph is shown in Fig. 2. Thermocouple is inexpensive and has a fast response time but it is less accurate and least stable and sensitive. RTD is best but has slow response time. Thermistor has a limited range.

2.4 Bimetallic Strip

Bimetallic sensor follows two basic principles of operation. The volumetric dimension of a metal changes proportional to the change in temperature. The change in dimension may be resulted into expansion or contraction of metal's body. Since different metals have a different coefficient of temperature, the rate of change of volume depends on the coefficient of temperature [7]. The two different metals and their bonding to form a spiral or twisted helix make a bimetallic strip. By welding or riveting, the one end of these two metals is joined together. The procedure begins when the opposite end, i.e., free end starts changing their shape with change in temperature as per their coefficient of temperature. It is shown in Fig. 3. The movement is linear in temperature and deflection can be easily extracted with the help of attached pointer. Above 420 °C, bimetallic sensors have not been suggested for use.

2.5 CMOS Temperature Sensor

The evolution of CMOS technology brings revolution in the electronics industry. Since its inception, MOSFET has been consistently used in almost every field. Researchers are adopting the technology to design circuits using CMOS technology which allows consumption of ultra-low power instead of high power which was

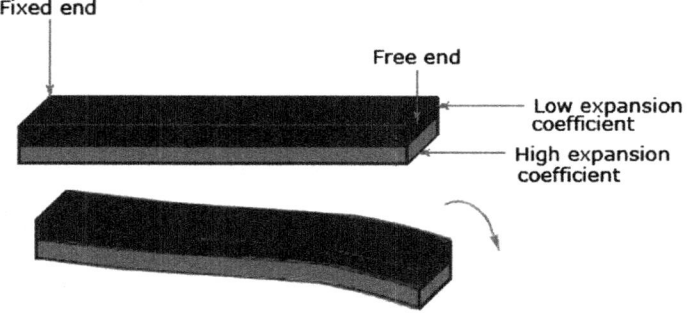

Fig. 3 Bimetallic strip

drawn earlier. CMOS technology-adopted temperature sensor has many advantages over traditional temperature sensors [8]. A linear is to be observed between the output voltage and temperature in saturation and sub-threshold region. The feature of low power consumption makes it more stable for the aerospace, automotive applications, etc. In comparison to bipolar junction transistor (BJT), the entire area of MOSFET is less than 5% of area required by BJT. Drain current of MOSFET depends on mobility and threshold voltage. The relationship between mobility and temperature is [9]

$$\mu \propto T^{-3/2} \tag{1}$$

On the other hand, drain current is proportional to mobility

$$I_D \propto T^{-3/2} \tag{2}$$

An increase in temperature leads to decrease in drain current. On the other hand, the threshold voltage is defined as minimum voltage where MOSFET enters into ON state. For better performance threshold voltage should be low. Low threshold voltage enables the device to operate with small supply voltage and makes MOSFET faster by reducing switching time.

The MOSFET threshold voltage is given by [10]

$$V_T = V_{FB} + 2V\phi_F + I\sqrt{2\phi_F} \tag{3}$$

$$V_{FB} = \phi_{MS} - (Q_{SS}/C_{OX}) \tag{4}$$

where V_{FB} is the flat band voltage

$$\phi_{MS} = \phi_T \ln\left(N_A N_G / n_i^2\right) \tag{5}$$

ϕ_{MS} is gate substrate contact potential, N_A and N_G are the substrate and gate doping concentrations, respectively, Q_{SS} is the surface charge density and C_{OX} is the oxide capacitance [11],

$$I = C_{OX}\sqrt{2q\epsilon_{si}N_A} \text{ is a body effect parameter,}$$

$\phi_F = \phi_T \ln(N_A/n_i)$ is the Fermi energy with the thermal voltage $\phi_T = kT/q$ and n_i is the intrinsic carrier concentration of Si.

Among a number of proposed CMOS temperature sensor, one of them is shown here where PTAT (Proportional to Absolute Temperature) and NTC (Negative Temperature Coefficient) voltages have been extracted. In Fig. 4, few resistors like M1, M2, M7, and M8 are working in the active region. To utilize the difference in current flow with respect to temperature, these transistors are connected in differential mode. Transistors pairs M3, M4 and M9, M10 make the current mirror circuit.

A supply voltage of ±0.5 V has been used to design the above circuit. Among extracted output of PTAT and NTC, NTC shows better result than PTAT. A temperature range of −60 to +150 °C can be sensed using above-proposed circuit [12] (Fig. 5 and Table 2).

Figure 6 depicts the power consumed by the proposed temperature sensor. At 150 °C and at room temperature, a power consumption of 862 and 216 nW is reported.

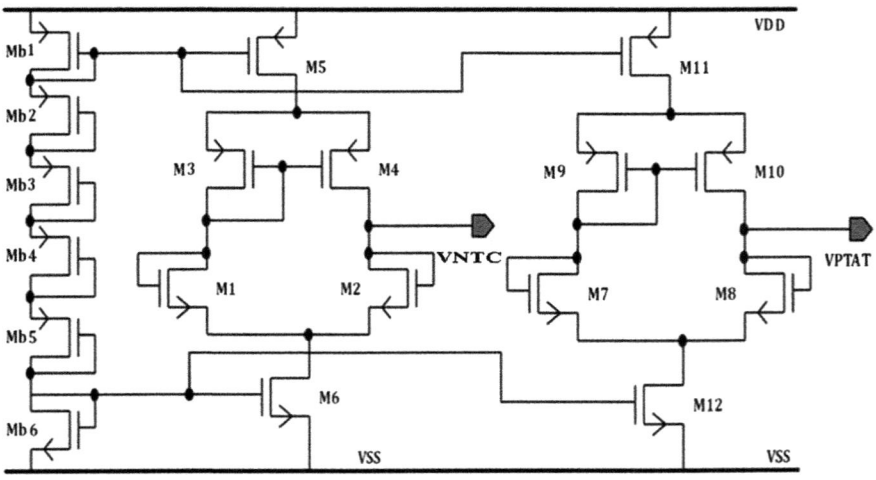

Fig. 4 CMOS temperature sensor

Fig. 5 Output voltage versus temperature

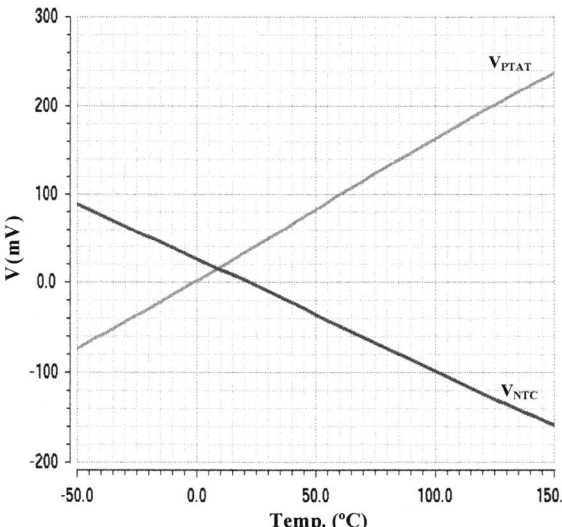

Table 2 Material and respective bandgap

MOSFET	W/L ratio	MOSFET	W/L ratio
M1	5 μm/90 nm	M10	2 μm/90 nm
M2	4 μm/90 nm	M11	260 nm/90 nm
M3	8 μm/90 nm	M12	120 nm/90 nm
M4	4 μm/90 nm	Mb1	120 nm/90 nm
M5	120 nm/90 nm	Mb2	120 nm/90 nm
M6	120 nm/90 nm	Mb3	120 nm/90 nm
M7	200 nm/90 nm	Mb4	120 nm/90 nm
M8	5 μm/90 nm	Mb5	120 nm/90 nm
M9	120 nm/90 nm	Mb6	120 nm/90 nm

Fig. 6 Total power consumption curve

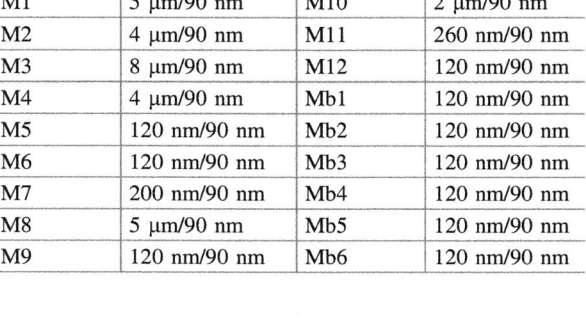

3 Conclusion

Though several years of engineering research offered a number of sensors to push technology one step ahead, yet the milestone is to be reached. CMOS temperature sensor has been used enormously nowadays due to its superior features.

The implementation of advanced CMOS temperature sensor and integrating with SOI chip will make the advanced industry technological sound, safer, beneficial, and economically viable.

Acknowledgments The authors are thankful to the Head of Department Dr. V. R. Gupta and Vice-Chancellor Dr. M. K. Mishra for their consistent support. The views expressed in this paper are of the authors and not necessarily of the organization they belong. We are thankful to RESPOND ISRO Ahmedabad for funding this project.

References

1. F. V. Thome and D. B. King, "Final Report on Results of a Survey and Workshop and High Temperature and Radiation-Hardened Electronics, SAND89-0975, 1989.
2. D. W. Palmer and R. C. Heckman, " Extreme temperature range electronics, IEEE Trans. Comp., Hybrids manuf. Technol, vol. CHMT-1, pp. 333–340, 1978.
3. R. K. Traeger and P. C. Lysne, "High temperature electronics application in well logging, "IEEE Trans. Nuc. Sci, vol. 35, pp. 852–854, 1988.
4. M. A. P. Pertijs and J. H. Huijsing, Precision temperature sensors in CMOS technology. Dordrecht, Netherland: Springer Science & Business Media, 2006.
5. R. Pallas-Areny and J. G. Webster, *Sensors and signal conditioning*.: Wiley, 2001.
6. A. Bakker and J. H. Huijsing, *High-accuracy CMOS smart temperature sensors*, ser. Int. Series in Engineering and Computer science. Boston, MA: Kluwer, 2000, vol. 595.
7. G. C. M. Meijer, R. V. Gelder, V. Nooder, J. V. Drecht, and H Kerkvliet, "A three-terminal integrated temperature transducer with microcomputer interfacing," *Sensors and Actuators*, vol. 18, no. 2, pp. 195–206, 1989.
8. A. Bakker and J. H. Huijsing, "Micropower CMOS temperature sensor with digital output," *IEEE Journal of Solid-State Circuits*, vol. 31, no. 7, pp. 933–937, 1996.
9. A. Bakker, "CMOS smart temperature sensors-an overview," in Proc. IEEE Sensors, 2002, pp. 1423–1427.
10. M. A. P. Pertijs et al., "A CMOS smart temperature sensor with a 3σ inaccuracy of ±0.5 °C from -50 to 120 °C," *IEEE Journal of solid-state circuits*, vol. 40, no. 2, pp. 454–461, 2005.
11. L. Ho-Yin, C. Shih-Lun, and L. Ching-Hsing, "A CMOS smart thermal sensor for biomedical application," *IEICE transactions on electronics*, vol. 91, no. 1, pp. 96–104, 2008.
12. Pandey. A and Nath. V, "A CMOS temperature sensor and auto-zeroing circuit with inaccuracy of $-1/+$ 0.7 °C between -30 and 150 °C," Microsystem Technologies, Vol. 23(9), pp. 4211–4219.

A 3.65 mW, Amplifier-Based Up-Conversion Mixer for Zigbee Application

Abhishek Pandey, Deepak Prasad and Vijay Nath

1 Introduction

With exploration and development of CMOS technology [1], it could have low cost, small size and low-voltage circuitry promising to integrate the whole system on a single chip. The challenges are continuous and imply attention in exploration of RF architectures [2–4]. The RF mixer is a critical component in RF system, because of its ability for frequency conversion. Basically, the frequency conversion can be done in two ways:

1. Up-Conversion Mixer (UCM)
2. Down-Conversion Mixer (DCM)

In up-conversion mixer, a mixer is used in the transmitter circuitry. It multiplies the low-frequency message signal with the local oscillator signal to convert low-frequency message signal to high-frequency message signal [5, 6].

Receiver circuitry is required for down-conversion mixer. It multiplies a high-frequency signal with a local oscillator signal to obtain low-frequency signal (IF signal) [6].

In an ideal situation, the output of the mixer is an exact replica of input signal. But in reality, due to the nonlinearity of the mixer some distortion occurs at the output. That's why linearity is one of the major parameters of mixer design. The mixer converts the frequency of the signal with some gain, this is called conversion

A. Pandey (✉) · D. Prasad · V. Nath
VLSI Design Group, Department of ECE, Birla Institute of Technology,
Mesra, Ranchi 835215, Jharkhand, India
e-mail: a.p.bitmesra@gmail.com

D. Prasad
e-mail: prasaddeepak007@gmail.com

V. Nath
e-mail: vijaynath@bitmesra.ac.in

© Springer Nature Singapore Pte Ltd. 2019
V. Nath and J. K. Mandal (eds.), *Proceeding of the Second International Conference on Microelectronics, Computing & Communication Systems (MCCS 2017)*, Lecture Notes in Electrical Engineering 476, https://doi.org/10.1007/978-981-10-8234-4_63

gain. In the design of mixer, attention should be given in the leakage of local oscillator to the output port [7, 8].

A good mixer should have the following qualities:

1. Large conversion gain
2. Good isolation
3. Small noise figure
4. High linearity

Since these parameters depend upon each other. Therefore, it is a tedious task to develop a suitable mixer topology, which can obtain a high conversion gain, and high linearity, low power, and noise figure, at the same time [2–5]. Zigbee IEEE 802.15.4 illustrates three frequency bands for operation: the 868-MHz, 915-MHz, and 2.4-GHz ISM bands in which the 2.4-GHz band is the most commonly used unlicensed band [2]. This standard is applicable in industrial, home automation, consumer electronics, and personal healthcare appliances [9].

The active mixer leads to a better conversion gain and low-noise figure. But the linearity and power consumption are not better [10]. Since linearity and conversion gain both oppose the enhancement of each other, therefore for proper value of conversion gains and IIP3, a novel architecture of mixer has been chosen. The two transconductance amplifiers have been used for boosting the conversion gain and derivative superposition theorem is used to enhance the linearity of the mixer.

This paper is organized in the following manner: The insight of the basics of Gilbert cell mixer design and operation are demonstrated in Sect. 2. A summarized analysis of proposed mixer topology is described in Sect. 3. Simulation results and discussion are demonstrated in Sect. 4. This consists of all the important data such as conversion gain, noise figure, linearity and 1 dB compression point. Finally, the conclusion is enunciated in Sect. 5.

2   Gilbert Cell Mixer Design

For constructing the active mixer, the Gilbert mixer cell is most commonly used topology [11–13]. Gilbert mixer has several salient features that are enough conversion gain with the proper load. A very good port-to-port isolation and low-noise figure is led by double-balanced Gilbert cell topology. It operates on the concept of translinear configuration. The drawbacks of this type of mixers are limited linearity and frequency, which depends on matching. The input transistors of these mixers should be in saturation region. The schematic diagram of Gilbert cell mixer has been depicted in Fig. 1.

I_{OUT} is tail current of transistor NM5. For governing the total current the tail current I_{OUT} should be in the saturation region. Transistors NM6 and NM7 are differential pairs, which operate in saturation region and change the input voltage to current. The linearity and gain of Gilbert cell mainly depend on these NM6 and NM7 transistors. Two pairs of switches are NM1, NM2, and NM3, NM4 operate in

Fig. 1 Schematic diagram of the Gilbert mixer

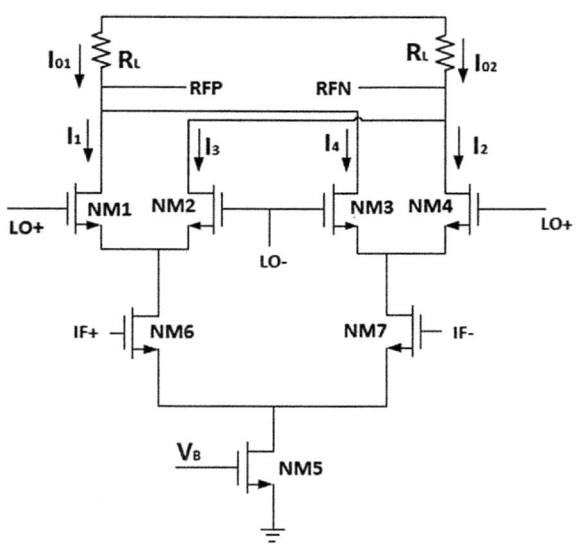

a saturation region. Mainly this transistor mixed the signal current from transconductors NM6 and NM7 [5] with the local oscillator (LOSC) signal current.

The Gilbert mixer output current can be expressed as follows [4]:

$$I_{\text{out}} = I_{\text{out1}} - I_{\text{out2}} = (I_1 - I_2) - (I_3 - I_4) \tag{1}$$

The transconductance of NM6 and NM7 is g_{m}:

$$g_{\text{m}_{\text{NM6}}} = g_{\text{m}_{\text{NM7}}} = g_{\text{m}} \tag{2}$$

The voltage conversion gain of Gilbert mixer is demonstrated as

$$G_{\text{CB}} = \frac{V_{\text{RF}}}{V_{\text{IF}}} = \frac{I_{\text{out}}R_{\text{L}}}{V_{\text{IF}}} = g_{\text{m}}R_{\text{L}} \sum_{n=1}^{\infty} \frac{\sin \frac{n\pi}{2}}{\frac{n\pi}{2}} [\cos(n\omega_{\text{LOSC}} + \omega_{\text{IF}})t + \cos(n\omega_{\text{LOSC}} - \omega_{\text{IF}})t]$$

$$\tag{3}$$

where R_{L} is the load resistance of the Gilbert cell mixer.

3 Transconductance Amplifier-Based Up-Conversion Mixer

The transconductance amplifier-based up-conversion mixer is shown in Fig. 2. The transconductance pair transistor M1 and M2 are used to convert the IF signal voltage to current. The transistors M3, M4 and M5, M6 behave as two pairs of ideal

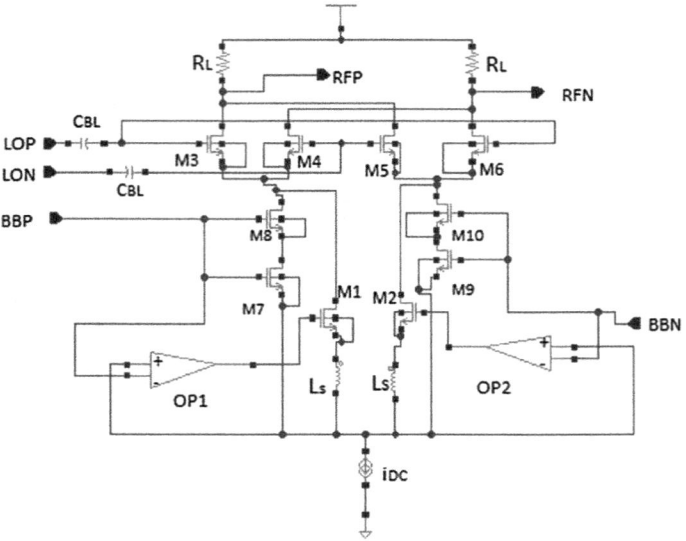

Fig. 2 Schematic diagram of the proposed up-conversion mixer circuit with transconductance amplifier

switches, which are biased in the saturation region. These transistors enhanced the current, supplied by transconductance pairs M1 and M2. To connect these four transistors to M1 and M2, the derivative superposition method is implied. The transistors M1 and M2 are connected to parallel with transistors M7, M8 and M9, M10. The transistors (M7–M10) are functioned in weak inversion region. The W/L ratio of these transistors is demonstrated as

$$R_{ON} = \frac{1}{\mu_n c_{ox} \frac{W}{L} (V_{GS} - V_T)} \quad (4)$$

If the aspect ratio (W/L ratio) of transistors M7 and M8 decreases, then R_{ON} resistance of these transistors is improved. With proper selection of W/L of transistors M7–M10 comes in weak inversion region. The parameter g_{m3} depends on V_{GS} [10]. The IIP3 is determined by the third-order coefficient of transconductance g_{m3}. If g_{m3} decreases, then the linearity will increase. For improving the circuit linearity, a source degeneration spiral inductor is used. The inductance is selected at the resonant frequency. The blocking capacitors C_{BL} behaves so as to isolate the input and output port from the DC sources.

The load resistors (R_L) connected on the top of LO switch transistors which is optimized to achieve a minimum power loss and better gain.

4   Simulation Results

The proposed circuit is simulated using Cadence analog and digital system design tools with PDK UMC 90 nm CMOS technology. The passive balun is used for the simulation of mixer circuit, which is depicted in Fig. 3.

The radio frequency (RF) of the mixer is designed at 2.4 GHz and the local oscillator is selected to operate at 2.3 GHz. The baseband signal is chosen at 100 MHz. The proposed circuit converts 100 MHz Baseband to 2.4 GHz radio frequency (RF) signal.

To find the transient response of the circuit, transient simulation techniques are required and the AC signals (RF and LO signals) must be applied to the mixer. The voltage against time (transient response) is presented in Fig. 4.

Figure 5 has presented the conversion gain of mixer, which is plotted against frequency is achieved at 21.4 dB. The 1 dB compression point and IIP3 of the mixer are found −5.61 and 4.24 dBm as shown in Figs. 6 and 7 respectively. The noise figure of the proposed mixer circuit produces 13 dB, which is shown in Fig. 8. The performance summary of the proposed mixer have been compared to other recent papers, and is summarized in Table 1.

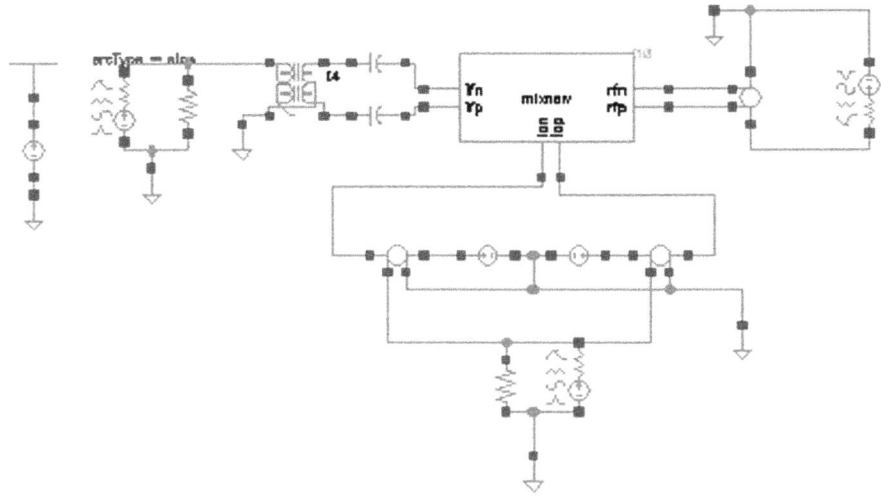

Fig. 3 Test bench used for mixer design

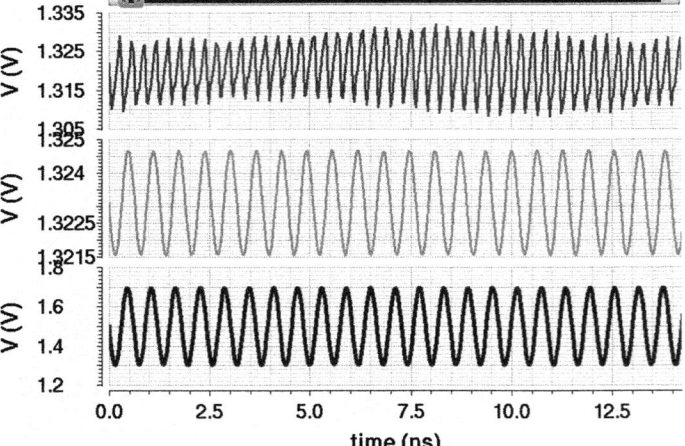

Fig. 4 Transient response of transconductance amplifier-based up-conversion mixer

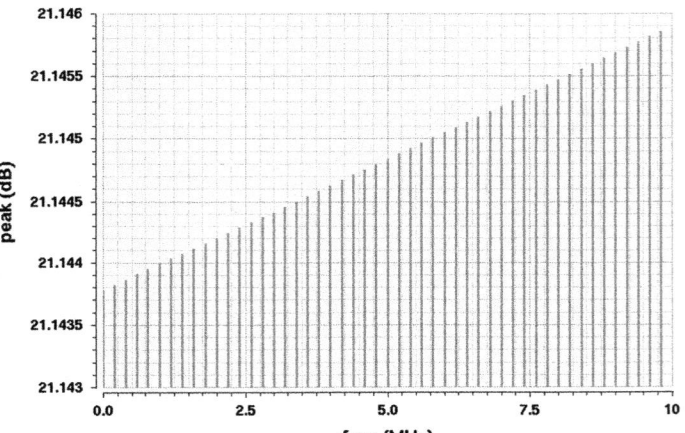

Fig. 5 Voltage conversion gain versus IF frequency of transconductance amplifier-based up-conversion mixer

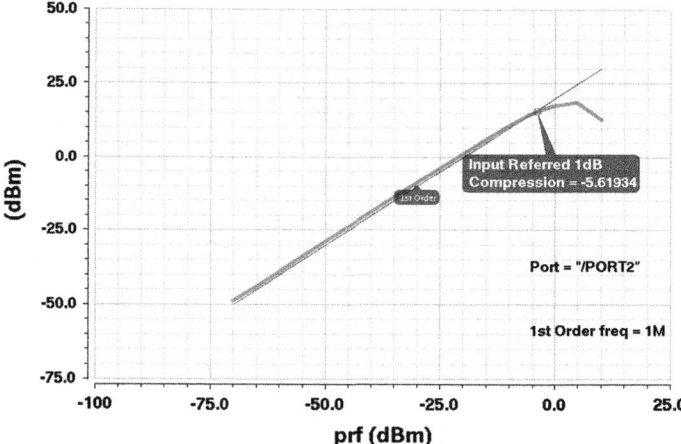

Fig. 6 1 dB compression point of transconductance amplifier-based up-conversion mixer

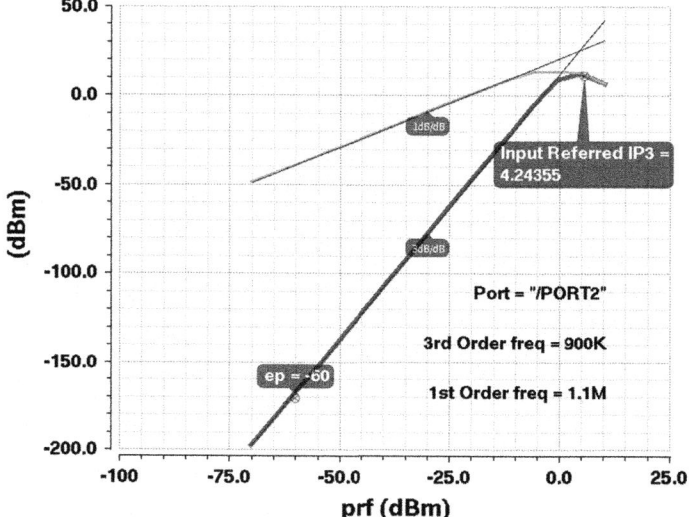

Fig. 7 IIP3 of transconductance amplifier-based up-conversion mixer

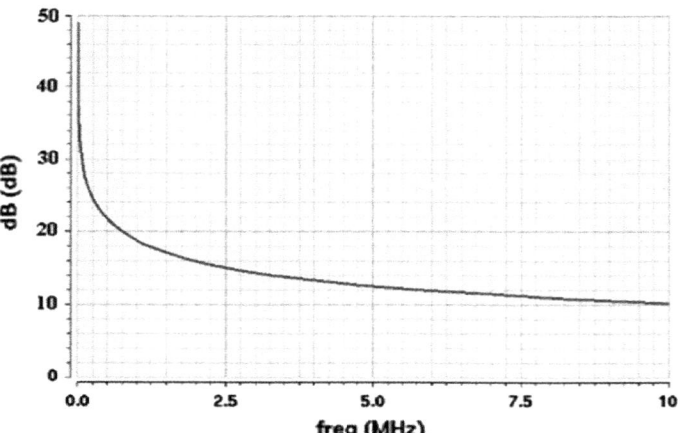

Fig. 8 Noise figure of transconductance amplifier-based up-conversion mixer

Table 1 Cross-platform comparative performance

Parameter	[3]	[4]	[5]	[6]	[7]	This work
Technology	0.18 μm	0.13 μm	0.18 μm	0.18 μm	0.18 μm	90 nm
Power supply (V)	1.2	1.2	1.8	1.8	1.8	1.8
IF freq (MHz)	100	10–400	1	100	100	100
LO freq (GHz)	1.8	1.8–2.6	2.419	5.1	5.1	2.3
RF freq (GHz)	1.9	1.810	2.42	5.2	5.2	2.4
Conversion gain (dB)	5	1.1	16.2	6	6	21.4
Input 1 dB compression point (dBm)			−20.7			−5.61
IIP3 (dBm)	14.68	6.45		8	15.7	4.26
Noise figure (dB)			18.6	26	24	13
Power consumption (mW)	9.45	10		8.6	7.5	3.65

5 Conclusion

The circuit is proposed to demonstrate the challenge of high conversion gain and low-noise figure with good linearity of the mixer for Zigbee band application. In this paper, two transconductance amplifiers with super derivative technique is used. It illustrates high forward gain, low-noise figure and linearity with power consumption of 3.65 mW. The circuit is compared with the newly designed mixers. So it is best suited for applying in energy efficient low-power Zigbee transmitter front-end.

References

1. Vijay Nath, http://www.ide.iitkgp.ernet.in/Pedagogy_view/example.jsp?USER_ID=210.
2. B. Razzavi, "Mixer Design Chapter 6," in *RF Micoelectronics 2nd edition.*
3. Yuan-Hao Shu, and Jeng-Rern Yang. "Low voltage, high linearity CMOS up-conversion mixer for LTE applications," *in Proc. IMFEDK –IEEE,* 2013, pp. 44–45.
4. Xiaopeng Sun et al., "A 1.8–2.6 GHz RF CMOS Up-Conversion Mixer for Wideband Applications," in Proc. IMWS-IEEE MTT-S International, 2012, pp. 1–4.
5. Wu Chenjian and Li Zhiqun, "A 0.18 μm CMOS Up-Conversion Mixer for Wireless Sensor Networks Application," *in Proc.WCSP IEEE,* 2011, pp. 1–4.
6. S.A.Z Murad et al., " A Design of 5.2 GHz CMOS Up-conversion Mixer with IF Input Active Balun," *in Proc. ISWTA IEEE,* 2011, pp. 1–4.
7. S.A.Z. Murad et al., "High Linearity 5.2 GHz CMOS up-conversion Mixer Using Super Derivative Superposition Method" *in Proc. TENCON IEEE,* 2010, pp. 1509–1512.
8. A. Saberkari et al., "A low voltage highly linear CMOS up-conversion mixer based on current conveyor," *IEICE Electronics Express,* vol. 6, pp. 930–935, 2009.
9. G. Sapone and G. Palmisano, "A 1.5 V 0.25 μm CMOS up converter for 3–5 GHz low power WPANs," *Microwave & Optical Technology Letter,* vol. 49, no. 9, pp. 2209–2212, Sept. 2007.
10. S. H-L. Tu and S.C-H. Chen, "A 5.26-GHz CMOS up-conversion mixer for IEEE 802.11a WLAN", *in Proc. of 4th IEEE International Conference on Circuits and Systems for Communications (ICCSC),* 2008, pp. 820–823.
11. Aparin, V. et al., "Modified Derivative superposition method for linearizing FET low –noise amplifiers," *IEEE Microwave Theory and Techniques,* vol. 53, pp. 571–581, Feb. 2005.
12. Ghulam Mehdi, "Highly Linear Mixer for On-Chip RF Test in 130 nm CMOS" Master thesis performed at *division of Electronic Devices at* Linköping Institute of Technology.
13. Shengchang Gao. "A High-Linearity Low-Noise Figure Active Mixer in 0.18 um CMOS", 2009 5th International Conference on Wireless communication Networking and Mobile computing, 09/2009.

Adaptive Compensation Algorithm for Flux Estimation of PM BLDC Motor Drives

Vijay Kumar Karan, P. R. Thakura and A. N. Thakur

1 Introduction

The back-emf derived through measurement of stator voltages $u_{s\alpha}, u_{s\beta}$ and currents $i_{s\alpha}, i_{s\beta}$ resolved in the stationary reference frame, on integration results into flux linkage as stated below [1]:

$$\Psi_{s\alpha} = \int \left(u_{s\alpha} - R i_{s\alpha} \right) \mathrm{d}t \tag{1}$$

$$\Psi_{s\beta} = \int \left(u_{s\beta} - R i_{s\beta} \right) \mathrm{d}t \tag{2}$$

$$\Psi_{s\beta} = u_{s\beta} \cdot t - R \int i_{s\beta} \mathrm{d}t + \Psi_{s\beta}(0) \tag{3}$$

$$\Psi_{s\alpha} = u_{s\alpha} \cdot t - R \int i_{s\alpha} \mathrm{d}t + \Psi_{s\alpha}(0) \tag{4}$$

where the initial values of stator flux are denoted as $\Psi_{s\alpha}(0)$ and $\Psi_{s\beta}(0)$ in the $\alpha - \beta$ axes and R is the stator winding resistance [2]. The magnitude and angular position of the stator flux linkage vector is obtained by

V. K. Karan (✉) · P. R. Thakura
Department of Electrical & Electronics Engineering, Birla Institute of Technology,
Mesra, Ranchi 835215, India
e-mail: vijaykaran1952@gmail.com

A. N. Thakur
Department of Electrical Engineering, National Institute of Technology, Jamshedpur,
Jamshedpur 834014, India
e-mail: anthakur@nitjsr.ac.in

© Springer Nature Singapore Pte Ltd. 2019
V. Nath and J. K. Mandal (eds.), *Proceeding of the Second International Conference on Microelectronics, Computing & Communication Systems (MCCS 2017)*, Lecture Notes in Electrical Engineering 476, https://doi.org/10.1007/978-981-10-8234-4_64

$$\Psi_s = \sqrt{\Psi_{s\alpha}^2 + \Psi_{s\beta}^2} \tag{5}$$

$$\theta_s = \tan^{-1} \frac{\Psi_{s\beta}}{\Psi_{s\alpha}} \tag{6}$$

The rotor flux components and angle can be derived using the stator flux as shown below:

$$\Psi_{r\alpha} = \Psi_{s\alpha} - L_s i_{s\alpha} \tag{7}$$

$$\Psi_{r\beta} = \Psi_{s\beta} - L_s i_{s\beta} \tag{8}$$

$$\theta_{re} = \tan^{-1} \frac{(\Psi_{s\beta} - L_s \cdot i_{s\beta})}{(\Psi_{s\alpha} - L_s \cdot i_{s\alpha})} \tag{9}$$

where L_s is the inductance of stator winding. The electromagnetic torque equation in it simple form has been derived below on the assumption of having a surface mounted permanent magnet brushless dc motor with a trapezoidal back-emf [3]:

$$T_e = \frac{3}{2} \frac{p}{2} \left[\frac{d\Psi_{r\alpha}}{d\theta_{re}} i_{s\alpha} + \frac{d\Psi_{r\beta}}{d\theta_{re}} i_{s\beta} \right] \tag{10}$$

In Direct Torque Control application, it is possible to control the torque efficiently by regulating rotating speed of stator flux while keeping its amplitude constant. This is done by selecting proper stator voltage vectors [4]. But to solve the common problems of integrators special integration algorithm for estimating stator flux is needed. Self-starting of the motor with no shaft position encoder has been suggested in [5]. Starting the motor in open loop to gain a speed of 200–300 rpm, to sense the induced emf tangibly in order to estimate the flux and torque for closed-loop control in an accurate manner.

In steady state, the stator flux linkage vector for PM BLDC motor attains a nearly circular shape about its origin. The sectoral kinks appear on larger load torque due to commutation effect. Figure 2a shows ideal stator flux achieved in steady-state which has been obtained without normal integration using an adaptive magnitude compensation method. The drift can be noticed in the stator flux graph achieved through the normal integration of the emf. Figure 3a–c shows the plots of $\Psi_{s\alpha}$ and $\Psi_{s\beta}$ and the stator flux linkage. This dc drift adds to the inaccuracy of stator flux angle and also the estimation of Torque in case of sensor-less shaft systems. To implement Direct Toque Control strategy, it becomes imminent then that flux linkage is estimated quite accurately in sensor-less drive.

2 Adaptive Magnitude Compensation Algorithm for Estimation of PM BLDC Motor Flux

Some new integration techniques have been suggested in [6] to eliminate the DC drift problem due to normal integration. To replace the pure integrator the expression is as follows:

$$y = \frac{1}{s}x; \quad \text{can be re-expressed as,}$$

$$y = \frac{\omega_c}{(s + \omega_c)}x + \frac{1}{(s + \omega_c)}y \tag{11}$$

where x and y are the input and output signals, and ω_c is the cut-off frequency. Out of three algorithms suggested, the Adaptive Magnitude Compensation Algorithm shown in Fig. 1, has been modelled in MATLAB for the simulation to replace the normal integrator in the PM BLDC drive simulation.

2.1 Adaptive Magnitude Compensation

This scheme, called Adaptive Magnitude Compensation was developed on the basis of the orthogonality of the back-emf with flux linkage. A quadrature detector was used to detect the orthogonal relation between the estimated flux and back-emf. A PI regulator was used for magnitude compensation. Using this model the stator flux has been estimated and the same has been depicted in Fig. 2a–f as ideal stator flux waveforms in steady state. Adaptive magnitude compensation has two

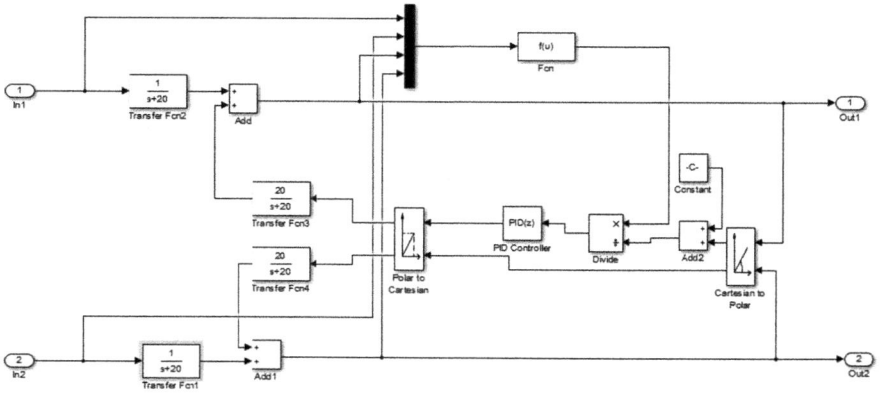

Fig. 1 Adaptive magnitude compensation block

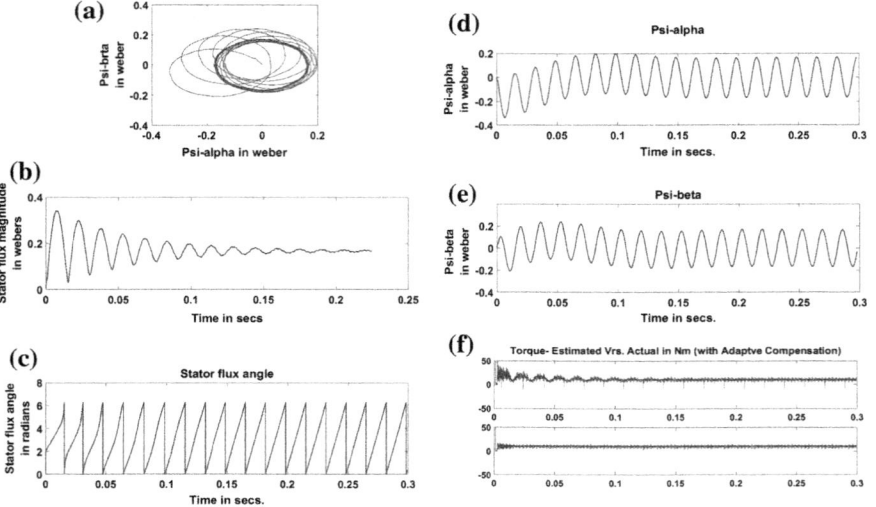

Fig. 2 a Stator flux using adaptive magnitude compensation. b Stator flux magnitude using adaptive magnitude compensation. c Stator flux angle using adaptive magnitude compensation. d Psi alpha using Adaptive magnitude compensation. e Psi beta using Adaptive magnitude compensation. f Estimated torque versus actual torque using adaptive magnitude compensation at 10 Nm Torque and 100 rad/s speed

components, feedforward and feedback. One from the output of the low-pass filter and second from the output of the magnitude compensation feedback. In the ideal case, the flux should be orthogonal to the back-emf and the output of quadrature detector should be zero.

Estimated torque follows closely the actual torque once the steady state is reached in adaptive magnitude compensation.

2.2 DC Offset Due to Normal Integration

Use of normal integration in estimating the stator flux from the emf in $\alpha - \beta$ axes results in dc drifted components and with the locus of stator flux having a dc offset. This happens due to the absence of initial values of flux at the time of start which was assumed to be zero. Figure 3a–f depicts the simulated output of a PM BLDC drive in closed-loop operation in 120° mode of PWM switching at 10 Nm load torque.

Estimated torque appears to be much higher than the actual torque when normal integration is used. The locus of the stator flux after normal integration has dc offset and its locus is displaced by an amount shown below: Though the simulation for estimating the above flux and torque has been based on closed-loop speed control

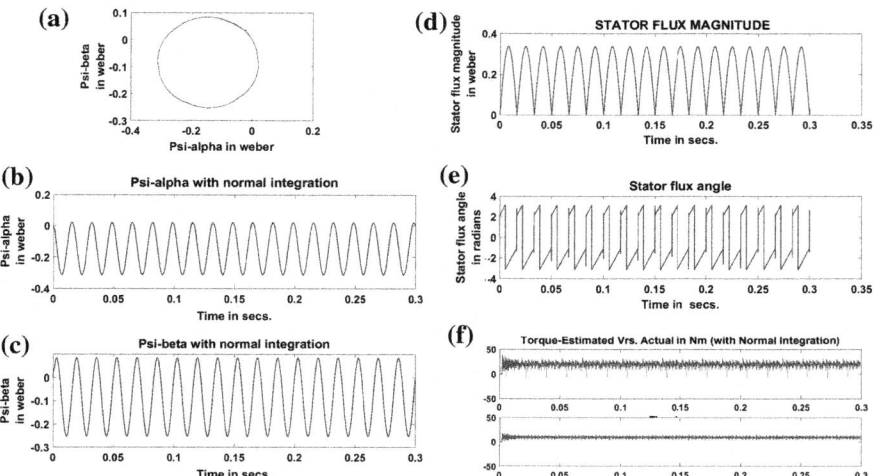

Fig. 3 **a** Stator flux using normal integration. **b** Psi alpha using normal integration. **c** Psi beta using normal integration. **d** Stator flux magnitude with normal integration. **e** Stator flux angle with normal integration. **f** Estimated torque versus actual torque using normal integration at 10 Nm Torque and 100 rad/s speed

using PWM, space vector method can be adopted for the direct control of the torque with improved switching strategies [7] to control torque ripple with faster torque response with this. With the correct estimation of stator flux and the derived rotor flux magnitude and angle, the rotor speed can be obtained using PLL scheme depicted in [8] to eliminate the enhancement of noise during differentiation as well (Fig. 4).

Fig. 4 Locus of drifted stator flux due to normal integration

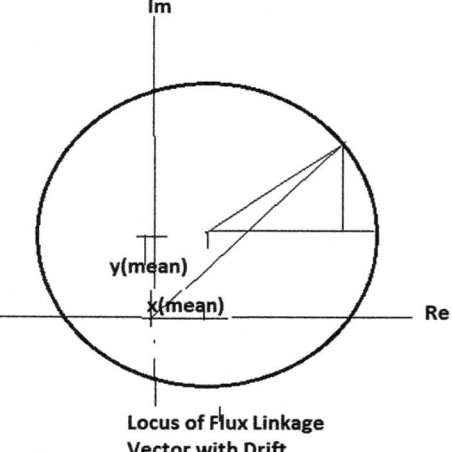

3   Sensor-less Drives Trend

Many sensor-less methods are discussed in [9, 10]. Back-emf sensing techniques from the terminal voltage of non-energized phase to determine rotor position by detection of zero crossing of line to line flux linkage is one, but they do not work satisfactorily at low speed with a noisy terminal voltage below 20% of the rated speed. The Third Harmonic Voltage Integrator has offset error due to integration. Free-wheeling Diodes conduction Detection (Terminal Current Sensing) has large hardware requirement, six independent power supplies for six operational amplifier–comparator circuits. Under this backdrop, it is suggested to use back-emf Integration which provides significant improvement using this new algorithm with adaptive control. This scheme eliminates the DC drift introduced in the estimation in successive corrective cycles.

4   Conclusion

The low-cost low-power Hybrid Electric Vehicle uses PM BLDC motor without shaft encoder. Therefore, the speed estimation has to be carried out through accurate estimation of flux and its angle. The adaptive magnitude compensation suggested for sinusoidal systems as a substitute for normal integration, works well for even non-sinusoidal emf of PM BLDC motor. The selection of cut-off frequency and tuning of PID has to be done in a manner to meet the wide speed range requirements. At lower speed PID tuning needs to be done optimally. However, the flux estimation can be carried out accurately up to speed down to 1 rad/s. The simulation has been run for 100 rad/s motor speed and at 10 Nm load torque. Results show the elimination of DC drift in the estimated stator flux.

References

1. Y. Liu, Z. Q. Zhu, and D. How, "Direct torque control of brushless dc drives with reduced torque ripple," IEEE Trans. Ind. Appl., vol. 41, no. 2, pp 599–608 Mar/Apr, 2005.
2. S.B. Ozturk and H.A. Toliyat, "Direct torque control of brushless dc motor with non-sinusoidal back-emf," in Proc. IEEE-IEMDC Biennial Meeting, Antalya, Turkey, May 3–5, 2007.
3. Z.Q. Zhu, Y. Liu, and D. Howe, "Comparison of performance of brushless DC drives under direct torque control and PWM current control," Proc. Eighth International Conference on Electrical Machines and Systems, 27–29 Sept., 2005, Nanjing, China, pp. 1486–1491.
4. L. Zhong, M.F. Rahman, W.Y. Hu, and K.W. Lim, "Analysis of direct torque control in permanent magnet synchronous motor drives," IEEE Trans. Power Electron., vol.12, no.3, pp. 528–536, May, 1997.
5. R. Krishnan and R. Ghosh, "Starting Algorithm and Performance of a PM DC Brushless Motor Drive System with No Position Sensor," IEEE-PESC Conf. Records, 1989, pp. 815–821.

6. J.Hu, B. Wu, "New Integration algorithms for estimating motor flux over a wide speed range," IEEE Trans. Power Electrons., vol. 13, pp. 967–977, Sep, 1998.
7. D. Talebi Kordkandi, H. Khanbabaie gardeshi, H. Torkaman, "An Improved Method to control the Speed and Flux of PM-BLDC Motors," The 6th International Power Electronics Drive Systems and Technologies Conference (EDSTC 2015), 3–4th Feb, 2015, pp. 638–644.
8. R. Sreepriya, Rajagopal Ragam, "Sensorless Control of Three Phase BLDC Motor Drive with Improved Flux Observer," IEEE Conference on Control, Communication and Computing (ICCC), 13–15 Dec, 2013, pp. 292–297.
9. Alin Stirban, Ion Boldea, Gheorghe-Daniel Andreescu, "Motion-Sensorless Control of BLDC-PM Motor With Offline FEM-Information- Assisted Position and Speed Observer," IEEE Trans. on Industry Appln., vol. 48, No. 6, Nov/Dec 2012, pp. 1950–1958.

Technical Reports

10. Jose'Carlos Gamazo-Real, Ernesto Va'zquez-Sanchez, and Jaime Go'mez-Gil, "Position and Speed Control of Brushless DC Motors using Sensorless Technologies and Application Trends," Open Access, www.mdpi.com/journals/sensors, 19th July, 2010.

Design and Analysis of 10-bit, 2 MS/s SAR ADC Using Nonredundant SAR and Split DAC

Kalmeshwar N. Hosur, Girish V. Attimarad, Harish M. Kittur, Gopalkrishna G. Mane and S. S. Kerur

1   Introduction

SAR ADC has attracted more attention because of the characteristics of digital in nature, scalability, excellent power efficiency and low power. In the literature survey, there are different kinds of SAR ADC architectures such as SAR ADC using charge redistribution DAC, SAR ADC using split DAC, SAR ADC using C-2C DAC, etc. In general, the working principle is dependent upon binary search algorithm to convert sinusoidal input into an equivalent discrete output. A good linearity performance is presented by SAR ADC's using binary-weighted DAC, however total capacitance exponentially depends upon resolution. Hence area occupied by SAR ADC is more and sampling speed becomes less. On the other hand, SAR ADC using C-2C DAC gives very high speed but the drawback is nonlinearity issue. SAR ADC using split DAC architectures maintaining speed advantage of C-2C SAR ADCs by decreasing the intermediary nodes whereas eliminate the resolution drawback. The split DAC consists of 2 capacitor arrays, which are segregated by LSB array and MSB array. To obtain the same scaling factor for 2 capacitor arrays the weight of the lower capacitor array of MSB should be identical to the weight of an array of LSB. The attenuation capacitor matching

K. N. Hosur (✉) · G. G. Mane · S. S. Kerur
Department of ECE, S.D.M. C.E.T., Dharwad, Karnataka, India
e-mail: kalmeshwar10@rediffmail.com

S. S. Kerur
e-mail: shreayagiri@rediffmail.com

G. V. Attimarad
Department of Electronics and Communication Engineering, K.S. School of Engineering and Management, Bangaluru, Karnataka, India

H. M. Kittur
School of Electrical Sciences, VIT University, Vellore, Tamil Nadu, India
e-mail: harish13579@rediffmail.com

© Springer Nature Singapore Pte Ltd. 2019
V. Nath and J. K. Mandal (eds.), *Proceeding of the Second International Conference on Microelectronics, Computing & Communication Systems (MCCS 2017)*, Lecture Notes in Electrical Engineering 476, https://doi.org/10.1007/978-981-10-8234-4_65

becomes difficult after realized design. Thus, the analog-to-digital converter linearity is substantially reduced [1–3].

Medium resolution SAR ADCs are increasingly used in high sampling rate applications. Ultra-low power and low-frequency SAR ADCs are being used in low-energy radios and biomedical applications. In a number of cases, capacitance DAC is the major power contributor of SAR ADC. Several researches have been carried out on DAC switching power reduction techniques [4]. More popular DAC architecture in SAR ADC is binary-weighted capacitive DAC. However, the exponential increase in the capacitance of the DAC array with the resolution, results in more settling time, larger area and larger consumption of switching energy. The split capacitive digital-to-analog converter is a precious alternative, which is useful for medium resolution applications. The main disadvantage of this architecture is the degradation of the conversion linearity [5].

In SAR ADC, mainly power is consumed by the comparator, SAR logic, and capacitive DAC. Due to technology innovation, the power consumption drastically reduced. The power consumption in capacitive DAC and comparator is reduced by matching of capacitance [6]. The split SAR ADC provides the excellent trade-off between area, power consumption, and speed [7]. The capacitive digital-to-analog converter dominates the overall power consumption compared with a digital control circuit and comparator in SAR analog-to-digital converter [8].

SAR ADC realizes the binary search algorithm with the help of SAR control logic. Basically, two methods are used for designing SAR logic namely (1) sequencer/code register SAR logic introduced by Anderson and in this type of SAR logic 2 N flip flops are used (2) SAR logic using MUX-D flip flop and some combinational logic and contains N flip flops [9, 10]. Because of power efficiency of SAR ADC, it plays a vital role as per the applications concerned and compared to other ADC configurations [11].

For optimized power, successive approximation register logic, respective solutions are investigated [10]. Switched capacitor-based binary search ADC has been used to solve the drawbacks of the standard SAR logic [8]. LFSR-based SAR Logic is presented in [12]. In this paper focused on designing and analysis of SAR ADC using nonredundant SAR and split DAC.

2 General SAR ADC Architecture

Figure 1 depicts a representation of SAR ADC block diagram. The basic theory of data conversion depends on binary search algorithm and is given in (1) [13].

$$V_{\text{analog}} - \left\{ V_{\text{ref}} \left[b_0 2^{-1} + b_1 2^{-2} + \cdots + b_{N-1} 2^{-N} \right] \right\} = 0 \tag{1}$$

Fig. 1 N-bits SAR ADC
structure

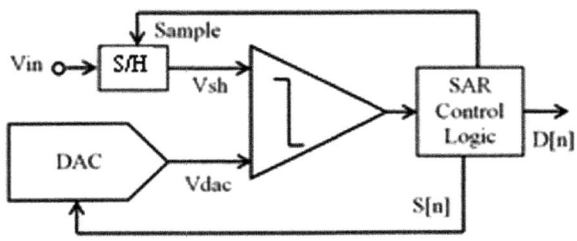

3 Main Internal Blocks of Proposed SAR ADC

3.1 Split Array DAC

The binary-weighted charge scaling DAC architecture is very much accepted since it is having comparative good accuracy and simple architecture. The most significant bit capacitor becomes a very large value for higher resolution. For instance 12-bit DAC is to be designed, if capacitor, $C = 0.5$ pF, then the value of MSB capacitor is given by

$$C_{\mathrm{MSB}} = 2^{N-1} \cdot 0.5 \text{ pF} = 1024 \text{ pF} \tag{2}$$

The size of the capacitors can be reduced using the above DAC configuration. Figure 2 shows a 6-bit split array DAC [14, 15].

The value of the attenuation capacitor can be found by

$$C_{\mathrm{att}} = \left(\frac{\text{Sum of the LSB array capacitors}}{\text{Sum of the MSB array capacitors}} \right) C \tag{3}$$

where the addition of the capacitors present at MSB side, is identical to the addition of the capacitors present at LSB side is lesser than $1C$. The output analog voltage is the attenuation factor times the sum of LSB and MSB bits multiplied by the reference voltage. The attenuation factor is defined as a capacitive divider between the sum of the LSB capacitors and the attenuation capacitor.

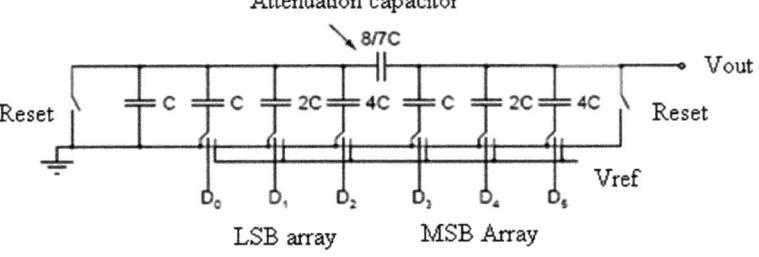

Fig. 2 Split array DAC

3.2 Nonredundant SAR Structure

This construction comprises less number of FFs and usually depends on the state of each bit with the other bit state. By activation of MSB, conversion algorithm starts considering all bits are zero, similarly the conversion of the remaining bits are sequentially activated, the value of one activated just prior is depends on the result of the comparator.

The fundamental architecture of SAR is consisting of many input 'N' bit shift registers. While starting step, activate start of conversion as all flip flops are made to initialization state that is zero, then for the consecutive state, by integrating multiplexer and decoder to all flip flops, every kth register will have the chance to select among the three data input lines from: output of flip flop (shift), the comparator output (cmp), the output of flip flop itself (k).

The assortment of the three lines depends on the present state and subsequent states of the following register. The structure with the complete realization of SAR Logic is depicted in Fig. 3 [3].

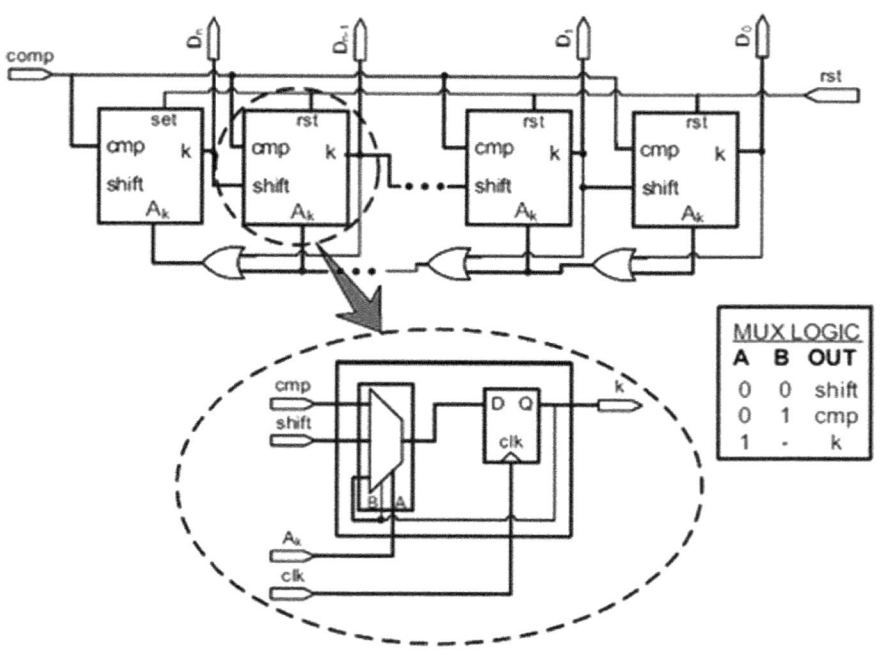

Fig. 3 Nonredundant SAR logic

4 Design and Implementation SAR ADC Using Nonredundant SAR and Split DAC

The 10-bit SAR ADC using nonredundant SAR and split DAC architecture is shown in Fig. 4. For this circuit analog signal of peak-to-peak amplitude of 1.2 V with an offset of 600 mV and frequency of 100 kHz is applied. The sampling signal of clock period is 1 us and high voltage of 1.2 V and low voltage of 0 V is given. The clock signal of SAR block is 50 ns. The results are shown in Fig. 5 and input samples and corresponding output in hex form are mentioned in Table 1 and power consumption of each block and total power is mentioned in Table 3. Performance comparison between this work and other recently published state-of-the-art results mentioned in Table 4. The comparative statement of power consumption and reduction are depicted in Table 2.

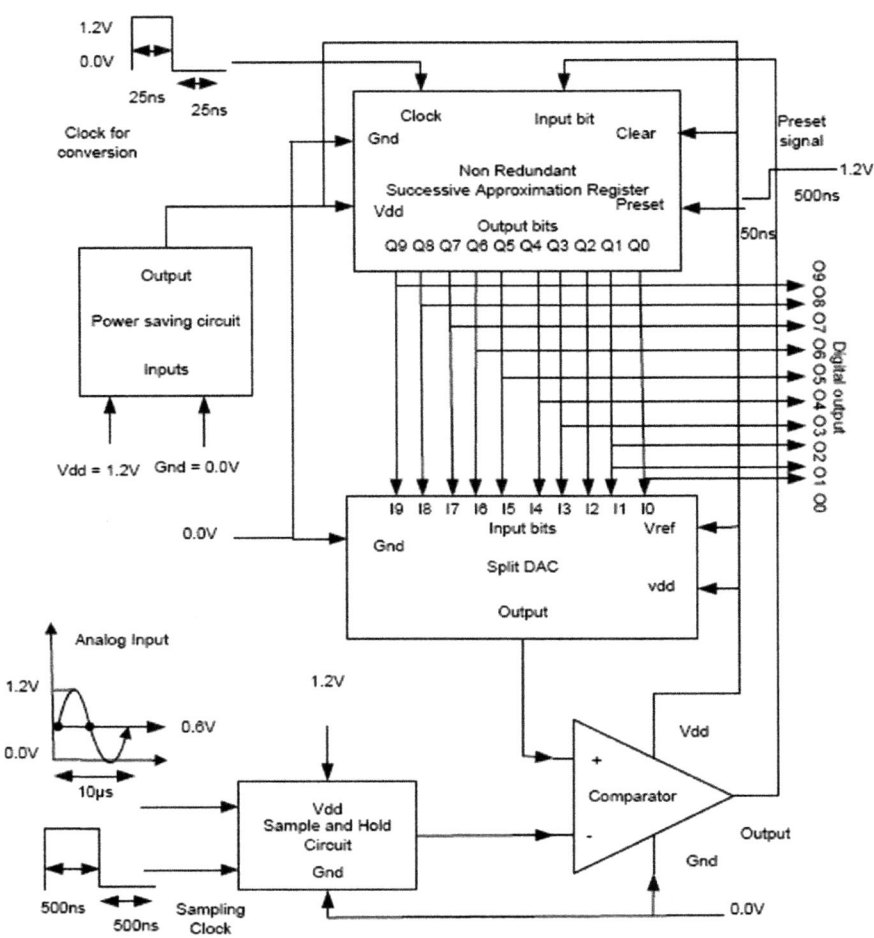

Fig. 4 SAR ADC schematic using nonredundant SAR and split DAC

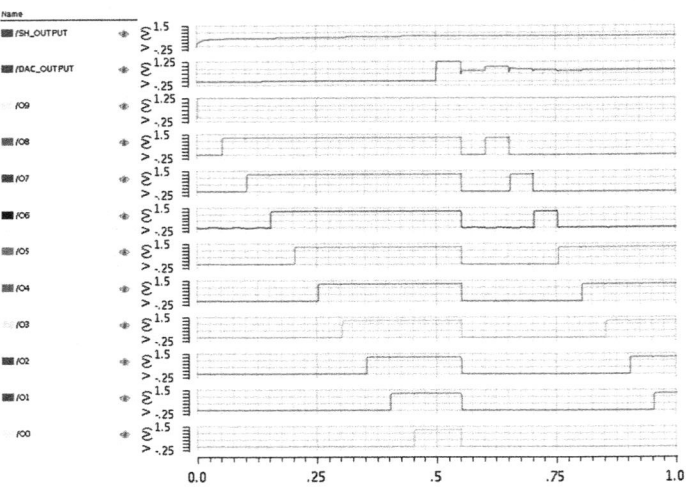

Fig. 5 Digital output = 1000111110 for an analog input = 754.3 mV

Table 1 Input sine wave samples and corresponding output in hex form

Time (μs)	Sample and hold circuit output	Split DAC output	SAR ADC output in HEX
1	754.3 mV	691.6 mV	23E
2	1.082 V	1.01 V	318
3	1.2 V	1.175 V	360
4	1.02 V	871.9 mV	280
5	777.0 mV	636.7 mV	200
6	375.3 mV	260.1 mV	05F
7	115.3 mV	235.1 mV	004
8	17.8 mV	198.7 mV	010
9	253.1 mV	179.4 mV	040
10	411.7 mV	444.8 mV	100

Table 2 Comparative statement of SAR ADC using the conventional method and nonredundant SAR and split DAC

Technique used	Power consumed (μW)	% of power reduction
SAR ADC using the conventional method	211.1908	100
Nonredundant SAR and split DAC	81.4721	38.57

Table 3 Power consumption of SAR ADC using nonredundant SAR and split DAC

Different blocks of SAR ADC	Power consumed	% of power consumed
Sample and hold circuit	21.4487 nW	0.0263
Comparator	10.4799 µW	12.8631
Nonredundant SAR	49.2569 µW	60.4586
Split DAC	17.4600 µW	21.4306
Power saving circuit	4.2539 µW	5.2212
Complete SAR ADC	81.4721 µW	100

Table 4 Comparison with the existing results

Specifications	[1]	[2]	[6]	This work
Supply voltage (V)	1.2	0.5	1.2	1.2
Technology (nm)	65	130	65	180
Resolution (Bits)	11	13	10	10
Sampling rate	50 MS/s	40 KS/s	70 MS/s	2 MS/s
Power consumption	2.48 mW	1.47 µW	0.96 mW	81.4721 µW

5   Conclusion

Designed 10 Bit, 2 MS/s Successive Approximation Register ADC using nonredundant SAR and split DAC and simulated using Cadence. Dynamic range for this architecture is 60.19 dB. The charge redistribution DAC in split capacitor structure has a total capacitance which is 96.87% lesser compared to a usual design. 10-bit SAR ADC constructed using two SAR configurations. The first one is sequencer code register successive approximation register configuration, which needs 20 FFs and thus desires more power consumption and occupies more area. The second one is nonredundant successive approximation register configuration, which needs 10 FFs and some combinational logic, thus requires less power consumption and occupies less area. Hence successive approximation registers ADC implementation using split array DAC and nonredundant SAR structure occupies a smaller area as well as consumes less power. The power consumed by SAR ADC using nonredundant SAR and split DAC obtained as 81.4721 µW and conventional SAR ADC using sequencer code register SAR and binary-weighted DAC are obtained as 211.1908 µW.

References

1. Anh Trong Huynh, Hoa Thai Duong, Hoang Viet Le, and Efstratios Skafidas, "Design and Implementation of an 11-bit 50-MS/s Split SAR ADC in 65 nm CMOS", 978-1-4799-3432-4/14/$31.00 ©2014 IEEE.
2. Hyunsoo Ha, Seon-Kyoo Lee, Byungsub Kim, Member, IEEE, Hong-June Park, Senior Member, IEEE, and Jae-Yoon Sim, Senior Member, IEEE, "A 0.5-V, 1.47-μW 40-kS/s 13-bit SAR ADC With Capacitor Error Compensation", IEEE TRANSACTIONS ON CIRCUITS AND SYSTEMS—II: EXPRESS BRIEFS, VOL. 61, NO. 11, NOVEMBER 2014.
3. Ji-Yong Um, Yoon-Jee Kim, Eun-Woo Song, Jae-Yoon Sim, Member, IEEE, and Hong-June Park, Member, IEEE, "A Digital-Domain Calibration of Split-Capacitor DAC for a Differential SAR ADC Without Additional Analog Circuits", IEEE TRANSACTIONS ON CIRCUITS AND SYSTEMS—I: REGULAR PAPERS, VOL. 60, NO. 11, NOVEMBER 2013.
4. Arindam Sanyal, Student Member, IEEE, and Nan Sun, Member, IEEE, "An Energy-Efficient Low Frequency-Dependence Switching Technique for SAR ADCs", IEEE TRANSACTIONS ON CIRCUITS AND SYSTEMS—II: EXPRESS BRIEFS, VOL. 61, NO. 5, MAY 2014.
5. Yan Zhu, Chi-Hang Chan, U-Fat Chio, Sai-Weng Sin, Seng-Pan U, Rui Paulo Martins, and Franco Maloberti, "Split-SAR ADCs: Improved Linearity With Power and Speed Optimization", IEEE TRANSACTIONS ON VERY LARGE SCALE INTEGRATION (VLSI) SYSTEMS, VOL. 22, NO. 2, FEBRUARY 2014.
6. Yue Wu, Xu Cheng and Xiaoyang Zeng, "A Split-capacitor Vcm-based Capacitor-switching Scheme for Low-power SAR ADCs", 978-1-4673-5762-3/13/$31.00 ©2013 IEEE.
7. Anh Trong Huynh, Hoa Thai Duong, Hoang Viet Le, and Efstratios Skafidas, "Design of a Capacitive DAC Mismatch Calibrator for Split SAR ADC in 65 nm CMOS", 2013 Asia-Pacific Microwave Conference Proceedings, 978-1-4799-1472-2/13/$31.00 ©2013 IEEE.
8. Liangbo Xie, Guangjun Wen, Jiaxin Liu and Yao Wang, "Energy-efficient hybrid capacitor switching scheme for SAR ADC", ELECTRONICS LETTERS 2nd January 2014 Vol. 50 No. 1 pp. 22–23.
9. Raheleh Hedayati, "A Study of Successive Approximation Registers and Implementation of an Ultra-Low Power 10-bit SAR ADC in 65 nm CMOS Technology", Master's thesis performed in Electronic Devices, LiTH-ISY-EX–11/4512—SE September 2011.
10. Dai Zhang, Ameya Bhide and Atila Alvandpour, A 53-nW 9.1-ENOB 1-kS/s SAR ADC in 0.13-m CMOS for Medical Implant Devices, 2012, IEEE Journal of Solid-State Circuits, (47),7, 1585-1593.
11. RVNR Suneel Krishna, Jyothirmayi, Geethanjali College of Engineering and Technology, Rangareddy Dist, A.P, India, "Low Power SAR-ADC in 0.18 µm Mixed-Mode CMOS Process for Biomedical Applications", IOSR Journal of VLSI and Signal Processing (IOSR-JVSP), Volume 3, Issue 1 (Sep.–Oct. 2013), PP 29–35, e-ISSN: 2319–4200, p-ISSN No.: 2319–4197.
12. ALGN Aditya, G. Rakesh Chowdary, J. Meenakshi, T. Praveen Blessington, M.S. Vamsi Krishna, "IMPLEMENTATION OF LOW POWER SUCCESSIVE APPROXIMATION ADC FOR MAV's", 2013 International Conference on Signal Processing, Image Processing and Pattern Recognition [ICSIPR].
13. P.E. Allen and D.R Holberg, CMOS Analog Circuit Design, Second Edition, Oxford University Press, 2002.
14. R. Jacob Baker, Harry W Li and David E. Boyce, CMOS Circuit Design, Layout, and Simulation, IEEE Press 2000.
15. Frank B. Boschker, "Design of a 12 bit 500 MS/s standalone charge redistribution Digital-to-Analog Converter", University of Twente, Msc. Thesis January 2008.

Performance Analysis of Inter-satellite Optical Wireless Communication Using 12 and 24 Transponders

Ikkurthy Kavya Sri and Avireni Srinivasulu

1 Introduction

At first, it is said that the satellites do not pose any competition for fiber in telecommunications, and are only used where fiber is simply diminishing from the field. This is known as "gap fillers", which is simply where the satellite is being used to fill the gaps in the coverage areas that not yet have access to fiber/cable. Satellites simply receive the radio signals and pass on them to its "gate way" at the other end. The gateway will have terrestrial fiber on to the final destination. As a result of the distance that has to be traveled, the satellite link(s) requires 240 ms to reach the satellite and above 240 ms there is high latency. This adds to 480 ms delay or latency in the data that flows over these links. Being so much closer to earth greatly reduces latency, but also limits the coverage area of each satellite. The LEO-based system uses laser communications between each of the satellites to create a mesh network covering the entire world [1, 2].

S-band, X-band, and Ku-band services presently support the trailing and knowledge acquisition desires of NASA's low Earth orbit from which knowledge is transferred on to earth stations via tracking and data relay satellite system (TDRSS), which is a component of NASA's space network (SN). The satellite link plays a major role in transferring data from the satellite to the ground station. In that process, we need to take care of different factors like uplink design, downlink design, link power budget, latency, system noise, satellite attitude control, and other

I. K. Sri
Department of Electronics & Communication Engineering, Vignan's University
(V.F.S.T.R University), Vadlamudi, Guntur 522213, India
e-mail: ikkurthy.kavyasri413@gmail.com

A. Srinivasulu (✉)
Department of Electronics & Communication Engineering, JECRC University,
Jaipur 303905, Rajasthan, India
e-mail: avireni_s@yahoo.com; avireni@ieee.org; avireni@gmail.com

© Springer Nature Singapore Pte Ltd. 2019
V. Nath and J. K. Mandal (eds.), *Proceeding of the Second International Conference on Microelectronics, Computing & Communication Systems (MCCS 2017)*, Lecture Notes in Electrical Engineering 476, https://doi.org/10.1007/978-981-10-8234-4_66

factors. Fiber optic-based technologies are primarily considered as unit comparatively involving unaccustomed satellite applications, and its area unit receiving
extended attention for planned applications in NASA, DOD, and industrial area
sectors. Initially, the satellites have not posed any competition for fiber in
telecommunications, and are only used where fiber simply does not exist. This is
known as "gap fillers", which is simply where the satellites are used to fill the gaps
in the coverage areas that not yet have access to fiber/cable. Now, as the technology
is advancing, we use the optical communication for inter-satellite links.

In satellite, the link plays an important role in communication. The link refers to
the transmission of the signal from the transmitter to the receiver and vice versa.
The communication ability depends upon the signal strength and the amount of
thermal noise that occurs. In the link, the important power calculation is drawn and
the same is received by the receiving station (P_R) (satellite receiver and the earth
station). The quality of satellite link is assessed by signal-to-noise ratio in the
receiving earth stations or in terms of energy per bit to noise density. Generally, the
satellites are placed in the LEO which is 160 km above (geocentric orbits). In
future, NASA has plans to launch several spacecraft in the LEO to support the
science machines and it requires the downlink in the order of several terabytes per
day. As the current systems are not capable of handling such large volumes of data,
two solutions are proposed. One is to establish the high data rate link between the
LEO spacecraft(s) via relay satellite to the ground and the other to establish a direct
link from LEO to the ground [3–8].

The European space program has launched the ARTEMIS satellite successfully
which is a data relay payload to provide satellite communication between the
ground and the low earth orbit space crafts. As such the analysis of the link
determines properties of satellite equipment (antennas, data rate, amplifiers, etc.).
Frequency is always specified as uplink/downlink frequencies.

2 Optical Inter-satellite Links Description and Analysis

In this paper, two new satellite communication systems using optical Inter-Satellite
Links (ISLs) with 12 and 24 transponders are proposed. The use of the ISL is that it
has the essential giant transmission capability, which helps it to have access to
supply ISDN subscriber channels among the mobile users. Therefore, the network
control station (NCS) is, however, placed on the earth [9].

The relocation of the broadband technology via the satellite communications has
been the first step in arriving at the satellite networks. The inter-satellite links are
playing an important role in transferring information between the satellites. So, the
number of earth stations will be getting decreased, and also where the ISLs
decrease, the amount of hops (uplink and downlink) too. Therefore, ISLs have
become a vital half of the area technology. The primary inter-satellite links using
microwave radio frequency (RF) is operated in two bands namely Ka-band and
V-band. The major optical ISL amid satellites victimize optical device with

lightweight was set between the European Space Agency (ESA), Cynthia satellite and French earth observation satellite SPOT_4, using Semiconductor Laser Inter-Satellite Link Experiment (SILEX). For optical ISLs, three specialized parameters are obligatory to determine fusion among the satellites. This can be the foremost fundamental thought for the signal transmission. Subsequent one is the multiple access that may have the capability for splitting the satellite electrical device to evade the involvement in the entering signals from alternative satellites to many ground stations, and also the data rate is given by $C = B \times \log2(1 + S/N)$. The last technique is that the modulation, and its key factor is to settle on the perfect modulation style in terms of a more theoretical capability. This process yields conjointly higher quality of safety precautions that are resistant to ECM and interception.

Optical inter-satellite link (OISL) is a leading telecommunication system and information transfer networks that are defined to transfer information and data with high security and high level of knowledge as a result of the slim beam breadth (typically five small radians). The concept of OISL is shown in Fig. 1. This is often considered a bonus for cover against interference between signals. OSIL has a high transmission capability and it does satisfy the rising demand of the users. OISL will outline two styles of optical systems. The foremost is the optical laser whose advantages are high capability, long communication distance, and lower transmitted power, and the other is a crystal rectifier. Several research and studies are done currently to use a crystal rectifier in ISL for a brief distance and to move less rates between nearest satellites. For instance, IrDA (Infrared Data Association) standardized the infrared for the device and low speed links and it is well known for the employment of the crystal rectifier in free area communications. There are 3 basic-level phases to ascertain an OISL [10–13].

Acquisition: The beam scans the regions of the house wherever the receiver is anticipated to be settled, this often is like a good beam to scale back the gaining

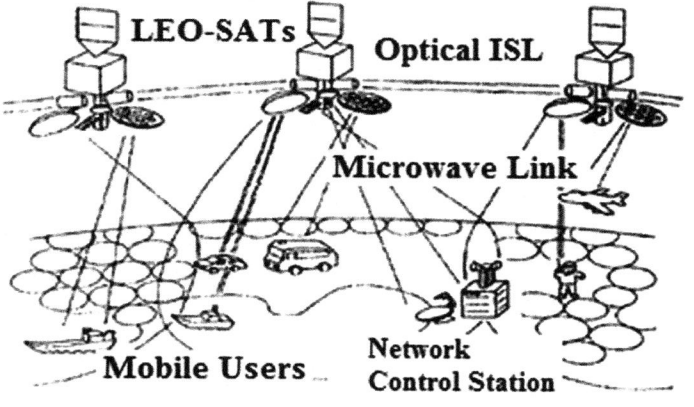

Fig. 1 The concept of optical inter-satellite links

time that needs a high power optical transmitter, the typical time for this section is regarded as 10 s.

Tracking: It occurs, once the receiver receives the signal, and then it transmits the same in an inward direction of the received signal. On receiving the signal from the receiver, the transmitter conjointly enters the section.

Communications: Exchange of information among two satellites is provided by a link. The major drawback facing the OISL between two satellites is vibration and also the relative speed between the 2 satellites.

2.1 Optical Transmitter

The optical transmitter is to come up with the optical signal, and imposes the information-bearing signal, then launches the modulated signal in the direction of the receiver. In this paper, the IsOWC for 12 and 24 transponders was designed using Optisystem software of version 14. The basic simplex model consists of a transmitter, channel, and receiver. There are four subsystems on the transmitter side. The pseudo-random bit sequence generator is the first subsystem. The required transmitted data or information is represented by the use of this subsystem. The data regularly comes from the satellite's TT&C system. In this work, to observe the performance of the system and the relationship between bit rate and distance the bit rate is varied. The NRZ pulse generator is the second subsystem that encodes the data from the pseudo-random bit sequence generator using the non-return zero encoding technique. Continuous Wave (CW) laser is the third subsystem which is on the transmitter. In CW laser the output signal of the laser is non-stoppable and unmodulated. Lasers are used as an alternative to LED in this system as it can transfer for long distances. The frequency of the light is selected to be 1550 nm or 353.0 THz with the key power of 64 and 12 dBm. Mach–Zehnder (MZ) modulator is the last subsystem whose function is to vary the intensity of the light source from the laser according to the NRZ pulse generator output.

2.2 Optical Wireless Channel

This optical wireless channel (OWC) acts as a medium in free space and is used to transfer the data from transmitter to receiver. It is present with 15 cm optical antenna at each end. Here, the gains of transmitter and receiver are equal to 0. There are no pointing errors and zero additional losses. Due to atmospheric effects, there will be no attenuation.

2.3 Optical Receiver

The receiver consists of an APD (Avalanche photodiode), low-pass Bessel filter, 3R regenerator, and a BER analyzer. The APD receives the optical signal and converts into electrical signals. To filter the high-frequency signals low-pass Bessel filter is used. The maximum attenuation of the filter is 100 dB. The 3R regenerator is used to restore the electrical signal of the original bit sequence and is used for BER analyzer.

3 Inter-satellite Links Equation

The proposed inter-satellite Optical Wireless Communication Model of 12 Tx/Rx is shown in Fig. 2. The received signal power P_R is equal to the transmitted power P_T, transmitter gain G_T, receiver gain G_R, free space loss L, transmitter efficiency τ_T, and receiver efficiency τ_R.

Received power

$$P_R = P_T \times G_T \times G_R \times L \times \tau_T \times \tau_R$$

Receiver antenna gain

$$G_R = \left(\frac{\pi D_R}{\lambda}\right)^2$$

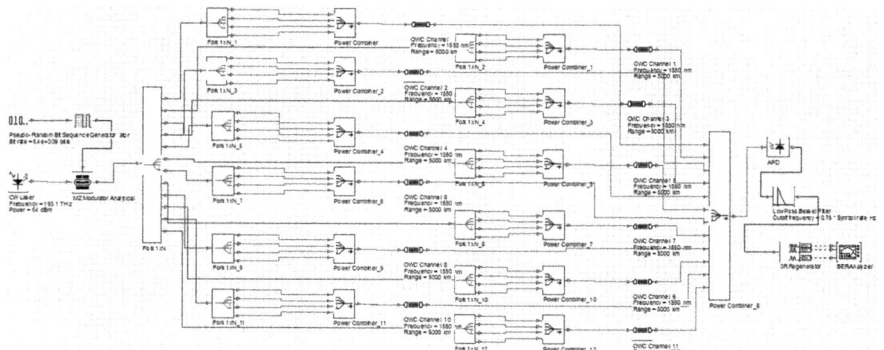

Fig. 2 Proposed inter-satellite optical wireless communication model of 12 Tx/Rx

Free space loss

$$L = \left(\frac{\lambda}{4\pi d}\right)^2$$

Transmitter antenna gain

$$G_T = \frac{32}{\theta_T^2}$$

where λ is the wavelength and θ_T is the transmitter divergence angle, it is thus given by $\theta_T = 1.22(\lambda/DT) = 18.91$ μ-radian. The system performance can be seen by analyzing the BER and Q-factor. The BER is defined as the ratio of the number of bit errors detected in the receiver and the number of bits transmitted as well. Whereas the Q is the quality factor which is the measurement of signal quality and is proportional to the systems S/N ratio. In the optical system the BER is very small to calculate so Q-factor stands well to be used.

4 Simulation Results

The circuit provided in Figs. 2 and 3 were simulated by the OPTISYSTEM-14 software which is a complete programming outline suite that empowers clients to arrange, test, and reenact optical connections in the transmission layer of present day optical systems. The CW laser at the transmitter frequency is taken as 193.1 THz and power is 30 dBm. The OWC channel frequency is 1550 nm and the range is 5000 km. The pseudo-random bit generator is taken as 1 Gbps and the Mach–Zehnder modulator extinction ratio is taken as 30 dB. Based on the changes in CW laser power and MZ extinction ratio the BER and Q-factor are shown in the figures mentioned as under. The impact of different parameters such as bit error rate, transmitted power, and Q-factor are analyzed. The simulation results for the Q-factor and BER is shown in Figs. 4 and 5 with 30 dBm power. Table 1 shows the comparison of the performance analysis of the candidate designs.

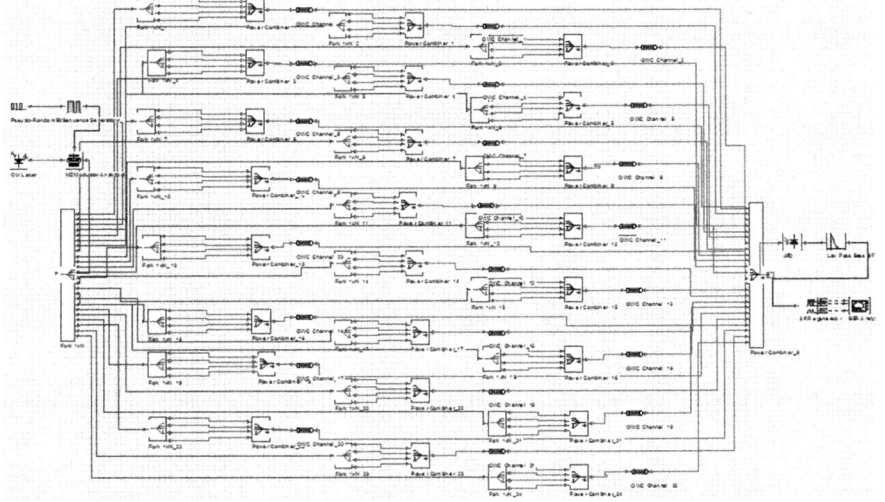

Fig. 3 Proposed inter-satellite optical wireless communication model of 24 Tx/Rx

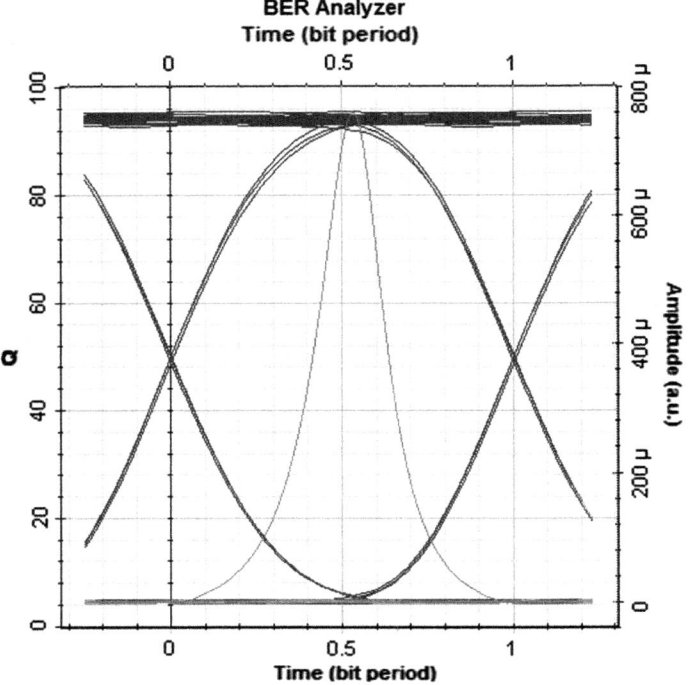

Fig. 4 Eye diagram of 12 Tx/Rx Q-factor at input power 30 dBm

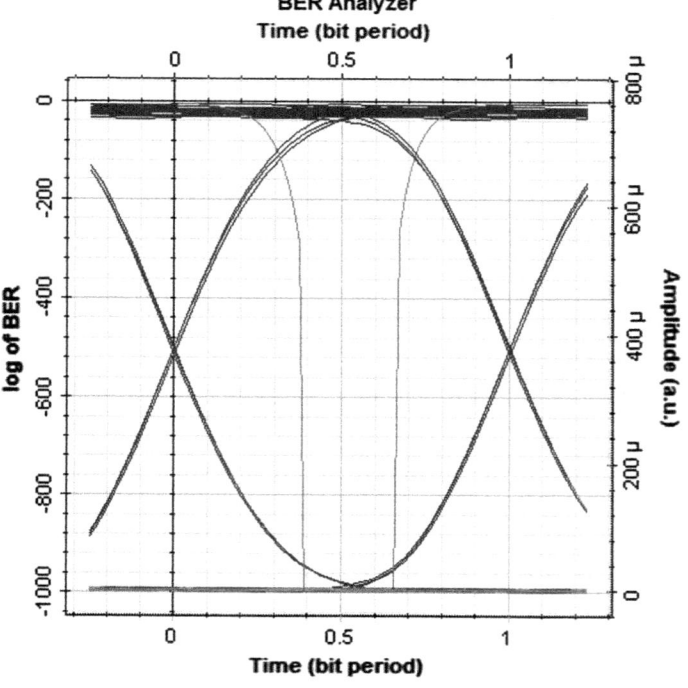

Fig. 5 Eye diagram of 12 Tx/Rx bit error rate at input power 30 dBm

Table 1 Simulated performance analysis of candidate designs

Reference	[10]	[11]	[12]	[13]	Proposed Fig. 2	Proposed Fig. 3
Distance (km)	1000–5000	5000	1000–5000	500–3000	5000	5000
Freq. (THz)	193.1	193.1	193.1	193.1	193.1	193.1
Q-factor	2.102	7.9582	525.17	–	99.62	164.36
BER	0.0177	8.722e−016	–	<10 to 9	0.01	0.001
Data rate (Gbps)	–	5.6	10	0.8	1	10
Power (dBm)	–	–	10	12	30	30
No. of Tx/Rx	8	4	–	–	12	24
Received power	−19	−25	–	–	−11.07	−6.24

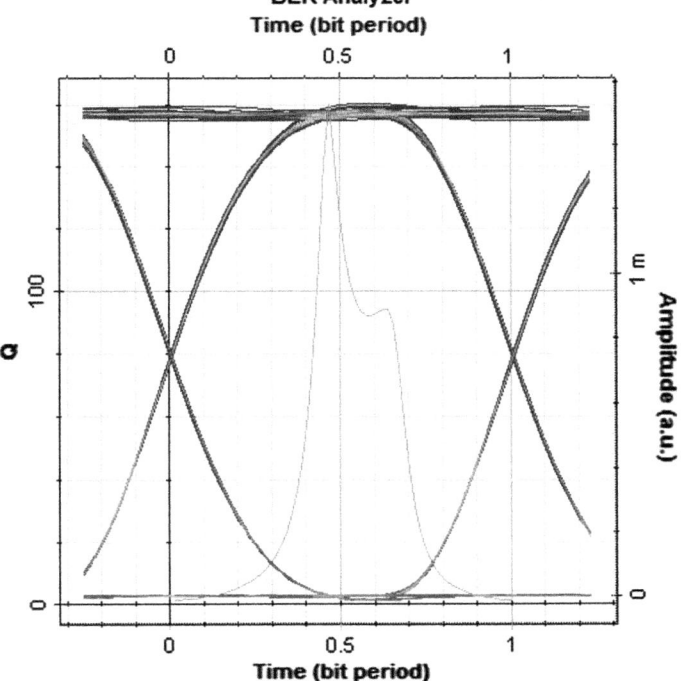

Fig. 6 Eye diagram of 24Tx/Rx Q-factor at input power 30 dBm

The simulation results of Fig. 3 for the Q-factor and BER is shown in Figs. 6 and 7 with 30 dBm power. The CW laser at the transmitter frequency is taken as 193.1 THz and power is 30 dBm. The OWC channel frequency is 1550 nm and the range is 5000 km. The pseudo-random bit generator is taken as 10 Gbps and the Mach–Zehnder modulator extinction ratio is taken as 30 dB.

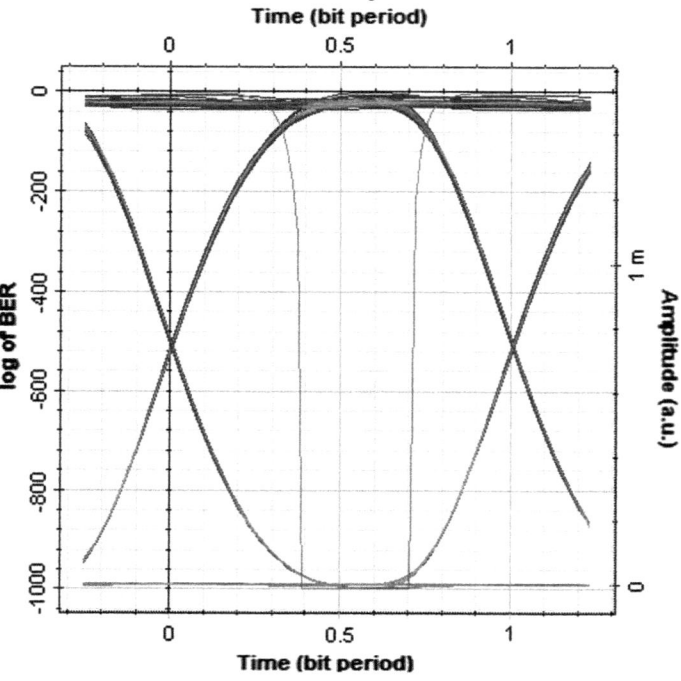

Fig. 7 Eye diagram of 24 Tx/Rx bit error rate at input power 30 dBm

5 Conclusion

IsOWC is the revolutionary technique which can establish the genuine communication. The link between satellites is established through lasers. With the
advancement of communication technology, unabated growth in information
technology is perceived in the field of top-speed knowledge transmissions. The
present RF satellite links will transmit knowledge with somewhat at same rate
around Mbps. So as to attain higher rate of knowledge optical links for houses
particularly exploitation optical laser is required. The light beam of an optical laser
provides a decent low chance for detective work, intercepting and reducing the
electronic jamming chance of the signal. Furthermore, it permits the utilization of
frequencies. OISL systems are quicker in information measures and in knowledge
transmission security.

References

1. X. Jia, Tao Lv, Feng He and Hejiao Huang, "Collaborative Data Downloading by Using Inter-Satellite Links in LEO Satellite Networks", IEEE Transactions on Wireless Communications, vol. 16, issue. 3, pp. 1523–1532, 2017, https://doi.org/10.1109/twc.2017.2647805.
2. J. Huang, Y. Su, W. Liu, F. Wang, "Adaptive modulation and coding techniques for global navigation satellite system inter-satellite communication based on the channel condition", IET Communications, vol. 10, issue. 16, pp. 2091–2095, 2016, https://doi.org/10.1049/iet-com.2016.0093.
3. W. Liang, Z. Naitong, W. Pingping, "Dynamic characteristics of inter-satellite links in LEO networks", Journal of Systems Engineering and Electronics, vol. 14, issue. 4, pp. 25–29, 2003.
4. Y. Tang, Y. Wang, J. Chen, "High sensitive acquisition of signals for inter-satellite links of navigation constellation based on two-dimension partitioned FFTs", IEEE International Conference on Signal and Image Processing (ICSIP), Beijing, China, 13–15 Aug. 2016, pp. 573-577, https://doi.org/10.1109/siprocess.2016.7888327.
5. B. A. Zaki M., A. Abd El Aziz, H. A. Fayed, M. H. Aly, "The Impact of Varying the Detector and Modulation Types on Inter Satellite Link (ISL) Realizing the Allowable High Data Rate", *2nd International Conference on Future Generation Communication Technology (FGCT)*, London, UK, 12–14 Nov. 2013, pp. 44-47, https://doi.org/10.1109/FGCT.2013.6767191.
6. Z. Yuan, Z. Liu and J. Zhang, "Inter-satellite Link Design for the LEO/MEO TwoLayered Satellite Network", *International Conference on Wireless Communications, Networking and Mobile Computing*, vol. 2, pp. 1072–1075, 2005, https://doi.org/10.1109/wcnm.2005.1544238.
7. Z. Yun-tao, F. Yong-xin, L. Fang, "An Improved Resistant to Destruction Routing Algorithm Based on LEO Satellite Network", *International Conference on Future Generation Communication and Networking*, Hainan Island, China, 13–15 Dec. 2008, vol. 1, pp. 230–233, https://doi.org/10.1109/fgcn.2008.44.
8. A. Nugrohoi, N. Jamalii, S. Tanuwijaya, "Introduction of the IiNUSAT inter-satellite link system", IEEE International Conference on Communication, Networks and Satellite (ComNetSat), Bali, Indonesia, 12–14 July 2012, pp. 192–195, https://doi.org/10.1109/comnetsat.2012.6380804.
9. L. Xu, Z. Xiaoxu, G. Lili, "An autonomous navigation study of Walker constellation based on reference satellite and inter-satellite distance measurement", Proceedings of 2014 IEEE Chinese Guidance, Navigation and Control Conference, Yantai, China, 8–10 Aug. 2014, pp. 2553–2557, https://doi.org/10.1109/cgncc.2014.7007568.
10. S. G. Dev, S. A. Thomas, "Performance Analysis Of Intersatellite Optical Wireless Link Using Multiple TX/RX And CO-OFDM Techniques", International Journal of Scientific & Engineering Research, vol. 6, issue 10, pp. 935–939, Oct. 2015.
11. I. Aggarwal, P. Chawla, R. Gupta, "Performance Evaluation of Intersatellite Free Space Optical Communication System with Varied Parameters and Transceiver Diversity", Advance in Electronic and Electric Engineering. vol. 3, no. 7, pp. 847–852, 2013. ISSN 2231-1297.
12. R. Gupta, S. Sharma, M. K. Sharma, "Modification in Parameter of Intersatellite Communication using Multiple Transmitter and Receiver", International Journal of Engineering Trends and Technology (IJETT), vol. 12 no. 10, pp. 495–498, Jun 2014.
13. Apurva Chaudhary, Sandeep Singh, Gagan Minocha, Haneet Rana, "Performance analysis of inter satellite optical link and the effect of transmitter and receiver aperture on its performance parameters", International Journal of Advanced Technology in Engineering and Science, vol. 02, issue. 05, pp. 139–144, May 2014.

Synthesis of High-Speed Multivalued ALU for ($2^p \pm q$) Radix

Arindam Banerjee, Swapan Bhattacharyya and Arpan Deyasi

1 Introduction

Arithmetic and logical circuits are the core components of all types of processors used in digital signal processing, image processing, artificial neural network, multimedia application, etc. [1–7]. But with an increase in market demand for increasing complexity and requirement of different functions inside a single computational unit, the problem of interconnection becomes severe; as edge connection increases linearly with edge length, whereas circuit space requirement increases with the square of the factor [8]. This forces to think people about design of on-chip bus and functional modularization [9]. This reduces the existing VLSI design problem to some extent as broadly usable functional modules help to design multivalued arithmetic logic unit with a reduction of high cost, to make it economically favorable. This represents the necessity of designing radix system with higher order.

In the past decade, researchers are working on the performance improvement of the processors by modifying the arithmetic and logical circuits of ALU structure. Many researchers have designed ALU in ternary and quaternary radix system [1, 2, 5–7]. But all the designs are dedicated for a particular radix system. The authors

A. Banerjee · S. Bhattacharyya
Department of Electronics and Communication Engineering, JIS College of Engineering, Kalyani, India
e-mail: banerjee.arindam1@gmail.com

S. Bhattacharyya
e-mail: swapanbhattacharyya@ieee.org

A. Deyasi (✉)
Department of Electronics and Communication Engineering, RCC Institute of Information Technology, Kolkata, India
e-mail: deyasi_arpan@yahoo.co.in

© Springer Nature Singapore Pte Ltd. 2019
V. Nath and J. K. Mandal (eds.), *Proceeding of the Second International Conference on Microelectronics, Computing & Communication Systems (MCCS 2017)*, Lecture Notes in Electrical Engineering 476, https://doi.org/10.1007/978-981-10-8234-4_67

have already proposed [10] a generalized ALU design scheme for $(2^p \pm q)$ radix system, and as per the knowledge of the authors, rare literature are available on the proposed generalized radix system. The present work is an extension of that where the circuit modules for arithmetic and logical operations have been modified with minimum circuit complexity such that the speed of operation of the processing elements can be improved. Detailed matrix representation has been carried out in tabular manner and pseudocode is also given following the algorithm. The result shows the importance of the present work in terms of reduced computational complexity. Results are verified by using Xilinx ISE 14.1 simulator using Vertex-7 XC7VX330t device.

2 Multivalued Arithmetic Operation

In multivalued logic arithmetic, the adder and subtractor circuits are based on the modulo operation for radix $(2^p \pm q)$. Also, the increment and decrement operations are performed which can be considered as arithmetic operations.

2.1 Addition Operation

The addition operation is shown in Table 1.

Table 1 shows the truth table for the addition of the radix $(2^p \pm q)$. The 'Sum' and 'Carry' operations and the corresponding matrices are represented below. Here 'a' and 'b' are the operands on which the arithmetic and logic operations like addition, subtraction, AND, OR, etc., are performed. For example, if $x = 3$ and

x	y	Carry	Sum
0	0	0	0
0	1	0	1
0	2	0	2
...	...	...	...
0	$2^p \pm q - 1$	0	$2^p \pm q - 1$
1	0	0	1
1	1	0	2
1	2	0	3
...	...	...	...
1	$2^p \pm q - 1$	1	0
...	...	...	...
$2^p \pm q - 1$	$2^p \pm q - 2$	1	$2^p \pm q - 3$
$2^p \pm q - 1$	$2^p \pm q - 1$	1	$2^p \pm q - 2$

Table 1 Truth table for addition operation

$y = 6$ and radix $= 7 = 2^3 - 1$, then Sum $= (x+y) \bmod 7 = 2$ and Carry $= (x+y)/7 = 1$. Here '2' and '1' are put in 'Sum' and 'Carry' columns respectively.

Tables 2 and 3 show the matrices for the 'Sum' and 'Carry' outputs. From Table 2, it is obvious that the elements of one row are rotated to the left by one position. Therefore, if we can compute the elements of the first row initially then no other arithmetic operations except rotation are performed. This approach definitely results in improvement of speed of operation of the processor. Similarly, from Table 3, it is obvious that it is simply a left-shift operation of the elements of each row and the rightmost cell is occupied by '1'. The first row must be initialized by '0's.

Based on these approaches, the following pseudocode has been developed for addition.

(1) *Pseudocode for addition*:

```
Input: integer x, y (∈ [0, 2^p ± q])
Output: integer sum, carry (∈ [0, 2^p ± q])
Variable: matrix1 [2^p ± q, 2^p ± q] , matrix2 [2^p ± q, 2^p ± q]
for i = 0 to 2^p ± q − 1
   matrix1[0,i] = i;
end;
for i = 1 to 2^p ± q − 1
for j = 0 to 2^p ± q − 1
   matrix1[i,j] = matrix1[i-1,((j+1) mod 4)];
end;
end;
for i = 0 to 2^p ± q − 1
for j = 0 to 2^p ± q − 1
   if(j >= (4-i)) then
        matrix2[i,j] = 1;
   else
        matrix2[i,j] = 0;
   end;
end;
end;
sum = matrix1[x,y];
carry = matrix2[x,y];
```

Table 2 Matrix for 'Sum' output

x \ y	0	1	2	...	$(2^p \pm q - 2)$	$(2^p \pm q - 1)$
0	0	1	2	...	$(2^p \pm q - 2)$	$(2^p \pm q - 1)$
1	1	2	...	$(2^p \pm q - 2)$	$(2^p \pm q - 1)$	0
2	2	...	$(2^p \pm q - 2)$	$(2^p \pm q - 1)$	0	1
...	...	...	...	...	...	...
$(2^p \pm q - 2)$	$(2^p \pm q - 2)$	$(2^p \pm q - 1)$	0	1	2	...
$(2^p \pm q - 1)$	$(2^p \pm q - 1)$	0	1	2	...	$(2^p \pm q - 2)$

Table 3 Matrix for 'Carry' output

x	y					
	0	1	2	...	$(2^p \pm q - 2)$	$(2^p \pm q - 1)$
0	0	0	0	...	0	0
1	0	0	0	...	0	1
2	0	0	0	...	1	1
...	...	...	...	...	...	...
$(2^p \pm q - 2)$	0	0	1	...	1	1
$(2^p \pm q - 1)$	0	1	1	...	1	1

2.2 Subtraction Operation

Subtraction can be performed similarly based on the following truth table.

Table 4 can be similarly represented in matrix form as shown. Table 5 shown is for 'Difference' output, and Table 6 is for 'Borrow' output. Now consider the previous example with $x = 3$ and $y = 6$ and radix $= 7 = 2^3 - 1$. Since $x < y$, then $(x - y) \bmod 7 = (3 - 6) \bmod 7 = (7 + 3 - 6) \bmod 7 = 4$. Therefore, Difference $= 4$ and Borrow $= 1$, since '7' is added to '3'.

From Table 5, it is observed that the elements of each row are rotated right by one position. The first row elements are computed and for the following rows, the elements are rotated by one position. From Table 6, it is obvious that the first row elements except the first position are loaded as '1' and the first position is loaded by '0'. For the remaining rows, the elements are shifted right and the leftmost position is loaded by '0'.

Table 4 Truth table for subtraction operation

x	y	Borrow	Difference
0	0	0	0
0	1	1	$2^p \pm q - 1$
0	2	1	$2^p \pm q - 2$
...	...	...	...
0	$2^p \pm q - 1$	1	1
1	0	0	1
1	1	0	0
1	2	1	$2^p \pm q - 1$
...	...	...	...
1	$2^p \pm q - 1$	1	2
...	...	...	...
$2^p \pm q - 1$	$2^p \pm q - 2$	0	1
$2^p \pm q - 1$	$2^p \pm q - 1$	0	0

Table 5 Matrix for 'Difference' output

x	y					
	0	1	2	...	$(2^p \pm q - 2)$	$(2^p \pm q - 1)$
0	0	$2^p \pm q - 1$	$2^p \pm q - 2$	...	2	1
1	1	0	$2^p \pm q - 1$	$2^p \pm q - 2$	...	2
2	2	1	0	$2^p \pm q - 1$	$2^p \pm q - 2$	...
...	...	...	...	...	...	...
$(2^p \pm q - 2)$	$2^p \pm q - 2$	$2^p \pm q - 3$	$2^p \pm q - 4$	...	0	$2^p \pm q - 1$
$(2^p \pm q - 1)$	$2^p \pm q - 1$	$2^p \pm q - 2$	$2^p \pm q - 3$	$2^p \pm q - 4$	...	0

Table 6 Matrix for 'Borrow' output

x	y					
	0	1	2	...	$(2^p \pm q - 2)$	$(2^p \pm q - 1)$
0	0	1	1	...	1	1
1	0	0	1	1	...	1
2	0	0	0	1	1	...
...	...	...	...	...	...	...
$(2^p \pm q - 2)$	0	0	0	...	0	1
$(2^p \pm q - 1)$	0	0	0	...	0	0

Based on Tables 5 and 6, the following pseudocode can be developed.

(1) *Pseudocode for subtraction*

```
Input: integer x, y (∈ [0, 2^p ± q])
Output: integer difference, borrow (∈ [0, 2^p ± q])
Variable: matrix1 [2^p ± q, 2^p ± q] , matrix2 [2^p ± q, 2^p ± q]
for i = 0 to 2^p ± q − 1
  matrix1[0,i] = (4-i) mod 4;
end;
for i = 1 to 2^p ± q − 1
for j = 0 to 2^p ± q − 1
  matrix1[i,j] = matrix1[i-1,((j-1) mod 4)];
end;
end;
for i in 0 to 2^p ± q − 1
for j in 0 to 2^p ± q − 1
  if(j > i) then
      matrix2[i,j] = 1;
```

```
    else
        matrix2[i,j] = 0;
    end;
end;
end;
difference = matrix1[x,y];
borrow = matrix2[x,y];
```

3 Increment–Decrement Operation

Increment or decrement operations are basically adding '1' to or subtracting '1' from the operand respectively. If the final result exceeds the radix or its multiple then the remainder of the result after division by the radix is the final result as per the definition of modulo number system. Since here the 'Sum' or the 'Difference' output is taken as the final output as requirement, then the 'Carry' or 'Borrow' output is redundant.

The increment–decrement operations, for operand 'x', are characterized by the following matrix.

From Table 7, it is obvious that for incrementing and decrementing the value of operand 'x', respectively rotate-left and rotate-right operations are performed. The following pseudocode describes the increment–decrement operation.

(1) Pseudocode for Increment–Decrement operation

```
Input: integer option, x (∈ [0, 2^p ± q])
Output: integer t (∈ [0, 2^p ± q])
Variable: matrix1, matrix2 [2^p ± q, 2^p ± q]
for i = 0 to 2^p ± q − 1
matrix1(i) = i;
end;
for i = 0 to 2^p ± q − 1
if(option = 0) then
matrix2(i) = matrix1((i + 1) mod (2^p ± q));
else if(option = 1) then
```

Table 7 Matrix for increment–decrement operation

op	x					
	0	1	2	...	$(2^p \pm q - 2)$	$(2^p \pm q - 1)$
0 (increment)	1	2	...	$(2^p \pm q - 2)$	$(2^p \pm q - 1)$	0
1 (decrement)	$(2^p \pm q - 1)$	0	1	...	$(2^p \pm q - 3)$	$(2^p \pm q - 2)$

```
matrix2((i + 1) mod (2^p ± q)) = matrix1(i);
end if;
end loop;
t = matrix2(x);
```

4 Multivalued Logic Operations

4.1 AND and NAND Operations

All the logical operations have already been defined in [10]. In this paper, the faster algorithms for their implementation have been presented. In the matrix representations for logic operations shown below, it can be observed that without repeated computation, the results can be obtained by some simple shift or rotate operations. The mathematical formula for logical AND operation is given below

$$x \operatorname{AND} y = \begin{cases} x, & \text{if } x \leq y \\ y, & \text{if } x > y \end{cases} \tag{1}$$

Similarly, the NAND operation is defined as follows.

$$x \operatorname{NAND} y = R - 1 - (x \operatorname{AND} y) \tag{2}$$

where

$$R = 2^p \pm q \tag{3}$$

The following matrices shown in Tables 8 and 9 are for the logical AND and NAND operations respectively. Here 'x' and 'y' are the operands as described earlier. For AND operation, minimum of the two operands are taken as the result. Similarly, for the NAND operation, the complement of the 'AND' result with respect to $(\text{Radix} - 1)$ is needed.

Table 8 Matrix for AND operation

| x | y | | | | | |
---	0	1	2	...	$(2^p \pm q - 2)$	$(2^p \pm q - 1)$
0	0	0	0	...	0	0
1	0	1	1	...	1	1
2	0	1	2	...	2	2
...	...	...	...	...	...	...
$(2^p \pm q - 2)$	0	1	2	...	$2^p \pm q - 2$	$2^p \pm q - 2$
$(2^p \pm q - 1)$	0	1	2	...	$2^p \pm q - 2$	$2^p \pm q - 1$

Table 9 Matrix for NAND operation

x	y					
	0	1	2	...	$(2^p \pm q - 2)$	$(2^p \pm q - 1)$
0	$2^p \pm q - 1$	$2^p \pm q - 1$	$2^p \pm q - 1$	...	$2^p \pm q - 1$	$2^p \pm q - 1$
1	$2^p \pm q - 1$	$2^p \pm q - 2$	$2^p \pm q - 2$	...	$2^p \pm q - 2$	$2^p \pm q - 2$
2	$2^p \pm q - 1$	$2^p \pm q - 2$	$2^p \pm q - 3$	...	$2^p \pm q - 3$	$2^p \pm q - 3$
...	...	...	...	...	...	...
$(2^p \pm q - 2)$	$2^p \pm q - 1$	$2^p \pm q - 2$	$2^p \pm q - 3$	...	1	1
$(2^p \pm q - 1)$	$2^p \pm q - 1$	$2^p \pm q - 2$	$2^p \pm q - 3$	...	1	0

Based on Tables 8 and 9, the following pseudocodes for 'AND' and 'NAND' operations have been developed.

(1) *Pseudocode for AND operation*

```
Input: integer x, y (∈ [0, 2^p ± q])
Output: integer t (∈ [0, 2^p ± q])
Variable: matrix1 [2^p ± q, 2^p ± q]
for i = 0 to 2^p ± q - 1
for j = i to 2^p ± q - 1
    matrix1[i,j] = i;
end;
for j = i+1 to 2^p ± q - 1
    matrix1[j,i] = i;
end;
end;
t = matrix1[x,y];
```

(2) *Pseudocode for NAND operation*

```
Input: integer x, y (∈ [0, 2^p ± q])
Output: integer t (∈ [0, 2^p ± q])
Variable: matrix1 [2^p ± q, 2^p ± q]
for i = 0 to 2^p ± q - 1
for j = i to 2^p ± q - 1
    matrix1[i,j] = 2^p ± q - 1) -i;
end;
for j = i+1 to 2^p ± q - 1
    matrix1[j,i] = (2^p ± q - 1) -i;
end;
end;
t = matrix1[x,y];
```

4.2 OR and NOR Operations

The mathematical formula for logical OR operation is given below as described in [10].

$$x \, \text{OR} \, y = \begin{cases} x, & \text{if } x > y \\ y, & \text{if } x \le y \end{cases} \tag{4}$$

Similarly the NOR operation is defined as follows.

$$x \, \text{NAND} \, y = R - 1 - (x \, \text{OR} \, y) \tag{5}$$

From the above mathematics, it is obvious that in OR operation, the maximum of the two operands is taken as the result. Similarly, for NOR operation, the complement of the 'OR' result is taken with respect to (Radix $-$ 1). The following matrices shown in Tables 10 and 11 define the OR and NOR operations respectively.

Based on Tables 10 and 11, the following pseudocodes for 'OR' and 'NOR' operations respectively have been developed.

(1) *Pseudocode for OR operation*

```
Input: integer x, y (∈ [0, 2ᵖ ± q])
Output: integer t (∈ [0, 2ᵖ ± q])
Variable: matrix1 [2ᵖ ± q, 2ᵖ ± q]
for i = 0 to 2ᵖ ± q − 1
for j = i to 2ᵖ ± q − 1
    matrix1[(2ᵖ ± q − 1) −i, (2ᵖ ± q − 1) −j] = (2ᵖ ± q − 1) −i;
end;
for j = i+1 to (2ᵖ ± q − 1)
```

Table 10 Matrix for OR operation

x	y					
	0	1	2	...	$(2^p \pm q - 2)$	$(2^p \pm q - 1)$
0	0	1	2	...	$(2^p \pm q - 2)$	$(2^p \pm q - 1)$
1	1	1	2	...	$(2^p \pm q - 2)$	$(2^p \pm q - 1)$
2	2	2	2	...	$(2^p \pm q - 2)$	$(2^p \pm q - 1)$
...	...	...	...	...	...	...
$(2^p \pm q - 2)$	$(2^p \pm q - 2)$	$(2^p \pm q - 2)$	$(2^p \pm q - 2)$	...	$(2^p \pm q - 2)$	$(2^p \pm q - 1)$
$(2^p \pm q - 1)$	$(2^p \pm q - 1)$	$(2^p \pm q - 1)$	$(2^p \pm q - 1)$	...	$(2^p \pm q - 1)$	$(2^p \pm q - 1)$

Table 11 Matrix for NOR operation

x	y					
	0	1	2	...	$(2^p \pm q - 2)$	$(2^p \pm q - 1)$
0	$(2^p \pm q - 1)$	$(2^p \pm q - 2)$	$(2^p \pm q - 3)$	...	1	0
1	$(2^p \pm q - 2)$	$(2^p \pm q - 2)$	$(2^p \pm q - 3)$	...	1	0
...	...	...	...	...	...	...
$(2^p \pm q - 2)$	1	1	1	...	1	0
$(2^p \pm q - 1)$	0	0	0	...	0	0

```
      matrix1[(2^p ± q - 1) -j,(2^p ± q - 1) -i] = (2^p ± q - 1) -i;
end;
end;
t = matrix1[x,y];
```

(2) *Pseudocode for NOR operation*

```
Input: integer x, y (∈ [0, 2^p ± q])
Output: integer t (∈ [0, 2^p ± q])
Variable: matrix1 [2^p ± q, 2^p ± q]
for i = 0 to 2^p ± q - 1
for j = i to 2^p ± q - 1
    matrix1[(2^p ± q - 1) -i,(2^p ± q - 1) -j] = i;
end;
for j = i+1 to 2^p ± q - 1
    matrix1[(2^p ± q - 1) -j,(2^p ± q - 1) -i] = i;
end;
end;
t = matrix1[x,y];
```

4.3 XOR and XNOR Operations

The mathematical formula for logical XOR operation is given below as described in [10].

$$x\,\text{XOR}\,y = \begin{cases} R - 1, & \text{if } x \neq y \\ 0, & \text{if } x = y \end{cases} \qquad (6)$$

Similarly the XNOR operation is defined as follows.

$$x \, \text{XNOR} \, y = R - 1 - (x \, \text{XOR} \, y) \tag{7}$$

From the above equations, it is obvious that if the two operands are identical then the 'XOR' output is '0' otherwise the output is $(\text{Radix} - 1)$. Similarly, for XNOR operation, the complement of the 'XOR' output is taken with respect to $(\text{Radix} - 1)$. Here also we need a comparison of the two operands. But the computations related to the comparison are not required. Using matrix manipulation technique shown below, it can be easier. Tables 12 and 13 show the matrices for generating XOR and XNOR gates which are characterized by Eqs. (6) and (7).

Based on Tables 12 and 13, the following pseudocodes for XOR and XNOR operations have been developed.

(1) *Pseudocode for XOR operation*

```
Input: integer x, y (∈ [0, 2ᵖ ± q])
Output: integer t (∈ [0, 2ᵖ ± q])
Variable: matrix1 [2ᵖ ± q, 2ᵖ ± q]
for i = 0 to 2ᵖ ± q - 1
for j = 0 to 2ᵖ ± q - 1
if(i = j) then
    matrix1[i,j] = 0;
else
    matrix1[i,j] = 2ᵖ ± q - 1 ;
end;
end;
end;
t = matrix1[x,y];
```

Table 12 Matrix for XOR operation

x	y					
	0	1	2	...	$(2^p \pm q - 2)$	$(2^p \pm q - 1)$
0	0	$(2^p \pm q - 1)$	$(2^p \pm q - 1)$	...	$(2^p \pm q - 1)$	$(2^p \pm q - 1)$
1	$(2^p \pm q - 1)$	0	$(2^p \pm q - 1)$	...	$(2^p \pm q - 1)$	$(2^p \pm q - 1)$
...	...	...	...	...	...	...
$(2^p \pm q - 2)$	$(2^p \pm q - 1)$	$(2^p \pm q - 1)$	$(2^p \pm q - 1)$	...	0	$(2^p \pm q - 1)$
$(2^p \pm q - 1)$	$(2^p \pm q - 1)$	$(2^p \pm q - 1)$	$(2^p \pm q - 1)$	...	$(2^p \pm q - 1)$	0

Table 13 Matrix for XNOR operation

x	y					
	0	1	2	...	$(2^p \pm q - 2)$	$(2^p \pm q - 1)$
0	$(2^p \pm q - 1)$	0	0	...	0	0
1	0	$(2^p \pm q - 1)$	0	...	0	0
...	...	...	...	...	...	...
$(2^p \pm q - 2)$	0	0	0	...	$(2^p \pm q - 1)$	0
$(2^p \pm q - 1)$	0	0	0	...	0	$(2^p \pm q - 1)$

5 Result Analysis

Tables 14 and 15 show the simulation results of all the operations designed and verified using FPGA. Here power has been computed using Xilinx Power Analyzer post implementation. Here it is revealed that a fixed portion of the temporary memory containing $(2^p \pm q) \times (2^p \pm q)$ array is required to compute all the operations. Using the control logic we can select the specified operation to optimize the memory usage.

Table 14 Simualted result for radix 4

Design	Radix-4		
	Area (slice no.)	Delay (ns)	Power (mW)
Adder	3	0.79	0.5
Subtractor	3	0.79	0.5
Adder/ Subtractor	8	1.4	0.62
AND	2	0.77	0.34
NAND	2	0.77	0.34
OR	2	0.77	0.34
NOR	2	0.77	0.34
XOR	1	0.77	0.33
XNOR	1	0.77	0.33
Inc/Dec	2	0.75	0.35

Table 15 Simaulted result for radix 6 and 8

Design	Radix-6			Radix-8		
	Area (slice no.)	Delay (ns)	Power (mw)	Area (slice no.)	Delay (ns)	Power (mw)
Adder	4	0.89	0.61	4	0.91	0.7
Subtractor	4	0.89	0.61	4	0.91	0.7
Adder/ Subtractor	17	1.8	0.72	25	2.1	0.83
AND	3	0.87	0.37	3	0.89	0.41
NAND	3	0.87	0.37	3	0.89	0.41
OR	3	0.87	0.37	3	0.89	0.41
NOR	3	0.87	0.37	3	0.89	0.41
XOR	1	0.87	0.36	1	0.89	0.39
XNOR	1	0.87	0.36	1	0.89	0.39
Inc/Dec	3	0.78	0.38	3	0.8	0.42

6 Conclusion

We have proposed a new and relatively fast synthesis technique for ALU for any radix system. Matrix element rotation or shifting technique, shown here, is very efficient in the computation of arithmetic and logical operations. Though we require few temporary memories but still it is helpful for faster operation (Table 16).

Table 15 shows the simulation results of all the operations designed and verified using FPGA Spartan 3 device. Here in [11], addition operation has been performed in Radix-4. Comparative analysis is made for radix 4, 6 and 8 systems using Xilinx simulator using FPGA board for area, delay and power calculations, which three together play a pivotal role in VLSI chip design.

Table 16 Comparison of quaternary adder using Spartan-3 FPGA device

No. of Qtrits	Area (no. of slices)		Delay (in ns)	
	[11]	Proposed	[11]	Proposed
2	15	12	6.21	4.54

References

1. C. Lazzari, P. Flores, J. Monteiro, L. Carro, "A new quaternary FPGA based on a voltage-mode multi-valued circuit", *Proc. of Int. Conf. on Design, Automation and Test in Europe*, March, 2010.

2. M. Eaton, "Design and Construction of a balanced ternary ALU with potential future cybernetic intelligent systems applications", *Int. Conf. on cybernetic Intelligent Systems (CIS)*, pp. 30–35, 2012.

3. N. Raad M. M. Mansour, "A Low Power 32-bit Quaternary Tree Adder", *Int. Conf. on Energy Aware Computing*, pp. 1–2, 2011.

4. H. Shirahama, A. Mochizuki, T. Hanyu, M. Nakajima K. Arimoto, "Design of a Processing Element based on Quaternary Differential Logic for a Multi-core SIMD processor", *Int. Symp. on Multiple Valued Logic*, pp. 43–45, 2007.

5. A. P. Dhande V. T. Ingole, "Design And Implementation Of 2 Bit Ternary ALU Slice", *Int. Conf. on Sciences of Electronic, Technologies of Information and Telecommunications*, Tunisia, March, 2005.

6. L. P. Nascimento, "An Automated Tool for Analysis and Design of MVL Digital Circuits", *Proc. of the 14th Symp. on Integrated circuits and systems design*, September, 2001.

7. Frank Buijs, "ALU synthesis from HDL descriptions to optimized multi-level logic", *Proc. of Int. Conf. on Euro design automation*, Oct 1992.

8. R. Keyes, "The Evolution of Digital Electronics towards VLSI", IEEE Jour. of solid state Circuits, vol. SC-14, pp. 193–201, 1979.

9. K. C. Smith, "The Prospects for Multi valued Logic: A Technology and Applications View", IEEE Trans on Comp, vol. C-30(9), pp. 619–634, 1981.

10. A. Banerjee, S. Bhattacharyya and A. Deyasi, "High Speed Reconfigurable ALU Design for Radix $(2^n \pm m)$", Advances In Industrial Engineering and Management (ASP), vol. 5(2), pp. 183–187, 2016.

11. S. A. Dakhane, A. M. Shah, "FPGA Implementation of Fast Arithmetic Unit Based on QSD", International Journal of Computer Science and Information Technologies, Vol. 5(3), pp. 3331–3334, 2014.

Author Index

© Springer Nature Singapore Pte Ltd. 2019
V. Nath and J. K. Mandal (eds.), *Proceeding of the Second International Conference on Microelectronics, Computing & Communication Systems (MCCS 2017)*, Lecture Notes in Electrical Engineering 476, https://doi.org/10.1007/978-981-10-8234-4

Printed by Printforce, the Netherlands